Space Sciences Series of ISSI

The Space Sciences Series of ISSI books are coherent reports of the findings, discussions, and ideas that result from international scientific workshops regularly held at the International Space Science Institute (ISSI) in Bern, Switzerland. ISSI's main task is to contribute to the achievement of a deeper understanding of the results from space-research missions, adding value to those results through multi-disciplinary research in an atmosphere of international cooperation. The books are reprints of special issues in the Space Science Reviews journal and occasionally of special issues in the Surveys in Geophysics journal.

More information about this series at https://link.springer.com/bookseries/6592

Manfred Schüssler · Robert Cameron ·
Paul Charbonneau · Mausumi Dikpati ·
Hideyuki Hotta · Leonid Kitchatinov
Editors

Solar and Stellar Dynamos

A New Era

The book is a spin-off from the Topical Collection "Solar and Stellar Dynamos: A New Era" of the journal Space Science Reviews

 Springer

Editors

Manfred Schüssler
Max Planck Institute for Solar System
 Research
Göttingen, Germany

Robert Cameron
Max Planck Institute for Solar System
 Research
Göttingen, Germany

Paul Charbonneau
Physics Department
Université de Montréal
Montréal, QC, Canada

Mausumi Dikpati
High Altitude Observatory
National Center of Atmospheric Research
Boulder, CO, USA

Hideyuki Hotta
Institute for Space-Earth Environmental
 Research
Nagoya University
Nagoya, Japan

Leonid Kitchatinov
Institute of Solar-Terrestrial Physics
Irkutsk, Russia

ISSN 1385-7525   Space Sciences Series of ISSI
ISBN 978-94-024-2261-0

Cover Image: Current-free extrapolations of the coronal magnetic fields of the Sun in 2019 August (left) and the rapidly-rotating K0 dwarf AB Doradus in 2007 December (right). Observational input data comprise, respectively, magnetograms from NASA Solar Dynamics Observatory (courtesy of NASA and the SDO/HMI science team), and inferred magnetic maps using Zeeman-Doppler Imaging (courtesy of G.A.J. Hussain – as used by O. Cohen, et al., Astrophys. J. 721, 80, 2010, following the method of G.A.J. Hussain et al., Astrophys. J. 575, 1078, 2002). The background image was downloaded from https://www.pexels.com/search/space%20background/.

This Springer imprint is published by the registered company Springer Nature B.V.
The registered company address is: Van Godewijckstraat 30, 3311 GX Dordrecht, The Netherlands

Paper in this product is recyclable.

Preface

Large-scale and small-scale dynamos generating magnetic flux are fundamental processes in astrophysics, for which solar and stellar dynamos provide crucial paradigms. During the last two decades, a wealth of new observational results, together with a new generation of large-scale numerical simulations, gave important insights in the complex interactions of turbulent convection, rotation, and magnetic fields. They permit tests and calibrations of mean-field models for the generation of large-scale magnetic flux, differential rotation, and meridional circulation in stellar convection zones. Understanding of the underlying processes and the possibility to compare with a broad base of observational data puts the necessary parametrizations in simplified dynamo models (e.g., "classical" α-effect models, flux-transport dynamos, or Babcock-Leighton models) on a firmer basis. At the same time, space-based observations of stars have dramatically increased the amount of information on rotation, age, magnetic activity, and cycles for a wide range of stellar parameters, thus opening the possibility to study dynamo action for a wide range of stellar parameters and evolutionary stages.

These developments led to substantial progress in our understanding of the generation of magnetic fields by self-excited dynamo action in the Sun and other stars. Dynamo models addressing the variation of the magnetic field in the solar interior and at its surface in the course of the 11-yr activity cycle have become more detailed and realistic. The important processes of formation, rise, and emergence of magnetic flux loops in the course of the solar dynamo process are now studied in simulations covering the whole range from the convection zone into the corona. The operation of small-scale dynamo processes is revealed by high-resolution solar observations, which can be directly compared with numerical simulations of radiative magneto-hydrodynamics. The ubiquitous magnetic fields generated by small-scale dynamo action are important for many astrophysical systems. They affect crucial aspects of convective turbulence, such as the transport of energy and angular momentum, as well as the driving of large-scale flows and differential rotation. 3D-MHD simulations show dynamo action with large-scale organization of the generated magnetic field and cyclic polarity reversals.

At this stage of affairs, it was the right time to review and put into perspective the wealth of results from observations, simulations, and simplified models. The one-week international workshop "Solar and Stellar Dynamos: a New Era" hosted by ISSI in Bern brought together more than 50 established as well as young researchers from altogether 13 countries working in the relevant areas: solar and stellar observations from space and ground, numerical simulations, turbulence theory, and dynamo models. This gave the unique opportunity to broadly review the state of the art, outline the open questions, and discuss approaches to make further progress.

This volume summarizes the outcome of the workshop in the form of 15 comprehensive review papers covering the whole range of topics discussed during the meeting. Charbonneau & Sokoloff set the stage for the subsequent papers with a historical and technical introduction, specifically focussing on a number of unresolved "tension points" and questions.

The reviews of Norton et al. and Biswas et al., respectively, address the basic observations of the solar cycle and its long-term development. Likewise, observations of magnetic fields and the activity cycles of cool stars are reviewed by Isik et al. and Jeffers et al., respectively. Weber et al. summarize the present understanding of the formation of magnetic structures in the convection zone and their emergence as active regions at the surfaces of the Sun and stars. The subsequent evolution of the emerged magnetic flux at the surface is the topic of the review by Yeates et al. Small-scale dynamo action and its reflection in observations are addressed by Rempel et al. Important ingredients relevant for models of large-scale dynamos are the structure and dynamics of the solar tachocline, the radial shear layer at the bottom of the convection zone (reviewed by Strugarek et al.), and the large-scale flows in the solar convective envelope (reviewed by Hotta et al.). The subsequent reviews cover a wide range of topics relevant for dynamo models: mean-field dynamos and turbulent processes (Brandenburg et al.), flux-transport dynamos and the generation of meridional flows (Hazra et al.), as well observationally guided models based upon the Babcock-Leighton scenario and the role of the surface field (Cameron & Schüssler). How such physical models could be used for the prediction of solar and stellar cycles is the topic of the review by Bhowmik et al. Finally, Käpylä et al. review the enormous progress that has been made by comprehensive 3D-MHD simulations of solar and stellar dynamos.

The editors would like to thank all authors and coordinators for their dedicated efforts to prepare the papers that are assembled in this collection. The friendly and always helpful assistance by the staff of ISSI before, during, and after the workshop was an invaluable support for the participants and the conveners. We also thank the editorial staff of Space Science Reviews for the smooth organization of the publishing process and efficient help whenever technical issues came up. Last not least, we are very grateful to ISSI for generous financial support in organizing the workshop and for the publication of the book.

Manfred Schüssler
Max Planck Institute for Solar System Research, Justus-von-Liebig-Weg 3, 37077 Göttingen, Germany

Robert Cameron
Max Planck Institute for Solar System Research, Justus-von-Liebig-Weg 3, 37077 Göttingen, Germany

Paul Charbonneau
Physics Department, Université de Montréal, C.P. 6128 Centre-Ville, Montréal, H3C-3J7, Qc, Canada

Mausumi Dikpati
High Altitude Observatory, NCAR, 3080 Center Green Drive, Boulder, 80310, CO, USA

Hideyuki Hotta
Institute for Space-Earth Environmental Research, Nagoya University, Chikusa-ku, Nagoya, 64-8601, Japan

Leonid Kitchatinov
Institute of Solar-Terrestrial Physics SB RAS, Lermontov Str. 126a, Irkutsk, 664033, Russia

Contents

Space Science Reviews (2023) 219:35
https://doi.org/10.1007/s11214-023-00980-0

Evolution of Solar and Stellar Dynamo Theory

Paul Charbonneau[1] · Dmitry Sokoloff[2,3,4]

Received: 9 March 2023 / Accepted: 31 May 2023 / Published online: 28 June 2023
© Crown 2023

Abstract

In this paper, written as a general historical and technical introduction to the various contributions of the collection "Solar and Stellar Dynamo: A New Era", we review the evolution and current state of dynamo theory and modelling, with emphasis on the solar dynamo. Starting with a historical survey, we then focus on a set of "tension points" that are still left unresolved despite the remarkable progress of the past century. In our discussion of these tension points we touch upon the physical well-posedness of mean-field electrodynamics; constraints imposed by magnetic helicity conservation; the troublesome role of differential rotation; meridional flows and flux transpost dynamos; competing inductive mechanisms and Babcock–Leighton dynamos; the ambiguous precursor properties of the solar dipole; cycle amplitude regulation and fluctuation through nonlinear backreaction and stochastic forcing, including Grand Minima; and the promises and puzzles offered by global magnetohydrodynamical numerical simulations of convection and dynamo action. We close by considering the potential bridges to be constructed between solar dynamo theory and modelling, and observations of magnetic activity in late-type stars.

Keywords Magnetohydrodynamics · Dynamo · Solar cycle · Stellar cycles

1 From First Ideas to Contemporary State

The history of solar dynamo studies can be said to begin with the famous talk delivered by J. Larmor (1919; the paper is also reprinted in Ruzmaikin et al. 1988). He attracted the

---

Solar and Stellar Dynamos: A New Era
Edited by Manfred Schüssler, Robert H. Cameron, Paul Charbonneau, Mausumi Dikpati,
Hideyuki Hotta and Leonid Kitchatinov

✉ P. Charbonneau
 paul.charbonneau@umontreal.ca

 D. Sokoloff
 sokoloff.dd@gmail.com

[1] Physics Department, Université de Montréal, C.P. 6128 Centre-Ville, Montréal, H3C-3J7, Qc, Canada

[2] Physics Department, Moscow State University, Leninskiye Gory, Moscow, 119991, Russia

[3] IZMIRAN, 4 Kaluzhskoe Shosse, Troitsk, Moscow, 108840, Russia

[4] Moscow Center of Fundamental and Applied Mathematics, Moscow, 119991, Russia

attention of the astronomical community to the fact that the only visible way to obtain solar magnetic fields, as observed a few years before by G.E. Hale and his colleagues seems to be electromagnetic induction in moving electrically conducting solar media. The car engine was at that time the latest impressive achievement of human civilization, and the idea became known as dynamo theory, after a part of this engine.

Formally speaking, the solar dynamo is an example of a wide variety of various instabilities interesting for astrophysics. Naively speaking, a century may seem as providing sufficient time to investigate an instability in all important details. Dynamos give an interesting and instructive impression here. Initially Larmor's idea proved to be very rich and to contain great potential for development. After various modifications, generalizations and improvements and being fruitfully combined with other scientific ideas, the dynamo still retains until now quite a lot of problems deserving clarification and development in the context of contemporary science. Our aim here is to describe some of these problems, and present the development of dynamo studies which lead to this new perspective. Of course, we can give here a number of instructive features only, and do not pretend to offer comprehensive review.

A quite obvious point is that Larmor spoke about the Sun rather than say about spiral galaxies just because the concept of galaxies was developed in the next decade only. Application of dynamo theory to the Earth, planets, stars, galaxies, clusters of galaxies, accretion discs, etc, has become an important part of astrophysics and has offered an attractive and important perspective for exoplanets studies (e.g. Rüdiger and Hollerbach 2004; Brandenburg and Subramanian 2005). These applications depart from solar dynamo studies in context of specific features of celestial bodies under consideration as well as contemporary observational abilities.

This extensive development of dynamo studies is very important for astronomy, however intensive investigation of the physical basis of solar dynamo looks instructive in a broader scientific context.

First of all, an attempt for straightforward realization of dynamo instability faces the fact that according to Lenz's Law, electromagnetic induction suppresses rather enhances the seed magnetic field. This is why rather simple spherically symmetric or planar 2D flows can not support dynamo action. A number of corresponding antidynamo theorems, from the initial idea of T.G. Cowling (1933) up to sophisticated 2D antidynamo theorem of Ya.B. Zeldovich (1957; see also a further discussion in Zeldovich and Ruzmaikin 1980) were suggested in following decades. Then Yu.B. Ponomarenko (1973) demonstrated that dynamo action is possible in a swirling jet and this idea remains the basis of contemporary experimental dynamo studies (e.g. Lathrop and Forest 2011; Sokoloff et al. 2014). This branch of dynamo studies was able to demonstrate, at the turn of the new millenium, that dynamo is far to be a purely theoretical speculation, but rather a real physical phenomenon which may have in a perspective even practical applications. Laboratory dynamo experiments are quite remote from direct astrophysical applications, however the very laboratory dynamo demonstration is important for solar physics. Contemporary authors do not need to demonstrate how a very weak solar magnetic field can be enhanced up to the kGs values and may admit, if desired, that solar matter was magnetized just in the very early Sun. The results obtained in these studies also contain an important impulse for theoretical mathematics as a fruitful example of problems for systems of parabolic differential equations with violation of the maximum principle.

Of course, we are interested here in dynamos in astrophysical context. The breakthrough here is associated with the famous migratory dynamo proposed by E.N. Parker (1955a). Parker demonstrated how one can overcome the problem with Lenz law using the idea of two magnetic circuits where the first circuit enhances magnetic field in the second one while the second circuit enhances magnetic field in the first. Conventional dipole magnetic field is associated with the first circuit and is transformed in the magnetic field of the second circuit by solar differential rotation. Magnetic field of the second circuit can be considered as a toroidal magnetic field hidden somewhere in the solar interior. A very important point of the scenario is that the recovery of the poloidal magnetic field from the toroidal one requires mirror asymmetry of the flow.

The point that mirror asymmetry of hydrodynamic flows plays a key role in astrophysical dynamo was suggested at the same time as the idea of effects of P-noninvariance was developed in particle physics (e.g. Lee and Yang 1956). The interplay between ideas of mirror-asymmetry effects in these two remote domains of physics emerged as a beautiful page of contemporary science.

Parker developed his idea based on his excellent physical intuition. A solid basis for the idea was suggested in a fully independent research ten year later by physicists from East Germany, namely M. Steenbeck, F. Krause and K.-H. Rädler (Steenbeck et al. 1966; Helmbold 2017; see also Krause and Rädler 1980) in the form of mean-field electrodynamics. The key effect of the scheme responsible for the mirror asymmetry became known as the α-effect (see also Sect. 2 below). The idea published initially in an East-German journal and written in German became accessible to the international community practically immediately due to the translation performed by P. Roberts and M. Stix (1971). Both are well-known experts in the field (e.g. Stix 2004; Glatzmaier and Roberts 1995) and Stix a native German speaker.

An interplay with ideas and experts from the West and East provides a part of the story very suitable for a novel. In fact Steenbeck, who was a very impressive person, leaves an instructive testimonial in a book about his life (Steenbeck 1971; see also Helmbold 2017), and some of its elements can be found in novels written later by H. Königsdorf in years of peaceful revolution in East Germany in the late 1980s. Regarding the Russian side of the story, one of the authors of this chapter keeps in his memory how he passed on to Paul Roberts a proof of the book (Zeldovich et al. 1983) on a street just nearby the Kremlin, in what looked like a scene in a spy movie.

Parker (1955a) as well as Steenbeck et al. (1966) considered the initially (almost) unmagnetized medium and associated the mirror asymmetry with Coriolis force action. The magnetic field creates the mirror asymmetry as well, and contemporary understanding of the solar situation is that this contribution is more important. One more point is the importance of meridional circulation as well as other physical effects which understandably were ignored in the early stages of scientific development (more on these in Sects. 5 and 6 below). Combined with the ideas of H.W. Babcock (1961) and R.B. Leighton (1964), this resulted in the contemporary flux-transport model of solar dynamo (e.g. Charbonneau 2007; Dikpati and Gilman 2009).

The first solar dynamo models dealing with the amplification of a weak seed magnetic field considered the prescribed flow properties (so-called kinematic models). A natural further step was to include a nonlinear dynamo suppression based on some balance arguments and conservation of energy looked as a natural idea for the balance. The situation occurred to be however much more complicated and after the very intensive scientific battle it becomes clear that the magnetic helicity conservation is more important for the problem (viz. Sect. 3 herein). Conservation of magnetic helicity was discovered as early as in the nineteenth century, however nobody considered it as something practically important until K. Moffatt (1978) reintroduced the idea in contemporary science. Mathematical aspects of the problem are that magnetic helicities (as well as various other helicities considered in dynamo studies) can be considered as instructive examples of topological invariants and its topological investigation belonged to activities of V.I. Arnol'd and his school (e.g. Arnol'd and Khesin 1992).

An important point here is that some crucial dynamo drivers including α-effect are associated with topological invariants and being inviscid integral of motions they redistribute in course of dynamo action between various layers in the solar convection shell. Presentation of this redistribution in various solar dynamo models still deserves investigation, however quite a rich bulk of ideas here is accumulated; we mention here as an early achievement the work of Ukrainian astronomer V.N. Krividubsky (e.g. Krivodubskij 2006).

Observational identification of dynamo drivers remains a part of astronomy which is still quite remote from its final stage. An important progress here was associated with the idea of N. Seehafer (1990) who suggested a method to observe magnetic helicity inside sunspots. Due to long-term observations undertaken by Chinese astronomers (Zhang et al. 2010) time-longitude distribution of magnetic helicity over several solar cycles was observed and the idea propagates on other relevant helicities in further studies by various groups.

Modelling of dynamo action in rotating turbulent spherical shells demonstrated that apart of solar equatorward propagating dynamo waves with polarities which follow the Hale laws, various less convenient magnetic field configurations may be excited (e.g. Jennings and Weiss 1991). This may be instructive to explain magnetic activity of some stars and exoplanets.

Helioseismological studies (e.g. Gough et al. 1996) are another (more obvious) way to know more about solar dynamo drivers. Development of helioseismology was associated with one more basic transformation in solar dynamo models (see Sect. 4 herein).

One more point in dynamo studies to be mentioned here is that the geodynamo models were one of the first cases where direct numerical simulations were able to reproduce very complicated physical processes in various fine details (Glatzmaier and Roberts 1995). Contemporary solar dynamo models successfully follow this way (see Sect. 10). These achievements impressively demonstrated direct numerical simulations and physical explanations of a phenomenon in terms of traditional theoretical physics in two separate problems. The point is that contemporary numerics are so powerful that they can mimic processes which theoreticians still can not explain in traditional terms. It looks as a general challenge for contemporary science to be addressed in its further development.

It is undeniable that in the past century our understanding of astrophysical dynamos has progressed remarkably, if somewhat non-linearly (in the geometrical sense of the word). Nonetheless, this progress has raised a host of new questions and puzzles, many still outstanding. The remainder of this review focuses on "tension points" left behind by this meandering path from early ideas to the present state, with emphasis on the solar dynamo.

2 Tension: Why Is Mean-Field Electrodynamics Working?

As just discussed, in the mid-1950s Parker argued that the cyclonicity imparted by the Coriolis force on convective updrafts and downdrafts could effectively break axisymmetry on small spatial scales, and in doing so bypass Cowling's anti-dynamo theorem (Parker 1955a). The basic idea is illustrated on Fig. 1. Parker showed that this mechanism could regenerate a poloidal magnetic field from an initially purely toroidal magnetic component, and, operating in conjunction with rotational shearing of the poloidal component so induced (the so-called $\alpha\Omega$ dynamo scenario), produce a working dynamo loop leading to regular polarity reversals.

Fig. 1 Twisting of an horizontal magnetic fieldline by a cyclonic fluid updraft. In this simple schematic depiction the fieldline is twisted outside of the plane of the page, forming a small loop in a plane perpendicular to the original fieldline. Under the right-hand rule, applying Ampère's Law to this small loops yields a current density pointing parallel to the undeformed magnetic field. Figure 1 in Parker (1970), reproduced with permission, copyright by AAS

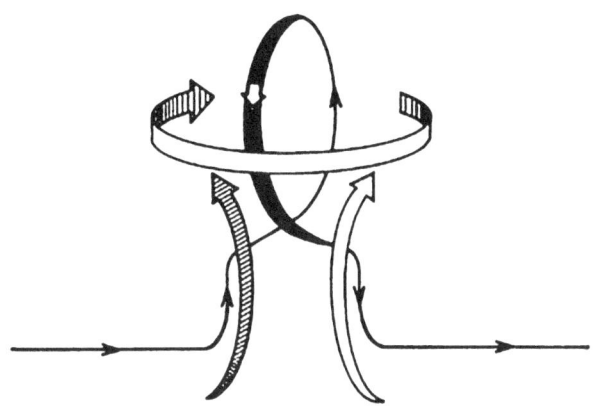

This groundbreaking idea was soon thereafter formalized through the development of mean-field electrodynamics (Steenbeck et al. 1966; Steenbeck and Krause 1969; see also Parker 1970; Moffatt 1978; Krause and Rädler 1980; Moffatt and Dormy 2019, and references therein). Separating the flow and magnetic field into large-scale, slowly varying "mean" component $\langle \mathbf{U} \rangle$, $\langle \mathbf{B} \rangle$ and small-scale rapidly varying "turbulent" components $\mathbf{u}'$, $\mathbf{b}'$, substitution into the induction equation and averaging yields an evolution equation for the mean magnetic field:

$$\frac{\partial \langle \mathbf{B} \rangle}{\partial t} = \nabla \times (\langle \mathbf{U} \rangle \times \langle \mathbf{B} \rangle + \boldsymbol{\xi} - \eta \nabla \times \langle \mathbf{B} \rangle) \, , \tag{1}$$

where the mean electromotive force $\boldsymbol{\xi}$ is given by the average of the small-scale flow-field cross-correlation:

$$\boldsymbol{\xi} = \langle \mathbf{u}' \times \mathbf{b}' \rangle \, . \tag{2}$$

Closure is achieved by expanding this turbulent electromotive force (emf) $\boldsymbol{\xi}$ in terms of $\langle \mathbf{B} \rangle$ and its derivatives:

$$\xi_i = a_{ij} \langle B_j \rangle + b_{ijk} \frac{\partial \langle B_j \rangle}{\partial x_k} + \cdots . \tag{3}$$

This latter expression highlights the fact that mean-field electrodynamics is fundamentally a linear theory, in the sense that the tensors $\mathbf{a}$, $\mathbf{b}$, etc, cannot themselves depend on $\langle \mathbf{B} \rangle$, but only on the statistical properties of the turbulent flow.

The symmetric part ($\boldsymbol{\alpha}$) of the $\mathbf{a}$ tensor captures the Parker mechanism of magnetic field deformation by non-mirror-symmetric turbulence, and is now known as the α-effect. The three components of its antisymmetric part can be recast in the form of a pseudo-velocity acting on the mean-magnetic field, called turbulent pumping. The antisymmetric part of the rank-3 tensor $\mathbf{b}$ can be recast as a rank-2 turbulent diffusivity tensor $\boldsymbol{\beta}$ (Schrinner et al. 2007).

The challenge is now to compute these tensorial quantities from known statistical properties of the turbulent flow, which turns out to be a tall order. There are three physical regimes under which this is tractable (see, e.g.,Sect. 6.3 in Brandenburg and Subramanian 2005; also Sect. 3.4.1 in Ossendrijver 2003; Rempel 2009 and Chap. 7 in Moffatt and Dormy 2019).

1. The energy density of the mean magnetic field is larger than the energy density of the small-scale field;
2. The magnetic Reynolds number is low;
3. The turbulent cyclonic eddies have a lifetime shorter than their characteristic turnover time.

These three physical regimes all amount to the mean magnetic field suffering little deformation by the small-scale turbulent flow either because magnetic tension kicks in and prevents large deformation (Regime 1), field/flow slippage occurs and prevents large deformation (Regime 2), or not enough time is available to induce a large deformation (Regime 3). Pictorially, going back to Fig. 1, the magnetic fieldline must be twisted out of the plane by an angle $\lesssim \pi/2$.

Regimes 1 and 2 are most certainly not applicable to solar interior conditions. Regime 3 is harder to assess, as it is notoriously difficult to predict the coherence time of a given turbulent flow, or even to extract it *a posteriori* from a numerical simulation. As we shall see presently, some circumstantial evidence exists suggesting that Regime 3 might be the key.

For turbulence that is isotropic and homogeneous, the α and β tensor reduce to diagonal forms $\alpha\mathbf{I}$, $\beta\mathbf{I}$, with $\mathbf{I}$ the identity tensor, and turbulent pumping vanishes. The second-order correlation approximation then leads to

$$\alpha = -\frac{1}{3}\tau_c \langle \mathbf{u}' \cdot (\nabla \times \mathbf{u}') \rangle , \qquad \beta = \frac{1}{3}\tau_c \langle (\mathbf{u}')^2 \rangle , \tag{4}$$

where τ_c is the coherence time of the small-scale turbulent flow. The α-effect is now simply proportional to the mean kinetic helicity of turbulence, and the turbulent diffusivity to its energy density. As shown by F. Krause in his Habilitation thesis (as cited in Steenbeck and Krause 1969), in the case of a solar/stellar stratified rotating convection zone:

$$\alpha = -\frac{16}{15}\tau_c^2 \langle (\mathbf{u}')^2 \rangle \mathbf{\Omega} \cdot \nabla (\log \rho u_{\mathrm{rms}}) , \qquad \text{[Northern hemisphere]} \tag{5}$$

where $u_{\mathrm{rms}} \equiv \sqrt{\langle (\mathbf{u}')^2 \rangle}$, $\mathbf{\Omega}$ is the solar angular velocity vector, and with a sign flip in the Southern solar hemisphere (see, e.g., Sect. 6.2 in Brandenburg and Subramanian 2005). Equation (5) implies that if the properties of turbulence are independent of latitude, then α is positive in the Northern solar hemisphere and proportional to $\cos\theta$, where θ is the polar angle (if in doubt, work through footnote 5 in Brandenburg and Subramanian 2005). Except for τ_c, the RHS of these expressions are readily extracted from numerical simulations upon suitable averaging. The **a** and **b** tensors can also be extracted using a variety of techniques (Brandenburg and Sokoloff 2002; Racine et al. 2011; Schrinner et al. 2007; Augustson et al. 2015; Simard et al. 2016; Warnecke et al. 2018, 2021; Shimada et al. 2022). The turbulent mean-field coefficients so extracted can then be used as input to a classical mean-field solar dynamo model, to ascertain whether the resulting large-scale magnetic field evolution resembles —or not— that characterizing the parent MHD simulation. Such tests of internal consistency have been carried out succesfully (Simard et al. 2013, 2016; Warnecke et al. 2021), with independent numerical simulations and extraction methods. This suggests that mean-field electrodynamics does properly capture the process of turbulent induction and resulting dynamo action, at least in these MHD numerical simulations, and by extension, hopefully, in the sun and stars as well.

Global MHD simulations of large-scale magnetic cycles can also be used to validate —or not— the analytical expressions (4). This exercise has been carried out by Simard et al. (2016) (among others), estimating τ_c by the common recipe consisting in equating τ_c with the convective turnover time ℓ/u_t, where ℓ and u_t are local measures of the density scale height and turbulent velocity. The spatial distributions they obtain for α and β reconstructed from Eq. (4) match tolerably well those directly extracted from their MHD simulation (cf. their Figs. 2 and 6), except for the global amplitude, which are larger by a factor of ≈ 5 in the reconstructions based on Eq. (4). The amplitudes can be reconciled provided one assumes that the coherence time τ_c is one fifth of the convective turnover time, which is consistent with low coherence time turbulence (Regime 3 above). The generality of this intriguing result remains to be established.

To sum up: Although tractable only in specific parameter regimes of dubious validity in the solar/stellar context, mean-field electrodynamics adequately captures turbulent induction in MHD simulations of solar convection and dynamo action, and leads to internally consistent spatiotemporal evolution of large-scale magnetic fields. Why it actually works so well remains an open question.

3 Tension: The Troublesome Magnetic Helicity

Magnetic helicity is a topological invariant measuring the linkage between magnetic flux systems (Moffatt 1978; Berger 1999; Moffatt and Dormy 2019). With the magnetic field expressible as $\mathbf{B} = \nabla \times \mathbf{A}$, the magnetic helicity content $\mathcal{H}_B$ of a volume V of magnetized fluid is given by:

$$\mathcal{H}_B = \int_V \mathbf{A} \cdot \mathbf{B}\, dV .$$ (6)

In the absence of a flux of helicity at the bounding surface of the volume V, $\mathcal{H}_B$ evolves according to:

$$\frac{d\mathcal{H}_B}{dt} = -2\mu_0\eta \int_V \mathbf{J} \cdot \mathbf{B}\, dV ,$$ (7)

where $\mathbf{J}$ is the current density and μ_0 the magnetic permeability. Magnetic helicity is clearly a conserved quantity in the ideal (dissipationless) limit $\eta \to 0$, expressing the (topological) fact that magnetic fieldlines cannot cross one another.

The solar large-scale magnetic field, associated with the magnetic cycle, is demonstrably helical. In the context of mean-field electrodynamics (Sect. 2), this helicity is imparted on the large-scale magnetic field by the α-effect, and is of a sign opposite to the kinetic helicity of the small-scale turbulent flow (viz. Eq. (4)). As the large-scale magnetic field is amplified, so must $\mathcal{H}_B$, in apparent violation of the above conservation argument. Note that polarity reversals of the large-scale magnetic field are irrelevant to the problem; reversing the magnetic polarity flips the sign of both $\mathbf{A}$ and $\mathbf{B}$, leaving the sign of $\mathcal{H}_B$ unchanged. How, then, can the solar large-scale magnetic field wax and wane in the course of the magnetic cycle?

Here again mean-field electrodynamic offers some useful insight. Applying scale separation to the vector potential $\mathbf{A}$ and current density $\mathbf{J}$, two evolution equation for the magnetic helicity associated with the large- and small-scale magnetic components can be obtained (Brandenburg and Subramanian 2005; Rempel 2009); in the ideal ($\eta \to 0$) limit:

$$\frac{d}{dt} \int_V \langle \mathbf{A} \rangle \cdot \langle \mathbf{B} \rangle\, dV = +2 \int_V \boldsymbol{\xi} \cdot \langle \mathbf{B} \rangle\, dV ,$$ (8)

$$\frac{d}{dt} \int_V \langle \mathbf{a}' \cdot \mathbf{b}' \rangle\, dV = -2 \int_V \boldsymbol{\xi} \cdot \langle \mathbf{B} \rangle\, dV$$ (9)

with the mean electromotive force $\boldsymbol{\xi}$ given by Eq. (2). Observe that the action of the turbulent emf on the large-scale magnetic field, i.e., the terms on the RHS of Eqs. (8)–(9), produces magnetic helicity of opposite signs at large and small scales, so that the *total* magnetic helicity produced by $\boldsymbol{\xi}$ acting on $\langle \mathbf{B} \rangle$ is thus nil. The large-scale field $\langle \mathbf{B} \rangle$ can now be amplified because magnetic helicity of opposite sign builds up at small scales. However, this turns out to oppose the α-effect, as originally demonstrated by Pouquet et al. (1976; see also Sect. 9 in Brandenburg and Subramanian 2005). The large-scale field is amplified, but at the cost of rapidly quenching the inductive part of the turbulent emf (see Kleeorin et al. 1995; Blackman and Brandenburg 2002; Brandenburg et al. 2009).

A way out of this quandary was identified in Brandenburg (2001). It consists in invoking a direct turbulent cascade of helicity towards even smaller scales than that at which $\boldsymbol{\xi}$ is operating, so that Ohmic dissipation sets in, as per Eq. (7), and dissipates the helicity produced at the inductive scale (RHS of Eq. (9)). At this dissipative scale, the magnetic Reynolds number $\mathrm{Rm} = u_t L / \eta$ is of order unity, but remains much larger at the inductive scale of $\boldsymbol{\xi}$, and even larger yet at the scale of $\langle \mathbf{B} \rangle$, so that Eqs. (8)–(9) effectively hold. Now the α-effect can operate, and a helical large-scale magnetic field can grow.

Another mechanism allowing to evade the constraint of magnetic helicity dissipation is to evacuate it through the volume boundaries. A star like the sun is not embedded in a true vaccum, and magnetic helicity can be evacuated through the corona. In particular, coronal mass ejections have been suggested to contribute significantly to the global magnetic helicity budget (Bieber and Rust 1995; Low 2001; Green et al. 2003; Lynch et al. 2005). Avoiding α-quenching via helicity flux across domain boundaries and/or cancellation across the equatorial plane has also found support in MHD numerical simulations (Brandenburg and Dobler 2001; Warnecke et al. 2011).

To sum up: Conservation of magnetic helicity in the high-Rm regimes poses a strong constraint on magnetic field amplification by turbulent induction, and can potentially quench the growth of the solar large-scale magnetic field. This constraints can be bypassed by a double turbulent cascade or expulsion of helicity from the region of dynamo action. Which of these mechanisms (if any or either) is regulating the overall solar magnetic helicity budget, remains an open question.

4 Tension: The Troublesome Solar Differential Rotation

Already in the nineteenth century, R.C. Carrington and G. Spörer independently noted that sunspots emerge closer and closer to the solar equator as the sunspot cycle unfolds. The first convincing explanation for this striking spatiotemporal pattern was proposed almost a century later by Parker, in the form of dynamo waves (Parker 1955a). In $\alpha\Omega$ mean-field dynamos, these waves propagate in a direction $\mathbf{s}$ given by

$$\mathbf{s} = \alpha \nabla \Omega \times \hat{\mathbf{e}}_\phi , \tag{10}$$

a result now known as the Parker-Stix-Yoshimura sign rule. Extending the observed surface latitudinal differential rotation pattern inwards along cylindrical isosurfaces yields a positive radial shear component at low latitudes, which then requires a negative α-effect in the Northern hemisphere[1] to produce equatorward propagation (Yoshimura 1975; Stix 1976).

---

[1]Or more precisely: a negative $\alpha_{\phi\phi}$ tensor component, the only component typically retained in classical $\alpha\Omega$ mean-field models.

Fig. 2 Northern hemisphere time-latitude diagrams of the large-scale toroidal magnetic component for three mean-field kinematic axisymmetric classical $\alpha\Omega$ dynamo solutions, all using the same solar-like parametrization of the solar internal differential rotation, but different latitudinal profile and sign for the α-effect, in all cases concentrated near the base of the convective envelope ($r/R = 0.7$, where the diagram are constructed). The toroidal magnetic field are normalized to their peak amplitude, and isocontours are equally spaced, with yellow→red (green→blue) indicating positive (negative) values. Figure 7 in Charbonneau (2020), used with permission

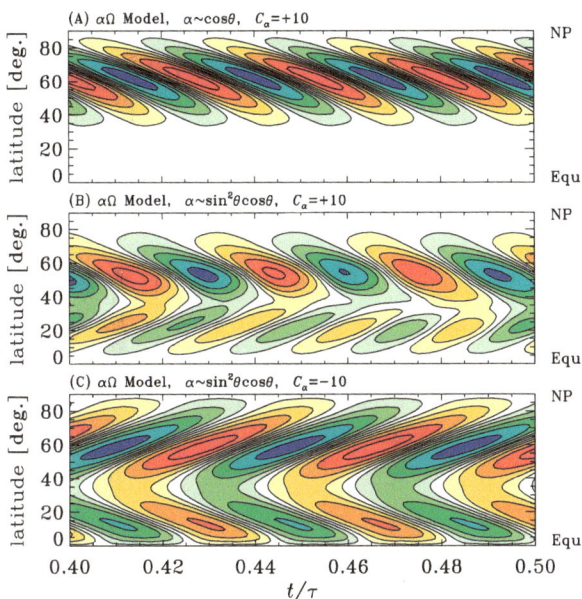

This neat picture was thrown into disarray by the first helioseismic inversions of the solar internal differential rotation (Brown et al. 1989; Dziembowski et al. 1989). Rather that cylindrical isocontours of angular velocity, these inversions revealed that the surface differential rotation remains constant along approximately radial segments, yielding a shear that is primarily latitudinal within the bulk of the convection zone, transiting beneath it to near-solid body rotation across a thin rotational shear layer since known as the tachocline (Spiegel and Zahn 1992; Tomczyk et al. 1995). As shown on Fig. 2, this complex form of the solar internal differential rotation yields very complex patterns of dynamo waves, even if the α-effect is artificially concentrated at the base of the convection zone, to suppress induction by the purely latitudinal shear above. The Figure shows Northern hemisphere time-latitude ("butterfly") diagrams for the toroidal magnetic component at the base of the convection zone, using different latitudinal dependency and sign for the α-effect in a classical $\alpha\Omega$ mean-field model (for more on these dynamo solutions, see Sect. 4.2 in Charbonneau 2020). Here, and even with the α-effect concentrated towards low latitude via a $\sin^2\theta\cos\theta$ dependency on polar angle (panels B and C), the strong positive radial shear in the high latitude regions of the tachocline dominate induction, leading to multiple branches and activity peaking at much higher latitudes than observed.

Within the standard $\alpha\Omega$ dynamo modelling framework, the only way to achieve equatorward dynamo wave propagation is to strongly concentrate a (negative) α effect not only radially at the base of the convection zone, but also latitudinally in its equatorial regions. Recall from Eq. (5) that the minimal latitudinal dependency expected from cyclonic convection leads to a positive α-effect concentrated at high latitudes, with $\alpha \propto \cos\theta$; this does lead to equatorward propagating dynamo waves (viz. Fig. 2A), but again peaking at far higher latitudes than observed on the Sun.

Helioseismology has also revealed the presence of a thin subsurface radial shear layer extending from the equator to mid-latitudes; equatorward propagating dynamo waves concentrated at low latitudes can then be produced in conjunction with a positive Northern hemisphere α-effect, but the small thickness of the layer sets the length scale of dynamo eigenmodes, typically leading to multiple overlapping magnetic flux systems even for weakly supercritical dynamos.

To sum up: in classical $\alpha\Omega$ mean-field dynamo models built using the solar internal differential rotation profile as inferred from helioseismic inversions, it is very hard to produce a sunspot butterfly diagram-like dynamo wave propagation pattern without making some very *ad hoc* assumptions regarding the spatial distribution and/or sign of the turbulent α-effect.

5 Tension: Flux Transport by Meridional Flows

Equatorward propagation of activity belts in the course of the cycle can also be achieved through advection by a bulk meridional flow acting as a "conveyor belt" displacing the internal toroidal magnetic field equatorward as it is amplified by rotational shearing (Wang et al. 1991; Choudhuri et al. 1995; Küker et al. 2001; Rempel 2006b; Pipin and Kosovichev 2011). Dynamo models achieving a solar-like butterfly diagram in this manner are known as *flux transport dynamos* (see Dikpati and Gilman 2009; Karak et al. 2014 for dedicated reviews). Adding a meridional flow to a classical $\alpha\Omega$ dynamo model of the type considered in Sect. 4, one finds that bulk transport of the magnetic field overwhelms the dynamo wave provided advection by the meridional flow dominates over diffusive transport. The ratio of these two transport mechanisms is quantified via a magnetic Reynolds number:

$$\mathrm{Rm} = \frac{u_0 L}{\eta_T} \, , \tag{11}$$

where u_0 is a typical speed for the meridional flow, η_T is a turbulent magnetic diffusivity, and L is a characteristic length scale for the meridional flow, usually taken as the solar radius. For magnetic flux transport to take place in the desired manner, this magnetic Reynolds number must be relatively high, i.e., $\sim 10^2$ or more. See Sect. 4.4 in Charbonneau (2020) for some representative dynamo solutions.

Powered by Reynolds stresses and pole-equator temperature differences caused by rotational influence on convective energy transport, meridional flows are as unavoidable as differential rotation in a rotating, stratified turbulent convective envelope (Kippenhahn 1963; Rüdiger 1989; Kitchatinov et al. 1994; Miesch and Toomre 2009; Balbus et al. 2012; Featherstone and Miesch 2015). This flow is observed at the solar surface, poleward-directed and with speeds peaking in the range 10–15 m s^{-1} at mid latitudes, with some variations in phase with the solar cycle (Hathaway 1996; Ulrich 2010; Cameron and Schüssler 2010). Mass conservation evidently requires an equatorward return flow somewhere within the convection zone. Helioseismic determinations of the internal meridional flow have yielded conflicting results, some inversions suggesting a very shallow equatorward return flow (Jackiewicz et al. 2015), others a complex flow pattern characterized by multiple flow cells stacked in radius and/or latitude (Schad et al. 2013; Zhao et al. 2013), while others yet are consistent with a single cell per meridional quadrant, with the equatorward return flow peaking near the base of the convection zone (Rajaguru and Antia 2015; Gizon et al. 2020).

Global numerical simulations of stratified rotating convection do provide additional insight on the matter. Solar-like differential rotation, in the sense of the rotation rate decreasing monotonically from the equator towards the poles, materializes when the Rossby number $\mathrm{Ro} = u_t/2\Omega L$ is sufficiently small, $\lesssim 0.3$, but in this regime the meridional flow is markedly multi-celled. A single meridional flow cell per meridional quadrant is achieved at higher Rossby number, but the differential rotation is no longer solar, with the equatorial regions rotating more slowly than the mid-latitudes[2] (Gastine et al. 2014; Brun et al. 2022, in particular their Fig. 10). Interestingly, given its rotation rate and luminosity, the sun appears to be characterized by a Rossby number near the tipping point between these two regimes, and some global MHD simulations indicate that magnetic stresses may turn an anti-solar differential rotation into a solar-like profile, while generating a solar-like single-cell meridional flow profile (see Karak et al. 2015; Hotta and Kusano 2021; Hotta et al. 2022, and references therein).

The implications of single vs multi-cell meridional flows for flux transport dynamos are profound. Multiple meridional flow cells can lead to a variety of time-latitude patterns departing significantly from the observed sunspot butterfly diagram (Jouve and Brun 2007; Pipin and Kosovichev 2013; Belucz et al. 2015). The dynamo simulations of Hazra et al. (2014a) do suggest that if the diffusivity is sufficiently high, an equatorward drift of the deep toroidal field can be achieved even in the presence multiple flow cells stacked in radius, provided the deeper cell has an equatorward return flow at the base of the convective envelope. The key parameter then becomes the magnetic Reynolds number (11), which is critically dependent on the assumed value for the (turbulent) magnetic diffusivity, a notoriously difficult quantity to compute from first principles.[3]

To sum up: flux transport dynamo can in principle produce solar-like "butterfly diagrams" even in cases where classical dynamo waves would do otherwise; however, their proper operation depends sensitively on the spatial form of the internal axisymmetric meridional flow, as well as on the value of turbulent magnetic diffusivity produced by solar convection.

6 Tension: Competing Inductive Mechanisms?

The turbulent α-effect is by no means the only way to evade Cowling's theorem. Originally proposed by Babcock (1961) and developed quantitatively by Leighton (1964, 1969), but largely eclipsed by the rise of mean-field electrodynamics until its vigorous revival a quarter of a century later (Wang et al. 1991; Choudhuri et al. 1995; Durney 1995; Dikpati and Charbonneau 1999; Nandy and Choudhuri 2001), what is now known as the Babcock–Leighton mechanism is arguably its most convincing alternative.

A little over a century ago Hale and collaborators established a number of empirical Laws describing the cycle-to-cycle variations in the hemispheric pattern of magnetic polarity measured in sunspots (Hale et al. 1919). They also established what is since known as *Joy's Law* namely the systematic inclinations of the line segment joining the poles of bipolar sunspot groups with respect to the E-W line, this tilt angle (γ) increasing with heliocentric latitude (λ). Leighton (1969) originally parametrized this variation as

$$\sin \gamma = 0.5 \sin \lambda \,, \tag{12}$$

---

[2] Note however that some mean-field turbulence models do produce a single-cell meridional flow in conjunction with a solar-like differential rotation profile; see, e.g., Kitchatinov and Olemskoy (2011).

[3] Alternate versions of flux transport dynamos can be constructed by relying on turbulent pumping (see §2) to achieve, in part or in its entirety, downward transport of the surface field (Jiang et al. 2013) and equatorward drift of the deep toroidal field (Guerrero and de Gouveia Dal Pino 2008; Hazra and Nandy 2016). Measurements of turbulent pumping in some MHD numerical simulations of solar convection do yield strong subsurface downward pumping as well as equatorward latitudinal pumping at mid- to low latitudes within the convecting fluid layers, with speeds of a few meters per second (Ossendrijver et al. 2002; Racine et al. 2011; Simard et al. 2016; Warnecke et al. 2018; Shimada et al. 2022), similar to the deep meridional flow speed.

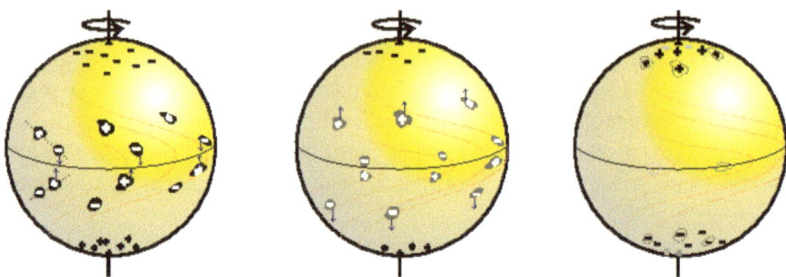

Fig. 3 Schematic illustration of the Babcock–Leighton mechanism in operation. At left, BMRs have just emerged, abiding to Hale's polarity Laws as well as Joy's Law. After some time (middle), the BMRs have decayed and spread diffusively, with preferential transequatorial dissipation of the leading polarities, and transport of the residual trailing polarity to high latitudes by surface flows. This eventually leads to the reversal of the pre-existing dipole (here negative), and buildup of a new (positive) dipole (at right). Diagram produced by D. Passos, used by permission

but other related forms fit the data equally well, in view of the large scatter of observed tilt angles about such mean relationships (see, e.g., McClintock and Norton 2013, and references therein).

Bipolar magnetic regions (BMRs) are believed to originate from magnetic flux ropes buoyantly rising through the convection zone and piercing the photosphere as "Ω-loops" (Parker 1955b). Modelling of this process in the thin flux tube approximation has allowed to identify the physical underpinning of Joy's Law, in the action of the Coriolis force on flows developing along the axis of buoyantly rising magnetic flux ropes (D'Silva and Choudhuri 1993; Fan et al. 1993; Caligari et al. 1995), and/or via the asymmetric buffeting imparted by cyclonic convection (Weber et al. 2011, 2013; see Fan 2021 for a comprehensive review).

The global dipole contribution δD associated with a BMR carrying a magnetic flux Φ with pole separation d and emerging at latitude λ (Petrovay et al. 2020) is given by:

$$\delta D = \frac{3\cos\lambda}{4\pi R^2}\Phi d \sin\gamma \ . \tag{13}$$

As BMRs decay and "release" this dipole moment, preferential cross-equatorial dissipation of the leading magnetic polarity and transport of the trailing polarity towards the poles leads to polarity reversal and subsequent buildup of a new global dipole moment. This surface transport of magnetic flux is observed at the solar surface, and has been modelled in detail (Wang et al. 1989; Jiang et al. 2014; Upton and Hathaway 2014; Lemerle et al. 2015; Whitbread et al. 2017), leaving little doubt to its role in reversing the surface dipole. Figure 3 illustrates schematically this sequence of events. Note that the tilt embodied in Joy's Law is crucial here; if BMRs emerge aligned with the E-W direction ($\gamma = 0$), then $\delta D \equiv 0$; both poles of the BMR then experience the same cross-equatorial diffusive cancellation, leaving behind no net hemispheric flux.

From end-to-end, the sequence of flux tube formation, destabilization, emergence as BMRs, surface decay and transport, thus converts a positive (negative) internal toroidal field into a positive (negative) dipole moment, in a manner analogous to a positive α-effect in mean-field electrodynamics.[4] Rotational shearing of this large-scale dipole can then regenerate the toroidal component and close the dynamo loop.

---

[4]Note that in both cases, the Coriolis force ultimately provides the break of axisymmetry needed to evade Cowling's theorem.

As with mean-field dynamos based on the turbulent α-effect, Babcock–Leighton dynamos must abide with helicity conservation. Magnetographic observations indicate that the magnetic flux ropes emerging as BMRs carry magnetic helicity in the form of internal twist about their axis (see, e.g., Seehafer 1990; López Fuentes et al. 2003), as expected if they form from a deep-seated dynamo-generated large-scale magnetic field that is itself helical. Ultimately, the large-scale twist of the flux rope itself (or writhe) associated with Joy's Law, acts as the global source of magnetic helicity in this class of dynamos. For more on these matters, see Pevtsov et al. (2014) and references therein.

Most contemporary versions of solar cycle models based on the Babcock–Leighton mechanism are formulated as flux transport dynamos, with the meridional flow carrying the surface dipole to the deep interior,[5] where rotational shearing takes place, and driving the equatorward propagation of emerging BMRs in the course of the cycle (viz. Sect. 5). It must be emphasized that to operate properly, *all* such solar dynamo models must invoke a strongly enhanced magnetic diffusivity, presumably of turbulent origin, as provided by mean-field theory. For more on such models, see Sects. 5.4 and 5.5 in Charbonneau (2020).

The polar cap (latitudes >60 degrees) magnetic flux amounts to $\sim 10^{22}$ Mx at times of peak surface dipole strength. The total unsigned magnetic flux emerging in the form of BMRs adds up to $\sim 10^{25}$ Mx in the course of a typical activity cycle. The toroidal-to-poloidal conversion efficiency of the Babcock–Leighton mechanism thus needs not be high, of order ~ 0.1 percent only. In fact it has been argued that the poloidal flux generated by the Babcock–Leighton mechanism is indeed sufficient, in conjunction with rotational shearing, to account for the emerging magnetic flux (Cameron and Schüssler 2015), although turbulent induction in the interior cannot be ruled out via this argument. Is the Babcock–Leighton mechanism then essential to the solar magnetic cycle? Answering this question on the basis of observations would require a detailed magnetic flux budget of the solar polar caps, i.e., accounting for flux emergence, submergence, transport from lower latitudes, as well as local generation.

To sum up: The Babcock–Leighton mechanism is observed operating at the solar surface, and in itself can account for the reversal of the surface dipole. Whether the surface dipole so generated feeds back into the dynamo loop, or is a mere side-effect of a deep-seated turbulent dynamo operating independently in the solar interior, remains an open question.

7 Tension: The Surface Dipole as Precursor

The surface dipole strength at activity minimum is long known to be a good precursor for the amplitude of the upcoming sunspot cycle (Schatten et al. 1978; Svalgaard et al. 2005; for reviews of solar cycle prediction schemes, see Pesnell 2016; Petrovay 2020). The dipole-as-precursor is also implemented in some dynamo model-based cycle forecasting schemes (Choudhuri et al. 2007; Jiang et al. 2007; Bhowmik and Nandy 2018). In such cases the details of the underlying flux transport dynamo model (Sect. 5) are secondary, as long as shearing by differential rotation is linear, i.e., there is no significant dynamical backreaction of the magnetic field on differential rotation, and the associated inductive shearing is not subjected to significant forcing by random fluctuations.

The good precursor value of the solar surface dipole is often matter-of-factly invoked as empirical support for the Babcock–Leighton "'picture"' of the solar dynamo, i.e., the large-

---

[5]Turbulent pumping has also been invoked as a flux transport mechanism in this context, in particular to ensure submergence of the surface magnetic field; see, e.g., Jiang et al. (2013).

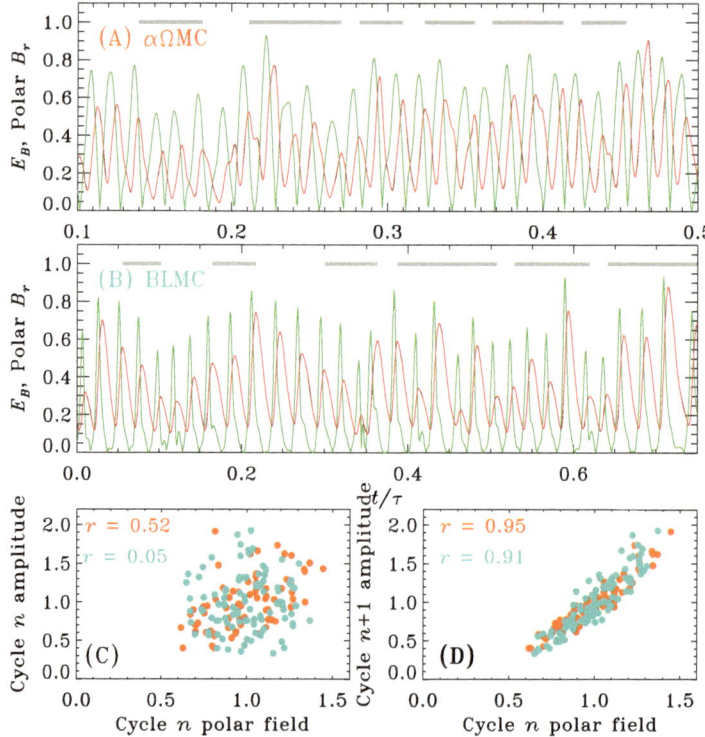

Fig. 4 Two solar cycle-like solutions in flux transport dynamo models differing only in their mechanism of poloidal field regeneration, and subjected to stochastic forcing of the latter (see text). Panel A and B are respectively an $\alpha\Omega$ and Babcock–Leighton solar cycle models, both including meridional circulation. Green lines are time series of the surface polar cap magnetic field, and red lines are time series of magnetic energy integrated over the solution domain, used here as a proxy of magnetic cycle amplitude. Neither model shows a significant correlation between cycle amplitude and dipole strength at the subsequent minimum (panel C), but both show a strong correlation between dipole strength at minimum and the amplitude of the subsequent cycle (panel D). Figure 19 in Charbonneau (2020), used with permission

scale poloidal magnetic component being regenerated by the surface decay of bipolar magnetic regions (viz. Sect. 6). Figure 4, adapted from Charbonneau and Barlet (2011), offers a specific counterexample to this claim. Panel (A) and (B) show time series of volumetric magnetic energy (red) and surface dipole (green; more precisely: Northern hemisphere polar cap unsigned magnetic field) produced by two dynamo models differing only in their poloidal source; the solution of panel (A) is a conventional mean-field $\alpha\Omega$ dynamo model, with the α-effect concentrated at the base of the convection zone, but includes a meridional flow and operates in the flux transport regime (Sect. 5). The solution of panel B is a mean-field-like Babcock–Leighton dynamo model using a non-local surface poloidal source term, as described in Dikpati and Charbonneau (1999). Both models are axisymmetic, kinematic, use the same solar-like parametrization of the solar internal differential rotation, and the quadrupolar meridional flow pattern of van Ballegooijen and Choudhuri (1988), characterized by a single flow cell per meridional quadrant spanning the full convection zone. In both cases zero-mean stochastic fluctuations are imposed on the dynamo number multiplying the poloidal source term, of amplitude corresponding to 50% of the mean and with coherence time of one month.

As shown on panel C, in either model no correlation is observed between the dipole strength at minimum and the amplitude of the cycle just ending, consistent with the fact that imposed random fluctuations affect the production of the poloidal component from the toroidal component. However, both models show a strong correlation between dipole strength and the amplitude of the subsequent cycle (panel D), as measured here via volumetric magnetic energy. In the case of the mean-field $\alpha\Omega$ model, this correlation vanishes altogether if the meridional flow is turned off and the model then operates as a classical $\alpha\Omega$ model, with equatorward propagation of activity belts driven by a dynamo wave. The surface dipole then becomes a side-effect of a dynamo operating in the deep interior, and does not feed back into the dynamo loop. Nonetheless, in the flux transport mean-field $\alpha\Omega$ model the surface dipole is as good a precursor of the next cycle as in the Babcock–Leighton model.

To sum up: for the surface dipole moment to act as a good precursor of the upcoming cycle's amplitude, two conditions must be met: (1) the primary source of fluctuation must reside in the regeneration of the large-scale poloidal magnetic component, and (2) the surface dipole must feed back into the dynamo loop. Many dynamo scenarios meet both constraints, and none can be favored over another only on the basis of the precursor value of the surface dipole.

8 Tension: Cycle Fluctuations: Stochastic or Nonlinear?

The sunspot numbers record reveals significant cycle-to-cycle variations in the amplitude, duration, shape, and hemispheric asymmetry of the solar cycle (see Hathaway 2015, and references therein). Reconstructions of solar activity based on cosmosgenic radioisotopes also reveals modulation patterns unfolding on centennial to millennial timescales (Usoskin 2023). What is the physical origin of these variability patterns?

Solar and stellar dynamos draws their energy from the kinetic energy of the participating inductive flows, through work done against Lorentz force associated with the dynamo-generated magnetic field. This is a nonlinear backreaction of the magnetic field on the flows, which most certainly is what prevents unbound growth of the cycle amplitude, and may also drive amplitude modulation on timescales longer than the cycle period.

Solar and stellar dynamos also operates in part or *in toto* in strongly turbulent convective envelopes. Turbulent convection acts as multiscale and spatiotemporally highly variable inductive component, which from the point of view of the large-scale magnetic field manifests itself as short coherence time stochastic "noise" superimposed on the mean electromotive force and induction by large-scale flows. Such stochastic noise can also cause significant cycle amplitude variability, Fig. 4 herein being a case in point.

Is stochastic forcing or deterministic nonlinear backreaction driving observed cycle fluctuations? This is a particularly complex question to answer because in the solar/stellar context, many flow components contribute to induction (and/or flux transport), and all are in principle impacted by the Lorentz force associated with the dynamo-generated large-scale magnetic field. Moreover, the magnetic field can also impact large-scale flows indirectly, via alterations of Reynolds stresses powering them (the so-called Λ-quenching; see Kitchatinov et al. 1994; Küker et al. 1996; Rempel 2006a), or via global constraints such as magnetic helicity conservation, as discussed in Sect. 3. To further complicate matters, in general the response of the dynamo to stochastic forcing depends on the nature of the nonlinearity regulating the average cycle amplitude (see, e.g., Biswas et al. 2022; Talafha et al. 2022 and references therein).

Assorted dynamo modelling work (see, e.g., Moss et al. 1992; Hoyng et al. 1994; Ossendrijver et al. 1996; Usoskin et al. 2009; Choudhuri and Karak 2012; Pipin et al. 2012; Olemskoy and Kitchatinov 2013; Hazra and Nandy 2019 for a representative subset) has amply demonstrated that in the presence of stochastic forcing, many solar-like behaviors, such as marked amplitude fluctuations, sustained mixed-parity modes, cross-equatorial activity, and intermittency, can materialize naturally in critical or very weakly supercritical dynamos. R. Cameron and M. Schüssler (2017) go further in arguing that a stochastically forced weakly supercritical dynamo is all that is needed to reproduce the observed spectral properties of solar activity, in a manner that is generic with respect to the amplitude-limiting nonlinearity. The key of their proposal is that the linear dynamo growth rate be much smaller than the cycle period —as one would indeed expect for very weakly supercritical dynamos. An ensemble average of their model runs yields a flat spectrum at low frequencies, but any single instance is characterized by spectral structure, of purely random origin (see their Fig. 3). Cosmogenic radioisotope reconstructions of solar activity on long timescale also show spectral structures at low frequencies, but also represent a single realization of a specific dynamo —the Sun! Can such reconstructions then actually prove or disprove the Cameron & Schüssler conjecture? Notwithstanding circumstantial evidence related to rotational evolution (Kitchatinov and Nepomnyashchikh 2017; Metcalfe et al. 2022), there are no *a priori* reasons to believe that the solar dynamo is only weakly supercritical. Moreover, long timescale modulation of cyclic behavior, as well as parity modulation, intermittency, etc, are also readily produced purely deterministically through various nonlinearities, notably magnetic backreaction on differential rotation (Tobias et al. 1995; Küker et al. 1999; Pipin 1999; Moss and Brooke 2000; Bushby 2006; Weiss and Tobias 2016; Simard and Charbonneau 2020; see also Sect. 4 in Charbonneau 2020). In a picture purely based on stochastic forcing, one would not expect any phase coherence in long term fluctuations. Cosmogenic reconstructions do suggest phase persistence and non-random clustering patterns for Grand Minima and Maxima (more on these further below), but even the most recent 9000 yr reconstructions (Usoskin et al. 2016; Wu et al. 2018) remain too short to yield strong statistical significance to confidently support or refute either class of explanation.

To sum up: The exact nature of the magnetic nonlinear backreaction mechanism(s) stabilizing the amplitude of the solar dynamos is not yet identified with confidence; nor are the mechanisms, whether of a stochastic or purely deterministic nature, driving apparently random cycle-to-cycle fluctuations in cycle characteristics such as amplitude and duration, as well as variability on timescale longer than the cycle period. These are absolutely fundamental gaps in our (lack of) understanding of solar and stellar dynamos.

9 Tension: Explaining Grand Minima

The most extreme pattern of solar fluctuation is arguably the Grand Minima of solar activity, epochs during which activity falls to very low levels over a time period much longer than the magnetic cycle period. First noted in the sunspot record independently by G. Spörer and E. W. Maunder in the late nineteenth century, and much later rediscovered (Eddy 1976), the 1645–1715 Maunder Minimum has become the archetype of such events, which have been found to recur aperiodically in the cosmogenic radioisototope record spanning the Holocene (see Usoskin et al. 2015, and references therein). Relatively recent similar events include the Spörer Minimum (ca. 1416–1534) and the Wolf Minimum (ca. 1282–1342). Other periods of sustained higher-than-average activity, "Grand Maxima", can also be identified (Usoskin 2023).

The Maunder Minimum remains unique in that it is the only Grand Minimum for which direct observations of sunspots are available. Extensive historical analyses have revealed that the Sun was not entirely devoid of sunspots during the Maunder Minimum, but that the few sunspots observed were almost all located in the Southern solar hemisphere (Ribes and Nesme-Ribes 1993; Hoyt and Schatten 1996; Usoskin et al. 2015); for a comprehensive review of historical sunspot observations, see Arlt and Vaquero (2020). Another intriguing pattern relates to a possible change in the surface latitudinal rotation, as revealed from analyses of sunspot drawings made before, during and after the Maunder Minimum (Eddy et al. 1977; Casas et al. 2006).

Dearth in production of sunspot does not necessarily mean a halt in cyclic regeneration of the solar large-scale magnetic field. The buoyant destabilization of magnetic flux rope formed in the deep solar interior (presumably), with subsequent rise through the convective envelope and emergence as bipolar magnetic regions, almost certainly involves a threshold in magnetic intensity (Schüssler et al. 1994). Consequently, the magnetic cycle may well continue unabated through Grand Minima, without reaching a magnetic amplitude sufficiently high to produce sunspots. Under this view, Grand Minima simply represent the low point of a large amplitude modulation. It is noteworthy that determinisitc, nonlinearly-driven backreaction on large-scale inductive flows (Sect. 8) can be accompanied by modulation of both flow and magnetic equatorial parity unfolding on long timescales, thus offering an attractive explanation for the strong hemispheric asymmetry of sunspot locations during the Maunder Minimum (Sokoloff and Nesme-Ribes 1994; Beer et al. 1998), as well as any variation in differential rotation. For a sample of dynamo models exhibiting this type of nonlinear modulation, see Küker et al. (1999), Bushby (2006), Weiss and Tobias (2016).

A distinct class of explanations for Grand Minima invokes *intermittency* (Platt et al. 1993), namely a transition between two distinct dynamo modes, the weaker one noncyclic and/or of very low magnetic amplitude. The switch between modes can be driven either stochastically or deterministically. For a sample of dynamo models generating Grand Minima-like episodes in this manner, see Ossendrijver (2000), Charbonneau et al. (2004), Moss et al. (2008), Usoskin et al. (2009), Choudhuri and Karak (2012), Olemskoy and Kitchatinov (2013), Inceoglu et al. (2017), Albert et al. (2021). Explaining Grand Minima via intermittency does pose a problem for dynamo models which are not self-excited, in the sense of being subjected to a lower operating threshold, e.g., dynamos relying on the Babcock–Leighton mechanism (Sect. 6). Jump-starting the primary dynamo out of a Grand Minimum then requires either a secondary self-excited dynamo (Passos et al. 2014; Hazra et al. 2014b; Ölçek et al. 2019; Saha et al. 2022), or some other source of magnetic fields acting as "magnetic noise" (see, e.g., Schmitt et al. 1996; Ossendrijver 2000; Charbonneau et al. 2004; Tripathi et al. 2021).

The observations of low amplitude cyclic activity in the ^{10}Be isotope record has been presented as evidence that the solar cycle was still running throughout the Maunder Minimum (Beer et al. 1998), the interplanetary magnetic field being a complex sampling of both active region magnetic fields as well as the global dipole. A turbulent $\alpha\Omega$ dynamo may well continue to reverse polarity, while failing to reach a magnetic amplitude sufficient to generate large emerging BMRs and associated sunspots. This type of behavior does materialize more naturally in dynamos undergoing large amplitude modulation through nonlinear backreaction by the Lorentz force. However, in some flux transport models undergoing intermittency, cyclic variations of the surface field can also take place as the meridional flow entrains the residual magnetic field (see Charbonneau et al. 2004; Saha et al. 2022 for specific examples). The persistence of cycle *phase* through Grand Minima is a potentially powerful discriminant, but quite challenging to harness in practice.

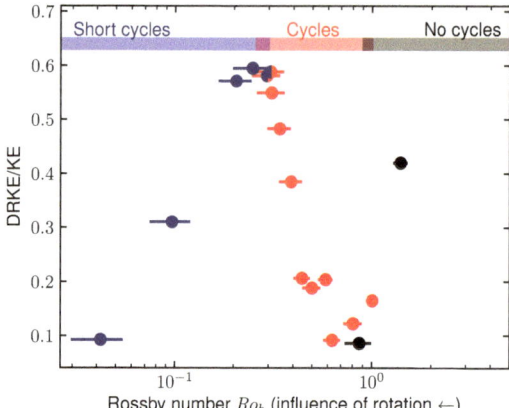

Fig. 5 A synthetic summary of a set of simulations from Strugarek et al. (2018), showing the ratio of kinetic energy contained in differential rotation to total kinetic energy, versus Rossby number. Deep seated, solar-like decadal magnetic cycles are plotted in red. The blue points refer to short (period of a few years) magnetic cycles unfolding in the upper third of the convecting fluid layers, while black point indicate simulations generating steady large-scale magnetic fields. Figure 2 in Strugarek et al. (2018), reproduced with permission, copyright by AAS

To sum up: A wide variety of potentially viable dynamo-based scenarios for Grand Minima have been proposed, but which (if any) actually applies to the Sun remains an open question.

10 Tension: Sensitive Cycles in MHD Simulations

For now more than a decade, many global magnetohydrodynamical simulations of solar convection have achieved the production of large-scale magnetic fields undergoing more or less regular polarity reversals, analogous to some extent to the solar magnetic cycle. For a representative sample of such simulations, see, e.g., Ghizaru et al. (2010), Masada et al. (2013), Nelson et al. (2013), Fan and Fang (2014), Mabuchi et al. (2015), Simitev et al. (2015), Hotta et al. (2016), Käpylä et al. (2017), Strugarek et al. (2018), Guerrero et al. (2019), Hotta et al. (2022). These simulations rely on markedly distinct computational approaches to the numerical solution of the MHD equations, in particular in the treatment of unresolved scales. While they often generate similar convective and large-scale flow patterns, the large-scale magnetic cycle they produce vary widely in their spatiotemporal evolution (see Sect. 3.2 in Charbonneau 2014 for a specific comparison). The origin of this "structural fragility" is multi-faceted and remains ill-understood.

Figure 5 summarizes cycle properties in a series of global MHD simulations from Strugarek et al. (2018), collectively spanning a factor of 10 in rotation rate and 3 in convective luminosity. The ratio of kinetic energy contained in differential rotation (DRKE) to total kinetic energy (KE) is plotted against Rossby number, with symbols colored according to the character of the large-scale magnetic cycle materializing in each simulation.

The sensitivity on the Rossby number is extreme; starting at $Ro \simeq 0.1$, increasing Ro by a mere a factor of 10 takes one from rapid subsurface cycles, through deep-seated decadal cycles in the range $0.3 \lesssim Ro \lesssim 1$, and on to steady large-scale magnetic field at $Ro \gtrsim 1$. This extreme sensitivity turns out to be robust, in the sense that it materializes in MHD simulations using entirely distinct numerical implementations and overall modelling frameworks;

consider the striking resemblance between Fig. 13 in Brun et al. (2022), working with the ASH LES code (Brun et al. 2004), with Fig. 5 herein, from Strugarek et al. (2018) using the EULAG-MHD ILES code (Smolarkiewicz and Charbonneau 2013).

There is much more to the sensitivity issue than the Rossby number, however. The large-scale flow and magnetic field emerging in MHD numerical simulations of solar convection are strongly influenced by the turbulent regime attained. At high Reynolds number, stresses associated with the turbulent magnetic field can have a strong impact on large-scale flows (Fan and Fang 2014; Hotta and Kusano 2021; Brun et al. 2022; Hotta et al. 2022), thus indirectly affecting global dynamo action. Likewise, the relative importance of the many dissipation channels available to the system, as measured by the viscous and magnetic Prandtl numbers, also influences significantly the large-scale flows and dynamo action (Käpylä et al. 2017; Tobias 2021).

Not surprisingly, the numerical treatment of small scales also impacts turbulent induction. Mean-field analyses of large-scale magnetic cycles in MHD simulations typically yield α-tensors that are full, with turbulent induction often contributing on par with large-scale shearing (Augustson et al. 2015; Simard et al. 2016; Warnecke et al. 2018; Viviani et al. 2019). In mean-field parlance, these simulations operate as $\alpha^2\Omega$ dynamos, or even α^2 dynamos, if turbulent induction dominates over shearing by differential rotation. Detailed analyses of various simulations also reveal that small-scale and large-scale inductive contribution sometimes counteract each other (Racine et al. 2011; Nelson et al. 2013; Brun et al. 2022; Shimada et al. 2022), which results in the total induction having a magnitude significantly smaller than its individual contributions. Relatively small changes in one contribution, for example in small-scale induction versus dissipation when distinct subgrid models are used or when the magnetic Prandtl number is varied, can have a much larger impact on total induction, and thus on the unfolding of the large-scale magnetic cycle.

Another complication arises in MHD simulations reaching very high Reynolds numbers (Hotta et al. 2016), namely small-scale dynamo action. Producing a strong small-scale magnetic field in this manner turns out to suppress the small-scale turbulent flow otherwise responsible for the turbulent diffusion of the large-scale magnetic component (Shimada et al. 2022). Somewhat counterintuitively, more strongly turbulent simulations end up sustaining large-scale magnetic fields better than less turbulent —and presumably less diffusive— simulations. Here again the computational treatment of subgrid effects can have a large impact, and the key to stable global magnetic cycles appears to be the minimization of dissipation at the larger scales (Strugarek et al. 2016, 2018).

Finally, the combination of multiple turbulent inductive contributions and relatively complex internal differential rotation profiles with distinct shearing regions can lead to the co-existence of multiple dynamo modes, spatially segregated but nonetheless interfering with one another, generating variability and modulation of magnetic cycle unfolding in MHD simulations (Beaudoin et al. 2016; Käpylä et al. 2016; Strugarek et al. 2018; Viviani et al. 2019). Interference between distinct dynamo modes can also yield occasional episodes of strongly reduced cyclic activity (Augustson et al. 2015; Käpylä et al. 2016), somewhat reminiscent of Grand Minima (viz. §9).

To sum up: Global MHD numerical simulations of solar convection can generate large-scale magnetic fields undergoing more or less regular polarity reversals. The unfolding of these magnetic cycles seems however far more sensitive to modelling details and physical parameter regimes that suggested by observations of magnetism and cycles in solar-type stars. The physical origin of this sensitivity remains an open question.

11 Tension(s): From Solar to Stellar Dynamos

The sun is but a star, yet its ease of observation makes it an essential springboard towards interpreting magnetic activity and cycles observed in other stars as resulting from the operation of dynamos. This is an immense topic, and we will close this review by only highlighting a few key points.

As revealed by the epoch-making Mt Wilson Ca H+K survey (Baliunas et al. 1995), all solar-type stars show evidence of magnetic activity, and a subset shows fairly well-defined cyclic variability on decadal timescales, presumably reflecting the presence of a dynamo-powered large-scale magnetic cycle analogous to the solar cycle. Observation of X-Ray emission, a tracer of coronal activity powered by magnetism, show a well-defined variation with the Rossby number inferred from the observed luminosity via mixing length theory. Particularly noteworthy is the fact that this trend is the same for solar-type stars (meaning, G and K dwarfs having a radiative core and overlying convective envelope), fully convective M dwarfs (Wright et al. 2018), and even evolved subgiants and giants (Lehtinen et al. 2020). This suggests —without strictly proving— convective turbulence-universality in the dynamo mechanism underlying stellar magnetic activity at large.

Again in late-type main-sequence stars, measurements of surface magnetism also show a fairly well-defined trend with Rossby number, also largely independent of spectral type (Reiners et al. 2022). This points again to a certain level of universality in stellar dynamo, which is not obvious to reconcile with the many detailed dynamo scenarios designed specifically to match solar observations (but do see Shulyak et al. 2015).

Even if complete knowledge of the solar dynamo were at hand —and it is not, as pointed out throughout this review,— going to the stars (even "only" to late-type main-sequence stars) demands answers to a number of crucial questions, including minimally:

1. Which is the primary polarity reversal mechanism: α-effect, or Babcock–Leighton, ... or something else?
2. How do differential rotation and meridional circulation vary with rotation rate, luminosity, and internal structure?
3. How do turbulent coefficients (α-effect, turbulent pumping, turbulent diffusion) vary with rotation rate, luminosity, and internal structure?
4. How do sunspots and BMRs form and decay in stars of varying structure (in particularly, depth of convective envelope), rotation rate and luminosity?

Harnessing knowledge acquired via solar dynamo modelling and MHD numerical simulations can in principle address many of these questions; for a thorough review see Brun and Browning (2017). The variation of differential rotation and meridional circulation is accessible via both semi-analytical turbulence models (e.g., Kitchatinov and Rüdiger 1993) and numerical simulations (Brun et al. 2022, and references therein), and there is now general agreement that (latitudinal) differential rotation does not vary much with rotation rate once in the solar-like (rapidly rotating equator) regime.

The behavior of large-scale magnetic cycles, on the other hand, shows greater disagreement between observations and theory; as a single example, consider the variation of the cycle period with Rossby number: the original dynamo analysis of Mt Wilson data by Noyes et al. (1984) suggested $P_{cyc}/P_{rot} \propto Ro^{0.25}$, while the distinct numerical simulations of Strugarek et al. (2018) and Warnecke (2018) indicate $P_{cyc}/P_{rot} \propto Ro^{-1.6}$ and $Ro^{-1.8}$, respectively. Reliably estimating the Rossby number is far from trivial, either from stellar data or from numerical simulations (see, e.g., Brun et al. 2022), but the fact that these two trends run in opposite direction is worth reflecting upon, to say the least.

To sum up: The rapidly growing body of high-quality observations of stellar activity and magnetism begs for the design of a unifying dynamo framework applicable to both the sun and solar type stars of varying spectral type, luminosity, and rotation rate. Which are the key elements on which to build such a framework remains an open question.

Acknowledgements This review was written following the workshop "Solar and Stellar Dynamos: A New Era", hosted and supported by the International Space Science Institute (ISSI) in Bern, Switzerland. The author wish to express their thanks to ISSI for their financial and logistical support.

Author Contribution Both authors contributed equally to this work.

Declarations

Competing Interests The authors have no relevant financial or non-financial interests to disclose.

References

Albert C, Ferriz-Mas A, Gaia F, Ulzega S (2021) Can stochastic resonance explain recurrence of grand minima? Astrophys J Lett 916(2):L9

Arlt R, Vaquero JM (2020) Historical sunspot records. Living Rev Sol Phys 17:1

Arnol'd VI, Khesin BA (1992) Topological methods in hydrodynamics. Annu Rev Fluid Mech 24:145–166

Augustson K, Brun AS, Miesch M, Toomre J (2015) Grand minima and equatorward propagation in a cycling stellar convective dynamo. Astrophys J 809:149

Babcock HW (1961) The topology of the Sun's magnetic field and the 22-year cycle. Astrophys J 133:572–589

Balbus SA, Latter H, Weiss N (2012) Global model of differential rotation in the Sun. Mon Not R Astron Soc 420(3):2457–2466

Baliunas SL, Donahue RA, Soon WH, Horne JH, Frazer J, Woodard-Eklund L, Bradford M, Rao LM, Wilson OC, Zhang Q, Bennett W, Briggs J, Carroll SM, Duncan DK, Figueroa D, Lanning HH, Misch T, Mueller J, Noyes RW, Poppe D, Porter AC, Robinson CR, Russell J, Shelton JC, Soyumer T, Vaughan AH, Whitney JH (1995) Chromospheric variations in main-sequence stars. II. Astrophys J 438:269

Beaudoin P, Simard C, Cossette J-F, Charbonneau P (2016) Double dynamo signatures in a global MHD simulation and mean-field dynamos. Astrophys J 826:138

Beer J, Tobias S, Weiss N (1998) An active Sun throughout the Maunder minimum. Sol Phys 181:237–249

Belucz B, Dikpati M, Forgács-Dajka E (2015) A Babcock–Leighton solar dynamo model with multi-cellular meridional circulation in advection- and diffusion-dominated regimes. Astrophys J 806(2):169

Berger MA (1999) Introduction to magnetic helicity. Plasma Phys Control Fusion 41(12B):B167–B175

Bhowmik P, Nandy D (2018) Prediction of the strength and timing of sunspot cycle 25 reveal decadal-scale space environmental conditions. Nat Commun 9:5209

Bieber JW, Rust DM (1995) The escape of magnetic flux from the Sun. Astrophys J 453:911

Biswas A, Karak BB, Cameron R (2022) Toroidal flux loss due to flux emergence explains why solar cycles rise differently but decay in a similar way. Phys Rev Lett 129(24):241102

Blackman EG, Brandenburg A (2002) Dynamic nonlinearity in large-scale dynamos with shear. Astrophys J 579(1):359–373

Brandenburg A (2001) The inverse cascade and nonlinear alpha-effect in simulations of isotropic helical hydromagnetic turbulence. Astrophys J 550(2):824–840

Brandenburg A, Dobler W (2001) Large scale dynamos with helicity loss through boundaries. Astron Astrophys 369:329–338

Brandenburg A, Sokoloff D (2002) Local and nonlocal magnetic diffusion and alpha-effect tensors in shear flow turbulence. Geophys Astrophys Fluid Dyn 96(4):319–344

Brandenburg A, Subramanian K (2005) Astrophysical magnetic fields and nonlinear dynamo theory. Phys Rep 417:1–209

Brandenburg A, Candelaresi S, Chatterjee P (2009) Small-scale magnetic helicity losses from a mean-field dynamo. Mon Not R Astron Soc 398:1414–1422

Brown TM, Christensen-Dalsgaard J, Dziembowski WA, Goode P, Gough DO, Morrow CA (1989) Inferring the Sun's internal angular velocity from observed p-mode frequency splittings. Astrophys J 343:526–546

Brun AS, Browning MK (2017) Magnetism, dynamo action and the solar-stellar connection. Living Rev Sol Phys 14:4

Brun AS, Miesch MS, Toomre J (2004) Global-scale turbulent convection and magnetic dynamo action in the solar envelope. Astrophys J 614:1073–1098

Brun AS, Strugarek A, Noraz Q, Perri B, Varela J, Augustson K, Charbonneau P, Toomre J (2022) Powering stellar magnetism: energy transfers in cyclic dynamos of Sun-like stars. Astrophys J 926(1):21

Bushby PJ (2006) Zonal flows and grand minima in a solar dynamo model. Mon Not R Astron Soc 371:772–780

Caligari P, Moreno-Insertis F, Schüssler M (1995) Emerging flux tubes in the solar convection zone. I. Asymmetry, tilt, and emergence latitudes. Astrophys J 441:886–902

Cameron RH, Schüssler M (2010) Changes of the solar meridional velocity profile during cycle 23 explained by flows toward the activity belts. Astrophys J 720:1030–1032

Cameron R, Schüssler M (2015) The crucial role of surface magnetic fields for the solar dynamo. Science 347:1333–1335

Cameron RH, Schüssler M (2017) Understanding solar cycle variability. Astrophys J 843:111

Casas R, Vaquero JM, Vazquez M (2006) Solar rotation in the 17th century. Sol Phys 234(2):379–392

Charbonneau P (2007) Babcock–Leighton models of the solar cycle: questions and issues. Adv Space Res 39(11):1661–1669

Charbonneau P (2014) Solar dynamo theory. Annu Rev Astron Astrophys 52:251–290

Charbonneau P (2020) Dynamo models of the solar cycle. Living Rev Sol Phys 17:4

Charbonneau P, Barlet G (2011) The dynamo basis of solar cycle precursor schemes. J Atmos Sol-Terr Phys 73:198–206

Charbonneau P, Blais-Laurier G, St-Jean C (2004) Intermittency and phase persistence in a Babcock–Leighton model of the solar cycle. Astrophys J Lett 616:L183–L186

Choudhuri AR, Karak BB (2012) Origin of grand minima in sunspot cycles. Phys Rev Lett 109(17):171103

Choudhuri AR, Schüssler M, Dikpati M (1995) The solar dynamo with meridional circulation. Astron Astrophys 303:L29

Choudhuri AR, Chatterjee P, Jiang J (2007) Predicting solar cycle 24 with a solar dynamo model. Phys Rev Lett 98(13):131103

Cowling TG (1933) The magnetic field of sunspots. Mon Not R Astron Soc 94:39–48

Dikpati M, Charbonneau P (1999) A Babcock–Leighton flux transport dynamo with solar-like differential rotation. Astrophys J 518:508–520

Dikpati M, Gilman PA (2009) Flux-transport solar dynamos. Space Sci Rev 144(1–4):67–75

D'Silva S, Choudhuri AR (1993) A theoretical model for tilts of bipolar magnetic regions. Astron Astrophys 272:621–633

Durney BR (1995) On a Babcock–Leighton dynamo model with a deep-seated generating layer for the toroidal magnetic field. Sol Phys 160:213–235

Dziembowski WA, Goode PR, Libbrecht KG (1989) The radial gradient in the Sun's rotation. Astrophys J Lett 337:L53

Eddy JA (1976) The Maunder minimum. Science 192(4245):1189–1202

Eddy JA, Gilman PA, Trotter DE (1977) Anomalous solar rotation in the early 17th century. Science 198(4319):824–829

Fan Y (2021) Magnetic fields in the solar convection zone. Living Rev Sol Phys 18:5

Fan Y, Fang F (2014) A simulation of convective dynamo in the solar convective envelope: maintenance of the solar-like differential rotation and emerging flux. Astrophys J 789:35

Fan Y, Fisher GH, Deluca EE (1993) The origin of morphological asymmetries in bipolar active regions. Astrophys J 405:390–401

Featherstone NA, Miesch MS (2015) Meridional circulation in solar and stellar convection zones. Astrophys J 804:67

Gastine T, Yadav RK, Morin J, Reiners A, Wicht J (2014) From solar-like to antisolar differential rotation in cool stars. Mon Not R Astron Soc 438:L76–L80

22    Springer

Ghizaru M, Charbonneau P, Smolarkiewicz PK (2010) Magnetic cycles in global large eddy simulations of solar convection. Astrophys J Lett 715:L133

Gizon L, Cameron RH, Pourabdian M, Liang Z-C, Fournier D, Birch AC, Hanson CS (2020) Meridional flow in the Sun's convection zone is a single cell in each hemisphere. Science 368(6498):1469–1472

Glatzmaier GA, Roberts PH (1995) A three-dimensional convective dynamo solution with rotating and finitely conducting inner core and mantle. Phys Earth Planet Inter 91(1):63–75

Gough DO, Kosovichev AG, Toomre J, Anderson E, Antia HM, Basu S, Chaboyer B, Chitre SM, Christensen-Dalsgaard J, Dziembowski WA, Eff-Darwich A, Elliott JR, Giles PM, Goode PR, Guzik JA, Harvey JW, Hill F, Leibacher JW, Monteiro MJPFG, Richard O, Sekii T, Shibahashi H, Takata M, Thompson MJ, Vauclair S, Vorontsov SV (1996) The seismic structure of the Sun. Science 272(5266):1296–1300

Green LM, López Fuentes MC, Mandrini CH, van Driel-Gesztelyi L, Démoulin P (2003) Active region helicity evolution and related coronal mass ejection activity. Adv Space Res 32(10):1959–1964

Guerrero G, de Gouveia Dal Pino EM (2008) Turbulent magnetic pumping in a Babcock–Leighton solar dynamo model. Astron Astrophys 485(1):267–273

Guerrero G, Zaire B, Smolarkiewicz PK, de Gouveia Dal Pino EM, Kosovichev AG, Mansour NN (2019) What sets the magnetic field strength and cycle period in solar-type stars? Astrophys J 880(1):6

Hale GE, Ellerman F, Nicholson SB, Joy AH (1919) The magnetic polarity of Sun-spots. Astrophys J 49:153

Hathaway DH (1996) Doppler measurements of the Sun's meridional flow. Astrophys J 460:1027–1033

Hathaway DH (2015) The solar cycle. Living Rev Sol Phys 12:4

Hazra S, Nandy D (2016) A proposed paradigm for solar cycle dynamics mediated via turbulent pumping of magnetic flux in babcock-leighton-type solar dynamos. Astrophys J 832(1):9

Hazra S, Nandy D (2019) The origin of parity changes in the solar cycle. Mon Not R Astron Soc 489(3):4329–4337

Hazra G, Karak BB, Choudhuri AR (2014a) Is a deep one-cell meridional circulation essential for the flux transport solar dynamo? Astrophys J 782:93

Hazra S, Passos D, Nandy D (2014b) A stochastically forced time delay solar dynamo model: self-consistent recovery from a Maunder-like grand minimum necessitates a mean-field alpha effect. Astrophys J 789:5

Helmbold B (2017) Wissenschaft und Politik im Leben von Max Steenbeck (1904-1981). Springer, Berlin

Hotta H, Kusano K (2021) Solar differential rotation reproduced with high-resolution simulation. Nat Astron 5:1100–1102

Hotta H, Rempel M, Yokoyama T (2016) Large-scale magnetic fields at high Reynolds numbers in magneto-hydrodynamic simulations. Science 351:1427–1430

Hotta H, Kusano K, Shimada R (2022) Generation of solar-like differential rotation. Astrophys J 933(2):199

Hoyng P, Schmitt D, Teuben LJW (1994) The effect of random alpha-fluctuations and the global properties of the solar magnetic field. Astron Astrophys 289:265–278

Hoyt DV, Schatten KH (1996) How well was the sun observed during the Maunder minimum? Sol Phys 165:181–192

Inceoglu F, Arlt R, Rempel M (2017) The nature of grand minima and maxima from fully nonlinear flux transport dynamos. Astrophys J 848(2):93

Jackiewicz J, Serebryanskiy A, Kholikov S (2015) Meridional flow in the solar convection zone. II. Helio-seismic inversions of GONG data. Astrophys J 805:133

Jennings RL, Weiss NO (1991) Symmetry breaking in stellar dynamos. Mon Not R Astron Soc 252:249–260

Jiang J, Chatterjee P, Choudhuri AR (2007) Solar activity forecast with a dynamo model. Mon Not R Astron Soc 381:1527–1542

Jiang J, Cameron RH, Schmitt D, Işık E (2013) Modeling solar cycles 15 to 21 using a flux transport dynamo. Astron Astrophys 553:A128

Jiang J, Hathaway DH, Cameron RH, Solanki SK, Gizon L, Upton L (2014) Magnetic flux transport at the solar surface. Space Sci Rev 186:491–523

Jouve L, Brun AS (2007) On the role of meridional flows in flux transport dynamo models. Astron Astrophys 474:239–250

Käpylä MJ, Käpylä PJ, Olspert N, Brandenburg A, Warnecke J, Karak BB, Pelt J (2016) Multiple dynamo modes as a mechanism for long-term solar activity variations. Astron Astrophys 589:A56

Käpylä PJ, Käpylä MJ, Olspert N, Warnecke J, Brandenburg A (2017) Convection-driven spherical shell dynamos at varying Prandtl numbers. Astron Astrophys 599:A4

Karak BB, Jiang J, Miesch MS, Charbonneau P, Choudhuri AR (2014) Flux transport dynamos: from kinematics to dynamics. Space Sci Rev 186(1–4):561–602

Karak BB, Käpylä PJ, Käpylä MJ, Brandenburg A, Olspert N, Pelt J (2015) Magnetically controlled stellar differential rotation near the transition from solar to anti-solar profiles. Astron Astrophys 576:A26

Kippenhahn R (1963) Differential rotation in stars with convective envelopes. Astrophys J 137:664

Kitchatinov L, Nepomnyashchikh A (2017) How supercritical are stellar dynamos, or why do old main-sequence dwarfs not obey gyrochronology? Mon Not R Astron Soc 470(3):3124–3130

Kitchatinov LL, Olemskoy SV (2011) Differential rotation of main-sequence dwarfs and its dynamo efficiency. Mon Not R Astron Soc 411(2):1059–1066

Kitchatinov LL, Rüdiger G (1993) λ-effect and differential rotation in stellar convection zones. Astron Astrophys 276:96–102

Kitchatinov LL, Rüdiger G, Küker M (1994) λ-quenching as the nonlinearity in stellar-turbulence dynamos. Astron Astrophys 292:125–132

Kleeorin N, Rogachevskii I, Ruzmaikin A (1995) Magnitude of the dynamo-generated magnetic field in solar-type convective zones. Astron Astrophys 297:159–167

Krause F, Rädler K-H (1980) Mean-field magnetohydrodynamics and dynamo theory. Pergamon Press, Oxford

Krivodubskij VN (2006) Dynamo-parameters of the convection zone of the Sun. Kinemat Phys Celest Bodies 22:1–20

Küker M, Rüdiger G, Pipin VV (1996) Solar torsional oscillations due to the magnetic quenching of the Reynolds stress. Astron Astrophys 312:615–623

Küker M, Arlt R, Rüdiger G (1999) The Maunder minimum as due to magnetic λ-quenching. Astron Astrophys 343:977–982

Küker M, Rüdiger G, Schulz M (2001) Circulation-dominated solar shell dynamo models with positive alpha effect. Astron Astrophys 374:301–308

Larmor J (1919) How could a rotating body such as the Sun become magnetic. Rep Brit Assoc Adv Sci, 159–160

Lathrop DP, Forest CB (2011) Magnetic dynamos in the lab. Phys Today 64(7):40

Lee TD, Yang CN (1956) Question of parity conservation in weak interactions. Phys Rev 104(1):254–258

Lehtinen JJ, Spada F, Käpylä MJ, Olspert N, Käpylä PJ (2020) Common dynamo scaling in slowly rotating young and evolved stars. Nat Astron 4:658–662

Leighton RB (1964) Transport of magnetic fields on the Sun. Astrophys J 140:1547–1562

Leighton RB (1969) A magneto-kinematic model of the solar cycle. Astrophys J 156:1–26

Lemerle A, Charbonneau P, Carignan-Dugas A (2015) A coupled 2×2D Babcock–Leighton solar dynamo model. I. Surface magnetic flux evolution. Astrophys J 810:78

López Fuentes MC, Démoulin P, Mandrini CH, Pevtsov AA, van Driel-Gesztelyi L (2003) Magnetic twist and writhe of active regions. On the origin of deformed flux tubes. Astron Astrophys 397:305–318

Low BC (2001) Coronal mass ejections, magnetic flux ropes, and solar magnetism. J Geophys Res 106(A11):25141–25164

Lynch BJ, Gruesbeck JR, Zurbuchen TH, Antiochos SK (2005) Solar cycle-dependent helicity transport by magnetic clouds. J Geophys Res Space Phys 110(A8):A08107

Mabuchi J, Masada Y, Kageyama A (2015) Differential rotation in magnetized and non-magnetized stars. Astrophys J 806:10

Masada Y, Yamada K, Kageyama A (2013) Effects of penetrative convection on solar dynamo. Astrophys J 778:11

McClintock BH, Norton AA (2013) Recovering Joy's law as a function of solar cycle, hemisphere, and longitude. Sol Phys 287:215–227

Metcalfe TS, Finley AJ, Kochukhov O, See V, Ayres TR, Stassun KG, van Saders JL, Clark CA, Godoy-Rivera D, Ilyin IV, Pinsonneault MH, Strassmeier KG, Petit P (2022) The origin of weakened magnetic braking in old solar analogs. Astrophys J Lett 933(1):L17

Miesch MS, Toomre J (2009) Turbulence, magnetism, and shear in stellar interiors. Annu Rev Fluid Mech 41:317–345

Moffatt HK (1978) Magnetic field generation in electrically conducting fluids. Cambridge monographs on mechanics and applied mathematics. Cambridge University Press, Cambridge

Moffatt KH, Dormy E (2019) Self-exciting fluid dynamos. Cambridge texts in applied mathematics. Cambridge University Press, Cambridge

Moss D, Brooke JM (2000) Towards a model of the solar dynamo. Mon Not R Astron Soc 315:521–533

Moss D, Brandenburg A, Tavakol R, Tuominen I (1992) Stochastic effects in mean-field dynamos. Astron Astrophys 265:843–849

Moss D, Sokoloff D, Usoskin I, Tutubalin V (2008) Solar grand minima and random fluctuations in dynamo parameters. Sol Phys 250:221–234

Nandy D, Choudhuri AR (2001) Toward a mean-field formulation of the Babcock–Leighton type solar dynamo. I. α-coefficient versus durney's double-ring approach. Astrophys J 551:576–585

Nelson NJ, Brown BP, Brun AS, Miesch MS, Toomre J (2013) Magnetic wreaths and cycles in convective dynamos. Astrophys J 762:73

Noyes RW, Weiss NO, Vaughan AH (1984) The relation between stellar rotation rate and activity cycle periods. Astrophys J 287:769–773

Ölçek D, Charbonneau P, Lemerle A, Longpré G, Boileau F (2019) Grand activity minima and maxima via dual dynamos. Sol Phys 294(7):99

Olemskoy SV, Grand LL (2013) Grand minima and north-south asymmetry of solar activity. Astrophys J 777:71

Ossendrijver MAJH (2000) Grand minima in a buoyancy-driven solar dynamo. Astron Astrophys 359:364–372

Ossendrijver M (2003) The solar dynamo. Astron Astrophys Rev 11:287–367

Ossendrijver AJH, Hoyng P, Schmitt D (1996) Stochastic excitation and memory of the solar dynamo. Astron Astrophys 313:938–948

Ossendrijver MAJH, Stix M, Brandenburg A, Rüdiger G (2002) Magnetoconvection and dynamo coefficients. II. Field-direction dependent pumping of magnetic field. Astron Astrophys 394:735–745

Parker EN (1955a) Hydromagnetic dynamo models. Astrophys J 122:293–314

Parker EN (1955b) The formation of sunspots from the solar toroidal field. Astrophys J 121:491

Parker EN (1970) The generation of magnetic fields in astrophysical bodies. I. The dynamo equations. Astrophys J 162:665

Passos D, Nandy D, Hazra S, Lopes I (2014) A solar dynamo model driven by mean-field alpha and Babcock–Leighton sources: fluctuations, grand-minima-maxima, and hemispheric asymmetry in sunspot cycles. Astron Astrophys 563:A18

Pesnell WD (2016) Predictions of solar cycle 24: how are we doing? Space Weather 14(1):10–21

Petrovay K (2020) Solar cycle prediction. Living Rev Sol Phys 17:2

Petrovay K, Nagy M, Yeates AR (2020) Towards an algebraic method of solar cycle prediction. I. Calculating the ultimate dipole contributions of individual active regions. J Space Weather Space Clim 10:50

Pevtsov AA, Berger MA, Nindos A, Norton AA, van Driel-Gesztelyi L (2014) Magnetic helicity, tilt, and twist. Space Sci Rev 186:285–324

Pipin VV (1999) The gleissberg cycle by a nonlinear $\alpha\lambda$ dynamo. Astron Astrophys 346:295–302

Pipin VV, Kosovichev AG (2011) Mean-field solar dynamo models with a strong meridional flow at the bottom of the convection zone. Astrophys J 738:104

Pipin VV, Kosovichev AG (2013) The mean-field solar dynamo with a double cell meridional circulation pattern. Astrophys J 776(1):36

Pipin VV, Sokoloff DD, Usoskin IG (2012) Variations of the solar cycle profile in a solar dynamo with fluctuating dynamo governing parameters. Astron Astrophys 542:A26

Platt N, Spiegel EA, Tresser C (1993) On-off intermittency: a mechanism for bursting. Phys Rev Lett 70:279–282

Ponomarenko YB (1973) Theory of the hydromagnetic generator. J Appl Mech Tech Phys 14(6):775–778

Pouquet A, Frish U, Leorat J (1976) Strong mhd helical turbulence and the nonlinear dynamo effect. J Fluid Mech 77:321–354

Racine É, Charbonneau P, Ghizaru M, Bouchat A, Smolarkiewicz PK (2011) On the mode of dynamo action in a global large-eddy simulation of solar convection. Astrophys J 735:46

Rajaguru SP, Antia HM (2015) Meridional circulation in the solar convection zone: time-distance helioseismic inferences from four years of HMI/SDO observations. Astrophys J 813:114

Reiners A, Shulyak D, Käpylä PJ, Ribas I, Nagel E, Zechmeister M, Caballero JA, Shan Y, Fuhrmeister B, Quirrenbach A, Amado PJ, Montes D, Jeffers SV, Azzaro M, Béjar VJS, Chaturvedi P, Henning T, Kürster M, Pallé E (2022) Magnetism, rotation, and nonthermal emission in cool stars. Average magnetic field measurements in 292 M dwarfs. Astron Astrophys 662:A41

Rempel M (2006a) Flux-transport dynamos with Lorentz force feedback on differential rotation and meridional flow: saturation mechanism and torsional oscillations. Astrophys J 647:662–675

Rempel M (2006b) Transport of toroidal magnetic field by the meridional flow at the base of the solar convection zone. Astrophys J 637:1135–1142

Rempel M (2009) Creation and destruction of magnetic field. In: Schrijver CJ, Siscoe GL (eds) Heliophysics: plasma physics of the local cosmos. Cambridge University Press, Cambridge, pp 42–76

Ribes JC, Nesme-Ribes E (1993) The solar sunspot cycle in the Maunder minimum AD1645 to AD1715. Astron Astrophys 276:549–563

Roberts P, Stix M (1971) The turbulent dynamo: a translation of a series of papers by F. Krause, K.-H. Radler, and M. Steenbeck (No. NCAR/TN-60+IA)

Rüdiger G, Hollerbach R (2004) The magnetic universe: geophysical and astrophysical dynamo theory. Wiley-VCH, Weinheim

Rüdiger G (1989) Differential rotation and stellar convection: Sun and the solar stars. De Gruyter, Berlin Boston

Ruzmaikin AA, Sokolov DD, Shukurov AM (1988) Magnetic fields of galaxies, vol 133. Kluwer Academic, Dordrecht

Saha C, Chandra S, Nandy D (2022) Evidence of persistence of weak magnetic cycles driven by meridional plasma flows during solar grand minima phases. Mon Not R Astron Soc 517(1):L36–L40

Schad A, Timmer J, Roth M (2013) Global helioseismic evidence for a deeply penetrating solar meridional flow consisting of multiple flow cells. Astrophys J Lett 778:L38

Schatten KH, Scherrer PH, Svalgaard L, Wilcox JM (1978) Using dynamo theory to predict the sunspot number during solar cycle 21. Geophys Res Lett 5:411–414

Schmitt D, Schüssler M, Ferriz-Mas A (1996) Intermittent solar activity by an on-off dynamo. Astron Astrophys 311:L1–L4

Schrinner M, Rädler K-H, Schmitt D, Rheinhardt M, Christensen UR (2007) Mean-field concept and direct numerical simulations of rotating magnetoconvection and the geodynamo. Geophys Astrophys Fluid Dyn 101:81–116

Schüssler M, Caligari P, Ferriz-Mas A, Moreno-Insertis F (1994) Instability and eruption of magnetic flux tubes in the solar convection zone. Astron Astrophys 281:L69–L72

Seehafer N (1990) Electric current helicity in the solar atmosphere. Sol Phys 125(2):219–232

Shimada R, Hotta H, Yokoyama T (2022) Mean-field analysis on large-scale magnetic fields at high Reynolds numbers. Astrophys J 935(1):55

Shulyak D, Sokoloff D, Kitchatinov L, Moss D (2015) Towards understanding dynamo action in M dwarfs. Mon Not R Astron Soc 449(4):3471–3478

Simard C, Charbonneau C (2020) Grand minima in a spherical non-kinematic $\alpha^2\Omega$ mean-field dynamo model. J Space Weather Space Clim 10:9

Simard C, Charbonneau P, Bouchat A (2013) Magnetohydrodynamic simulation-driven kinematic mean field model of the solar cycle. Astrophys J 768(1):16

Simard C, Charbonneau P, Dubé C (2016) Characterisation of the turbulent electromotive force and its magnetically-mediated quenching in a global EULAG-MHD simulation of solar convection. Adv Space Res 58:1522–1537

Simitev RD, Kosovichev AG, Busse FH (2015) Dynamo effects near the transition from solar to anti-solar differential rotation. Astrophys J 810:80

Smolarkiewicz PK, Charbonneau P (2013) EULAG, a computational model for multiscale flows: an MHD extension. J Comput Phys 236:608–623

Sokoloff D, Nesme-Ribes E (1994) The Maunder minimum: a mixed-parity dynamo mode? Astron Astrophys 288:293–298

Sokoloff DD, Stepanov RA, Frick PG (2014) Dynamos: from an astrophysical model to laboratory experiments. Phys Usp 57(3):292–311

Spiegel EA, Zahn J-P (1992) The solar tachocline. Astron Astrophys 265:106–114

Steenbeck M (1971) Impulse und Impulse und Wirkungen: Schritte auf meinem Lebensweg. Verlag der Nation, Berlin

Steenbeck M, Krause F (1969) On the dynamo theory of stellar and planetary magnetic fields. I. AC dynamos of solar type. Astron Nachr 291:49–84

Steenbeck M, Krause F, Rädler KH (1966) Berechnung der mittleren Lorentz-Feldstärke $\overline{\mathbf{v} \times \mathbf{B}}$ für ein elektrisch leitendes Medium in turbulenter, durch Coriolis-Kräfte beeinflußter Bewegung. Z Naturforsch Teil A 21:369

Stix M (1976) Differential rotation and the solar dynamo. Astron Astrophys 47:243–254

Stix M (2004) The Sun: an introduction. Springer, Berlin

Strugarek A, Beaudoin P, Brun AS, Charbonneau P, Mathis S, Smolarkiewicz PK (2016) Modeling turbulent stellar convection zones: sub-grid scales effects. Adv Space Res 58(8):1538–1553

Strugarek A, Beaudoin P, Charbonneau P, Brun AS (2018) On the sensitivity of magnetic cycles in global simulations of solar-like stars. Astrophys J 863(1):35

Svalgaard L, Cliver EW, Kamide Y (2005) Sunspot cycle 24: smallest cycle in 100 years? Geophys Res Lett 32:L01104

Talafha M, Nagy M, Lemerle A, Petrovay K (2022) Role of observable nonlinearities in solar cycle modulation. Astron Astrophys 660:A92

Tobias SM (2021) The turbulent dynamo. J Fluid Mech 912:P1

Tobias SM, Weiss NO, Kirk V (1995) Chaotically modulated stellar dynamos. Mon Not R Astron Soc 273:1150–1166

Tomczyk S, Schou J, Thompson MJ (1995) Measurement of the rotation rate in the deep solar interior. Astrophys J Lett 448:L57–L60

Tripathi B, Nandy D, Banerjee S (2021) Stellar mid-life crisis: subcritical magnetic dynamos of solar-like stars and the breakdown of gyrochronology. Mon Not R Astron Soc 506(1):L50–L54

Ulrich RK (2010) Solar meridional circulation from Doppler shifts of the Fe I line at 5250 Å as measured by the 150-foot solar tower telescope at the Mt. Wilson Observatory. Astrophys J 725:658–669

Upton L, Hathaway DH (2014) Predicting the Sun's polar magnetic fields with a surface flux transport model. Astrophys J 780:5

Usoskin IG (2023) A history of solar activity over millennia. Living Rev Sol Phys 20:2

Usoskin IG, Sokoloff D, Moss D (2009) Grand minima of solar activity and the mean-field dynamo. Sol Phys 254:345–355

Usoskin IG, Arlt R, Asvestari E, Hawkins E, Käpylä M, Kovaltsov GA, Krivova N, Lockwood M, Mursula K, O'Reilly J, Owens M, Scott CJ, Sokoloff DD, Solanki SK, Soon W, Vaquero JM (2015) The Maunder minimum (1645-1715) was indeed a grand minimum: a reassessment of multiple datasets. Astron Astrophys 581:A95

Usoskin IG, Gallet Y, Lopes F, Kovaltsov GA, Hulot G (2016) Solar activity during the Holocene: the Hallstatt cycle and its consequence for grand minima and maxima. Astron Astrophys 587:A150

van Ballegooijen AA, Choudhuri AR (1988) The possible role of meridional circulation in suppressing magnetic buoyancy. Astrophys J 333:965–977

Viviani M, Käpylä MJ, Warnecke J, Käpylä PJ, Rheinhardt M (2019) Stellar dynamos in the transition regime: multiple dynamo modes and antisolar differential rotation. Astrophys J 886(1):21

Wang Y-M, Nash AG, Sheeley NR Jr (1989) Magnetic flux transport on the Sun. Science 245:712–718

Wang Y-M, Sheeley NR Jr, Nash AG (1991) A new cycle model including meridional circulation. Astrophys J 383:431–442

Warnecke J (2018) Dynamo cycles in global convection simulations of solar-like stars. Astron Astrophys 616:A72

Warnecke J, Brandenburg A, Mitra D (2011) Dynamo-driven plasmoid ejections above a spherical surface. Astron Astrophys 534:A11

Warnecke J, Rheinhardt M, Tuomisto S, Käpylä PJ, Käpylä MJ, Brandenburg A (2018) Turbulent transport coefficients in spherical wedge dynamo simulations of solar-like stars. Astron Astrophys 609:A51

Warnecke J, Rheinhardt M, Viviani M, Gent FA, Tuomisto S, Käpylä MJ (2021) Investigating global convective dynamos with mean-field models: full spectrum of turbulent effects required. Astrophys J Lett 919(2):L13

Weber MA, Fan Y, Miesch MS (2011) The rise of active region flux tubes in the turbulent solar convective envelope. Astrophys J 741(1):11

Weber MA, Fan Y, Miesch MS (2013) Comparing simulations of rising flux tubes through the solar convection zone with observations of solar active regions: constraining the dynamo field strength. Sol Phys 287:239–263

Weiss NO, Tobias SM (2016) Supermodulation of the Sun's magnetic activity: the effects of symmetry changes. Mon Not R Astron Soc 456(3):2654–2661

Whitbread T, Yeates AR, Muñoz-Jaramillo A, Petrie GJD (2017) Parameter optimization for surface flux transport models. Astron Astrophys 607:A76

Wright NJ, Newton ER, Williams PKG, Drake JJ, Yadav RK (2018) The stellar rotation-activity relationship in fully convective M dwarfs. Mon Not R Astron Soc 479(2):2351–2360

Wu CJ, Usoskin IG, Krivova N, Kovaltsov GA, Baroni M, Bard E, Solanki SK (2018) Solar activity over nine millennia: a consistent multi-proxy reconstruction. Astron Astrophys 615:A93

Yoshimura H (1975) Solar-cycle dynamo wave propagation. Astrophys J 201:740–748

Zeldovich YB (1957) The magnetic field in the two-dimensional motion of a conducting turbulent liquid. Sov Phys JETP 4:460–462

Zeldovich YB, Ruzmaikin AA (1980) Magnetic field of a conducting fluid in two-dimensional motion. Zh Eksp Teor Fiz 78:980–986

Zeldovich Y, Ruzmaikin A, Sokoloff D (1983) Magnetic fields in astrophysics, vol 3. Gordon & Breach, New York

Zhang H, Sakurai T, Pevtsov A, Gao Y, Xu H, Sokoloff DD, Kuzanyan K (2010) A new dynamo pattern revealed by solar helical magnetic fields. Mon Not R Astron Soc 402(1):L30–L33

Zhao J, Bogart RS, Kosovichev AG, Duvall TL Jr, Hartlep T (2013) Detection of equatorward meridional flow and evidence of double-cell meridional circulation inside the Sun. Astrophys J Lett 774:L29

Publisher's Note Springer Nature remains neutral with regard to jurisdictional claims in published maps and institutional affiliations.

Space Science Reviews (2023) 219:64
https://doi.org/10.1007/s11214-023-01008-3

Solar Cycle Observations

Aimee Norton[1] · Rachel Howe[2] · Lisa Upton[3] · Ilya Usoskin[4]

Received: 30 May 2023 / Accepted: 26 September 2023 / Published online: 17 October 2023
© The Author(s) 2023

Abstract

We describe the defining observations of the solar cycle that provide constraints for the dynamo processes operating within the Sun. Specifically, we report on the following topics: historical sunspot numbers and revisions; active region (AR) flux ranges and lifetimes; bipolar magnetic region tilt angles; Hale and Joy's law; the impact of rogue ARs on cycle progression and the amplitude of the following cycle; the spatio-temporal emergence of ARs that creates the butterfly diagram; polar fields; large-scale flows including zonal, meridional, and AR in-flows; short-term cycle variability; and helioseismic results including mode parameter changes.

Keywords Sunspots – 1653 · Solar cycle – 1487 · Dynamo – 2001 · Helioseismology – 709

1 Introduction

In the mid 1800s, Schwabe (1844) discovered the solar cycle by observing that sunspot numbers rise and fall over the course of roughly 11 years. This discovery likely inspired Wolfe to take daily observations of the Sun, thus beginning the crucial, historical recording

Solar and Stellar Dynamos: A New Era
Edited by Manfred Schüssler, Robert H. Cameron, Paul Charbonneau, Mausumi Dikpati, Hideyuki Hotta and Leonid Kitchatinov

✉ R. Howe
r.howe@bham.ac.uk

A. Norton
aanorton@stanford.edu

L. Upton
lisa.upton@swri.org

I. Usoskin
ilya.usoskin@oulu.fi

[1] HEPL Solar Physics, Stanford University, Stanford, CA, 94305-4085, USA

[2] School of Physics and Astronomy, University of Birmingham, Edgbaston, Birmingham, B15 2TT, UK

[3] Southwest Research Institute, Boulder, CO, 80302, USA

[4] Sodankylä Geophysical Observatory and Space Physics and Astronomy Research Unit, University of Oulu, Oulu, 90014, Finland

of the sunspot number. See Sect. 2 for the history, and recent revision, of the sunspot numbers. Studies of individual ARs and their properties began with sunspot drawings and daily observations, but AR research was revolutionized by Hale (1908) who demonstrated their magnetic nature, as discussed in Sect. 3.

Larmor (1919) proposed that the solar magnetic fields observed by Hale and his colleagues were generated through the process of electromagnetic induction in the electrically conducting solar plasma. This idea became known as dynamo theory and its evolution as applied to the Sun and stars is discussed extensively in Charbonneau and Sokoloff (2023).

In addition to records of sunspot numbers, the area and position of sunspots has been recorded since 1874, beginning at the Royal Observatory, Greenwich and later continuing via the National Oceanic and Atmospheric Administration (NOAA). The distribution of sunspot area as a function of latitude and time revealed that two sunspot bands existed on either side of the heliographic equator and that these bands moved equatorward during the course of the solar cycle. This pattern is known as the butterfly diagram and is discussed in Sect. 4. One criteria of success for any dynamo model is its ability to reproduce the features of the sunspot bands, including the observed equatorward migration.

While sunspots are distinctive, visible features containing strong magnetic fields, sunspots account for less than 1% of the solar surface area even at solar cycle maximum. Determining the larger, global-scale magnetic structure of the Sun required measurements of weaker, more spatially distributed fields (Stenflo 1970). There is a large-scale dipole field that dominates at cycle minimum. The amplitude of this dipole is a reliable precursor for the next cycle amplitude, which is best-studied through the accumulation of small-scale flux at the poles, as discussed in Sect. 5.

The magnetic nature of the solar cycle is only a part of the story. The behavior of large-scale flows informs us of the variations associated with the dynamo. These plasma motions include differential rotation (radial and latitudinal), torsional oscillations, meridional flow and AR in-flows, as discussed in Sect. 6. Finally, short-term cycle variations and helioseismic mode parameter changes are mentioned in Sects. 7 and 8, respectively.

This review paper is a result of an International Space Science Institute Workshop titled "Solar and Stellar Dynamos: A New Era" held in June, 2022. Discussions in this review paper are intentionally brief and may not be comprehensive, since the purpose is to introduce the observations that have inspired the up-to-date research summarized in the other papers that are part of this collection.

2 Sunspot Number

The sunspot number (SN) is a synthetic (not physical) quantitative index of solar activity, which is historically widely used because of its simplicity and long (more than 400 years) available dataset. The SN is not equal to the *number of sunspots* (denoted as s below) but includes also the weighted number of sunspot groups g, using the formula introduced by Rudolf Wolf in the middle of the 19th century:

$$SN = k \cdot (10 \cdot g + s), \tag{1}$$

where k is a scaling factor reducing the data quality (related, e.g., to the quality of the instrumentation used) of individual observers to that of the reference one (usually Rudolf Wolf or Alfred Wolfer are considered as the reference observers). A single spot on the Sun ($s = 1$) is counted as a single sunspot group leading thus to SN = 11. The classical Wolf's method

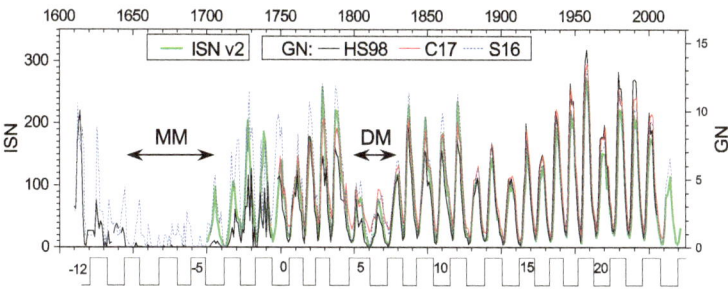

Fig. 1 Annual sunspot activity for the last centuries according to different recent reconstructions: International sunspot number (ISN) series version 2 (green ISN v2 curve, left axis) obtained from SILSO; Sunspot group number (GN, right axis), according to HS98 – Hoyt and Schatten (1998); C17 – Chatzistergos et al. (2017); S16 – Svalgaard and Schatten (2016). Standard (Zürich) cycle numbering is shown between the panels. Approximate dates of the Maunder minimum (MM) and Dalton minimum (DM) are labeled within the figure. Modified after Usoskin (2023)

uses observations of only one, so-called primary observer for each day. If the primary observer's data was not available for a day, secondary, tertiary, etc. observers were used, but always only one per day (see Waldmeier 1961). This makes the SN series easy to calculate but leaves no way to verify it nor to estimate its uncertainty. This forms the so-called *Wolf* or Zürich SN series (WSN or *Z*). The WSN was continuously produced by Zürich Observatory using roughly nearly the same, reproducible, techniques. The main shortcoming of the WSN is that it is not transparent and cannot be presently revisited, corrected or verified, since only the final product has been published while raw data were hand-written in log books. These old log books are being digitized now making it potentially possible to revise the WSN in the future (Friedli 2020).

The production of the SN series was ceased in the 1980s in Zürich and smoothly transmitted to Brussels (Sunspot Index and Long-term Solar Observations project, SILSO – http://sidc.be/silso), where it is continued in the form of the International Sunspot Number, ISN (Clette et al. 2007). SILSO continues using the same formula (Eq. (1)) for ISN but changed the methodology so that not only a single primary observer's data, but a weighted sum of all available data are used for each day.

For more than a century, the WSN was the "gold standard" in solar studies, but then several problems were identified, including difficulties maintaining consistency with new data. Hoyt and Schatten (1998) developed a new index, called *Group sunspot number*, GSN (or GN), based on a weighted sum of the number of sunspot groups reported for each day by all possible observers. They neglected the number of individual spots as less reliably detected. Hoyt and Schatten (1998) added a lot of new data not known to R. Wolf and his successors and, most importantly, published the entire database of raw data, making it possible to assess the uncertainties and add or revise the data if needed. This also allows evaluation of the related uncertainties. The database of individual historical observations is continuously updated at http://haso.unex.es/?q=content/data (Vaquero et al. 2016).

After a careful study, several issues have been found in the WSN/ISN dataset, such as discontinuous transitions between different observers or changed methodology (e.g., Leussu et al. 2013; Clette et al. 2014). These obvious discontinuities have been ad-hoc corrected in the revised ISN series, which have been also normalized to A. Wolfer as the reference observer – the latter leads to a scaling factor of 1.667 with respect to the classical WSN. This forms the ISN version 2 dataset which is considered as a current version (Clette and Lefèvre 2016), as shown by the green curve in Fig. 1. A new revision of the ISN, version 3, is pending in the near future as the first consensus dataset using the best of our present-day knowledge.

Independently of the WSN/ISN, the methodology has been revisited for the GSN series, starting from the raw-data database. Several new approaches have been developed in this direction. One was made by Svalgaard and Schatten (2016) who performed a daisy-chain "backbone" GSN composition following the classical scheme of linearly scaling individual observers between each other (blue dotted curve in Fig. 1). The daisy-chain approach was further improved by Chatzistergos et al. (2017) who accounted for non-linear relations between data from overlapping observers and composed a new GN series (red curve in Fig. 1). A new approach has been developed recently (Usoskin et al. 2016; Willamo et al. 2017) that uses the active-day (days with at least one sunspot observed) fraction as the metrics of the minimum size of sunspot group that could be detected by an observer due to instrumentation and seeing conditions.

Of special interest is the level of solar activity during the Maunder minimum of 1645–1715 (Eddy 1976): while the present paradigm is that it was nearly sunspot free (e.g., Usoskin et al. 2015; Carrasco et al. 2021), some estimates predict low but still significant sunspot activity (Svalgaard and Schatten 2016; Zolotova and Ponyavin 2015). However, a consistent analysis of the multitude of other data, such as cosmogenic isotopes, auroral records, solar eclipse observations, confirms the very low level of solar activity during the Maunder minimum (e.g., Usoskin et al. 2015; Asvestari et al. 2017; Carrasco et al. 2021; Hayakawa et al. 2021), implying particularly that the reconstruction by Svalgaard and Schatten (2016) is too high in the 18th century.

Thus, at present there is a zoo of SN and GN reconstructions, as shown in Fig. 1. Generally, they are all consistent after about 1870 but somewhat disagree for the period between 1749–1870, with the difference being indicative of the systematic uncertainties. The GSN series by Hoyt and Schatten (1998) and by Svalgaard and Schatten (2016) can be considered as conservative lower and upper bounds, respectively, while other models lie between them. A consensus-based SN reconstruction is presently not available but it is under consideration by the research community.

3 Active Regions (ARs)

An AR is identified as a dark feature observed in the solar photosphere in white light observations. ARs contain strong, bipolar magnetic fields and are associated with sunspots. In this section, we discuss observational aspects of ARs that contribute to our understanding of the solar cycle. For a review of the origins of ARs and their emergence process, see Weber et al. (2023).

3.1 Hale's Law

Hale (1908) realized that magnetic fields were the cause of sunspots after observing the Zeeman splitting of a spectral line from the light originating in a sunspot. He also noted that sunspots appeared in pairs of positive and negative magnetic polarity and that the leading polarity (with respect to rotation) in each hemisphere changes from one sunspot cycle to the next. This is known as Hale's polarity law, see Fig. 2. While Hale's law is straightforward, it has profound implications for the solar dynamo. It implies that the large-scale organization of the magnetic field in the interior is mostly toroidal (East-West) in orientation and oppositely directed on either side of the equator. ARs adhere to Hale's law ~92–95% of the time (Wang and Sheeley 1989; McClintock et al. 2014; Li 2018; Muñoz-Jaramillo et al. 2021).

Fig. 2 (Top) Hale's law describes how bipolar magnetic regions in one hemisphere tend to have the same leading magnetic polarity while those in the other hemisphere have the opposite leading polarity. This leading polarity switches from one Solar Cycle to the next, i.e., cycle N (N + 1) shows the expected polarity for Cycle 24 (25). The average tilt angles between the magnetic polarities, depicted by red lines, increase with increasing latitudes. (Bottom) HMI magnetograms from Cycles 24 (2012.04.21) and 25 (2022.05.15) show the manifestation of Hale's and Joy's law on any given day with orange-red (green-blue) colors identifying the location of negative (positive) polarity

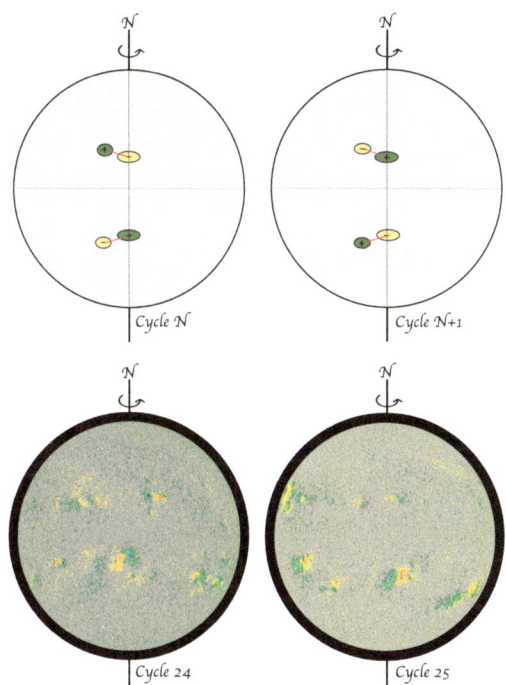

3.2 Flux Ranges and Lifetimes

ARs are part of a spectrum of magnetic bipoles that emerge into the photosphere and have a smooth, distribution function in regards to size and total absolute flux values ranging from 10^{18} to 10^{23} Mx. Ephemeral regions are smaller, short-lived regions with flux less than 1×10^{20} Mx and have a lifetime of hours, i.e., shorter than a day (Hagenaar et al. 2003). Small regions appear as pores with flux in the range of 1×10^{20} to 5×10^{21} Mx (van Driel-Gesztelyi and Green 2015). Larger sunspots develop well-defined penumbra and have 5×10^{21} to several $\times 10^{23}$ Mx. They live on the order of several weeks to several months. Typically, the flux emergence period is 15 – 30% of the total lifetime (van Driel-Gesztelyi and Green 2015), with most ARs fully emerged within 3–5 days (Harvey-Angle 1993; Norton et al. 2017) and an average emergence time of ~2 days (Weber et al. 2023). After flux emergence, there is a plateau of stability before the flux begins to decay.

3.3 Tilt Angles

Bipolar sunspot pairs are, on average, oriented so that the leading sunspot is closer to the equator than the following sunspot, see Fig. 3 for an example, and also the red lines connecting the bipolar sunspot pairs in Fig. 2, with the angle being a measure of the orientation of the bipolar magnetic region's axis with respect to the East–West direction. On average, the tilt angles increase with latitude, and this trend was named "Joy's Law" by Zirin (1988), see Fig. 3. Tilt angles are crucial in flux-transport dynamo models where it plays a role in the formation and evolution of polar fields (see, e.g., Wang and Sheeley 1991; Dikpati and Charbonneau 1999a). Tilts serve as an observable feature of the conversion of toroidal magnetic field into poloidal, i.e., the α-effect, and the reversal of axial dipole between cycles

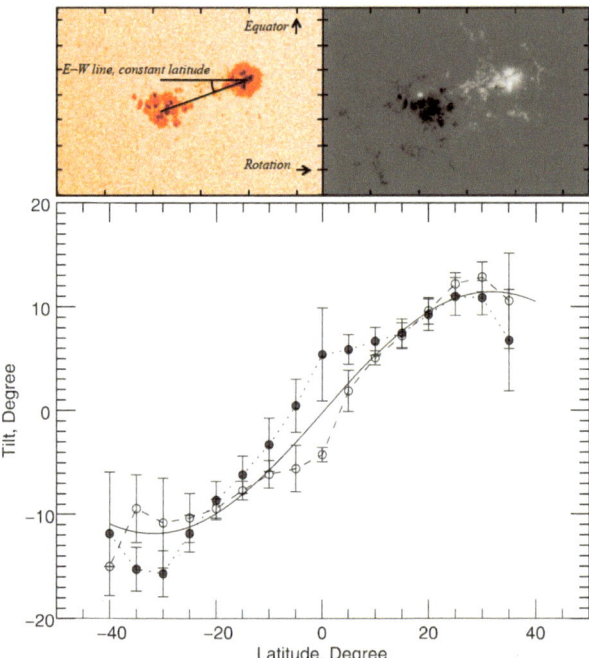

Fig. 3 The majority of ARs emerge in nearly an E-W orientation with a tendency for the leading spot (with respect to rotation) to be closer to the equator. A southern hemisphere AR from Cycle 24 is shown (top) as observed in SDO/HMI intensity and magnetogram. (Bottom) The binned, mean tilt of magnetic bipoles as a function of latitude from Cycles 15−24 from Mt. Wilson Observatory daily sunspot drawings with magnetic polarity indicated. RMS of mean tilt shown as error bars with even (open circles, dashed line) and odd cycles (filled circles, dotted line) as well as a fit to all data (solid line) of the form $\gamma = 0.2 \sin(2.8\theta)$ where γ is the tilt angle and θ is latitude. This form of Joy's law captures the inflection point in tilt values at mid-latitudes. Lower image reproduced with permission from Tlatova et al. (2018), copyright from Springer Nature

(Cameron et al. 2018). There are two dominant, physical explanations for the origin of Joy's law. First, as proposed by Babcock (1961), the tilt angle observed in the photosphere reflects the directional components of the global magnetic field at depth and is a direct consequence of the "winding up" of the poloidal field in the solar interior. Second, Wang and Sheeley (1991) propose that Joy's law is a result of the Coriolis force acting on flows within the flux tube as it rises through the convection zone.

Joy's law is a statistical law and only becomes obvious after much averaging. A study by Wang and Sheeley (1989) with over 2500 bipolar magnetic regions reported that 16.6% had no measurable tilts, 19% were anti-Joy, 4.4% were anti-Hale. That is 39.9% of regions that were not obeying Joy's law. The data are so noisy that Joy's law cannot be recovered for Cycle 17 (Cycle 19) in the northern (southern) hemisphere, respectively (McClintock and Norton 2013) The scatter is thought to have a physical origin, the buffeting of flux tubes by convective motions (Fisher et al. 1995; Weber et al. 2011). In addition to the high scatter of tilt angles, the expansion of the sunspot group along its major axis is observed with the possibility of differential rotation acting on the poleward and equatorward spots accordingly (Gilman and Howard 1986; Schunker et al. 2020).

Simulations of thin flux tubes rising through the convection zone with the Coriolis force acting on flows within the flux tube have been able to recreate both Joy's law and its scatter

(D'Silva and Choudhuri 1993; Fan 2009; Weber et al. 2011) with scatter increasing for flux tubes that spend a longer time rising. Results from 3D dynamo models show some promise of producing bipolar magnetic regions that adhere to Hale and Joy's law (Nelson et al. 2013); there is certainly no consensus as to which models most accurately represent solar conditions and recreate tilt angle distributions.

Various forms of Joy's law are reported in the literature including the following where γ is the tilt angle and θ the latitude:

$$\sin \gamma = m \cdot \sin \theta + c, \tag{2}$$

$$\gamma = m \cdot \theta + c, \tag{3}$$

$$\gamma = m \cdot \sin \theta + c, \tag{4}$$

$$\gamma = m \cdot \sin(k \cdot \theta), \tag{5}$$

The reported best-fit values for slope, m, and intercept, c, depend on the solar cycle, instrument, type of data (white-light or magnetogram), and sampling techniques used for the determination of the tilt angles. Note that some researchers force the fits through zero, $c = 0$, while others allow a y-intercept that is non-zero. For a few examples, Wang and Sheeley (1991) report Joy's law in the form of Eq. (2) with $m = 0.48$ and $c = 0.03$, Norton and Gilman (2005) use Eq. (3) with $m = 0.2$ and $c = 0.2$, Tlatova et al. (2018) uses Eq. (5) with $m = 0.2$ and $k = 2.8$. Li (2018) comprehensively reports fits to the forms of Eqs. (2)–(4). These are only a few examples, as it is beyond the scope of this review to report an all-inclusive list of Joy's law fits.

A list of observational aspects of tilt angles are as follows:

- the dependence on latitude differs between cycles and hemispheres (Dasi-Espuig et al. 2010; McClintock and Norton 2013; Tlatova et al. 2018);
- there is evolution as the AR emerges and decays so the time of measuring a tilt angle matters (McClintock and Norton 2016; Schunker et al. 2020);
- scatter is higher during the first day of emergence (Schunker et al. 2020),
- the value tends to settle near the end of emergence (Stenflo and Kosovichev 2012; Schunker et al. 2020);
- there are conflicting reports as to whether the tilts show a dependence on magnetic flux as predicted in thin flux tube modeling (Fisher et al. 1995; Stenflo and Kosovichev 2012; Jiang et al. 2014; McClintock and Norton 2016; Schunker et al. 2020);
- the scatter in the tilt values has a dependence on flux but not latitude (Fisher et al. 1995);
- the mean and median tilts of regions near the equator are not zero indicating that forcing a fit for Joy's law through the origin may be unphysical,
- an inflection point in the fit of tilts as a function of latitude occurs around 30° in both hemispheres (Tlatova et al. 2018);
- the smallest bipoles appear to have negative tilts (Tlatov et al. 2013);
- and the anti-Hale regions may not simply be the tail of the distribution of tilt angles as Muñoz-Jaramillo et al. (2021) reports they prefer an east-west orientation and have a distribution distinct from the ARs that follow Joy's law.

Improvements in tilt angle measurements and databases is ongoing work. Traditional determinations of Joy's law have been based on white-light images because magnetograms only became routinely available in the mid-1960s. White-light studies yield median tilt angles that are smaller and increase less steeply with latitude (lower slopes) than those obtained from magnetic data as shown by Wang et al. (2015), who also pointed out that a substantial fraction of tilts determined from white-light data were erroneous since they were from sunspots of the same polarity. In addition, if plage is included in the calculation, the tilt angle is usually higher. Given the errors in tilt angle determinations prior to routine magnetograms, and inconsistent methodologies (i.e., tilt angles determined including only umbra versus those determined using umbra, penumbra and plage), it is not clear if long-term trends of tilt angles using only white-light data are valid.

An anti-correlation between area-weighted mean tilt angles (normalized by latitude) and cycle strength was shown by Dasi-Espuig et al. (2010) for cycles 15–21, indicating that the surface source for the poloidal field becomes weaker for stronger cycles, potentially limiting the strength of the next cycle, and providing a feedback mechanism ("tilt-quenching") that prevents runaway solutions to the cycle amplitude. However, McClintock and Norton (2013) could only recover the Dasi-Espuig et al. (2010) result for the Southern hemisphere, not the Northern, and the Cycle 19 outlier value dominated the fit for the Southern hemispheric data. Nevertheless, non-linear feedback mechanisms that affect average tilt angles appear effective. Surface flux-transport modeling by Cameron and Schüssler (2012) and Jiang et al. (2010) incorporated the effect of AR inflows into surface flux transport models and found that strong cycles produce strong in-flows which result in a lower tilt angle and decreased resulting axial dipole moment.

3.4 Rogue Active Regions

The progression of any solar cycle, including the polarity reversal and gradual strengthening of the polar caps responsible for the axial dipole moment, is punctuated by the appearance of unusually influential, or rogue, ARs. The term "rogue AR" was coined by Nagy et al. (2017) who reported that a single rogue bipolar magnetic region in their simulations was found to have a major effect on the development of subsequent solar cycles, either increasing or decreasing the amplitudes, and in extreme cases, triggering a grand minimum. Nagy et al. (2020) then proposed the AR Degree of Rogueness (ARDoR) quantity that is the difference between the final contribution to the axial dipole moment from an individual AR and an ideal contribution from a region at the same latitude that has an expected tilt angle prescribed by Joy's Law and separation of opposite polarity footpoints typical for an AR of similar flux. Meaning, a region is defined as rogue when its contribution to the final axial dipole moment is significantly different from an average active region emerging at the same latitude.

Petrovay et al. (2020) formulated an algebraic method that consists of summing the ultimate contributions of individual ARs to the solar axial dipole moment at the end of the cycle. Nagy et al. (2020) performed a statistical analysis of a large number of simulated activity cycles and ranked the ARs from most to least influential depending on their contribution to the final axial dipole moment. The model by Lemerle and Charbonneau (2017) used in the simulation couples a conventional surface flux transport (SFT) 2D simulation defined over a spherical surface with a 2D axisymmetric flux transport dynamo (FTD) simulation defined in a meridional plane (Charbonneau et al. 2005). In this hybrid 2×2D Babcock–Leighton dynamo model, the SFT component provides the surface poloidal source term for the FTD simulation, while the FTD component provides the magnetic emergence events input to the SFT simulation (Charbonneau et al. 2005; Nagy et al. 2017). They showed that the top 50 influential ARs of any given cycle are sufficient to reproduce the final dipole moment of that cycle. Rogue ARs have a variety of characteristics but their rogueness is commonly determined by having one or more of the following characteristics: a very large amount of flux ($\Phi > 1 \times 10^{22}$ Mx); an abnormal tilt angle such as one that is anti-Joy or

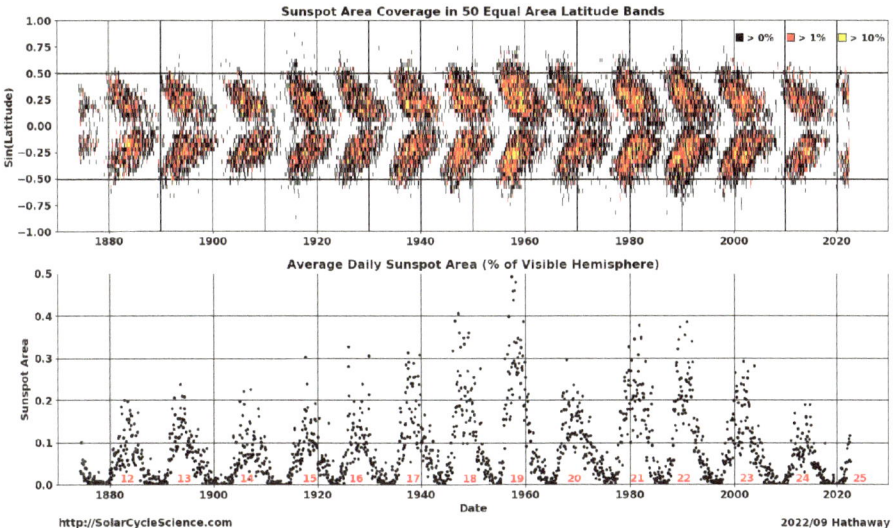

Fig. 4 Butterfly diagram shows sunspot area from the Royal Greenwich Observatory, color coded as a percentage of the solar disk, plotted as a function of time and sine latitude (top panel). For reference, the total sunspot area is also plotted as a function of time (lower panel). Figure courtesy D.H. Hathaway via www.solarcyclescience.com

anti-Hale (90–180° away from Joy's law); an unusually large separation distance between the polarities (>70 Mm); or being very close to the equator.

4 The Butterfly Diagram

When the location of sunspots or ARs are plotted as a function of latitude and time, a striking pattern emerges that resembles butterfly wings. This so called "Butterfly Diagram", first depicted by Maunder (1904), shows bands of sunspot activity in both the Northern and Southern hemispheres. Modern depictions include a third dimension, the fractional area of the Sun covered by sunspots, see Fig. 4. Inspection of the butterfly diagram reveals that early in the cycle, ARs begin emerging at mid-latitudes (approximately 30 degrees) and as the cycle progresses the emergence moves closer to the equator. This equatorward progression of AR emergence is known as Spörer's Law (Maunder 1903). Stronger cycles tend to begin emergence at higher latitudes than weaker cycles. The latitudinal width of the "Butterfly wings" also changes over the course of the cycle and is proportional to the strength of the solar cycle (Ivanov and Miletsky 2011), producing a tapering of the wings at both the start and the end of the cycle. Typically the cycles overlap in time by about one–two years, with the new cycle beginning at mid-latitudes before the previous cycle has finished. However, this overlap is proportional to the strength of the following cycle such that the weakest cycles have little to no overlap with the previous cycle. Asymmetry between the northern and southern hemispheric cycle progression was noted by Spörer (1894) and Maunder (1904) with unequal sunspot activity persisting for several years. Norton et al. (2014) showed that the hemispheric asymmetry never had more than a 20% difference in the amplitude of sunspot number or sunspot area. Nor was the time lag (measured by time of polar reversal, cycle

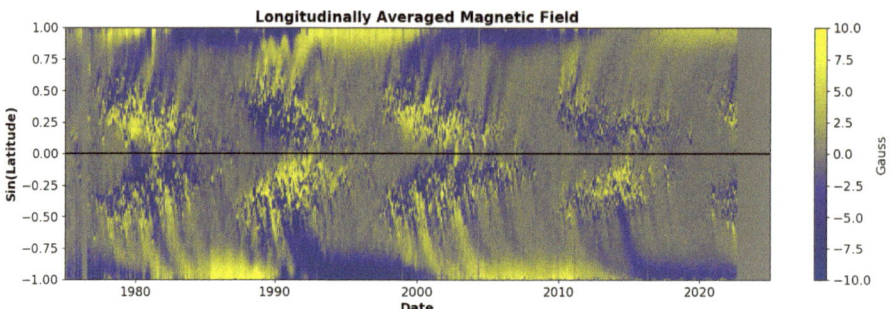

Fig. 5 The magnetic butterfly diagram shows the latitudinal distribution of the magnetic field as a function of time. The magnetic field is averaged over all available longitudes and over each Carrington Rotation using data from SOLIS/MDI/HMI. The color indicated the sign of the polarity, with yellow (blue) for positive (negative) radial magnetic fields. Figure courtesy D.H. Hathaway via www.solarcyclescience.com

maximum or minimum in each hemisphere) more than 20% of the entire cycle length. In other words, the hemispheres are strongly coupled – to within ≈80%.

While the traditional Butterfly Diagram is created in statistical manner by plotting the sunspot area as a function of latitude and time, a similar plot can be created by plotting the longitudinally averaged magnetic field instead (Harvey 1994). This "Magnetic Butterfly Diagram", see Fig. 5 reveals several other characteristics of the solar cycle. Most notable are the appearance of Joy's Law and Hale's law. Each wing displays predominantly leading polarity on the southern edge and the opposite following polarity predominantly on the northern edge (Joy's Law). The wing polarity is opposite across hemispheres and switches from one cycle to the next (Hale's law). In addition to the butterfly wings, streams of flux can be seen emerging from the wings and moving towards the poles. They are most prominent during solar cycle maximum and are dominated by the following sunspot polarity flux for that cycle (though intermittent leading-sunspot polarity streams are also present). These streams are a signature of the pole-ward meridional flow transporting residual AR flux to the polar regions. This process forms strong flux concentrations at the poles, i.e. the polar fields, that reverse polarity near the time of solar maximum.

5 Polar Fields

During solar minimum, the Sun's magnetic field resembles that of a dipole, with opposite polarity magnetic field concentrations at the poles. This dipolar magnetic field acts as the seed field for the solar cycle described in Charbonneau and Sokoloff (2023), Cameron and Schüssler (2023).

The Sun's polar magnetic fields can be measured by averaging the magnetic field strength over the polar cap to get the flux density over each polar region or by calculating the axial dipole moment of the magnetic field configuration. The latter provides a single value for the state of the Sun's global magnetic field as a whole, while the former provides additional information about the differences between the North and South hemispheric polar magnetic field.

The Wilcox Solar Observatory (WSO) has been measuring the Sun's line-of-sight magnetic field daily since 1976 and has provided measurements of both the polar field strength

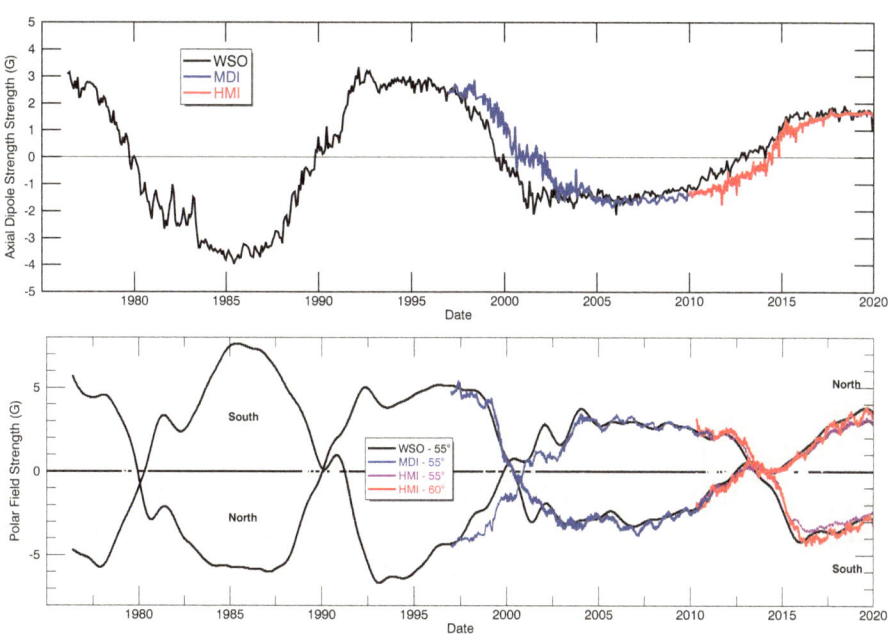

Fig. 6 The Sun's polar fields measured by the axial dipole moment (top) and by the average field strength over the polar caps (bottom). Data is shown from WSO in black, SOHO/MDI in blue, and SDO/HMI in purple and red. WSO and MDI averaged polar fields are measured from 55° and above, while HMI is shown for both 55° and 60° (purple and red) and above. The polar fields are smoothed over 13 Carrington Rotations (for reference, the unsmoothed WSO and HMI measurements are shown in grey, highlighting the annual signal caused by the changing inclination of the Sun over Earth's yearly orbit)

and the axial dipole since that time[1] (Svalgaard et al. 1978; Hoeksema 1995), and the axial dipole component is shown as a black line in Fig. 6 (top panel). The axial dipole moment is an integrated quantity that measures the axisymmetric component of the large-scale photospheric magnetic field. The polar field strength, bottom panel in Fig. 6, is defined as the flux density of the magnetic field above a specific latitude. For WSO this is limited by the spatial resolution and taken to be the line of sight field strength measured in the highest-latitude pixel, which is taken to be between 55° and the poles (but the actual latitude range varies with the Earth's orbit). Space based missions have a better resolution and in the case of HMI[2] calculate the polar fields as the inferred radial component of the magnetic field measured at 60° and above (Sun et al. 2015). While the improved resolution does mitigate the projection effects, a residual annual oscillation (gray lines in the bottom panel in Fig. 6) is evidence that there is still uncertainty in these measurements due to the poor viewing angle. While the flux density over each polar region offers insight into hemispheric asymmetries, the innate ambiguity associated with this measurement may make the axial component of the Sun's magnetic dipole a better metric for solar cycle prediction (Upton and Hathaway 2014b).

The polar fields are out of phase with the sunspot number, with the reversal occurring near the time of solar cycle maximum. The peak in the polar field strength typically occurs at or just before solar cycle minimum. The amplitude of the Sun's polar fields (as measured by the

---

[1] http://wso.stanford.edu/Polar.html courtesy of J.T. Hoeksema.

[2] http://jsoc.stanford.edu/data/hmi/polarfield/ courtesy of Xudong Sun.

Fig. 7 Top: Differential rotation measured by feature tracking (from Snodgrass and Ulrich 1990, © AAS, reproduced by permission). Bottom: Solar rotation profile from 2D inversions of HMI helioseismic observations, averaged from 2010 to 2022, prepared by the authors for this review

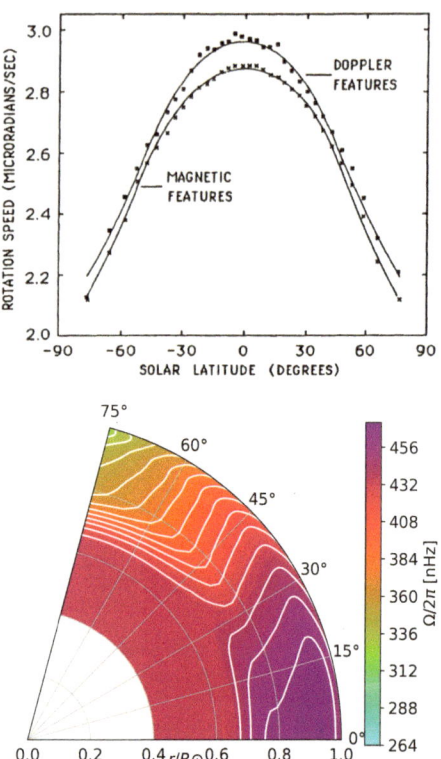

axial dipole or by the field strength over the polar cap) at the time of solar cycle minimum are proportional to the amplitude of the next solar cycle. Consequently, the amplitude of the polar fields at the time of cycle minimum have proven to be successful predictors of solar cycle amplitude (Schatten et al. 1978; Svalgaard et al. 2005; Petrovay 2010; Muñoz-Jaramillo et al. 2013; Bhowmik et al. 2023). This can be understood in terms of the dynamo model proposed by Babcock (1961) and extended by Leighton (1964). For a detailed account of the Babcock–Leighton model, see Charbonneau and Sokoloff (2023) and Cameron and Schüssler (2023).

6 Flows

The observed large-scale flows of the Sun — differential rotation, torsional oscillations, meridional circulation, large-scale convection and the recently observed inertial modes — provide a set of measurements that characterize the solar convective processes. For a detailed account of the plasma flows in the Sun, including Rossby waves and inertial modes, see Hotta et al. (2023). Herein, we introduce the fundamental observations of these flows.

6.1 Solar Rotation Profile

The mean solar rotation profile is well known. At the surface the latitudinal differential rotation can be measured by tracking features such as sunspots, revealing that the rotation rate is highest at the solar equator and decreases towards the poles (Fig. 7, top). A comprehensive

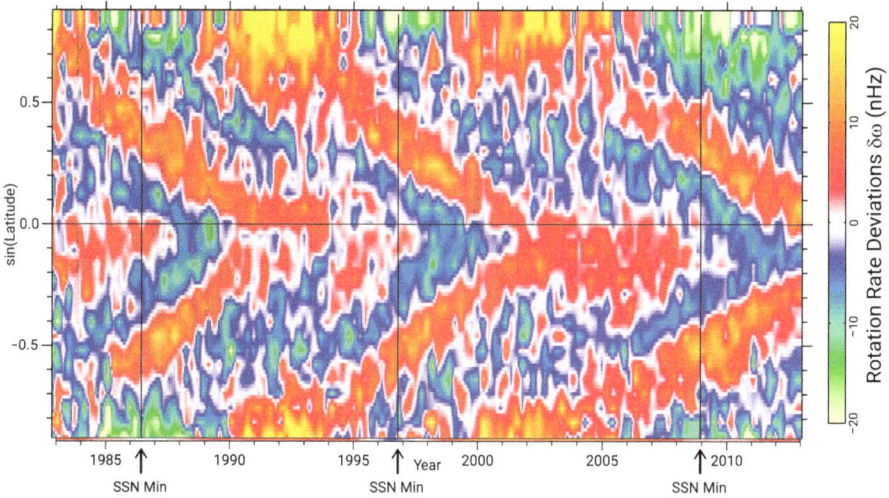

Fig. 8 The torsional oscillation in Mount Wilson surface Doppler observations, adapted from Ulrich et al. (2022) under the CC BY 4.0 license

review of these measurements has been made by Beck (2000). Helioseismology (see, for example, Thompson et al. 1996; Schou et al. 1998; Larson and Schou 2018) has revealed the interior rotation profile (Fig. 7 bottom). It features a near-surface shear layer (sometimes abbreviated as NSSL) where the rotation rate increases with depth down to about $0.95\,R_\odot$. Below this layer, latitudinal differential rotation persists through the bulk of the convection zone, approximately constant on radial lines although the isorotation contours tend to lie at about a 25-degree angle to the rotation axis over a wide range of latitudes (Gilman and Howe 2003). There is another shear layer or "tachocline" at the $0.71\,R_\odot$ base of the convection zone, which is narrower in reality than it appears in most helioseismic profiles due to the finite resolution of the inversions; the consensus (see Table 2 of Howe 2009, and references therein) is that the thickness is around $0.05\,R_\odot$, but at least one estimate (Corbard et al. 1999) puts it as low as $0.01\,R_\odot$. Below the tachocline, in the radiative interior, there is roughly rigid rotation down to the limits of reliable measurement at around $0.2\,R_\odot$ (e.g., Eff-Darwich and Korzennik 1998; Couvidat et al. 2003, although the former authors note that it is possible that the core is rotating somewhat faster than the bulk of the radiative interior).

6.2 Zonal Flows: Torsional Oscillations

The solar rotation profile is modulated by a pattern of bands of faster- and slower-than-average rotation, which can be considered respectively as prograde or retrograde flows, and which migrate in latitude in synchrony with the solar cycle (see Figs. 8 and 9). This pattern, revealed when a temporal average is subtracted from the rotation rate at each latitude, was first observed, and dubbed the "torsional oscillation", by Howard and LaBonte (1980) in surface Doppler observations from the 150 ft tower at the Mount Wilson Observatory. The Mount Wilson observations continued until 2013, and Fig. 8 from Ulrich et al. (2022) shows the pattern over three solar cycles. The main feature is the band of faster rotation in each hemisphere that moves from mid-latitudes towards the equator between one solar minimum and the next; as pointed out by Howard and LaBonte (1980), the latitude of maximum flux

Fig. 9 Zonal flow map from helioseismic inversions of GONG (1995–2022), MDI (1996–2010), and HMI (2011–2022) data, at a target depth of $0.99\,R_\odot$, with a temporal mean over the whole dataset subtracted at each latitude. Note that $\delta\Omega$ here is the same quantity as $\delta\omega$ in Fig. 8. Reproduced from Howe et al. (2022) under the CC BY 4.0 license

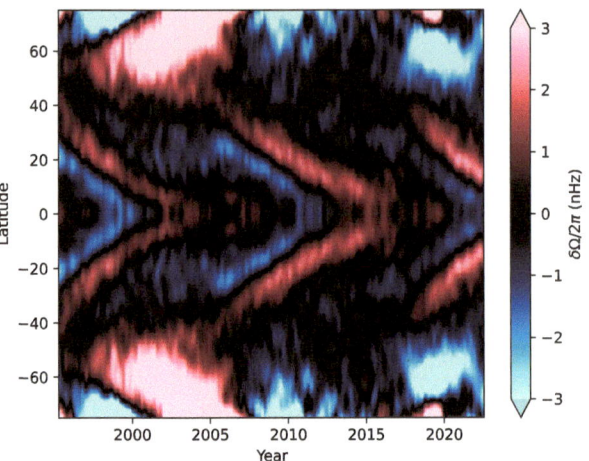

falls close to the edge of this belt. These flows are relatively weak compared to the mean solar rotation, with amplitudes close to the surface of less than ten meters per second, or a fraction of a per cent of the equatorial rotation rate.

The flow patterns were seen in helioseismic data in the rising phase of Solar Cycle 23 by Schou (1999), and Howe et al. (2000) found that the patterns penetrated at least 0.08 solar radii into the convection zone; subsequent work, for example by Vorontsov et al. (2002), suggested that the variation in rotation involves most of the bulk of the convection zone. A strong band of faster flow migrating from mid-latitudes towards the poles early in Solar Cycle 23 was reported by Antia and Basu (2001).

Because the mid-latitude rotation begins to speed up before significant surface activity is seen, the flow pattern towards the end of one solar cycle can give some indication of the timing of the onset of the following one, as reported by Howe et al. (2009) for Cycle 24 and Howe et al. (2018) for Cycle 25. In particular, the time at which the main belt of faster rotation reaches a latitude of around 25 degrees seems to coincide with solar activity becoming widespread in a new cycle. The strong poleward branch seen in Cycle 23 was not repeated in Cycle 24 (Howe et al. 2013). This seems to be associated with small but significant deceleration at higher latitudes, possibly related to the weaker polar fields in Cycle 24 (Rempel 2012). Figure 9 shows the flow residuals from inversions of GONG, MDI, and HMI data, as reported by Howe et al. (2022). We note that the global helioseismic inversions can only show the North–South symmetric part of the flow pattern, while the surface measurements and those from local helioseismology (e.g., Komm et al. 2018; Lekshmi et al. 2018) can distinguish the two hemispheres. The relationship between the flow pattern and magnetic butterfly diagram is complex, but Lekshmi et al. (2018) found that the hemispheric asymmetry of the flows is related to, and is a leading indicator of, the magnetic asymmetry; asymmetry in the flows is seen in advance of the corresponding asymmetry in the magnetic activity.

6.3 Meridional Flows

The solar meridional flow is the North-South motion of the plasma. At the surface this flow plays a critical role in the solar dynamo by transporting residual flux from ARs to the poles in order to generate the magnetic field to initialize the next solar cycle. This plasma flow moves

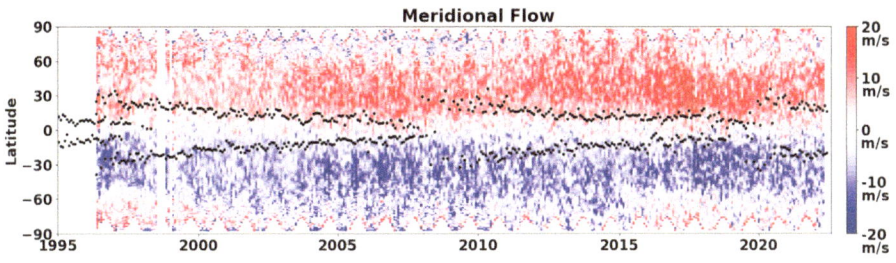

Fig. 10 The evolution of the meridional flow measured by magnetic pattern tracking of MDI/HMI data over the last two cycles (adapted from Hathaway et al. 2022)

from the equator to the poles in each hemisphere with an amplitude of $\sim 10\text{--}20 \text{ ms}^{-1}$. The meridional flow is 1-2 orders of magnitude weaker than the differential rotation (relative velocities of $\sim 200\text{--}250 \text{ ms}^{-1}$) and the convective flows (velocities of $\sim 500 \text{ ms}^{-1}$ for supergranules and $\sim 3000 \text{ ms}^{-1}$ for granules), making it the most challenging plasma flow to measure. The meridional flow is typically measured in the same manner as (and along with) the differential rotation (e.g., Doppler imaging, helioseismology, tracking techniques, etc.). Characterizing this flow is particularly challenging because independent measurement techniques can often give very different measurements, thought to be a consequence of the different depths sampled by each technique. For an in depth review, see Hanasoge (2022) and references therein.

High resolution continuous magnetic data from space-based observatories (i.e., SOHO/ MDI and SDO/HMI) have ushered in a new era, paving the way for meridional flow measurements with unprecedented spatial and temporal resolution, revealing that the amplitude and structure vary with the solar cycle (Gizon 2004; González Hernández et al. 2008; Hathaway and Rightmire 2010). The meridional flow measured by magnetic pattern tracking (Hathaway et al. 2022) for the last two solar cycles (see Fig. 10) shows that the meridional flow is the strongest at solar cycle minimum and weakens during solar minimum. This weakening of the meridional flow was more pronounced during the stronger Solar Cycle 23 than it was for the weak Solar Cycle 24. The relative magnitude of this cycle dependent change in the flow speed is illustrated in the left panel of Fig. 11. This modulation of the meridional flow by the presence of ARs may serve as a nonlinear feedback mechanism for regulating the solar cycle, as described in the next section.

Another aspect of the meridional flow is the quite contentious existence of the high-latitude equatorward flows, sometimes referred to as polar counter-cells. If present, these flows would have implications for the build up of the polar fields and thus the strength of the solar cycle (Jiang et al. 2009; Upton and Hathaway 2014a). The possibility of these flows was suggested by Ulrich (2010) as well as Hathaway and Rightmire (2010), but later dismissed (Rightmire-Upton et al. 2012) as an instrumental artifact because the counter-cells were not originally present in high-resolution HMI data. However, more recent analysis (Hathaway et al. 2022) now suggests these flows may have returned and are now observed in the HMI measurements (see Fig. 10). As of yet, their appearance does not seem to have a solar-cycle dependence but rather to occur somewhat sporadically. Resolving these structures unambiguously remains a challenge for several reasons. First and foremost, these flows only appear to be $\sim 1\text{--}2 \text{ ms}^{-1}$, an order of magnitude weaker than the already difficult to measure standard meridional flow. Secondly, they appear at latitudes above $60°$, where the radial component of the magnetic field is not well resolved and signal to noise is small. While advancement in the measurement techniques may eventually shed some light on this

Fig. 11 The average meridional flow profiles for different time periods measured by magnetic pattern tracking of MDI/HMI data are shown in the top panel. The red (blue) represents the flow during Cycle 23 (24) maximum. The purple line represents cycle minimum during Solar Cycle 23 and 24. The black line represents an average over both cycles. Two possible, idealized meridional circulation patterns are shown in the bottom panel, adapted from Stejko et al. (2021) (© AAS, reproduced by permission): the classical single-cell with a deep return flow (labeled K1) and a double-cell circulation profile with a stronger return flow (labeled K2)

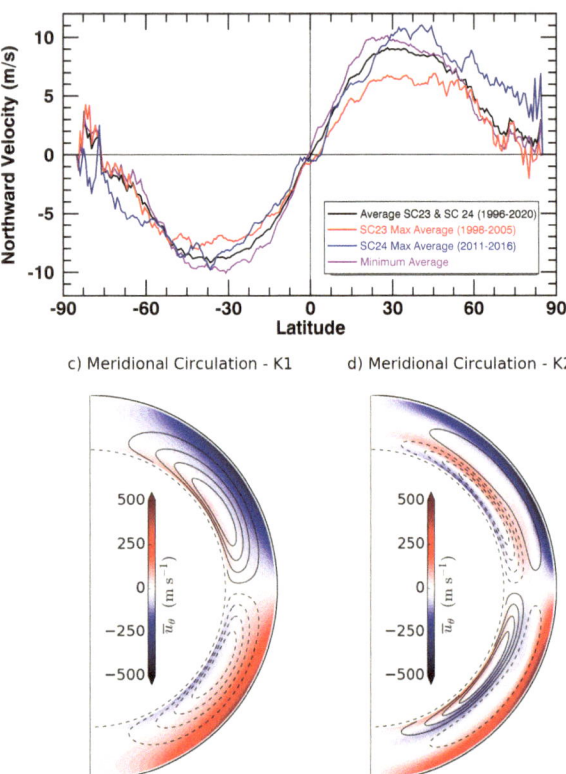

ambiguous aspect of the meridional flow, a mission to directly observe the poles with a Doppler-magnetograph may ultimately be needed to fully resolve these controversial flows.

In order to satisfy mass conservation, the meridional flow must have an equatorward return flow at some depth, and thus it is also referred to as the meridional circulation. In addition to generating the polar fields at the surface to initialize the solar cycle, the meridional circulation in the interior is believed to play an important role in setting the period of the cycle (Dikpati and Charbonneau 1999b). Long thought to be a single circulating cell in each hemisphere, modern observations are challenging that notion (Hathaway 2012; Zhao et al. 2013) with indications that a double cell may exist at times in each hemisphere. However, Gizon et al. (2020) finds evidence of a single meridional circulation cell using recent observations, so a discrepancy exists. Understanding the implications of different possible configurations (e.g., see the lower panel of Fig. 11) of the meridional return flow in the solar interior has become an integral focus of dynamo modelers (Bekki and Yokoyama 2017; Stejko et al. 2021). For a more in depth discussion on this, refer to Hazra et al. (2023), Hotta et al. (2023).

6.4 Active Region Inflows

Inflows towards AR belts are observed by local helioseismic techniques (e.g., Gizon et al. 2001; Zhao and Kosovichev 2004; Haber et al. 2004; González Hernández et al. 2008) and these flows are observed from approximately $10°$ from the AR with amplitudes up to $50\ \mathrm{ms^{-1}}$ of horizontal velocities. The AR inflows modulate the N-S meridional flow, which is on the same order of magnitude.

The explanation for the inflows is a geostrophic flow caused by increased radiative loss in the AR belt (Gizon and Rempel 2008). Gottschling et al. (2022) studied the evolution of the AR inflows and reports that converging flows are present one day prior to emergence and that these pre-emergence flows do not depend on latitude or flux. A prograde flow of about 40 ms^{-1} is found at the leading polarity during emergence (Birch et al. 2019; Gottschling et al. 2022) with the increase in amplitudes of the inflows occurring between 1–4 days after emergence.

One important consequence of AR inflows is that they slow the flux diffusion, advection, and cancellation (De Rosa and Schrijver 2006). Surface flux transport modeling highlights how AR inflows may modulate the amplitude of the global magnetic field in several ways. First, AR inflows can limit the latitudinal separation of the AR polarities, thus weakening the contribution of a given bipolar region to the axial dipole field (Jiang et al. 2010), or second, by increasing the cross-equatorial transport of magnetic flux in weaker cycles, when sunspots emerge at lower latitudes, which ultimately strengthens the axial dipole field (Cameron and Schüssler 2012). For further discussion of the implementation of AR inflows into surface flux transport models, and the results thereof, see Yeates et al. (2023) discussion on fluctuating large-scale flows.

7 Short-Term Solar Cycle Variability

There are two significant variations seen in solar-cycle data (i.e., sunspot number and area, 10.7 cm radio emission, mean solar magnetic field, coronal green line, H-alpha flare number, solar neutrino flux, p-mode frequencies) on a time period shorter than the sunspot cycle: the quasi-biennial oscillation (QBO) (1–4 years) and Rieger-type variations (50–200 days). The QBO was observed as a roughly 2-year period (Benevolenskaya 1995) in polar field components and also manifests as the double peak in sunspot numbers observed most easily near the maximum of the sunspot cycle. This double-peak is also known as the Gnevyshev gap (Gnevyshev 1967). A thorough look at QBOs can be found in Bazilevskaya et al. (2014) in which the following characteristics are listed: the QBO timescales change within the range 1–4 years with no dominant frequency; they develop in each solar hemisphere independently, but are synchronous within one hemisphere with signatures in the atmosphere and beneath the photosphere based on helioseismology (Bazilevskaya et al. 2014); they are observed in the photospheric magnetic field in phase with other solar activity indices; the QBOs are transferred into the interplanetary medium by the Sun's open magnetic flux. The Rieger type variations were first observed in gamma-ray flare activity in the 1980s with a 154 d periodicity (Rieger et al. 1984). They were subsequently shown to have many shorter periodicities and be present in sunspot number and area and photospheric magnetic field indices (Bai 2003).

Other stars show secondary, shorter cycles with smaller amplitudes than their primary cycle (Böhm-Vitense 2007), and one explanation is that the dynamo is fed by the deep-seated and near-surface shear layer. Another explanation is that the QBOs are caused by the interaction between the dipole and quadrupole terms of the solar dynamo (Wang and Sheeley 2003). The physical mechanisms responsible for the Rieger type variability may be as simple as AR evolution (Vecchio et al. 2012) or a harmonic of the QBO (Krivova and Solanki 2002). One compelling mechanism that can produce a range of short-term variability is found in MHD shallow-water modeling of an instability involving the differential rotation and toroidal field bands in the solar tachocline (Dikpati and Gilman 2005). This instability generates quasi-periodic tachocline nonlinear oscillations (TNOs) with periodicities of 2-20 months that can be correlated with the formation of persistent active longitudes (de Toma et al. 2000) seen in photospheric magnetic field data. (Dikpati et al. 2018, 2021). The mechanism involves the production of upward bulges at selected longitudes in the overshoot tachocline that contain significant toroidal fields. For a review of long-term modulation of the solar cycle, see Biswas et al. (2023).

8 Helioseismic Mode Parameter Changes

The frequency (e.g., Woodard and Noyes 1985; Libbrecht and Woodard 1990; Elsworth et al. 1990), amplitude (e.g., Elsworth et al. 1993), and lifetime (Chaplin et al. 2000; Komm et al. 2002) of the acoustic modes used in helioseismology all vary with the solar cycle, and they are spatially and temporally correlated with magnetic activity on a wide range of scales, with changes in low-degree modes following global activity measures such as the sunspot number and 10.7 cm radio flux (RF), while in local helioseismology we can see changes down to the scale of ARs (Hindman et al. 2000; Rajaguru et al. 2001; Howe et al. 2004). Most of these changes are believed (e.g. Libbrecht and Woodard 1990) to arise quite close to the surface, where the cavity in which the modes propagate is modified directly or indirectly by the presence of activity and the excitation and damping of the modes influenced by magnetic fields. The exact interpretation of these changes is difficult, but, for example Basu and Mandel (2004), Verner et al. (2006), and more recently Watson and Basu (2020) found evidence of solar-cycle changes in the signature that the helium ionization zone at $0.98R_\odot$ makes in helioseismic frequencies. These near-surface effects dominate the changes and make it difficult to use helioseismology to infer changes in the internal solar structure or magnetic fields.

While short-lived, high-degree modes can be used to study local, near-surface effects on timescales as short as a day, global helioseismology requires integration times of at least a few solar rotations to obtain the necessary precision to resolve the interior structure and dynamics, and this precludes the possibility, for example, of using helioseismology to follow the rise of an individual flux tube through the convection zone. On the timescale of a solar cycle, some marginal effects have been reported. For example, Baldner and Basu (2008) and more recently Basu (2021) found small changes in the sound speed at the base of the convection zone. The latter work reports a change of about 2×10^{-5} in the squared sound-speed at the base of the convection zone between solar maximum and minimum, anticorrelated with the activity level; this is just below the 3×10^{-5} upper limit found by Eff-Darwich et al. (2002). Small changes in the sound-speed near the base of the convection zone were also seen by Chou and Serebryanskiy (2005) using a different technique. Because these effects are so difficult to measure, they have not been widely studied. Any effects of solar-cycle changes in the magnetic fields near the base of the convection zone, which could be valuable to help in understanding the solar dynamo, remain close to or below the limits of detection.

9 Discussion

The traditional index of solar activity is the (group) sunspot number, which however, is not robustly defined before the middle of the 19th century and particularly poor in the first half of the 18th century. The research community is working hard to reconcile the sunspot dataset.

While the solar-cycle phenomena in the Sun's surface and outer atmosphere can be studied in great detail using a variety of observing techniques, helioseismology reveals motions – and to a limited extent structural changes related to the solar cycle – far below the photosphere. The torsional oscillation and the meridional circulation penetrate throughout the convection zone and play a crucial role in the solar dynamo. Therefore, it is imperative that we achieve an unambiguous inference of the structure of the meridional circulation at depth and its evolution over the solar cycle. If, for example, the observations completely ruled out a single-cell meridional circulation flow in each hemisphere, it would rule out dynamo models that rely on such a configuration.

We are able to infer flows at depths, yet the great majority of solar magnetic fields remain unobserved in the solar interior. The bipolar magnetic regions observed in the photosphere represent only the "tip of the iceberg". Until we can reliably infer magnetic field strengths and dynamics in the interior, we must rely on observations at the surface coupled with simulations to infer dynamics, amplitude and structure of the magnetic fields at depth.

To be considered successful, solar dynamo theories and simulations must be able to reproduce, to some degree, key observations. This includes the modulation of cycle amplitude as measured in the sunspot number, the observed large-scale flows, the adherence to Hale's law, the trends and inherent scatter in the tilt angles, the equator-ward migration of the active latitudes that produces the butterfly diagram, the evolution of the polar fields.

Acknowledgements All authors thank the International Space Science Institute for supporting the workshop where this review originated.

Author Contribution I.U. contributed §2 and Fig. 1. A.N. contributed §1, 3, 6.4, 7 and Figs. 2 and 3. L.U. contributed §4, 5 and Figs. 4, 5, 6, 10 and 11. R.H. contributed §6.1-3, 8 and Figs. 7, 8, and 9. All authors reviewed the full manuscript and contributed to §9.

Funding I.U. acknowledges partial support from the Academy of Finland (project 321882 ESPERA). L.U. was supported by NASA Heliophysics Living With a Star grants 80HQTR18T0023, 80NSSC20K0220, and 80NSSC23K0048 and NASA grant 80NSSC22M0162 to the COFFIES DRIVE Center managed by Stanford University. A.N. acknowledges NASA DRIVE Center COFFIES grant 80NSSC20K0602. R.H. acknowledges the support of the UK Science and Technology Facilities Council (STFC) through grant ST/V000500/1.

Declarations

References

Antia HM, Basu S (2001) Temporal variations of the solar rotation rate at high latitudes. Astrophys J Lett 559(1):L67–L70. https://doi.org/10.1086/323701. arXiv:astro-ph/0108226 [astro-ph]

Asvestari E, Usoskin IG, Kovaltsov GA et al (2017) Assessment of different sunspot number series using the cosmogenic isotope [44]Ti in meteorites. Mon Not R Astron Soc 467:1608–1613. https://doi.org/10.1093/mnras/stx190

Babcock HW (1961) The topology of the Sun's magnetic field and the 22-year cycle. Astrophys J 133:572. https://doi.org/10.1086/147060

Bai T (2003) Periodicities in solar flare occurrence: analysis of cycles 19-23. Astrophys J 591(1):406–415. https://doi.org/10.1086/375295

Baldner CS, Basu S (2008) Solar cycle related changes at the base of the convection zone. Astrophys J 686(2):1349–1361. https://doi.org/10.1086/591514. arXiv:0807.0442 [astro-ph]

Basu S (2021) Evidence of solar-cycle-related structural changes in the solar convection zone. Astrophys J 917(1):45. https://doi.org/10.3847/1538-4357/ac0c11. arXiv:2106.08383 [astro-ph.SR]

Basu S, Mandel A (2004) Does solar structure vary with solar magnetic activity? Astrophys J Lett 617(2):L155–L158. https://doi.org/10.1086/427435. arXiv:astro-ph/0411427 [astro-ph]

Bazilevskaya G, Broomhall AM, Elsworth Y et al (2014) A combined analysis of the observational aspects of the quasi-biennial oscillation in solar magnetic activity. Space Sci Rev 186(1–4):359–386. https://doi.org/10.1007/s11214-014-0068-0

Beck JG (2000) A comparison of differential rotation measurements - (invited review). Sol Phys 191(1):47–70. https://doi.org/10.1023/A:1005226402796

Bekki Y, Yokoyama T (2017) Double-cell-type solar meridional circulation based on a mean-field hydrodynamic model. Astrophys J 835(1):9. https://doi.org/10.3847/1538-4357/835/1/9. arXiv:1612.00174 [astro-ph.SR]

Benevolenskaya EE (1995) Double magnetic cycle of solar activity. Sol Phys 161(1):1–8. https://doi.org/10.1007/BF00732080

Bhowmik P, Jiang J, Upton L et al (2023) Physical models for solar cycle predictions. Space Sci Rev 219(5):40. https://doi.org/10.1007/s11214-023-00983-x. arXiv:2303.12648 [astro-ph.SR]

Birch AC, Schunker H, Braun DC et al (2019) Average surface flows before the formation of solar active regions and their relationship to the supergranulation pattern. Astron Astrophys 628:A37. https://doi.org/10.1051/0004-6361/201935591

Biswas A, Karak BB, Usoskin I et al (2023) Long-term modulation of solar cycles. Space Sci Rev 219(3):19. https://doi.org/10.1007/s11214-023-00968-w. arXiv:2302.14845 [astro-ph.SR]

Böhm-Vitense E (2007) Chromospheric activity in G and K main-sequence stars, and what it tells us about stellar dynamos. Astrophys J 657(1):486–493. https://doi.org/10.1086/510482

Cameron RH, Schüssler M (2012) Are the strengths of solar cycles determined by converging flows towards the activity belts? Astron Astrophys 548:A57. https://doi.org/10.1051/0004-6361/201219914. arXiv:1210.7644 [astro-ph.SR]

Cameron R, Schüssler M (2023) Observationally guided models for the solar dynamo and the role of the surface field. Space Sci Rev 219. https://doi.org/10.1007/s11214-023-01004-7. arXiv:2305.02253 [astro-ph.SR]

Cameron RH, Duvall TL, Schüssler M et al (2018) Observing and modeling the poloidal and toroidal fields of the solar dynamo. Astron Astrophys 609:A56. https://doi.org/10.1051/0004-6361/201731481. arXiv:1710.07126 [astro-ph.SR]

Carrasco VMS, Hayakawa H, Kuroyanagi C et al (2021) Strong evidence of low levels of solar activity during the Maunder minimum. Mon Not R Astron Soc 504(4):5199–5204. https://doi.org/10.1093/mnras/stab1155.

Chaplin WJ, Elsworth Y, Isaak GR et al (2000) Variations in the excitation and damping of low-l solar p modes over the solar activity cycle. Mon Not R Astron Soc 313(1):32–42. https://doi.org/10.1046/j.1365-8711.2000.03176.x

Charbonneau P, Sokoloff D (2023) Evolution of solar and stellar dynamo theory. Space Sci Rev 219(5):35. https://doi.org/10.1007/s11214-023-00980-0. arXiv:2305.16553 [astro-ph.SR]

Charbonneau P, St-Jean C, Zacharias P (2005) Fluctuations in Babcock-Leighton dynamos. I. Period doubling and transition to chaos. Astrophys J 619(1):613–622. https://doi.org/10.1086/426385

Chatzistergos T, Usoskin IG, Kovaltsov GA et al (2017) New reconstruction of the sunspot group numbers since 1739 using direct calibration and "backbone" methods. Astron Astrophys 602:A69. https://doi.org/10.1051/0004-6361/201630045. arXiv:1702.06183 [astro-ph.SR]

Chou DY, Serebryanskiy A (2005) In search of the solar cycle variations of p-mode frequencies generated by perturbations in the solar interior. Astrophys J 624:420–427. https://doi.org/10.1086/428925. arXiv:astro-ph/0405175

Clette F, Lefèvre L (2016) The new sunspot number: assembling all corrections. Sol Phys 291:2629–2651. https://doi.org/10.1007/s11207-016-1014-y

Clette F, Berghmans D, Vanlommel P et al (2007) From the Wolf number to the international sunspot index: 25 years of SIDC. Adv Space Res 40:919–928. https://doi.org/10.1016/j.asr.2006.12.045

Clette F, Svalgaard L, Vaquero J et al (2014) Revisiting the sunspot number: a 400-year perspective on the solar cycle. Space Sci Rev 186:35. https://doi.org/10.1007/s11214-014-0074-2

Corbard T, Blanc-Féraud L, Berthomieu G et al (1999) Non linear regularization for helioseismic inversions. Application for the study of the solar tachocline. Astron Astrophys 344:696–708. arXiv:astro-ph/9901112 [astro-ph]

Couvidat S, García RA, Turck-Chièze S et al (2003) The rotation of the deep solar layers. Astrophys J Lett 597(1):L77–L79. https://doi.org/10.1086/379698. arXiv:astro-ph/0309806 [astro-ph]

Dasi-Espuig M, Solanki SK, Krivova NA et al (2010) Sunspot group tilt angles and the strength of the solar cycle. Astron Astrophys 518:A7. https://doi.org/10.1051/0004-6361/201014301. arXiv:1005.1774 [astro-ph.SR]

De Rosa ML, Schrijver CJ (2006) Consequences of large-scale flows around active regions on the dispersal of magnetic field across the solar surface. In: Fletcher K, Thompson M (eds) Proceedings of SOHO 18/GONG 2006/HELAS I. Beyond the spherical Sun. ESA Special Publication, vol SP-624. ESA, Nordwijk, p 12

de Toma G, White OR, Harvey KL (2000) A picture of solar minimum and the onset of solar Cycle 23. I. Global magnetic field evolution. Astrophys J 529(2):1101–1114. https://doi.org/10.1086/308299

Dikpati M, Charbonneau P (1999a) A Babcock-Leighton flux transport dynamo with solar-like differential rotation. Astrophys J 518(1):508–520. https://doi.org/10.1086/307269

Dikpati M, Charbonneau P (1999b) Intermittency in solar cycle caused by stochastic fluctuation in meridional circulation. In: American astronomical society meeting abstracts, vol #194, p 92.04

Dikpati M, Gilman PA (2005) A shallow-water theory for the Sun's active longitudes. Astrophys J Lett 635(2):L193–L196. https://doi.org/10.1086/499626

Dikpati M, McIntosh SW, Bothun G et al (2018) Role of interaction between magnetic Rossby waves and tachocline differential rotation in producing solar seasons. Astrophys J 853(2):144. https://doi.org/10.3847/1538-4357/aaa70d

Dikpati M, McIntosh SW, Chatterjee S et al (2021) Deciphering the deep origin of active regions via analysis of magnetograms. Astrophys J 910(2):91. https://doi.org/10.3847/1538-4357/abe043

D'Silva S, Choudhuri AR (1993) A theoretical model for tilts of bipolar magnetic regions. Astron Astrophys 272:621

Eddy J (1976) The Maunder minimum. Science 192:1189–1202. https://doi.org/10.1126/science.192.4245.1189

Eff-Darwich A, Korzennik SG (1998) The rotation of the solar core: compatibility of the different data sets available. In: Korzennik S (ed) Structure and dynamics of the interior of the sun and sun-like stars. ESA, Nordwijk, p 685

Eff-Darwich A, Korzennik SG, Jiménez-Reyes SJ et al (2002) An upper limit on the temporal variations of the solar interior stratification. Astrophys J 580(1):574–578. https://doi.org/10.1086/343063. arXiv:astro-ph/0207402 [astro-ph]

Elsworth Y, Howe R, Isaak GR et al (1990) Variation of low-order acoustic solar oscillations over the solar cycle. Nature 345:322–324. https://doi.org/10.1038/345322a0

Elsworth Y, Howe R, Isaak GR et al (1993) The variation in the strength of low-l solar p-modes - 1981-92. Mon Not R Astron Soc 265:888–898

Fan Y (2009) Magnetic fields in the solar convection zone. Living Rev Sol Phys 6(1):4. https://doi.org/10.12942/lrsp-2009-4

Fisher GH, Fan Y, Howard RF (1995) Comparisons between theory and observation of active region tilts. Astrophys J 438:463. https://doi.org/10.1086/175090

Friedli TK (2020) Recalculation of the Wolf series from 1877 to 1893. Sol Phys 295(6):72. https://doi.org/10.1007/s11207-020-01637-9

Gilman PA, Howard R (1986) Rotation and expansion within sunspot groups. Astrophys J 303:480. https://doi.org/10.1086/164093

Gilman PA, Howe R (2003) Meridional motion and the slope of isorotation contours. In: Sawaya-Lacoste H (ed) GONG+ 2002. Local and global helioseismology: the present and future. ESA Special Publication, vol SP-517. ESA, Nordwijk, pp 283–285

Gizon L (2004) Helioseismology of time-varying flows through the solar cycle. Sol Phys 224(1–2):217–228. https://doi.org/10.1007/s11207-005-4983-9

Gizon L, Rempel M (2008) Observation and modeling of the solar-cycle variation of the meridional flow. Sol Phys 251(1–2):241–250. https://doi.org/10.1007/s11207-008-9162-3. arXiv:0803.0950 [astro-ph]

Gizon L, Duvall JTL, Larsen RM (2001) Probing surface flows and magnetic activity with time-distance helioseismology. In: Brekke P, Fleck B, Gurman JB (eds) Recent insights into the physics of the sun and heliosphere: highlights from SOHO and other space missions, p 189

Gizon L, Cameron RH, Pourabdian M et al (2020) Meridional flow in the Sun's convection zone is a single cell in each hemisphere. Science 368(6498):1469–1472. https://doi.org/10.1126/science.aaz7119

Gnevyshev MN (1967) On the 11-years cycle of solar activity. Sol Phys 1(1):107–120. https://doi.org/10.1007/BF00150306

González Hernández I, Kholikov S, Hill F et al (2008) Subsurface meridional circulation in the active belts. Sol Phys 252(2):235–245. https://doi.org/10.1007/s11207-008-9264-y. arXiv:0808.3606 [astro-ph]

Gottschling N, Schunker H, Birch AC et al (2022) Testing solar surface flux transport models in the first days after active region emergence. Astron Astrophys 660:A6. https://doi.org/10.1051/0004-6361/202142071. arXiv:2111.01896 [astro-ph.SR]

Haber DA, Hindman BW, Toomre J et al (2004) Organized subsurface flows near active regions. Sol Phys 220(2):371–380. https://doi.org/10.1023/B:SOLA.0000031405.52911.08

Hagenaar HJ, Schrijver CJ, Title AM (2003) The properties of small magnetic regions on the solar surface and the implications for the solar dynamo(s). Astrophys J 584(2):1107–1119. https://doi.org/10.1086/345792

Hale GE (1908) On the probable existence of a magnetic field in sun-spots. Astrophys J 28:315. https://doi.org/10.1086/141602

Hanasoge SM (2022) Surface and interior meridional circulation in the sun. Living Rev Sol Phys 19(1):3. https://doi.org/10.1007/s41116-022-00034-7

Harvey KL (1994) The solar magnetic cycle. In: Rutten RJ, Schrijver CJ (eds) Solar surface magnetism. NATO ASI Series, vol 433. Springer, Dordrecht, pp 347–363. https://doi.org/10.1007/978-94-011-1188-1_30

Harvey-Angle KL (1993) Magnetic bipoles on the sun. PhD thesis. Utrecht University

Hathaway DH (2012) Supergranules as probes of the Sun's meridional circulation. Astrophys J 760(1):84. https://doi.org/10.1088/0004-637X/760/1/84. arXiv:1210.3343 [astro-ph.SR]

Hathaway DH, Rightmire L (2010) Variations in the Sun's meridional flow over a solar cycle. Science 327(5971):1350. https://doi.org/10.1126/science.1181990

Hathaway DH, Upton LA, Mahajan SS (2022) Variations in differential rotation and meridional flow within the sun's surface shear layer 1996–2022. Front Astron Space Sci 9:419. https://doi.org/10.3389/fspas.2022.1007290. arXiv:2212.10619 [astro-ph.SR]

Hayakawa H, Lockwood M, Owens M et al (2021) Graphical evidence for the solar coronal structure during the Maunder minimum: comparative study of the total eclipse drawings in 1706 and 1715. J Space Weather Space Clim 11:1. https://doi.org/10.1051/swsc/2020035

Hazra G, Nandy D, Kitchatinov L et al (2023) Mean field models of flux transport dynamo and meridional circulation in the sun and stars. Space Sci Rev 219(5):39. https://doi.org/10.1007/s11214-023-00982-y. arXiv:2302.09390 [astro-ph.SR]

Hindman B, Haber D, Toomre J et al (2000) Local fractional frequency shifts used as tracers of magnetic activity. Sol Phys 192:363–372. https://doi.org/10.1023/A:1005283302728

Hoeksema JT (1995) The large-scale structure of the heliospheric current sheet during the Ulysses epoch. Space Sci Rev 72(1–2):137–148. https://doi.org/10.1007/BF00768770

Hotta H, Bekki Y, Gizon L et al (2023) Dynamics of solar large-scale flows. https://doi.org/10.48550/arXiv.2307.06481. arXiv:2307.06481 [astro-ph.SR]

Howard R, LaBonte BJ (1980) The sun is observed to be a torsional oscillator with a period of 11 years. Astrophys J Lett 239:L33–L36. https://doi.org/10.1086/183286

Howe R (2009) Solar interior rotation and its variation. Living Rev Sol Phys 6(1):1. https://doi.org/10.12942/lrsp-2009-1. arXiv:0902.2406 [astro-ph.SR]

Howe R, Christensen-Dalsgaard J, Hill F et al (2000) Deeply penetrating banded zonal flows in the solar convection zone. Astrophys J Lett 533:L163–L166. https://doi.org/10.1086/312623. arXiv:astro-ph/0003121

Howe R, Komm RW, Hill F et al (2004) Activity-related changes in local solar acoustic mode parameters from Michelson Doppler imager and global oscillations network group. Astrophys J 608(1):562–579. https://doi.org/10.1086/392525

Howe R, Christensen-Dalsgaard J, Hill F et al (2009) A note on the torsional oscillation at solar minimum. Astrophys J Lett 701(2):L87–L90. https://doi.org/10.1088/0004-637X/701/2/L87. arXiv:0907.2965 [astro-ph.SR]

Howe R, Christensen-Dalsgaard J, Hill F et al (2013) The high-latitude branch of the solar torsional oscillation in the rising phase of cycle 24. Astrophys J Lett 767(1):L20. https://doi.org/10.1088/2041-8205/767/1/L20

Howe R, Hill F, Komm R et al (2018) Signatures of solar cycle 25 in subsurface zonal flows. Astrophys J Lett 862(1):L5. https://doi.org/10.3847/2041-8213/aad1ed. arXiv:1807.02398 [astro-ph.SR]

Howe R, Chaplin WJ, Christensen-Dalsgaard J et al (2022) Update on global helioseismic observations of the solar torsional oscillation. Res Notes AAS 6(12):261. https://doi.org/10.3847/2515-5172/aca97d

Hoyt DV, Schatten KH (1998) Group sunspot numbers: a new solar activity reconstruction. Sol Phys 181:491–512. https://doi.org/10.1023/A:1005056326158

Ivanov VG, Miletsky EV (2011) Width of sunspot generating zone and reconstruction of butterfly. Sol Phys 268(1):231–242. https://doi.org/10.1007/s11207-010-9665-6. arXiv:1011.4800 [astro-ph.SR]

Jiang J, Cameron R, Schmitt D et al (2009) Countercell meridional flow and latitudinal distribution of the solar polar magnetic field. Astrophys J Lett 693(2):L96–L99. https://doi.org/10.1088/0004-637X/693/2/L96

Jiang J, Işik E, Cameron RH et al (2010) The effect of activity-related meridional flow modulation on the strength of the solar polar magnetic field. Astrophys J 717(1):597–602. https://doi.org/10.1088/0004-637X/717/1/597. arXiv:1005.5317 [astro-ph.SR]

Jiang J, Cameron RH, Schüssler M (2014) Effects of the scatter in sunspot group tilt angles on the large-scale magnetic field at the solar surface. Astrophys J 791(1):5. https://doi.org/10.1088/0004-637X/791/1/5. arXiv:1406.5564 [astro-ph.SR]

Komm R, Howe R, Hill F (2002) Localizing width and energy of solar global p-modes. Astrophys J 572:663–673. https://doi.org/10.1086/340196

Komm R, Howe R, Hill F (2018) Subsurface zonal and meridional flow during cycles 23 and 24. Sol Phys 293(10):145. https://doi.org/10.1007/s11207-018-1365-7

Krivova NA, Solanki SK (2002) The 1.3-year and 156-day periodicities in sunspot data: wavelet analysis suggests a common origin. Astron Astrophys 394:701–706. https://doi.org/10.1051/0004-6361:20021063

Larmor J (1919) How could a rotating body such as the sun become magnetic. Rep Br Assoc Adv Sci 159:160

Larson TP, Schou J (2018) Global-mode analysis of full-disk data from the Michelson Doppler Imager and the Helioseismic and Magnetic Imager. Sol Phys 293(2):29. https://doi.org/10.1007/s11207-017-1201-5

Leighton RB (1964) Transport of magnetic fields on the sun. Astrophys J 140:1547. https://doi.org/10.1086/148058

Lekshmi B, Nandy D, Antia HM (2018) Asymmetry in solar torsional oscillation and the sunspot cycle. Astrophys J 861(2):121. https://doi.org/10.3847/1538-4357/aacbd5. arXiv:1807.03588 [astro-ph.SR]

Lemerle A, Charbonneau P (2017) A coupled 2 × 2D Babcock-Leighton solar dynamo model. II. Reference dynamo solutions. Astrophys J 834(2):133. https://doi.org/10.3847/1538-4357/834/2/133. arXiv:1606.07375 [astro-ph.SR]

Leussu R, Usoskin IG, Arlt R et al (2013) Inconsistency of the Wolf sunspot number series around 1848. Astron Astrophys 559:A28. https://doi.org/10.1051/0004-6361/201322373. arXiv:1310.8443 [astro-ph.SR]

Li J (2018) A systematic study of hale and anti-hale sunspot physical parameters. Astrophys J 867(2):89. https://doi.org/10.3847/1538-4357/aae31a. arXiv:1809.08980 [astro-ph.SR]

Libbrecht KG, Woodard MF (1990) Solar-cycle effects on solar oscillation frequencies. Nature 345:779–782. https://doi.org/10.1038/345779a0

Maunder EW (1903) Spoerer's law of zones. Observatory 26:329–330

Maunder EW (1904) Note on the distribution of sun-spots in heliographic latitude, 1874-1902. Mon Not R Astron Soc 64:747–761. https://doi.org/10.1093/mnras/64.8.747

McClintock BH, Norton AA (2013) Recovering Joy's law as a function of solar cycle, hemisphere, and longitude. Sol Phys 287(1–2):215–227. https://doi.org/10.1007/s11207-013-0338-0. arXiv:1305.3205 [astro-ph.SR]

McClintock BH, Norton AA (2016) Tilt angle and footpoint separation of small and large bipolar sunspot regions observed with HMI. Astrophys J 818(1):7. https://doi.org/10.3847/0004-637X/818/1/7. arXiv:1602.04154 [astro-ph.SR]

McClintock BH, Norton AA, Li J (2014) Re-examining sunspot tilt angle to include anti-hale statistics. Astrophys J 797(2):130. https://doi.org/10.1088/0004-637X/797/2/130. arXiv:1412.5094 [astro-ph.SR]

Muñoz-Jaramillo A, Dasi-Espuig M, Balmaceda LA et al (2013) Solar cycle propagation, memory, and prediction: insights from a century of magnetic proxies. Astrophys J Lett 767(2):L25. https://doi.org/10.1088/2041-8205/767/2/L25. arXiv:1304.3151 [astro-ph.SR]

Muñoz-Jaramillo A, Navarrete B, Campusano LE (2021) Solar anti-hale bipolar magnetic regions: a distinct population with systematic properties. Astrophys J 920(1):31. https://doi.org/10.3847/1538-4357/ac133b. arXiv:2203.11898 [astro-ph.SR]

Nagy M, Lemerle A, Labonville F et al (2017) The effect of "rogue" active regions on the solar cycle. Sol Phys 292(11):167. https://doi.org/10.1007/s11207-017-1194-0. arXiv:1712.02185 [astro-ph.SR]

Nagy M, Lemerle A, Charbonneau P (2020) Impact of nonlinear surface inflows into activity belts on the solar dynamo. J Space Weather Space Clim 10:62. https://doi.org/10.1051/swsc/2020064

Nelson NJ, Brown BP, Brun AS et al (2013) Magnetic wreaths and cycles in convective dynamos. Astrophys J 762(2):73. https://doi.org/10.1088/0004-637X/762/2/73. arXiv:1211.3129 [astro-ph.SR]

Norton AA, Gilman PA (2005) Recovering solar toroidal field dynamics from sunspot location patterns. Astrophys J 630(2):1194–1205. https://doi.org/10.1086/431961. arXiv:astro-ph/0506025 [astro-ph]

Norton AA, Charbonneau P, Passos D (2014) Hemispheric coupling: comparing dynamo simulations and observations. Space Sci Rev 186(1–4):251–283. https://doi.org/10.1007/s11214-014-0100-4. arXiv:1411.7052 [astro-ph.SR]

 Springer

Norton AA, Jones EH, Linton MG et al (2017) Magnetic flux emergence and decay rates for preceder and follower sunspots observed with HMI. Astrophys J 842(1):3. https://doi.org/10.3847/1538-4357/aa7052. arXiv:1705.02053 [astro-ph.SR]

Petrovay K (2010) Solar cycle prediction. Living Rev Sol Phys 7:6. https://doi.org/10.12942/lrsp-2010-6. arXiv:1012.5513 [astro-ph.SR]

Petrovay K, Nagy M, Yeates AR (2020) Towards an algebraic method of solar cycle prediction. I. Calculating the ultimate dipole contributions of individual active regions. J Space Weather Space Clim 10:50. https://doi.org/10.1051/swsc/2020050. arXiv:2009.02299 [astro-ph.SR]

Rajaguru SP, Basu S, Antia HM (2001) Ring diagram analysis of the characteristics of solar oscillation modes in active regions. Astrophys J 563:410–418. https://doi.org/10.1086/323780. arXiv:astro-ph/0108227

Rempel M (2012) High latitude solar torsional oscillations during phases of changing magnetic cycle amplitude. Astrophys J Lett 750(1):L8. https://doi.org/10.1088/2041-8205/750/1/L8

Rieger E, Share GH, Forrest DJ et al (1984) A 154-day periodicity in the occurrence of hard solar flares? Nature 312(5995):623–625. https://doi.org/10.1038/312623a0

Rightmire-Upton L, Hathaway DH, Kosak K (2012) Measurements of the Sun's high-latitude meridional circulation. Astrophys J Lett 761(1):L14. https://doi.org/10.1088/2041-8205/761/1/L14. arXiv:1211.0944 [astro-ph.SR]

Schatten KH, Scherrer PH, Svalgaard L et al (1978) Using dynamo theory to predict the sunspot number during Solar Cycle 21. Geophys Res Lett 5(5):411–414. https://doi.org/10.1029/GL005i005p00411

Schou J (1999) Migration of zonal flows detected using Michelson Doppler Imager F-mode frequency splittings. Astrophys J Lett 523:L181–L184. https://doi.org/10.1086/312279

Schou J, Antia HM, Basu S et al (1998) Helioseismic studies of differential rotation in the solar envelope by the solar oscillations investigation using the Michelson Doppler Imager. Astrophys J 505(1):390–417. https://doi.org/10.1086/306146

Schunker H, Baumgartner C, Birch AC et al (2020) Average motion of emerging solar active region polarities. II. Joy's law. Astron Astrophys 640:A116. https://doi.org/10.1051/0004-6361/201937322. arXiv:2006.05565 [astro-ph.SR]

Schwabe H (1844) Sonnenbeobachtungen im Jahre 1843. Von Herrn Hofrath Schwabe in Dessau. Astron Nachr 21(15):233. https://doi.org/10.1002/asna.18440211505

Snodgrass HB, Ulrich RK (1990) Rotation of Doppler features in the solar photosphere. Astrophys J 351:309. https://doi.org/10.1086/168467

Spörer G (1894) Beobachtungen von Sonnenflecken in den Jahren 1885 bis 1893. Publ Astrophys Obs Potsdam 10:3–147

Stejko AM, Kosovichev AG, Pipin VV (2021) Forward modeling helioseismic signatures of one- and two-cell meridional circulation. Astrophys J 911(2):90. https://doi.org/10.3847/1538-4357/abec70. arXiv:2101.01220 [astro-ph.SR]

Stenflo JO (1970) Hale's attempts to determine the Sun's general magnetic field. Sol Phys 14(2):263–273. https://doi.org/10.1007/BF00221312

Stenflo JO, Kosovichev AG (2012) Bipolar magnetic regions on the sun: global analysis of the SOHO/MDI data set. Astrophys J 745(2):129. https://doi.org/10.1088/0004-637X/745/2/129. arXiv:1112.5226 [astro-ph.SR]

Sun X, Hoeksema JT, Liu Y et al (2015) On polar magnetic field reversal and surface flux transport during solar cycle 24. Astrophys J 798(2):114. https://doi.org/10.1088/0004-637X/798/2/114. arXiv:1410.8867 [astro-ph.SR]

Svalgaard L, Schatten KH (2016) Reconstruction of the sunspot group number: the backbone method. Sol Phys 291:2653–2684. https://doi.org/10.1007/s11207-015-0815-8. arXiv:1506.00755 [astro-ph.SR]

Svalgaard L, Duvall JTL, Scherrer PH (1978) The strength of the Sun's polar fields. Sol Phys 58(2):225–239. https://doi.org/10.1007/BF00157268

Svalgaard L, Cliver EW, Kamide Y (2005) Sunspot cycle 24: smallest cycle in 100 years? Geophys Res Lett 32(1):L01104. https://doi.org/10.1029/2004GL021664

Thompson MJ, Toomre J, Anderson ER et al (1996) Differential rotation and dynamics of the solar interior. Science 272(5266):1300–1305. https://doi.org/10.1126/science.272.5266.1300

Tlatov A, Illarionov E, Sokoloff D et al (2013) A new dynamo pattern revealed by the tilt angle of bipolar sunspot groups. Mon Not R Astron Soc 432(4):2975–2984. https://doi.org/10.1093/mnras/stt659. arXiv:1302.2715 [astro-ph.SR]

Tlatova K, Tlatov A, Pevtsov A et al (2018) Tilt of sunspot bipoles in solar cycles 15 to 24. Sol Phys 293(8):118. https://doi.org/10.1007/s11207-018-1337-y. arXiv:1807.07913 [astro-ph.SR]

Ulrich RK (2010) Solar meridional circulation from Doppler shifts of the Fe I line at 5250 a as measured by the 150-foot solar tower telescope at the mt. Wilson observatory. Astrophys J 725(1):658. https://doi.org/10.1088/0004-637X/725/1/658

52

Ulrich RK, Tran T, Boyden JE (2022) Polar upwelling at three sunspot minima. Res Notes AAS 6(9):181. https://doi.org/10.3847/2515-5172/ac905f

Upton L, Hathaway DH (2014a) Effects of meridional flow variations on solar Cycles 23 and 24. Astrophys J 792(2):142. https://doi.org/10.1088/0004-637X/792/2/142. arXiv:1408.0035 [astro-ph.SR]

Upton LA, Hathaway DH (2014b) Predicting the sun's polar magnetic fields with a surface flux transport model. Astrophys J 780:5. https://doi.org/10.1088/0004-637X/780/1/5. arXiv:1311.0844 [astro-ph.SR]

Usoskin IG (2023) A history of solar activity over millennia. Living Rev Sol Phys 20:2. https://doi.org/10.1007/s41116-023-00036-z

Usoskin IG, Arlt R, Asvestari E et al (2015) The Maunder minimum (1645-1715) was indeed a grand minimum: a reassessment of multiple datasets. Astron Astrophys 581:A95. https://doi.org/10.1051/0004-6361/201526652. arXiv:1507.05191 [astro-ph.SR]

Usoskin IG, Kovaltsov GA, Lockwood M et al (2016) A new calibrated sunspot group series since 1749: statistics of active day fractions. Sol Phys 291:2685–2708. https://doi.org/10.1007/s11207-015-0838-1. arXiv:1512.06421 [astro-ph.SR]

van Driel-Gesztelyi L, Green LM (2015) Evolution of active regions. Living Rev Sol Phys 12(1):1. https://doi.org/10.1007/lrsp-2015-1

Vaquero J, Svalgaard L, Carrasco V et al (2016) A revised collection of sunspot group numbers. Sol Phys 291:3061–3074. https://doi.org/10.1007/s11207-016-0982-2

Vecchio A, Laurenza M, Meduri D et al (2012) The dynamics of the solar magnetic field: polarity reversals, butterfly diagram, and quasi-biennial oscillations. Astrophys J 749(1):27. https://doi.org/10.1088/0004-637X/749/1/27

Verner GA, Chaplin WJ, Elsworth Y (2006) BiSON data show change in solar structure with magnetic activity. Astrophys J Lett 640(1):L95–L98. https://doi.org/10.1086/503101

Vorontsov S, Christensen-Dalsgaard J, Schou J et al (2002) Helioseismic measurement of solar torsional oscillations. Science 296:101–103. https://doi.org/10.1126/science.1069190

Waldmeier M (1961) The sunspot activity in the years 1610-1960. Schulthess, Zürich

Wang YM, Sheeley NR Jr (1989) Average properties of bipolar magnetic regions during sunspot cycle-21. Sol Phys 124(1):81–100. https://doi.org/10.1007/BF00146521

Wang YM, Sheeley NR Jr (1991) Magnetic flux transport and the Sun's dipole moment: new twists to the Babcock-Leighton model. Astrophys J 375:761. https://doi.org/10.1086/170240

Wang YM, Sheeley NR Jr (2003) On the fluctuating component of the Sun's large-scale magnetic field. Astrophys J 590(2):1111–1120. https://doi.org/10.1086/375026

Wang YM, Colaninno RC, Baranyi T et al (2015) Active-region tilt angles: magnetic versus white-light determinations of Joy's law. Astrophys J 798(1):50. https://doi.org/10.1088/0004-637X/798/1/50. arXiv:1412.2329 [astro-ph.SR]

Watson CB, Basu S (2020) Solar-cycle-related changes in the helium ionization zones of the sun. Astrophys J Lett 903(2):L29. https://doi.org/10.3847/2041-8213/abc348. arXiv:2010.11215 [astro-ph.SR]

Weber MA, Fan Y, Miesch MS (2011) The rise of active region flux tubes in the turbulent solar convective envelope. Astrophys J 741(1):11. https://doi.org/10.1088/0004-637X/741/1/11. arXiv:1109.0240 [astro-ph.SR]

Weber MA, Schunker H, Jouve L et al (2023) Understanding active region emergence and origins on the sun and other cool stars. https://doi.org/10.48550/arXiv.2306.06536. arXiv:2306.06536 [astro-ph.SR]

Willamo T, Usoskin IG, Kovaltsov GA (2017) Updated sunspot group number reconstruction for 1749-1996 using the active day fraction method. Astron Astrophys 601:A109. https://doi.org/10.1051/0004-6361/201629839. arXiv:1705.05109 [astro-ph.SR]

Woodard MF, Noyes RW (1985) Change of solar oscillation eigenfrequencies with the solar cycle. Nature 318(6045):449–450. https://doi.org/10.1038/318449a0

Yeates AR, Cheung MCM, Jiang J et al (2023) Surface Flux Transport on the Sun. https://doi.org/10.48550/arXiv.2303.01209, arXiv:2303.01209 [astro-ph.SR]

Zhao J, Kosovichev AG (2004) Torsional oscillation, meridional flows, and vorticity inferred in the upper convection zone of the sun by time-distance helioseismology. Astrophys J 603(2):776–784. https://doi.org/10.1086/381489

Zhao J, Bogart RS, Kosovichev AG et al (2013) Detection of equatorward meridional flow and evidence of double-cell meridional circulation inside the sun. Astrophys J Lett 774(2):L29. https://doi.org/10.1088/2041-8205/774/2/L29. arXiv:1307.8422 [astro-ph.SR]

Zirin H (1988) Astrophysics of the sun. Cambridge University Press, Cambridge

Zolotova NV, Ponyavin DI (2015) The Maunder minimum is not as grand as it seemed to be. Astrophys J 800:42. https://doi.org/10.1088/0004-637X/800/1/42

Publisher's Note Springer Nature remains neutral with regard to jurisdictional claims in published maps and institutional affiliations.

Space Science Reviews (2023) 219:19
https://doi.org/10.1007/s11214-023-00968-w

Long-Term Modulation of Solar Cycles

Akash Biswas[1] · Bidya Binay Karak[1] · Ilya Usoskin[2] · Eckhard Weisshaar[3]

Received: 13 December 2022 / Accepted: 1 March 2023 / Published online: 17 March 2023
© The Author(s) 2023

Abstract

Solar activity has a cyclic nature with the ≈ 11-year Schwabe cycle dominating its variability on the interannual timescale. However, solar cycles are significantly modulated in length, shape and magnitude, from near-spotless grand minima to very active grand maxima. The ≈ 400-year-long direct sunspot-number series is inhomogeneous in quality and too short to study robust parameters of long-term solar variability. The cosmogenic-isotope proxy extends the timescale to twelve millennia and provides crucial observational constraints of the long-term solar dynamo modulation. Here, we present a brief up-to-date overview of the long-term variability of solar activity at centennial–millennial timescales. The occurrence of grand minima and maxima is discussed as well as the existing quasi-periodicities such as centennial Gleissberg, 210-year Suess/de Vries and 2400-year Hallstatt cycles. It is shown that the solar cycles contain an important random component and have no clock-like phase locking implying a lack of long-term memory. A brief yet comprehensive review of the theoretical perspectives to explain the observed features in the framework of the dynamo models is presented, including the nonlinearity and stochastic fluctuations in the dynamo. We keep gaining knowledge of the processes driving solar variability with the new data acquainted and new models developed.

Keywords Solar activity · Solar cycle · Cosmogenic isotopes

1 Introduction

Sun is a magnetically active star whose activity is a result of the magnetic dynamo process operating in the Sun's convection zone (see, e.g., Karak et al. 2014; Charbonneau 2020).

Solar and Stellar Dynamos: A New Era
Edited by Manfred Schüssler, Robert H. Cameron, Paul Charbonneau, Mausumi Dikpati, Hideyuki Hotta and Leonid Kitchatinov

✉ I. Usoskin
   ilya.usoskin@oulu.fi

1   Department of Physics, Indian Institute of Technology (Banaras Hindu University), Varanasi, 221005, UP, India

2   Space Physics and Astronomy Research Unit and Sodankylä Geophysical Observatory, University of Oulu, Oulu, 90014, Finland

3   Company, Software & Automation, Brunnenstr. 58, Bad Nauheim, 61231, Germany

Solar surface magnetic activity varies cyclicly with the main period of about 11 years (called the Schwabe cycle) or, considering inversion of the sign of its magnetic polarity, the 22-year Hale cycle. More details can be found in an extensive review by Hathaway (2015). The physics of the dynamo mechanism is currently believed to be reasonably well understood. However, solar cyclicity is far from being a regularly ticking clock and experiences essential long-term variability at timescales longer than the Schwabe cycle. The solar cycles are not perfectly regular and vary in length, shape, and strength/intensity, or even can enter periods of almost inactive state, called grand minima of solar activity (e.g., Usoskin 2017).

The standard index quantifying solar activity is related to sunspot numbers which are available from 1610 AD onward with the quality degrading backwards in time, as discussed in Sect. 2. On one hand, this 410-year-long series exhibits a great deal of variability covering the range from an almost spotless period of the Maunder minimum between 1645 – 1715 AD (Eddy 1976) to an epoch of very active Sun between 1940 – 2009 called the Modern grand maximum (Solanki et al. 2004; Usoskin et al. 2007). This great variability raises important questions, answers to which can put crucial observational constraints on the solar/stellar dynamo theory:

- Do the changes between the Maunder minimum and the Modern grand maximum cover the full possible range of solar variability?
- Do the grand minima and maxima represent special states of the solar dynamo or simply represent the tails of the distribution?
- How typical are these changes?
- Do the grand minima episodes appear periodically or randomly?
- What physical processes drive such changes?

The four-century-long sunspot number series is not sufficiently long to answer these questions, and a much longer dataset is needed to form a basis for the answers. Fortunately, solar activity can be reliably reconstructed from indirect natural proxy data (cosmogenic radioisotopes) on the timescale of 10 – 12 millennia, during the period of the Holocene with a stable warm climate on Earth, as discussed in Sect. 3. This reconstruction extends the solar-activity dataset by a factor of about 25 making it possible to perform a thorough statistical analysis of solar variability as discussed in Sect. 4, while statistical properties of the solar-cycle modulation are summarized in Sect. 5. In Sect. 6, we discuss the implications of the long-term solar variability for the solar dynamo theory and our present level of understanding of the related physics.

2 Direct Sunspot Number Series Since 1610

Sunspots have been more or less systematically studied since 1610, soon after the invention of the telescope. Thousands of observational records and drawings exist in archives as being continuously recovered and analyzed (e.g., Vaquero and Vázquez 2009; Arlt and Vaquero 2020). The most recent and continuously updated database of raw sunspot-group observation is collected at the HASO (Historical Archive of Sunspot Observations, http://haso.unex.es/haso – Vaquero et al. 2016).

Despite numerous observational records, it was noticed only in the middle of the 18th century by the Danish astronomer Christian Horrebow and finally confirmed in the early 19th century by the German observer Heinrich Schwabe, that the number of sunspots varies cyclicly with about 10-year period. This cycle was later shown to be of about 11 years mean length and appears to be a fundamental feature of solar activity and is now called the

Schwabe cycle. More details of the sunspot number measurements and reconstructions can be found elsewhere in this volume or in comprehensive reviews by Hathaway (2015) and Usoskin (2017).

2.1 Wolf Sunspot Series R_W and International Sunspot Number R_I

Following the discovery of the solar cycle, Rudolf Wolf from Zürich Observatory founded a synthetic index called the sunspot number presently known as Wolf or Zürich sunspot number R_W (WSN) defined as

$$R_W = k \cdot (10 \cdot G + S), \tag{1}$$

where G and S are the numbers of sunspot groups and all sunspots, including those in groups, respectively, visible on the solar disc during a given day by the primary observer whose quality scaling factor k is set to reduce his/her counts to the reference observer with $k \equiv 1$. Obviously, the sunspot number is not the same as the number of spots, and for a single sunspot, $R_W = 11$ assuming $k = 1$. This series, constructed by R. Wolf in 1861 using his own and recovered earlier observations, formally covered the period since 1749 (solar cycle SC #1 in Wolf's numbering), but in fact, it was more or less reliable only since the 1820's when H. Schwabe started his observations. Later it was extended back to 1700 with unreliable data. The compilation of the R_W was continued at Zürich by Wolf's successors Wolfer, Brunner, Waldmeier and Koeckelenbergh until 1981 when the formation of the sunspot series was transferred to the Royal Observatory of Belgium (Clette et al. 2007).

Until 1981, the R_W was constructed considering the observation of only one primary observer for each day, all other observations were discarded. This series could not, till now, be revisited or redone because of the lack of original raw data. Accordingly, when several apparent inhomogeneities were found in the standard Wolf sunspot series (Leussu et al. 2013; Clette et al. 2014; Lockwood et al. 2014), only step-wise corrections to the old series could be done (Clette et al. 2014; Clette and Lefèvre 2016). This 'corrected' sunspot series is known as the International sunspot number series version 2.0, $R_I(2.0)$, and is available at the SILSO (Sunspot Index and Long-term Solar Observations, https://www.sidc.be/silso/datafiles) formally since 1700. The $R_I(2.0)$ is shown in Fig. 1a along with the standard Zürich sunspot cycle numbering.

Although the update of the series was through several adjustments of scaling jumps, an important effort is currently done by the community to restore and digitize old raw data (Clette et al. 2021) so that it will be possible to redo the sunspot number series from scratch increasing its reliability and assessing realistic uncertainties.

2.2 Group Sunspot Number Series G_N

Since the sunspot number (Equation 1) includes both numbers of sunspot groups (weighted by a factor of 10) and individual sunspots, it is sensitive to the quality of observations. This was addressed by Hoyt and Schatten (1998) who noticed that sunspot groups are defined more reliably than individual spots and created the group sunspot number series G_N which is simply the number of sunspot groups G on the solar disc corrected for the observer's quality. This series is shown in Fig. 1b. Sometimes it is scaled up to match the values typical for R_W. However, contrary to R_W, G_N is based on the average of all available observations for each day, not only the primary ones. Another principal difference between R_W and G_N is that Hoyt and Schatten (1998) created and published a full database of raw data they used

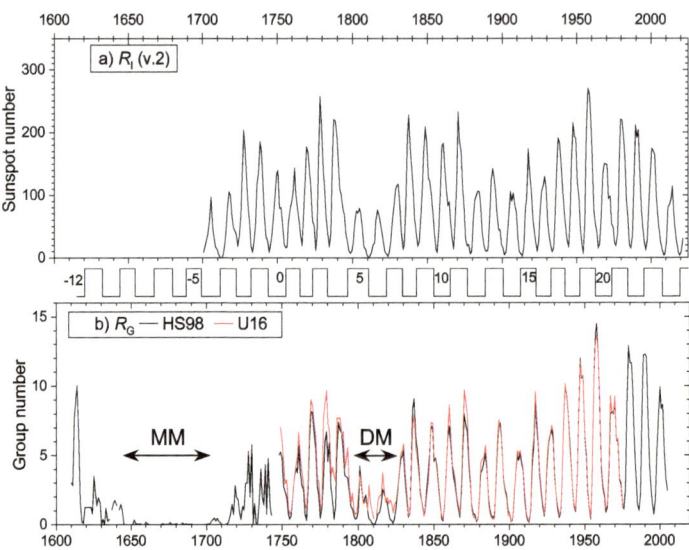

Fig. 1 Annual sunspot activity for the last centuries based on direct sunspot observations: a) International sunspot number series version 2 from SILSO (http://sidc.be/silso/datafiles). b) Number of sunspot groups according to Hoyt and Schatten (1998, – HS98) and Usoskin et al. (2016b, – U16). Approximate dates of the Maunder minimum (MM) and Dalton minimum (DM) are shown in the lower panel. Standard (Zürich) cycle numbering is shown between the panels. Cycles during the MM are only indicative as provided by Usoskin et al. (2000)

to construct the G_N series. Accordingly, this series can be completely redone as a whole, without limitation to the 'correction factors'.

It was recognized that the original G_N underestimated solar activity during the 19-th century (Clette et al. 2014), and several efforts have been made to revisit it using different methodologies and inter-calibrations (e.g., Svalgaard and Schatten 2016; Usoskin et al. 2016b; Chatzistergos et al. 2017; Willamo et al. 2017). One of the reconstructions is also shown in Fig. 1b. However, these new series often moderately disagree with each other illustrating the problem of compiling a homogeneous series from individual raw datasets (Muñoz-Jaramillo and Vaquero 2019). It is presently impossible to decide between different reconstructions of the group sunspot series, but the zoo of those gives a clue of what the related uncertainties are, and presently they are bounded by the series of Svalgaard and Schatten (2016) from the top and from below by Hoyt and Schatten (1998).

3 Cosmogenic-Isotope-Based Reconstructions of Long-Term Solar Variability

The sunspot number series covers ca. 410 years in the past with the quality degrading back in time (Muñoz-Jaramillo and Vaquero 2019) and principally cannot be extended before the 17-th century because of the lack of instrumental data. Unaided (naked-eye) observations of sunspots do not provide systematic quantitative information on solar activity (Usoskin 2017). There are some other proxy-based indices of solar activity, such as geomagnetic or heliospheric activity, and radio-emission of the Sun, but they all are based on scientific measurements and typically do not go beyond the middle of the 19-th century. Fortunately, there is one solar-activity proxy which can help in reconstructing solar variability on the multi-millennial timescale. This is related to cosmogenic radioisotopes which are produced and preserved in dateable archives in a natural way.

3.1 Method of Cosmogenic Isotopes

Solar surface magnetic activity and hot corona create the solar wind which is a supersonic outflow of solar coronal plasma permanently emitted from the Sun (see, e.g., Vidotto 2021). Because of its high conductivity, solar wind drags away the solar magnetic field which appears 'frozen' in the solar-wind plasma. This wind radially expands forming the heliosphere, a region of about 200 astronomical units across which is totally controlled (in the magneto-hydrodynamical sense) by the solar wind and magnetic field (e.g., Owens and Forsyth 2013). The heliosphere makes an obstacle for charged highly energetic particles of galactic cosmic rays (GCRs) which permanently bombard it isotropically with nearly constant flux. Inside the heliosphere, cosmic rays are affected by four major processes, viz. scattering and diffusion on magnetic irregularities, convection by expanding solar wind, adiabatic cooling, and large-scale drifts. All these processes are ultimately driven by solar activity leading to the solar modulation of cosmic-ray flux near Earth so that the cosmic-ray flux is stronger when solar activity is weak and vice-versa (e.g., Potgieter 2013). Thus, knowing the modulated flux of GCRs at a moment in time, one can assess the level of solar activity slightly before that (within one year – Koldobskiy et al. 2022). Of course, there were no scientific cosmic-ray detectors in the distant past, but there is a natural cosmic-ray monitor – cosmogenic radioisotopes.

Cosmogenic radioisotopes are unstable nuclides, which cannot survive from the time of the solar-system formation, and whose main source is related to nuclear reactions caused by cosmic rays in the Earth's atmosphere (Beer et al. 2012). After production in the atmosphere by GCR, nuclides can be stored in natural independently dateable archives, such as tree trunks, polar ice cores, lake/marine sediments, etc. Accordingly, the flux of GCR can be estimated in the past by measuring the abundance of such isotopes in the archives, forming the only quantitative proxy of solar activity over long timescales (see more details in Beer 2000; Usoskin 2017). The most important cosmogenic isotopes are ^{14}C 'radiocarbon' (half-life 5730 years) measured in dendrochronologically dated tree rings and ^{10}Be ($\approx 1.4 \cdot 10^6$ years) measured in glaciologically dated ice cores.

Conversion between the measured isotope concentration and production by cosmic rays requires a knowledge of the isotope's transport and deposition processes which are currently well modelled (e.g., Roth and Joos 2013; Heikkilä et al. 2013; Golubenko et al. 2021). Additionally, it needs to be corrected for the changing geomagnetic field (e.g., Pavón-Carrasco et al. 2018), and the resulting variability can be attributed to solar activity. The conversion from the cosmic-ray modulation to the heliospheric properties (open solar flux) and then to the pseudo-sunspot numbers is done via a chain of physics-based models making it possible to reconstruct solar activity and the related uncertainties (see, e.g., Usoskin 2017; Wu et al. 2018).

3.2 Holocene ($\approx$12 kyr) Decadal Reconstruction

While the idea of the use of cosmogenic-isotope data as a proxy to solar activity has been discussed since long (Stuiver 1961; Lal and Peters 1962), first approaches were empirical as based on timescale separation of the cosmogenic data: timescales longer than 500 years were thought to be caused by changes in the large-scale geomagnetic field, while shorter

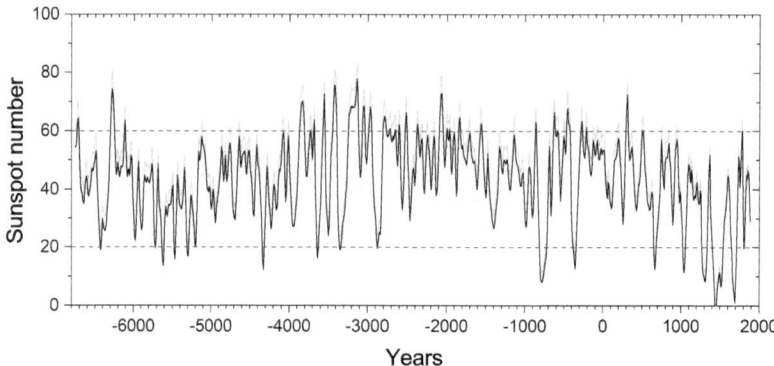

Fig. 2 Multi-proxy reconstruction of the decadal sunspot numbers (in the classical Wolf's definition) over the last nine millennia, along with the 1σ uncertainties (Wu et al. 2018). The blue and red dashed lines approximately denote the low (Grand minimum) and high states of solar activity

time scales – by solar activity (Damon and Sonett 1991). That approach made it possible to identify grand solar minima (Eddy 1976; Stuiver and Braziunas 1989) but was unable to provide a quantitative reconstruction of solar activity because both factors are important at the centennial timescales. A full reconstruction of solar activity from cosmogenic-isotope data became possible only after the development of models of cosmic-ray-induced atmospheric cascades (Masarik and Beer 1999). The first quantitative reconstruction of solar activity using a physics-based approach was made by Usoskin et al. (2003) on the millennial time scale (see also Solanki et al. 2004). Later the reconstructions were extended to the Holocene (the present period of stable warm climate lasting for about 12 millennia) using different cosmogenic isotopes (e.g., Vonmoos et al. 2006; Steinhilber et al. 2012; Usoskin et al. 2016b). The most recent and accurate multi-millennial solar-activity reconstruction by Wu et al. (2018) is based on a multi-proxy Bayesian approach providing also realistic uncertainties. It is shown in Fig. 2. One can see that solar activity varies essentially between the grand minima, visible at sharp dips down to $10-20$ (in sunspot number, SN), and grand maxima when SN exceeds 60, while most of the time the solar-activity level remains moderate at SN= 40 ± 10 (see more detail in Usoskin et al. 2014). The results of an analysis of the solar-activity variability are reviewed in Sect. 4.

Because of the low time resolution of the cosmogenic-isotope throughout the Holocene (typically decadal – see, e.g., Reimer et al. 2020), reconstructions of solar activity are also usually limited to the 10-year resolution being thus unable to resolve individual solar cycles. Long-term reconstructions of solar activity are limited to the Holocene timescale because of the stable climate so that the standard models of the isotope atmospheric transport and deposition can apply. However, for the ice-age-type of climate, the properties of the atmospheric transport are quite uncertain including the large-scale atmospheric and ocean circulation, which prevents quantitative assessment of solar activity. At present, there is no model which is able to handle this in a satisfactory manner, but progress is expected in the future.

3.3 ≈100 Solar Cycles Reconstructed

Thanks to the recent technological progress, high-precision measurements of annual ^{14}C concentrations have been performed with the annual resolution for the last millennium (Brehm et al. 2021). It allowed us to make, by applying the physics-based model, the first

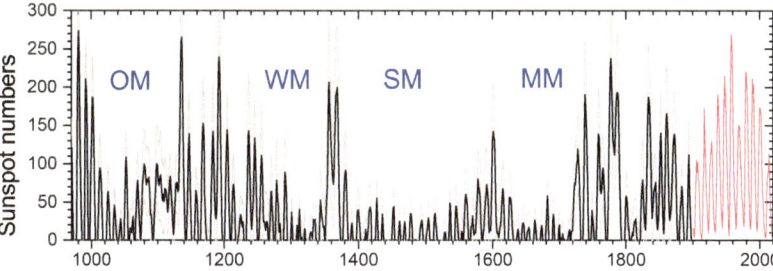

Fig. 3 Annual reconstruction, based on high-precision ^{14}C data, of the sunspot numbers over the last millennium (970 – 1900), along with the 1σ uncertainties (Usoskin et al. 2021). The red curve presents the ISN (v.2) since 1900. Approximate periods of the Oort (OM), Wolf (WM), Spörer (SM) and Maunder (MM) grand minima are indicated in blue letters

reliable reconstruction of individual solar cycles beyond the epoch of telescopic observations (Usoskin et al. 2021) as shown in Fig. 3. Four known grand minima are seen – Oort, Wolf, Spörer and Maunder minima, and between the minima, there are clear solar cycles of variable amplitude. In this way, 85 individual solar cycles have been reconstructed from ^{14}C of which 35 cycles are reasonably and well resolved, 21 are poorly and 29 are not reliably resolved, mostly during the grand minima of activity. Overall, including both direct solar observations and proxy-based reconstructions, we now have information on 96 solar cycles of which 50 are well resolved, thus nearly tripling the extent of the solar-cycle knowledge and doubling the number of well-defined cycles.

The extended statistic made it possible to perform a primary analysis of the solar-cycle parameters. The length of the well-defined cycles was 10.8 ± 1.4 years which is in good agreement with 11.0 ± 1.1 years known for the ISN dataset. The statistical significance of the Waldmeier rule (solar-cycle height is inversely correlated with the length of the ascending phase – high cycles rise fast) has been confirmed with the extended dataset, implying its robust nature (Usoskin et al. 2021). However, the Gnevyshev-Ohl rule of even–odd cycle pairing (Gnevyshev and Ohl 1948; Usoskin et al. 2001) has not been confirmed, nor rejected with the extended data. A more detailed analysis of this new dataset is still pending.

4 Long-Term Solar Activity

With the reconstructed long series, one can investigate properties of solar variability which pose observational constraints crucially important for solar physics but cannot be set by the too short-ranging conventional direct telescopic observations of the Sun. While the 11-year solar cycle forms the main feature of solar activity, the cycles are far from being perfect clock ticks – they vary by both duration and amplitude including periods of greatly suppressed activity, grand minima (see Fig. 3). Here we review the most important features of long-term solar variability.

4.1 Long Quasi-Periodic Variations (Gleissberg, Suess/de Vries, Hallstatt Cycles)

It is hardly possible to distinguish whether solar variability on a long-term scale (Fig. 2) is stochastic/chaotic or (quasi)periodic. Power-spectrum analyses are controversial but generally agree that there are three period ranges with apparent and barely significant variability. An example of the global wavelet power spectrum is shown in Fig. 4.

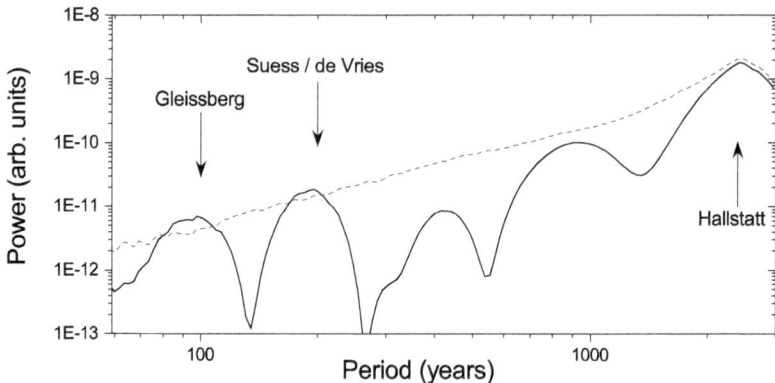

Fig. 4 Global wavelet (Morlet basis) power spectrum (black curve) of the long-term sunspot-number series shown in Fig. 2. Blue-dashed line denotes the 90% confidence level estimated using the AR1 auto-regressive noise, following the methodology of Grinsted et al. (2004). Approximate locations of the discussed quasi-periodic variations (Sect. 4.1) are indicated by vertical arrows

One is the centennial variability, called the *Gleissberg* cycle, which is not a strict periodicity but a characteristic period range between 60 – 140 years (e.g., Peristykh and Damon 2003; Ogurtsov 2004). The Gleissberg cycle is clearly seen in the direct sunspot data but is less pronounced throughout the Holocene.

Another important periodicity is the *Suess* cycle (called also *de Vries* cycle in the literature), which has a narrow period range between 200 – 210 years and an intermittent occurrence. It is typically seen as a recurrence of grand minima within clusters of reduced solar activity (Usoskin et al. 2014) as seen, e.g., in Fig. 3, but is not readily observed during the epochs of moderate solar activity.

Sometimes, the so-called *Eddy* millennial cycle is claimed to exist (Steinhilber et al. 2012), but it is unstable and cannot be identified in a significant way (see Fig. 4).

Additionally, there exists a very-long cycle with a timescale of 2000 – 2400 years called the *Hallstatt* cycle (Damon and Sonett 1991; Vasiliev and Dergachev 2002; Usoskin et al. 2016a). Because of its length, it cannot be robustly defined in the ≈10-kyr time series (see Fig. 4). The nature of the Hallstatt cycle is still unclear: it is likely to be ascribed to the Sun (Usoskin et al. 2016a) but geomagnetic or climatic origin cannot be excluded. Longer-scale variability cannot be reliably assessed from the cosmogenic-isotope data, in particular, because of the unresolved discrepancy between ^{14}C and ^{10}Be datasets on the multi-millennial timescale as probably related to the effect of deglaciation (e.g., Vonmoos et al. 2006; Usoskin et al. 2016a; Wu et al. 2018).

4.2 Grand Minima and Maxima

As seen, e.g., in Figs. 2 and 3, solar activity sometimes drops fast, within one–two solar cycles, to the very quiet level with almost no sunspots on the solar surface. These drops are called grand minima of activity. Until the 1970s, the existence of such minima was debated, but Eddy (1976) had convincingly proved that the sunspot activity indeed dropped to almost no sunspots between 1645 – 1715 as confirmed also by other proxies such as auroral displays at mid-latitudes. That grand minimum was called the *Maunder* minimum. More grand minima have been found later using the cosmogenic-isotope data (e.g., Usoskin et al. 2007; Inceoglu et al. 2015). At present, about 30 grand minima of duration ranging between

40 – 70 (Maunder-type minima) and 100 – 140 years (Spörer-type) each, have been identified during the Holocene occupying about 1/6 of the time. It has been shown that the grand minima correspond to a special state of the solar dynamo (e.g., Usoskin et al. 2014).

Solar activity was abnormally high in the second half of the 20th century compared to the 19th or 21st centuries (Lockwood et al. 1999) but it was unknown whether this high level is unique or typical. Using the cosmogenic-isotope data, it was discovered that the period from the 1940s to 2010 was not unique and there are other similarly high but very rare episodes, that forms the concept of a *grand solar maximum* (Usoskin et al. 2003; Solanki et al. 2004). Grand maxima represent periods of enhanced solar activity covering at least a few solar cycles. There were about 20 grand maxima over the Holocene which cover ≈10% of the time (Usoskin et al. 2007; Inceoglu et al. 2015), but they are defined not as robustly as grand minima. No apparent clustering in the grand-maxima occurrence or duration has been found, nor do they form a special distribution of solar cycles (Usoskin et al. 2014, 2016a). It is still unknown whether grand maxima make a special mode of the dynamo, similar to grand minima, or just represent a rare tail of the solar-cycle-strength distribution.

5 Statistical Properties of the Long-Term Modulation of Solar Cycles

As historical records show, solar cycles are highly variable in amplitude and length. The validity of theoretical models that attempt to predict this variability depends heavily on whether the cycle exhibits long-term phase stability or whether the phase is subject to a random walk, or a mixture of these. In the first of the two extreme cases, the system has infinite phase memory and in the second case no phase memory at all. Phase stability could be achieved through synchronization processes, such as high-quality torsional oscillations in the solar interior (Dicke 1970) or the weak tidal forces of planets (e.g., Stefani et al. 2021). Dynamo models generally predict phase progression without memory. An insightful summary of the use of historical observations to explain solar phenomena was given by Vaquero and Vázquez (2009).

The question of the regularities and randomness of solar activity variability has been studied for a long time. For example, statistical methods including those based on the Lyapunov and Hurst exponents or Kolmogorov entropy (e.g., Ostriakov and Usoskin 1990; Mundt et al. 1991; Carbonell et al. 1994; Ruzmaikin et al. 1994; Lepreti et al. 2021) were inconclusive, implying that a mixture of different components is likely (see more details in Usoskin 2017; Petrovay 2020).

Various publications (e.g., Lomb 2013; Russell et al. 2019; Stefani et al. 2020) claim that the solar cycle is phase stable. However, to answer the question of whether the phase is stable or not, one needs a clear definition of phase stability, an appropriate statistical analysis as well as reliable data on which to apply the analysis. Dicke (1978) and Gough (1978) were among the first to perform a systematic statistical analysis based on telescopic sunspot records. Independently, but using similar concepts, they concluded that the time span of the available data was too small for a clear distinction between the two cases. Later, Gough (1981, 1983, 1988) corrected and modified his earlier analysis without altering the conclusion.

Interestingly, Eddington and Plakidis (1929) analyzed the light-curve variations of long-period variable stars, a problem close to the variability of the solar cycle. By deriving a statistical function to which the processed observational data were fitted, they were able to determine two indicators for the composition of clock-synchronised phase perturbations and random phase perturbations of the light signal.

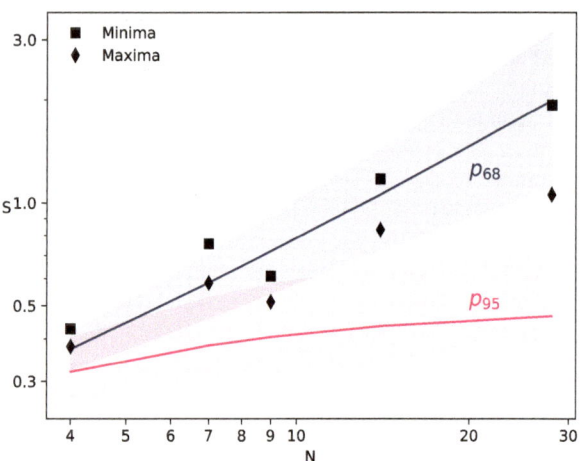

Fig. 5 Modified Gough test S applied to the epochs of sunspot minima and maxima of 28 activity cycles between 1712 and 2019. Symbols correspond to the solar cycle maxima and minima, as denoted in the legend. The black line with the shaded 68% confidence interval depicts the random phase hypothesis (Eq. 4). The red curve with the shaded 95% c.i. depicts the clock phase hypothesis (Eq. 3)

Weisshaar et al. (2023) have revisited Gough's analysis based on newly available data. For clarity, a brief outline of Gough's test is given here: From the arithmetic mean of the individual cycle lengths (Gough 1981), the regular minima or maxima of the hypothetical dynamo or clock cycles and thus the corresponding phase deviations can be determined as the difference to the observed minima or maxima. The basic statistics are the expectation values of the variances of cycle period, $E(\sigma_P^2)$, and phase, $E(\sigma_\phi^2)$. The final statistics is defined as the ratio of the two variances to cancel out the unknown fluctuation amplitude:

$$S = \frac{E(\sigma_\phi^2)}{E(\sigma_P^2)} \tag{2}$$

Later, Gough (1983) modified the method by replacing the arithmetic mean of the cycle period with a value that minimizes the variance of the phase deviations, resulting in a more sensitive distinction between the clock regime and the random phase regime. Calculating the expectation values of the variances for the two cases, one obtains the following expressions for S_c (clock) and S_r (random phase) using the modified method:

$$S_c = \frac{E(\sigma_\phi^2)}{E(\sigma_P^2)} = \frac{N^2}{2(N+1)^2} \tag{3}$$

which asymptotically reaches $N \to \infty$, $S_c \to \frac{1}{2}$;

$$S_r = \frac{E(\sigma_\phi^2)}{E(\sigma_P^2)} = \frac{N(N+3)}{15(N+1)} \tag{4}$$

which asymptotically reaches $N \to \infty$, $S_r \to \frac{N}{15}$.

The procedure to apply Gough's test to an observed data set is as follows: The data set is divided into contiguous segments of N cycles each. Then the ratio of the averages of the empirical variances is calculated and compared with the ratio of the expectation values, plotted as functions of N in Fig. 6.

Weisshaar et al. (2023) augmented the method by determining suitable confidence intervals through Monte Carlo simulations for the clock and the random phase cases, assuming normally distributed variations in cycle length. They applied the test to the extended sunspot

Fig. 6 Modified Gough test (notations are similar to those in Fig. 5) applied to the series of 84 cycles covering the period between 976 and 1999 as reconstructed from ^{14}C data by (Usoskin et al. 2021). The data agree with a random phase shift, while synchronization with the "clock" is rejected at the confidence level much higher than 99% due to the longer data set

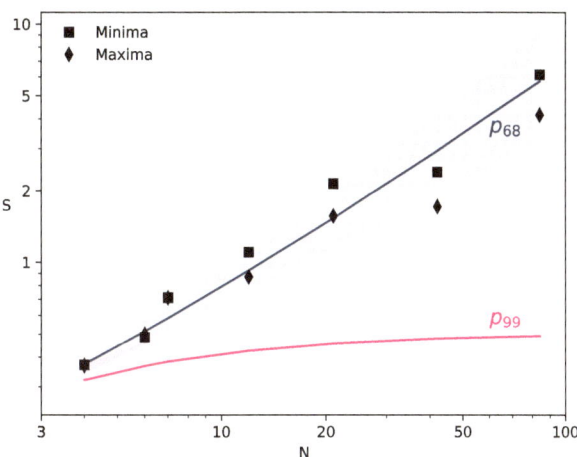

record of now 28 cycles, four more than available to Gough. The main improvement is narrower confidence intervals, rejecting the synchronization hypothesis on a 2σ level (Fig. 5).

Recently, a reconstruction of yearly sunspot numbers from the record of cosmogenic ^{14}C in tree rings for the years 976 until 1888 (Brehm et al. 2021; Usoskin et al. 2021) has extended the number of contiguous cycles available for the analysis to 84. The Gough test confirms the previous result based on the direct sunspot record, in fact strengthening it significantly, since now the synchronization hypothesis can be rejected even on a $> 3\sigma$ level (Fig. 6).

Weisshaar et al. (2023) also applied the method of Eddington and Plakidis (1929) mentioned above to these new data and found, consistent with the analysis discussed here, that the fraction of clock-synchronised perturbations is negligible.

The question may arise how misidentifications of the observed solar cycles can affect the results. If this happens not too common, the nature of the fluctuations (phase stability or migration) is not expected to be changed by this bias. As a test, a lost cycle between more distant minima was "restored" by placing a minimum in between. This did not cause the S-values to leave the phase migration confidence interval.

Furthermore, the above-mentioned method of the phase evolution of empirical cycle data is therefore consistent with a random walk (such as provided by a memory-less dynamo process). External synchronization by a 'clock' is clearly excluded at a high significance.

6 Implications for the Dynamo Theory

The solar magnetic cycle is maintained by a dynamo process, operating in the solar convection zone (SCZ). Thus, it is natural to expect that the variations in the solar cycle are caused by some mechanisms in the solar dynamo. Here we identify the causes of the variations in the solar cycle and demonstrate them by presenting results from some illustrative models. Let us first summarise the mechanism of the solar dynamo.

6.1 Introduction to the Solar Dynamo

There is enough evidence that the solar dynamo is a mechanism in which toroidal and poloidal fields sustain each other through a cyclic loop (e.g., Parker 1955; Cameron and

Schüssler 2015). In this loop, the toroidal field is generated due to the shearing of the poloidal field by the differential rotation in the deeper CZ. The toroidal field rises to the surface due to magnetic buoyancy to give rise to sunspots or more generally bipolar magnetic regions (BMRs). These BMRs are systematically tilted with respect to their East-West orientations. Due to these tilts, after their decay, BMRs produce a poloidal field. This, the so-called Babcock–Leighton process is clearly identified in the observed magnetic field data on the solar surface (e.g., Mordvinov et al. 2022). The observed correlation between the polar field (or its proxy) at the solar minima and the amplitude of the next cycle (Wang and Sheeley 2009; Kitchatinov and Olemskoy 2011; Muñoz-Jaramillo et al. 2013; Priyal et al. 2014) and the flux budgets of the observed and the generated poloidal and toroidal fields (Cameron and Schüssler 2015) suggest that the Babcock–Leighton process is possibly the main source of the poloidal field in the Sun.

There is however another mechanism through which the poloidal field in the sun can be produced and that is the classical α effect as originally proposed by Parker (1955) and mathematically formulated by Steenbeck et al. (1966). In this mechanism, the toroidal field is twisted by the helically rising blobs of plasma in the SCZ. However, this process of lifting and twisting of the field by the convective flow experiences catastrophic quenching due to helicity conservation and thus this process operates when the energy density of the toroidal field is less than the energy density of the convective motion (Sect. 8.7 of Brandenburg and Subramanian 2005). Therefore, this α effect is unfavourable in the solar convection zone and the obvious option is to consider the observationally supported Babcock–Leighton process for the generation of the poloidal field in the sun.

To study the dynamo action, we need to begin with at least following two fundamental equations of magnetohydrodynamics (MHD).

$$\frac{\partial \boldsymbol{B}}{\partial t} = \nabla \times (\boldsymbol{v} \times \boldsymbol{B} - \eta \nabla \times \boldsymbol{B}), \tag{5}$$

$$\rho \left[\frac{\partial \boldsymbol{v}}{\partial t} + (\boldsymbol{v} \cdot \nabla)\boldsymbol{v} \right] = -\nabla P + \boldsymbol{J} \times \boldsymbol{B} + \nabla \cdot (2\nu\rho S) + \boldsymbol{F}, \tag{6}$$

where $\boldsymbol{B}$ and $\boldsymbol{v}$ are the magnetic and velocity fields, respectively, η is the magnetic diffusivity, ρ is the density, P is the pressure, $\boldsymbol{J} = \nabla \times \boldsymbol{B}/\mu_0$, the current density, ν is the kinetic viscosity, $S_{ij} = \frac{1}{2}(\nabla_i v_j + \nabla_j v_i) - \frac{1}{3}\delta_{ij} \nabla \cdot \boldsymbol{v}$ is the rate-of-strain tensor, and the term $\boldsymbol{F}$ includes gravitational, Coriolis and any other body forces acting on the fluid. These equations along with the mass continuity and energy equations and equation of state are numerically solved with appropriate boundary conditions in the solar CZ to study the dynamo problem. Broadly there are two approaches for doing this, namely, the global MHD simulations and mean-field modellings. In global MHD simulations, we solve the above MHD equations numerically to resolve the full spectrum of turbulent convection. In mean-field models, we study the evolution of the mean/large-scale quantities by parameterizing the small-scale/fluctuating quantities using suitable approximations.

Global MHD simulations for the Sun are challenging due to extreme parameter regimes, such as high fluid and magnetic Reynolds numbers and large stratification. Despite these, simulations have begun to produce some solar-like features; see Sect. 6 of Charbonneau (2020). However, due to their computationally expensive nature, these simulations were rarely run for many cycles so that the cycle variabilities can be studied. Passos and Charbonneau (2014) have produced simulations for several cycles and shown long-term modulations (also see Karak et al. 2015, for a simulation at solar rotation rate although ran for not many cycles). Augustson et al. (2015) and Käpylä et al. (2016) performed MHD convection simulations for the cases of three and five times the solar rotation rate, respectively. They both found an episode of suppressed surface activity, somewhat resembling the solar grand minimum. Although these results of cycle modulations are encouraging, simulations face serious issues when matching with observations, for example, concerning solar observations, simulations (i) produce higher power at the largest length scale, (ii) do not produce BMRs, and (iii) do not produce correct large-scale flows, particularly, they produce a large variation in the differential rotation.

On the other hand, mean-field models are computationally less expensive and easy to analyse their results. Probably due to these reasons, long-term modulations are studied using mean-field dynamo models. Due to the observational facts that the magnetic field at the solar minima and the large-scale velocity field are largely axisymmetric, historically the mean-field models are constructed under axisymmetric approximation. With this approximation, the equations for the poloidal and toroidal fields are written as

$$
\frac{\partial A}{\partial t} + \frac{1}{s}(\boldsymbol{v_m} \cdot \boldsymbol{\nabla})(s A) = \eta_t \left(\nabla^2 - \frac{1}{s^2} \right) A + \alpha B, \tag{7}
$$

$$
\frac{\partial B}{\partial t} + \frac{1}{r} \left[\frac{\partial (r v_r B)}{\partial r} + \frac{\partial (v_\theta B)}{\partial \theta} \right] = \eta_t \left(\nabla^2 - \frac{1}{s^2} \right) B + s(\boldsymbol{B_p} \cdot \boldsymbol{\nabla})\Omega + \frac{1}{r} \frac{d\eta_t}{dr} \frac{\partial (r B)}{\partial r}, \tag{8}
$$

where A is the potential for the poloidal field ($\boldsymbol{B_p} = \boldsymbol{\nabla} \times (A\hat{\boldsymbol{\phi}})$, B is the toroidal field, $s = r \sin\theta$, $\boldsymbol{v_m}(= v_r \hat{\boldsymbol{r}} + v_\theta \hat{\boldsymbol{\theta}})$ represents the meridional circulation, η_t is the turbulent diffusivity which is assumed to depend only on r, α is the α effect, and Ω is the angular frequency.

The term αB in Equation (7) is the source for the poloidal field through the α effect. The generation of the poloidal field through the Babcock–Leighton process is also parameterised in the 2D (axisymmetric models) through the same αB term. However, this α operates near the surface of the sun and it has a completely different origin than the α effect which operates in the whole convection zone due to helical convection. In comprehensive 3D dynamo models (Yeates and Muñoz-Jaramillo 2013; Miesch and Dikpati 2014; Miesch and Tewelde-birhan 2016; Kumar et al. 2019; Bekki and Cameron 2023), this αB term is not added in Equation (7), instead, explicit BMRs are deposited whose decay produces a poloidal field. The source for the toroidal field in Equation (8) is due to the Ω-effect which is represented by the term: $s(\boldsymbol{B_p} \cdot \boldsymbol{\nabla})\Omega$. The above equations technically represent the equations for the $\alpha\Omega$ dynamo model, in which the generation of the toroidal field through the α effect is assumed to be much less than the generation due to Ω effect, which is true in the sun; see e.g., Cameron and Schüssler (2015).

6.2 Causes for Long-Term Variations in the Solar Activity

With the above discussion of the solar dynamo, we now identify the causes of the cycle modulation. As the solar dynamo is nonlinear, it is natural to expect that the modulation in the solar cycle is caused by the back reaction of the flow on the magnetic field. Therefore, we first identify the nonlinearities in the dynamo models and check if they can lead to cycle modulations.

6.2.1 Nonlinearities in the Dynamo

As we can see from Equation (6), the magnetic field can alter the flow directly through the Lorentz force. The Lorentz force can come from the mean magnetic field and the mean current (which is popularly known as the Malkus-Proctor effect (Malkus and Proctor 1975) in the mean-field context) and from the fluctuating magnetic field and the current. The mean magnetic field can also alter the anisotropic convection which is responsible for transporting angular momentum and maintaining differential rotation and meridional flow in the Sun (Kitchatinov et al. 1994b). This effect is also called micro-feedback. When these Lorentz feedbacks of the magnetic fields are included in the flow, we expect a long-term modulation in the flow and the magnetic cycle.

In mean-field models, the magnetic feedback is captured by considering a direct Lorentz force of the mean magnetic field in the zonal flow (e.g., Bushby 2006) and/or by a quenching term in the Λ effect (e.g., Küker et al. 1999). Cycle modulations in these systems can generally happen in two ways. In the first one, the magnetic energy of the primary mode (the equatorial symmetry or antisymmetric) can oscillate due to the energy exchange between the flow and the magnetic field via the nonlinear Lorentz feedback. In this case, a considerable amount of modulation in the differential rotation is observed. In the second case, a small magnetic perturbation on the differential rotation can slowly change one dominant dynamo mode into another. In this case, the magnetic field parity can change (between equatorially symmetric (quadrupole) and antisymmetric (dipole)) without producing a large change in the differential rotation. These two mechanisms are respectively coined as Type II and I modulations. Mean-field models have demonstrated that nonlinear back reaction of magnetic field on large-scale flow through these types of modulations can induce a variety of modulation patterns in the cycle amplitude, including grand minima and parity modulations which do not leave a strong imprint in differential rotation (e.g., Beer et al. 1998; Knobloch et al. 1998; Bushby 2006; Weiss and Tobias 2016). Both types of modulation can arise in a model, however, as the observed differential rotation shows a tiny variation over the solar cycle, we expect the Type II modulation is less likely to occur in the Sun. Even for Type I modulation, a detailed comparison of the magnetic field and the flows in these models with the observations is missing (also see Sect. 7 of Charbonneau 2020, for a discussion on this topic).

Next is the meridional flow, which is the second important large-scale flow in the Sun. As it arises due to a slight imbalance between the non-conservative centrifugal and buoyancy forces, we expect its large variation. In fact, the global simulations find a large variation in the meridional flow despite a small variation in the differential rotation (Karak et al. 2015). In Babcock–Leighton type dynamo models, meridional circulation plays a crucial role in transporting the field on the surface from low to high latitudes and down to the deeper CZ where the shear produces a toroidal field. The toroidal field is transported to the low latitudes via the equatorward return flow and possibly causes the equatorward migration of the sunspot belt. Thus, in these models, meridional circulation largely regulates the cycle period (Dikpati and Charbonneau 1999; Karak and Choudhuri 2011). It also affects the strength of the field as a weak meridional circulation allows the field to advect slowly and gives more time for diffusion (Yeates et al. 2008). Karak (2010) showed that when a variable meridional flow is used in a high diffusivity dynamo model to match the observed solar cycle periods, the amplitudes of the cycles are also modelled up to some extent (also see Karak and Choudhuri 2011; Hazra et al. 2015, for modelling various aspects of solar cycle using variable meridional flow). In an extreme case, a largely reduced meridional circulation can trigger a Maunder-like grand minimum. In reality, how large the variation in the meridional flow occurred in the past remains uncertain. However, it is obvious that any changes in the flow can lead to modulation in the solar cycle.

Turbulent transport as parameterized by, for example, the turbulent diffusivity, Λ effect, and heat diffusion are also nonlinear because the Lorentz force of the small-scale as well

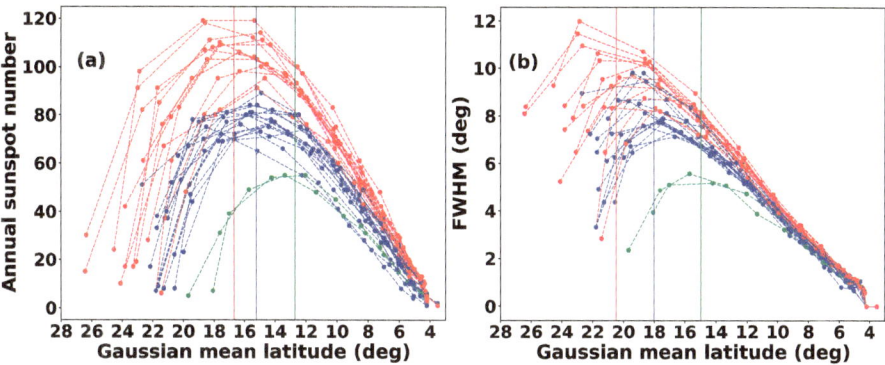

Fig. 7 The trajectories of (a) annual sunspot number and (b) FWHM vs the central latitude of the annual spot distribution obtained from a dynamo simulation with buoyancy-induced flux loss (Biswas et al. 2022). Curves clearly show that the beginning phases of the cycles differ widely depending on their strengths but they decline in the same way irrespective of their strengths. This property closely matches with the observations of Cameron and Schüssler (2016)

as the large-scale dynamo-generated fields act on the small-scale turbulent flows. However, due to limited knowledge in the turbulence theory for solar parameter regions, we do not have a satisfactory model for the magnetic field-dependent form of the turbulent transport parameters; however, see Ruediger and Kichatinov (1993) and Kitchatinov et al. (1994a) respectively, for the magnetic field-dependent forms of α and η based on the quasi-linear approximation.

Finally, the toroidal to poloidal part of the dynamo loop involves some nonlinearities. When the generation of poloidal field is due to the classical α effect, there is a well-known α quenching of the form $1/\left(1 + (B/B_{eq})^2\right)$ with B_{eq} being the equipartition field strength. However, this type of α quenching tries to make a stable cycle rather than producing irregularity in the cycle. In the Babcock–Leighton dynamo, the generation of the poloidal field from the toroidal one also involves several nonlinearities. Here we discuss the following three potential candidates for these.

- Flux loss due to magnetic buoyancy

The magnetic buoyancy as proposed by Parker (1955) plays a critical role in the emergence of BMRs on the solar surface. As the shearing of the poloidal field due to differential rotation intensifies the strength of the toroidal field, there comes a point where the magnetic energy density of the toroidal flux tubes becomes greater than the kinetic energy of the local convective plasma inside the CZ, as a result, the flux tubes become buoyant and start rising through the CZ, eventually giving birth to the sunspots. Following this process, the strength of the magnetic field gets locally reduced as a part of it rises due to buoyancy and the flux tube becomes inefficient to produce further sunspots for some time (however see a counter-argument by Rempel and Schüssler 2001). The sharp rise in the flux loss once the toroidal field strength exceeds a certain value clearly indicates a nonlinear mechanism in the solar dynamo. Incorporating this mechanism of toroidal flux loss due to buoyancy in a simple manner, Biswas et al. (2022) showed that this nonlinear process plays a critical role in limiting the growth of the solar dynamo which is a potential mechanism to explain why different solar cycles rise differently depending on their strength but all the solar cycles decay with similar statistical properties (see Fig. 7). They found that introducing the flux

Fig. 8 Demonstration of latitude quenching: Temporal evolution of the net polar flux generated from two BMRs deposited symmetrically in two hemispheres at different latitudes. The solid and dashed lines are the polar fields produced from the BMRs deposited at 25° and 5° latitudes, respectively

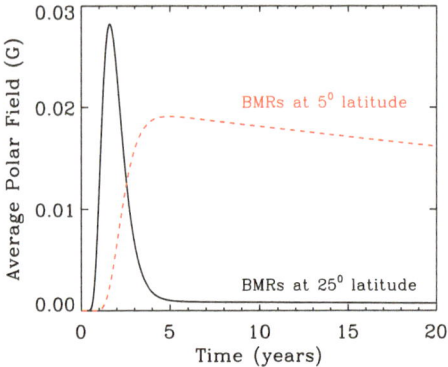

loss in the dynamo simulations was critical to reproduce the long-term features of the latitudinal distribution of the sunspots (Waldmeier 1955; Cameron and Schüssler 2016); also see Cameron and Schüssler (2016) and Talafha et al. (2022) for an alternative explanation of the universal decay of the solar cycle using cross-equatorial diffusion.

• Latitude quenching

It has been found that when BMRs appear in low latitudes, the leading polarities from both hemispheres get efficiently cancelled at the equator. This leads to the following polarities of the BMRs efficiently getting carried to the poles and contributing to the polar field, see Fig. 8. On the other hand, BMRs appearing in the high latitudes do not exhibit efficient cross-hemisphere cancellation and thus do not contribute significantly to the polar field (Jiang et al. 2014; Karak and Miesch 2018). It is seen that strong cycles produce more BMRs at high latitudes. In other words, the average latitude of the BMRs is high for the strong cycles (Solanki et al. 2008; Mandal et al. 2017). Hence for a strong cycle, most of its BMRs emerging at high latitudes would be less efficient in polar field production and vice versa for the weak cycles. This mechanism, so-called the *latitude quenching* (Petrovay 2020) may help to stabilize the growth of the magnetic field in the Sun (Jiang 2020).

Introducing a latitude-dependent threshold on the BMR emergence condition into a 3D Babcock–Leighton dynamo simulation, Karak (2020) showed that latitude quenching can regulate the growth of a magnetic field when the dynamo is not too supercritical.

• Tilt quenching

The tilt angle of BMR plays a crucial role in generating poloidal field in the Sun. For a given latitude, the amount of generated poloidal field increases with the increase of tilt. The thin flux tube model for the sunspot formation suggests that the tilt of the BMR is produced due to a torque acting on the diverging flows produced from the apex of the rising flux tube which forms the BMR (D'Silva and Choudhuri 1993; Fan et al. 1994). Thus, if the magnetic field of the sunspot-forming flux tube is strong, then it will rise quickly and the Coriolis force will get less time to induce tilt. In a strong cycle, the toroidal magnetic field is strong and the number of BMRs with strong magnetic field tends to be high (Jha et al. 2020). Thus, we expect the mean tilt in that cycle to be smaller. A lesser tilt will produce less poloidal field and the next cycle will be weak. Hence, this may be a potential mechanism for stabilizing the growth of the magnetic cycle through the reduction of tilt which is known as the *tilt quenching*.

The observational evidence of tilt quenching is limited. Dasi-Espuig et al. (2010), Jiao et al. (2021) showed that there is a statistical anti-correlation between the cycle-average

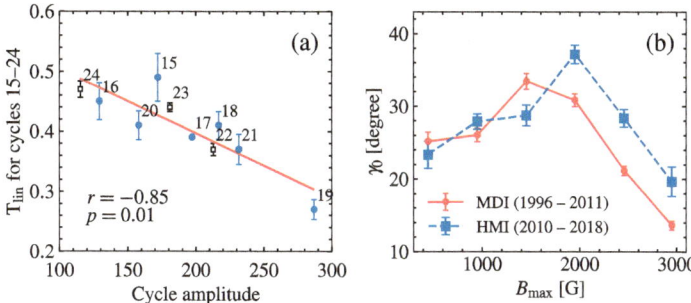

Fig. 9 Demonstration of tilt quenching: (a) Tilt coefficient (mean tilt normalized by the mean latitude) vs the cycle strength (Jiao et al. 2021); also see Dasi-Espuig et al. (2010). (b) The slope of Joy's law vs the maximum field strength in the BMR (Jha et al. 2020)

tilt of the sunspots with the cycle strength (Fig. 9a). On the other hand, Jha et al. (2020) examined the variation of BMR tilt with the strength of its magnetic field within a cycle. They found a non-monotonous dependence of the tilt with the BMR field strength as seen in Fig. 9(b). For weak field strengths, the tilt first increases, however at sufficiently strong field strengths, the BMR tilt starts to decrease.

6.2.2 Stochastic Effects in the Dynamo

The solar convection zone is turbulent and thus the turbulent quantities (such as α effect) are subject to fluctuate around their means. Hoyng (1993) showed that as there are finite numbers of convection eddies along the longitudes in the sun, the fluctuations of the turbulent transport coefficients can be larger than their means. There is a long history including the stochastic noise in the α effect in the mean-field dynamo models. Most of these studies find long-term modulations in the cycle and grand minima in a certain parameter range of the dynamo number (Choudhuri 1992; Ossendrijver and Hoyng 1996; Ossendrijver et al. 1996; Gómez and Mininni 2006; Brandenburg and Spiegel 2008; Moss et al. 2008).

In Babcock–Leighton dynamo also stochastic fluctuations are unavoidable. The toroidal to poloidal part of this model primarily involves stochastic fluctuations due to the following effects.

- Scatter around Joy's law

Observations find that the tilt "statistically" increases with the increase of latitude, which is known as Joy's law. However, a large number of BMRs do not follow this relation (so-called non-Joy), as seen by a huge scatter around the mean trend in Fig. 10. In fact, there are many BMRs which are of anti-Hale type. These anti-Hale and non-Joy BMRs, having opposite tilts (negative in the northern hemisphere) are responsible for generating opposite polarity field (with respect to the expected polarity) and lead to large fluctuations in the polar field (Jiang et al. 2014; Hazra et al. 2017; Nagy et al. 2017; Mordvinov et al. 2022).

- Variations in the BMR eruption rates

There are spatial and temporal variations in the BMR eruptions. BMRs near the equator are much more efficient in generating poloidal field in the Sun because for them the leading polarity can easily connect with the opposite polarity flux from the opposite hemisphere (Cameron et al. 2013; Jiang et al. 2014; Karak and Miesch 2018; Karak 2020; Mordvinov

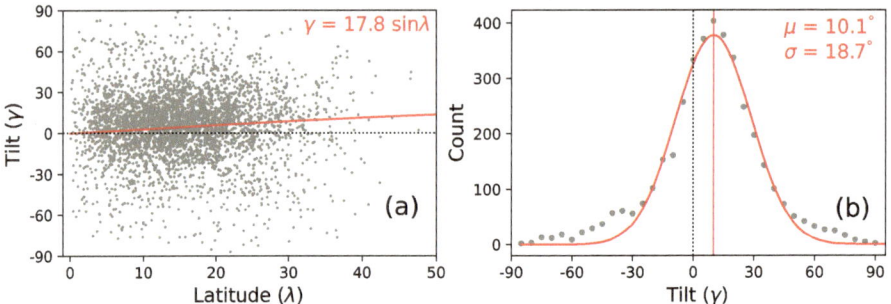

Fig. 10 (a) Scatter of BMR tilt around Joy's law (solid line). (b) The tilt distribution with fitted Gaussian (solid line). Here the tilt angles of BMRs are computed by tracking the MDI line-of-sight magnetograms for September 1996–December 2008 (A. Sreedevi personal communication, 2023)

et al. 2022). Thus variation in the latitudinal position can produce variation in the generated poloidal field. Next, the rate of BMR eruption is not the same—there is a distribution. Thus, the rate of generation of the poloidal field is not the same (Karak and Miesch 2017). Furthermore, the flux contents of the BMR has also a distribution and thus a wrongly tilted BMR with *high flux* can disturb the polar field in the sun considerably (Nagy et al. 2017).

In summary, the randomness involved in the BMR properties (originated due to the turbulent nature of the convection) produces variation in the poloidal field. Although the sun produces thousands of spots in a cycle, only a few spots are produced (on average) per day. This leads to variations in the polar field comparable to its mean value. In the next section, we shall demonstrate some illustrative results from stochastically driven Babcock–Leighton dynamo models.

6.3 Babcock–Leighton Dynamo Models for the Long-Term Variation

As discussed above, the generation of the poloidal field in the Babcock–Leighton dynamo models involves some randomness. Thus, in axisymmetric dynamo models, these randomnesses were captured by adding a noise term in the poloidal source (e.g., Charbonneau and Dikpati 2000). Long-term modulations, including Gnevyshev-Ohl/Odd-Even rule (Charbonneau 2001; Charbonneau et al. 2007) and grand minima (Charbonneau et al. 2004; Choudhuri and Karak 2009; Passos et al. 2012, 2014) are naturally produced in these models. Variations within the cycle, like the amplitude-period anti-correlation (Charbonneau and Dikpati 2000; Karak 2010) and Waldmeier effect (Karak and Choudhuri 2011; Biswas et al. 2022) are also reproduced. Karak et al. (2018) showed that a large variation in the Babcock–Leighton process can change the polar field abruptly and this can lead to double peaks in the following cycle. While in most of the studies, the level of fluctuations was tuned to produce the observed variation of the solar cycle including a reasonable number of grand minima, Choudhuri and Karak (2012) and Olemskoy and Kitchatinov (2013) made some estimate of the fluctuations in the Babcock–Leighton process from observations. Choudhuri and Karak (2012) found the correct frequency of grand minima as observed in the cosmogenic data for the last 11,000 years. Olemskoy and Kitchatinov (2013) showed that the statistics of grand minima are consistent with the Poisson random process, indicating the initiation of grand minima to be independent of the history of the past minima.

In recent years, cycle modulations were, in particular, produced by including the variations in the BMR properties in two comprehensive models, namely, 2×2D (Lemerle and

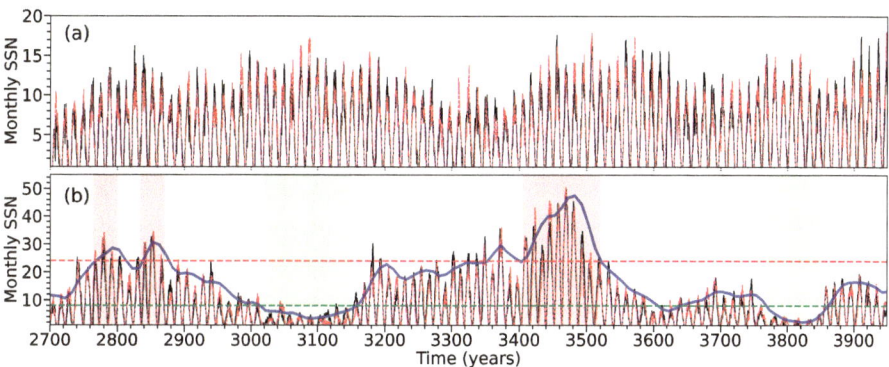

Fig. 11 Time series of the monthly BMR number from a 3D dynamo model of Karak and Miesch (2017) (a) without tilt scatter around Joy's law and (b) with scatter of $\sigma_\delta = 18°$ (close to the observed value). The black/red curves indicate the north/south hemispheres. The blue curve in panel (b) is the smoothed curve of the cycle trajectories, and the green and red dashed horizontal lines indicate the thresholds for the grand minima and grand maxima, respectively. The green and red shaded regions indicate the grand minima and grand maxima episodes, respectively

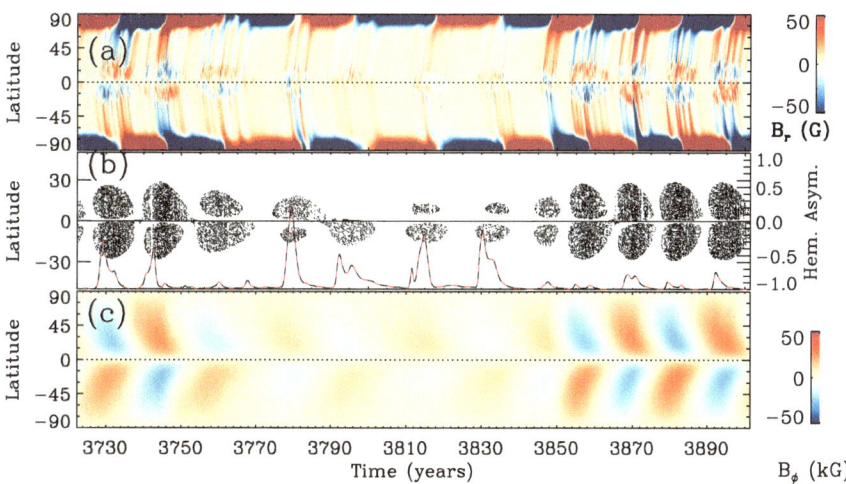

Fig. 12 Zoomed-in view of a grand minimum presented in Fig. 11. Evolution of (a) the surface radial field (b) BMR eruptions and hemispheric asymmetry of the toroidal field (black/red curve), and (c) the toroidal field at the bottom of the convection zone

Charbonneau 2017) and 3D dynamo models (Karak and Miesch 2017). In Fig. 11, we show cycles from the 3D dynamo model presented by Karak and Miesch (2017). As seen in Fig. 11(a), the variation in the BMR emergence rate and the flux distribution produce little variation in the solar cycle. When the variation around Joy's law tilt is included, it produces a large variation, including suppressed magnetic activity like the one seen during Dalton minimum and Maunder minimum as shown in Fig. 11b (the regions shaded in green). Here, the grand minima are identified in the same manner as done in the observed data (Usoskin et al. 2007), i.e., the modelled-sunspot data are first binned in 10 years window and smoothed and then a grand minimum is considered when the smoothed data fall below 50% of the average at least for two cycles.

In Fig. 12, we present a detailed view of a grand minimum. We find that some of the observed features of the Maunder minimum (hemispheric asymmetry, gradual recovery, slightly longer cycle) are reproduced in this figure. We note that during this grand minimum, some BMRs are still produced, the number of which is a bit larger than that was observed during Maunder minimum (Usoskin et al. 2015; Vaquero et al. 2015; Zolotova and Ponyavin 2016; Carrasco et al. 2021). However, we should keep in mind that the observations during Maunder minimum were limited (due to the poor resolving power of the 17th-century telescopes) to detect the small BMRs (e.g., Vaquero and Vázquez 2009); only big sunspots were detected. In our Babcock–Leighton dynamo model, few BMRs erupt which produces a poloidal field at a slow rate through the Babcock–Leighton process and the model emerges from the grand minimum episode. It is the downward magnetic pumping included in our model which helps to reduce the magnetic flux loss through the surface and recovers the model from grand minima (Cameron et al. 2013; Karak and Cameron 2016).

There have been suggestions that during Maunder-like extended grand minima, the Babcock–Leighton process may not operate due to few observed sunspots, and α effect (Parker 1955) is the best candidate for this as it efficiently operates in sub-equipartition field strength (Karak and Choudhuri 2013; Passos et al. 2014; Ölçek et al. 2019). We observe that our model also fails to recover when it enters a deep grand minimum and stops producing BMRs due to the fall of the toroidal field below the threshold for BMR formation. However, this happens very rarely. While it is a critical question to answer what mechanism dominates in recovering the Sun from an extended grand minimum, it is expected that Babcock–Leighton process becomes less efficient during this phase and the α effect certainly helps in recovering the Sun from grand minima.

Dynamo models with stochastic fluctuations also produce grand maxima. Our model presented in Fig. 11b also produces a few grand maxima shown by the regions shaded in red. Similar to the grand minima, grand maxima are also computed based on the smoothed sunspot number, but here the threshold is taken as 150% of the long-term mean. Systematic studies of grand maxima using dynamo models are limited (however, see Karak and Choudhuri 2013; Olemskoy and Kitchatinov 2013; Inceoglu et al. 2017). Kitchatinov and Olemskoy (2016) showed that at the beginning of the cycle, if the generation of the poloidal field is reversed (say due to the emergence of some wrongly tilted BMRs), then it will amplify the existing polar field, instead of reversing it. This increase in the magnetic field can lead to a grand maximum. Another mechanism of grand maxima was given by Ölçek et al. (2019), who showed that when the deep-seated α effect is coupled with the surface Babcock–Leighton source, then these two sources more or less contribute equally to generate a strong poloidal field through a sort of constructive interference.

Finally, for the secular and supersecular modulations (modulations beyond 11-year periodicity, e.g., Gleissberg cycle, Suess/de Vries cycle, Eddy cycle, and 2400-year Hallstatt cycle; Beer et al. 2018), there are limited studies available in the literature. In a simplified $\alpha\Omega$ dynamo model coupled with the angular momentum equation, Pipin (1999) found the Gleissberg cycle as a result of the re-establishment of differential rotation after the magnetic feedback on the angular momentum transport. Cameron and Schüssler (2017) modelled the overall power spectrum of solar activity using a generic normal form model for a noisy and weakly nonlinear limit cycle, and Cameron and Schüssler (2019) showed that the long-term modulations beyond the 11-year cycle are consistent with the realization noise, thus casting doubt whether secular and supersecular modulations are connected to the intrinsic periodicities of the solar dynamo.

7 Summary

Herewith, a brief overview is presented of the long-term variability of solar activity at centennial–millennial timescales. The main feature of solar variability is the 11-year quasi-periodic Schwabe cycles, which is however variable *per se* in both magnitudes, duration and phase. While the direct telescopic observations of the Sun cover roughly four centuries since 1610 and cover a full range of solar-activity levels from the Maunder minimum in the 17th century to the Modern grand maximum in the late 20th century, the quality of the sunspot-number dataset is inhomogeneous and greatly degrades back in time, being quite imprecise before ≈1820s. Moreover, it is too short to study the statistical properties of the solar-cycle modulation on a long timescale.

The cosmogenic-isotope method provides quantitative reconstructions of solar activity on the multi-millennial timescale with stable quality throughout ages making it possible to study long-term solar-cycle modulation. Using the decadal data for the Holocene (the last twelve millennia), it is possible to identify specific observed properties of solar variability beyond the Schwabe cycle:

- The Sun spends about $1/6$ of its time in the grand minimum state, grand minima tend to cluster with a ≈210-year recurrence time;
- The Sun spends about $1/10$ of its time in the grand maximum state, grand maxima appear without any regular pattern;
- During the major fraction of time, the Sun is in the cyclic moderate activity state;
- Several quasi-periodicities can be found in long-term solar variability, but they are intermittent and barely significant: Centennial *Gleissberg* cycle which is an oscillation with the characteristic time of 60–140 years; 210-year *Suess/de Vries* cycles manifesting itself through intermittent recurrence of grand minima; About 2400-year *Hallstatt* cycle whose nature is still unclear; Other long-term cycles, including the millennial Eddy cycle, are insignificant.

A recent reconstruction of the annual sunspot numbers from high-precision radiocarbon data for the last millennium makes greatly extended, nearly tripling, the statistic of solar cycles to 96 individually resolved cycles. In particular, the Waldmeier rule (high cycles rise faster) is statistically confirmed on a larger statistical basis, while the Genvyshev-Ohl rule of the even-odd cycle pairing is not confirmed. The extended statistic of solar cycles has made it possible, for the first time, to answer the question principle to the solar dynamo theory: is the solar cycle phase-locked, implying an intrinsic synchronisation process as proposed by some external clocking mechanisms, or is random and incoherent. The new analysis excludes the phase-locking hypothesis at a high significance level, implying that solar cycles vary randomly.

A brief review of the theoretical perspectives to explain the observed features in the framework of the dynamo models is presented. It is discussed that the nonlinearities in the dynamo, including the effects of the flux loss due to magnetic buoyancy as well as latitude and tilt quenching, help to stabilize the solar dynamo, rather than producing variability in the solar cycle. Primary causes of the solar cycle variability are the stochastic fluctuations in the dynamo which are inherent in different processes such as a large scatter of the BMR's tilts around Joy's law, and variability in the BMR eruption rates and locations. On one hand, while modern dynamo models are able to reproduce, with a reasonable ad-hoc tuning of the parameters, the observed features of solar variability, the exact role of those factors is not clear, and some discrepancies between the model results and the data still remain. On the other hand, the progress in the accuracy of models is significant, and we keep gaining knowledge of the processes driving solar variability with the new data acquainted and new models developed.

Acknowledgements Authors thank Anu B Sreedevi for providing Fig. 10 and Bibhuti Jha and Jie Jiang for their help in preparing Fig. 9. IU acknowledges the Academy of Finland (project ESPERA No. 321882). AB and BBK gratefully acknowledge the financial support provided by ISRO/RESPOND (project No. ISRO/RES/2/430/19-20), the Department of Science and Technology (SERB/DST), India through the Ramanujan Fellowship (project No. SB/S2/RJN-017/2018), the International Space Science Institute (ISSI, Team 474), and the computational resources of the PARAM Shivay Facility under the National Supercomputing Mission, the Government of India, at the Indian Institute of Technology Varanasi. This work was performed in the framework of the ISSI workshop "Solar and Stellar Dynamos: A New Era".

Funding Note Open Access funding provided by University of Oulu including Oulu University Hospital.

Declarations

Competing Interests The authors declare they have no conflicts of interest.

References

Arlt R, Vaquero JM (2020) Historical sunspot records. Living Rev Sol Phys 17:1. https://doi.org/10.1007/s41116-020-0023-y

Augustson K, Brun AS, Miesch M et al (2015) Grand minima and equatorward propagation in a cycling stellar convective dynamo. Astrophys J 809:149. https://doi.org/10.1088/0004-637X/809/2/149. arXiv: 1410.6547 [astro-ph.SR]

Beer J (2000) Neutron monitor records in broader historical context. Space Sci Rev 93:107–119. https://doi.org/10.1023/A:1026536226656

Beer J, Tobias S, Weiss N (1998) An active sun throughout the Maunder minimum. Sol Phys 181:237–249. https://doi.org/10.1023/A:1005026001784

Beer J, McCracken K, von Steiger R (2012) Cosmogenic radionuclides: theory and applications in the terrestrial and space environments. Physics of Earth and space environments. Springer, Berlin. https://doi.org/10.1007/978-3-642-14651-0

Beer J, Tobias SM, Weiss NO (2018) On long-term modulation of the Sun's magnetic cycle. Mon Not R Astron Soc 473(2):1596–1602. https://doi.org/10.1093/mnras/stx2337

Bekki Y, Cameron R (2023) Three-dimensional non-kinematic simulation of post-emergence evolution of bipolar magnetic regions and Babcock-Leighton dynamo of the Sun. Astron Astrophys. 670:A101. https://doi.org/10.1051/0004-6361/202244990

Biswas A, Karak BB, Cameron R (2022) Toroidal flux loss due to flux emergence explains why solar cycles rise differently but decay in a similar way. Phys Rev Lett 129(24):241102. https://doi.org/10.1103/PhysRevLett.129.241102

Brandenburg A, Spiegel EA (2008) Modeling a Maunder minimum. Astron Nachr 329:351

Brandenburg A, Subramanian K (2005) Astrophysical magnetic fields and nonlinear dynamo theory. Phys Rep 417:1–209. https://doi.org/10.1016/j.physrep.2005.06.005. https://arxiv.org/abs/astro-ph/0405052

Brehm N, Bayliss A, Christl M et al (2021) Eleven-year solar cycles over the last millennium revealed by radiocarbon in tree rings. Nat Geosci 14:10–15. https://doi.org/10.1038/s41561-020-00674-0

Bushby PJ (2006) Zonal flows and grand minima in a solar dynamo model. Mon Not R Astron Soc 371(2):772–780. https://doi.org/10.1111/j.1365-2966.2006.10706.x

Cameron R, Schüssler M (2015) The crucial role of surface magnetic fields for the solar dynamo. Science 347:1333–1335. https://doi.org/10.1126/science.1261470. arXiv:1503.08469 [astro-ph.SR]

Cameron RH, Schüssler M (2016) The turbulent diffusion of toroidal magnetic flux as inferred from properties of the sunspot butterfly diagram. Astron Astrophys 591:A46. https://doi.org/10.1051/0004-6361/201527284. arXiv:1604.07340 [astro-ph.SR]

Cameron RH, Schüssler M (2017) Understanding Solar cycle variability. Astrophys J 843(2):111. https://doi.org/10.3847/1538-4357/aa767a. arXiv:1705.10746 [astro-ph.SR]

Cameron RH, Schüssler M (2019) Solar activity: periodicities beyond 11 years are consistent with random forcing. Astron Astrophys 625:A28. https://doi.org/10.1051/0004-6361/201935290

Cameron RH, Dasi-Espuig M, Jiang J et al (2013) Limits to solar cycle predictability: cross-equatorial flux plumes. Astron Astrophys 557:A141. https://doi.org/10.1051/0004-6361/201321981. arXiv:1308.2827 [astro-ph.SR]

Carbonell M, Oliver R, Ballester J (1994) A search for chaotic behaviour in solar activity. Astron Astrophys 290:983–994

Carrasco VMS, Hayakawa H, Kuroyanagi C et al (2021) Strong evidence of low levels of solar activity during the Maunder minimum. Mon Not R Astron Soc 504(4):5199–5204. https://doi.org/10.1093/mnras/stab1155.

Charbonneau P (2001) Multiperiodicity, chaos, and intermittency in a reduced model of the solar cycle. Sol Phys 199(2):385–404. https://doi.org/10.1023/A:1010387509792

Charbonneau P (2020) Dynamo models of the solar cycle. Living Rev Sol Phys 17:4. https://doi.org/10.1007/s41116-020-00025-6

Charbonneau P, Dikpati M (2000) Stochastic fluctuations in a Babcock-Leighton model of the solar cycle. Astrophys J 543(2):1027–1043. https://doi.org/10.1086/317142

Charbonneau P, Blais-Laurier G, St-Jean C (2004) Intermittency and phase persistence in a Babcock-Leighton model of the Solar Cycle. Astrophys J Lett 616:L183–L186. https://doi.org/10.1086/426897

Charbonneau P, Beaubien G, St-Jean C (2007) Fluctuations in Babcock-Leighton dynamos. II. Revisiting the Gnevyshev-Ohl rule. Astrophys J 658(1):657–662. https://doi.org/10.1086/511177

Chatzistergos T, Usoskin IG, Kovaltsov GA et al (2017) New reconstruction of the sunspot group numbers since 1739 using direct calibration and "backbone" methods. Astron Astrophys 602:A69. https://doi.org/10.1051/0004-6361/201630045. arXiv:1702.06183 [astro-ph.SR]

Choudhuri AR (1992) Stochastic fluctuations of the solar dynamo. Astron Astrophys 253:277–285

Choudhuri AR, Karak BB (2009) A possible explanation of the Maunder minimum from a flux transport dynamo model. Res Astron Astrophys 9:953–958. https://doi.org/10.1088/1674-4527/9/9/001. arXiv:0907.3106 [astro-ph.SR]

Choudhuri AR, Karak BB (2012) Origin of grand minima in sunspot cycles. Phys Rev Lett 109(17):171103

Clette F, Lefèvre L (2016) The new sunspot number: assembling all corrections. Sol Phys 291:2629–2651. https://doi.org/10.1007/s11207-016-1014-y

Clette F, Berghmans D, Vanlommel P et al (2007) From the Wolf number to the international sunspot index: 25 years of SIDC. Adv Space Res 40:919–928. https://doi.org/10.1016/j.asr.2006.12.045

Clette F, Svalgaard L, Vaquero J et al (2014) Revisiting the sunspot number: a 400-year perspective on the solar cycle. Space Sci Rev 186:35. https://doi.org/10.1007/s11214-014-0074-2

Clette F, Lefèvre L, Bechet S et al (2021) Reconstruction of the sunspot number source database and the 1947 Zurich discontinuity. Sol Phys 296(9):137. https://doi.org/10.1007/s11207-021-01882-6

Damon P, Sonett C (1991) Solar and terrestrial components of the atmospheric c-14 variation spectrum. In: Sonett C, Giampapa M, Matthews M (eds) The sun in time. University of Arizona Press, Tucson, U.S.A., pp 360–388

Dasi-Espuig M, Solanki SK, Krivova NA et al (2010) Sunspot group tilt angles and the strength of the solar cycle. Astron Astrophys 518:A7. https://doi.org/10.1051/0004-6361/201014301. arXiv:1005.1774 [astro-ph.SR]

Dicke RH (1970) The rotation of the sun. In: Slettebak A (ed) IAU Colloq. 4: stellar rotation. Reidel, Dordrecht, p 289

Dicke RH (1978) Is there a chronometer hidden deep in the Sun? Nature 276(5689):676–680. https://doi.org/10.1038/276676b0

Dikpati M, Charbonneau P (1999) A Babcock-Leighton flux transport dynamo with solar-like differential rotation. Astrophys J 518:508–520. https://doi.org/10.1086/307269

D'Silva S, Choudhuri AR (1993) A theoretical model for tilts of bipolar magnetic regions. Astron Astrophys 272:621

Eddington AS, Plakidis S (1929) Irregularities of period of long-period variable stars. Mon Not R Astron Soc 90:65–71. https://doi.org/10.1093/mnras/90.1.65.

Eddy J (1976) The maunder minimum. Science 192:1189–1202. https://doi.org/10.1126/science.192.4245.1189.

Fan Y, Fisher GH, McClymont AN (1994) Dynamics of emerging active region flux loops. Astrophys J 436:907–928. https://doi.org/10.1086/174967.

77

Gnevyshev MN, Ohl A (1948) On the 22-year cycle of solar activity. Astron Ž 25:18–20

Golubenko K, Rozanov E, Kovaltsov G et al (2021) Application of CCM SOCOL-AERv2-BE to cosmogenic beryllium isotopes: description and validation for polar regions. Geosci Model Dev 14(12):7605–7620. https://doi.org/10.5194/gmd-14-7605-2021

Gómez DO, Mininni PD (2006) Description of Maunder-like events from a stochastic Alpha–Omega model. Adv Space Res 38(5):856–861. https://doi.org/10.1016/j.asr.2005.07.032

Gough D (1978) The significance of solar oscillations. In: Dumont S, Roesch J (eds) Pleins Feux sur la Physique Solaire, p 81

Gough D (1981) On the seat of the solar cycle. In: NASA conference publication, pp 185–206

Gough DO (1983) Temporal solar variations. ESA J 7(4):325–339

Gough DO (1988) Theory of solar variation. In: Cini Castagnoli G (ed) Solar-terrestrial relationships and the Earth environment in the last Millennia, p 90. North-Holland, Amsterdam

Grinsted A, Moore JC, Jevrejeva S (2004) Application of the cross wavelet transform and wavelet coherence to geophysical time series. Nonlinear Process Geophys 11:561–566

Hathaway DH (2015) The solar cycle. Living Rev Sol Phys 12:4. https://doi.org/10.1007/lrsp-2015-4

Hazra G, Karak BB, Banerjee D et al (2015) Correlation between decay rate and amplitude of Solar Cycles as revealed from observations and dynamo theory. Sol Phys 290:1851–1870. https://doi.org/10.1007/s11207-015-0718-8. arXiv:1410.8641 [astro-ph.SR]

Hazra G, Choudhuri AR, Miesch MS (2017) A theoretical study of the build-up of the Sun's polar magnetic field by using a 3D kinematic dynamo model. Astrophys J 835:39. https://doi.org/10.3847/1538-4357/835/1/39. arXiv:1610.02726 [astro-ph.SR]

Heikkilä U, Beer J, Abreu JA et al (2013) On the atmospheric transport and deposition of the cosmogenic radionuclides (^{10}Be): a review. Space Sci Rev 176:321–332. https://doi.org/10.1007/s11214-011-9838-0

Hoyng P (1993) Helicity fluctuations in mean field theory: an explanation for the variability of the solar cycle? Astron Astrophys 272:321

Hoyt DV, Schatten KH (1998) Group sunspot numbers: a new solar activity reconstruction. Sol Phys 179:189–219. https://doi.org/10.1023/A:1005007527816

Inceoglu F, Simoniello R, Knudsen VF et al (2015) Grand solar minima and maxima deduced from ^{10}Be and ^{14}C: magnetic dynamo configuration and polarity reversal. Astron Astrophys 577:A20. https://doi.org/10.1051/0004-6361/201424212

Inceoglu F, Arlt R, Rempel M (2017) The nature of grand minima and maxima from fully nonlinear flux transport dynamos. Astrophys J 848(2):93. https://doi.org/10.3847/1538-4357/aa8d68. arXiv:1710.08644 [astro-ph.SR]

Jha BK, Karak BB, Mandal S et al (2020) Magnetic field dependence of bipolar magnetic region tilts on the Sun: indication of tilt quenching. Astrophys J Lett 889(1):L19. https://doi.org/10.3847/2041-8213/ab665c. arXiv:1912.13223 [astro-ph.SR]

Jiang J (2020) Nonlinear mechanisms that regulate the solar cycle amplitude. Astrophys J 900(1):19. https://doi.org/10.3847/1538-4357/abaa4b. arXiv:2007.07069 [astro-ph.SR]

Jiang J, Cameron RH, Schüssler M (2014) Effects of the scatter in sunspot group tilt angles on the large-scale magnetic field at the solar surface. Astrophys J 791:5. https://doi.org/10.1088/0004-637X/791/1/5. arXiv:1406.5564 [astro-ph.SR]

Jiao Q, Jiang J, Wang ZF (2021) Sunspot tilt angles revisited: dependence on the solar cycle strength. Astron Astrophys 653:A27. https://doi.org/10.1051/0004-6361/202141215. arXiv:2106.11615 [astro-ph.SR]

Käpylä MJ, Käpylä PJ, Olspert N et al (2016) Multiple dynamo modes as a mechanism for long-term solar activity variations. Astron Astrophys 589:A56. https://doi.org/10.1051/0004-6361/201527002. arXiv:1507.05417 [astro-ph.SR]

Karak BB (2010) Importance of meridional circulation in flux transport dynamo: the possibility of a Maunder-like grand minimum. Astrophys J 724:1021. https://doi.org/10.1088/0004-637X/724/2/1021. arXiv:1009.2479 [astro-ph.SR]

Karak BB (2020) Dynamo saturation through the latitudinal variation of bipolar magnetic regions in the Sun. Astrophys J Lett 901(2):L35. https://doi.org/10.3847/2041-8213/abb93f. arXiv:2009.06969 [astro-ph.SR]

Karak BB, Cameron R (2016) Babcock-Leighton solar dynamo: the role of downward pumping and the equatorward propagation of activity. Astrophys J 832:94. https://doi.org/10.3847/0004-637X/832/1/94. arXiv:1605.06224 [astro-ph.SR]

Karak BB, Choudhuri AR (2011) The Waldmeier effect and the flux transport solar dynamo. Mon Not R Astron Soc 410:1503–1512. https://doi.org/10.1111/j.1365-2966.2010.17531.x. arXiv:1008.0824 [astro-ph.SR]

Karak BB, Choudhuri AR (2013) Studies of grand minima in sunspot cycles by using a flux transport solar dynamo model. Res Astron Astrophys 13:1339. https://doi.org/10.1088/1674-4527/13/11/005. arXiv:1306.5438 [astro-ph.SR]

Karak BB, Miesch M (2017) Solar cycle variability induced by tilt angle scatter in a Babcock-Leighton Solar dynamo model. Astrophys J 847:69. https://doi.org/10.3847/1538-4357/aa8636. arXiv:1706.08933 [astro-ph.SR]

Karak BB, Miesch M (2018) Recovery from Maunder-like grand minima in a Babcock–Leighton solar dynamo model. Astrophys J Lett 860:L26. https://doi.org/10.3847/2041-8213/aaca97. arXiv:1712.10130 [astro-ph.SR]

Karak BB, Jiang J, Miesch MS et al (2014) Flux transport dynamos: from kinematics to dynamics. Space Sci Rev 186:561–602. https://doi.org/10.1007/s11214-014-0099-6

Karak BB, Käpylä PJ, Käpylä MJ et al (2015) Magnetically controlled stellar differential rotation near the transition from solar to anti-solar profiles. Astron Astrophys 576:A26. https://doi.org/10.1051/0004-6361/201424521. arXiv:1407.0984 [astro-ph.SR]

Karak BB, Mandal S, Banerjee D (2018) Double Peaks of the Solar Cycle: an explanation from a dynamo model. Astrophys J 866(1):17. https://doi.org/10.3847/1538-4357/aada0d. arXiv:1808.03922 [astro-ph.SR]

Kitchatinov LL, Olemskoy SV (2011) Does the Babcock-Leighton mechanism operate on the Sun? Astron Lett 37:656–658. https://doi.org/10.1134/S0320010811080031. arXiv:1109.1351 [astro-ph.SR]

Kitchatinov LL, Olemskoy SV (2016) Dynamo model for grand maxima of solar activity: can superflares occur on the Sun? Mon Not R Astron Soc 459(4):4353–4359. https://doi.org/10.1093/mnras/stw875. arXiv:1602.08840 [astro-ph.SR]

Kitchatinov LL, Pipin VV, Ruediger G (1994a) Turbulent viscosity, magnetic diffusivity, and heat conductivity under the influence of rotation and magnetic field. Astron Nachr 315:157–170. https://doi.org/10.1002/asna.2103150205

Kitchatinov LL, Rüdiger G, Küker M (1994b) Lambda-quenching as the nonlinearity in stellar-turbulence dynamos. Astron Astrophys 292:125–132

Knobloch E, Tobias SM, Weiss NO (1998) Modulation and symmetry changes in stellar dynamos. Mon Not R Astron Soc 297(4):1123–1138. https://doi.org/10.1046/j.1365-8711.1998.01572.x

Koldobskiy SA, Kähkönen R, Hofer B et al (2022) Time lag between cosmic-ray and solar variability: sunspot numbers and open solar magnetic flux. Sol Phys 297(3):38. https://doi.org/10.1007/s11207-022-01970-1

Küker M, Arlt R, Rüdiger G (1999) The Maunder minimum as due to magnetic Λ-quenching. Astron Astrophys 343:977–982

Kumar R, Jouve L, Nandy D (2019) A 3D kinematic Babcock Leighton solar dynamo model sustained by dynamic magnetic buoyancy and flux transport processes. Astron Astrophys 623:A54. https://doi.org/10.1051/0004-6361/201834705. arXiv:1901.04251 [astro-ph.SR]

Lal D, Peters B (1962) Cosmic ray produced isotopes and their application to problems in geophysics. In: Wilson J, Wouthuysen S (eds) Progress in elementary particle and cosmic ray physics, vol 6. North Holland, Amsterdam, pp 77–243.

Lemerle A, Charbonneau P (2017) A coupled 2 × 2D Babcock-Leighton Solar dynamo model. II. Reference dynamo solutions. Astrophys J 834:133. https://doi.org/10.3847/1538-4357/834/2/133. arXiv:1606.07375 [astro-ph.SR]

Lepreti F, Carbone V, Vecchio A (2021) Scaling properties and persistence of long-term solar activity. Atmosphere 12(6):733. https://doi.org/10.3390/atmos12060733

Leussu R, Usoskin IG, Arlt R et al (2013) Inconsistency of the Wolf sunspot number series around 1848. Astron Astrophys 559:A28. https://doi.org/10.1051/0004-6361/201322373. arXiv:1310.8443 [astro-ph.SR]

Lockwood M, Stamper R, Wild MN (1999) A doubling of the Sun's coronal magnetic field during the past 100 years. Nature 399:437–439. https://doi.org/10.1038/20867

Lockwood M, Owens MJ, Barnard L (2014) Centennial variations in sunspot number, open solar flux, and streamer belt width: 1. Correction of the sunspot number record since 1874. J Geophys Res Space Phys 119:5172–5182. https://doi.org/10.1002/2014JA019970

Lomb N (2013) The sunspot cycle revisited. J Phys Conf Ser 440:012042. https://doi.org/10.1088/1742-6596/440/1/012042

Malkus WVR, Proctor MRE (1975) The macrodynamics of α-effect dynamos in rotating fluids. J Fluid Mech 67:417–443. https://doi.org/10.1017/S0022112075000390

Mandal S, Karak BB, Banerjee D (2017) Latitude distribution of sunspots: analysis using sunspot data and a dynamo model. Astrophys J 851:70. https://doi.org/10.3847/1538-4357/aa97dc. arXiv:1711.00222 [astro-ph.SR]

Masarik J, Beer J (1999) Simulation of particle fluxes and cosmogenic nuclide production in the Earth's atmosphere. J Geophys Res 104:12,099–12,111

Miesch MS, Dikpati M (2014) A three-dimensional Babcock-Leighton solar dynamo model. Astrophys J Lett 785:L8. https://doi.org/10.1088/2041-8205/785/1/L8. arXiv:1401.6557 [astro-ph.SR]

Miesch MS, Teweldebirhan K (2016) A three-dimensional Babcock-Leighton solar dynamo model: Initial results with axisymmetric flows. Adv Space Res 58(8):1571–1588. https://doi.org/10.1016/j.asr.2016. 02.018. arXiv:1511.03613 [astro-ph.SR]

Mordvinov AV, Karak BB, Banerjee D et al (2022) Evolution of the Sun's activity and the poleward transport of remnant magnetic flux in Cycles 21-24. Mon Not R Astron Soc 510(1):1331–1339. https://doi.org/ 10.1093/mnras/stab3528. arXiv:2111.15585 [astro-ph.SR]

Moss D, Sokoloff D, Usoskin I et al (2008) Solar grand minima and random fluctuations in dynamo parameters. Sol Phys 250:221–234. https://doi.org/10.1007/s11207-008-9202-z

Muñoz-Jaramillo A, Vaquero J (2019) Visualization of the challenges and limitations of the long-term sunspot number record. Nat Astron 3:205–211. https://doi.org/10.1038/s41550-018-0638-2

Muñoz-Jaramillo A, Dasi-Espuig M, Balmaceda LA et al (2013) Solar cycle propagation, memory, and prediction: insights from a century of magnetic proxies. Astrophys J Lett 767:L25. https://doi.org/10.1088/ 2041-8205/767/2/L25. arXiv:1304.3151 [astro-ph.SR]

Mundt M, Maguire W II, Chase R (1991) Chaos in the sunspot cycle: analysis and prediction. J Geophys Res 96:1705–1716

Nagy M, Lemerle A, Labonville F et al (2017) The effect of "rogue" active regions on the solar cycle. Sol Phys 292:167. https://doi.org/10.1007/s11207-017-1194-0. arXiv:1712.02185 [astro-ph.SR]

Ogurtsov M (2004) New evidence for long-term persistence in the Sun's activity. Sol Phys 220:93–105. https://doi.org/10.1023/B:sola.0000023439.59453.e5

Ölçek D, Charbonneau P, Lemerle A et al (2019) Grand activity minima and maxima via dual dynamos. Sol Phys 294(7):99. https://doi.org/10.1007/s11207-019-1492-9

Olemskoy SV, Kitchatinov LL (2013) Grand minima and North-South asymmetry of solar activity. Astrophys J 777:71

Ossendrijver AJH, Hoyng P (1996) Stochastic and nonlinear fluctuations in a mean field dynamo. Astron Astrophys 313:959–970

Ossendrijver AJH, Hoyng P, Schmitt D (1996) Stochastic excitation and memory of the solar dynamo. Astron Astrophys 313:938–948

Ostriakov V, Usoskin I (1990) On the dimension of solar attractor. Sol Phys 127:405–412. https://doi.org/10. 1007/BF00152177

Owens MJ, Forsyth RJ (2013) The heliospheric magnetic field. Living Rev Sol Phys 10:5. https://doi.org/10. 12942/lrsp-2013-5

Parker EN (1955) Hydromagnetic dynamo models. Astrophys J 122:293. https://doi.org/10.1086/146087

Passos D, Charbonneau P (2014) Characteristics of magnetic solar-like cycles in a 3D MHD simulation of solar convection. Astron Astrophys 568:A113. https://doi.org/10.1051/0004-6361/201423700

Passos D, Charbonneau P, Beaudoin P (2012) An exploration of non-kinematic effects in flux transport dynamos. Sol Phys 279(1):1–22. https://doi.org/10.1007/s11207-012-9971-2

Passos D, Nandy D, Hazra S et al (2014) A solar dynamo model driven by mean-field alpha and Babcock-Leighton sources: fluctuations, grand-minima-maxima, and hemispheric asymmetry in sunspot cycles. Astron Astrophys 563:A18. https://doi.org/10.1051/0004-6361/201322635. arXiv:1309.2186 [astro-ph.SR]

Pavón-Carrasco FJ, Gómez-Paccard M, Campuzano S et al (2018) Multi-centennial fluctuations of radionuclide production rates are modulated by the Earth's magnetic field. Sci Rep 8:9820. https://doi.org/10. 1038/s41598-018-28115-4

Peristykh A, Damon P (2003) Persistence of the gleissberg 88-year solar cycle over the last ∼12,000 years: evidence from cosmogenic isotopes. J Geophys Res 108:1003. https://doi.org/10.1029/2002JA009390

Petrovay K (2020) Solar cycle prediction. Living Rev Sol Phys 17:2. https://doi.org/10.1007/s41116-020-0022-z. arXiv:1907.02107 [astro-ph.SR]

Pipin VV (1999) The Gleissberg cycle by a nonlinear $\alpha\Lambda$ dynamo. Astron Astrophys 346:295–302

Potgieter M (2013) Solar modulation of cosmic rays. Living Rev Sol Phys 10:3. https://doi.org/10.12942/ lrsp-2013-3. arXiv:1306.4421 [physics.space-ph]

Priyal M, Banerjee D, Karak BB et al (2014) Polar network index as a magnetic proxy for the solar cycle studies. Astrophys J Lett 793:L4. https://doi.org/10.1088/2041-8205/793/1/L4. arXiv:1407.4944 [astro-ph.SR]

Reimer P, Austin W, Bard E et al (2020) The INTCAL20 Northern hemisphere radiocarbon age calibration curve (0-55 CAL KBP). Radiocarbon 62(4):725–757. https://doi.org/10.1017/RDC.2020.41

Rempel M, Schüssler M (2001) Intensification of magnetic fields by conversion of potential energy. Astrophys J Lett 552(2):L171–L174. https://doi.org/10.1086/320346

Roth R, Joos F (2013) A reconstruction of radiocarbon production and total solar irradiance from the Holocene ^{14}C and CO_2 records: implications of data and model uncertainties. Clim Past 9:1879–1909. https://doi.org/10.5194/cp-9-1879-2013

Ruediger G, Kichatinov LL (1993) Alpha-effect and alpha-quenching. Astron Astrophys 269(1–2):581–588

Russell CT, Jian LK, Luhmann JG (2019) The solar clock. Rev Geophys 57:1129

Ruzmaikin A, Feynman J, Robinson P (1994) Long-term persistence of solar activity. Sol Phys 149:395–403

Solanki SK, Usoskin IG, Kromer B et al (2004) Unusual activity of the Sun during recent decades compared to the previous 11,000 years. Nature 431:1084–1087. https://doi.org/10.1038/nature02995

Solanki SK, Wenzler T, Schmitt D (2008) Moments of the latitudinal dependence of the sunspot cycle: a new diagnostic of dynamo models. Astron Astrophys 483:623–632. https://doi.org/10.1051/0004-6361: 20054282.

Steenbeck M, Krause F, Rädler KH (1966) Berechnung der mittleren Lorentz-Feldstärke für ein elektrisch leitendes Medium in turbulenter, durch Coriolis-Kräfte beeinflußter Bewegung. Z Naturforsch A 21:369. https://doi.org/10.1515/zna-1966-0401

Stefani F, Beer J, Giesecke A et al (2020) Phase coherence and phase jumps in the Schwabe cycle. Astron Nachr 341(600):600–615. https://doi.org/10.1002/asna.202013809. arXiv:2004.10028 [astro-ph.SR]

Stefani F, Stepanov R, Weier T (2021) Shaken and stirred: when bond meets Suess-de Vries and Gnevyshev-Ohl. Sol Phys 296(6):88. https://doi.org/10.1007/s11207-021-01822-4. arXiv:2006.08320 [astro-ph.SR]

Steinhilber F, Abreu J, Beer J et al (2012) 9,400 years of cosmic radiation and solar activity from ice cores and tree rings. Proc Natl Acad Sci USA 109(16):5967–5971. https://doi.org/10.1073/pnas.1118965109

Stuiver M (1961) Variations in radiocarbon concentration and sunspot activity. J Geophys Res 66:273–276

Stuiver M, Braziunas T (1989) Atmospheric ^{14}C and century-scale solar oscillations. Nature 338:405–408. https://doi.org/10.1038/338405a0

Svalgaard L, Schatten KH (2016) Reconstruction of the sunspot group number: the Backbone method. Sol Phys 291:2653–2684. https://doi.org/10.1007/s11207-015-0815-8. arXiv:1506.00755 [astro-ph.SR]

Talafha M, Nagy M, Lemerle A et al (2022) Role of observable nonlinearities in solar cycle modulation. Astron Astrophys 660:A92. https://doi.org/10.1051/0004-6361/202142572. arXiv:2112.14465 [astro-ph.SR]

Usoskin I, Mursula K, Kovaltsov G (2000) Cyclic behaviour of sunspot activity during the Maunder minimum. Astron Astrophys 354:L33–L36

Usoskin IG (2017) A history of solar activity over millennia. Living Rev Sol Phys 14:3. https://doi.org/10.1007/s41116-017-0006-9

Usoskin IG, Mursula K, Kovaltsov GA (2001) Was one sunspot cycle lost in late XVIII century? Astron Astrophys 370:L31–L34. https://doi.org/10.1051/0004-6361:20010319

Usoskin IG, Solanki SK, Schüssler M et al (2003) Millennium-scale sunspot number reconstruction: evidence for an unusually active Sun since the 1940s. Phys Rev Lett 91:211101. https://doi.org/10.1103/PhysRevLett.91.211101. arXiv:astro-ph/0310823

Usoskin IG, Solanki SK, Kovaltsov GA (2007) Grand minima and maxima of solar activity: new observational constraints. Astron Astrophys 471:301–309. https://doi.org/10.1051/0004-6361:20077704. arXiv:0706.0385

Usoskin IG, Hulot G, Gallet Y et al (2014) Evidence for distinct modes of solar activity. Astron Astrophys 562:L10. https://doi.org/10.1051/0004-6361/201423391. arXiv:1402.4720 [astro-ph.SR]

Usoskin IG, Arlt R, Asvestari E et al (2015) The Maunder minimum (1645-1715) was indeed a grand minimum: a reassessment of multiple datasets. Astron Astrophys 581:A95. https://doi.org/10.1051/0004-6361/201526652. arXiv:1507.05191 [astro-ph.SR]

Usoskin IG, Gallet Y, Lopes F et al (2016a) Solar activity during the Holocene: the Hallstatt cycle and its consequence for grand minima and maxim. Astron Astrophys 587:A150. https://doi.org/10.1051/0004-6361/201527295

Usoskin IG, Kovaltsov GA, Lockwood M et al (2016b) A new calibrated sunspot group series since 1749: statistics of active day fractions. Sol Phys 291:2685–2708. https://doi.org/10.1007/s11207-015-0838-1. arXiv:1512.06421 [astro-ph.SR]

Usoskin IG, Solanki SK, Krivova N et al (2021) Solar cyclic activity over the last millennium reconstructed from annual ^{14}C data. Astron Astrophys 649:A141. https://doi.org/10.1051/0004-6361/202140711

Vaquero J, Svalgaard L, Carrasco V et al (2016) A revised collection of sunspot group numbers. Sol Phys 291:3061–3074. https://doi.org/10.1007/s11207-016-0982-2.

Vaquero JM, Vázquez M (2009) The Sun recorded through history. Astrophysics and space science library, vol 361. Springer, New York. https://doi.org/10.1007/978-0-387-92790-9

Vaquero JM, Kovaltsov GA, Usoskin IG et al (2015) Level and length of cyclic solar activity during the Maunder minimum as deduced from the active-day statistics. Astron Astrophys 577:A71. https://doi.org/10.1051/0004-6361/201525962. arXiv:1503.07664 [astro-ph.SR]

Vasiliev S, Dergachev V (2002) The ~2400-year cycle in atmospheric radiocarbon concentration: bispectrum of ^{14}C data over the last 8000 years. Ann Geophys 20:115–120

Vidotto AA (2021) The evolution of the solar wind. Living Rev Sol Phys 18:3. https://doi.org/10.1007/s41116-021-00029-w. arXiv:2103.15748 [astro-ph.SR]

 Springer

Reprinted from the journal

Vonmoos M, Beer J, Muscheler R (2006) Large variations in Holocene solar activity: constraints from [10]Be in the Greenland Ice Core Project ice core. J Geophys Res A10:105. https://doi.org/10.1029/2005JA011500

Waldmeier M (1955) Ergebnisse und Probleme der Sonnenforschung. Geest & Portig, Leipzig

Wang YM, Sheeley NR (2009) Understanding the geomagnetic precursor of the Solar Cycle. Astrophys J Lett 694:L11–L15. https://doi.org/10.1088/0004-637X/694/1/L11

Weiss NO, Tobias SM (2016) Supermodulation of the Sun's magnetic activity: the effects of symmetry changes. Mon Not R Astron Soc 456(3):2654–2661. https://doi.org/10.1093/mnras/stv2769

Weisshaar E, Cameron RH, Schüssler M (2023) No evidence for synchronization of the solar cycle by a "clock". Astron Astrophys 671:A87. https://doi.org/10.1051/0004-6361/202244997. arXiv:2301.07469 [astro-ph.SR]

Willamo T, Usoskin IG, Kovaltsov GA (2017) Updated sunspot group number reconstruction for 1749-1996 using the active day fraction method. Astron Astrophys 601:A109. https://doi.org/10.1051/0004-6361/201629839. arXiv:1705.05109 [astro-ph.SR]

Wu CJ, Usoskin IG, Krivova N et al (2018) Solar activity over nine millennia: a consistent multi-proxy reconstruction. Astron Astrophys 615:A93. https://doi.org/10.1051/0004-6361/201731892. arXiv:1804.01302 [astro-ph.SR]

Yeates AR, Muñoz-Jaramillo A (2013) Kinematic active region formation in a three-dimensional solar dynamo model. Mon Not R Astron Soc 436(4):3366–3379. https://doi.org/10.1093/mnras/stt1818. arXiv:1309.6342 [astro-ph.SR]

Yeates AR, Nandy D, Mackay DH (2008) Exploring the physical basis of solar cycle predictions: flux transport dynamics and persistence of memory in advection- versus diffusion-dominated solar convection zones. Astrophys J J.673:544–556. https://doi.org/10.1086/524352. arXiv:0709.1046

Zolotova NV, Ponyavin DI (2016) How deep was the Maunder minimum? Sol Phys 291:2869–2890. https://doi.org/10.1007/s11207-016-0908-z

Space Science Reviews (2023) 219:70
https://doi.org/10.1007/s11214-023-01016-3

Scaling and Evolution of Stellar Magnetic Activity

Emre Işık[1,2] · Jennifer L. van Saders[3] · Ansgar Reiners[4] · Travis S. Metcalfe[5]

Received: 7 July 2023 / Accepted: 12 October 2023 / Published online: 30 October 2023

Abstract

Magnetic activity is a ubiquitous feature of stars with convective outer layers, with implications from stellar evolution to planetary atmospheres. Investigating the mechanisms responsible for the observed stellar activity signals from days to billions of years is important in deepening our understanding of the spatial configurations and temporal patterns of stellar dynamos, including that of the Sun. In this paper, we focus on three problems and their possible solutions. We start with direct field measurements and show how they probe the dependence of magnetic flux and its density on stellar properties and activity indicators. Next, we review the current state-of-the-art in physics-based models of photospheric activity patterns and their variation from rotational to activity-cycle timescales. We then outline the current state of understanding in the long-term evolution of stellar dynamos, first by using chromospheric and coronal activity diagnostics, then with model-based implications on magnetic braking, which is the key mechanism by which stars spin down and become inactive as they age. We conclude by discussing possible directions to improve the modeling and analysis of stellar magnetic fields.

Solar and Stellar Dynamos: A New Era
Edited by Manfred Schüssler, Robert H. Cameron, Paul Charbonneau, Mausumi Dikpati, Hideyuki Hotta and Leonid Kitchatinov

✉ A. Reiners
Ansgar.Reiners@phys.uni-goettingen.de

E. Işık
isik@mps.mpg.de

J.L. van Saders
jlvs@hawaii.edu

T.S. Metcalfe
travis@wdrc.org

[1] Max-Planck-Institut für Sonnensystemforschung, Justus-von-Liebig-Weg 3, Göttingen, 37077, Germany

[2] Department of Computer Science, Turkish-German University, Şahinkaya Cd. 94, Beykoz, 34820, Istanbul, Turkey

[3] Institute for Astronomy, University of Hawaii, 2680 Woodlawn Dr., Honolulu, HI 96822, USA

[4] Institut für Astrophysik und Geophysik, Georg-August-Universität Göttingen, Friedrich-Hund-Platz 1, Göttingen, 37077, Germany

[5] White Dwarf Research Corporation, 9020 Brumm Trl, Golden, CO 80403, USA

Keywords Cool stars · Stellar magnetism · Stellar activity · Angular momentum loss

1 Overview

Magnetism is ubiquitous in stars and yet it is relatively poorly understood, even in our closest neighbor. We necessarily rely on observational constraints—direct measurements or proxies for magnetism—to probe magnetic behavior across stellar types and lifetimes, and to connect these observations to underlying theoretical descriptions (Schrijver and Zwaan 2000). In this paper, we highlight ways in which magnetism on stars reveals itself, and the insights those physical manifestations provide about the underlying physics of magnetic fields in stellar systems. The purpose of this paper is to report on recent progress in the following particular problems.

- How does magnetic flux and its density scale with rotation and the fractional depth of the convection-zone?
- How can physics-based diagnostic modeling help us to constrain surface patterns and their evolution?
- What is responsible for the spin-down and the weakening of outer-atmospheric activity indicators with age?

Following a summary of our recent attempts to find answers to these questions, we present an outlook on possible avenues to better understand the scaling relations of stellar magnetic activity. More extensive reviews can be found in the literature (e.g., Donati and Landstreet 2009; Strassmeier 2009; Reiners 2012; Engvold et al. 2019; Basri 2021).

The structure and dynamics of magnetic fields threading the atmospheres of stars other than the Sun are observed mostly indirectly. Magnetic field measurements from Zeeman splitting of photospheric spectral lines are becoming more accessible owing to instrumentation at optical and near-infrared wavelengths, such as CRIRES$^+$ (Dorn et al. 2014), PEPSI (Strassmeier et al. 2015), ESPaDOnS (Donati et al. 2006), NARVAL (Aurière 2003), HARP-Spol (Snik et al. 2008; Piskunov et al. 2011), SPIRou (Donati et al. 2020), and CARMENES (Quirrenbach et al. 2014) and HPF (Mahadevan et al. 2012). This promotes reliable quantification of the magnetic flux and its heating mechanisms observable in indirect activity indicators, and their scaling laws for different types of stars. We cover recent work on direct magnetic field measurements and their use in constraining the rotation-activity scalings in Sect. 2.

While our understanding is often driven by observations, numerical simulation frameworks are essential tools to better evaluate observational trends of stellar magnetic activity on cool stars. Forward modeling of observational diagnostics are mostly based on physical models developed originally for the Sun. Scaling laws are often used to extend the solar paradigm to younger and more active suns as well as for cooler stars with deeper convection zones. More physically motivated applications involve dynamo models of the global magnetic field and the flux emergence process as a function of stellar properties. We present some important recent developments in modeling photospheric diagnostics of stellar magnetism in Sect. 3.

The indirect diagnostics—also called proxies—of magnetic activity include: (1) disk-integrated brightness in intermediate and broad bandpasses for the effects on the photosphere; (2) narrow-band radiative fluxes centered on spectral lines such as Hα and singly ionized Ca and Mg that probe the chromosphere; and (3) outer-atmospheric indicators of non-thermal heating by magnetic fields, at X-ray and radio wavelengths. We review recent advances in chromospheric and coronal proxies in relation to the long-term evolution of stellar activity in Sect. 4.

Stars with convective outer layers have dynamos that support large-scale fields from their interiors to their magnetospheric environments, also called astrospheres. Following star formation and disk dispersal, rapid stellar rotation coupled with convection leads to strong magnetic activity, which is responsible for strong magnetized winds removing angular momentum from the star. Rotational evolution of cool stars can be used as a proxy for the integrated large-scale field behavior. We discuss the connection between magnetic braking and global field properties in Sect. 5.

We propose strategies for further progress in the problems introduced above in Sect. 6.

2 Scaling of Magnetic Flux and Non-thermal Emission

Our picture of stellar magnetism and its influence on stellar evolution and activity is anchored in the detailed observational data from the Sun. Among other examples, the spatial correspondence between active regions and magnetic flux concentrations, the occurrence of faculae and dark spots along latitudinal bands, differential rotation, the thermal and magnetic structure of active regions, and the temporal variability including magnetic cycle(s), are phenomena that can be observed in the Sun (see, e.g., Schrijver 1987; Solanki 2003; Solanki et al. 2006). In analogy, they are assumed to occur on other stars (e.g., Berdyugina 2005).

As a *direct* measurement of magnetic fields, we understand determination of the immediate influence of the field, e.g., the signatures of the Zeeman effect on line profiles. This is in contrast to *indirect* measurements, which include the measurement of proxies of magnetic activity, for example non-thermal emission. The direct measurement of magnetic fields is hampered by the fact that other stars are very far away and cannot be spatially resolved. One of the consequences is that spectroscopy can be obtained from the spatially integrated stars but not from individual areas, which leads to blending of spectroscopic effects caused by magnetism with those from other atmospheric effects and partial cancellation of polarization (see, e.g., Donati and Landstreet 2009; Reiners 2012; Kochukhov 2021). Direct observations of magnetic fields therefore typically require substantial spectral resolution and signal-to-noise (S/N), and sometimes recurrent observations of the same star over different rotational phases. Therefore, other *indirect* indicators of stellar magnetism are often employed to characterize stellar magnetic activity. These are usually indicators of non-thermal emission that is generated by the stellar magnetic field in analogy to the solar example (see, e.g., Reiners et al. 2022, and references therein).

Direct measurements of magnetism are usually those that investigate an immediate observable of the magnetic fields, i.e., spectroscopic signatures like the Zeeman effect. In solar-type and low-mass stars, the Zeeman effect causes the most often used *direct* signatures that are Zeeman broadening (observed in integrated light, Stokes I; Saar 1988) and polarization (see, e.g., Landi Degl'Innocenti and Landolfi 2004). Observational biases can be rather strong leading to large uncertainties in the measured field strengths and/or the amount of magnetism unseen by the observations (see, e.g., Kochukhov and Reiners 2020, and references therein) of several directions of polarization, and monitoring projects delivering very high S/N data have provided a wealth of information from direct field measurements.

Solar-type and low-mass stars show a clear relation between rotation and non-thermal emission observed in several activity indicators, e.g., X-rays (Skumanich 1972; Noyes et al. 1984; Pizzolato et al. 2003; Wright et al. 2011; Reiners et al. 2014). The causal connection

Fig. 1 Average surface magnetic field measurements from Stokes I in sun-like and low-mass stars (adapted from Reiners et al. 2022). Grey dashed lines indicate the relation between field strength and Rossby number, $Ro = P/\tau$, for two groups of "slow" and "fast" rotators. Stellar mass is indicated with symbol size and color, downward triangles indicate upper limits in $\langle B \rangle$

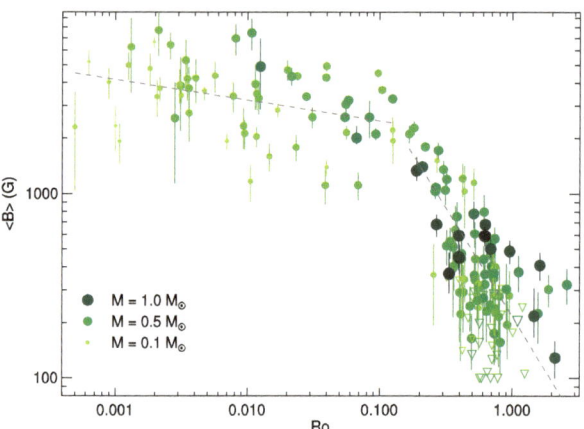

between stellar rotation, stellar activity, and rotational evolution is supposed to be (1) a mechanism providing more surface magnetic flux at high rotation rates in any given star, and (2) the generation of non-thermal emission in proportion to magnetic energy (or magnetic flux) at the stellar surface. Currently, there is no physical model that explains any of the two phenomena from first principles.

The relation between average surface magnetic flux density (or field strength, $\langle B \rangle$) and rotation is shown in Fig. 1. Similar to the rotation-activity relation, the Rossby number, $Ro = P/\tau$ with τ the convective turnover time, is used as a normalized proxy of rotation. Figure 1 shows two regimes of magnetic fields: a group of rapid rotators with $Ro \leq 0.1$ and $\langle B \rangle > 1$ kG, and the slower rotators with significantly weaker fields. Both groups show a statistically significant relation between $\langle B \rangle$ and Ro. The dependence of $\langle B \rangle$ on Ro is a lot stronger among the slow rotators than within the more rapidly rotating group. The dashed lines show the relations:

$$\langle B \rangle = 199\,\text{G} \times Ro^{-1.26\pm0.10} \text{ for slow rotation} \tag{1}$$

$$\langle B \rangle = 2050\,\text{G} \times Ro^{-0.11\pm0.03} \text{ for rapid rotation} \tag{2}$$

The magnetic field-rotation relation (Fig. 1) closely resembles the rotation-activity relation mentioned above. This suggests that there is also a tight relation between magnetism and non-thermal emission. A relation between X-ray luminosity and magnetic flux was reported by Pevtsov et al. (2003) and re-investigated for a sample of stars with measured surface fields by Feiden and Chaboyer (2013). Figure 2 shows relations between X-ray, Hα, and Ca H&K luminosity as a function of surface magnetic flux, Φ_B. All three luminosities significantly grow with Φ_B. It should be noted that both luminosity and magnetic flux scale with radius squared, implying that at least parts of the relations seen in Fig. 2 (and in Pevtsov et al. 2003) could be caused by the different radii. An analysis of magnetic flux density and normalized luminosities reveals that higher magnetic field strengths in fact cause stronger emission (see Reiners et al. 2022).

Figures 1 and 2 are adapted from Reiners et al. (2022) and mainly cover M dwarf stars plus several young Suns from the literature. A broad range of stellar masses and rotation rates are included. The more rapidly rotating stars are observed to show stronger surface fields, and stellar surface magnetic flux leads to a proportional amount of non-thermal chromospheric and coronal emission. The Ca H&K lines show emission already at relatively low

Fig. 2 Relations between
magnetic flux, Φ_B, and total
non-thermal emission as
observed in X-rays (top panel),
Hα (center panel), and Ca H&K
(bottom panel). Stellar mass is
indicated with symbol size and
color, leftward triangles indicate
upper limits in $\langle B \rangle$. A maximum
field strength of 800 G was used
for the calculation of Φ_B in the
bottom panel (see text and
discussion in Reiners et al. 2022)

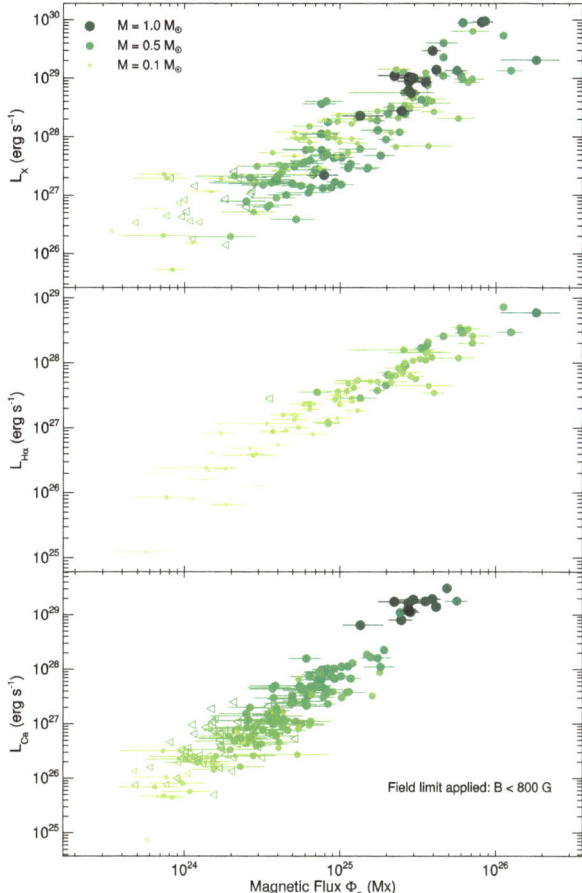

magnetic flux levels, and the field strength adopted for the calculation of Φ_B was limited to $\langle B \rangle < 800$ G indicating saturation of Ca H&K emission in very active stars (see Reiners et al. 2022, for more details). Coronal X-ray emission requires a somewhat higher magnetic flux level to generate observable emission rates, and Hα becomes visible in emission at similar or higher magnetic flux levels. The observed relations between magnetism and rotation, and between magnetism and non-thermal emission provide a link between the wealth of information on stellar rotation and stellar activity. This should help to further constrain the processes of magnetic field generation and flux emergence in sun-like and low-mass stars.

An important property of stellar magnetic fields is its distribution across the stellar surface. Measurements of Zeeman broadening are sensitive to the integrated field across the stellar surface, but they are less sensitive to the distribution of individual field components. Measurements of polarized light in addition with observations taken at different times, with the goal of sampling the star at different rotational phases, can provide important information about the geometry of the field. The field measured in polarized light, in particular in the case of circular polarization only (Stokes V), resembles the distribution of the so-called large-scale field. This is because magnetic fields of opposite polarity can cancel each other and remain invisible to this type of observation. In general, results about the scaling of large-scale fields with stellar mass and rotation are consistent with observations of average surface fields and non-thermal emission (Vidotto et al. 2014).

Observations of large-scale fields have provided insight into magnetic geometries in very different stars (Donati and Landstreet 2009; Marsden et al. 2014; Kochukhov 2021). Additional information can be derived from the ratio between the large-scale field from circular polarization and the average field from Zeeman broadening. The ratio between Stokes V and Stokes I field strengths is typically on the order of 10% with individual stars showing up to 40% or less than 1% of their magnetic fields on large-scales. There is a slight trend of larger ratios $\langle B \rangle_V / \langle B \rangle_I$ occuring among the lower mass / smaller stars, which may indicate different modes of magnetic field generation but could also be influenced by biases in the observational methods. In addition to gathering more data on magnetic fields from Stokes V and I observations, it is very important to understand the systematic bias introduced by both methods. These include effects like unknown velocity fields and line profile distortions caused by non-thermal emission in Stokes I measurements, and the consequences of incomplete phase coverage, uncertainties of the inclination angle, and flux cancellation in Stokes V measurements.

3 Modeling Photospheric Magnetism

Magnetic features on stars other than the Sun can only be observed indirectly (except in a few cases with interferometry), as mentioned in Sect. 2. In essence, filling factors and distributions of starspots can only be inferred from disc- integrated diagnostics, often through data-driven modeling. Forward modeling (i.e., via numerical simulations) of surface magnetic features come into stage here, as they often bring physical insight that helps us interpret observations. This section is devoted to recent attempts in numerical simulations of observational diagnostics, based on physical models of magnetic flux generation and transport.

3.1 Distribution of Surface Magnetic Flux

Attempts to model surface magnetic activity patterns on cool stars range from star-in-a-box simulations of convectively driven dynamos in M stars (e.g., Yadav et al. 2015) to solar-like models applied to stars rotating faster than the Sun, which we will focus on here.

The existence of polar or high-latitude spots on young solar-type stars rotating much faster than the Sun was revealed by Doppler imaging studies (e.g., Strassmeier and Rice 1998). The formation and structure of near-polar spots is one of the problems in cool-star research. The main question is whether they emerge at near-polar latitudes, or are transported there by surface flows. There are two non-mutually exclusive explanations for such spot patterns in the literature. Firstly, any radially rising buoyant concentration of toroidal magnetic flux is expected to be deflected towards the rotation axis, owing to angular momentum conservation (or Coriolis force in the co-rotating frame), leading to poleward deflection that increases with the stellar rotation rate (Schüssler and Solanki 1992; Weber et al. 2023). Numerical simulations of flux-tube emergence through the convection zone as a function of the stellar rotation rate provided further support for this hypothesis (Schüssler et al. 1996). According to the hypothesis, poleward deflection of rising tubes is controlled not only by the rotation rate, but also by the fractional depth of the convection zone (in turn, the Rossby number). A flux loop that starts from a low latitude and rises parallel to the rotation axis would thus have an emergence latitude that increases towards later spectral types as shown in simulations by Granzer et al. (2000). The second explanation involves the transport of emerging flux by differential rotation, meridional flow and supergranulation (Schrijver and Title 2001). Surface flux transport (SFT) simulations for various differential rotation and stellar radius configurations by Işık et al. (2007) have shown that, mid-latitude emergence of highly tilted bipolar regions sustain polar spots, mainly by diffusion and meridional flow. Faster-than-solar meridional flow speeds were invoked by Holzwarth et al. (2006), who showed formation of possible starspots with intermingled polarities around the rotational pole. All these studies showed various possibilities for the maintenance of long-lived polar spots by SFT processes.

Stellar dynamo models were incorporated to simulate stellar cycle characteristics in addition to surface distributions. In a model that integrates a deep-seated dynamo, flux-tube rise and surface transport, Işık et al. (2011) estimated the evolving surface distribution of large-scale radial magnetic flux through several dynamo cycles. They found that for intermediate rotators ($P_{rot} \sim 10$ d), the combined effects of enhanced cycle overlap and large tilt angles of emerging bipoles can lead to an unsigned-flux balance between polar caps and low latitudes that are modulated in anti-phase. This was suggested as an explanation for the existence of moderately active but non-cycling stars that were observed in S-index time series (Hall and Lockwood 2004). For more extensive reviews of modeling work on stellar activity cycles, see Biswas et al. (2023) and Hazra et al. (2023).

Aiming to forward-model brightness variability in rotational time scales, Işık et al. (2018) constructed a Flux Emergence and Transport (FEAT) simulation framework for solar-type stars with rotation rates and activity levels from the solar reference levels up to 8 times higher values. Assuming a solar-like latitudinal distribution of magnetic flux at the base of the convection zone, the authors modeled the surface emergence patterns by using simulations of buoyant flux tubes rising through the convection zone. The latitudes and tilt angles of emerging bipolar regions are used as input to a surface flux transport model, which calculates the time-dependent distribution of surface magnetic flux with a daily cadence, for a decade. The FEAT model also features active-region nesting at the time of emergence, with the probability that an active region emerges near the previous one being an adjustable parameter. They found that polar spots start to form between 4 and 8 times the solar rotation rate and activity level, by accumulation of trailing-polarity flux from tilted bipoles emerging at mid-latitudes. The tilt angles and thus the dipole contributions of emerging active regions increase in average, along with their variance around the mean (Weber et al. 2023). An enhanced nesting tendency locally increases the flux density reached in certain regions as well as in the formation of the spotted polar caps.

3.2 Line Profile Modeling

Despite the currently inaccessible detailed structure of stellar active regions, solar observations indicate that the magnetic field should be mostly radial upon emergence into the photosphere, owing to the steep density gradient in the atmosphere. Along with Gauss's law, this enforces a bipolar distribution of radial magnetic field throughout an active region. However, Zeeman-Doppler imaging (ZDI) studies indicate that, as the activity level increases, some rapidly rotating, young cool stars manifest strong azimuthal fields that can form axisymmetric bands (e.g., Folsom et al. 2018). The azimuthal field component tends to follow power laws with two different exponents on both sides of a stellar mass of about $0.5\,M_\odot$, and it becomes more axisymmetric and confined to higher latitudes for more rapid rotators (See et al. 2015). Strong azimuthal fields on some stars much more active than the Sun was interpreted by Solanki (2002) as the effect of differential rotation acting on strong fields that accumulate or emerge near the rotational poles, resulting in amplification of azimuthal fields at mid-latitudes. Motivated by ZDI results, Lehmann et al. (2017) decomposed simulated surface flux distributions into spherical harmonics, to obtain relative fractions of magnetic energy in radial, azimuthal, and meridional components (see Vidotto 2016, for a similar analysis for the Sun). They found that magnetic multipoles beyond the quadrupole (mainly contributed by azimuthal fields) host poloidal and toroidal field components following a fixed ratio. Later, Lehmann et al. (2019) synthesised Stokes profiles from surface flux transport simulations for various activity levels and concluded that ZDI overestimates the relative contributions from axisymmetric and toroidal fields, particularly as the axial inclination decreases towards pole-on configurations.

As a first application of the FEAT model described in Sect. 3.1, Şenavcı et al. (2021) synthesised Doppler images of the young solar analogue EK Draconis, using SFT snapshots from the FEAT model, and compared them with the Doppler images they generated from observed spectra. The overall latitudinal distribution of spots in the simulations were consistent with the observations in the case of strong differential rotation, which was previously reported for EK Dra. The simulations also showed that low-latitude spots in observed Doppler images can result from mid-latitude activity in the partially visible rotational hemisphere, owing to the axial inclination of 63°. This study has shown the importance of forward modeling of the observed signals from physics-based models, in the interpretation of observations.

3.3 Brightness Variability

Broad-band variability of cool stars on time scales ranging from the rotation period to the activity cycle result from photospheric manifestations of magnetic flux: mainly spots in active regions (ARs) and faculae resulting from AR network fields. Intriguingly, some solar-type stars with near-solar rotation rates and temperatures display unexpectedly large photometric variability amplitudes on the rotational time scale, in comparison to the Sun (Reinhold et al. 2020). By numerical simulations of light curves with different modes and degrees of active-region nesting, Işık et al. (2020) showed that such high-amplitude light curves can be explained by a moderate increase of the emergence frequency and a high degree of nesting (up to 90%). The same study showed that active-longitude-type nesting can reproduce light-curve morphology of some stars with sinusoidal-like light curves in the same sample better than free nesting, where nests are allowed to form within active latitudes and a random longitude.

The Sun's brightness variability on the cycle timescale is dominated by bright faculae, because they cover larger areas despite their visibility being confined to near-limb regions in visible wavelengths. However, more active stars are known to get dimmer as they reach peak activity levels in their S-index time series (e.g., Radick et al. 1998). On the Sun, the area fractions of spots and faculae depend quadratically and linearly on the S-index, respectively. These dependencies can be used to predict that in more active stars, spots would dominate over faculae in terms of disc-area coverages and hence of brightness (Shapiro et al. 2014). The physical mechanism underlying this transition from facula-dominated to spot-dominated variability has been explained by Nèmec et al. (2022), who carried out surface flux transport (SFT) simulations of a single solar-like activity cycle with gradually higher activity levels. The simulations were carried out for equator-on and pole-on inclinations. They showed qualitative agreement with the observed transition between facula- and spot-dominated cyclic variations of the solar-type sub-sample of Radick et al. (1998). Figure 3 shows the Strömgren $b + y$ brightening in units of the change (increase) in S-index as a function of $\log R'_{HK}$ for the stellar sample. Also shown are brightening estimates from a series

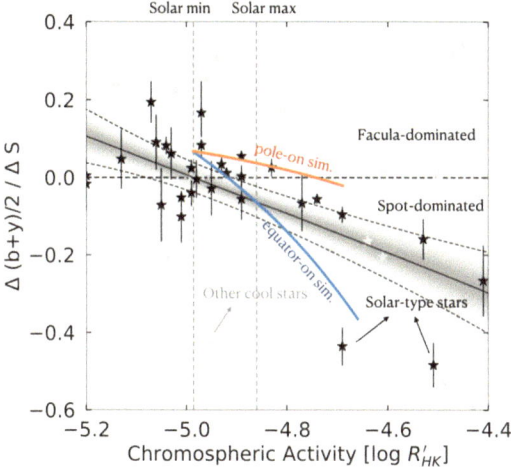

Fig. 3 Activity-related brightening (negative for dimming) as a function of the mean level of chromospheric activity in the HK bands. The stellar observations are shown by black stars (T_{eff} to within 200 K of the solar value and relative brightness uncertainty below 0.01) and gray stars corresponding to other stars in the sample of Radick et al. (1998). The gray-shaded region shows the posterior distribution to 2σ of Bayesian linear regression to the near-solar sample, using Gaussian priors for a quadratic function. Blue and orange curves show the calculated brightening functions using SFT simulations, for equator-on and pole-on views, respectively (Nèmec et al. 2022).

of SFT simulations for two extreme axial inclinations. The expected transition from spot- to facula-dominated regime is not far from the current solar activity level during maxima. Here, the suggested mechanism for the transition is that for more active stars with presumably higher flux emergence rates, the active-region network field spanning much larger areas than sunspots finds more chance to undergo flux cancellation, owing to random superpositions of opposite magnetic polarities. On the Sun, the activity level is low enough to avoid such an overall flux cancellation effect, and the facular state remains the dominant counterpart, leading to a slight brightening of the Sun with increasing activity.

Physics-based forward modeling of light curves can help constrain the parameter space of stellar surface brightness distributions and the axial inclination. In such an attempt, the FEAT model (Sect. 3.1) was used to calculate synthetic light curves of G2V stars with rotation rates between $\Omega_\odot$ and $8\Omega_\odot$ (Nèmec et al. 2023). The method involved integration of facular and spot disc coverages weighted by the *Kepler* transmission function, evaluated at various axial inclinations. The results reproduced several observed characteristics of *Kepler* light curves of stars with different activity levels and variability patterns. However, for a better match to observations and for an improved understanding of the observed change in stellar variability patterns as a function of the rotation rate, empirical relationships involving the magnetic flux emergence rate should be established and incorporated into models of flux emergence and transport.

3.4 Astrometric and Radial Velocity Jitter

One promising method to infer activity-pattern characteristics is to make use of time-resolved high-precision astrometric measurements, exploiting the spatial symmetry breaking around the line of sight through the disc centre (Lanza et al. 2008). This lack of axial symmetry results from the fact that different portions of the disc have varying surface brightness.

Based on a method developed by Shapiro et al. (2021), numerical simulations of activity-induced astrometric jitter of solar-type stars were carried out by Sowmya et al. (2021a) as a function of axial inclination, metallicity, and active-region nesting, finding that activity cycles can be inferred from systematic changes in the photocenter positions (see also Meunier et al. 2020). Moreover, when the degree of active-region nesting is high enough, the cyclic changes in the photocenter jitter can be detected by Gaia. Simulations for more active and rapidly rotating solar-type stars have shown that the jitter becomes spot-dominated and could be observed even on monthly timescales (Sowmya et al. 2022).

The radial velocity (RV) time series derived from photospheric absorption lines includes information on magnetic features as they transit the visible stellar disc, often hampering high-precision exoplanet detection with the same method. As part of a physics-based modeling framework of several activity indicators as a function of several stellar properties (Meunier et al. 2019), the radial velocity time series were also modelled by Meunier and Lagrange (2019). The RV variations led by magnetic features and their spatial distributions on the stellar disc showed general agreement with observations. The main features strongly affecting RV amplitudes were found to be the latitude coverage of active regions, the level of activity, and axial inclination.

4 Chromospheric and Coronal Activity

The most widely used indicator of chromospheric activity in solar-type stars is known as the S-index, which was devised in the late-1960s at Mount Wilson Observatory (Wilson 1968). The S-index is a measurement of the stellar flux in the cores of the Ca II H and K spectral lines (N_H, N_K) relative to the flux in two neighboring pseudo-continuum bands (N_V, N_R).

$$S = \alpha \cdot \frac{N_\mathrm{H} + N_\mathrm{K}}{N_\mathrm{V} + N_\mathrm{R}}, \tag{3}$$

where α is a calibration constant that can be determined for any instrument using observations of standard stars monitored by the Mount Wilson survey (Vaughan et al. 1978). Time series measurements of the S-index for any given star can reveal variability due to stellar rotation and magnetic cycles (Jeffers et al. 2023). However, a meaningful comparison of stars with different spectral types requires a small correction for the photospheric contribution to the H and K emission, as well as a normalization by the bolometric luminosity (Noyes et al. 1984). This normalized activity indicator is known as R'_HK, and has been measured for thousands of solar-type stars (Boro Saikia et al. 2018).

The evolution of R'_HK over stellar lifetimes provides one method of estimating ages for isolated solar-type stars (Mamajek and Hillenbrand 2008; Lorenzo-Oliveira et al. 2018). During the first half of their main-sequence lifetimes, the rotation rates and activity levels of solar-type stars appear to decline together roughly with the square-root of the age (Skumanich 1972). At some critical value of the Rossby number (Ro $\equiv P_\mathrm{rot}/\tau_\mathrm{c}$) when the rotation period becomes comparable to the convective turnover time, rotation and activity become decoupled and the subsequent evolution of activity appears to be dominated by slow changes in the mechanical energy available from convection (Böhm-Vitense 2007; Metcalfe et al. 2016). Figure 4 shows the rotation-activity relation for solar-type stars in the Mount Wilson survey (gray points; Baliunas et al. 1996) and for Kepler asteroseismic targets (colored points; Metcalfe et al. 2016). For the Mount Wilson stars, Rossby numbers have been calculated from $B - V$ colors (Noyes et al. 1984), and show a large scatter particularly at low activity levels (log $R'_\mathrm{HK} < -5$). For the Kepler targets, Rossby numbers have been calculated

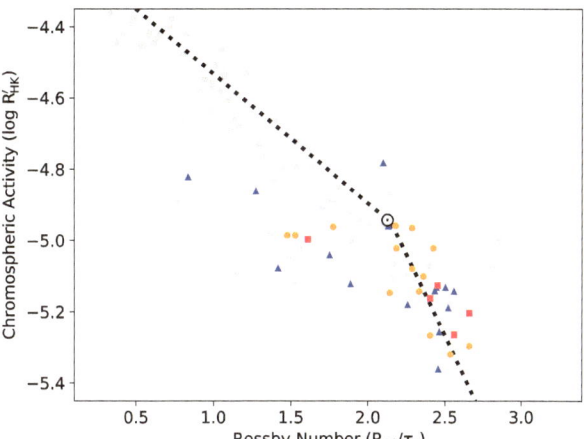

Fig. 4 The evolution of chromospheric activity with Rossby number for solar-type stars in the Mount Wilson survey (gray points) and for Kepler asteroseismic targets with precise Rossby numbers (colored points). Note the change in slope near the solar activity level

using turnover times near the base of the convective envelope from an asteroseismic model for each star (Metcalfe et al. 2014), including hotter F-type (blue triangles), sun-like G-type (yellow circles), and cooler K-type stars (red squares). Most of this sample appears to follow a common relation at low activity levels, possibly due to the higher precision of their Rossby numbers. Some outliers remain at lower Rossby number and higher activity level, perhaps from activity variations within an unknown magnetic cycle. Note that the rotation-activity relation is consistent with a change in slope near the solar activity level, which corresponds to the apparent onset of weakened magnetic braking (van Saders et al. 2016).

Modeling the chromospheric emission due to magnetic activity is important to better understand the physical characteristics of magnetism as a function of stellar properties. Such models would also be useful in disentangling various mechanisms that are likely responsible for generating the observed distributions of chromospheric emissions. Recently, Sowmya et al. (2021b) developed a physics-based approach to forward-model S-index variations of the Sun as a star, from rotational to century-scale variations, but for the full range of the inclination of the rotation axis with respect to the line of sight. Based on their calculated S-index time series of the Sun for cycles 1-23, they found that the variability amplitude in the chromospheric emission of the Sun as seen from the full range of inclinations was representative of the variability distribution of other solar-type stars with similar $\langle R'_{HK} \rangle$, adapted from Radick et al. (1998).

While magnetic heating of the chromosphere is evident in optical and ultraviolet spectral lines, similar processes heat the stellar corona to $\sim$1 million K, emitting at X-ray wavelengths. A measurement of the X-ray luminosity is the coronal equivalent of the chromospheric S-index, and normalizing by the bolometric luminosity facilitates the comparison of stars with different spectral types like R'_{HK}. Depending on their evolutionary state, solar-type stars typically have a fractional X-ray luminosity $R_x \equiv L_x/L_{bol} \sim 10^{-3}$ to 10^{-8} (Schmitt and Liefke 2004). Rather than decrease monotonically with rotation rate or Rossby number, there appear to be different regimes in the rotation-activity relation for X-rays. For the youngest and most rapidly rotating stars, there is a "saturated" regime in which R_x appears relatively constant for Ro <0.1 (Pizzolato et al. 2003). It was initially unclear whether this saturation represented active regions filling the entire stellar surface, or a saturation of the underlying dynamo mechanism (Vilhu 1984), but recent evidence suggests that it is a feature of the dynamo (Reiners et al. 2022). For stars with Rossby numbers between $\sim$0.1 and the

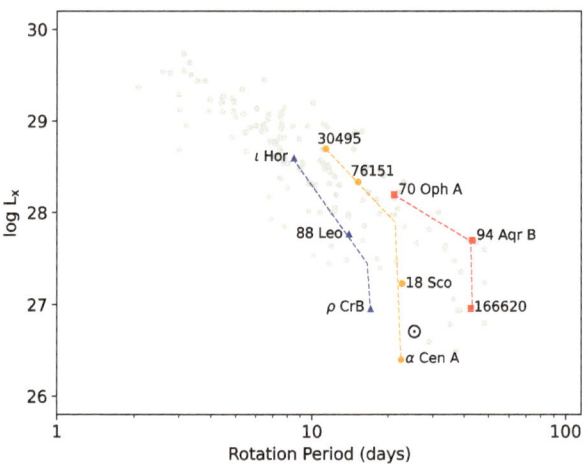

Fig. 5 The evolution of coronal activity with rotation for solar-type stars in the unsaturated regime (gray points; Wright et al. 2011), with evolutionary sequences of F-type (blue), G-type (yellow), and K-type stars (red) for comparison. The vertical tracks for older stars suggest a previously unrecognized "decoupled" regime in the rotation-activity relation

solar value, there is an "unsaturated" regime in which R_x decreases linearly with log Ro, following a power law with an index $\beta \sim 2$ (Wright et al. 2011).

The evolution of X-ray luminosity for Rossby numbers greater than the solar value has been largely unexplored, with only a few measurements available at $R_x < 10^{-6}$. Considering the rotational and magnetic transitions that have recently been identified near the solar Rossby number (van Saders et al. 2016; Metcalfe et al. 2016), we can evaluate the currently available data from a new perspective. Rotation periods and X-ray luminosities are shown in Fig. 5 for solar-type stars in the unsaturated regime (gray points; Wright et al. 2011), with evolutionary sequences of hotter F-type (blue triangles), sun-like G-type (yellow circles), and cooler K-type stars (red squares) for comparison. When stars reach the critical rotation period that corresponds to the onset of weakened magnetic braking for a given spectral type, the X-ray luminosity continues to decline at roughly constant rotation period. For example, despite having very similar rotation periods, the X-ray luminosities of the solar analogs 18 Sco (3.7 Gyr; Li et al. 2012) and α Cen A (5.4 Gyr; Bazot et al. 2016) differ by nearly an order of magnitude. This behavior suggests a previously unrecognized "decoupled" regime in the rotation-activity relation, where the X-ray luminosity is no longer determined by rotation.

5 Magnetic Braking

Magnetic braking is the loss of angular momentum (AM) through the interaction of stellar mass loss and magnetic fields. Stars with convective outer envelopes ($T_{\mathrm{eff}} < 6250$ K) and magnetic dynamos spin down over time due to angular momentum loss from magnetized winds (Kraft 1967; Skumanich 1972). It represents a unique window into the magnetism of the Sun and stars because operates in a feedback loop: slower rotation leads to weaker magnetic fields, which lead to less AM loss via magnetized winds. It is therefore both a consequence and a driver of magnetic evolution. Rotational evolution can occur over billion-year timescales, and represents one the best tests of the integrated behavior of the large-scale magnetic fields of stars over stellar lifetimes. Spin-down is most directly a probe of the strength of the large-scale dipole (Weber and Leverett 1967; Kawaler 1988), with only minor contributions from higher-order fields under specific conditions (See et al. 2019). However, observed rotational evolution depends on many interacting ingredients—the magnetic field, details of the mass loss and wind flow, initial rotation rates, and internal angular momentum redistribution—making the interpretation of rotational evolution in the context of magnetism a subtle task.

5.1 Modeling Magnetic Braking

A model of rotational evolution has three ingredients: 1) the braking law, which most directly probes the magnetic field behavior, 2) some assumption of the initial AM, and 3) a prescription for internal AM redistribution. Together, they predict rotational evolution as a function of time.

5.1.1 Braking Laws

The main sequence (MS) is where magnetic braking has the largest impact on rotational evolution for single stars. Over the MS lifetime, the braking appears to be a strong function of the rotation velocity— $\frac{dJ}{dt} \sim \omega^3$ (Skumanich 1972), which asymptotically forces convergence to a narrow range of rotation periods at a given time for a given stellar mass. This has the benefit of making the rotation rates of old stars insensitive to the significant (and for purposes here, uninteresting) spread in birth rotation periods. Observationally, this convergence happens first in the more massive stars and later in the low-mass stars, with open clusters showing tight, converged rotation sequences by a few hundred Myr around solar temperatures.

The standard $\frac{dJ}{dt} \propto \omega^3$ spin down results in a period-age relation that goes roughly as $P_{rot} \propto \sqrt{t}$— so-called "Skumanich" spin down (Skumanich 1972). Much of the literature on braking seeks to empirically constrain this period-age relation while being largely agnostic to the underlying physical mechanisms (Barnes 2007, 2010; Mamajek and Hillenbrand 2008; Angus et al. 2015). To use braking as a constraint on magnetic field behavior, we turn to physically motivated magnetic braking laws that follow the basic formalism of (Weber and Leverett 1967; Mestel 1968) for angular momentum loss from magnetized stellar winds. The torque on a star is most simply written as:

$$\tau = \dot{M}\Omega\langle r_A \rangle^2, \tag{4}$$

where r_A is the Alfvén radius where magnetic Alfvén velocity and wind velocity are equivalent. The average radius defines the effective "lever arm" of the torque, and depends both on the strength and morphology of the magnetic field (Réville et al. 2015; Garraffo et al. 2016; Finley and Matt 2018). In practice, the challenge in defining a braking law is in prescribing $\dot{M}$ and B (as it enters in r_A). Many authors choose to fix a magnetic field morphology and make some assumption about how the magnetic field scales with stellar properties, often in the form of a Rossby scaling (e.g. Kawaler 1988; Krishnamurthi et al. 1997; van Saders and Pinsonneault 2013; Matt et al. 2015). Modern braking laws increasingly draw their forms from > 1D MHD simulations of mass loss entrained in a magnetized wind (e.g. Matt and Pudritz 2008), but still fundamentally require some additional input of how B scales with stellar properties, and necessarily make strong assumptions about the nature of wind launching. For mass loss, authors commonly adopt either empirical (e.g. Wood et al. 2005, 2021) or theoretical scalings (Cranmer and Saar 2011), but the choice remains a significant uncertainty.

5.1.2 Additional Ingredients

Stars are born with roughly two orders of magnitude of spread in their initial rotation periods (Irwin and Bouvier 2009; Herbst et al. 2002), due both to stochasticity in the birth angular momentum itself and star-disk interactions that occur in the first few Myr of the star's life (Matt and Pudritz 2005; Shu et al. 1994; Koenigl 1991). While rotational evolution in solar mass stars becomes insensitive to choices of initial conditions within a few hundred Myrs, lower mass stars may retain sensitivity for Gyrs (Gallet and Bouvier 2015), complicating the interpretation of their rotation periods (Epstein and Pinsonneault 2014; Roquette et al. 2021). Authors generally either consider a range of initial rotation periods motivated by those observed in the youngest open clusters, or "launch" their braking simulations from initial conditions defined by a benchmark cluster (see Somers et al. 2017; Epstein and Pinsonneault 2014). There is some evidence that stellar environments may alter the distribution of initial periods (Coker et al. 2016; Roquette et al. 2021).

Many braking prescriptions make the simplest (and often reasonably correct) assumption that internal AM transport is instantaneous compared to evolutionary timescales, resulting in solid body rotation. However, the efficiency of internal AM transport does affect the evolution of the surface rotation period. Literature braking laws that allow for interior AM transport generally use either 1) simple "two-zone" models (MacGregor and Brenner 1991) allow rotation of the core and convective envelope to evolve separately, coupled by AM transport over some characteristic timescale τ_{ce}, or 2) allow for extra AM transport via a diffusion term (Denissenkov et al. 2010; Somers and Pinsonneault 2016; Spada and Lanzafame 2020) in interiors models. There is no accepted first-principles mechanism for interior AM transport that fully reproduces observations. Magnetically mediated transport at the tachocline has been proposed (Oglethorpe and Garaud 2013), as have internal gravity waves (Denissenkov et al. 2008; Fuller et al. 2014; Somers et al. 2017; Cao et al. 2023), but no conclusive mechanism has been identified.

5.2 Recent Modifications to Braking Laws

Because of the uncertainties in both the braking process and the fundamental underlying stellar processes of mass loss and magnetic field generation, all modern braking laws are tuned via fitting parameters to observations. Improvements in the underlying physical models come largely from examining whether the braking prescription captures the observed mass and time dependence of the spin down.

Observational benchmark systems must have both well-known ages and rotation periods. The most impactful class of calibrating systems to date has been the open clusters, which represent coeval stellar populations that span a range of masses at uniform composition. However, cluster calibrators have historically been confined to young (< 1Gyr) ages and solar composition stars; old open clusters are rare and tend to be distant and challenging to study. Although the lack of calibrating sources at low masses, old ages, and non-solar compositions remains a persistent roadblock in the testing and validation of magnetic braking models, two classes of new calibrator sources have fueled recent refinements of magnetic braking laws: intermediate-age open clusters observed with space- and ground-based photometric missions, and bright field stars with precise, asteroseismically measured ages. Both classes have suggested significant alterations to the Skumanich-type spin-down that we will discuss below.

We show a sample of calibrator stars selected to lie in a relatively narrow slice between $1.0 < M_\odot < 1.1$ and $0.0 <$ [Fe/H] < 0.2 in Fig. 6. These calibrators are shown against

Fig. 6 Calibrator stars (points) of known rotation period and age shown against calibrated braking laws that include a range of initial conditions (teal tracks), internal AM redistribution, and weakened magnetic braking. Cluster stars (circles) are drawn from the compilation in Curtis et al. (2020), and asteroseismic targets (stars) from van Saders et al. (2016) and Hall et al. (2021)

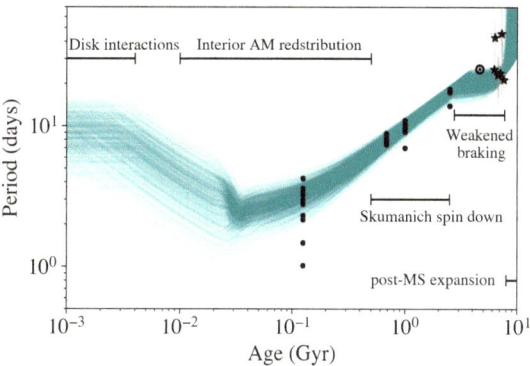

a braking law that includes 1) a range of initial conditions (gaussian centered at 8 days, with a 1σ width of 4 days), 2) disk-locking timescales drawn uniformly from 1-5 Myr, 3) the braking law form from van Saders et al. (2016) with the addition of core-envelope decoupling following the Somers and Pinsonneault (2016) 2-zone model. Rough regions where different processes discussed in this section are important for interpreting the rotation period are noted, but readers should be aware that the exact boundaries of these regions are both mass and metallicity dependent.

5.2.1 Young Stars

Young open clusters retain a subset of rapidly rotating stars for hundreds of millions of years, longer than Skumanich-type spin-down law would predict given the simultaneous existence of slow rotators. This observation motivates the inclusion of an epoch of "saturated" spin down in nearly all magnetic braking models, in which the spin down instead goes as $\frac{dJ}{dt} \propto \omega_{crit}^2 \omega$ (Krishnamurthi et al. 1997). This saturated regime is the analog of saturation in other magnetic proxies: the observation that beyond some ω_{crit} more rapid rotation no longer results in a stronger magnetic response, in this case reflected in the angular momentum loss. In practice, the saturation threshold is often included as a Rossby scaling, with $\mathrm{Ro}_{sat} \propto (\omega_{crit}\tau_{cz})^{-1}$. A saturation threshold of $\mathrm{Ro}_{sat} \sim 0.1$ is broadly consistent both with the saturation threshold in other magnetic proxies, and with the observed rotational behavior in clusters.

5.2.2 Intermediate-Age Stars

In detailed cluster observations, spin-down rates depart from those expected for unsaturated, Skumanich-type spin-down in young and intermediate-age stars. Stars initially spin-down faster-than-expected (Denissenkov et al. 2010), and then "stalled out" at intermediate ages with what appears to be minimal evolution in the surface rotation rate (Agüeros et al. 2018; Curtis et al. 2020). Although we cannot entirely rule out changes to the magnetic braking itself as the cause, internal AM transport is emerging as a leading explanation. While the epoch of "core-envelope" decoupling is brief and subtle for solar mass stars (Denissenkov et al. 2010; Gallet and Bouvier 2015), the apparently long timescales for transport in low-mass stars produces a stark pileup in the cluster sequences of 1 Gyr old open clusters (Curtis et al. 2020). There are also puzzling gaps in the distribution of cool field stars at periods just long-ward of feature, which may suggest a similar phenomenon in the population at large (Lu et al. 2022). Cao et al. (2023) identified a magnetic counterpart to the observed rotational stalling, and showed that stars putatively undergoing core-envelope recoupling in the Praesepe open cluster had starkly elevated surface spot-filling fractions, which the authors argued is the consequence of the radial shears present during this epoch of AM redistribution. At a minimum, the spin-down evolution of intermediate-age stars is more complex than Skumanich; the exact mechanism (and magnetic involvement) remain areas of active inquiry.

5.2.3 Old Stars

Recent observations of bright field stars with precisely measured asteroseismic ages provide insight into rotation at a wider range of ages and compositions in solar mass stars. The seismic sample also displays non-Skumanich behavior: it traces the standard magnetic braking patterns at young and intermediate ages, but appears to undergo dramatically reduced angular momentum loss in the latter half of the main sequence lifetime (van Saders et al. 2016). Followup work showed that the observed long-period edge in the field star rotation distribution (McQuillan et al. 2014; Matt et al. 2015) was also consistent with weakened braking (van Saders et al. 2019; David et al. 2022), and that the weakened braking was apparent even when rotation rates were measured methods other than spot modulation (Hall et al. 2021; Masuda et al. 2022). These braking models allow for standard evolution until some critical Rossby number is reached—Ro_{crit}—after which the braking is severely reduced or AM loss truncated entirely. The mechanism again remains uncertain: van Saders et al. (2016) and Metcalfe et al. (2016) suggested that an overall weakening of the magnetic field strength and shift to higher-order morphologies could be responsible, and detailed studies of stars on either side of the transition have thus far supported the picture of weakening dipole fields as a driver for the change in spin-down (see Metcalfe et al. 2023, and references therein). However, changes in the mass loss rates, and details of the wind launching and flow are not ruled out.

6 Outlook

Because the underlying mechanisms are still poorly understood, progress in understanding stellar magnetism is very much driven by observations, and then theoretical efforts to reproduce the patterns we see. Larger, more complete, and more comprehensive datasets drive this progress. Those data are challenging to obtain—whether they be subtle spot signatures, direct measurements of weak fields, precise ages and rotation rates, or decades-long activity cycles—with significant progress in the last decade. There are two classes of observational benchmarks that enable progress: small, but exquisitely studied samples of stars with multiple precision measurements of magnetic proxies that allow us to build a complete picture of their behavior, and truly large but comparatively more poorly constrained samples that allow us to probe the properties and patterns in populations.

The various attempts at forward modeling of photospheric activity diagnostics are helpful in deepening our understanding of the physics of magnetic activity, along with qualitative comparisons with the available observations. Extending magnetic flux emergence and subsequent surface evolution models of spots and faculae to wide ranges of stellar properties (e.g., T_{eff}, differential rotation, metallicity) will be important in evaluating observational data. We note, however, that care should be taken when interpreting observations with the matching simulations. The inverse problem of recovering the surface magnetic patterns is often ill-posed, involving several parameter degeneracies. Simulation-based inference frameworks have the potential to illuminate parameter ranges that optimally simulate observed data, allowing a Bayesian way of inverting observations (e.g., Cranmer et al. 2020; Asensio Ramos et al. 2022).

Models of magnetic braking have seen frequent, large revisions in recent years as new data become available. Because the number of calibrating sources still spans a relatively narrow range of masses, ages, and compositions, there is likely much to learn as new corners of parameter space become accessible. Observational improvements are likely to come from a combination of many datasets, with two paths for growth: 1) small samples with exceptional observational coverage (rotation, asteroseismology, magnetic activity cycle measurements, direct magnetic field mapping for all targets), and 2) very large but more poorly characterized stellar samples, enabled by large photometric, spectroscopic, and astrometric surveys. Pushing to lower masses, older ages, and less solar-like compositions are the most critical directions for new observational insight into braking behavior. On the theoretical side, progress can be made by increasingly ground braking models in first-principles or semi-empirical models of magnetized winds, and gradually phasing out the more purely empirical scalings currently in use (e.g., Chebly et al. 2023; Evensberget et al. 2023). As has been the case since the beginning of this field, the interplay between improved observational calibrator sets and more sophisticated physical models will drive progress.

Acknowledgements This review was written following the workshop "Solar and Stellar Dynamos: A New Era", hosted and supported by the International Space Science Institute (ISSI) in Bern, Switzerland. The authors wish to express their thanks to ISSI for their financial and logistical support.

Author Contribution All authors contributed equally to this work.

Funding Open Access funding enabled and organized by Projekt DEAL.

Declarations

References

Agüeros MA, Bowsher EC, Bochanski JJ et al (2018) A new look at an old cluster: the membership, rotation, and magnetic activity of low-mass stars in the 1.3 Gyr old open cluster NGC 752. Astrophys J 862(1):33. https://doi.org/10.3847/1538-4357/aac6ed. arXiv:1804.02016 [astro-ph.SR]

Angus R, Aigrain S, Foreman-Mackey D et al (2015) Calibrating gyrochronology using Kepler asteroseismic targets. Mon Not R Astron Soc 450(2):1787–1798. https://doi.org/10.1093/mnras/stv423. arXiv:1502.06965 [astro-ph.EP]

Asensio Ramos A, Díaz Baso CJ, Kochukhov O et al (2022) Approximate Bayesian neural Doppler imaging. Astron Astrophys 658:A162. https://doi.org/10.1051/0004-6361/202142027. arXiv:2108.09266 [astro-ph.IM]

Aurière M (2003) Stellar Polarimetry with NARVAL. In: Arnaud J, Meunier N (eds) Magnetism and Activity of the Sun and Stars. EAS Publications Series, p 105

Baliunas S, Sokoloff D, Soon W (1996) Magnetic field and rotation in lower main-sequence stars: an empirical time-dependent magnetic Bode's relation? Astrophys J Lett 457:L99. https://doi.org/10.1086/309891

Barnes SA (2007) Ages for illustrative field stars using gyrochronology: viability, limitations, and errors. Astrophys J 669(2):1167–1189. https://doi.org/10.1086/519295. arXiv:0704.3068 [astro-ph]

Barnes SA (2010) A simple nonlinear model for the rotation of main-sequence cool stars. I. Introduction, implications for gyrochronology, and color-period diagrams. Astrophys J 722(1):222–234. https://doi.org/10.1088/0004-637X/722/1/222

Basri G (2021) An introduction to stellar magnetic activity. IOP Publishing, Bristol. https://doi.org/10.1088/2514-3433/ac2956

Bazot M, Christensen-Dalsgaard J, Gizon L et al (2016) On the uncertain nature of the core of α Cen A. Mon Not R Astron Soc 460(2):1254–1269. https://doi.org/10.1093/mnras/stw921. arXiv:1603.07583 [astro-ph.SR]

Berdyugina SV (2005) Starspots: a key to the stellar dynamo. Living Rev Sol Phys 2(1):8. https://doi.org/10.12942/lrsp-2005-8

Biswas A, Karak BB, Usoskin I et al (2023) Long-term modulation of solar cycles. Space Sci Rev 219(3):19. https://doi.org/10.1007/s11214-023-00968-w. arXiv:2302.14845 [astro-ph.SR]

Böhm-Vitense E (2007) Chromospheric activity in G and K main-sequence stars, and what it tells us about stellar dynamos. Astrophys J 657(1):486–493. https://doi.org/10.1086/510482

Boro Saikia S, Marvin CJ, Jeffers SV et al (2018) Chromospheric activity catalogue of 4454 cool stars. Questioning the active branch of stellar activity cycles. Astron Astrophys 616:A108. https://doi.org/10.1051/0004-6361/201629518. arXiv:1803.11123 [astro-ph.SR]

Cao L, Pinsonneault MH, van Saders JL (2023) Core-envelope decoupling drives radial shear dynamos in cool stars. Astrophys J Lett 951:L49. https://doi.org/10.3847/2041-8213/acd780. arXiv:2301.07716 [astro-ph.SR]

Chebly JJ, Alvarado-Gómez JD, Poppenhäger K et al (2023) Numerical quantification of the wind properties of cool main sequence stars. Mon Not R Astron Soc 524(4):5060–5079. https://doi.org/10.1093/mnras/stad2100. arXiv:2307.04615 [astro-ph.SR]

Coker CT, Pinsonneault M, Terndrup DM (2016) Evidence for cluster to cluster variations in low-mass stellar rotational evolution. Astrophys J 833(1):122. https://doi.org/10.3847/1538-4357/833/1/122. arXiv:1604.05729 [astro-ph.SR]

Cranmer K, Brehmer J, Louppe G (2020) The frontier of simulation-based inference. Proc Natl Acad Sci 117(48):30,055–30,062. https://doi.org/10.1073/pnas.1912789117. arXiv:1911.01429 [stat.ML]

Cranmer SR, Saar SH (2011) Testing a predictive theoretical model for the mass loss rates of cool stars. Astrophys J 741(1):54. https://doi.org/10.1088/0004-637X/741/1/54. arXiv:1108.4369 [astro-ph.SR]

Curtis JL, Agüeros MA, Matt SP et al (2020) When do stalled stars resume spinning down? Advancing gyrochronology with Ruprecht 147. Astrophys J 904(2):140. https://doi.org/10.3847/1538-4357/abbf58. arXiv:2010.02272 [astro-ph.SR]

David TJ, Angus R, Curtis JL et al (2022) Further evidence of modified spin-down in sun-like stars: pileups in the temperature-period distribution. Astrophys J 933(1):114. https://doi.org/10.3847/1538-4357/ac6dd3. arXiv:2203.08920 [astro-ph.SR]

Denissenkov PA, Pinsonneault M, MacGregor KB (2008) What prevents internal gravity waves from disturbing the solar uniform rotation? Astrophys J 684(1):757–769. https://doi.org/10.1086/589502. arXiv:0801.3622 [astro-ph]

Denissenkov PA, Pinsonneault M, Terndrup DM et al (2010) Angular momentum transport in solar-type stars: testing the timescale for core-envelope coupling. Astrophys J 716(2):1269–1287. https://doi.org/10.1088/0004-637X/716/2/1269. arXiv:0911.1121 [astro-ph.SR]

Donati JF, Landstreet JD (2009) Magnetic fields of nondegenerate stars. Annu Rev Astron Astrophys 47:333–370. https://doi.org/10.1146/annurev-astro-082708-101833. arXiv:0904.1938 [astro-ph.SR]

Donati JF, Catala C, Landstreet JD et al (2006) ESPaDOnS: the new generation stellar spectro-polarimeter. Performances and first results. In: Casini R, Lites BW (eds) Solar Polarization 4. ASP Conference Series, vol. 358. Astronomical Society of the Pacific, p 362

Donati JF, Kouach D, Moutou C et al (2020) SPIRou: NIR velocimetry and spectropolarimetry at the CFHT. Mon Not R Astron Soc 498(4):5684–5703. https://doi.org/10.1093/mnras/staa2569. arXiv:2008.08949 [astro-ph.IM]

Dorn RJ, Anglada-Escude G, Baade D et al (2014) CRIRES+: exploring the cold universe at high spectral resolution. Messenger 156:7–11

Engvold O, Vial JC, Skumanich A (2019) The sun as a guide to stellar physics. Elsevier, Amsterdam. https://doi.org/10.1016/C2017-0-01365-4

Epstein CR, Pinsonneault MH (2014) How good a clock is rotation? The stellar rotation-mass-age relationship for old field stars. Astrophys J 780(2):159. https://doi.org/10.1088/0004-637X/780/2/159. arXiv:1203.1618 [astro-ph.SR]

Evensberget D, Marsden SC, Carter BD et al (2023) The winds of young solar-type stars in the Pleiades, AB Doradus, Columba, and β Pictoris. Mon Not R Astron Soc 524(2):2042–2063. https://doi.org/10.1093/mnras/stad1650. arXiv:2305.17427 [astro-ph.SR]

Feiden GA, Chaboyer B (2013) Magnetic inhibition of convection and the fundamental properties of low-mass stars. I. Stars with a radiative core. Astrophys J 779(2):183. https://doi.org/10.1088/0004-637X/779/2/183. arXiv:1309.0033 [astro-ph.SR]

Finley AJ, Matt SP (2018) The effect of combined magnetic geometries on thermally driven winds. II. Dipolar, quadrupolar, and octupolar topologies. Astrophys J 854(2):78. https://doi.org/10.3847/1538-4357/aaaab5. arXiv:1801.07662 [astro-ph.SR]

Folsom CP, Bouvier J, Petit P et al (2018) The evolution of surface magnetic fields in young solar-type stars II: the early main sequence (250-650 Myr). Mon Not R Astron Soc 474(4):4956–4987. https://doi.org/10.1093/mnras/stx3021. arXiv:1711.08636 [astro-ph.SR]

Fuller J, Lecoanet D, Cantiello M et al (2014) Angular momentum transport via internal gravity waves in evolving stars. Astrophys J 796(1):17. https://doi.org/10.1088/0004-637X/796/1/17. arXiv:1409.6835 [astro-ph.SR]

Gallet F, Bouvier J (2015) Improved angular momentum evolution model for solar-like stars. II. Exploring the mass dependence. Astron Astrophys 577:A98. https://doi.org/10.1051/0004-6361/201525660. arXiv:1502.05801 [astro-ph.SR]

Garraffo C, Drake JJ, Cohen O (2016) The missing magnetic morphology term in stellar rotation evolution. Astron Astrophys 595:A110. https://doi.org/10.1051/0004-6361/201628367. arXiv:1607.06096 [astro-ph.SR]

Granzer T, Schüssler M, Caligari P et al (2000) Distribution of starspots on cool stars. II. Pre-main-sequence and ZAMS stars between 0.4 M_{sun} and 1.7 M_{sun}. Astron Astrophys 355:1087–1097

Hall JC, Lockwood GW (2004) The chromospheric activity and variability of cycling and flat activity solar-analog stars. Astrophys J 614(2):942–946. https://doi.org/10.1086/423926

Hall OJ, Davies GR, van Saders J et al (2021) Weakened magnetic braking supported by asteroseismic rotation rates of Kepler dwarfs. Nat Astron 5:707–714. https://doi.org/10.1038/s41550-021-01335-x. arXiv:2104.10919 [astro-ph.SR]

Hazra G, Nandy D, Kitchatinov L et al (2023) Mean field models of flux transport dynamo and meridional circulation in the sun and stars. Space Sci Rev 219(5):39. https://doi.org/10.1007/s11214-023-00982-y. arXiv:2302.09390 [astro-ph.SR]

Herbst W, Bailer-Jones CAL, Mundt R et al (2002) Stellar rotation and variability in the Orion Nebula Cluster. Astron Astrophys 396:513–532. https://doi.org/10.1051/0004-6361:20021362

Holzwarth V, Mackay DH, Jardine M (2006) The impact of meridional circulation on stellar butterfly diagrams and polar caps. Mon Not R Astron Soc 369(4):1703–1718. https://doi.org/10.1111/j.1365-2966.2006.10407.x. arXiv:astro-ph/0604102 [astro-ph]

Işık E, Schüssler M, Solanki SK (2007) Magnetic flux transport on active cool stars and starspot lifetimes. Astron Astrophys 464(3):1049–1057. https://doi.org/10.1051/0004-6361:20066623. arXiv:astro-ph/0612399 [astro-ph]

Işık E, Schmitt D, Schüssler M (2011) Magnetic flux generation and transport in cool stars. Astron Astrophys 528:A135. https://doi.org/10.1051/0004-6361/201014501. arXiv:1102.0569 [astro-ph.SR]

Işık E, Solanki SK, Krivova NA et al (2018) Forward modelling of brightness variations in Sun-like stars. I. Emergence and surface transport of magnetic flux. Astron Astrophys 620:A177. https://doi.org/10.1051/0004-6361/201833393. arXiv:1810.06728 [astro-ph.SR]

Işık E, Shapiro AI, Solanki SK et al (2020) Amplification of brightness variability by active-region nesting in solar-like stars. Astrophys J Lett 901(1):L12. https://doi.org/10.3847/2041-8213/abb409. arXiv:2009.00692 [astro-ph.SR]

Irwin J, Bouvier J (2009) The rotational evolution of low-mass stars. In: Mamajek EE, Soderblom DR, Wyse RFG (eds) The ages of stars, Proc IAU, vol S258. Cambridge University Press, pp 363–374. https://doi.org/10.1017/S1743921309032025. arxiv:0901.3342

Jeffers SV, Kiefer R, Metcalfe TS (2023) Stellar Activity Cycles. Space Sci Rev 219:54. https://doi.org/10.1007/s11214-023-01000-x. arXiv:2309.14138 [astro-ph.SR]

Kawaler SD (1988) Angular momentum loss in low-mass stars. Astrophys J 333:236. https://doi.org/10.1086/166740

Kochukhov O (2021) Magnetic fields of M dwarfs. Astron Astrophys Rev 29(1):1. https://doi.org/10.1007/s00159-020-00130-3. arXiv:2011.01781 [astro-ph.SR]

Kochukhov O, Reiners A (2020) The magnetic field of the active planet-hosting M Dwarf AU Mic. Astrophys J 902(1):43. https://doi.org/10.3847/1538-4357/abb2a2. arXiv:2008.10668 [astro-ph.SR]

Koenigl A (1991) Disk accretion onto magnetic T Tauri stars. Astrophys J Lett 370:L39. https://doi.org/10.1086/185972

Kraft RP (1967) Studies of stellar rotation. V. The dependence of rotation on age among solar-type stars. Astrophys J 150:551. https://doi.org/10.1086/149359

Krishnamurthi A, Pinsonneault MH, Barnes S et al (1997) Theoretical models of the angular momentum evolution of solar-type stars. Astrophys J 480(1):303–323. https://doi.org/10.1086/303958

Landi Degl'Innocenti E, Landolfi M (2004) Polarization in spectral lines. Astrophysics and Space Science Library, vol 307. Springer, Dordrecht. https://doi.org/10.1007/1-4020-2415-0

Lanza AF, De Martino C, Rodonò M (2008) Astrometric effects of solar-like magnetic activity in late-type stars and their relevance for the detection of extrasolar planets. New Astron 13(2):77–84. https://doi.org/10.1016/j.newast.2007.06.009. arXiv:0706.2942 [astro-ph]

Lehmann LT, Jardine MM, Vidotto AA et al (2017) The energy budget of stellar magnetic fields: comparing non-potential simulations and observations. Mon Not R Astron Soc 466(1):L24–L28. https://doi.org/10.1093/mnrasl/slw225. arXiv:1610.08314 [astro-ph.SR]

Lehmann LT, Hussain GAJ, Jardine MM et al (2019) Observing the simulations: applying ZDI to 3D non-potential magnetic field simulations. Mon Not R Astron Soc 483(4):5246–5266. https://doi.org/10.1093/mnras/sty3362. arXiv:1811.03703 [astro-ph.SR]

Li TD, Bi SL, Liu K et al (2012) Stellar parameters and seismological analysis of the star 18 Scorpii. Astron Astrophys 546:A83. https://doi.org/10.1051/0004-6361/201219063

Lorenzo-Oliveira D, Freitas FC, Meléndez J et al (2018) The solar twin planet search. The age-chromospheric activity relation. Astron Astrophys 619:A73. https://doi.org/10.1051/0004-6361/201629294. arXiv:1806.08014 [astro-ph.SR]

Lu YL, Curtis JL, Angus R et al (2022) Bridging the gap-the disappearance of the intermediate period gap for fully convective stars, uncovered by new ZTF rotation periods. Astron J 164(6):251. https://doi.org/10.3847/1538-3881/ac9bee. arXiv:2210.06604 [astro-ph.SR]

MacGregor KB, Brenner M (1991) Rotational evolution of solar-type stars. I. Main-sequence evolution. Astrophys J 376:204. https://doi.org/10.1086/170269

Mahadevan S, Ramsey L, Bender C et al (2012) The habitable-zone planet finder: a stabilized fiber-fed NIR spectrograph for the Hobby-Eberly telescope. In: McLean IS, Ramsay SK, Takami H (eds) Ground-based and airborne instrumentation for Astronomy IV, p 84461S. https://doi.org/10.1117/12.926102. arxiv:1209.1686

Mamajek EE, Hillenbrand LA (2008) Improved age estimation for solar-type dwarfs using activity-rotation diagnostics. Astrophys J 687(2):1264–1293. https://doi.org/10.1086/591785. arXiv:0807.1686 [astro-ph]

Marsden SC, Petit P, Jeffers SV et al (2014) A BCool magnetic snapshot survey of solar-type stars. Mon Not R Astron Soc 444(4):3517–3536. https://doi.org/10.1093/mnras/stu1663. arXiv:1311.3374 [astro-ph.SR]

Masuda K, Petigura EA, Hall OJ (2022) Inferring the rotation period distribution of stars from their projected rotation velocities and radii: application to late-F/early-G Kepler stars. Mon Not R Astron Soc 510(4):5623–5638. https://doi.org/10.1093/mnras/stab3650. arXiv:2112.07162 [astro-ph.SR]

Matt S, Pudritz RE (2005) Accretion-powered stellar winds as a solution to the stellar angular momentum problem. Astrophys J Lett 632(2):L135–L138. https://doi.org/10.1086/498066. arXiv:astro-ph/0510060 [astro-ph]

Matt S, Pudritz RE (2008) Accretion-powered stellar winds. II. Numerical solutions for stellar wind torques. Astrophys J 678(2):1109–1118. https://doi.org/10.1086/533428. arXiv:0801.0436 [astro-ph]

Matt SP, Brun AS, Baraffe I et al (2015) The mass-dependence of angular momentum evolution in sun-like stars. Astrophys J Lett 799(2):L23. https://doi.org/10.1088/2041-8205/799/2/L23. arXiv:1412.4786 [astro-ph.SR]

McQuillan A, Mazeh T, Aigrain S (2014) Rotation periods of 34,030 Kepler main-sequence stars: the full autocorrelation sample. Astrophys J Suppl Ser 211(2):24. https://doi.org/10.1088/0067-0049/211/2/24. arXiv:1402.5694 [astro-ph.SR]

Mestel L (1968) Magnetic braking by a stellar wind-I. Mon Not R Astron Soc 138:359. https://doi.org/10.1093/mnras/138.3.359

Metcalfe TS, Creevey OL, Doğan G et al (2014) Properties of 42 solar-type Kepler targets from the asteroseismic modeling portal. Astrophys J Suppl Ser 214(2):27. https://doi.org/10.1088/0067-0049/214/2/27. arXiv:1402.3614 [astro-ph.SR]

Metcalfe TS, Egeland R, van Saders J (2016) Stellar evidence that the solar dynamo may be in transition. Astrophys J Lett 826(1):L2. https://doi.org/10.3847/2041-8205/826/1/L2. arXiv:1606.01926 [astro-ph.SR]

Metcalfe TS, Strassmeier KG, Ilyin IV et al (2023) Constraints on magnetic braking from the G8 Dwarf stars 61 UMa and τ Cet. Astrophys J Lett 948(1):L6. https://doi.org/10.3847/2041-8213/acce38. arXiv:2304.09896 [astro-ph.SR]

Meunier N, Lagrange AM (2019) Activity time series of old stars from late F to early K. II. Radial velocity jitter and exoplanet detectability. Astron Astrophys 628:A125. https://doi.org/10.1051/0004-6361/201935347. arXiv:1909.02969 [astro-ph.SR]

Meunier N, Lagrange AM, Boulet T et al (2019) Activity time series of old stars from late F to early K. I. simulating radial velocity, astrometry, photometry, and chromospheric emission. Astron Astrophys 627:A56. https://doi.org/10.1051/0004-6361/201834796. arXiv:1904.01437 [astro-ph.SR]

Meunier N, Lagrange AM, Borgniet S (2020) Activity time series of old stars from late F to early K. V. effect on exoplanet detectability with high-precision astrometry. Astron Astrophys 644:A77. https://doi.org/10.1051/0004-6361/202038710. arXiv:2011.02158 [astro-ph.SR]

Nèmec NE, Shapiro AI, Işık E et al (2022) Faculae cancel out on the surfaces of active suns. Astrophys J Lett 934(2):L23. https://doi.org/10.3847/2041-8213/ac8155. arXiv:2207.06816 [astro-ph.SR]

Nèmec NE, Shapiro AI, Işık E et al (2023) Forward modelling of brightness variations in sun-like stars. II. Light curves and variability. Astron Astrophys 672:A138. https://doi.org/10.1051/0004-6361/202244412. arXiv:2303.03040 [astro-ph.SR]

Noyes RW, Hartmann LW, Baliunas SL et al (1984) Rotation, convection, and magnetic activity in lower main-sequence stars. Astrophys J 279:763–777. https://doi.org/10.1086/161945

Oglethorpe RLF, Garaud P (2013) Spin-down dynamics of magnetized solar-type stars. Astrophys J 778(2):166. https://doi.org/10.1088/0004-637X/778/2/166. arXiv:1401.0932 [astro-ph.SR]

Pevtsov AA, Fisher GH, Acton LW et al (2003) The relationship between X-ray radiance and magnetic flux. Astrophys J 598:1387–1391. https://doi.org/10.1086/378944

Piskunov N, Snik F, Dolgopolov A et al (2011) HARPSpol — the new polarimetric mode for HARPS. Messenger 143:7–10

Pizzolato N, Maggio A, Micela G et al (2003) The stellar activity-rotation relationship revisited: dependence of saturated and non-saturated X-ray emission regimes on stellar mass for late-type dwarfs. Astron Astrophys 397:147–157. https://doi.org/10.1051/0004-6361:20021560

Quirrenbach A, Amado PJ, Caballero JA et al (2014) CARMENES instrument overview. In: Ramsay SK, McLean IS, Takami H (eds) Ground-based and airborne instrumentation for Astronomy V, p 91471F. https://doi.org/10.1117/12.2056453

Radick RR, Lockwood GW, Skiff BA et al (1998) Patterns of variation among sun-like stars. Astrophys J Suppl Ser 118(1):239–258. https://doi.org/10.1086/313135

Reiners A (2012) Observations of cool-star magnetic fields. Living Reviews in Sol Phys 9(1). arXiv:1203.0241 [astro-ph.SR]

Reiners A, Schüssler M, Passegger VM (2014) Generalized investigation of the rotation-activity relation: favoring rotation period instead of Rossby number. Astrophys J 794(2):144. https://doi.org/10.1088/0004-637X/794/2/144. arXiv:1408.6175 [astro-ph.SR]

Reiners A, Shulyak D, Käpylä PJ et al (2022) Magnetism, rotation, and nonthermal emission in cool stars. Average magnetic field measurements in 292 M dwarfs. Astron Astrophys 662:A41. https://doi.org/10.1051/0004-6361/202243251. arXiv:2204.00342 [astro-ph.SR]

Reinhold T, Shapiro AI, Solanki SK et al (2020) The sun is less active than other solar-like stars. Science 368(6490):518–521. https://doi.org/10.1126/science.aay3821. arXiv:2005.01401 [astro-ph.SR]

Réville V, Brun AS, Matt SP et al (2015) The effect of magnetic topology on thermally driven wind: toward a general formulation of the braking law. Astrophys J 798(2):116. https://doi.org/10.1088/0004-637X/798/2/116. arXiv:1410.8746 [astro-ph.SR]

Roquette J, Matt SP, Winter AJ et al (2021) The influence of the environment on the spin evolution of low-mass stars - I. External photoevaporation of circumstellar discs. Mon Not R Astron Soc 508(3):3710–3729. https://doi.org/10.1093/mnras/stab2772. arXiv:2109.10296

Saar SH (1988) Improved methods for the measurement and analysis of stellar magnetic fields. Astrophys J 324:441. https://doi.org/10.1086/165907

Schmitt JHMM, Liefke C (2004) NEXXUS: a comprehensive ROSAT survey of coronal X-ray emission among nearby solar-like stars. Astron Astrophys 417:651–665. https://doi.org/10.1051/0004-6361:20030495. arXiv:astro-ph/0308510 [astro-ph]

Schrijver CJ (1987) Solar active regions - radiative intensities and large-scale parameters of the magnetic field. Astron Astrophys 180(1–2):241–252

Schrijver CJ, Title AM (2001) On the formation of polar spots in sun-like stars. Astrophys J 551(2):1099–1106. https://doi.org/10.1086/320237

Schrijver CJ, Zwaan C (2000) Solar and stellar magnetic activity. Cambridge University Press, Cambridge

Schüssler M, Solanki SK (1992) Why rapid rotators have polar spots. Astron Astrophys 264:L13–L16

Schüssler M, Caligari P, Ferriz-Mas A et al (1996) Distribution of starspots on cool stars. I. Young and main sequence stars of $1M_{sun}$. Astron Astrophys 314:503–512

See V, Jardine M, Vidotto AA et al (2015) The energy budget of stellar magnetic fields. Mon Not R Astron Soc 453(4):4301–4310. https://doi.org/10.1093/mnras/stv1925, arXiv:1508.01403 [astro-ph.SR]

See V, Matt SP, Finley AJ et al (2019) Do non-dipolar magnetic fields contribute to spin-down torques? Astrophys J 886(2):120. https://doi.org/10.3847/1538-4357/ab46b2. arXiv:1910.02129 [astro-ph.SR]

Şenavcı HV, Kılıçoğlu T, Işık E et al (2021) Observing and modelling the young solar analogue EK Draconis: starspot distribution, elemental abundances, and evolutionary status. Mon Not R Astron Soc 502(3):3343–3356. https://doi.org/10.1093/mnras/stab199. arXiv:2101.07248 [astro-ph.SR]

Shapiro AI, Solanki SK, Krivova NA et al (2014) Variability of sun-like stars: reproducing observed photometric trends. Astron Astrophys 569:A38. https://doi.org/10.1051/0004-6361/201323086. arXiv:1406.2383 [astro-ph.SR]

Shapiro AI, Solanki SK, Krivova NA (2021) Predictions of astrometric jitter for sun-like stars. I. The model and its application to the sun as seen from the ecliptic. Astrophys J 908(2):223. https://doi.org/10.3847/1538-4357/abd630. arXiv:2012.12312 [astro-ph.SR]

Shu F, Najita J, Ostriker E et al (1994) Magnetocentrifugally driven flows from young stars and disks. I. A generalized model. Astrophys J 429:781. https://doi.org/10.1086/174363

Skumanich A (1972) Time scales for ca II emission decay, rotational braking, and lithium depletion. Astrophys J 171:565. https://doi.org/10.1086/151310

Snik F, Jeffers S, Keller C et al (2008) The upgrade of HARPS to a full-Stokes high-resolution spectropolarimeter. In: McLean IS, Casali MM (eds) Ground-based and airborne instrumentation for Astronomy II, p 70140O. https://doi.org/10.1117/12.787393

Solanki SK (2002) The magnetic structure of sunspots and starspots. Astron Nachr 323:165–177. https://doi.org/10.1002/1521-3994(200208)323:3/4<165::AID-ASNA165>3.0.CO;2-U

Solanki SK (2003) Sunspots: an overview. Astron Astrophys Rev 11(2–3):153–286. https://doi.org/10.1007/s00159-003-0018-4

Solanki SK, Inhester B, Schüssler M (2006) The solar magnetic field. Rep Prog Phys 69(3):563–668. https://doi.org/10.1088/0034-4885/69/3/R02. arXiv:1008.0771 [astro-ph.SR]

Somers G, Pinsonneault MH (2016) Lithium depletion is a strong test of core-envelope recoupling. Astrophys J 829(1):32. https://doi.org/10.3847/0004-637X/829/1/32. arXiv:1606.00004 [astro-ph.SR]

Somers G, Stauffer J, Rebull L et al (2017) M dwarf rotation from the K2 Young clusters to the field. I. A mass-rotation correlation at 10 Myr. Astrophys J 850(2):134. https://doi.org/10.3847/1538-4357/aa93ed. arXiv:1710.07638 [astro-ph.SR]

Sowmya K, Nèmec NE, Shapiro AI et al (2021a) Predictions of astrometric jitter for sun-like stars. II. Dependence on inclination, metallicity, and active-region nesting. Astrophys J 919(2):94. https://doi.org/10.3847/1538-4357/ac111b. arXiv:2107.01493 [astro-ph.SR]

Sowmya K, Shapiro AI, Witzke V et al (2021b) Modeling stellar Ca II H and K emission variations. I. Effect of inclination on the S-index. Astrophys J 914(1):21. https://doi.org/10.3847/1538-4357/abf247. arXiv:2103.13893 [astro-ph.SR]

Sowmya K, Nèmec NE, Shapiro AI et al (2022) Predictions of astrometric jitter for sun-like stars. III. Fast rotators. Astrophys J 934(2):146. https://doi.org/10.3847/1538-4357/ac79b3. arXiv:2206.07702 [astro-ph.SR]

Spada F, Lanzafame AC (2020) Competing effect of wind braking and interior coupling in the rotational evolution of solar-like stars. Astron Astrophys 636:A76. https://doi.org/10.1051/0004-6361/201936384. arXiv:1908.00345 [astro-ph.SR]

Strassmeier KG (2009) Starspots. Astron Astrophys Rev 17(3):251–308. https://doi.org/10.1007/s00159-009-0020-6

Strassmeier KG, Rice JB (1998) Doppler imaging of stellar surface structure. VI. HD 129333 = EK Draconis: a stellar analog of the active young Sun. Astron Astrophys 330:685–695

Strassmeier KG, Ilyin I, Järvinen A et al (2015) PEPSI: the high-resolution échelle spectrograph and polarimeter for the large binocular telescope. Astron Nachr 336(4):324. https://doi.org/10.1002/asna.201512172. arXiv:1505.06492 [astro-ph.IM]

van Saders JL, Pinsonneault MH (2013) Fast star, slow star; old star, young star: subgiant rotation as a population and stellar physics diagnostic. Astrophys J 776(2):67. https://doi.org/10.1088/0004-637X/776/2/67. arXiv:1306.3701 [astro-ph.SR]

van Saders JL, Ceillier T, Metcalfe TS et al (2016) Weakened magnetic braking as the origin of anomalously rapid rotation in old field stars. Nature 529(7585):181–184. https://doi.org/10.1038/nature16168. arXiv:1601.02631 [astro-ph.SR]

van Saders JL, Pinsonneault MH, Barbieri M (2019) Forward modeling of the Kepler stellar rotation period distribution: interpreting periods from mixed and biased stellar populations. Astrophys J 872(2):128. https://doi.org/10.3847/1538-4357/aafafe. arXiv:1803.04971 [astro-ph.SR]

Vaughan AH, Preston GW, Wilson OC (1978) Flux measurements of Ca II and K emission. Publ Astron Soc Pac 90:267–274. https://doi.org/10.1086/130324.

Vidotto AA (2016) The magnetic field vector of the Sun-as-a-star. Mon Not R Astron Soc 459(2):1533 1542. https://doi.org/10.1093/mnras/stw758. arXiv:1603.09226 [astro-ph.SR]

Vidotto AA, Gregory SG, Jardine M et al (2014) Stellar magnetism: empirical trends with age and rotation. Mon Not R Astron Soc 441(3):2361–2374. https://doi.org/10.1093/mnras/stu728. arXiv:1404.2733 [astro-ph.SR]

Vilhu O (1984) The nature of magnetic activity in lower main sequence stars. Astron Astrophys 133:117–126

Weber EJ, Leverett Jr D (1967) The angular momentum of the solar wind. Astrophys J 148:217–227. https://doi.org/10.1086/149138

Weber MA, Schunker H, Jouve L et al (2023) Understanding active region emergence and origins on the Sun and other cool stars. Space Sci Rev 219:63. https://doi.org/10.1007/s11214-023-01006-5. arXiv:2306.06536 [astro-ph.SR]

Wilson OC (1968) Flux measurements at the centers of stellar H- and K-lines. Astrophys J 153:221. https://doi.org/10.1086/149652

Wood BE, Müller HR, Zank GP et al (2005) New mass-loss measurements from astrospheric Lyα absorption. Astrophys J Lett 628(2):L143–L146. https://doi.org/10.1086/432716. arXiv:astro-ph/0506401 [astro-ph]

Wood BE, Müller HR, Redfield S et al (2021) New observational constraints on the winds of M dwarf stars. Astrophys J 915(1):37. https://doi.org/10.3847/1538-4357/abfda5. arXiv:2105.00019 [astro-ph.SR]

Wright NJ, Drake JJ, Mamajek EE et al (2011) The stellar-activity-rotation relationship and the evolution of stellar dynamos. Astrophys J 743(1):48. https://doi.org/10.1088/0004-637X/743/1/48. arXiv:1109.4634 [astro-ph.SR]

Yadav RK, Christensen UR, Morin J et al (2015) Explaining the coexistence of large-scale and small-scale magnetic fields in fully convective stars. Astrophys J Lett 813(2):L31. https://doi.org/10.1088/2041-8205/813/2/L31. arXiv:1510.05541 [astro-ph.SR]

Publisher's Note Springer Nature remains neutral with regard to jurisdictional claims in published maps and institutional affiliations.

Space Science Reviews (2023) 219:54
https://doi.org/10.1007/s11214-023-01000-x

Stellar Activity Cycles

Sandra V. Jeffers[1] · René Kiefer[2] · Travis S. Metcalfe[3]

Received: 20 February 2023 / Accepted: 6 September 2023 / Published online: 28 September 2023
© The Author(s) 2023

Abstract

The magnetic field of the Sun is generated by internal dynamo process with a cyclic period of 11 years or a 22 year magnetic cycle. The signatures of the Sun's magnetic cycle are observed in the different layers of its atmosphere and in its internal layers. In this review, we use the same diagnostics to understand the magnetic cycles of other stars with the same internal structure as the Sun. We review what is currently known about mapping the surface magnetic fields, chromospheric and coronal indicators, cycles in photometry and asteroseismology. We conclude our review with an outlook for the future.

Keywords Stars: activity cycles · Stars: photospheres · Stars: chromospheres · Stars: corona · Stars: interiors

1 Introduction

Magnetic fields on the Sun are well characterised with observations at both high spatial and temporal resolutions. The signatures of the solar dynamo are observable in the different layers of the Sun's atmosphere using multi-wavelength observations and through seismology that allows us to directly probe beneath the Sun's surface. For a comprehensive review of the solar dynamo we refer to the recent review of Charbonneau (2020). For stars with masses ranging from slightly higher than the Sun down to about one-third of the Sun's mass, they are known to have a comparable internal structure as the Sun, in the form of an internal radiative zone and an convective envelope. Since the presence of an outer convective envelope

Solar and Stellar Dynamos: A New Era
Edited by Manfred Schüssler, Robert H. Cameron, Paul Charbonneau, Mausumi Dikpati, Hideyuki Hotta and Leonid Kitchatinov

✉ S.V. Jeffers
jeffers@mps.mpg.de

R. Kiefer
kiefer@leibniz-kis.de

T.S. Metcalfe
travis@wdrc.org

[1] Max Planck Institut für Sonnensystemforschung, Justus-von-Liebig-Weg 3, Göttingen, 37077, Germany

[2] Leibniz Institute for Solar Physics (KIS), Schöneckstraße 6, Freiburg, 79104, Germany

[3] White Dwarf Research Corporation, 9020 Brumm Trail, Golden, Colorado 80403, USA

is a key ingredient in the generation of the Solar dynamo, we also expect to see comparable signatures of magnetic activity on other solar-type stars. While it is unfortunately not possible to observe other stars with the same spatial resolution or temporal cadence as the Sun, they do allow us to understand how key components of the solar dynamo, such as stellar mass and rotation rate, impact the internal dynamo processes.

A key component of the solar dynamo is that the dynamo mechanism produces a balance between the amounts of magnetic flux generated and lost over its 11-year activity cycle. At the beginning of the cycle, magnetic flux emerges at mid-latitudes on the Sun's surface. Over the course of the cycle, the latitude of flux emergence decreases, reflecting the changing nature of the Sun's internal magnetic field, and finally reaches the Sun's equator. This winged pattern of flux emergence is commonly depicted in a solar butterfly diagram (Maunder 1904). The importance of the surface magnetic fields in the dynamo process has been recently reviewed by Cameron and Schüssler (2023). The first detection of a magnetic field in sunspots was reported by Hale (1908). The evolution of the Sun's large-scale magnetic field over its cycle has been monitored by regular polarimetric observations since the 1970s (Cameron et al. 2018). For each 11-year cycle, the magnetic field changes polarity, leading to a 22-year magnetic cycle (Hathaway 2015). Higher up in the Sun's atmosphere, chromospheric diagnostics are commonly used such as the S index. The S index is a measure of the emission in the cores of the Ca II H and K lines, at 396.6 nm and 393.4 nm respectively, relative to nearby spectral continuum regions (Wilson 1968). Another key feature of the solar dynamo is that the Sun's S index is co-incident with the evolution of the geometry of the solar large-scale magnetic field, where a complex geometry occurs at S index activity maximum and a more simple dipole at S index activity minimum. Similarly, the Sun's coronal emission variations, as observed in X-rays, are also co-incident with the evolution of the Sun's large-scale magnetic field (Ayres 2020). While Radick et al. (2018) reported that the photometric variations of the Sun are co-incident with variations in its Ca II emission and the total solar irradiance.

Probing the internal structure of the Sun using the technique of helioseismology allows us to understand how the internal layers of the Sun are impacted by its magnetic cycle. The primary diagnostic is via globally resonant acoustic waves, or p-modes. The specific way in which the p-modes' parameters vary as a consequence of the changing levels of magnetic activity and magnetic field strength, in particular the p-mode frequencies, encodes information about the perturbation causing these changes. In the Sun, the frequencies of p-modes are correlated with the level of magnetic activity, whereas p-mode amplitudes are anti-correlated with the level of magnetic activity. While there are many more indicators to characterise the magnetic cycle of the Sun, these are the main seismic diagnostics that can be searched for in the photometric time series of other stars and can yield insights into their magnetic cycles.

One of the challenges in observing and characterising stellar cycles is the long timespans that are needed to acquire definitive results. The longest data sets are those of the S index for the Sun and for other stars. The longest continuous sets of S index chromospheric activity measurements were obtained over 35 years at the Mount Wilson observatory[1] between 1968 and 2003 (Oláh et al. 2016), although most of the published datasets end in 1992 (Baliunas et al. 1995). Additional long-term observations are available from Lowell Observatory (1992–2020; Hall et al. 2007), Keck Observatory (1996–present; Baum et al. 2022), the TIGRE telescope (2013–present; González-Pérez et al. 2022), ESPaDOnS (Echelle SpectroPolarimetric Device for the Observation of STARS) (2006–present; Brown et al. 2022)

---

[1]For publicly available data, see https://dataverse.harvard.edu/dataverse/mwo_hk_project.

and NARVAL (2007–present; Brown et al. 2022) in the northern hemisphere, and from ESO (2003–present; Lovis et al. 2011) and SMARTS (2007–2013; Metcalfe et al. 2009) in the southern hemisphere.

In terms of understanding stellar cycles, there has recently been significant progress in understanding how the geometry of a star's large-scale magnetic field varies over its magnetic cycle. This is due to the development of instrumentation to reconstruct, and dedicated telescopes to monitor, the large-scale magnetic fields of many stars. The main instruments used to observe the magnetic fields of stars are: (1) ESPaDOnS at the 3.6 m Canada France Hawaii telescope (Donati et al. 2006), its twin (2) NARVAL at the 2.2 m Telescope Bernard Lyot (TBL; Aurière 2003), and (3) HARPSpol (Snik et al. 2008; Piskunov et al. 2011). The advantage of ESPaDOnS and NARVAL is that they were specifically designed with the purpose of monitoring the magnetic fields of stars. For example, their long wavelength ranges cover wavelengths from the far-UV to the nIR and are capable of observing many thousands of lines that can be combined to increase the information content using techniques such as least-squares deconvolution (LSD; Donati et al. 1997; Kochukhov et al. 2010). Additionally, NARVAL is the only instrument at the TBL allowing detailed monitoring of the large-scale magnetic field of stars on short to long timescales. Even though these instrumental developments were commissioned more than 15 years ago, it is only now that we have a sufficiently long time span of data to understand the intrinsic variability of these stars on timescales of the order of the solar cycle.

Furthermore, the field of asteroseismology has rapidly advanced over the last 15 years with the advent of high-precision photometric missions such as CoRoT (Baglin et al. 2006; Auvergne et al. 2009), *Kepler* (Borucki et al. 2010; Koch et al. 2010) and currently TESS (Ricker et al. 2014). Data from these space missions secured with a high cadence has allowed us to put the Sun in the context of other stars and to monitor the impact of stellar magnetic fields on the stars' internal layers.

In this review article, we focus on the observational diagnostics of stellar magnetic cycles of F, G and K dwarfs. We first address photospheric signatures of stellar cycles in Sect. 2, and then observations of cycles in the stellar chromosphere and corona in Sect. 3. Finally, we present cycles in terms of the internal layers of stars via astroseismology in Sect. 4. For completeness we also briefly summarise what is known about photometric cycles in Sect. 5. We then discuss the future prospects in Sect. 6.

2 Photospheric Diagnostics

To reconstruct the large-scale magnetic field of stars other than the Sun, a commonly used method is Zeeman-Doppler Imaging (ZDI). This technique can measure the net magnetic field from a non-homogeneous distribution of circularly polarised light as a function of stellar rotation phase. As previously mentioned, the information content of thousands of spectral lines is combined using the technique of LSD. From a time series of LSD Stokes V observations covering not more than a few stellar rotation periods, the ZDI technique inverts the observed circularly polarised profiles into the strength, polarisation and distribution on the stellar surface in terms of the poloidal, toroidal and meridional magnetic fields (via spherical harmonics expansion Donati et al. 2006b; Folsom et al. 2018). An example of the time series of Stokes V LSD profiles and the corresponding ZDI map is shown in Fig. 1 for the K dwarf ϵ Eri in October 2015. In this review we specifically focus on the results from ZDI as these are many times more numerous than the tentative results from Doppler

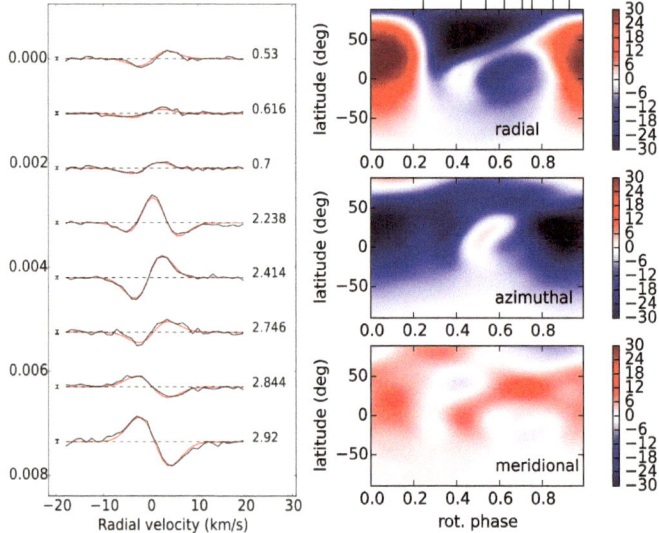

Fig. 1 Time-series of Stokes V profiles for ϵ Eri in October 2015 (*left panel*) with the reconstructed large-scale magnetic field geometry shown in the three panels on the right. For the Stokes V profiles, the black lines represent the data and red dotted lines correspond to synthetic profiles of our magnetic model and where successive profiles are shifted vertically for display clarity. Rotational phases of observations are indicated in the right part of the Stokes V LSD profiles and as vertical ticks at the top of the ZDI map. Data is from the NARVAL spectropolarimeter at the 2.2 m Telescope Bernard Lyot secured as part of the BCool collaboration

imaging alone (for example Jeffers et al. 2007, Hackman et al. 2012, for the long-term brightness monitoring of AB Dor and II Peg).

A key diagnostic of the solar magnetic cycle is that the Sun's chromospheric cycle, or S index cycle, is in phase with its magnetic cycle meaning that its large-scale magnetic field switches polarity at activity maximum. The first star reported to show a solar-like magnetic cycle is the K dwarf 61 Cyg A (Boro Saikia et al. 2016, 2018). 61 Cyg A is an old K5 dwarf, with a very slow rotation rate and a S index cycle of 7.3 yr. It shows polarity switches of its large-scale magnetic field in phase with its S index activity maximum. Similar to the Sun, the large-scale field geometry is complex at activity maximum and dipolar at activity minimum. A summary of 61 Cyg A's magnetic cycle is shown in Fig. 2. Another K dwarf with an extensive time span of magnetic maps is the young, rapidly rotating K dwarf, ϵ Eri which has been shown by Metcalfe et al. (2013) to have two S index cycles of 2.95 yr and 12.7 yr. In contrast to 61 Cyg A and the Sun, ϵ Eri's large-scale magnetic field geometry shows a high degree of complexity at the minimum of the shorter $\sim$ 3-year S index cycle (Jeffers et al. 2022). It also does not show a change in polarity with every S index maximum (Jeffers et al. 2014; Petit et al. 2021). However, Jeffers et al. (2022) recently showed that this could be explained if ϵ Eri's shorter $\sim$ 3-year cycle is a modulation of its longer $\sim$ 13-year cycle and that a polarity switch in its large-scale magnetic field should occur in phase with the longer S index cycle (Fig. 3). Indications of third cycle period has been recently been presented by Fuhrmeister et al. (2023), though long-term observations of ϵ Eri's S-index are needed to confirm this additional cycle. Other young K dwarfs that have been monitored using multi-epoch ZDI observations include the very young stars AB Dor and LQ Hya (Donati et al. 2003), though no clear cyclic behaviour was identified. More recently, Lehtinen et al.

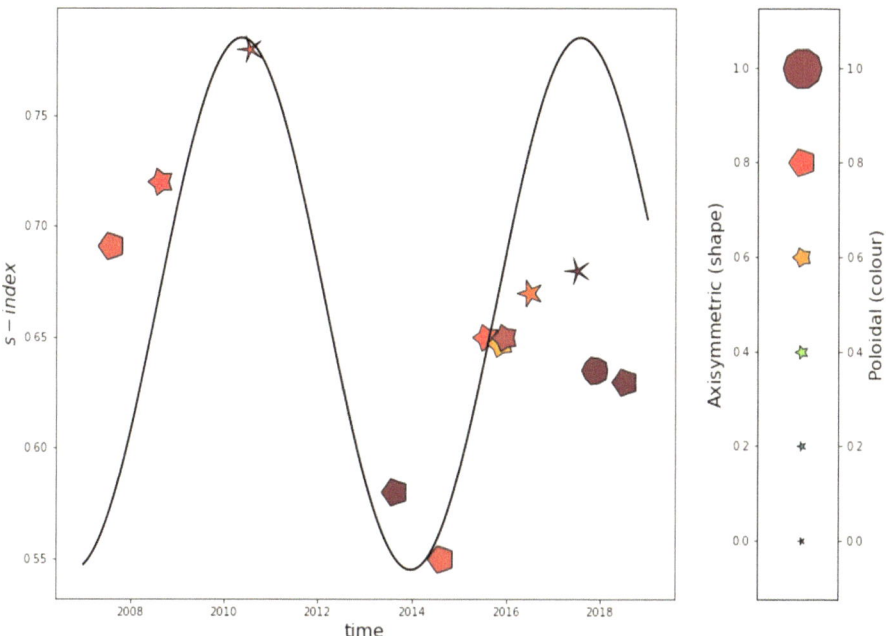

Fig. 2 Magnetic and S index cycle of 61 Cyg A as a function of time. The solid black line is a sinusoid with the period 61 Cyg A's S index cycle of 7.3 yr. Each of the coloured data points summarises the magnetic field geometry of one magnetic map. Data points are coloured depending on whether the large-scale field is poloidal (dark red) or toroidal (blue). The shape of the points indicates the dipolar fraction of the poloidal field, with decagon shaped symbols having a high degree of axisymmetry. Data is from the NARVAL spectropolarimeter at the 2.2 m Telescope Bernard Lyot secured as part of the BCool collaboration

(2022) reported a polarity reversal in LQ Hya, that is coincident with a possible S-index activity minimum.

For G dwarfs that are approximately within 10% of the Sun's mass, we have a long timespan of magnetic maps for HD 171488 (Marsden et al. 2006; Jeffers and Donati 2008; Jeffers et al. 2011; Willamo et al. 2022), HN Peg (Boro Saikia et al. 2015), EK Dra (Waite et al. 2017), HD 190771 (Petit et al. 2009; Morgenthaler et al. 2011) and κ Cet (Boro Saikia et al. 2022). However, we are yet to observe a solar-like magnetic cycle where the large-scale magnetic field switches polarity at S index maximum on another early-G dwarf. This is mainly because the G dwarfs that have been investigated with multi-epoch maps all are much younger than the Sun. For the three young stars EK Dra, HD 171488, and HN Peg the large-scale magnetic field evolves rapidly with little evidence of polarity switches in either the poloidal or toroidal fields. In particular, for HN Peg, the toroidal field appears and disappears again without any correlation with other activity indicators such as the S index. Polarity reversals have been observed in the G2 dwarf HD 190771 (Petit et al. 2009; Morgenthaler et al. 2011), however, they appear first in the azimuthal field (Petit et al. 2009) but are not observed in subsequent observations (Morgenthaler et al. 2011). Somewhat surprisingly, Morgenthaler et al. (2011) report a polarity reversal in subsequent epochs of HD 190771's radial field. Recently, Boro Saikia et al. (2022) reported a potential $\sim$ 10-year magnetic cycle for the G5 V star κ Cet, a moderately active star with a rotation period of 9.2 d. κ Cet is also slightly older than EK Dra, HD 171488, and HN Peg with an age of $\sim$ 750 Myrs. Similar to the K2 dwarf ϵ Eri, κ Cet shows evidence for having two chromospheric cycle periods of

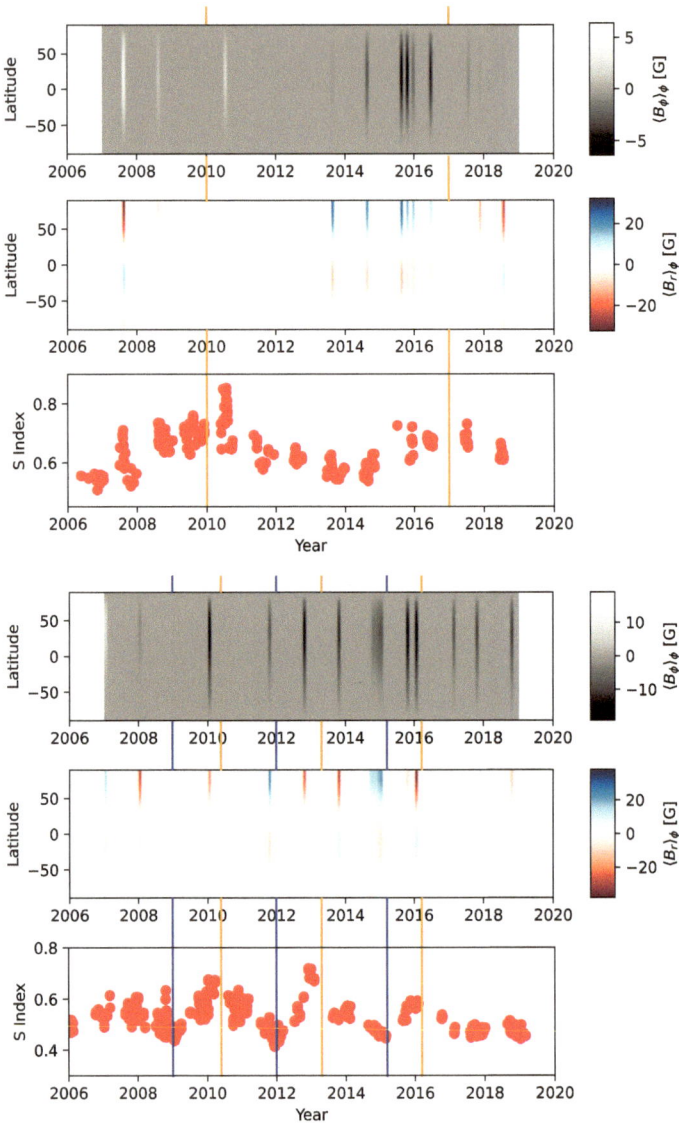

Fig. 3 Butterfly diagrams for 61 Cyg A (upper three panels) and ϵ Eri (lower three panels). The vertical blue and orange lines indicate the S index minima and maxima, respectively. Taken from Jeffers et al. (2022)

3.1 yr and 6 yr, respectively. The longer period dominates for most of the data set analysed by Boro Saikia et al. (2022). Interestingly, the shorter period is present at the beginning and end of the dataset, when the longer period disappears. Additional spectropolarimetric observations that densely sample κ Cet's S index cycle periods will help to resolve the intriguing case of the cyclic evolution of κ Cet's magnetic field.

The first star to show successive polarity switches in its large-scale magnetic field geometry is the planet-hosting late-F dwarf τ Boo (Catala et al. 2007; Donati et al. 2008; Fares et al. 2009, 2013). With a rotation period of only 3.3 d, τ Boo is more massive and has a much faster rotation than 61 Cyg A. The first results indicated a polarity switch occurring between June 2006 and June 2007, and then again between June 2007 and July 2008 from which Fares et al. (2013) concluded that the magnetic cycle was 2 yr long. Subsequent work by Mengel et al. (2016) presented additional epochs of τ Boo's large-scale magnetic field and reported that τ Boo has an S index cycle length of 120 d. More recently, the densely sampled observations by Jeffers et al. (2018) over τ Boo's S index cycle showed that τ Boo's magnetic cycle is indeed co-incident with its S index cycle and with a polarity reversal at S index activity maximum. This makes τ Boo one of the shortest magnetic cycles with a length of 240 d. τ Boo's giant ~ 6 Jupiter-mass planet, which orbits at a distance of ~ 0.5 AU, has been considered to play a role in its internal dynamo processes via tidal locking, similar to the increased activity of stars in binary systems (see Fares et al. 2013). Recent work by Brown et al. (2021) showed that similar levels of activity and polarity reversals also exist on HD 75332 a star with a similar mass and rotation rate as τ Boo. Using 12 epochs of magnetic maps, Brown et al. (2021) showed that HD 75332 has a rapid 1.06 yr cycle and that a polarity reversal at activity maximum is consistent with polarity switches at activity maximum but a repeated polarity switch is required to confirm the cyclic nature of HD 75332's large-scale magnetic field. Another late-F star that shows evidence for a potential 3-year cycle is HD 78366 (Morgenthaler et al. 2011), where the radial field shows polarity reversals with a possible 3-year cycle. In contrast, the recently work of Marsden et al. (2023) showed that the large-scale magnetic field of the old F7 dwarf Dra is stable over a 5 year time span and does not show indications of cyclic behaviour. The results from these stars show the that rapid cyclic nature of the large-scale magnetic field of late F-stars could be an intrinsic feature of young to middle aged stars with a shallow convective zone, and that more stable patterns emerge as these stars evolve off the main-sequence.

While the observational data is not yet conclusive on several targets, the dense phase coverage of 61 Cyg A and ϵ Eri allow a more detailed comparison with the workings of the solar dynamo processes. In the case of the Sun, it is well established that the axisymmetric component of the toroidal field $\langle B_\phi \rangle$ (Cameron et al. 2018) is a proxy for flux emergence of the global dynamo and follows the Sun's S index. Recent work by Jeffers et al. (2022) shows that this relation still holds at the resolution of the magnetic maps reconstructed with ZDI. Applying this to 61 Cyg A shows that the flux emergence also follows its S index, while for ϵ Eri it shows two cycles and the potential onset of an extended inactive period. The work of Jeffers et al. (2022) concludes that surface magnetic fields play a crucial role in the dynamos of 61 Cyg A, ϵ Eri, and the Sun. For further discussion on the nature of the stellar dynamo from a modelling perspective we refer to Brun et al. (2022), Käpylä et al. (2023) and references therein.

3 Chromospheric and Coronal Diagnostics

Most of the available data on stellar activity cycles come from observations of the Ca II H (396.6 nm) and K (393.4 nm) spectral lines (hereafter Ca HK). Emission in the cores of these lines is a well-established proxy for magnetic heating in the chromosphere (Leighton 1959). Time series observations with an appropriate cadence can probe both the long-term variations due to magnetic activity cycles (Wilson 1978), as well as shorter-term modulation due to stellar rotation (Baliunas et al. 1983). Considering the 11-year sunspot cycle, decades of observations are typically required to measure activity cycles in other stars. A list of stars with currently known chromospheric cycles was recently compiled by Mittag et al. (2023) (see their Table 1).

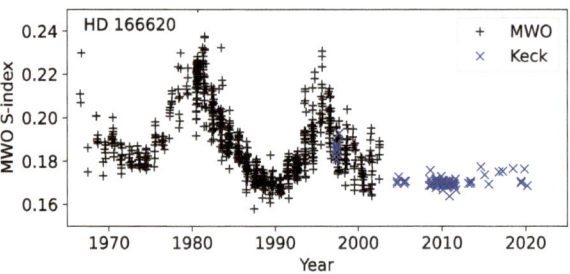

Fig. 4 Stellar activity measurements for the old K-type star HD 166620 spanning more than 50 years, including observations from Mount Wilson (black points) and Keck (blue points). This is the first unambiguous example of a Sun-like star entering a grand magnetic minimum (data from Oláh et al. 2016; Baum et al. 2022)

As an illustration of what we can learn from multi-decadal time series measurements, the combined datasets from Mount Wilson and Keck for the old K-type star HD 166620 are shown in Fig. 4. The Mount Wilson Observatory (MWO) S index is one of the standard proxies for magnetic activity, measuring emission in the Ca HK line cores relative to nearby pseudo-continuum bands. The first few decades of observations from Mount Wilson (black points in Fig. 4; Oláh et al. 2016) reveal a regular activity cycle with a period $P_{\rm cyc} \sim 15$ yr. Higher-cadence observations beginning in the 1980s reveal rotational modulation with a mean period $P_{\rm rot} \sim 42$ d, and significant differential rotation from seasonal variations (33.4–50.8 d), presumably as active regions migrate to different latitudes through the cycle (Donahue et al. 1996). In addition, the apparent correlation between cycle amplitude and the rise time for individual cycles has been used to examine whether the Waldmeier Effect in the Sun (Waldmeier 1935) is also observed for other Sun-like stars (Garg et al. 2019; Willamo et al. 2020). Most notably, the continued observations from Keck (blue points in Fig. 4; Baum et al. 2022) reveal a smooth transition from cycling to constant activity, the first unambiguous example of a Sun-like star entering a grand magnetic minimum (Luhn et al. 2022). The high-cadence observations around 2010 coincide with the next expected maximum of the cycle, which is clearly absent. This may support the idea that stellar cycles can become intermittent as stars evolve through the critical activity level where weakened magnetic braking appears to begin (van Saders et al. 2016; Metcalfe et al. 2022).

Some stellar cycles have also been observed at X-ray wavelengths. The magnetic processes that heat the chromosphere also heat the corona, which has a much higher temperature ($\sim 10^6$ K, emitting at X-ray wavelengths) and fills a larger volume around the star. Variations around the mean X-ray luminosity are substantially larger than variations around the mean Ca HK emission, providing a higher contrast for the detection of magnetic cycles. For example, the young solar-type star ι Hor shows fractional variations that are 3–4 times larger in X-ray luminosity than in Ca HK emission (Sanz-Forcada et al. 2013). The main challenge is that X-ray measurements must be obtained from above the Earth's atmosphere, so the cadence and duration of the observations are limited by competition for time on space telescopes and the longevity of X-ray missions (for instrument stability). These challenges have effectively forced studies of coronal activity cycles to rely on measurements from two long-lived X-ray missions, *XMM-Newton* and *Chandra*.

Most observations of coronal activity cycles have focused on stars with previously known activity cycles from Ca HK measurements. The earliest detections included the 7.3-year cycle in 61 Cyg A (Hempelmann et al. 2006; Robrade et al. 2012) and the 8.2-year cycle in HD 81809 (Favata et al. 2008; Orlando et al. 2017), both of which appeared to be approximately in phase with chromospheric variations. The discovery of substantially shorter

Fig. 5 Dependence of activity cycle period on rotation, showing two distinct sequences (solid lines). Points are coloured by effective temperature, indicating F-type (blue triangles), G-type (yellow and orange circles), and K-type stars (red squares). Schematic evolutionary tracks are shown as dashed lines, leading to stars with constant activity that appear to have shut down their global dynamos (arrows along top). Several notable stars are labelled

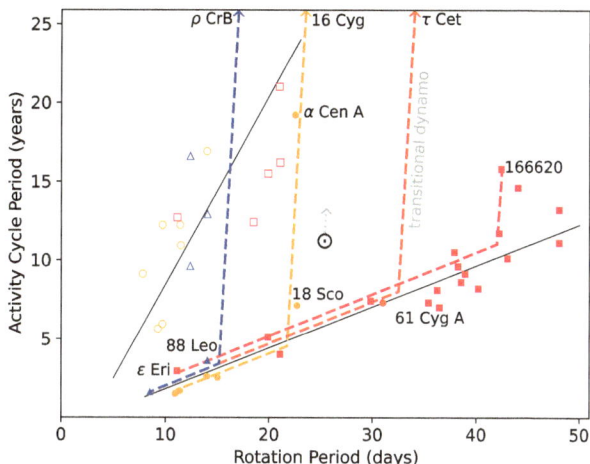

chromospheric activity cycles in the young solar-type stars ι Hor (Metcalfe et al. 2010; Alvarado-Gómez et al. 2018) and ϵ Eri (Metcalfe et al. 2013) provided new opportunities to study coronal activity cycles on shorter timescales (Sanz-Forcada et al. 2013; Coffaro et al. 2020), and revealed some fascinating incongruities between chromospheric and coronal variations. Discoveries of previously unknown activity cycles are currently limited to the southern hemisphere stars α Cen A & B (Robrade et al. 2012), which were inaccessible to the Mount Wilson survey. Characterization of the coronal activity cycle in the 5.4 Gyr solar analog α Cen A is of particular interest as a constraint on the future of the 11-year solar cycle. The amplitude of its 19.2-year X-ray cycle is about one-third that of the Sun (Ayres 2020), suggesting that the solar cycle may be growing longer and weaker (Metcalfe and van Saders 2017).

With high-quality measurements of stellar activity cycles and rotation periods, it is natural to ask whether there is any discernible relationship between these observables, as expected from dynamo theory. This question was examined empirically by Erika Böhm-Vitense in 2007 (Böhm-Vitense 2007). In the first figure of her thought-provoking paper, she simply plotted P_{cyc} against P_{rot} for stars in the Mount Wilson survey with the most reliable measurements (Saar and Brandenburg 1999). An updated version of this plot is shown in Fig. 5, which reveals two distinct relationships between these two observables (solid lines). There is an upper sequence of long-period cycles (open points; but see Boro Saikia et al. 2018), and a lower sequence of short-period cycles (solid points), with the solar cycle falling curiously in between. Some of the stars exhibit cycles on both branches simultaneously, leading Böhm-Vitense to suggest that the two branches may represent two distinct dynamos operating in different regions of the star. Considering other properties of the stellar sample, she suggested that cycles on the long-period sequence may be driven in the near-surface shear layer, while the cycles on the short-period sequence may be driven at the base of the convection zone.

The first explanation for the peculiar position of the Sun in Fig. 5 came ten years later (Metcalfe and van Saders 2017). The updated version of the diagram coloured the points by spectral type, indicating hotter F-type (blue triangles), Sun-like G-type (yellow circles), and cooler K-type stars (red squares). The authors added several stars with measured rotation periods but no activity cycles (arrows along top) and included schematic evolutionary tracks (dashed lines) to indicate where rotation periods became nearly constant for different spectral types, apparently due to weakened magnetic braking (van Saders et al. 2016;

Metcalfe et al. 2016). According to this interpretation, activity cycles initially grow longer along each sequence as the stellar rotation period slows over time. However, when stars reach a critical Rossby number (the rotation period normalised by the convective turnover time), the rotation period remains nearly constant while the cycle gradually grows longer and weaker before disappearing entirely. This explains why hotter stars are confined to the left side of the diagram, while progressively cooler stars (with longer convective turnover times) continue to evolve further towards the right side. It also suggests that the Sun may be in a transitional evolutionary phase and that the solar cycle may represent a special case of dynamo theory. These conclusions still hold even with the more recent results of Boro Saikia et al. (2018) who confirm the lower sequence but question the presence of the upper sequence in Fig. 5. Linking chromospheric cycles with the photospheric large-scale magnetic field (see Sect. 2), See et al. (2016) reported that there are indications that stars on the upper sequence have highly variable toroidal fields, while stars on the lower branch have stable poloidal fields. However, more stars with both chromospheric cycles and ZDI maps are needed to confirm this conclusion at short rotation periods.

An analysis contemporaneous to Metcalfe and van Saders (2017) by Axel Brandenburg and collaborators relied on a different representation of the measurements, more closely connected to dynamo theory (Brandenburg et al. 2017). Rather than plot P_{cyc} against P_{rot}, the authors plotted $\log(P_{rot}/P_{cyc})$ (related to the strength of the α effect) against $\log R'_{HK}$ (related to the strength of the magnetic field; Brandenburg et al. 1998). The latter is the chromospheric emission from the MWO S index, corrected for a small photospheric contribution and normalised by the bolometric luminosity of the star, allowing meaningful comparisons of stars with different spectral types. In this representation, a constant slope in a plot of P_{cyc} against P_{rot} becomes a horizontal line. However, the stellar data actually show a slope, indicating a weaker α effect as the magnetic field grows weaker. The solar cycle and HD 166620 both appear closer to the short-period sequence in this analysis, but the old solar analog α Cen A (Judge et al. 2017) and the K-type subgiant 94 Aqr A (Metcalfe et al. 2020) remain significant outliers.

4 Seismology: Insight into the Internal Structure

As the Sun and the stars pass through their activity cycles, the physical conditions in the regions in which the magnetic concentrations are located change over time. Solar and stellar oscillations propagating in these regions[2] are sensitive to these changes. The specific way in which the modes parameters consequently vary, in particular the mode frequencies, contains valuable information about the perturbation causing these changes, i.e., the varying magnetic field. Therefore, helio- and asteroseismology enable us to probe the interior and atmospheric magnetic structure of the Sun and the stars.

4.1 On the Sun

For the Sun, essentially all fundamental p-mode parameters are observed to vary over the solar activity cycle. The first parameter for which this was noticed, was the mode frequencies of low-degree modes (Woodard and Noyes 1985). This detection has been confirmed and expanded over the following decades (Elsworth et al. 1990; Libbrecht and Woodard 1990;

---

[2]p-, g-, and mixed modes; consult, e.g., the review by Hekker and Christensen-Dalsgaard 2017 for more details about the different types of modes.

Jiménez-Reyes et al. 1998; Howe et al. 1999; Chaplin et al. 2001; Salabert et al. 2015; Tripathy et al. 2015). Now, the cyclic shift of p-mode frequencies, which is tightly in phase with the activity cycle for p-modes below the acoustic cut-off frequency, has been confirmed for a wide range of frequencies and harmonic degrees (e.g., Broomhall 2017). In the context of stellar cycles, it is important to note that – in contrast to p-mode frequencies, which are, as mentioned, correlated with the level of magnetic activity – solar p-mode amplitudes are indeed anti-correlated with the level of magnetic activity (e.g., Komm et al. 2000; Jiménez et al. 2002; Jiménez-Reyes et al. 2003, 2004; Salabert et al. 2004; Burtseva et al. 2009; Broomhall et al. 2014, 2015; Kiefer et al. 2018). Also mode linewidths, which are related to mode damping (e.g., Jefferies et al. 1991; Chaplin et al. 2000), mode energies (e.g., Komm et al. 2000b; Kiefer et al. 2018), mode energy supply rates (Kiefer and Broomhall 2021), and mode parameters of pseudomodes above the acoustic cut-off frequency (Kosak et al. 2022) vary through the solar cycle.

The sensitivity of p-modes to perturbations depends on their frequency as well as on their harmonic degree, increasing with both. This behaviour is largely due to the modes' inertia decreasing with both frequency and harmonic degree (see, e.g., Christensen-Dalsgaard and Berthomieu 1991; Komm et al. 2000b; Chaplin et al. 2001 and for a more in-depth discussion and more references consult the review article by Basu 2016). In contrast to the Sun, for stars, only the lowest harmonic degrees $l = 0, 1, 2$ of p-modes can be measured, as the stellar photometric time series integrate the light of the full stellar disk. For these low harmonic degrees, the mode inertia does not differ very much between them. Any detected variation in the frequency shifts between modes of different harmonic degrees can be utilised to infer the latitudinal distribution of magnetic activity (Moreno-Insertis and Solanki 2000; Chaplin et al. 2007; Thomas et al. 2021). Further, mode frequency shifts, which increase with mode frequency, can be attributed to magnetic perturbations that are located very close to the surface, as higher frequency modes are concentrated to shallower layers (e.g., Basu et al. 2012; Salabert et al. 2015; Broomhall 2017).

4.2 Asteroseismic Detections of Stellar Magnetic Activity (Cycles)

It is the tight anti-correlation between the activity-related variations in p-mode frequencies and p-mode amplitudes that is a tell-tale seismic signature for varying levels of stellar magnetic activity in solar-like oscillators. This signature can be searched for in high-quality photometric time series that were delivered to us by the satellite missions CoRoT (Baglin et al. 2006; Auvergne et al. 2009), *Kepler* (Borucki et al. 2010; Koch et al. 2010), and currently TESS (Ricker et al. 2014). If a stellar activity cycle is not fully covered by the data, the measured p-mode parameter variation generally presents a lower boundary on each star's cycle variability. In contrast, short activity cycles with periods of weeks or a few months may be missed by seismology: The length of the time series segments, that are needed to achieve the required frequency resolution to detect p-mode frequency variation on the order of a few tenths of μHz, is typically around 100 d.

Chaplin et al. (2007) and Karoff et al. (2009) investigated which types of stars ought to be observed and what characteristics the data must have if stellar activity cycles are to be detected and characterised seismically. Karoff et al. (2009) also include ground-based observations of chromospheric activity in their considerations. They found that, most importantly, the photometric time series as well as the ground-based observations need to be sufficiently long – at least several consecutive months – and the amplitude of the acoustic modes of the observed stars should be large enough so they protrude the noise (also see Chaplin et al. 2011a,b; Campante et al. 2016b; Ball et al. 2018; Schofield et al. 2019). An in-depth review

Fig. 6 First detection of the asteroseismic fingerprint of stellar magnetic activity on the CoRoT target HD 49933 by García et al. (2010). *Top panel:* p-mode frequency shifts $\delta\nu$ as measured via the cross-correlation (red diamonds) and the peak bagging approaches (black circles). *Middle panel:* Mean p-mode amplitudes A obtained from peak-bagging. *Bottom panel:* Starspot proxy S_{ph} measured from the standard deviation of segments of the time series. This figure is a reproduction of Fig. 1 from García et al. (2010) based on their original data

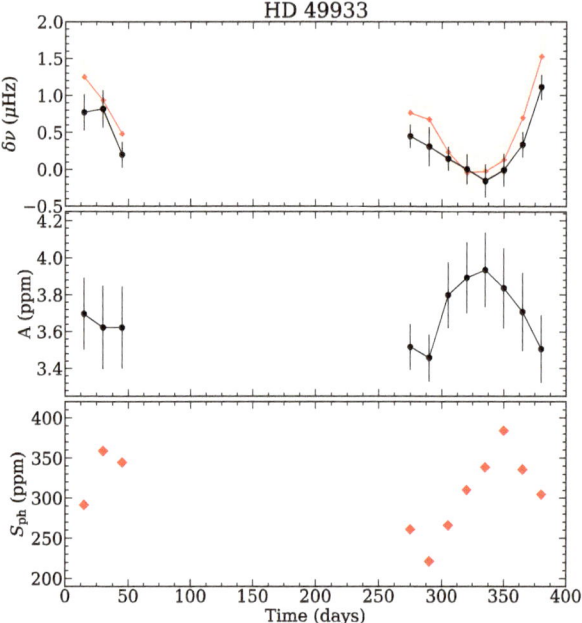

of the inferences asteroseismology can yield on stellar activity and activity cycles was also provided by Chaplin and Basu (2014).

The first detection of activity-related p-mode parameter variations was achieved by García et al. (2010) for the F-type dwarf star HD 49933 with CoRoT data. Their main results are reproduced in Fig. 6. The p-mode frequency shifts are depicted in the top panel as measured with two different methods (p-mode peak bagging in black circles, cross-correlation of the periodogram as red diamonds). The middle panel shows the p-mode amplitudes and the bottom panel shows a "starspot proxy", which is the standard deviation of segments of the photometric time series. Indeed, the temporal changes in mode frequencies and mode amplitudes are clearly anti-correlated, pointing towards magnetic activity being the cause of these variations. As for the Sun, also HD 49933's frequency shifts increase with mode frequency, as Salabert et al. (2011) found. This indicates that the magnetic perturbation is located close to the star's surface.

Using *Kepler* data of 24 solar-like stars with a length of at least 960 d, Kiefer et al. (2017a) found significant frequency shifts ($> 1\sigma$) on 23 stars, showing that p-mode frequency variations are a very widespread phenomenon in solar-like oscillators. For six of these stars, the variation of p-mode amplitudes is also strongly anti-correlated (Spearman rank correlation coefficient $\rho < -0.5$) with the observed shifts of their frequencies. Shortly before, Salabert et al. (2016b) already found that the young solar analog KIC 10644253 exhibits activity-related p-mode frequency shifts. Based on spectroscopic observations with the HERMES spectrograph, they also demonstrated that this star is more active than the Sun.

The search for seismically detected magnetic activity (cycles) was then further expanded by Santos et al. (2018) to 87 solar-like stars, including the 66 stars from the *Kepler* LEGACY sample (Lund et al. 2017; Aguirre et al. 2017) and 25 solar-like KOI targets (Campante et al. 2016a). They used a Bayesian peak-bagging technique on 90-day segments of the photometric time series to measure the shifts of individual p-mode peaks in the periodogram enabling

Fig. 7 Asteroseismic detection of magnetic activity in the two solar-like *Kepler* targets KIC 8006161 and KIC 5184732. *Top panel for each star:* Mean frequency shifts $\delta\nu$ averaged over all available p-modes as a function of time as measured by Bayesian peak-bagging (black circles) and as measured with the cross-correlation technique (red diamonds). *Bottom panel for each star:* Logarithmic mode height $\ln S$ of the p-modes obtained from peak-bagging. This figure was produced based on the original data from Santos et al. (2018)

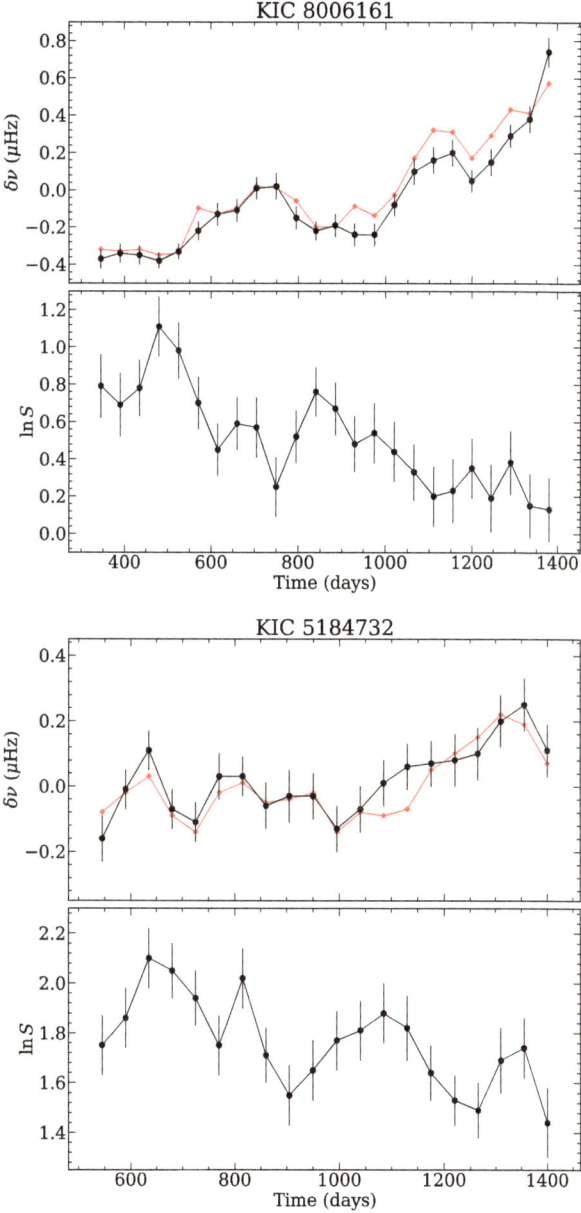

them to analyse different harmonic degrees and even azimuthal orders for variations in their parameters. The results of two of the stars in their sample are shown in Fig. 7. The two top panels show the frequency shifts of KIC 8006161 in the top panel and the logarithm of the mode amplitudes obtained from Bayesian peak-bagging in the bottom panel. The results of KIC 5184732, a more typical example from the analysed sample regarding the magnitude of the uncertainties and amplitude of the variations, are shown in the two bottom panels. In both cases, there are significant and systematic variations in p-mode frequencies as well as in mode amplitudes. What is more, these variations are anti-correlated (as measured by Pearson's correlation coefficient r) with one another, at a level of $r = -0.791$ for KIC 8006161 and $r = -0.482$ for KIC 5184732, signifying magnetic activity is likely causing these variations.

Indeed, KIC 8006161 is probably one of the most intensively studied stars from the *Kepler* seismic sample due to its very significant asteroseismic signature of magnetic activity. This star is very similar to the Sun in mass and radius but has a metallicity which is about twice the solar value. Karoff et al. (2018) analysed spectroscopic observations from the MWO program of KIC 8006161 spanning almost 20 years. This uncovered an activity cycle with a period of approximately 7.4 yr. Through its cycle, KIC 8006161 has a significantly higher variability in its photospheric activity proxy S_{ph} and its S index than the Sun. Karoff et al. (2018) postulated the star's higher metallicity brings about a deeper convective envelope compared to the Sun. This, in turn, then causes stronger levels of activity. The authors also show that the *Kepler* era of observations is coincident with the rising period of magnetic activity for KIC 8006161. This lends further support to activity being the root cause of the measured p-mode parameter variations. Due to the exquisite quality of the *Kepler* data and the very good signal-to-noise ratio of the p-mode peaks in the periodogram of KIC 8006161, Thomas et al. (2019) were able to constrain its latitudinal distribution of active regions by remapping the observed frequency shift to the stellar surface. Their technique utilises that modes of different harmonic degrees differ in their sensitivity to the latitudinal distribution of the perturbation causing the frequencies to shift (Moreno-Insertis and Solanki 2000; Chaplin et al. 2007). The authors determined that KIC 8006161's active regions are distributed over a wider band of latitudes and are located at higher latitudes than for the Sun. Based on a model of the rotation profile and the rotational modulation of this star's *Kepler* time series, Bazot et al. (2018) constructed a butterfly diagram. In their result, KIC 8006161 exhibits spots at both low latitudes close to the equator and, during some periods, at higher latitudes around 40°.

Following the advice of Karoff et al. (2009), Karoff et al. (2013) observed 20 Sun-like stars in the *Kepler* field-of-view with the Nordic Optical Telescope (NOT) and determined their excess flux (surface flux arising from magnetic sources) and S index. From the stars' *Kepler* light curves, they measured the rotation periods and the small frequency separation, which they used to guide the target selection for their program. The stellar fundamental parameters were obtained using an asteroseismic modelling code. Karoff et al. (2013) found that the ten stars from their sample which have independent measurements of asteroseismic ages, rotation periods and excess flux follow the Skumanich relations (Skumanich 1972) reasonably well. Further and interestingly, they obtained a much stronger relation between asteroseismically determined stellar properties and the stars' excess flux than with their S index. Karoff et al. (2019) subsequently analysed the full four years of NOT spectroscopic data (covering the complete *Kepler* main mission 2009–2013) as well as the photometric variability and p-mode frequency shifts of these 20 stars. They detected a strong correlation between the different activity proxies only for a few targets, most notably for KIC 8006161. The authors attribute this to the rather sparse sampling of spectroscopic data and the relative shortness of the photometric time series compared to the expected length of activity cycles. While Karoff et al. (2013, 2019) did not specifically look for or find new asteroseismic detections of stellar activity cycles, they showed how asteroseismology – in conjunction with ground-based spectroscopic data – can usefully inform research on stellar activity, activity cycles, and the investigation of age–rotation–activity relations of solar-like oscillators.

Fig. 8 Frequency shift amplitudes $\delta\nu_{max}$ of 75 *Kepler* stars as a function of different stellar parameters: chromospheric activity level $\log R'_{HK}$ (*first panel*), effective Temperature T_{eff} (*second panel*), rotation period P_{rot} (*third panel*), and age (*fourth panel*). The colours of the data points indicate the stars' ages, except for the bottom panel, where they encode rotation period. KIC 8006161 and the Sun are highlighted by the light green square and yellow star, respectively. This figure was produced based on the original data from Santos et al. (2019)

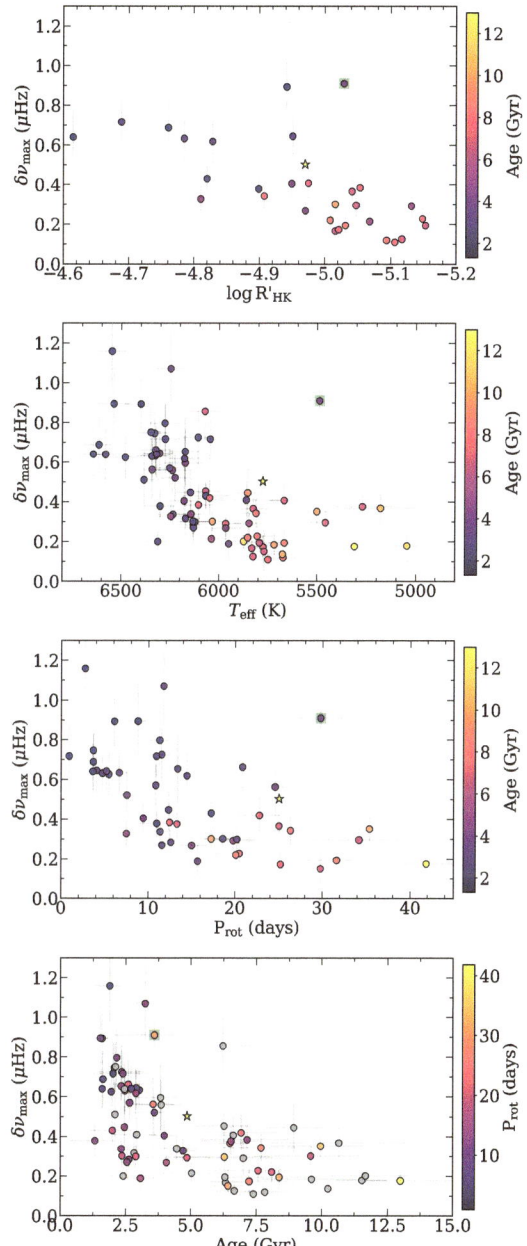

4.3 P-Mode Frequency Variations and Their Relations to Fundamental Stellar Parameters

Santos et al. (2019) investigated whether the amplitudes of the mean frequency shifts $\delta\nu_{max}$ given by Santos et al. (2018) depend on fundamental stellar parameters. Parts of their results are reproduced in Fig. 8. They found that there is a strong correlation between the amplitude of the frequency shifts and the chromospheric activity level as measured by the log R'_{HK} index with a Spearman rank correlation coefficient of $\rho = 0.69$ (first panel of Fig. 8), supporting magnetic activity as the root cause of the measured shifts. As a function of effective temperature T_{eff}, $\delta\nu_{max}$ increases. This is not necessarily caused by an increase in the strength of magnetic activity with T_{eff} but is most likely caused by the increased mode sensitivity. Both, Metcalfe et al. (2007) (based on work done by Dziembowski and Goode 2004) and Kiefer et al. (2019) used a simple scaling model for the frequency shifts, assuming that they are proportional to the depth of the perturbation causing them (magnetic activity), the p-modes' inertia, and the strength of the activity. The relations which they deducted reproduce the course of $\delta\nu_{max}$ seen in the second panel of Fig. 8 reasonably well. As can be seen in the bottom two panels of Fig. 8, $\delta\nu_{max}$ decreases as rotation period P_{rot} increases and as stars age. This is expected, as stellar activity as well as stellar rotation is known to decay as stars get older (see, e.g, Skumanich 1972; Vaughan et al. 1981; Noyes et al. 1984).

As the frequencies of p-modes of unequal harmonic degrees are variably susceptible to perturbations, also the frequency separation ratios are subject to magnetic activity-induced variations. Thomas et al. (2021) investigated the bias caused by activity-perturbed separation ratios on the estimates for fundamental stellar parameters through stellar modelling pipelines. They determined that, for solar-like stars with activity levels similar to the Sun, the bias is typically less than 0.5% for mass, but can affect estimates of stellar age by up to 5% and core hydrogen content by up to 3%. Stronger than solar activity levels consequently increases these errors, as will extreme inclination angles, because the separation ratios are more strongly perturbed in such a scenario. Similarly, Pérez Hernández et al. (2019) found that the activity-induced variation of the small frequency separation, i.e., the frequency separation of consecutive quadrupole and radial modes, can cause misdetermination of stellar age by up to 10% and of mass and radius by a few percent. However, they also find this variation can often be masked by filtering out surface effects from the mode frequencies.

As all p-mode frequencies shift with activity, the frequency of maximum oscillation amplitude ν_{max} follows suit. In their study of the temporal variation of the solar $\nu_{max,\odot}$, Howe et al. (2020) indeed found that it is positively correlated with the level of solar magnetic activity and that it changes by as much as $\simeq 25$ µHz between solar activity minimum and maximum. As $\nu_{max,\odot}$ is used in asteroseismic scaling relations, this shift can incur an error of up to 0.8% and 2.4% in the estimates of stellar radius and mass, respectively.

4.4 Theoretical Groundwork and Recent Detection of Interior Magnetic Fields

Before the advent of asteroseismology, there have been a number of theoretical developments in helioseismology dealing with the effects of internal magnetic field on the oscillation frequencies of the Sun. Gough and Thompson (1990) used perturbation theory for the calculation of the frequency shifts of p-mode multiplets caused by buried magnetic field distributions. Their framework was later expanded by Antia et al. (2000) and Baldner et al. (2009) for the investigation of long time series of solar p-mode splitting coefficients to yield information on magnetic field concentrations in the solar convective envelope. Utilizing another flavour of perturbation theory, Lavely and Ritzwoller (1992) applied a quasi-degenerate perturbational ansatz to deduct a theoretical framework for the calculation of the effect of convection and elastic-gravitational asphericities on p-modes. This was later expanded to include the effect of sub-surface magnetic field concentrations by Kiefer et al. (2017b) and Kiefer and Roth (2018). Similarly, Hanasoge (2017) and Das et al. (2020) deduct expressions describing the effect of Lorentz stresses on the coupling of solar oscillations.

These theories can also be applied to other solar-like stars – albeit in a more limited fashion due to the spatially unresolved nature of stellar observations – and as long as the physical boundary conditions with which the theory has been developed are respected. Using stellar structural and oscillation models, these theories can emulate the fingerprint of cyclic variations of magnetic field strengths and configurations in the stellar oscillation frequencies.

Over the last few years, several groups have pushed forward the theoretical description of and, consequently, the search for signatures of the interplay of internal magnetic fields and stellar oscillation. These recent studies largely focused on the impact of internal magnetic field on gravity or mixed-mode frequencies: Van Beeck et al. (2020) calculated frequency shifts in gravity-mode pulsators at the end of the main-sequence and find that axisymmetric poloidal-toroidal fields stronger than 10^6 G should be detectable from *Kepler* time series, if these fields exist. Prat et al. (2020) predicted that oblique dipolar magnetic fields leave detectable signatures in the gravity mode periods by applying their theory to a magnetic, rapidly rotating and slowly pulsating B-type star. Several recent studies investigated the effects of magnetic field distributions of various configurations on mixed modes in red giants (Gomes and Lopes 2020; Mathis et al. 2021; Bugnet et al. 2021; Bugnet 2022; Loi 2020, 2021) and γ Dor and SPB stars (Dhouib et al. 2022).

In the wake of this flurry of recent theoretical studies, the existing data sets were reanalysed, looking for the predicted signatures of magnetic fields in the oscillation frequencies. Using *Kepler* data, Li et al. (2022) detected strong magnetic fields in the cores of giant stars. For three hydrogen-shell burning giants, they measured asymmetries in the (mixed-mode) oscillation multiplets, which translate into magnetic field strengths of 102 ± 12 kG for KIC 8684542, 98 ± 24 kG for KIC 7518143, and an upper limit of 41 kG for KIC 11515377. Recently, building on the technique presented by Li et al. (2022), Deheuvels et al. (2023) seismically detected strong magnetic fields in the cores of 11 red giants, again using *Kepler* data. These observational studies are thus far restricted to static magnetic fields, i.e., one data point in time without the detection of cyclic activity. However, they clearly show that – given long enough data sets – variable or cyclic magnetic field concentrations can be detected in the stellar interiors using asteroseismology.

5 Activity Cycles in Photometric Time Series

Stellar activity cycles can also reveal themselves as periodic variations of stellar brightness (see, e.g., Baliunas and Vaughan 1985), just as the Sun's total irradiance varies slightly over the solar 11-year cycle (see, e.g., Yeo et al. 2014). Using long-term observations of the V-band magnitude, Oláh et al. (2000) were able to detect starspot cycles for nine out of the ten stars in their sample of rapidly-rotating active stars (a mix of K and G stars, including main-sequence stars, subgiants, and giants) with data lengths between 11–30 yr. They found that the detected photometric cycle lengths agree with those from other activity proxies found by other authors. Later, Oláh and Strassmeier (2002) expanded this study to a baseline of up to 34 yr and could confirm that cycle period depends on rotation rate.

Evaluating four-year-long photometric light curves from the *Kepler* satellite, Vida et al. (2014) presented evidence for activity cycles for nine of the 39 fast-rotating late-type active stars they investigated. The cycles they detected have periods between 300–900 d. Vida et al. (2014) used the temporal variation of the stars' rotation period as an indicator for the cycles. Using multi-decadal ground-based photometric data, this approach enabled Oláh et al. (2009) to detect activity cycles in at least 15 of the 20 active stars in their sample. Ferreira Lopes et al. (2015) found evidence for stellar cycles in the photometric data of 16

CoRoT FGK main sequence stars. These cycles follow the earlier-found relations between the length of the activity cycles and the stars' rotation periods. Further, in addition to the active and inactive branches in the P_{rot}-P_{cyc} diagram proposed by Böhm-Vitense (2007), they detected hints for a possible third branch for short cycles.

With between 16 and 27 years of Johnson B- and V-band photometry from the Automatic Photoelectric Telescope (APT) at the Fairborn Observatory in Arizona, Lehtinen et al. (2016) investigated differential photometry from 21 young solar-type stars. They detected photometric activity cycles in nearly all of the targeted stars. Populating the $\log \frac{P_{rot}}{P_{cyc}}$-$\log Ro^{-1}$ diagram as defined in Saar and Brandenburg (1999), they could confirm the active and transitional activity branches and found that the transitional branch merges with the active branch at $\approx \log Ro^{-1} = 1.42$, similar to what was reported by Boro Saikia et al. (2018). Using time-frequency analysis, Soon et al. (2019) investigated the temporal variations in chromospheric (S index) and photometric (differential photometry in b- and y-bands) of decades-long time series. They found that activity of the young rapidly rotating solar analogue HD 30495 (also see Egeland et al. 2015 for a detailed analysis of this star), as measured with these two time series are strongly correlated. They detected activity cycles in the 'mid-term' regime with a length of 1.6–1.8 yr and confirmed a longer cycle with a period of $\approx$11 yr.

Mathur et al. (2014) defined two simple measures of photometric activity levels: S_{ph} is the standard deviation of the complete time series and thus reflects an average level of activity. With $\langle S_{ph,k} \rangle$, the standard deviation of the time series is calculated over $k \times P_{rot}$. They identified $k = 5$ to be a good value, which smoothes out variations by rotation sufficiently, while still leaving longer-cycle variations intact. These measures are both often-used in the investigation of space-photometry for activity (e.g., Salabert et al. 2016a; Karoff et al. 2018; Mathur et al. 2019). Distefano and Lanzafame (2020) show that stellar activity cycles can also be detected in Gaia photometric time series. For two Gaia targets, they present evidence of cyclic photometric variations with cycle lengths of $P_{cyc} \approx 500$ d for Gaia DR2 2925085041699059712 and of $P_{cyc} = 3262 \pm 125$ d for Gaia DR2 3246069594362282752.

The as-of-yet largest number of detections of photometric activity cycles was achieved by Reinhold et al. (2017). They analysed the long-cadence *Kepler* times series of 23601 stars. As a signature for photometric variability, and hence for varying levels of magnetic activity, they measured the variability amplitude within each *Kepler* quarter (≈ 90 d) as the difference between the 5th and 95th percentiles of the light curve. They found amplitude periodicities in 3203 stars with cycle periods between 0.5 yr $< P_{cyc} <$ 6 yr with stellar rotation periods between 1 d $< P_{rot} <$ 40 d. Interestingly, they confirmed, by folding all of the detected cycles, that the average shape of the stellar activity cycle deviates from a perfect sine, in particular during epochs of maximal and minimal activity. No dependence on stellar effective temperature was detected for this behaviour. The detections are scattered around the inactive (I) branch in the P_{cyc}-P_{rot} diagram (cf., Saar and Brandenburg 1999; Böhm-Vitense 2007) with only few detections on the active (A) and short-cycle (S) branches. The authors propose that this may be due to the strong sensitivity of *Kepler* photometry to spots and plages in the photosphere, while other studies, which have detected the A and S branches, used chromospheric activity indicators.

The *Kepler* satellite recorded 53 full-frame images (FFIs) over the course of its main mission. Montet et al. (2017) used these FFIs to investigate a set of 3845 stars (F7 to G4) for signs of long-term photometric variability. For approximately 10% of their targets, 463 stars, Montet et al. (2017) observed significant ($> 3\sigma$) brightness variations over the *Kepler* mission. By eye, they detected apparently complete cycles for 28 stars. Further, they identified the range of rotation periods during which the transition from spot- to facula-dominated variability occurs, to lie between $15\,\text{d} < P_{\text{rot}} < 25\,\text{d}$. Also, the detected cycles appear to follow the A and I branches in the P_{cyc}-P_{rot} diagram.

Basri and Shah (2020) generated a large number of light curves based on starspot models in order to understand degeneracies in these light curves affected by starspots with varying lifetimes and distributions as well as underlying global and differential rotation. In light of their study, caution must be taken when interpreting short-term cyclic behavior in photometric light curves as the signature of possible activity cycles: such cycle-like behaviour can be the result of random fluctuations. The authors urge to reconsider past identifications of short-term activity cycles, such as those reported by, e.g., Vida et al. (2014), Reinhold et al. (2017), and encourage to scrutinize detections of cycles, which are solely based on variations in photometric time series, more carefully in future analyses.

6 Summary and Future Prospects

In this review, we have presented the latest results on understanding stellar magnetic cycles on stars other than the Sun using the same diagnostics as are commonly used to quantify the solar cycle. Over the 11-year solar cycle, changes in the internal structure of the Sun are observed in acoustic oscillations (p-modes), while on the Sun's surface or the photosphere, the 11-year evolution of activity is evident from the patterns of spot emergence with ever decreasing latitudes as the cycle progresses. At solar activity maximum there is the largest number of starspots, and the Sun's large-scale magnetic field geometry is complex. In the Sun's outer atmospheric layers, for example in the chromosphere and corona, variations in its S index and X-rays are co-incident with activity maximum.

While the spatial resolution and time cadence of stellar observations is several orders of magnitude lower than observations of the Sun, we can use the same multi-wavelength diagnostics to quantify stellar magnetic activity. These are namely Doppler and Zeeman-Doppler imaging to map photospheric features such as dark, bright and magnetic spots, the S index and X-ray observations as diagnostics of the stellar chromosphere and corona. In particular, we now have the capability to observe stars over time bases that are comparable to the length of the solar cycle.

Observations of solar-like cycles have been observed on other stars, where the large-scale magnetic field geometry, including polarity reversals, is co-incident with S index activity minimum and X-ray variations. For stars that have the same mass but are very much younger than the Sun, having just arrived on the zero-age main-sequence, they show more extreme levels of activity without any cyclic behaviour. As the stars age, the variations in magnetic activity start to become more cyclic like what we observe on the Sun, and could even have superimposed cycles. Long timespans of S-index monitoring show that cycles on solar-mass stars older than the Sun start to become lower in amplitude before eventually disappearing as the stars evolve off the main-sequence. This is consistent with the recent results of Brown et al. (2022) where chromospheric activity and variability is shown to decrease together with the toroidal field strength as stars evolve through their main-sequence lifetimes. This is also in agreement with the results of Radick et al. (2018) where they report that young stars have an inverse correlation between photometric brightness and Ca II emission, while more evolved main-sequence stars tend to show a direct correlation.

Looking to the future, more long-term monitoring of the large-scale magnetic field geometry and S-index of stars that are both younger and older than the Sun, but with a similar mass, will provide us with an important insight into exactly when activity patterns start to become more regular and when they disappear. Similarly, the systematic observations of young, middle-aged and old stars with masses lower than the Sun will also provide us with an insight into how crucial stellar mass is for the stars' internal dynamo generation mechanisms.

Asteroseismology of stellar activity and activity cycles will take the next step forward with ESA's exoplanet-finding PLATO mission (Rauer et al. 2014). PLATO will observe two Long-duration Observation Phase (LOP) fields for two years each. The exact position of these fields will be fixed two years before the targeted 2026 launch (Nascimbeni et al. 2022). At least 15000 dwarf and subgiant stars in the spectral type range F5–K7 with magnitudes $V \leq 11$ will be observed in the stellar sample with the highest priority at a time cadence of 25 s (sample P1, see Goupil 2017; Montalto et al. 2021). Another at least 1000 stars (dwarfs and subgiants, F5–K7) with $V \leq 8.5$ will be observed during a LOP (sample P2). This sample's higher brightness increases the feasibility of ground-based follow-up observations. Additional $\geq 245{,}000$ stars (sample P5, dwarfs and subgiants, F5–K7) with $V \leq 13$ will be observed with a lower signal-to-noise ratio than those in P1. The duration of two consecutive years is not optimal for seismic studies of complete activity cycles, as shown in Sect. 4. PLATO will still most likely expand the number of seismic targets with detected signatures of magnetic activity considerably and thus help improve our understanding of stellar dynamos, cycles, and how these depend on fundamental stellar parameters. If the PLATO mission should be extended and the same LOP be revisited, this would substantially increase the potential of the seismic detection and probing of activity cycles.

Funding Open Access funding enabled and organized by Projekt DEAL. SVJ acknowledges the support of the German Science Foundation (DFG) priority program SPP 1992 'Exploring the Diversity of Extrasolar Planets' (JE 701/5-1).

Declarations

Competing Interests There are no conflicting interests to declare.

References

Aguirre VS, Lund MN, Antia HM et al (2017) Standing on the shoulders of dwarfs: the Kepler asteroseismic LEGACY sample II – radii, masses, and ages. Astrophys J 835(2):173. https://doi.org/10.3847/1538-4357/835/2/173. arXiv:1611.08776

Alvarado-Gómez JD, Hussain GAJ, Drake JJ et al (2018) Far beyond the Sun – I. The beating magnetic heart in Horologium. Mon Not R Astron Soc 473(4):4326–4338. https://doi.org/10.1093/mnras/stx2642, arXiv:1710.02438 [astro-ph.SR]

Antia HM, Chitre SM, Thompson MJ (2000) The sun's acoustic asphericity and magnetic fields in the solar convection zone. Astron Astrophys 360:335–344

Aurière M (2003) Stellar polarimetry with NARVAL. Arnaud J, Meunier N (eds) EAS Publications Series, p 105

Auvergne M, Bodin P, Boisnard L et al (2009) The CoRoT satellite in flight: description and performance. Astron Astrophys 506(1):411–424. https://doi.org/10.1051/0004-6361/200810860

Ayres TR (2020) In the trenches of the solar-stellar connection. I. Ultraviolet and X-ray flux-flux correlations across the activity cycles of the Sun and Alpha Centauri AB. Astrophys J Suppl Ser 250(1):16. https://doi.org/10.3847/1538-4365/aba3c6

Baglin A, Auvergne M, Barge P et al (2006) Scientific objectives for a minisat: CoRoT. In: Fridlund M, Baglin A, Lochard J et al (eds) The CoRoT mission pre-launch status – stellar seismology and planet finding, Noordwijk, p 33

Baldner CS, Antia HM, Basu S et al (2009) Solar magnetic field signatures in helioseismic splitting coefficients. Astrophys J 705(2):1704–1713. https://doi.org/10.1088/0004-637X/705/2/1704

Baliunas SL, Vaughan AH (1985) Stellar activity cycles. Annu Rev Astron Astrophys 23:379–412. https://doi.org/10.1146/annurev.aa.23.090185.002115

Baliunas SL, Hartmann L, Noyes RW et al (1983) Stellar rotation in lower main-sequence stars measured from time variations in H and K emission-line fluxes. II. Detailed analysis of the 1980 observing season data. Astrophys J 275:752–772. https://doi.org/10.1086/161572

Baliunas SL, Donahue RA, Soon WH et al (1995) Chromospheric variations in main-sequence stars. II. Astrophys J 438:269. https://doi.org/10.1086/175072

Ball WH, Chaplin WJ, Schofield M et al (2018) A synthetic sample of short-cadence solar-like oscillators for TESS. Astrophys J Suppl Ser 239(2):34. https://doi.org/10.3847/1538-4365/aaedbc

Basri G, Shah R (2020) The information content in analytic spot models of broadband precision light curves. II. Spot distributions and lifetimes and global and differential rotation. Astrophys J 901(1):14. https://doi.org/10.3847/1538-4357/abae5d. arXiv:2008.04969 [astro-ph.SR]

Basu S (2016) Global seismology of the Sun. Living Rev Sol Phys 13:2. https://doi.org/10.1007/s41116-016-0003-4

Basu S, Broomhall AM, Chaplin WJ et al (2012) Thinning of the Sun's magnetic layer: the peculiar solar minimum could have been predicted. Astrophys J 758(1):43. https://doi.org/10.1088/0004-637X/758/1/43

Baum AC, Wright JT, Luhn JK et al (2022) Five decades of chromospheric activity in 59 Sun-like stars and new Maunder minimum candidate HD 166620. Astron J 163(4):183. https://doi.org/10.3847/1538-3881/ac5683. arXiv:2203.13376 [astro-ph.SR]

Bazot M, Nielsen MB, Mary D et al (2018) Butterfly diagram of a Sun-like star observed using asteroseismology. Astron Astrophys 619:L9. https://doi.org/10.1051/0004-6361/201834251. arXiv:1810.08630 [astro-ph.SR]

Böhm-Vitense E (2007) Chromospheric activity in G and K main-sequence stars, and what it tells us about stellar dynamos. Astrophys J 657(1):486–493. https://doi.org/10.1086/510482

Boro Saikia S, Jeffers SV, Petit P et al (2015) Variable magnetic field geometry of the young sun HN Pegasi (HD 206860). Astron Astrophys 573:A17. https://doi.org/10.1051/0004-6361/201424096. arXiv:1410.8307 [astro-ph.SR]

Boro Saikia S, Jeffers SV, Morin J et al (2016) A solar-like magnetic cycle on the mature K-dwarf 61 Cygni A (HD 201091). Astron Astrophys 594:A29. https://doi.org/10.1051/0004-6361/201628262. arXiv:1606.01032 [astro-ph.SR]

Boro Saikia S, Lueftinger T, Jeffers SV et al (2018) Direct evidence of a full dipole flip during the magnetic cycle of a Sun-like star. Astron Astrophys 620:L11. https://doi.org/10.1051/0004-6361/201834347. arXiv:1811.11671 [astro-ph.SR]

Boro Saikia S, Lüftinger T, Folsom CP et al (2022) Time evolution of magnetic activity cycles in young suns: the curious case of κ ceti. Astron Astrophys 658:A16. https://doi.org/10.1051/0004-6361/202141525. arXiv:2110.06000 [astro-ph.SR]

Borucki WJ, Koch D, Basri G et al (2010) Kepler planet-detection mission: introduction and first results. Science 327(5968):977–980. https://doi.org/10.1126/science.1185402

Brandenburg A, Saar SH, Turpin CR (1998) Time evolution of the magnetic activity cycle period. Astrophys J Lett 498(1):L51–L54. https://doi.org/10.1086/311297

Brandenburg A, Mathur S, Metcalfe TS (2017) Evolution of co-existing long and short period stellar activity cycles. Astrophys J 845(1):79. https://doi.org/10.3847/1538-4357/aa7cfa. arXiv:1704.09009 [astro-ph.SR]

Broomhall AM (2017) A helioseismic perspective on the depth of the minimum between solar cycles 23 and 24. Sol Phys 292:67. https://doi.org/10.1007/s11207-017-1068-5

Broomhall AM, Chatterjee P, Howe R et al (2014) The Sun's interior structure and dynamics, and the solar cycle. Space Sci Rev 186:191–225. https://doi.org/10.1007/s11214-014-0101-3

Broomhall AM, Pugh C, Nakariakov V (2015) Solar cycle variations in the powers and damping rates of low-degree solar acoustic oscillations. Adv Space Res 56(12):2706–2712. https://doi.org/10.1016/j.asr.2015.04.018

Brown EL, Marsden SC, Mengel MW et al (2021) Magnetic field and chromospheric activity evolution of HD 75332: a rapid magnetic cycle in an F star without a hot Jupiter. Mon Not R Astron Soc 501(3):3981–4003. https://doi.org/10.1093/mnras/staa3878, arXiv:2012.05407 [astro-ph.SR]

Brown EL, Jeffers SV, Marsden SC et al (2022) Linking chromospheric activity and magnetic field properties for late-type dwarf stars. Mon Not R Astron Soc 514(3):4300–4319. https://doi.org/10.1093/mnras/stac1291, arXiv:2205.03108 [astro-ph.SR]

Brun AS, Strugarek A, Noraz Q et al (2022) Powering stellar magnetism: energy transfers in cyclic dynamos of Sun-like stars. Astrophys J 926(1):21. https://doi.org/10.3847/1538-4357/ac469b. arXiv:2201.13218 [astro-ph.SR]

Bugnet L (2022) Magnetic signatures on mixed-mode frequencies. II. Period spacings as a probe of the internal magnetism of red giants. Astron Astrophys 667:A68. https://doi.org/10.1051/0004-6361/202243167. arXiv:2208.14954 [astro-ph.SR]

Bugnet L, Prat V, Mathis S et al (2021) Magnetic signatures on mixed-mode frequencies. I. An axisymmetric fossil field inside the core of red giants. Astron Astrophys 650:A53. https://doi.org/10.1051/0004-6361/202039159. arXiv:2102.01216 [astro-ph.SR]

Burtseva O, Hill F, Kholikov S et al (2009) Lifetimes of high-degree p modes in the quiet and active sun. Sol Phys 258(1):1–11. https://doi.org/10.1007/s11207-009-9399-5. https://arxiv.org/abs/0902.2016

Cameron R, Schüssler M (2023) Observationally guided models for the solar dynamo and the role of the surface field. Space Sci Rev 219. arXiv:2305.02253 [astro-ph.SR]

Cameron RH, Duvall TL, Schüssler M et al (2018) Observing and modeling the poloidal and toroidal fields of the solar dynamo. Astron Astrophys 609:A56. https://doi.org/10.1051/0004-6361/201731481. arXiv:1710.07126 [astro-ph.SR]

Campante TL, Lund MN, Kuszlewicz JS et al (2016a) Spin-orbit alignment of exoplanet systems: ensemble analysis using asteroseismology. Astrophys J 819(1):85. https://doi.org/10.3847/0004-637x/819/1/85. arXiv:1601.06052

Campante TL, Schofield M, Kuszlewicz JS et al (2016b) The asteroseismic potential of TESS: exoplanet-host stars. Astrophys J 830(2):138. https://doi.org/10.3847/0004-637X/830/2/138

Catala C, Donati JF, Shkolnik E et al (2007) The magnetic field of the planet-hosting star τ Bootis. Mon Not R Astron Soc 374(1):L42–L46. https://doi.org/10.1111/j.1745-3933.2006.00261.x. arXiv:astro-ph/0610758 [astro-ph]

Chaplin WJ, Basu S (2014) Inferences on stellar activity and stellar cycles from asteroseismology. Space Sci Rev 186(1–4):437–456. https://doi.org/10.1007/s11214-014-0090-2

Chaplin WJ, Elsworth Y, Isaak GR et al (2000) Variations in the excitation and damping of low- solar p modes over the solar activity cycle. Mon Not R Astron Soc 313(1):32–42. https://doi.org/10.1046/j.1365-8711.2000.03176.x

Chaplin W, Appourchaux T, Elsworth Y et al (2001) The phenomenology of solar-cycle-induced acoustic eigenfrequency variations: a comparative and complementary analysis of GONG, BiSON and VIRGO/LOI data. Mon Not R Astron Soc 324(4):910–916. https://doi.org/10.1046/j.1365-8711.2001.04357.x

Chaplin WJ, Elsworth Y, Houdek G et al (2007) On prospects for sounding activity cycles of Sun-like stars with acoustic modes. Mon Not R Astron Soc 377(1):17–29. https://doi.org/10.1111/j.1365-2966.2007.11581.x

Chaplin WJ, Bedding TR, Bonanno A et al (2011a) Evidence for the impact of stellar activity on the detectability of solar-like oscillations observed by Kepler. Astrophys J Lett 732(1):L5. https://doi.org/10.1088/2041-8205/732/1/L5

Chaplin WJ, Kjeldsen H, Bedding TR et al (2011b) Predicting the detectability of oscillations in solar-type stars observed by Kepler. Astrophys J 732(1):54. https://doi.org/10.1088/0004-637X/732/1/54

Charbonneau P (2020) Dynamo models of the solar cycle. Living Rev Sol Phys 17(1):4. https://doi.org/10.1007/s41116-020-00025-6

Christensen-Dalsgaard J, Berthomieu G (1991) Theory of solar oscillations. In: Cox AN, Livingston WC, Matthews M (eds) Solar interior and atmosphere. University of Arizona Press, Tucson, pp 401–478

Coffaro M, Stelzer B, Orlando S et al (2020) An X-ray activity cycle on the young solar-like star ϵ Eridani. Astron Astrophys 636:A49. https://doi.org/10.1051/0004-6361/201936479. arXiv:2002.11009 [astro-ph.SR]

Das SB, Chakraborty T, Hanasoge S et al (2020) Sensitivity kernels for inferring Lorentz stresses from normal-mode frequency splittings in the Sun. Astrophys J 897(1):38. https://doi.org/10.3847/1538-4357/ab8e3a

Deheuvels S, Li G, Ballot J et al (2023) Strong magnetic fields detected in the cores of 11 red giant stars using gravity-mode period spacings. Astron Astrophys 670:L16. https://doi.org/10.1051/0004-6361/202245282, arXiv:2301.01308 [astro-ph.SR]

 Springer

Dhouib H, Mathis S, Bugnet L et al (2022) Detecting deep axisymmetric toroidal magnetic fields in stars – the traditional approximation of rotation for differentially rotating deep spherical shells with a general azimuthal magnetic field. Astron Astrophys 661:A133. https://doi.org/10.1051/0004-6361/202142956. arXiv:2202.10026

Distefano E, Lanzafame A (2020) Detection and characterization of stellar magnetic activity with Gaia. Astron Nachr 341(5):508–512. https://doi.org/10.1002/asna.202013732

Donahue RA, Saar SH, Baliunas SL (1996) A relationship between mean rotation period in lower main-sequence stars and its observed range. Astrophys J 466:384. https://doi.org/10.1086/177517

Donati JF, Semel M, Carter BD et al (1997) Spectropolarimetric observations of active stars. Mon Not R Astron Soc 291(4):658–682. https://doi.org/10.1093/mnras/291.4.658

Donati JF, Collier Cameron A, Semel M et al (2003) Dynamo processes and activity cycles of the active stars AB Doradus, LQ Hydrae and HR 1099. Mon Not R Astron Soc 345(4):1145–1186. https://doi.org/10.1046/j.1365-2966.2003.07031.x

Donati JF, Catala C, Landstreet JD et al (2006) ESPaDOnS: the new generation stellar spectro-polarimeter. Performances and first results. In: Casini R, Lites BW (eds) Solar polarization 4. ASP Conference Series, vol 358. Astronomical Society of the Pacific., San Francisco, pp 362–368

Donati JF, Howarth ID, Jardine MM et al (2006b) The surprising magnetic topology of τ Sco: fossil remnant or dynamo output? Mon Not R Astron Soc 370(2):629–644. https://doi.org/10.1111/j.1365-2966.2006.10558.x. arXiv:astro-ph/0606156 [astro-ph]

Donati JF, Moutou C, Farès R et al (2008) Magnetic cycles of the planet-hosting star τ Bootis. Mon Not R Astron Soc 385(3):1179–1185. https://doi.org/10.1111/j.1365-2966.2008.12946.x. arXiv:0802.1584 [astro-ph]

Dziembowski WA, Goode PR (2004) Helioseismic probing of solar variability: the formalism and simple assessments. Astrophys J 600(1):464–479. https://doi.org/10.1086/379708, arXiv:astro-ph/0310095

Egeland R, Metcalfe TS, Hall JC et al (2015) Sun-like magnetic cycles in the rapidly-rotating young solar analog HD 30495. Astrophys J 812(1):12. https://doi.org/10.1088/0004-637X/812/1/12. arXiv:1507.03611 [astro-ph.SR]

Elsworth Y, Howe R, Isaak GR et al (1990) Variation of low-order acoustic solar oscillations over the solar cycle. Nature 345(6273):322–324. https://doi.org/10.1038/345322a0

Fares R, Donati JF, Moutou C et al (2009) Magnetic cycles of the planet-hosting star τ Bootis – II. a second magnetic polarity reversal. Mon Not R Astron Soc 398(3):1383–1391. https://doi.org/10.1111/j.1365-2966.2009.15303.x. arXiv:0906.4515 [astro-ph.SR]

Fares R, Moutou C, Donati JF et al (2013) A small survey of the magnetic fields of planet-host stars. Mon Not R Astron Soc 435(2):1451–1462. https://doi.org/10.1093/mnras/stt1386, arXiv:1307.6091 [astro-ph.SR]

Favata F, Micela G, Orlando S et al (2008) The X-ray cycle in the solar-type star HD 81809. XMM-Newton observations and implications for the coronal structure. Astron Astrophys 490(3):1121–1126. https://doi.org/10.1051/0004-6361:200809694. arXiv:0806.2279 [astro-ph]

Ferreira Lopes CE, Leão IC, de Freitas DB et al (2015) Stellar cycles from photometric data: CoRoT stars. Astron Astrophys 583:A134. https://doi.org/10.1051/0004-6361/201424900. arXiv:1508.06194 [astro-ph.SR]

Folsom CP, Bouvier J, Petit P et al (2018) The evolution of surface magnetic fields in young solar-type stars II: the early main sequence (250-650 Myr). Mon Not R Astron Soc 474(4):4956–4987. https://doi.org/10.1093/mnras/stx3021 arXiv:1711.08636 [astro-ph.SR]

Fuhrmeister B, Coffaro M, Stelzer B et al (2023) A multi-wavelength view of the multiple activity cycles of ε Eridani. Astron Astrophys 672:A149. https://doi.org/10.1051/0004-6361/202245201. arXiv:2303.08487 [astro-ph.SR]

García RA, Mathur S, Salabert D et al (2010) CoRoT reveals a magnetic activity cycle in a Sun-like star. Science 329(5995):1032. https://doi.org/10.1126/science.1191064

Garg S, Karak BB, Egeland R et al (2019) Waldmeier effect in stellar cycles. Astrophys J 886(2):132. https://doi.org/10.3847/1538-4357/ab4a17. arXiv:1909.12148 [astro-ph.SR]

Gomes P, Lopes I (2020) Core magnetic field imprint in the non-radial oscillations of red giant stars. Mon Not R Astron Soc 496(1):620–628. https://doi.org/10.1093/mnras/staa1585, arXiv:2007.09632 [astro-ph.SR]

González-Pérez JN, Mittag M, Schmitt JHMM et al (2022) Eight years of TIGRE robotic spectroscopy: operational experience and selected scientific results. Front Astron Space Sci 9:912546. https://doi.org/10.3389/fspas.2022.912546, arXiv:2206.02832 [astro-ph.IM]

Gough DO, Thompson MJ (1990) The effect of rotation and a buried magnetic field on stellar oscillations. Mon Not R Astron Soc 242(1):25–55. https://doi.org/10.1093/mnras/242.1.25

Goupil M (2017) Expected asteroseismic performances with the space project PLATO. EPJ Web Conf 160:01003. https://doi.org/10.1051/epjconf/201716001003

Hackman T, Mantere MJ, Lindborg M et al (2012) Doppler images of II Pegasi for 2004-2010. Astron Astrophys 538:A126. https://doi.org/10.1051/0004-6361/201117603. arXiv:1106.6237 [astro-ph.SR]

Hale GE (1908) On the probable existence of a magnetic field in sun-spots. Astrophys J 28:315. https://doi.org/10.1086/141602

Hall JC, Lockwood GW, Skiff BA (2007) The activity and variability of the Sun and Sun-like stars. I. Synoptic Ca II H and K observations. Astron J 133(3):862–881. https://doi.org/10.1086/510356

Hanasoge SM (2017) Seismic sensitivity of normal-mode coupling to Lorentz stresses in the Sun. Mon Not R Astron Soc 470(3):2780–2790. https://doi.org/10.1093/mnras/stx1342, arXiv:1705.09431

Hathaway DH (2015) The solar cycle. Living Rev Sol Phys 12(1):4. https://doi.org/10.1007/lrsp-2015-4. arXiv:1502.07020 [astro-ph.SR]

Hekker S, Christensen-Dalsgaard J (2017) Giant star seismology. Astron Astrophys Rev 25:1. https://doi.org/10.1007/s00159-017-0101-x

Hempelmann A, Robrade J, Schmitt JHMM et al (2006) Coronal activity cycles in 61 Cygni. Astron Astrophys 460(1):261–267. https://doi.org/10.1051/0004-6361:20065459

Howe R, Komm R, Hill F (1999) Solar cycle changes in GONG P-mode frequencies, 1995-1998. Astrophys J 524(2):1084–1095. https://doi.org/10.1086/307851

Howe R, Chaplin WJ, Basu S et al (2020) Solar cycle variation of ν_{max} in helioseismic data and its implications for asteroseismology. Mon Not R Astron Soc Lett 493(1):L49–L53. https://doi.org/10.1093/MNRASL/SLAA006

Jefferies SM, Pomerantz MA, Duvall TLJ et al (1991) Characteristics of intermediate-degree solar p-mode line widths. Astrophys J 377:330–336. https://doi.org/10.1086/170362

Jeffers SV, Donati JF (2008) High levels of surface differential rotation on the young G0 dwarf HD171488. Mon Not R Astron Soc 390(2):635–644. https://doi.org/10.1111/j.1365-2966.2008.13695.x

Jeffers SV, Donati J-F, Collier Cameron A (2007) Magnetic activity on AB Doradus: temporal evolution of star-spots and differential rotation from 1988 to 1994. Mon Not R Astron Soc 375(2):567–583. https://doi.org/10.1111/j.1365-2966.2006.11154.x

Jeffers SV, Donati JF, Alecian E et al (2011) Observations of non-solar-type dynamo processes in stars with shallow convective zones. Mon Not R Astron Soc 411(2):1301–1312. https://doi.org/10.1111/j.1365-2966.2010.17762.x

Jeffers SV, Petit P, Marsden SC et al (2014) ϵ Eridani: an active K dwarf and a planet hosting star? The variability of its large-scale magnetic field topology. Astron Astrophys 569:A79. https://doi.org/10.1051/0004-6361/201423725

Jeffers SV, Mengel M, Moutou C et al (2018) The relation between stellar magnetic field geometry and chromospheric activity cycles – II the rapid 120-day magnetic cycle of τ Bootis. Mon Not R Astron Soc 479(4):5266–5271. https://doi.org/10.1093/mnras/sty1717 arXiv:1805.09769 [astro-ph.SR]

Jeffers SV, Cameron RH, Marsden SC et al (2022) The crucial role of surface magnetic fields for stellar dynamos: ϵ Eridani, 61 Cygni A, and the Sun. Astron Astrophys 661:A152. https://doi.org/10.1051/0004-6361/202142202. arXiv:2201.07530 [astro-ph.SR]

Jiménez A, Roca Cortés TR, Jiménez-Reyes J (2002) Variation of the low-degree solar acoustic mode parameters over the solar cycle. Sol Phys 209(2):247–263. https://doi.org/10.1023/A:1021226503589

Jiménez-Reyes SJ, Régulo C, Pallé PL et al (1998) Solar activity cycle frequency shifts of low-degree p-modes. Astron Astrophys 329:1119–1124

Jiménez-Reyes SJ, García RA, Jiménez A et al (2003) Excitation and damping of low-degree solar p-modes during activity cycle 23: analysis of GOLF and VIRGO sun photometer data. Astrophys J 595(1):446–457. https://doi.org/10.1086/377304

Jiménez-Reyes SJ, Chaplin WJ, Elsworth Y et al (2004) Tracing the "acoustic" solar cycle: a direct comparison of BiSON and GOLF low-l p-mode variations. Astrophys J 604(2):969–976. https://doi.org/10.1086/381936

Judge PG, Egeland R, Metcalfe TS et al (2017) The magnetic future of the Sun. Astrophys J 848(1):43. https://doi.org/10.3847/1538-4357/aa8d6a. arXiv:1710.05088 [astro-ph.SR]

Käpylä PJ, Browning MK, Brun AS et al (2023) Simulations of solar and stellar dynamos and their theoretical interpretation. Space Sci Rev 219. arXiv:2305.16790 [astro-ph.SR]

Karoff C, Metcalfe TS, Chaplin WJ et al (2009) Sounding stellar cycles with Kepler – I. Strategy for selecting targets. Mon Not R Astron Soc 399(2):914–923. https://doi.org/10.1111/j.1365-2966.2009.15323.x

Karoff C, Metcalfe TS, Chaplin WJ et al (2013) Sounding stellar cycles with Kepler – II. Ground-based observations. Mon Not R Astron Soc 433(4):3227–3238. https://doi.org/10.1093/mnras/stt964

Karoff C, Metcalfe TS, Santos ÂRG et al (2018) The influence of metallicity on stellar differential rotation and magnetic activity. Astrophys J 852(1):46. https://doi.org/10.3847/1538-4357/aaa026

Karoff C, Metcalfe TS, Montet BT et al (2019) Sounding stellar cycles with Kepler – III. Comparative analysis of chromospheric, photometric and asteroseismic variability. Mon Not R Astron Soc 485(4):5096–5104. https://doi.org/10.1093/mnras/stz782. arXiv:1902.02172

Kiefer R, Broomhall AM (2021) They do change after all: 25 yr of GONG data reveal variation of p-mode energy supply rates. Mon Not R Astron Soc 500(3):3095–3110. https://doi.org/10.1093/mnras/staa3198, arXiv:2010.06287 [astro-ph.SR]

Kiefer R, Roth M (2018) The effect of toroidal magnetic fields on solar oscillation frequencies. Astrophys J 854(1):74. https://doi.org/10.3847/1538-4357/aaa3f7

Kiefer R, Schad A, Davies G et al (2017a) Stellar magnetic activity and variability of oscillation parameters: an investigation of 24 solar-like stars observed by Kepler. Astron Astrophys 598:A77. https://doi.org/10.1051/0004-6361/201628469

Kiefer R, Schad A, Roth M (2017b) The direct effect of toroidal magnetic fields on stellar oscillations: an analytical expression for the general matrix element. Astrophys J 846(2):162. https://doi.org/10.3847/1538-4357/aa8634

Kiefer R, Komm R, Hill F et al (2018) GONG p-mode parameters through two solar cycles. Sol Phys 293(11):151. https://doi.org/10.1007/s11207-018-1370-x

Kiefer R, Broomhall AM, Ball WH (2019) Seismic signatures of stellar magnetic activity—what can we expect from TESS? Front Astron Space Sci 6:52. https://doi.org/10.3389/fspas.2019.00052

Koch DG, Borucki WJ, Basri G et al (2010) Kepler mission design, realized photometric performance, and early science. Astrophys J Lett 713(2):L79–L86. https://doi.org/10.1088/2041-8205/713/2/L79

Kochukhov O, Makaganiuk V, Piskunov N (2010) Least-squares deconvolution of the stellar intensity and polarization spectra. Astron Astrophys 524:A5. https://doi.org/10.1051/0004-6361/201015429. arXiv:1008.5115 [astro-ph.SR]

Komm RW, Howe R, Hill F (2000) Solar-cycle changes in gong p-mode widths and amplitudes 1995–1998. Astrophys J 531(2):1094–1108. https://doi.org/10.1086/308518

Komm RW, Howe R, Hill F (2000b) Width and energy of solar p-modes observed by global oscillation network group. Astrophys J 543(1):472–485. https://doi.org/10.1086/317101

Kosak K, Kiefer R, Broomhall AM (2022) A multi-instrument investigation of the frequency stability of oscillations above the acoustic cut-off frequency with solar activity. Mon Not R Astron Soc 512(4):5743–5754. https://doi.org/10.1093/mnras/stac647 arXiv:2203.03685 [astro-ph.SR]

Lavely EM, Ritzwoller MH (1992) The effect of global-scale, steady-state convection and elastic-gravitational asphericities on helioseismic oscillations. Philos Trans R Soc Lond A, Math Phys Eng Sci 339(1655):431–496. https://doi.org/10.1098/rsta.1992.0048

Lehtinen J, Jetsu L, Hackman T et al (2016) Activity trends in young solar-type stars. Astron Astrophys 588:A38. https://doi.org/10.1051/0004-6361/201527420. arXiv:1509.06606 [astro-ph.SR]

Lehtinen JJ, Käpylä MJ, Hackman T et al (2022) Topological changes in the magnetic field of LQ Hya during an activity minimum. Astron Astrophys 660:A141. https://doi.org/10.1051/0004-6361/201936780. arXiv:1909.11028 [astro-ph.SR]

Leighton RB (1959) Observations of solar magnetic fields in plage regions. Astrophys J 130:366. https://doi.org/10.1086/146727

Li G, Deheuvels S, Ballot J et al (2022) Magnetic fields of 30 to 100 kG in the cores of red giant stars. Nature 610(7930):43–46. https://doi.org/10.1038/s41586-022-05176-0. arXiv:2208.09487 [astro-ph.SR]

Libbrecht KG, Woodard MF (1990) Solar-cycle effects on solar oscillation frequencies. Nature 345(6278):779–782. https://doi.org/10.1038/345779a0

Loi ST (2020) Effect of a strong magnetic field on gravity-mode period spacings in red giant stars. Mon Not R Astron Soc 496(3):3829–3840. https://doi.org/10.1093/MNRAS/STAA1823, arXiv:2006.08635

Loi ST (2021) Topology and obliquity of core magnetic fields in shaping seismic properties of slowly rotating evolved stars. Mon Not R Astron Soc 504(3):3711–3729. https://doi.org/10.1093/mnras/stab991, arXiv:2104.03112 [astro-ph.SR]

Lovis C, Dumusque X, Santos NC et al (2011) The HARPS search for southern extra-solar planets. XXXI. Magnetic activity cycles in solar-type stars: statistics and impact on precise radial velocities. arXiv e-prints arXiv:1107.5325 [astro-ph.SR]

Luhn JK, Wright JT, Henry GW et al (2022) HD 166620: portrait of a star entering a grand magnetic minimum. Astrophys J Lett 936(2):L23. https://doi.org/10.3847/2041-8213/ac8b13. arXiv:2207.00612 [astro-ph.SR]

Lund MN, Aguirre VS, Davies GR et al (2017) Standing on the shoulders of dwarfs: the Kepler asteroseismic LEGACY sample. I. Oscillation mode parameters. Astrophys J 835(2):172. https://doi.org/10.3847/1538-4357/835/2/172

Marsden SC, Donati JF, Semel M et al (2006) Surface differential rotation and photospheric magnetic field of the young solar-type star HD 171488 (V889 Her). Mon Not R Astron Soc 370(1):468–476. https://doi.org/10.1111/j.1365-2966.2006.10503.x

Marsden SC, Evensberget D, Brown EL et al (2023) The magnetic field and stellar wind of the mature late-F star χ Draconis A. Mon Not R Astron Soc 522(1):792–810. https://doi.org/10.1093/mnras/stad925

Mathis S, Bugnet L, Prat V et al (2021) Probing the internal magnetism of stars using asymptotic magneto-asteroseismology. Astron Astrophys 647:A122. https://doi.org/10.1051/0004-6361/202039180. arXiv:2012.11050 [astro-ph.SR]

Mathur S, Salabert D, García RA et al (2014) Photometric magnetic-activity metrics tested with the Sun: application to Kepler M dwarfs. J Space Weather Space Clim 4:A15. https://doi.org/10.1051/swsc/2014011, arXiv:1404.3076 [astro-ph.SR]

Mathur S, García RA, Bugnet L et al (2019) Revisiting the impact of stellar magnetic activity on the detectability of solar-like oscillations by Kepler. Front Astron Space Sci 6:46. https://doi.org/10.3389/fspas.2019.00046, arXiv:1907.01415

Maunder EW (1904) Note on the distribution of Sun-spots in Heliographic Latitude, 1874-1902. Mon Not R Astron Soc 64:747–761. https://doi.org/10.1093/mnras/64.8.747

Mengel MW, Fares R, Marsden SC et al (2016) The evolving magnetic topology of τ Boötis. Mon Not R Astron Soc 459(4):4325–4342. https://doi.org/10.1093/mnras/stw828, arXiv:1604.02501 [astro-ph.SR]

Metcalfe TS, van Saders J (2017) Magnetic evolution and the disappearance of Sun-like activity cycles. Sol Phys 292(9):126. https://doi.org/10.1007/s11207-017-1157-5. arXiv:1705.09668 [astro-ph.SR]

Metcalfe TS, Dziembowski WA, Judge PG et al (2007) Asteroseismic signatures of stellar magnetic activity cycles. Mon Not R Astron Soc Lett 379(1):L16–L20. https://doi.org/10.1111/j.1745-3933.2007.00325.x

Metcalfe TS, Judge PG, Basu S et al (2009) Activity cycles of Southern Asteroseismic Targets. arXiv e-prints arXiv:0909.5464 [astro-ph.SR]

Metcalfe TS, Basu S, Henry TJ et al (2010) Discovery of a 1.6 year magnetic activity cycle in the exoplanet host star ι horologii. Astrophys J Lett 723(2):L213–L217. https://doi.org/10.1088/2041-8205/723/2/L213. arXiv:1009.5399 [astro-ph.SR]

Metcalfe TS, Buccino AP, Brown BP et al (2013) Magnetic activity cycles in the exoplanet host star epsilon Eridani. Astrophys J Lett 763(2):L26. https://doi.org/10.1088/2041-8205/763/2/L26. arXiv:1212.4425 [astro-ph.SR]

Metcalfe TS, Egeland R, van Saders J (2016) Stellar evidence that the solar dynamo may be in transition. Astrophys J Lett 826(1):L2. https://doi.org/10.3847/2041-8205/826/1/L2. arXiv:1606.01926 [astro-ph.SR]

Metcalfe TS, van Saders JL, Basu S et al (2020) The evolution of rotation and magnetic activity in 94 Aqr Aa from asteroseismology with TESS. Astrophys J 900(2):154. https://doi.org/10.3847/1538-4357/aba963. arXiv:2007.12755 [astro-ph.SR]

Metcalfe TS, Finley AJ, Kochukhov O et al (2022) The origin of weakened magnetic braking in old solar analogs. Astrophys J Lett 933(1):L17. https://doi.org/10.3847/2041-8213/ac794d. arXiv:2206.08540 [astro-ph.SR]

Mittag M, Schmitt JHMM, Schröder KP (2023) Revisiting the cycle-rotation connection for late-type stars. Astron Astrophys 674:A116. https://doi.org/10.1051/0004-6361/202245060. arXiv:2306.05866 [astro-ph.SR]

Montalto M, Piotto G, Marrese PM et al (2021) The all-sky PLATO input catalogue. Astron Astrophys 653:A98. https://doi.org/10.1051/0004-6361/202140717. arXiv:2108.13712 [astro-ph.EP]

Montet BT, Tovar G, Foreman-Mackey D (2017) Long-term photometric variability in Kepler full-frame images: magnetic cycles of Sun-like stars. Astrophys J 851(2):116. https://doi.org/10.3847/1538-4357/aa9e00. arXiv:1705.07928 [astro-ph.SR]

Moreno-Insertis F, Solanki SK (2000) Distribution of magnetic flux on the solar surface and low-degree p-modes. Mon Not R Astron Soc 313(2):411–422. https://doi.org/10.1046/j.1365-8711.2000.03246.x

Morgenthaler A, Petit P, Morin J et al (2011) Direct observation of magnetic cycles in Sun-like stars. Astron Nachr 332:866. https://doi.org/10.1002/asna.201111592, arXiv:1109.3982 [astro-ph.SR]

Nascimbeni V, Piotto G, Börner A et al (2022) The PLATO field selection process. I. Identification and content of the long-pointing fields. Astron Astrophys 658:A31. https://doi.org/10.1051/0004-6361/202142256. arXiv:2110.13924 [astro-ph.EP]

Noyes RW, Hartmann LW, Baliunas SL et al (1984) Rotation, convection, and magnetic activity in lower main-sequence stars. Astrophys J 279(1):763–777. https://doi.org/10.1086/161945

Oláh K, Strassmeier KG (2002) Starspot cycles from long-term photometry. Astron Nachr 323:361–366. https://doi.org/10.1002/1521-3994(200208)323:3/4<361::AID-ASNA361>3.0.CO;2-1

Oláh K, Kolláth Z, Strassmeier KG (2000) Multiperiodic light variations of active stars. Astron Astrophys 356:643–653

Oláh K, Kolláth Z, Granzer T et al (2009) Multiple and changing cycles of active stars. II. Results. Astron Astrophys 501(2):703–713. https://doi.org/10.1051/0004-6361/200811304. arXiv:0904.1747 [astro-ph.SR]

Oláh K, Kővári Z, Petrovay K et al (2016) Magnetic cycles at different ages of stars. Astron Astrophys 590:A133. https://doi.org/10.1051/0004-6361/201628479. arXiv:1604.06701 [astro-ph.SR]

Orlando S, Favata F, Micela G et al (2017) Fifteen years in the high-energy life of the solar-type star HD 81809. XMM-Newton observations of a stellar activity cycle. Astron Astrophys 605:A19. https://doi. org/10.1051/0004-6361/201731301. arXiv:1707.06437 [astro-ph.SR]

Pérez Hernández F, García RA, Mathur S et al (2019) Influence of magnetic activity on the determination of stellar parameters through asteroseismology. Front Astron Space Sci 6:41. https://doi.org/10.3389/ fspas.2019.00041, arXiv:1906.10569

Petit P, Dintrans B, Morgenthaler A et al (2009) A polarity reversal in the large-scale magnetic field of the rapidly rotating sun HD 190771. Astron Astrophys 508(1):L9–L12. https://doi.org/10.1051/0004-6361/ 200913285. arXiv:0909.2200 [astro-ph.SR]

Petit P, Folsom CP, Donati JF et al (2021) Multi-instrumental view of magnetic fields and activity of ϵ Eridani with SPIRou, NARVAL, and TESS. Astron Astrophys 648:A55. https://doi.org/10.1051/0004-6361/202040027. arXiv:2101.02643 [astro-ph.SR]

Piskunov N, Snik F, Dolgopolov A et al (2011) HARPSpol — the new polarimetric mode for HARPS. Messenger 143:7–10

Prat V, Mathis S, Neiner C et al (2020) Period spacings of gravity modes in rapidly rotating magnetic stars. Astron Astrophys 636:A100. https://doi.org/10.1051/0004-6361/201937398

Radick RR, Lockwood GW, Henry GW et al (2018) Patterns of variation for the Sun and Sun-like stars. Astrophys J 855(2):75. https://doi.org/10.3847/1538-4357/aaaae3

Rauer H, Catala C, Aerts C et al (2014) The PLATO 2.0 mission. Exp Astron 38(1-2):249–330. https://doi. org/10.1007/s10686-014-9383-4. arXiv:1310.0696 [astro-ph.EP]

Reinhold T, Cameron RH, Gizon L (2017) Evidence for photometric activity cycles in 3203 Kepler stars. Astron Astrophys 603:A52. https://doi.org/10.1051/0004-6361/201730599. arXiv:1705.03312 [astro-ph.SR]

Ricker GR, Winn JN, Vanderspek R et al (2014) Transiting Exoplanet Survey Satellite. Proc SPIE 9143:914320. https://doi.org/10.1117/1.JATIS.1.1.014003, arXiv:1406.0151

Robrade J, Schmitt JHMM, Favata F (2012) Coronal activity cycles in nearby G and K stars. XMM-Newton monitoring of 61 Cygni and α Centauri. Astron Astrophys 543:A84. https://doi.org/10.1051/0004-6361/ 201219046. arXiv:1205.3627 [astro-ph.SR]

Saar SH, Brandenburg A (1999) Time evolution of the magnetic activity cycle period. II. Results for an expanded stellar sample. Astrophys J 524(1):295–310. https://doi.org/10.1086/307794

Salabert D, Fossat E, Gelly B et al (2004) Solar p modes in 10 years of the IRIS network. Astron Astrophys 413(3):1135–1142. https://doi.org/10.1051/0004-6361:20031541

Salabert D, Régulo C, Ballot J et al (2011) About the p-mode frequency shifts in HD 49933. Astron Astrophys 530:A127. https://doi.org/10.1051/0004-6361/201116633

Salabert D, García RA, Turck-Chièze S (2015) Seismic sensitivity to sub-surface solar activity from 18 yr of GOLF/SoHO observations. Astron Astrophys 578:A137. https://doi.org/10.1051/0004-6361/ 201425236

Salabert D, García RA, Beck PG et al (2016a) Photospheric and chromospheric magnetic activity of seismic solar analogs. Observational inputs on the solar-stellar connection from Kepler and Hermes. Astron Astrophys 596:A31. https://doi.org/10.1051/0004-6361/201628583. arXiv:1608.01489 [astro-ph.SR]

Salabert D, Régulo C, García RA et al (2016b) Magnetic variability in the young solar analog KIC 10644253. Astron Astrophys 589:A118. https://doi.org/10.1051/0004-6361/201527978

Santos ARG, Campante TL, Chaplin WJ et al (2018) Signatures of magnetic activity in the seismic data of solar-type stars observed by Kepler. Astrophys J Suppl Ser 237(1):17. https://doi.org/10.3847/1538-4365/aac9b6

Santos ÂRG, Campante TL, Chaplin WJ et al (2019) Signatures of magnetic activity: on the relation between stellar properties and p-mode frequency variations. Astrophys J 883(1):65. https://doi.org/10. 3847/1538-4357/ab397a

Sanz-Forcada J, Stelzer B, Metcalfe TS (2013) ι Horologi, the first coronal activity cycle in a young solar-like star. Astron Astrophys 553:L6. https://doi.org/10.1051/0004-6361/201321388. arXiv:1305.1132 [astro-ph.SR]

Schofield M, Chaplin WJ, Huber D et al (2019) The Asteroseismic Target List (ATL) for solar-like oscillators observed in 2-minute cadence with the Transiting Exoplanet Survey Satellite (TESS). Astrophys J Suppl Ser 241(1):12. https://doi.org/10.3847/1538-4365/ab04f5. arXiv:1901.10148

See V, Jardine M, Vidotto AA et al (2016) The connection between stellar activity cycles and magnetic field topology. Mon Not R Astron Soc 462(4):4442–4450. https://doi.org/10.1093/mnras/stw2010, arXiv: 1610.03737 [astro-ph.SR]

Skumanich A (1972) Time scales for CaII emission decay, rotational braking, and lithium depletion. Astrophys J 171:565–567. https://doi.org/10.1086/151310

Snik F, Jeffers S, Keller C et al (2008) The upgrade of HARPS to a full-Stokes high-resolution spectropolarimeter. In: McLean IS, Casali MM (eds) Ground-based and airborne instrumentation for astronomy II, p 70140O. https://doi.org/10.1117/12.787393

Soon W, Velasco Herrera VM, Cionco RG et al (2019) Covariations of chromospheric and photometric variability of the young Sun analogue HD 30495: evidence for and interpretation of mid-term periodicities. Mon Not R Astron Soc 483(2):2748–2757. https://doi.org/10.1093/mnras/sty3290

Thomas AEL, Chaplin WJ, Davies GR et al (2019) Asteroseismic constraints on active latitudes of solar-type stars: HD 173701 has active bands at higher latitudes than the Sun. Mon Not R Astron Soc 485(3):3857–3868. https://doi.org/10.1093/mnras/stz672, arXiv:1903.04998 [astro-ph.SR]

Thomas AE, Chaplin WJ, Basu S et al (2021) Impact of magnetic activity on inferred stellar properties of main-sequence Sun-like stars. Mon Not R Astron Soc 502(4):5808–5820. https://doi.org/10.1093/MNRAS/STAB354, https://academic.oup.com/mnras/article/502/4/5808/6131852, arXiv:2102.02566

Tripathy SC, Jain K, Hill F (2015) Variations in high degree acoustic mode frequencies of the Sun during solar cycle 23 and 24. Astrophys J 812(1):20. https://doi.org/10.1088/0004-637X/812/1/20. https://arxiv.org/abs/1509.05474

Van Beeck J, Prat V, Van Reeth T et al (2020) Detecting axisymmetric magnetic fields using gravity modes in intermediate-mass stars. Astron Astrophys 638:A149. https://doi.org/10.1051/0004-6361/201937363. arXiv:2005.02411 [astro-ph.SR]

van Saders JL, Ceillier T, Metcalfe TS et al (2016) Weakened magnetic braking as the origin of anomalously rapid rotation in old field stars. Nature 529(7585):181–184. https://doi.org/10.1038/nature16168, arXiv:1601.02631 [astro-ph.SR]

Vaughan AH, Baliunas SL, Middelkoop F et al (1981) Stellar rotation in lower main-sequence stars measured from time variations in H and K emission-line fluxes. I. Initial results. Astrophys J 250:276–283. https://doi.org/10.1086/159372

Vida K, Oláh K, Szabó R (2014) Looking for activity cycles in late-type Kepler stars using time-frequency analysis. Mon Not R Astron Soc 441(3):2744–2753. https://doi.org/10.1093/mnras/stu760 arXiv:1404.4359 [astro-ph.SR]

Waite IA, Marsden SC, Carter BD et al (2017) Magnetic fields on young, moderately rotating Sun-like stars – II. EK Draconis (HD 129333). Mon Not R Astron Soc 465(2):2076–2091. https://doi.org/10.1093/mnras/stw2731, arXiv:1611.07751 [astro-ph.SR]

Waldmeier M (1935) Neue Eigenschaften der Sonnenfleckenkurve. Astron Mitt Eidgenöss Sternwarte Zür 14:105–136

Willamo T, Hackman T, Lehtinen JJ et al (2020) Shapes of stellar activity cycles. Astron Astrophys 638:A69. https://doi.org/10.1051/0004-6361/202037666. arXiv:2002.04300 [astro-ph.SR]

Willamo T, Hackman T, Lehtinen JJ et al (2022) V889 Her: abrupt changes in the magnetic field or differential rotation? Open J Astrophys 5. https://doi.org/10.21105/astro.2203.13398, arXiv:2203.13398 [astro-ph.SR]

Wilson OC (1968) Flux measurements at the centers of stellar H- and K-lines. Astrophys J 153:221. https://doi.org/10.1086/149652

Wilson OC (1978) Chromospheric variations in main-sequence stars. Astrophys J 226:379–396. https://doi.org/10.1086/156618

Woodard MF, Noyes RW (1985) Change of solar oscillation eigenfrequencies with the solar cycle. Nature 318(6045):449–450. https://doi.org/10.1038/318449a0

Yeo KL, Krivova NA, Solanki SK et al (2014) Reconstruction of total and spectral solar irradiance from 1974 to 2013 based on KPVT, SoHO/MDI, and SDO/HMI observations. Astron Astrophys 570:A85. https://doi.org/10.1051/0004-6361/201423628. arXiv:1408.1229 [astro-ph.SR]

Publisher's Note Springer Nature remains neutral with regard to jurisdictional claims in published maps and institutional affiliations.

Space Science Reviews (2023) 219:77
https://doi.org/10.1007/s11214-023-01021-6

Dynamics of Large-Scale Solar Flows

Hideyuki Hotta[1] · Yuto Bekki[2] · Laurent Gizon[2,3] · Quentin Noraz[4,5] · Mark Rast[6]

Received: 29 June 2023 / Accepted: 18 October 2023 / Published online: 17 November 2023
© The Author(s) 2023

Abstract

The Sun's axisymmetric large-scale flows, differential rotation and meridional circulation, are thought to be maintained by the influence of rotation on the thermal-convective motions in the solar convection zone. These large-scale flows are crucial for maintaining the Sun's global magnetic field. Over the last several decades, our understanding of large-scale motions in the Sun has significantly improved, both through observational and theoretical efforts. Helioseismology has constrained the flow topology in the solar interior, and the growth of supercomputers has enabled simulations that can self-consistently generate large-scale flows in rotating spherical convective shells. In this article, we review our current understanding of solar convection and the large-scale flows present in the Sun, including those associated with the recently discovered inertial modes of oscillation. We discuss some issues still outstanding, and provide an outline of future efforts needed to address these.

Keywords Convection · Differential rotation · Meridional flow · Helioseismology · Numerical simulation

1 Observations of Large-Scale Flows in the Sun

1.1 Solar Differential Rotation and Meridional Circulation

Differential Rotation Due to the solar rotation, axisymmetry (about its rotational axis) of the large-scale flows is established to some degree in the Sun's interior. *Differential rotation* denotes the longitudinal component of the axisymmetric (longitudinally-averaged) flow which varies with radius and latitude. It arises from the nonlinear interaction of the rotationally-influenced solar magneto-convection (e.g., Miesch 2005). Differential rotation represents a shear in the rotation rate and is thought to play a significant role in the solar dynamo by stretching and amplifying the magnetic field lines (Charbonneau 2020).

The solar differential rotation profile can be measured by *global helioseismology*, which analyzes small frequency splittings of resonant acoustic oscillations (global standing acoustic modes) (Duvall et al. 1984; Thompson et al. 1996; Schou et al. 1998; Howe et al. 2000). Figure 1 shows the observationally-inferred profile of the internal differential rotation of the Sun (Howe et al. 2005; Larson and Schou 2018). We summarize striking features of the solar differential rotation as follows:

Solar and Stellar Dynamos: A New Era
Edited by Manfred Schüssler, Robert H. Cameron, Paul Charbonneau, Mausumi Dikpati, Hideyuki Hotta and Leonid Kitchatinov

Extended author information available on the last page of the article

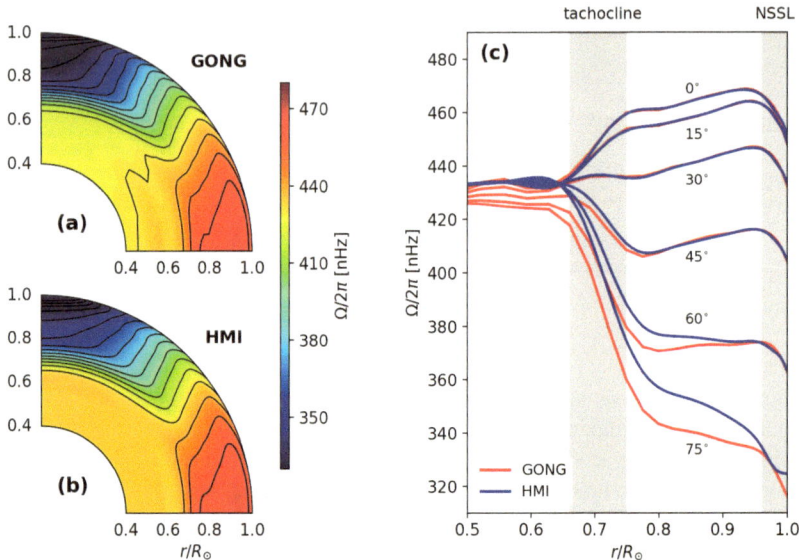

Fig. 1 Internal profile of the solar differential rotation deduced from global helioseismology and averaged from April 2010 to February 2021. Panels (a) and (b) show the results obtained by the Global Oscillation Network Group (GONG) (data courtesy of R. Howe, using the method of Howe et al. 2005) and the Helioseismic and Magnetic Imager (HMI) onboard the Solar Dynamics Observatory (SDO) (Larson and Schou 2018). Panel (c) shows the radial differential rotation at selected latitudes. Grey shades denote the layers of strong radial rotational shear known as the tachocline and the near-surface shear layer (NSSL)

- The radiative interior rotates almost rigidly.
- In the convection zone, the equator rotates about 30% faster than the poles.
- The transition from uniformly-rotating radiation zone to differentially-rotating convection zone occurs in a thin layer from 0.68 $R_\odot$ to 0.73 $R_\odot$. This layer is called the *tachocline*.
- In the bulk of the convection zone (0.73 $R_\odot < r < 0.96 R_\odot$), the rotation rate varies strongly with latitude and much more weakly with radius.
- In a shallow surface layer ($r \gtrsim 0.96 R_\odot$), the rotation rate decreases by about 5% at all latitudes. This layer is called the *near surface shear layer*.
- The contours of constant angular velocity are inclined by about 25° with respect to the rotational axis over a wide range of latitude. In other words, the differential rotation does not follow the Taylor-Proudman theorem.

These observational facts need to be explained by theoretical and numerical models of rotating solar magneto-convection.

Meridional Circulation *Meridional circulation* represents radial and latitudinal components of the large-scale axisymmetric flow in the Sun, i.e., a poloidal flow in a meridional plane. Meridional circulation, as well as the differential rotation, is believed to play a significant role in the solar dynamo by advecting the magnetic flux in both radial and latitudinal directions (e.g., Charbonneau 2020).

The meridional circulation is much weaker than the differential rotation (two orders of magnitudes smaller in flow amplitude) and cannot be inferred using conventional global-mode helioseismology. Therefore, it is an extremely difficult task to measure the meridional flow in the Sun. Near the solar surface, the meridional flow is poleward in both hemispheres with typical amplitudes of $\sim$10–20 m s^{-1}. This was first measured by Duvall (1979) using Doppler measurements and then robustly confirmed in follow-up studies by a variety of methods (e.g., Patron et al. 1995; Giles et al. 1997; Hathaway 1996; Braun and Fan 1998; Haber et al. 2002; Ulrich 2010; Basu and Antia 2010).

Local helioseismology can extend these measurements into the deeper convection zone (e.g., Gizon and Birch 2005). In particular, time-distance helioseismology (Duvall et al. 1993) and ring-diagram analysis (Hill 1988) can be used to measure the effects of the meridional flow on north-south propagating waves. In time-distance helioseismology, wave travel times are extracted from the cross-covariance of the Doppler signals between points along meridians. The technique was applied first by Giles et al. (1997) to the Michelson Doppler Imager (MDI) data onboard SOHO (Scherrer et al. 1995). However, this is an extremely difficult measurement to make because the deep meridional circulation is very weak (no greater than 3–5 m s^{-1}) and the sensitivity of the travel times to flows also decreases with depth. The analysis of very long time series is required to reduce noise. Furthermore, for accurate measurements it is critically important to apply corrections to the measurements, especially a center-to-limb correction (Duvall and Hanasoge 2009; Zhao et al. 2012), and corrections for the P-angle and the Carrington elements (e.g. Giles 2000; Hathaway and Rightmire 2010; Liang et al. 2017). Often pixels that are in regions of very strong magnetic fields (e.g. in sunspots) are excluded from the cross-covariances (e.g. Liang and Chou 2015). The center-to-limb effect is, in general, very significant and depends strongly on time and instrument (Liang et al. 2018). It may have both an instrumental and physical component. When Giles et al. (1997) made their measurements, no center-to-limb correction was applied to the MDI travel times, as it was very small during the year 1996 (Liang et al. 2018). We refer the reader to the review by Hanasoge (2022) for additional information.

Many different inferences of the solar meridional circulation using time-distance helioseismology have been published. They are not consistent. Using travel time measurements obtained from SDO/HMI in 2010–2012, Zhao et al. (2013) reported an equatorward return flow in the middle of the convection zone (0.82–0.91 $R_\odot$) and a poleward flow below 0.82 $R_\odot$, indicating a double-cell structure in radius (Fig. 2a). A similar result was obtained by Chen and Zhao (2017) who used seven years of data from HMI and active regions were masked out to remove the contribution from pixels with strong magnetic fields. Note, however, that Zhao et al. (2013) and Chen and Zhao (2017) did not invert the radial component of the meridional flow which is required by the local mass conservation. An attempt to test a mass conservation constraint was made by Jackiewicz et al. (2015) who found that the equation of continuity is poorly satisfied for the inverted flows from GONG and HMI (only two years of data were used). Under the constraint of mass conservation in terms of the stream function, Rajaguru and Antia (2015), using four years of HMI data report a single-cell meridional circulation in each hemisphere, with an equatorial return flow near the base of the convection zone (below 0.77 $R_\odot$). A similar result was reported by Mandal et al. (2018).

Liang et al. (2017) have compared the north-south travel-time data from SOHO/MDI and those from SDO/HMI over the 1-year overlap period of 2010. After correcting for the center-to-limb effect (Zhao et al. 2012), it is found that the travel-times are consistent within the error bars, although an overlap period of only 288 days is far from enough for such consistency test. When comparing the travel-times with different forward models of meridional flows, Liang et al. (2018) found that the MDI/Cycle 23 data point to a single cell meridional flow in both hemispheres, while the HMI/Cycle 24 data point to a double cell in the north and a single cell in the south.

Following this work, Gizon et al. (2020a) carried out comprehensive measurements of the north-south travel-time measurements using all available data sets from GONG

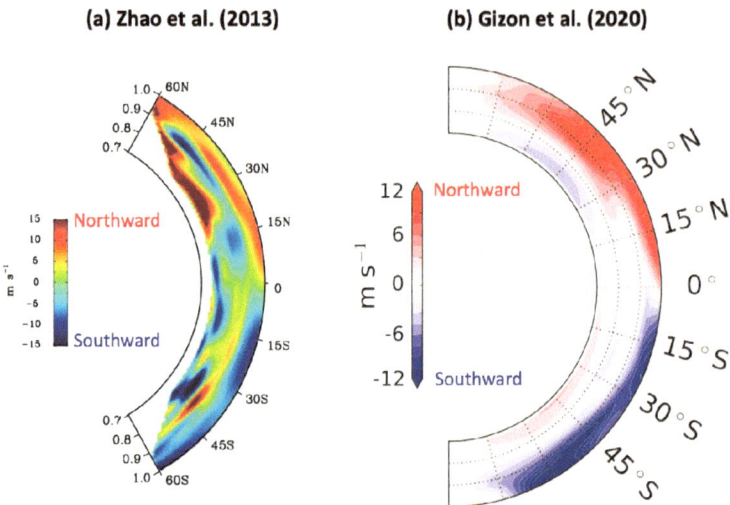

(a) Zhao et al. (2013)

(b) Gizon et al. (2020)

Fig. 2 Latitudinal component of the meridional flow inferred by time-distance local helioseismology. The red and blue shades correspond to the northward and southward directions respectively. (a) The result obtained by Zhao et al. (2013) using SDO/HMI data (2010–2012), reproduced with permission. (b) The result obtained by Gizon et al. (2020a) using GONG data (2008–2019), reprinted with permission from AAAS

(2001–2019), SOHO/MDI (1996–2011), and SDO/HMI (2010–2019). After correcting for all known systematic errors (CCD orientation, center-to-limb effect, Carrington elements, pixels in sunspots and active regions), it was found that the north-south travel times measured from GONG and MDI are in good agreement over the overlap period of 1996–2011. However, a small travel-time offset is apparent in the HMI data compared to the GONG data over the period 2010–2019. No explanation was found for this HMI offset and the data were set aside: The HMI measurements imply a strikingly different meridional flow pattern between the northern and southern hemispheres, as hinted at by Liang et al. (2018) and confirmed by Gizon et al. (2020a) and Braun et al. (2021). Gizon et al. (2020a) find that the mass-conserving meridional flows during cycle 23 (from MDI/GONG data) and cycle 24 (from GONG data) display a single-cell pattern in each hemisphere, as shown in Fig. 2b.

Work is needed to resolve the remaining issues that affect helioseismic inferences of the deep meridional flow. Efforts are ongoing to identify the source of the HMI travel-time offset, which is extremely small (0.2 s) but significant. The other pressing issue is to gain an understanding of the origin of the center-to-limb effect (e.g. Chen and Zhao 2018). The analysis of travel-times in different frequency bands is a promising avenue to make progress in this area (Chen and Zhao 2018; Rajaguru and Antia 2020).

Observational determination of the solar meridional circulation is crucial, not only for constraining solar dynamo models (e.g., Hazra et al. 2014), but also for properly understanding the angular momentum flux balance in the Sun's convection zone (e.g., Featherstone and Miesch 2015).

Cycle Variations of Large-Scale Mean Flows Solar Doppler observations and global helioseismology have revealed that the Sun's differential rotation exhibits a temporal variation pattern consisting of multiple bands of faster- and slower-than-average zonal flows which migrate in latitude with the phase of the solar dynamo cycle (e.g., Howe 2009). The so-called *torsional oscillation* (coined by Howard and Labonte 1980) has two distinct branches, one at active-region latitudes and one at latitudes above $50°$. The low-latitude branch migrates equatorward together with the magnetic activity belts and shows a clear 11-yr periodicity. On the other hand, the poleward-migrating high-latitude branch tends to show a rather irregular behavior: it was clearly seen in Cycle 23 but did not appear in Cycle 24 (e.g., Howe et al. 2013). For more details on the observational aspects of the Sun's torsional oscillations, see Norton et al. (2023) (their Sect. 6.2). The amplitude of the solar torsional oscillation is ≈ 3–5 nHz and it is important to understand its physical origin as it likely reflects the nonlinear feedback of the dynamo-generated magnetic fields onto the large-scale flows (e.g., Rempel 2007; Beaudoin et al. 2013; Pipin and Kosovichev 2019; Brun et al. 2022).

As is the case of differential rotation, the poleward meridional circulation shows a temporal variation associated with the solar magnetic cycle (Beck et al. 2002; Gizon 2004; Gizon et al. 2010; Mahajan et al. 2023); see also Sect. 6.3 in Norton et al. (2023). This variation is of order ≈ 5 m s^{-1} or more, which is a significant fraction of the maximum meridional flow amplitude. It may be explained, at least in part, by the north-south component of the inflows around active regions (Gizon et al. 2010; Mahajan et al. 2023). In addition to this component, Mahajan et al. (2023) find that there is a residual solar-cycle component as small as ≈ 2 m s^{-1}, which is seen around cycle minima.

1.2 Solar Convective Flows

Convection on the Sun occurs over a wide range of spatial scales, and while the spectrum is continuous, apparent characteristic scales are commonly cited: granulation, mesogranulation, supergranulation, and giant cells. Granulation (Herschel 1801) is readily apparent in high-resolution images of the solar photosphere, as a pattern of bright upflowing regions separated by darker downflowing lanes. The characteristic upflow cells have diameters of ~ 1000 km, lifetimes of about 0.2 hr, and vertical flow speeds of ~ 1 km s^{-1}. The upflow velocity often peaks near the granular boundaries (e.g., Nesis et al. 1992; Rast 1995; Hirzberger 2002; Nordlund et al. 2009; Falco et al. 2017, and reference therein). These properties reflect the compressible flow dynamics of a strongly cooled radiative boundary layer, with observations confirming the convective nature of the flow via measurement of the correlation between the vertical velocity and plasma temperature (e.g., Canfield and Mehltretter 1973). Granulation is well observed and robustly modeled (se e.g., Nordlund et al. 2009), even in quite shallow domains, by codes that capture the rapid change in radiative opacity in the solar photosphere and implement an open lower boundary condition to minimize bottom-up influences on the top-down dynamics of the radiative boundary layer.

Mesogranulation (November et al. 1981), on the other hand, is observationally elusive. With a reported length scale of about 5–10 Mm, ~ 60 m s^{-1} vertical flow speeds, and ~ 2–3 hr lifetime, its identification as a convective feature is still debated. Most recent studies suggest that no distinct mesogranular scale is present in the broad range of convective scales observed (e.g., Rincon and Rieutord 2018, and references therein). One possibility is that there is weak advective self-organization of the granular flows, a process first proposed in the context of supergranulation (Rieutord et al. 2000; Rast 2003) but likely more relevant on mesogranular scales (Cattaneo et al. 2001; Berrilli et al. 2005; Leitzinger et al. 2005; Duvall and Birch 2010). However, the absence of a mesogranular scale in the clustering of magnetic elements in high resolution magnetograms suggests that this mechanism too leads to a continuous exponential distribution of scales between 2 and 10 Mm, with no distinctive characteristic peak (Berrilli et al. 2013).

Supergranulation (Hart 1954; Leighton et al. 1962) is the largest likely-convective scale of motion readily visible in the solar photosphere. It is observed directly in spectral Doppler shifts away from disk center (due to horizontal motions) and is traced by network magnetic elements which are prominent in magnetograms and in emission in low chromospheric lines such as Ca II K. There is good correlation between Ca II K emission and magnetic flux density (Ortiz and Rast 2005, and references therein). Supergranular cells have diameter of $\sim$30 Mm, horizontal flow velocities of $\sim$100 m s^{-1}, and lifetimes of $\sim$20 hr. After the intensity contributions of the small scale magnetic field elements has been removed, they show an average continuum intensity contrast across the cells of about 0.1%, corresponding to about one degree Kelvin in brightness temperature (Goldbaum et al. 2009).

The origin of the supergranular motions has been widely debated (see Rincon and Rieutord 2018, and references therein). It has recently been proposed that the scale of supergranulation reflects not a selected convective scale, but is instead defined by the scale above which convective power declines (Lord et al. 2014; Cossette and Rast 2016). This interpretation, and the reasons underlying the power reduction, links the well observed phenomenon of supergranulation to the *convective conundrum*, an outstanding discrepancy between models and observations (see this section below, and Sects. 2.2 and 3.1). We note that the early suggestion that helium ionization plays a role in determining the mesogranular and supergranular scales (Leighton et al. 1962; Simon and Leighton 1964) is not supported by numerical simulations or simplified models base on them (Rast and Toomre 1993; Lord et al. 2014). Additionally, the presence of the network magnetic elements themselves (Crouch et al. 2007; Thibault et al. 2012) or the enhanced radiative loses through them (Rast 2003) do not seem to play a role in scale selection role (Lord 2014). Finally, it is important to note that supergranulation shows peculiar unexplained wave-like properties (Gizon et al. 2003; Schou 2003; Langfellner et al. 2018).

In contrast with supergranulation, which is readily observed but not captured by any local-area or global spherical-shell simulation, solar giant cells (Simon and Weiss 1968), motions on the scale of the solar convection zone depth ($\sim$200 Mm), dominate global spherical-shell simulations but are very difficult to observe (Hathaway et al. 2013; Hathaway and Upton 2021, and references therein). If, in the Sun, giant-cells had the amplitude they do in simulations, they would be easily observed in the solar photosphere. This is the simplest manifestation of the convective conundrum: that supergranulation, rather than giant-cell scale motions, are the largest readily observed motions in the solar photosphere. The implication for solar differential rotation is fundamental. The enhanced amplitude of the large-scale convective motions in global numerical simulations tend to place those simulations in a Rossby-number regime that favors anti-solar differential rotation profiles.

These issues are critical to our understanding of large-scale motions on the Sun. As can be seen in Fig. 3 (from Hathaway et al. 2015), only two of the components described above are evident as distinct features in the observed spectrum of motions in the solar photosphere. Granulation is responsible for the most pronounced peak at high spherical-harmonic degree and supergranulation for the smaller peak near spherical-harmonic degree 120. Added to the plot are vertical fiducial lines indicating the approximate scale of supergranulation and giant cells. Additionally, a *blue dotted* line has been added to schematically indicate the monotonic increase of power to low wavenumbers seen in all realistic numerical simulations up until the most recent of Hotta and Kusano (2021). In the Hotta and Kusano (2021) simulations in the power rolls over at spherical-harmonic degree $\sim$10. It is the discrepancy in low spatial-frequency power between simulations and observations that has come to be known as the convective conundrum.

It is important to note that the spectrum plotted in Fig. 3 is a composite, with vertical velocities dominating at high spatial wavenumbers (granular scales) and horizontal motions most important at supergranular scales. The vertical velocity contribution decreases from the

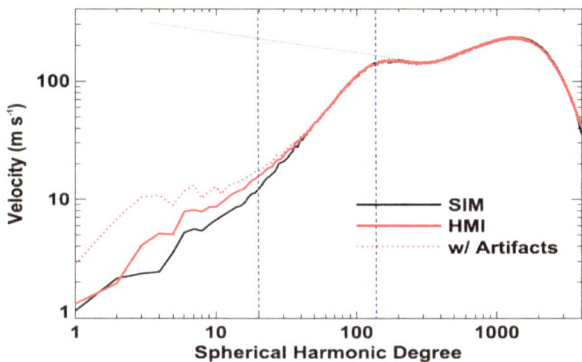

Fig. 3 From Hathaway et al. (2015): Solar Doppler velocity spectrum as determined from Helioseismic and Magnetic Imager (HMI) observations (*red* curves, with and without removal of image artifacts), along with a random-phase synthetic spectrum (*black* curve). Vertical *blue dashed* fiducial lines have been added indicating the approximate scale of supergranulation and giant-cells. The *blue dotted* line approximately over-plots the spectrum seen in numerical simulations. Figure without added *dashed* and *dotted blue* lines used with permission of Hathaway et al. (2015)

granular peak towards lower wavenumbers, with horizontal velocity contribution increasing to spherical-harmonic degree ∼120 before rolling over beyond that. The supergranular peak results from this decrease in the power beyond spherical-harmonic degree ∼120. Thus, with respect to photospheric flow observations, the convective conundrum refers to the scale and amplitude of the horizontal-flows in the photosphere. No global-spherical-shell or local-area simulation of solar convection yet captures the supergranular scale maximum in photospheric horizontal-flow power.

1.3 Solar Inertial Modes

On scales much larger than the supergranules, the non-axisymmetric flows have long timescales and are thus strongly influenced by solar rotation via the Coriolis force. A component of these flows has recently been observed as retrograde waves of radial vorticity at the surface and are known as inertial modes. In this section, we give an overview of their characteristics and properties.

Inertial modes owe their existence to rotation (Greenspan et al. 1968). Their restoring force is the Coriolis force. In a uniformly rotating sphere, the frequencies of inertial modes are limited to a range of $|\omega| < 2\Omega_0$ in the co-rotating frame, where Ω_0 denotes the angular velocity. The traditional Rossby modes (or r modes) correspond to a variety of inertial modes that have quasi-toroidal motions. Although these modes have been expected to exist in the Sun and stars since the late 1970's (e.g., Papaloizou and Pringle 1978; Saio 1982; Unno et al. 1989), they were not observed on the Sun until very recently. Inertial modes on the Sun have very long oscillation periods (of the order of months) and very small velocity amplitudes (of the order of 1 m s^{-1}). Therefore, long-term and high-precision observations of horizontal flows over many years are required to detect them.

Löptien et al. (2018) discovered the solar equatorial Rossby modes using both a granulation-tracking method and a ring-diagram analysis applied to six years of SDO/HMI data. They detected excess power in the radial vorticity along the dispersion relation of the sectoral Rossby modes, $\omega = -2\Omega_{\text{eq}}/(m + 1)$ in the Carrington frame, for azimuthal orders $3 \leq m \leq 15$. In this formula, Ω_{eq} is the equatorial rotation rate at the Sun's surface. For

$m \gtrsim 5$, the latitudinal eigenfunctions of the equatorial Rossby modes significantly deviate from the sectoral spherical harmonics: the radial vorticity peaks at the equator but changes sign at middle latitudes (this is due to differential rotation, see below). The detection of solar equatorial Rossby modes has been confirmed in follow-up studies using various other observational datasets and methods (Liang et al. 2019; Proxauf et al. 2020; Mandal and Hanasoge 2020; Hanson et al. 2020; Hathaway and Upton 2021).

The observed equatorial Rossby modes exist only at very large scales with $m \leq m_{\text{crit}} \approx 15$. Löptien et al. (2018) speculated that this critical azimuthal order m_{crit} might correspond to the Rhines scale $l_{\text{Rhines}} = \sqrt{R_{\odot} v_c / \Omega_{\odot}}$ above which rotation strongly affects turbulent convection (Rhines 1975). Assuming $m_{\text{crit}} \approx R_{\odot} / l_{\text{Rhines}}$, a typical speed of turbulent convection can be roughly estimated as $v_c \approx 10 \text{ m s}^{-1}$, which is about one order magnitude smaller than the typical mixing-length estimate (Böhm-Vitense 1958; Stix 2002). This may add another piece of information with regard to the solar convective conundrum (see Sects. 2.2 and 3.1 below).

Recently, Gizon et al. (2021) analyzed more than 10 years of data from both SDO/HMI and GONG and detected many additional quasi-toroidal inertial modes with $1 \leq m \leq 10$. With the help of a 2.5D linear eigenvalue solver applied to a model of the differentially rotating convection zone (Bekki et al. 2022b), they identified modes at middle and high latitudes, in addition to the equatorial Rossby modes. The observational power spectra and the measured eigenfunctions of three selected inertial modes with $m = 1, 2$, and 3 are shown in Fig. 4. These modes are very sensitive to the solar latitudinal differential rotation (see Gizon et al. 2020b, for a discussion in the β plane). For many of the inertial modes, there exists latitudes at which the phase speed is equal the local differential rotation speed; such latitudes are called critical latitudes (see Fig. 4, middle column). The $m = 1$ high-latitude mode has a large amplitude ($v_{\phi} \approx 15 \text{ m s}^{-1}$) at high latitudes (above $\approx 50°$) and its surface eigenfunction exhibits a spiral pattern in the polar regions. The horizontal flow associated with the $m = 1$ high-latitude mode was first observed by Hathaway et al. (2013), but misidentified at the time as giant cell convection. It had also been observed by Bogart et al. (2015) and Howe et al. (2015). A linear model suggests that the $m = 1$ high-latitude mode is self-excited when a large enough latitudinal entropy gradient exists in the convection zone (Bekki et al. 2022b).

More recently, Hanson et al. (2022) reported the detection of another class of inertial modes with north-south anti-symmetric radial vorticity across the equator. These modes propagate in a retrograde direction with the phase speed roughly three times faster than the that of the equatorial Rossby modes. Compared with the equatorial Rossby modes and the high-latitude modes reported in Gizon et al. (2021), these modes have lower velocity amplitudes and are thus much harder to distinguish from the background noise in the power spectra. According to simplified linear eigenmode calculations of a uniformly-rotating Sun (Triana et al. 2022; Bhattacharya and Hanasoge 2023), these modes are not quasi-toroidal, i.e., substantial radial motions are involved.

Though it remains unclear how important the solar inertial modes are to the overall convection zone dynamics, they play an important diagnostic role. Gizon et al. (2021) and Bekki et al. (2022b) have shown that the properties of some inertial modes (i.e., frequencies, linewidths, and surface eigenfunctions) are sensitive to the turbulent viscous diffusivity ν_t and to the superadiabaticity δ of the convection zone. Gizon et al. (2021) inferred that, on average, $\nu_t \lesssim 10^{12} \text{ cm}^2 \text{ s}^{-1}$ and $\delta < 2 \times 10^{-7}$. These values are an order of magnitude smaller than the theoretical estimates from the local mixing length model (Christensen-Dalsgaard et al. 1996; Muñoz-Jaramillo et al. 2011). It is noteworthy that both of these parameters cannot be constrained by conventional p-mode helioseismology, and are important in discussions of the convective conundrum (Sects. 2.2 and 3.1 below) and the solar

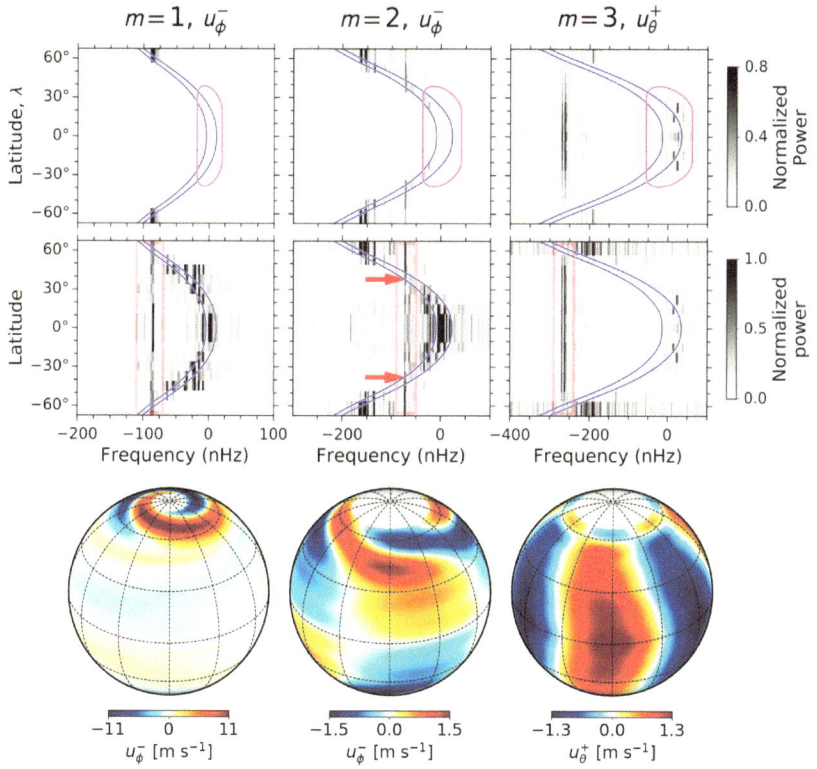

Fig. 4 Observational power spectra in the Carrington frame and eigenfunctions of three selected inertial modes of the Sun. *Top row*: Power spectra of the longitudinal component of velocity u_ϕ for $m = 1$ (left column) and $m = 2$ (middle column), and power spectrum of the colatitudinal component of velocity u_θ for $m = 3$. The blue curves show the differential rotation rate at $r = 0.96\ R_\odot$ and $R_\odot$. The purple contour indicates the region affected by active-region flows. *Middle row*: The same power spectra but normalized at each latitude by their average value over the frequency range between the orange bars. Excess power is seen at a specific frequency at all latitudes in each of the three cases. The red arrows point to critical latitudes at the surface for the case $m = 2$. *Bottom row*: Observed horizontal velocity eigenfunctions at the surface for the $m = 1$ high-latitude mode, the $m = 2$ critical-latitude mode, and the $m = 3$ equatorial Rossby mode. These figures are courtesy of Gizon et al. (2021)

dynamo. The amplitudes of the linearly-stable modes might provide additional constraints on the turbulent viscosity (Philidet and Gizon 2023). On the other hand, to better understand the amplitudes of the linearly-unstable inertial modes, nonlinear numerical simulations will be required (Bekki et al. 2022a; Matilsky et al. 2022; Bekki and Cameron 2023).

1.4 Observations of Large-Scale Flows on Solar-Type Stars

It is possible to obtain some information about the rotation of distant stars using observations of photometric variability, radial velocity measurements, and asteroseismology. Such methods are challenging, however recent results are setting new constraints on stellar differential rotation.

One possibility is to measure the effects of rotation on stellar oscillations. Rotation lifts the degeneracy of the frequencies of the modes of oscillation with different azimuthal orders.

Latitudinal and radial differential rotation may leave a signature in the fine structure of the oscillation power spectra. See, e.g., Aerts et al. (2010) and García and Ballot (2019) for reviews of *asteroseismology*. In the last years it has become possible to probe the deep interior of some evolved stars, including red giants (Beck et al. 2012; Deheuvels et al. 2012) and, more recently, subgiants (Deheuvels et al. 2020). Probing the slowly-rotating Sun-like stars is especially difficult because rotational splitting frequencies are often too small compared to the mode linewidths. Constraints on the mean angular velocity and the inclination angle of the rotation axis have been obtained in a few cases (e.g., Gizon et al. 2013; Chaplin et al. 2013). Strong latitudinal differential rotation has recently been observed on a few selected main-sequence solar-type stars (Benomar et al. 2018). Solar-like differential rotation profile (equator faster than the poles) is detected on 13 stars with a much stronger amplitude than on the Sun (on average twice the solar value). Subsequently, Bazot et al. (2019) reported latitudinal rotation contrasts closer to the solar value for the two solar analogs 16 Cyg A and B. The equator-to-pole difference in angular velocity, $\Delta\Omega$, may depend on the mean rotation rate of the stars, with the majority of stars in Benomar et al. (2018)'s study rotating faster than the Sun. Only upper limits have been set on the radial differential rotation of Sun-like stars (Nielsen et al. 2014). To understand differential rotation in stars, it is important to study its relationship to the mean rotation rate Ω_*, the stellar mass M_*, the stellar age, and the stellar composition.

Surface rotation and differential rotation can also be studied using observations of stellar photometric variability. The presence of starspots and faculae on rotating stars modulate the photometric signal at the rotation period (e.g., Nielsen et al. 2013; Reinhold et al. 2022). Some information about latitudinal differential rotation can be retrieved when magnetic active regions are present over a range of latitudes on the stellar surface (e.g., Reinhold et al. 2013; Nielsen et al. 2019). The stellar differential rotational versus mean rotation rate may be described by a power law, $\Delta\Omega \propto \Omega_*^n$. Figure 3 of Barnes et al. (2005) summarizes the different studies that were available at the time. The long-term monitoring of photometric modulations point toward a weak rotational dependence, $n = 0.24$ (Henry et al. 1995), while similar observations in H & K bands of Ca II emission suggest a significantly higher value, $n = 0.7$ (Donahue et al. 1996). The spectroscopic analysis of Reiners and Schmitt (2003) gave a power law index of $n = 0.66$. Barnes et al. (2005) proposes an index of $n = 0.15$, but also show that this index is very sensitive to the spectral-type diversity of the target sample considered (see also Reiners and Schmitt 2003). For instance, Reinhold et al. (2013) find a value of $n = 0.3$ for cool stars and Balona and Abedigamba (2016) report $n = 0.2$ for G-stars. Asteroseismic studies tend to find values for n between 0.3 and 0.45 (García and Ballot 2019), and underline its sensitivity to the effective temperature T_{eff} range of the stars considered (see Reinhold and Gizon 2015, and references therein).

The dependence of the latitudinal contrast $\Delta\Omega$ on spectral type also appears to be significant for fast main-sequence rotators. Collier Cameron (2007) studies fast M, K, G, and F main-sequence rotators and reports a strong $\Delta\Omega \propto T_{\text{eff}}^{8.6}$ dependence on the effective temperature, consistent with Barnes et al. (2005). Large error bars are associated with the hottest spectral types, however the models of Kueker and Ruediger (2011) tend to confirm a strong dependence (see Reinhold and Gizon 2015, for a detailed discussion).

To take into account both rotational and spectral-type aspects, Saar (2010) proposed to consider the global shear $\Delta\Omega$ as a function of the stellar Rossby number $Ro_s = \tau_c / \Omega_*$ (see also Brun et al. 2017 and Noraz et al. 2022a for definitions and prescriptions). Indeed, it is possible to parameterize the convective turnover time τ_c as a function of T_{eff} (Cranmer and Saar 2011). In particular, he finds that $\Delta\Omega \propto Ro_s^{-1}$ for unsaturated rotators ($\Omega \leq 12\Omega_\odot$), pointing then toward $n = 1$ when fixing T_{eff} and the composition. Numerical studies of

144

the last decade have confirmed that the Rossby number is a major parameter to consider regarding the characterization of large-scale flows along stellar evolution (see Sect. 2.6). However, it has to be mentioned here that the high-Rossby regime still needs observations to constrain theoretical results. A difficulty lies here in the sensitivity of the aforementioned techniques to the rotation rate, which decreases signals significantly when considering slow-rotators.

Finally, recent observational studies have started to investigate the impact of the metallicity, for instance with the solar analog HD 173701 studied by Karoff et al. (2018). Its parameters are indeed close to the solar ones, while having a significantly higher metallicity ([Fe/H] = 0.3 ± 0.1). Using different methods previously mentioned, the authors report a solar-like differential rotation (fast equator) with a latitudinal contrast being twice the solar one. Monitoring of its chromospheric and photometric emissions also show a cyclic activity shorter than the one observed on the Sun ($P_{cyc} = 7.4$ against 11 years), while having a higher amplitude of variation, which underlines the entanglement between composition, large-scale flows, and magnetism (Brun and Browning 2017; See et al. 2021).

To summarize, differential rotation is a characteristic quantity of stellar convective envelopes. Apart from the Sun, the exact quantification of the surface rotational contrast remains difficult because it requires a high degree of precision of the instruments used and long acquisition periods for the different targets. New observations by the upcoming PLATO mission should allow major advances in this direction (Rauer et al. 2014). A particular focus for future prospects will lie on the monitoring and characterization of slow-rotators, which appear to be challenging targets for current techniques (Noraz et al. 2022a; Donati et al. 2023, see also Sect. 2.6). In the meantime, theoretical modeling of these flows are hence crucial for the understanding and support of these observations, and in order to guide those to come.

2 Models of Large-Scale Flows

Large-scale flows in the Sun are generated and maintained by thermal convection. In this section we briefly summarize the current theoretical understanding of how that occurs.

2.1 Mixing Length Theory and Energy Budget

Mixing-length theory (MLT; Böhm-Vitense 1958) remains a remarkably useful way to describe the mean energy transport by convection even when the actual dynamics is far from localized eddy motions. In the mixing length formulation, we can relate energy flux to the convective velocity and the stratification through a parameter called the mixing-length parameter $\alpha_{MLT} = L_{MLT}/H_p$, where L_{MLT} and $H_p = -(d \log p/dr)^{-1}$ are the mixing length and the pressure scale height in the convecting fluid. The mixing-length parameter, α_{MLT} is a parameter of order unity which mainly affects the stellar radius (Demarque and Percy 1964) when used in a stellar structure model. Here we take $\alpha_{MLT} = 1$ (i.e., $L_{MLT} = H_p$) for the simplest mixing-length formulation. The typical temperature perturbation in the convection ΔT is then evaluated as:

$$\Delta T = \left[\left(\frac{dT}{dr} \right)_{ad} - \frac{dT}{dr} \right] L_{MLT} = T \left[\left(\frac{d \log T}{d \log p} \right)_{ad} - \left(\frac{d \log T}{d \log p} \right) \right]$$

$$= -T\delta, \tag{1}$$

where $\delta = (d\log T/d\log p) - (d\log T/d\log p)_{\text{ad}}$ is the superadiabaticity, the deviation from the adiabatic stratification. When the speed of sound in the medium is much faster than convection, the sound wave instantaneously relaxes any pressure perturbation Δp, and the density perturbation $\Delta\rho$ can be estimated from the linearized equation of state as

$$\frac{\Delta\rho}{\rho} = -\frac{\Delta T}{T} = \delta. \tag{2}$$

A simplified equation of motion is converted with rough dimensional analysis as

$$\rho\frac{Dv}{Dt} = -\Delta\rho g \rightarrow \frac{v_c}{\tau_c} = -\frac{\Delta\rho}{\rho}g, \tag{3}$$

where v_c is the typical convection velocity. Here we can evaluate the typical time for the convection as $\tau_c = H_p/v_c$. Then the convection velocity can be written as

$$v_c^2 = \delta\frac{p}{\rho} \sim \delta c_s^2, \tag{4}$$

where c_s is the speed of sound. Together, these approximations yield several important approximations for the superadiabaticity:

$$\delta \sim \left|\frac{\Delta T}{T}\right| \sim \left|\frac{\Delta\rho}{\rho}\right| \sim \left(\frac{v_c}{c_s}\right)^2. \tag{5}$$

While some more careful treatments include additional physical effects, such as variation in the mean molecular weight of the plasma or the fluid drag force, in the practical application of MLT, the relationships captured by Equation (5) change little. What is important is that these relations lead to a ratio of the kinetic energy E_{kin} to the internal energy E_{int} that scales with superadiabaticity δ:

$$\frac{E_{\text{kin}}}{E_{\text{int}}} \sim \frac{\rho v_c^2/2}{\rho c_v T} \sim \left(\frac{v_c}{c_s}\right)^2 \sim \delta. \tag{6}$$

In the convection zone, heat is mainly transported by the convection, with the enthalpy flux given by

$$F_e = \rho c_p \Delta T v_c \sim \rho v_c^3. \tag{7}$$

Normalizing this equation yields

$$\frac{F_e}{\rho c_s^3} \sim \left(\frac{v_c}{c_s}\right)^3 \sim \delta^{3/2}. \tag{8}$$

Over the depth of the convection zone, with the superadiabaticity being 10^{-6} and 10^{-1} at the base and surface respectively (e.g., Ossendrijver 2003), the convective velocity amplitudes should change by a factor of a few hundred. At the base of the solar convection zone, the flow is subsonic (Mach number $\sim 10^{-3}$) and the internal energy of the fluid is much larger (10^6 times larger) than the kinetic energy of the flows.

2.2 The Effect of Stratification on Observed Horizontal Convective Flow Amplitudes

As is clear from the mixing-length calculation above, one of the most important aspect of stellar envelope convection is the steep stratification of the mean state. The solar convection zone is about 210 Mm deep, over which the density changes by a factor of about one million and the pressure by about 800 million. The density scale height in the photosphere is about 150 km, while at the convection zone base it is equal to nearly half the depth. This has profound influence on the convective dynamics.

By mass conservation, only a very small fraction of the upwelling fluid from the deep convection zone makes it into the photosphere. The rest must overturn. Over each scale height, the density decreases by a factor of $1/e$, so that $1 - 1/e$ of the mass must overturn. Similarly, the downwelling fluid must entrain mass at this rate. For simple assumptions about the flow geometry, this implies a characteristic horizontal flow scale at each depth $d = 4H_\rho$, where H_ρ is the density scale height (Nordlund et al. 2009). Taking this to be the integral (driving) scale of the motions (Stein et al. 2009) allows a simple two component model that can reproduce the observed spectrum of horizontal motions in the solar photosphere (Lord et al. 2014).

For statistically steady motions and small horizontal density gradients compared to the mean vertical stratification, the equation of mass continuity takes the form

$$\nabla_{\rm h} \cdot \boldsymbol{v}_{\rm h} = -\frac{\partial v_z}{\partial z} - \frac{v_z}{H_\rho}. \tag{9}$$

This suggest two flow regimes. For $\partial v_z/\partial z \gg v_z/H_\rho$ the motions are nearly divergenceless and for $\partial v_z/\partial z \ll v_z/H_\rho$ the vertical stratification dominates. In these two limits, Equation (9) can be used to determine the horizontal velocity power spectrum given that of the vertical velocity. For small scale motions, high horizontal wavenumbers $k_{\rm h}$, the motions are nearly isotropic with

$$\widetilde{\boldsymbol{v}}_{\rm h}^* \cdot \widetilde{\boldsymbol{v}}_{\rm h} = \widetilde{v}_z^* \widetilde{v}_z, \tag{10}$$

where $\widetilde{\boldsymbol{v}}_{\rm h}^* \cdot \widetilde{\boldsymbol{v}}_{\rm h}$ is the power spectrum of the horizontal flows and $\widetilde{v}_z^* \widetilde{v}_z$ is that of the vertical flows. For low wavenumber components of the flow, on the other hand, stratification is important and

$$\widetilde{\boldsymbol{v}}_{\rm h}^* \cdot \widetilde{\boldsymbol{v}}_{\rm h} = \frac{2}{k_{\rm h}^2 H_\rho^2} \widetilde{v}_z^* \widetilde{v}_z. \tag{11}$$

The cross over between these two regimes occurs at the integral scale $4H_\rho$ at each depth. Figure 1 of Lord et al. (2014) confirms that this two-component continuity balance determines the relationship between the vertical and horizontal-velocity power spectra in radiative hydrodynamic simulations.

Using this balance, a model of the horizontal motions observed in the solar photosphere can be constructed from the vertical velocity spectrum at each depth. For example, the high-wavenumber vertical-velocity spectrum can be taken to be Kolmogorov above the integral scale $4H_\rho$, with a mixing-length approximation to determine the total integrated power. Since no scales larger than that are driven at any given depth, the amplitudes of modes with scales larger than the driving scale at any depth can be determined by their decay with height from the depth at which they were last driven. A potential flow approximation on these scales can be used to model that amplitude decrease with height. With these ingredients, working from the bottom of the convection zone upwards, the power spectrum of the horizontal velocity at each depth can be determined. It matches that seen in three-dimensional radiative hydrodynamic simulations (Lord et al. 2014). The broader and critical take away from this simplified approach is that the horizontal-velocity spectrum in the photosphere depends on the vertical-velocity flow amplitudes at depth. The larger the horizontal flow scale observed in the photosphere, the deeper it originates. This suggests that the dramatic decrease in the observed horizontal-velocity power above the supergranular scale on the Sun reflects weak convective driving at depth (at depths below ~ 10 Mm) and that supergranulation represents the largest buoyantly driven scale of motion (Lord et al. 2014; Cossette and Rast 2016).

A number of reasons for the low-convective amplitudes at depth are possible, including highly non-local dynamics (maintenance of small scale downflowing plumes generated in the photosphere with little horizontal diffusion) that ensures that the mean stratification of the solar convection zone is closer to adiabatic than numerical models can achieve (Cossette and Rast 2016; Rast 2020), this possibly due to the presence of small scale magnetic field (O'Mara et al. 2016; Bekki et al. 2017); subadiabatic stratification in the deep solar convection zone due to internal heating by radiation; reduced convective amplitudes due to the stabilizing influence of rotation in the lower convection zone (Featherstone and Hindman 2016; Vasil et al. 2021); or a combination of these. Any mechanism that leads to reduced convective amplitudes in the deep layers of the solar convection zone will be reflected in reduced low wavenumber power in the horizontal-flows at the surface.

In radiative hydrodynamic simulations, a factor of ~ 2.5 reduction in convective amplitudes below ~ 10 Mm depth is sufficient to resolve the convective conundrum (Lord et al. 2014). Though such a reduction has not been yet achieved in a first principles model of convection, we note that, by simple mixing-length scaling arguments (Equation (5)), it is consistent with the reduction in superadiabaticity ($\delta < 2 \times 10^{-7}$) suggested by the Gizon et al. (2021) analysis of solar inertial modes.

2.3 Gyroscopic Pumping and Thermal Wind Balance

Given the convective motions, it is important to understand how, in combination with solar rotation, they generate and maintain large-scale solar differential rotation and meridional circulation. For that purpose, the concepts of gyroscopic pumping and the thermal wind balance are useful (McIntyre 1999; Miesch et al. 2008; Miesch and Hindman 2011). We discuss those in this section. For simplicity, the magnetic field is ignored, though it may be critical in some cases (Hotta et al. 2022). We use a notation where a quantity Q is divided into a mean (longitudinal average) $\langle Q \rangle$ and perturbation Q', i.e., $Q = \langle Q \rangle + Q'$.

Gyroscopic pumping is a reflection of angular momentum conservation. The longitudinally averaged longitudinal equation of motion under the anelastic approximation can be written as

$$\rho_0 \frac{\partial \langle \mathcal{L} \rangle}{\partial t} = -\nabla \cdot (\rho_0 \langle \boldsymbol{v}_m \mathcal{L} \rangle) , \tag{12}$$

where $\mathcal{L} = \varpi v_\phi$ and $\boldsymbol{v}_m = v_r \boldsymbol{e}_r + v_\theta \boldsymbol{e}_\theta$ are the specific angular momentum and the meridional flow velocity, respectively, with $\varpi = r \sin\theta$ in the spherical geometry (r, θ, ϕ). We also note ρ_0 the density background profile (spherical average). The flow velocity is then divided into components, as above, $\boldsymbol{v} = \langle \boldsymbol{v} \rangle + \boldsymbol{v}'$ and, assume a steady state $\partial/\partial t = 0$ balance, the equation for gyroscopic pumping can be written

$$\rho_0 \langle \boldsymbol{v}_m \rangle \cdot \nabla \langle \mathcal{L} \rangle = -\nabla \cdot \left(\rho \varpi \langle \boldsymbol{v}'_m v'_\phi \rangle \right) . \tag{13}$$

While the angular momentum conservation equation (12) determines the temporal evolution of the differential rotation, the gyroscopic pumping balance (Equation (13)) mainly determines the meridional flow $\langle v_m \rangle$ in a steady state. The distribution of the specific angular momentum $\langle \mathcal{L} \rangle$ in the Sun is known to be mostly cylindrical, i.e., $\partial \langle \mathcal{L} \rangle / \partial z \sim 0$, where z denotes the direction of the rotational axis. Thus, gyroscopic pumping can be rewritten as

$$\rho \langle v_\varpi \rangle \frac{\partial \mathcal{L}}{\partial \varpi} \approx -\nabla \cdot \left(\rho_0 \varpi \langle v_m' v_\phi' \rangle \right). \tag{14}$$

Since we know that the sign of $\partial \langle \mathcal{L} \rangle / \partial \varpi$ is greater than zero, the sign of the axial torque $\mathcal{T} = -\nabla \cdot \left(\rho_0 \varpi \langle v_m' v_\phi' \rangle \right)$ directly determines the direction of the meridional flow.

The thermal wind balance equation is derived from the longitudinal vorticity equation,

$$\frac{\partial \langle \zeta_\phi \rangle}{\partial t} = \langle \nabla \times (v \times \zeta) \rangle_\phi + \varpi \frac{\partial \langle \Omega \rangle^2}{\partial z} - \frac{g}{r c_p} \frac{\partial \langle s \rangle}{\partial \theta}, \tag{15}$$

where $\zeta = \nabla \times v$ is the vorticity. This equation is for the evolution of the meridional flow (v_r and v_θ) in terms of the longitudinal vorticity ζ_ϕ. In a steady state ($\partial / \partial t = 0$),

$$\varpi \frac{\partial \langle \Omega \rangle^2}{\partial z} = -\langle \nabla \times (v \times \zeta) \rangle_\phi + \frac{g}{r c_p} \frac{\partial \langle s \rangle}{\partial \theta}, \tag{16}$$

which reduces to the Taylor-Proudman theorem $\partial \langle \Omega \rangle / \partial z = 0$ when advection $\nabla \times (v \times \zeta)$ and the latitudinal entropy gradient $\partial \langle s \rangle / \partial \theta$ are ignored. Under the Taylor-Proudman constraint, contour lines of the angular velocity must be parallel to the rotational axis. As shown in Fig. 1 the solar differential rotation does not follow this configuration, showing instead the prominent tachocline and the near surface shear layer at its boundaries and a more conical profile in the interior. The latitudinal entropy gradient plays a dominant role in forcing the rotation profile away from the Taylor-Proudman state in the deep convection zone where the rotational influence is important. The Reynolds stresses arising from vigorous surface convection (and the magnetic field which we ignored here) are also crucial in the near surface layer (e.g., Hotta et al. 2022) (see Sect. 2.5 for more details).

2.4 Governing Equations for Numerical Simulations

Simulating the global scale motions seen on the Sun directly requires simulating the convective motions in a spherical domain over many scale heights. This in turn requires running efficient numerical code on high-performance supercomputers. For tractability, a number of approximations must be made during formulation.

The solar convection zone is fully ionized below about 20 Mm, and, for global spherical shell magnetohydrodynamic models of convection below that depth, we can reliably assume the equation of state is close to that of a perfect gas. Simulations of the deep solar convection zone must include rotation along with gravitational stratification, and since the superadiabaticity in the solar convection zone is tiny ($\delta \lesssim 10^{-6}$), solving an entropy equation is preferable to formulations in terms of the total energy or internal energy. The set of equations to be solved can thus be written as

$$\frac{\partial \rho}{\partial t} = -\nabla \cdot (\rho v),$$

$$\frac{\partial}{\partial t} (\rho v) = -\nabla \cdot (\rho v v) - \nabla p + \rho g + \frac{1}{4\pi} (\nabla \times B) \times B + 2\rho (v \times \Omega_0),$$

$$\rho T \frac{\partial s}{\partial t} = -\rho T \left(\boldsymbol{v} \cdot \nabla \right) s + Q_{\text{rad}},$$

$$\frac{\partial \boldsymbol{B}}{\partial t} = \nabla \times \left(\boldsymbol{v} \times \boldsymbol{B} \right),$$

$$s = c_{\text{v}} \log \left(\frac{p}{\rho^\gamma} \right).$$

Here, we ignore the viscosity, the magnetic diffusivity, and the thermal conductivity due to the large fluid/magnetic Reynolds and Peclet numbers in the Sun. Moreover, since perturbations in the thermodynamic variables scale with the superadiabicity (Equation (2)) a linearized equation of state is appropriate,

$$\frac{s_1}{c_{\text{v}}} = \frac{p_1}{p_0} - \gamma \frac{\rho_1}{\rho_0}, \tag{17}$$

where γ is the specific heat ratio and subscripts 0 and 1 denote the background and perturbed variables, such that $Q = Q_0 + Q_1$, and we typically assume the hydrostatic balance for the background stratification ρ_0, p_0, and T_0:

$$\frac{dp_0}{dr}(r) = -\rho_0(r)g(r).$$

Pressure gradient and the gravitational forces in the momentum equation are then due to perturbations about this background state,

$$-\nabla p + \rho \boldsymbol{g} \rightarrow -\nabla p_1 + \rho_1 \boldsymbol{g}. \tag{18}$$

Due to a large optical depth in the deep convection zone, the diffusion approximation can be used for radiation energy transfer Q_{rad},

$$Q_{\text{rad}} = -\kappa_{\text{rad}} \nabla T, \tag{19}$$

where κ_{rad} is the radiation diffusion coefficient estimated from the local opacity.

Employing an anelastic approximation is important for deep solar convection simulations since the speed of sound c_s is much faster than the convection velocity v_c. Explicitly solving for sound waves in the domain severely restricts the CFL condition on the time stepping Δt, and a huge number of the time steps would be required to evolve the convective motions and larger-scale flows over dynamical time scales. To avoid this difficulty, the anelastic approximation (Gough 1969) is widely used (Clune et al. 1999), which simplifies the equation of continuity,

$$\nabla \cdot (\rho_0 \boldsymbol{v}) = 0, \tag{20}$$

filtering out sound waves by taking the sound speed to be infinite. In the context of MHD, the anelastic approximation eliminates the fast magneto-sonic waves, while preserving the Alfvén waves and the slow magneto-sonic waves.

Other sonic-filter formulations have been developed including the *Lantz-Braginsky-Roberts* (LBR) method, in which a reduce pressure is introduced and interactions between fluctuating pressure and stratification are neglected (Lantz 1992; Braginsky and Roberts 1995). The LBR method has the advantage of conserving energy well in both unstable convective zones and stable radiative interiors, critical for simulating gravity wave excitation and propagation (Brown et al. 2012; Vasil et al. 2013).

Recently, a method that has come to be known as the Reduced Speed of Sound Technique (RSST) has also found extensive use. With this, the continuity equation is altered,

$$\frac{\partial \rho_1}{\partial t} = -\frac{1}{\xi^2} \nabla \cdot (\rho \boldsymbol{v}), \tag{21}$$

so that the effective speed of sound is reduced by a factor of ξ (Rempel 2005; Hotta et al. 2012). An advantage of the RSST method is that it does not require global communications in a parallel computing environment, in contrast to the anelastic approximation which requires frequent global communication to solve the elliptic equation (20). Thus, the RSST is very useful when solving the MHD equations with massively parallel supercomputers. Additionally, an inhomogeneous ξ can be employed, taking ξ large in the deep layers where the sound speed is fast and $\xi = 1$ in the near-surface layer where the anelastic approximation is not valid. This enables simulation over a continuous domain that covers the whole convection zone (Hotta et al. 2019).

2.5 Modeling of the Solar Large-Scale Flows

The last fifty years have brought tremendous advances in the simulation of solar and stellar convection and our understanding of the global flows that result in rotating domains. Initial analytic analysis of convection in a rotating sphere (Chandrasekhar 1961; Roberts 1968; Busse 1970) were extended to numerical linear and nonlinear numerical studies aimed at understanding the behavior of rotating turbulent astrophysical bodies (Gilman 1975, 1977). In particular, the aim was to understand the importance of nonlinear process in both the solar (Gilman 1979) and terrestrial (Cuong and Busse 1981) contexts. In the solar case, this was strongly motivated by the need to explain solar differential rotation, which had at the time been observed for more than one century (Carrington 1860).

The earliest numerical calculations were done using the Boussinesq approximation, but quickly extended to include stratification using the anelastic approximation (Gilman and Glatzmaier 1981; Glatzmaier and Gilman 1982). At the same time, dynamo calculations were solving for the evolution of a magnetic field in these simulations, adding new theoretical constraints on the understanding of its generation within the Sun (Gilman and Miller 1981; Glatzmaier 1984, 1985). In parallel, stellar evolution models became sophisticated enough to model in detail the solar stratification (i.e., density, pressure, temperature, equation-of-state, opacity as functions of depth) and these models were found to be highly consistent with deductions based on helioseismic measurements (Model S: Christensen-Dalsgaard et al. 1996). With the stratification well modeled, notable advances could be made in a more realistic reproduction of thermal convection (e.g., Miesch et al. 2000; Brun and Toomre 2002). Although spatial resolutions were moderate ($N_\theta < 128$, $\ell_{max} < 85$), the solar-like differential rotation profile (with faster equator and slower poles) was reproduced in these anelastic studies.

However, the differential-rotation profiles found in numerical solutions tended to obey the Taylor-Proudman theorem, i.e., $\partial \langle \Omega \rangle / \partial z = 0$, in contrast to the solar observation (Fig. 1). Motivated by a thorough assessment of mean-field dynamics (Rempel 2005), Miesch et al. (2006) adopted latitudinal entropy gradient at the bottom boundary and achieved non-Taylor-Proudman differential rotation, as indeed is suggested by Equation (16). This somewhat ad-hoc method was adopted by several follow-up studies (e.g. Miesch et al. 2008; Fan and Fang 2014). In particular, Brun et al. (2011) included the overshoot layer between the convection zone and the deeper radiative layer below in a self-consistent model, suggesting that the interaction between the two layers is responsible for the crucial latitudinal entropy gradient, the tachocline, and the conical profile observed in the bulk of the convection zone.

This conclusion has not gone without debate. Earlier, Miesch et al. (2000) concluded that anisotropic entropy transport by the overshooting downflows may not be enough to maintain a solar-like differential rotation profile, though Hotta (2018) point out that an efficient small-scale dynamo can amplify the effect (see also Hotta et al. 2015a), and help produce required non-Taylor-Proudman state. Recently Matilsky et al. (2020) have pointed out that the latitudinal entropy gradient achieved is sensitive to the radial boundary condition imposed. Resolution of these uncertainties is critical to our understanding of the origin of the observed radial profile of the solar differential rotation.

The other non-Taylor-Proudman feature in the solar convection zone is the near-surface shear layer (NSSL). The NSSL is thought to be generated by the radially-inward angular momentum transport by near-surface convection which is not strongly influenced by rotation (e.g., Foukal and Jokipii 1975). Numerical simulations have succeeded in reproducing part of the observed feature (Guerrero et al. 2013; Hotta et al. 2015b; Matilsky et al. 2019), with turbulent viscosity playing an important role (Hotta et al. 2015b), however, the models fail to reproduce key aspects of the NSSL, especially at mid-latitude. Matilsky et al. (2019) argue that the role of large-scale columnar convection (banana cells) and meridional circulation and their balance differ with latitude, but are unable to produce a solution that captures the NSSL in both the equatorial and high-latitude regions. Overall, a consensus has not been reached on the maintenance mechanism for the NSSL.

2.6 Towards Modeling Other Stars

Using models such as the ones presented previously, numerical simulations are a powerful tool to study the formation and dynamics of large-scale flows in the astrophysical context. In particular, full sphere simulations, resolving a broad range of turbulent convective scales, are suitable tools to probe the large-scale dynamics of distant stars. As examples, numerical studies of solar-type stars recently provided new constraints on the trends we can expect for differential rotation ($\Delta\Omega \propto \Omega_*^{0.46}$) on G and K stars (Brun et al. 2022), and highlight the different large-scale flows regimes possible. These regimes can be characterized by the Rossby number (Gastine et al. 2014; Brun and Browning 2017; Hindman et al. 2020).

Thanks to numerical simulations (Matt et al. 2011; Guerrero et al. 2013; Käpylä et al. 2014; Simitev et al. 2015; Brun et al. 2017; Karak et al. 2018), three regimes are currently acknowledged in global rotating models of main-sequence solar-type stars: 1. At low Rossby numbers, differential rotation profiles are highly constrained (Taylor-Proudman) with flows that become cylindrical (Gilman 1975; Elliott et al. 2000; Miesch et al. 2006; Brown et al. 2008). In extreme cases, such profiles show alternating prograde and retrograde zonal jets, also called Jupiter-like jets (Rhines 1975; Heimpel et al. 2016). The meridional circulations under these conditions is typically multicellular in each hemisphere and aligned along the vertical axis (see for example Brun et al. 2017). When magnetic fields are included in the calculation, the Lorentz force feedback can quench the flows, (Brun 2004; Yadav et al. 2015), resulting in a significant decrease of the differential rotation contrast (Ω-quenching) and the appearance of trans-equatorial meridional-circulation cells (Brun et al. 2022). 2. At intermediate Rossby numbers, the differential rotation adopts a typical solar-like conical profile, with a fast equator and slow poles (see for instance Miesch et al. 2006; Hotta et al. 2022). In this regime the power sustaining the differential rotation contrast can reach tens of percent of the stellar luminosity (Brun et al. 2022). 3. At high Rossby number (typically over the unity), differential-rotation profiles in simulations become "anti-solar" (Gastine et al. 2014), showing then a slow equator and fast poles (Gilman 1977).

As stars spin down along the main sequence (Skumanich 1972; Gallet and Bouvier 2013; Ahuir et al. 2021), its rotational influence changes and so does by definition its Rossby number which will increase. The resulting transitions between these large-scale flow regimes may then influence the evolution of the star (Metcalfe et al. 2022), and in particular play important roles in global dynamo processes (Strugarek et al. 2018; Warnecke 2018; Guerrero et al. 2019). Large-scale flows indeed strongly influence the nature of such dynamo processes, however the Lorentz force feedback on these flows are likely to play an important role, both in the construction of the solar-type differential rotation profile (Hotta et al. 2022) and magnetic cycle emergence (Gilman 1983; Augustson et al. 2015). Recent studies highlighted that the latter behavior may arise in specific stellar parameter ranges (Strugarek et al. 2018), where the back-reaction of the Maxwell stress modulates the latitudinal differential rotation contrast, and thus the large-scale magnetic field generation. Probing of these modulations (torsional oscillations) have recently been constrained on the Sun (see Sect. 1.1) and quantified for other stars in numerical parametric studies (Brun et al. 2022).

Recent observations show that the composition of the star too plays a role in the convective dynamo (See et al. 2021). One-dimensional stellar models show varying sensitivities to metallicity (Amard and Matt (2020), see also Sect. 4 of Noraz et al. (2022a) for a discussion). Typically, when metallicity is increased at a given stellar mass and age, the opacity also increases, and thus so too does the temperature gradient within the star. The convective zone becomes deeper in proportion to the stellar radius, with longer convective turnover times at its base because of the higher inertia induced by a higher density. This then likely modifies the convective scale distribution and also decrease the Rossby number of the star, which subsequently impacts the dynamics of the large-scale flows (Bessolaz and Brun 2011). Such metallicity effects may lie at the origin of the observed differences in the magnetism of HD 173701 (mentioned in Sect. 1.4). Karoff et al. (2018) suggest that the higher metallicity of that star could either enhance the magnetic field generated or the observed facular contrast, yielding then the large amplitude brightness variations observed during the activity cycle. Comparisons between solar twins of different metallicities are needed to either confirm or decipher such mechanisms. To guide those observations, local-area numerical studies of the metalicity impact on photospheric convection have already started (Witzke et al. 2022), and studies using global simulations are currently being investigated (Noraz 2022).

A major point to mention here, is that clear detections of anti-solar rotators are still pending for solar-type stars on the main sequence. In addition to the decrease of sensitivity regarding slow-rotation for current observational techniques, recent results highlight that a change in nature of the dynamo may be induced by a rotational transition toward anti-solar rotation (Karak et al. 2015; Viviani et al. 2019; Noraz et al. 2022b; Brun et al. 2022). In that context, a possible disappearance of star-spots is currently discussed in the community, which would make rotational characterization of the high-Rossby regime even more difficult via photometric techniques. An active search is currently underway to better constrain this regime for solar-type stars (Reiners 2007; Reinhold and Arlt 2015; Benomar et al. 2018; Noraz et al. 2022a), and its implications for the solar case.

3 Discrepancies Between Observations and Models and Possible Solutions

Significant difficulties remain when making direct comparison of the observed large-scale motions on the Sun and those produced in numerical simulations still exist. Most fundamentally, until recently (Hotta and Kusano 2021), all global spherical shell models of global

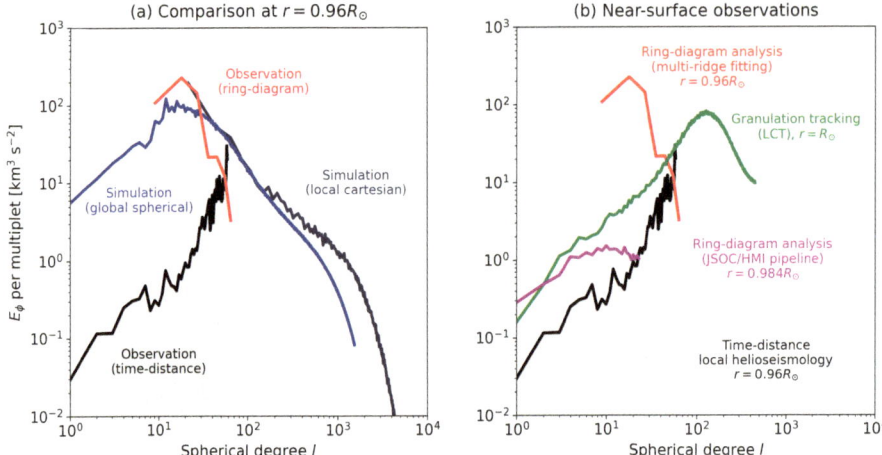

Fig. 5 Horizontal velocity power spectra near the solar surface. (a) Comparison of the spectra between numerical simulations and the observations at $r = 0.96 \, R_\odot$ (Birch 2023). *Blue*: A global full-spherical simulation of rotating magneto-convection by Hotta and Kusano (2021). *Navy*: A local cartesian box simulation of solar convection by Hotta et al. (2019). *Black (gray area)*: Observational upper limits inferred by deep-focusing time-distance helioseismic measurement of Hanasoge et al. (2012), revised recently by Proxauf (2020). (b) Comparison of the spectra obtained by various observational measurements at various depths. *Red*: Multi-ridge fitting ring-diagram analysis by Greer et al. (2015), revised recently by Nagashima et al. (2020). *Green*: Local correlation tracking of surface granulation (Proxauf 2020). *Magenta*: The SDO/HMI ring-diagram pipeline (Bogart et al. 2011a,b; Proxauf 2020). All observations reported in the above plots are available online (Birch 2023)

convection and circulation produced anti-solar differential rotation profiles (a slow equator and a fast pole) at solar rotation rates and at the solar luminosity, and no local area models of solar convection produces a peak in the photospheric horizontal velocity a supergranular scales, as observed. These discrepancies likely reflect a mismatch in convective amplitudes at depth, that has been come to be known as the 'convective conundrum' (O'Mara et al. 2016). The global implications were first recognized when increased numerical resolution in simulations resulted in lower diffusivity, faster convective flows, and consequent difficulty in achieving a solar-like differential rotation profile (Miesch et al. 2008).

3.1 The Convection Conundrum

There is an observation mismatch between flow velocities on the Sun and those found in numerical simulations. In the surface layers this is dramatically illustrated by the absence of supergranulation in local area models, compared to its importance in photospheric observations. That aspect of the convective conundrum is discussed in Sect. 1.2 and 2.2 above. In summary, these disparities suggest reduced convective amplitudes with depth, by a factor of about 2.5 below 10 Mm, with several causes for this reduction suggested in the literature, any one or more of which would suffice.

Convective amplitudes in the near solar surface region of the solar convection zone can be measured using a variety of local helioseismic techniques. As illustrated by Fig. 5 these do not agree with each other or with numerical simulations (Hanasoge et al. 2012; Gizon and Birch 2012; Greer et al. 2015). In general models produce flow with significant power at low wavenumbers, monotonically increasing to low wavenumber, while observations indicate reduced power there. A modeling exception is the recent Hotta and Kusano (2021) simulation, in which the horizontal velocity power rolls over at large scales, though at scales somewhat beyond that of supergranulation. An observational exception is the ring-diagram result of Greer et al. (2015) which shows significant power at scales larger than supergranulation at a depth of 0.96 $R_\odot$.

While refinement and revisions of the techniques employed have brought measurements and models closer together (Nagashima et al. 2020; Proxauf 2020), there are significant and important remaining discrepancies. It is imperative to resolve these, not just to understand the Sun and its dynamo, but because the solar case serves as the touchstone for stellar modeling.

3.2 Columnar Convective Modes

Numerical simulations of rotating convection have repeatedly found that the giant-cell convection tends to exist as banana cells in a strongly rotationally-constrained regime (e.g., Miesch et al. 2000). These banana cells are located outside the tangential cylinder and can be seen as north-south aligned downflow lanes across the equator. They are known to propagate in a prograde direction with frequencies higher than the local differential rotation rate (Miesch et al. 2008).

The prograde propagation of banana cells can be understood in terms of a special class of inertial modes called *columnar convective modes* or *thermal Rossby modes* (e.g., Busse 2002; Miesch et al. 2008; Bekki et al. 2022a). They are z-vorticity waves arising from the compressional β-effect due to the strong background density stratification. A linear dispersion relation of the columnar convective modes was first derived by Glatzmaier and Gilman (1981) using a cylindrical model and later by Hindman and Jain (2022) in Cartesian geometry. The most realistic spherical-shell model has recently been presented by Bekki et al. (2022b). They have shown that the dispersion relation of the columnar convective modes is very sensitive to the superadiabaticity δ, and that the modes become convectively-unstable (exponentially growing) when the background is slightly superadiabatic ($\delta > 0$). These modes have the associated Reynolds stress $\langle v_r v_\phi \rangle > 0$ and $\langle v_\theta v_\phi \rangle > 0$ (< 0) in the northern (southern) hemisphere, implying that they transport the angular momentum radially upward and equatorward outside the tangential cylinder. The great importance of *banana cells* on the establishment of the solar-like differential rotation has been repeatedly appreciated in previous literature (Käpylä et al. 2011; Gastine et al. 2013; Hotta et al. 2015b; Matilsky et al. 2020; Camisassa and Featherstone 2022).

Despite their prominence in numerical simulations, columnar convective modes have not been observed on the surface of the Sun, though very large-scale flows have been deduced from correlation tracking of the solar supergranulation (Hathaway et al. 2013; Hathaway and Upton 2021) and possible indirect evidence may come from the alignment of the solar supergranulation (Lisle et al. 2004; Nagashima et al. 2011). It remains a mystery why the columnar convective modes are not directly detected in the solar surface observations. One possible scenario is that they indeed exist hidden in the deep convection zone with substantial amplitudes but are concealed by surface small-scale convective feature (Guerrero et al. 2013). The other scenario is that they are simply absent in the Sun or too weak to be detected (van Ballegooijen 1986; Lord et al. 2014, Sect. 2.2 above). If the latter is the case, the angular momentum needs to be transported by something other than the large-scale columnar convective modes. For instance, the equatorward angular momentum transport can be achieved in the Sun by the Reynolds and/or Maxwell stresses at much smaller spatial scales.

3.3 Differential Rotation

An important aspect of the convective conundrum is its implications for differential rotation. In early phase of the solar differential rotation research, the solar differential rotation profile could be reproduced relatively easily. As supercomputer power grew, allowing simulation of higher resolution, the problem became apparent. Simulations began to fail to reproduce the solar-like differential rotation profile at solar rotation rates.

A well-known feature of differential rotation is that a fast equator (poles) is obtained with strong (weak) rotational influence (e.g., Gastine et al. 2013; Featherstone and Miesch 2015; Karak et al. 2015). The rotational influence is measured by the Rossby number $Ro = v/(2\Omega H)$, where v is a characteristic convective velocity and H a characteristic spatial scale. Although we believed that the Sun is in a low Rossby-number regime, high resolution simulations most often produce anti-solar differential rotation profiles (fast poles). High resolution simulations introduce the small-scale turbulence which is important for heat transport but is only very weakly rotationally constrained. This has been recently confirmed with scale-dependence analyses of the angular momentum flux (Mori and Hotta 2023). The small-scale turbulence tends to transport the angular momentum radially inward which leads one-cell meridional flow, which transports angular momentum and accelerate the poles (see also Featherstone and Miesch 2015). In other words, high-resolution decreases the convective spatial scale H placing the simulation in a high Rossby number regime, resulting an anti-solar differential rotation.

Three mechanisms have been employed in global simulations in order to maintain a solar-like differential rotation profile, i.e, to maintain the low Rossby number, in the face of vigorous smaller-scale convective flows.

- Increasing the rotation rate Ω_0 (Brown et al. 2008; Nelson et al. 2013; Hotta 2018).
- Reducing the luminosity L (Hotta et al. 2015b), which leads to a reduction of v.
- Increasing viscosity v and/or thermal conductivity κ (Miesch et al. 2000, 2008; Fan and Fang 2014; Hotta et al. 2016), which leads to a reduction of v and an increase of H.
- Increasing radiative conductivity κ_{rad} (Käpylä et al. 2019; Noraz 2022), which leads to a reduction of v while conserving L.

None of these reflect a possible physical mechanism operating on the Sun but not captured in simulations. They cannot provide an answer to the question: Why does the Sun have such a low Rossby number?

Recently, Hotta and Kusano (2021) reproduced the solar-like differential rotation in extremely high resolution simulation without using these manipulations. Hotta et al. (2022) showed that in that simulation the strong magnetic field maintained by the convection when the magnetic diffusivity is very low, as it is in their simulation, causes angular momentum to be transported outward. A double-cell meridional flow is generated and that flow transports angular momentum equatorward, resulting in the solar-like differential rotation profile achieved. An important point is that the Hotta and Kusano (2021) simulation convection remains in a high Rossby number regime but the columnar convective modes no longer play an dominant role in the angular momentum transport. The Rossby number range in which a solar-like differential rotation regime is possible is under investigation, and recent studies suggest that this range depends sensitively on the Prandtl number (Käpylä 2023; Noraz 2022).

4 Future Prospects

From a theoretical view point, some more significant progress can be made in the near future. Hotta and Kusano (2021) show that the high-resolution simulations remain a promising approach to understand the solar large-scale flows. Simulations have not yet reached *numerical convergence*, where by numerical convergence we mean that when we double the resolution the large-scale structure does not change. The solar turbulent convection in the simulations has a spatial spectrum which has more power at low wavenumbers than is observed on the Sun. With the injection scale at 200 Mm and the dissipation scale around 1 cm, it is impossible to directly resolve all the energy containing scales with numerical simulations. Although higher-resolution numerical simulations are required to better understand the nature of turbulent convection, it is also crucial to construct reliable turbulence models which can mimic the essential features of unresolved flows.

Such models require a deep understanding of the key physical components of the convection. Important progress has been made in understanding the essential role of rotation (e.g., Vasil et al. 2021, and references therein) and preliminary work has begun to characterize the highly nonlocal convective flows that result from radiative cooling of the photosphere which may play an important role in heat transport and allow a mean gradient much closer to isentropic than that achieved by current simulations (Brandenburg 2016; Cossette and Rast 2016). Moreover, radiative heating of the lower convection zone likely plays an important role, one that has only begun to be examined. Brun et al. (2011) have undertaken an important study of the interaction between the convection and radiation zone, but, as Käpylä (2019) pointed out, a fully consistent treatment of the overshoot layer requires huge numerical resources, which are currently not available (see also Hotta 2017). Solving these problems on convection is essential to understanding the large-scale flow dynamics.

Another direct approach is to combine in one simulation the radiative magnetohydrodynamics of the photosphere and the global scale convective motions. Including the photosphere may have a significant impact on the deep large-scale convection (Spruit 1997). Hotta et al. (2019) found only weak (or no) influence from the photosphere on the deep convection, however their result may not reflect a lack of importance of these flows, but the difficulties faced in maintaining them with depth given the diffusivities required in global models. Even when the resolution of the simulation is increased (and numerical or explicit diffusivities are reduced), the problem persists because the horizontal scales of the flow structures are also decreased. What is required to maintain nonlocal transport in a simulation is that the diffusion time across a downflowing plume be greater than the transit time of that fluid across the fluid layer. This is extremely difficult to achieve. Because of the steep stratification, local area radiative magnetohydrodynamic simulations, and ambitious global models which include an upper radiative boundary, may be able to produce granular scale downflow structures but not the low thermal diffusivity required for them to maintain their role in transport at depth.

Over the past fifty years, tremendous progress has been made in understanding and simulating solar convection and the global scale flows that result in a rotating domain. The Sun allows us to directly confront that progress with observations that, with the advent and development of helioseismology over the same time period, have also made previously unimaginable advances. New diagnostics based on the study of the solar inertial modes are expected to provide additional constraints on the physics of the deep convection zone. Some gaps in the observations and some fundamental issues in our understanding however remain. Resolving those will not only advance our understanding of the origin of large-scale global flows, but will also allow us to more robustly model the solar dynamo, perhaps predict solar behavior critical to human activities in space and on Earth, and extend our understanding to other stars for which comparison with data will remain less constraining. Understanding the Sun is thus a touchstone activity.

Acknowledgements We acknowledge the support from ISSI Bern for our participation in the workshop. We thank Zhi-Chao Liang for helpful comments on the manuscript.

Funding Open Access funding enabled and organized by Projekt DEAL. HH is supported by JSPS KAK-ENHI grant Nos. JP20K14510, JP21H04492, JP21H01124, JP21H04497, and MEXT as a Program for Promoting Researches on the Supercomputer Fugaku ("Toward a unified view of the universe: from large-scale structures to planets," grant No. 20351188). YB, LG and QN acknowledge financial support from ERC Synergy Grant WHOLE SUN 810218. QN was funded in part by an INSU/PNST grant and CNES (Solar Orbiter). MPR acknowledges partial support for this work by the National Science Foundation under award number NSF 1841100.

Declarations

Competing Interests The authors declare they have no conflicts of interest.

References

Aerts C, Christensen-Dalsgaard J, Kurtz DW (2010) Asteroseismology. Springer, Dordrecht. https://doi.org/10.1007/978-1-4020-5803-5

Ahuir J, Strugarek A, Brun AS et al (2021) Magnetic and tidal migration of close-in planets: influence of secular evolution on their population. Astron Astrophys 650:A126 https://doi.org/10.1051/0004-6361/202040173

Amard L, Matt SP (2020) The impact of metallicity on the evolution of the rotation and magnetic activity of Sun-like stars. Astrophys J 889(2):108. https://doi.org/10.3847/1538-4357/ab6173

Augustson K, Brun AS, Miesch M et al (2015) Grand minima and equatorward propagation in a cycling stellar convective dynamo. Astrophys J 809(2):149. https://doi.org/10.1088/0004-637X/809/2/149

Balona LA, Abedigamba OP (2016) Differential rotation in K, G, F and A stars. Mon Not R Astron Soc 461(1):497–506 https://doi.org/10.1093/mnras/stw1443

Barnes JR, Cameron AC, Donati JF et al (2005) The dependence of differential rotation on temperature and rotation. Mon Not R Astron Soc 357(1):L1–L5 https://doi.org/10.1111/j.1745-3933.2005.08587.x.

Basu S, Antia HM (2010) Characteristics of solar meridional flows during solar cycle 23. Astrophys J 717(1):488–495 https://doi.org/10.1088/0004-637X/717/1/488. arXiv:1005.3031 [astro-ph.SR]

Bazot M, Benomar O, Christensen-Dalsgaard J et al (2019) Latitudinal differential rotation in the solar analogues 16 Cygni A and B. Astron Astrophys 623:A125. https://doi.org/10.1051/0004-6361/201834594

Beaudoin P, Charbonneau P, Racine E et al (2013) Torsional oscillations in a global solar dynamo. Sol Phys 282(2):335–360. https://doi.org/10.1007/s11207-012-0150-2. arXiv:1210.1209 [astro-ph.SR]

Beck JG, Gizon L, Duvall Jr TL (2002) A new component of solar dynamics: North-south diverging flows migrating toward the equator with an 11 year period. Astrophys J 575(1):L47–L50. https://doi.org/10.1086/342636.

Beck PG, Montalban J, Kallinger T et al (2012) Fast core rotation in red-giant stars as revealed by gravity-dominated mixed modes. Nature 481(7379):55–57. https://doi.org/10.1038/nature10612. arXiv:1112.2825 [astro-ph.SR]

Bekki Y, Cameron RH (2023) Three-dimensional non-kinematic simulation of the post-emergence evolution of bipolar magnetic regions and the Babcock-Leighton dynamo of the Sun. Astron Astrophys 670:A101. https://doi.org/10.1051/0004-6361/202244990. arXiv:2209.08178 [astro-ph.SR]

Bekki Y, Hotta H, Yokoyama T (2017) Convective velocity suppression via the enhancement of the subadiabatic layer: role of the effective Prandtl number. Astrophys J 851(2):74. https://doi.org/10.3847/1538-4357/aa9b7f. arXiv:1711.05960 [astro-ph.SR]

Bekki Y, Cameron RH, Gizon L (2022a) Theory of solar oscillations in the inertial frequency range: amplitudes of equatorial modes from a nonlinear rotating convection simulation. Astron Astrophys 666:A135. https://doi.org/10.1051/0004-6361/202244150. arXiv:2208.11081 [astro-ph.SR]

Bekki Y, Cameron RH, Gizon L (2022b) Theory of solar oscillations in the inertial frequency range: linear modes of the convection zone. Astron Astrophys 662:A16. https://doi.org/10.1051/0004-6361/202243164. arXiv:2203.04442 [astro-ph.SR]

Benomar O, Bazot M, Nielsen MB et al (2018) Asteroseismic detection of latitudinal differential rotation in 13 Sun-like stars. Science 361(6408):1231–1234. https://doi.org/10.1126/science.aao6571

Berrilli F, Del Moro D, Russo S et al (2005) Spatial clustering of photospheric structures. Astrophys J 632(1):677–683. https://doi.org/10.1086/432708.

Berrilli F, Scardigli S, Giordano S (2013) Multiscale magnetic underdense regions on the solar surface: granular and mesogranular scales. Sol Phys 282(2):379–387. https://doi.org/10.1007/s11207-012-0179-2. arXiv:1208.2669 [astro-ph.SR]

Bessolaz N, Brun AS (2011) Hunting for giant cells in deep stellar convective zones using wavelet analysis. Astrophys J 728(2):115. https://doi.org/10.1088/0004-637X/728/2/115

Bhattacharya J, Hanasoge SM (2023) A spectral solver for solar inertial waves. Astrophys J Suppl Ser 264(1):21. https://doi.org/10.3847/1538-4365/aca09a. arXiv:2211.03323 [astro-ph.SR]

Birch A (2023) Convection spectra from the thesis of B. Proxauf. https://doi.org/10.17617/3.DFU3SQ

Bogart RS, Baldner C, Basu S et al (2011a) HMI ring diagram analysis I. The processing pipeline. In: GONG-SoHO 24: a new era of seismology of the Sun and solar-like stars, p 012008. https://doi.org/10.1088/1742-6596/271/1/012008

Bogart RS, Baldner C, Basu S et al (2011b) HMI ring diagram analysis II. Data products. In: GONG-SoHO 24: a new era of seismology of the Sun and solar-like stars, p 012009. https://doi.org/10.1088/1742-6596/271/1/012009

Bogart RS, Baldner CS, Basu S (2015) Evolution of near-surface flows inferred from high-resolution ring-diagram analysis. Astrophys J 807(2):125. https://doi.org/10.1088/0004-637X/807/2/125. arXiv:1506.01733 [astro-ph.SR]

Böhm-Vitense E (1958) Über die Wasserstoffkonvektionszone in Sternen verschiedener Effektivtemperaturen und Leuchtkräfte. Mit 5 Textabbildungen. Z Astrophys 46:108

Braginsky SI, Roberts PH (1995) Equations governing convection in Earth's core and the geodynamo. Geophys Astrophys Fluid Dyn 79(1–4):1–97. https://doi.org/10.1080/03091929508228992

Brandenburg A (2016) Stellar mixing length theory with entropy rain. Astrophys J 832(1):6. https://doi.org/10.3847/0004-637X/832/1/6. arXiv:1504.03189 [astro-ph.SR]

Braun DC, Fan Y (1998) Helioseismic measurements of the subsurface meridional flow. Astrophys J Lett 508:L105–L108. https://doi.org/10.1086/311727

Braun DC, Birch AC, Fan Y (2021) Probing the solar meridional circulation using Fourier Legendre decomposition. Astrophys J 911(1):54. https://doi.org/10.3847/1538-4357/abe7e4. arXiv:2103.02499 [astro-ph.SR]

Brown BP, Browning MK, Brun AS et al (2008) Rapidly rotating suns and active nests of convection. Astrophys J 689(2):1354–1372. https://doi.org/10.1086/592397

Brown BP, Vasil GM, Zweibel EG (2012) Energy conservation and gravity waves in sound-proof treatments of stellar interiors. Part I. Anelastic approximations. Astrophys J 756(2):109. https://doi.org/10.1088/0004-637X/756/2/109

Brun AS (2004) On the interaction between differential rotation and magnetic fields in the Sun. Sol Phys 220(2):333–344. https://doi.org/10.1023/B:SOLA.0000031384.75850.68

Brun AS, Browning MK (2017) Magnetism, dynamo action and the solar-stellar connection. Living Rev Sol Phys 14(1):4. https://doi.org/10.1007/s41116-017-0007-8

Brun AS, Toomre J (2002) Turbulent convection under the influence of rotation: sustaining a strong differential rotation. Astrophys J 570(2):865–885. https://doi.org/10.1086/339228

Brun AS, Miesch MS, Toomre J (2011) Modeling the dynamical coupling of solar convection with the radiative interior. Astrophys J 742(2):79. https://doi.org/10.1088/0004-637X/742/2/79

Brun AS, Strugarek A, Varela J et al (2017) On differential rotation and overshooting in solar-like stars. Astrophys J 836(2):192. https://doi.org/10.3847/1538-4357/aa5c40

Brun AS, Strugarek A, Noraz Q et al (2022) Powering stellar magnetism: energy transfers in cyclic dynamos of Sun-like stars. Astrophys J 926(1):21. https://doi.org/10.3847/1538-4357/ac469b

Busse FH (1970) Thermal instabilities in rapidly rotating systems. J Fluid Mech 44(3):441–460. https://doi.org/10.1017/S0022112070001921

Busse FH (2002) Convective flows in rapidly rotating spheres and their dynamo action. Phys Fluids 14(4):1301–1314. https://doi.org/10.1063/1.1455626

Camisassa ME, Featherstone NA (2022) Solar-like to antisolar differential rotation: a geometric interpretation. Astrophys J 938(1):65. https://doi.org/10.3847/1538-4357/ac879f. arXiv:2208.05591 [astro-ph.SR]

Canfield RC, Mehltretter JP (1973) Fluctuations of brightness and vertical velocity at various heights in the photosphere. Sol Phys 33(1):33–48. https://doi.org/10.1007/BF00152375

Carrington RC (1860) On two cases of solar spots in high latitudes, and on the surface currents indicated by the observations. Mon Not R Astron Soc 20:254. https://doi.org/10.1093/mnras/20.6.254

Cattaneo F, Lenz D, Weiss N (2001) On the origin of the solar mesogranulation. Astrophys J 563(1):L91–L94. https://doi.org/10.1086/338355

Chandrasekhar S (1961) Hydrodynamic and hydromagnetic stability. Dover, New York

Chaplin WJ, Sanchis-Ojeda R, Campante TL et al (2013) Asteroseismic determination of obliquities of the exoplanet systems Kepler-50 and Kepler-65. Astrophys J 766(2):101. https://doi.org/10.1088/0004-637X/766/2/101. arXiv:1302.3728 [astro-ph.EP]

Charbonneau P (2020) Dynamo models of the solar cycle. Living Rev Sol Phys 17(1):4. https://doi.org/10.1007/s41116-020-00025-6

Chen R, Zhao J (2017) A comprehensive method to measure solar meridional circulation and the center-to-limb effect using time-distance helioseismology. Astrophys J 849(2):144. https://doi.org/10.3847/1538-4357/aa8eec. arXiv:1709.07905 [astro-ph.SR]

Chen R, Zhao J (2018) Frequency dependence of helioseismic measurements of the center-to-limb effect and flow-induced travel-time shifts. Astrophys J 853(2):161. https://doi.org/10.3847/1538-4357/aaa3e3

Christensen-Dalsgaard J, Däppen W, Ajukov SV et al (1996) The current state of solar modeling. Science 272(5266):1286–1292. https://doi.org/10.1126/science.272.5266.1286

Clune T, Elliott J, Miesch M et al (1999) Computational aspects of a code to study rotating turbulent convection in spherical shells. Parallel Comput 25(4):361–380. https://doi.org/10.1016/S0167-8191(99)00009-5

Collier Cameron A (2007) Differential rotation on rapidly rotating stars. Astron Nachr 328(10):1030–1033. https://doi.org/10.1002/asna.200710880

Cossette JF, Rast MP (2016) Supergranulation as the largest buoyantly driven convective scale of the Sun. Astrophys J 829(1):L17. https://doi.org/10.3847/2041-8205/829/1/L17. arXiv:1606.04041 [astro-ph.SR]

Cranmer SR, Saar SH (2011) Testing a predictive theoretical model for the mass loss rates of cool stars. Astrophys J 741(1):54. https://doi.org/10.1088/0004-637X/741/1/54

Crouch AD, Charbonneau P, Thibault K (2007) Supergranulation as an emergent length scale. Astrophys J 662(1):715–729. https://doi.org/10.1086/515564

Cuong P, Busse F (1981) Generation of magnetic fields by convection in a rotating sphere, I. Phys Earth Planet Inter 24(4):272–283. https://doi.org/10.1016/0031-9201(81)90114-X

Deheuvels S, García RA, Chaplin WJ et al (2012) Seismic evidence for a rapidly rotating core in a lower-giant-branch star observed with KEPLER. Astrophys J 756(1):19. https://doi.org/10.1088/0004-637X/756/1/19

Deheuvels S, Ballot J, Eggenberger P et al (2020) Seismic evidence for near solid-body rotation in two Kepler subgiants and implications for angular momentum transport. Astron Astrophys 641:A117. https://doi.org/10.1051/0004-6361/202038578

Demarque PR, Percy JR (1964) A series of solar models. Astrophys J 140:541. https://doi.org/10.1086/147947

Donahue RA, Saar SH, Baliunas SL (1996) A relationship between mean rotation period in lower main-sequence stars and its observed range. Astrophys J 466:384. https://doi.org/10.1086/177517

Donati JF, Lehmann LT, Cristofari PI et al (2023) Magnetic fields & rotation periods of M dwarfs from SPIRou spectra. https://doi.org/10.48550/arXiv.2307.14190. arXiv:2307.14190 [astro-ph.SR]

Duvall Jr TL (1979) Large-scale solar velocity fields. Sol Phys 63(1):3–15. https://doi.org/10.1007/BF00155690

Duvall Jr TL, Birch AC (2010) The vertical component of the supergranular motion. Astrophys J 725(1):L47–L51. https://doi.org/10.1088/2041-8205/725/1/L47

Duvall Jr TL, Dziembowski WA, Goode PR et al (1984) Internal rotation of the Sun. Nature 310(5972):22–25. https://doi.org/10.1038/310022a0

Duvall Jr TL, Jefferies SM, Harvey JW et al (1993) Time-distance helioseismology. Nature 362:430–432. https://doi.org/10.1038/362430a0

Duvall Jr TL, Hanasoge SM (2009) Measuring meridional circulation in the Sun. In: Dikpati M, Arentoft T, González Hernández I et al (eds) Solar-stellar dynamos as revealed by helio- and asteroseismology: GONG 2008/SOHO 21, p 103. arXiv:0905.3132

Elliott JR, Miesch MS, Toomre J (2000) Turbulent solar convection and its coupling with rotation: the effect of Prandtl number and thermal boundary conditions on the resulting differential rotation. Astrophys J 533(1):546–556. https://doi.org/10.1086/308643

Falco M, Puglisi G, Guglielmino SL et al (2017) Comparison of different populations of granular features in the solar photosphere. Astron Astrophys 605:A87. https://doi.org/10.1051/0004-6361/201629881

Fan Y, Fang F (2014) A simulation of convective dynamo in the solar convective envelope: maintenance of the solar-like differential rotation and emerging flux. Astrophys J 789(1):35. https://doi.org/10.1088/0004-637X/789/1/35. arXiv:1405.3926 [astro-ph.SR]

Featherstone NA, Hindman BW (2016) The emergence of solar supergranulation as a natural consequence of rotationally constrained interior convection. Astrophys J 830(1):L15. https://doi.org/10.3847/2041-8205/830/1/L15. arXiv:1609.05153 [astro-ph.SR]

Featherstone NA, Miesch MS (2015) Meridional circulation in solar and stellar convection zones. Astrophys J 804(1):67. https://doi.org/10.1088/0004-637X/804/1/67. arXiv:1501.06501 [astro-ph.SR]

Foukal P, Jokipii JR (1975) On the rotation of gas and magnetic fields at the solar photosphere. Astrophys J 199:L71–L73. https://doi.org/10.1086/181851

Gallet F, Bouvier J (2013) Improved angular momentum evolution model for solar-like stars. Astron Astrophys 556:A36. https://doi.org/10.1051/0004-6361/201321302

García RA, Ballot J (2019) Asteroseismology of solar-type stars. Living Rev Sol Phys 16(1):4. https://doi.org/10.1007/s41116-019-0020-1

Gastine T, Wicht J, Aurnou JM (2013) Zonal flow regimes in rotating anelastic spherical shells: an application to giant planets. Icarus 225(1):156–172. https://doi.org/10.1016/j.icarus.2013.02.031. arXiv:1211.3246 [astro-ph.EP]

Gastine T, Yadav RK, Morin J et al (2014) From solar-like to antisolar differential rotation in cool stars. Mon Not R Astron Soc 438(1):L76–L80. https://doi.org/10.1093/mnrasl/slt162

Giles PM (2000) Time-distance measurements of large-scale flows in the solar convection zone. PhD thesis, Stanford University, California

Giles PM, Duvall TL, Scherrer PH et al (1997) A subsurface flow of material from the Sun's equator to its poles. Nature 390:52–54. https://doi.org/10.1038/36294

Gilman PA (1975) Linear simulations of Boussinesq convection in a deep rotating spherical shell. J Atmos Sci 32(7):1331–1352

Gilman PA (1977) Nonlinear dynamics of Boussinesq convection in a deep rotating spherical shell. I. Geophys Astrophys Fluid Dyn 8:93–135. https://doi.org/10.1080/03091927708240373

Gilman PA (1979) Model calculations concerning rotation at high solar latitudes and the depth of the solar convection zone. Astrophys J 231:284. https://doi.org/10.1086/157191

Gilman PA (1983) Dynamically consistent nonlinear dynamos driven by convection in a rotating spherical shell. II – dynamos with cycles and strong feedbacks. Astrophys J Suppl Ser 53:243. https://doi.org/10.1086/190891

Gilman PA, Glatzmaier GA (1981) Compressible convection in a rotating spherical shell. I. Anelastic equations. Astrophys J Suppl Ser 45(2):53

Gilman PA, Miller J (1981) Dynamically consistent nonlinear dynamos driven by convection in a rotating spherical shell. Astrophys J Suppl Ser 46:211. https://doi.org/10.1086/190743

Gizon L (2004) Helioseismology of time-varying flows through the solar cycle. Sol Phys 224(1–2):217–228. https://doi.org/10.1007/s11207-005-4983-9

Gizon L, Birch AC (2005) Local helioseismology. Living Rev Sol Phys 2(1):6. https://doi.org/10.12942/lrsp-2005-6

Gizon L, Birch AC (2012) Helioseismology challenges models of solar convection. Proc Natl Acad Sci 109(30):11896–11897. https://doi.org/10.1073/pnas.1208875109

Gizon L, Duvall TL, Schou J (2003) Wave-like properties of solar supergranulation. Nature 421(6918):43–44. https://doi.org/10.1038/nature01287. arXiv:astro-ph/0208343 [astro-ph]

Gizon L, Birch AC, Spruit HC (2010) Local helioseismology: three-dimensional imaging of the solar interior. Annu Rev Astron Astrophys 48:289–338. https://doi.org/10.1146/annurev-astro-082708-101722. arXiv:1001.0930 [astro-ph.SR]

Gizon L, Ballot J, Michel E et al (2013) Seismic constraints on rotation of Sun-like star and mass of exoplanet. Proc Natl Acad Sci 110(33):13267–13271. https://doi.org/10.1073/pnas.1303291110. arXiv:1308.4352 [astro-ph.SR]

Gizon L, Cameron RH, Pourabdian M et al (2020a) Meridional flow in the Sun's convection zone is a single cell in each hemisphere. Science 368(6498):1469–1472. https://doi.org/10.1126/science.aaz7119

Gizon L, Fournier D, Albekioni M (2020b) Effect of latitudinal differential rotation on solar Rossby waves: critical layers, eigenfunctions, and momentum fluxes in the equatorial β plane. Astron Astrophys 642:A178. https://doi.org/10.1051/0004-6361/202038525. arXiv:2008.02185 [astro-ph.SR]

Gizon L, Cameron RH, Bekki Y et al (2021) Solar inertial modes: observations, identification, and diagnostic promise. Astron Astrophys 652:L6. https://doi.org/10.1051/0004-6361/202141462. arXiv:2107.09499 [astro-ph.SR]

Glatzmaier GA (1984) Numerical simulations of stellar convective dynamos. I. The model and method. J Comput Phys 55(3):461–484. https://doi.org/10.1016/0021-9991(84)90033-0

Glatzmaier GA (1985) Numerical simulations of stellar convective dynamos. II – field propagation in the convection zone. Astrophys J 291:300–307. https://doi.org/10.1086/163069

Glatzmaier GA, Gilman PA (1981) Compressible convection in a rotating spherical shell – part three – analytic model for compressible vorticity waves. Astrophys J Suppl Ser 45:381. https://doi.org/10.1086/190716

Glatzmaier GA, Gilman PA (1982) Compressible convection in a rotating spherical shell. V – induced differential rotation and meridional circulation. Astrophys J 256:316–330. https://doi.org/10.1086/159909

Goldbaum N, Rast MP, Ermolli I et al (2009) The intensity profile of the solar supergranulation. Astrophys J 707(1):67–73. https://doi.org/10.1088/0004-637X/707/1/67. arXiv:0909.3310 [astro-ph.SR]

Gough DO (1969) The anelastic approximation for thermal convection. J Atmos Sci 26(3):448–456. https://doi.org/10.1175/1520-0469(1969)026<0448:TAAFTC>2.0.CO;2

Greenspan H, Batchelor C, Ablowitz M et al (1968) The theory of rotating fluids. Cambridge monographs on mechanics. Cambridge University Press, Cambridge. https://books.google.de/books?id=2R47AAAAIAAJ

Greer BJ, Hindman BW, Featherstone NA et al (2015) Helioseismic imaging of fast convective flows throughout the near-surface shear layer. Astrophys J 803(2):L17. https://doi.org/10.1088/2041-8205/803/2/L17. arXiv:1504.00699 [astro-ph.SR]

Guerrero G, Smolarkiewicz PK, Kosovichev AG et al (2013) Differential rotation in solar-like stars from global simulations. Astrophys J 779:176. https://doi.org/10.1088/0004-637X/779/2/176

Guerrero G, Zaire B, Smolarkiewicz PK et al (2019) What sets the magnetic field strength and cycle period in solar-type stars? Astrophys J 880(1):6. https://doi.org/10.3847/1538-4357/ab224a

Haber DA, Hindman BW, Toomre J et al (2002) Evolving submerged meridional circulation cells within the upper convection zone revealed by ring-diagram analysis. Astrophys J 570(2):855–864. https://doi.org/10.1086/339631

Hanasoge SM (2022) Surface and interior meridional circulation in the Sun. Living Rev Sol Phys 19(1):3. https://doi.org/10.1007/s41116-022-00034-7

Hanasoge SM, Duvall TL, Sreenivasan KR (2012) Anomalously weak solar convection. Proc Natl Acad Sci 109(30):11928–11932. https://doi.org/10.1073/pnas.1206570109. arXiv:1206.3173 [astro-ph.SR]

Hanson CS, Gizon L, Liang ZC (2020) Solar Rossby waves observed in GONG++ ring-diagram flow maps. Astron Astrophys 635:A109. https://doi.org/10.1051/0004-6361/201937321. arXiv:2002.01194 [astro-ph.SR]

Hanson CS, Hanasoge S, Sreenivasan KR (2022) Discovery of high-frequency retrograde vorticity waves in the Sun. Nat Astron 6:708–714. https://doi.org/10.1038/s41550-022-01632-z

Hart AB (1954) Motions in the Sun at the photospheric level. IV. The equatorial rotation and possible velocity fields in the photosphere. Mon Not R Astron Soc 114:17. https://doi.org/10.1093/mnras/114.1.17

Hathaway DH (1996) Doppler measurements of the Sun's meridional flow. Astrophys J 460:1027. https://doi.org/10.1086/177029

Hathaway DH, Rightmire L (2010) Variations in the Sun's meridional flow over a solar cycle. Science 327(5971):1350. https://doi.org/10.1126/science.1181990

Hathaway DH, Upton LA (2021) Hydrodynamic properties of the Sun's giant cellular flows. Astrophys J 908(2):160. https://doi.org/10.3847/1538-4357/abcbfa. arXiv:2006.06084 [astro-ph.SR]

Hathaway DH, Upton L, Colegrove O (2013) Giant convection cells found on the Sun. Science 342(6163):1217–1219. https://doi.org/10.1126/science.1244682

Hathaway DH, Teil T, Norton AA et al (2015) The Sun's photospheric convection spectrum. Astrophys J 811(2):105. https://doi.org/10.1088/0004-637X/811/2/105. arXiv:1508.03022 [astro-ph.SR]

Hazra G, Karak BB, Choudhuri AR (2014) Is a deep one-cell meridional circulation essential for the flux transport solar dynamo? Astrophys J 782(2):93. https://doi.org/10.1088/0004-637X/782/2/93. arXiv:1309.2838 [astro-ph.SR]

Heimpel M, Gastine T, Wicht J (2016) Simulation of deep-seated zonal jets and shallow vortices in gas giant atmospheres. Nat Geosci 9(1):19–23. https://doi.org/10.1038/ngeo2601

Henry GW, Eaton JA, Hamer J et al (1995) Starspot evolution, differential rotation, and magnetic cycles in the chromospherically active binaries lambda Andromedae, sigma Geminorum, II Pegasi, and V711 Tauri. Astrophys J Suppl Ser 97:513. https://doi.org/10.1086/192149

Herschel W (1801) Observations tending to investigate the nature of the Sun, in order to find the causes or symptoms of its variable emission of light and heat; with remarks on the use that may possibly be drawn from solar observations. Philos Trans R Soc Lond Ser A 91:265–318

Hill F (1988) Rings and trumpets – three-dimensional power spectra of solar oscillations. Astrophys J 333:996. https://doi.org/10.1086/166807

Hindman BW, Jain R (2022) Radial trapping of thermal Rossby waves within the convection zones of low-mass stars. Astrophys J 932(1):68. https://doi.org/10.3847/1538-4357/ac6d64. arXiv:2205.02346 [astro-ph.SR]

Hindman BW, Featherstone NA, Julien K (2020) Morphological classification of the convective regimes in rotating stars. Astrophys J 898(2):120. https://doi.org/10.3847/1538-4357/ab9ec2

Hirzberger J (2002) On the brightness and velocity structure of solar granulation. Astron Astrophys 392:1105–1118. https://doi.org/10.1051/0004-6361:20020902

Hotta H (2017) Solar overshoot region and small-scale dynamo with realistic energy flux. Astrophys J 843(1):52. https://doi.org/10.3847/1538-4357/aa784b. arXiv:1706.06413 [astro-ph.SR]

Hotta H (2018) Breaking Taylor-Proudman balance by magnetic fields in stellar convection zones. Astrophys J 860(2):L24. https://doi.org/10.3847/2041-8213/aacafb. arXiv:1806.01452 [astro-ph.SR]

Hotta H, Kusano K (2021) Solar differential rotation reproduced with high-resolution simulation. Nat Astron 5:1100–1102. https://doi.org/10.1038/s41550-021-01459-0. arXiv:2109.06280 [astro-ph.SR]

Hotta H, Rempel M, Yokoyama T et al (2012) Numerical calculation of convection with reduced speed of sound technique. Astron Astrophys 539:A30. https://doi.org/10.1051/0004-6361/201118268. arXiv:1201.1061 [astro-ph.SR]

Hotta H, Rempel M, Yokoyama T (2015a) Efficient small-scale dynamo in the solar convection zone. Astrophys J 803(1):42. https://doi.org/10.1088/0004-637X/803/1/42. arXiv:1502.03846 [astro-ph.SR]

Hotta H, Rempel M, Yokoyama T (2015b) High-resolution calculation of the solar global convection with the reduced speed of sound technique. II. Near surface shear layer with the rotation. Astrophys J 798(1):51. https://doi.org/10.1088/0004-637X/798/1/51. arXiv:1410.7093 [astro-ph.SR]

Hotta H, Rempel M, Yokoyama T (2016) Large-scale magnetic fields at high Reynolds numbers in magneto-hydrodynamic simulations. Science 351(6280):1427–1430. https://doi.org/10.1126/science.aad1893

Hotta H, Iijima H, Kusano K (2019) Weak influence of near-surface layer on solar deep convection zone revealed by comprehensive simulation from base to surface. Sci Adv 5(1):2307. https://doi.org/10.1126/sciadv.aau2307

Hotta H, Kusano K, Shimada R (2022) Generation of solar-like differential rotation. Astrophys J 933(2):199. https://doi.org/10.3847/1538-4357/ac7395. arXiv:2202.04183 [astro-ph.SR]

Howard R, Labonte BJ (1980) The sun is observed to be a torsional oscillator with a period of 11 years. Astrophys J 239:L33–L36. https://doi.org/10.1086/183286

Howe R (2009) Solar interior rotation and its variation. Living Rev Sol Phys 6(1):1. https://doi.org/10.12942/lrsp-2009-1. arXiv:0902.2406 [astro-ph.SR]

Howe R, Christensen-Dalsgaard J, Hill F et al (2000) Dynamic variations at the base of the solar convection zone. Science 287(5462):2456–2460. https://doi.org/10.1126/science.287.5462.2456

Howe R, Christensen-Dalsgaard J, Hill F et al (2005) Solar convection-zone dynamics 1995-2004. Astrophys J 634(2):1405–1415. https://doi.org/10.1086/497107

Howe R, Christensen-Dalsgaard J, Hill F et al (2013) The high-latitude branch of the solar torsional oscillation in the rising phase of cycle 24. Astrophys J 767(1):L20. https://doi.org/10.1088/2041-8205/767/1/L20

Howe R, Komm RW, Baker D et al (2015) Persistent near-surface flow structures from local helioseismology. Sol Phys 290(11):3137–3149. https://doi.org/10.1007/s11207-015-0747-3. arXiv:1507.06525 [astro-ph.SR]

Jackiewicz J, Serebryanskiy A, Kholikov S (2015) Meridional flow in the solar convection zone. II. Helioseismic inversions of GONG data. Astrophys J 805(2):133. https://doi.org/10.1088/0004-637X/805/2/133. arXiv:1504.08071 [astro-ph.SR]

Käpylä PJ (2019) Overshooting in simulations of compressible convection. Astron Astrophys 631:A122. https://doi.org/10.1051/0004-6361/201834921. arXiv:1812.07916 [astro-ph.SR]

Käpylä PJ (2023) Transition from anti-solar to solar-like differential rotation: dependence on Prandtl number. Astron Astrophys 669:A98. https://doi.org/10.1051/0004-6361/202244395

Käpylä PJ, Mantere MJ, Guerrero G et al (2011) Reynolds stress and heat flux in spherical shell convection. Astron Astrophys 531:A162. https://doi.org/10.1051/0004-6361/201015884. arXiv:1010.1250 [astro-ph.SR]

Käpylä PJ, Käpylä MJ, Brandenburg A (2014) Confirmation of bistable stellar differential rotation profiles. Astron Astrophys 570:A43. https://doi.org/10.1051/0004-6361/201423412

Käpylä PJ, Viviani M, Käpylä MJ et al (2019) Effects of a subadiabatic layer on convection and dynamos in spherical wedge simulations. Geophys. Astrophys. Fluid Dyn. 113(1–2):149–183. https://doi.org/10.1080/03091929.2019.1571584. arxiv:1803.05898

Karak BB, Käpylä PJ, Käpylä MJ et al (2015) Magnetically controlled stellar differential rotation near the transition from solar to anti-solar profiles. Astron Astrophys 576:A26. https://doi.org/10.1051/0004-6361/201424521

Karak BB, Miesch M, Bekki Y (2018) Consequences of high effective Prandtl number on solar differential rotation and convective velocity. Phys Fluids 30(4):046602. https://doi.org/10.1063/1.5022034. arXiv: 1801.00560

Karoff C, Metcalfe TS, Santos ARG et al (2018) The influence of metallicity on stellar differential rotation and magnetic activity. Astrophys J 852(1):46. https://doi.org/10.3847/1538-4357/aaa026

Kueker M, Ruediger G (2011) Differential rotation and meridional flow on the lower zero age main sequence: Reynolds stress versus baroclinic flow. Astron Nachr 332(9–10):933–938. https://doi.org/10.1002/asna. 201111628. arXiv:1110.4757

Langfellner J, Birch AC, Gizon L (2018) Evolution and wave-like properties of the average solar supergranule. Astron Astrophys 617:A97. https://doi.org/10.1051/0004-6361/201732471. arXiv:1805.12522 [astro-ph.SR]

Lantz SR (1992) Dynamical behavior of magnetic fields in a stratified, convecting fluid layer. PhD thesis, Cornell University, New York

Larson TP, Schou J (2018) Global-mode analysis of full-disk data from the Michelson Doppler Imager and the Helioseismic and Magnetic Imager. Sol Phys 293(2):29. https://doi.org/10.1007/s11207-017-1201-5

Leighton RB, Noyes RW, Simon GW (1962) Velocity fields in the solar atmosphere. I. Preliminary report. Astrophys J 135:474. https://doi.org/10.1086/147285

Leitzinger M, Brandt PN, Hanslmeier A et al (2005) Dynamics of solar mesogranulation. Astron Astrophys 444(1):245–255. https://doi.org/10.1051/0004-6361:20053152

Liang ZC, Chou DY (2015) Effects of solar surface magnetic fields on the time-distance analysis of solar subsurface meridional flows. Astrophys J 805(2):165. https://doi.org/10.1088/0004-637X/805/2/165

Liang ZC, Birch AC, Duvall Jr TL et al (2017) Comparison of acoustic travel-time measurements of solar meridional circulation from SDO/HMI and SOHO/MDI. Astron Astrophys 601:A46. https://doi.org/10. 1051/0004-6361/201730416. arXiv:1704.00475 [astro-ph.SR]

Liang ZC, Gizon L, Birch AC et al (2018) Solar meridional circulation from twenty-one years of SOHO/MDI and SDO/HMI observations. Helioseismic travel times and forward modeling in the ray approximation. Astron Astrophys 619:A99. https://doi.org/10.1051/0004-6361/201833673. arXiv:1808.08874 [astro-ph.SR]

Liang ZC, Gizon L, Birch AC et al (2019) Time-distance helioseismology of solar Rossby waves. Astron Astrophys 626:A3. https://doi.org/10.1051/0004-6361/201834849. arXiv:1812.07413 [astro-ph.SR]

Lisle JP, Rast MP, Toomre J (2004) Persistent North-south alignment of the solar supergranulation. Astrophys J 608(2):1167–1174. https://doi.org/10.1086/420691

Löptien B, Gizon L, Birch AC et al (2018) Global-scale equatorial Rossby waves as an essential component of solar internal dynamics. Nat Astron 2:568–573. https://doi.org/10.1038/s41550-018-0460-x. arXiv: 1805.07244 [astro-ph.SR]

Lord JW (2014) Deep convection, magnetism and solar supergranulation. PhD thesis, University of Colorado at Boulder

Lord JW, Cameron RH, Rast MP et al (2014) The role of subsurface flows in solar surface convection: modeling the spectrum of supergranular and larger scale flows. Astrophys J 793(1):24. https://doi.org/ 10.1088/0004-637X/793/1/24. arXiv:1407.2209 [astro-ph.SR]

Mahajan SS, Sun X, Zhao J (2023) Removal of active region inflows reveals a weak solar cycle scale trend in the near-surface meridional flow. Astrophys J 950(1):63. https://doi.org/10.3847/1538-4357/acc839. arXiv:2304.02158 [astro-ph.SR]

Mandal K, Hanasoge S (2020) Properties of solar Rossby waves from normal mode coupling and characterizing its systematics. Astrophys J 891(2):125. https://doi.org/10.3847/1538-4357/ab7227. arXiv:1908. 05890 [astro-ph.SR]

Mandal K, Hanasoge SM, Rajaguru SP et al (2018) Helioseismic inversion to infer the depth profile of solar meridional flow using spherical born kernels. Astrophys J 863(1):39. https://doi.org/10.3847/1538-4357/aacea2. arXiv:1807.00314 [astro-ph.SR]

Matilsky LI, Hindman BW, Toomre J (2019) The role of downflows in establishing solar near-surface shear. Astrophys J 871(2):217. https://doi.org/10.3847/1538-4357/aaf647. arXiv:1810.00115 [astro-ph.SR]

Matilsky LI, Hindman BW, Toomre J (2020) Revisiting the Sun's strong differential rotation along radial lines. Astrophys J 898(2):111. https://doi.org/10.3847/1538-4357/ab9ca0. arXiv:2004.00208 [astro-ph.SR]

Matilsky LI, Hindman BW, Featherstone NA et al (2022) Confinement of the solar tachocline by dynamo action in the radiative interior. Astrophys J 940(2):L50. https://doi.org/10.3847/2041-8213/ac93ef. arXiv: 2206.12920 [astro-ph.SR]

Matt S, Do Cao O, Brown B et al (2011) Convection and differential rotation properties of G and K stars computed with the ASH code. Astron Nachr 332(9–10):897–906. https://doi.org/10.1002/asna.201111624

McIntyre ME (1999) Breaking waves and global-scale chemical transport in the Earth's atmosphere, with spinoffs for the Sun's interior. Prog Theor Phys 101(1):189. https://doi.org/10.1143/PTP.101.189

Metcalfe TS, Finley AJ, Kochukhov O et al (2022) The origin of weakened magnetic braking in old solar analogs. Astrophys J 933(1):L17. https://doi.org/10.3847/2041-8213/ac794d

Miesch MS (2005) Large-scale dynamics of the convection zone and tachocline. Living Rev Sol Phys 2(1):1. https://doi.org/10.12942/lrsp-2005-1

Miesch MS, Hindman BW (2011) Gyroscopic pumping in the solar near-surface shear layer. Astrophys J 743(1):79. https://doi.org/10.1088/0004-637X/743/1/79. arXiv:1106.4107 [astro-ph.SR]

Miesch MS, Elliott JR, Toomre J et al (2000) Three-dimensional spherical simulations of solar convection. I. Differential rotation and pattern evolution achieved with laminar and turbulent states. Astrophys J 532(1):593–615. https://doi.org/10.1086/308555

Miesch MS, Brun AS, Toomre J (2006) Solar differential rotation influenced by latitudinal entropy variations in the tachocline. Astrophys J 641(1):618–625. https://doi.org/10.1086/499621

Miesch MS, Brun AS, DeRosa ML et al (2008) Structure and evolution of giant cells in global models of solar convection. Astrophys J 673(1):557–575. https://doi.org/10.1086/523838. arXiv:0707.1460 [astro-ph]

Mori K, Hotta H (2023) Investigation of the dependence of angular momentum transport on spatial scales for construction of differential rotation. Mon Not R Astron Soc 519(2):3091–3097. https://doi.org/10.1093/mnras/stac3804. arXiv:2212.11502 [astro-ph.SR]

Muñoz-Jaramillo A, Nandy D, Martens PCH (2011) Magnetic quenching of turbulent diffusivity: reconciling mixing-length theory estimates with kinematic dynamo models of the solar cycle. Astrophys J 727(1):L23. https://doi.org/10.1088/2041-8205/727/1/L23. arXiv:1007.1262 [astro-ph.SR]

Nagashima K, Zhao J, Kosovichev AG et al (2011) Detection of supergranulation alignment in polar regions of the Sun by helioseismology. Astrophys J 726(2):L17. https://doi.org/10.1088/2041-8205/726/2/L17. arXiv:1011.1025 [astro-ph.SR]

Nagashima K, Birch AC, Schou J et al (2020) An improved multi-ridge fitting method for ring-diagram helioseismic analysis. Astron Astrophys 633:A109. https://doi.org/10.1051/0004-6361/201936662. arXiv:1911.07772 [astro-ph.SR]

Nelson NJ, Brown BP, Brun AS et al (2013) Magnetic wreaths and cycles in convective dynamos. Astrophys J 762(2):73. https://doi.org/10.1088/0004-637X/762/2/73. arXiv:1211.3129 [astro-ph.SR]

Nesis A, Hanslmeier A, Hammer R et al (1992) Dynamics of the solar granulation. I – a phenomenological approach. Astron Astrophys 253(2):561–566

Nielsen MB, Gizon L, Schunker H et al (2013) Rotation periods of 12 000 main-sequence Kepler stars: dependence on stellar spectral type and comparison with v sin i observations. Astron Astrophys 557:L10. https://doi.org/10.1051/0004-6361/201321912. arXiv:1305.5721 [astro-ph.SR]

Nielsen MB, Gizon L, Schunker H et al (2014) Rotational splitting as a function of mode frequency for six Sun-like stars. Astron Astrophys 568:L12. https://doi.org/10.1051/0004-6361/201424525. arXiv:1408.4307 [astro-ph.SR]

Nielsen MB, Gizon L, Cameron RH et al (2019) Starspot rotation rates versus activity cycle phase: butterfly diagrams of Kepler stars are unlike that of the Sun. Astron Astrophys 622:A85. https://doi.org/10.1051/0004-6361/201834373. arXiv:1812.06414 [astro-ph.SR]

Noraz Q (2022) Magnétisme et dynamique des étoiles de type solaire. PhD thesis, Université Paris-Cité

Noraz Q, Breton SN, Brun AS et al (2022a) Hunting for anti-solar differentially rotating stars using the Rossby number: an application to the Kepler field. Astron Astrophys 667:A50. https://doi.org/10.1051/0004-6361/202243890

Noraz Q, Brun AS, Strugarek A et al (2022b) Impact of anti-solar differential rotation in mean-field solar-type dynamos: exploring possible magnetic cycles in slowly rotating stars. Astron Astrophys 658:A144. https://doi.org/10.1051/0004-6361/202141946

Nordlund Å, Stein RF, Asplund M (2009) Solar surface convection. Living Rev Sol Phys 6(1):2. https://doi.org/10.12942/lrsp-2009-2

Norton A, Howe R, Upton L et al (2023) Solar cycle observations. Space Sci Rev 219:64. https://doi.org/10.1007/s11214-023-01008-3. arXiv:2305.19803 [astro-ph.SR]

November LJ, Toomre J, Gebbie KB et al (1981) The detection of mesogranulation on the Sun. Astrophys J 245:L123–L126. https://doi.org/10.1086/183539

O'Mara B, Miesch MS, Featherstone NA et al (2016) Velocity amplitudes in global convection simulations: the role of the Prandtl number and near-surface driving. Adv Space Res 58(8):1475–1489. https://doi.org/10.1016/j.asr.2016.03.038. arXiv:1603.06107 [astro-ph.SR]

Ortiz A, Rast M (2005) How good is the Ca II K as a proxy for the magnetic flux? Mem Soc Astron Ital 76:1018

Ossendrijver M (2003) The solar dynamo. Astron Astrophys Rev 11(4):287–367. https://doi.org/10.1007/s00159-003-0019-3

Papaloizou J, Pringle JE (1978) Non-radial oscillations of rotating stars and their relevance to the short-period oscillations of cataclysmic variables. Mon Not R Astron Soc 182:423–442. https://doi.org/10.1093/mnras/182.3.423

Patron J, Hill F, Rhodes JEJ et al (1995) Velocity fields within the solar convection zone: evidence from oscillation ring diagram analysis of mount Wilson dopplergrams. Astrophys J 455:746. https://doi.org/10.1086/176620

Philidet J, Gizon L (2023) Interaction of solar inertial modes with turbulent convection. A 2D model for the excitation of linearly stable modes. Astron Astrophys 673:A124. https://doi.org/10.1051/0004-6361/202245666. arXiv:2304.05926 [astro-ph.SR]

Pipin VV, Kosovichev AG (2019) On the origin of solar torsional oscillations and extended solar cycle. Astrophys J 887(2):215. https://doi.org/10.3847/1538-4357/ab5952. arXiv:1908.04525 [astro-ph.SR]

Proxauf B (2020) Observations of large-scale solar flows. PhD thesis, Georg August University of Gottingen, Germany

Proxauf B, Gizon L, Löptien B et al (2020) Exploring the latitude and depth dependence of solar Rossby waves using ring-diagram analysis. Astron Astrophys 634:A44. https://doi.org/10.1051/0004-6361/201937007. arXiv:1912.02056 [astro-ph.SR]

Rajaguru SP, Antia HM (2015) Meridional circulation in the solar convection zone: time-distance helioseismic inferences from four years of HMI/SDO observations. Astrophys J 813(2):114. https://doi.org/10.1088/0004-637X/813/2/114. arXiv:1510.01843 [astro-ph.SR]

Rajaguru SP, Antia HM (2020) Time-distance helioseismology of deep meridional circulation. In: Monteiro MJPFG, García RA, Christensen-Dalsgaard J et al (eds) Dynamics of the Sun and stars; honoring the life and work of Michael J. Thompson, pp 107–113. https://doi.org/10.1007/978-3-030-55336-4_11. arXiv:2004.12708

Rast MP (1995) On the nature of "exploding" granules and granule fragmentation. Astrophys J 443:863. https://doi.org/10.1086/175576

Rast MP (2003) The scales of granulation, mesogranulation, and supergranulation. Astrophys J 597(2):1200–1210. https://doi.org/10.1086/381221

Rast MP, Toomre J (1993) Compressible convection with ionization. I. Stability, flow asymmetries, and energy transport. Astrophys J 419:224. https://doi.org/10.1086/173477

Rast MP (2020) Deciphering solar convection. In: Monteiro MJPFG, García RA, Christensen-Dalsgaard J et al (eds) Dynamics of the Sun and stars; honoring the life and work of Michael J. Thompson. Astrophysics and Space Science Proceedings, vol 57. Springer, Cham, pp 149–161. https://doi.org/10.1007/978-3-030-55336-4_23

Rauer H, Catala C, Aerts C et al (2014) The PLATO 2.0 mission. Exp Astron 38(1–2):249–330. https://doi.org/10.1007/s10686-014-9383-4

Reiners A (2007) Differential rotation in F stars. Astron Nachr 328(10):1034–1036. https://doi.org/10.1002/asna.200710853

Reiners A, Schmitt JHMM (2003) Rotation and differential rotation in field F- and G-type stars. Astron Astrophys 398(2):647–661. https://doi.org/10.1051/0004-6361:20021642

Reinhold T, Arlt R (2015) Discriminating solar and antisolar differential rotation in high-precision light curves. Astron Astrophys 576:A15. https://doi.org/10.1051/0004-6361/201425337

Reinhold T, Gizon L (2015) Rotation, differential rotation, and gyrochronology of active Kepler stars. Astron Astrophys 583:A65. https://doi.org/10.1051/0004-6361/201526216

Reinhold T, Reiners A, Basri G (2013) Rotation and differential rotation of active Kepler stars. Astron Astrophys 560:A4. https://doi.org/10.1051/0004-6361/201321970. arXiv:1308.1508 [astro-ph.SR]

Reinhold T, Shapiro AI, Solanki SK et al (2022) Measuring periods in aperiodic light curves-applying the GPS method to infer the rotation periods of solar-like stars. Astrophys J 938(1):L1. https://doi.org/10.3847/2041-8213/ac937a. arXiv:2209.12593 [astro-ph.SR]

Rempel M (2005) Solar differential rotation and meridional flow: the role of a subadiabatic tachocline for the Taylor-Proudman balance. Astrophys J 622(2):1320–1332. https://doi.org/10.1086/428282. arXiv:astro-ph/0604451 [astro-ph]

Rempel M (2007) Origin of solar torsional oscillations. Astrophys J 655(1):651–659. https://doi.org/10.1086/509866. arXiv:astro-ph/0610221 [astro-ph]

Rhines PB (1975) Waves and turbulence on a beta-plane. J Fluid Mech 69(3):417–443. https://doi.org/10.1017/S0022112075001504

Rieutord M, Roudier T, Malherbe JM et al (2000) On mesogranulation, network formation and supergranulation. Astron Astrophys 357:1063–1072

Rincon F, Rieutord M (2018) The Sun's supergranulation. Living Rev Sol Phys 15(1):6. https://doi.org/10.1007/s41116-018-0013-5

Roberts PH (1968) On the thermal instability of a rotating-fluid sphere containing heat sources. Philos Trans R Soc Lond Ser A 263(1136):93–117. https://doi.org/10.1098/rsta.1968.0007

Saar SH (2010) Starspots, cycles, and magnetic fields. Proc Int Astron Union 6(S273):61–67. https://doi.org/10.1017/S1743921311015018

Saio H (1982) R-mode oscillations in uniformly rotating stars. Astrophys J 256:717–735. https://doi.org/10.1086/159945

Scherrer PH, Bogart RS, Bush RI et al (1995) The solar oscillations investigation – Michelson Doppler Imager. Sol Phys 162(1–2):129–188. https://doi.org/10.1007/BF00733429

Schou J (2003) Wavelike properties of solar supergranulation detected in Doppler shift data. Astrophys J 596(2):L259–L262. https://doi.org/10.1086/379529

Schou J, Antia HM, Basu S et al (1998) Helioseismic studies of differential rotation in the solar envelope by the solar oscillations investigation using the Michelson Doppler Imager. Astrophys J 505(1):390–417. https://doi.org/10.1086/306146

See V, Roquette J, Amard L et al (2021) Photometric variability as a proxy for magnetic activity and its dependence on metallicity. Astrophys J 912:127. https://doi.org/10.3847/1538-4357/abed47 arXiv:2103.05675 [astro-ph]

Simitev RD, Kosovichev AG, Busse FH (2015) Dynamo effects near the transition from solar to anti-solar differential rotation. Astrophys J 810(1):80. https://doi.org/10.1088/0004-637X/810/1/80

Simon GW, Leighton RB (1964) Velocity fields in the solar atmosphere. III. Large-scale motions, the chromospheric network, and magnetic fields. Astrophys J 140:1120. https://doi.org/10.1086/148010

Simon GW, Weiss NO (1968) Supergranules and the hydrogen convection zone. Z Astrophys 69:435

Skumanich A (1972) Time scales for Ca II emission decay, rotational braking, and lithium depletion. Astrophys J 171:565

Spruit HC (1997) Convection in stellar envelopes: a changing paradigm. Mem Soc Astron Ital 68:397–413. arXiv:astro-ph/9605020 [astro-ph]

Stein RF, Georgobiani D, Schafenberger W et al (2009) Supergranulation scale convection simulations. In: Stempels E (ed) 15th Cambridge workshop on cool stars, stellar systems, and the Sun, pp 764–767. https://doi.org/10.1063/1.3099227

Stix M (2002) The sun: an introduction. Springer, Berlin

Strugarek A, Beaudoin P, Charbonneau P et al (2018) On the sensitivity of magnetic cycles in global simulations of solar-like stars. Astrophys J 863(1):35. https://doi.org/10.3847/1538-4357/aacf9e

Thibault K, Charbonneau P, Crouch AD (2012) The buildup of a scale-free photospheric magnetic network. Astrophys J 757(2):187. https://doi.org/10.1088/0004-637X/757/2/187

Thompson MJ, Toomre J, Anderson ER et al (1996) Differential rotation and dynamics of the solar interior. Science 272(5266):1300–1305. https://doi.org/10.1126/science.272.5266.1300

Triana SA, Guerrero G, Barik A et al (2022) Identification of inertial modes in the solar convection zone. Astrophys J 934(1):L4. https://doi.org/10.3847/2041-8213/ac7dac. arXiv:2204.13007 [astro-ph.SR]

Ulrich RK (2010) Solar meridional circulation from Doppler shifts of the Fe I line at 5250 Å as measured by the 150-foot Solar Tower Telescope at the Mt. Wilson Observatory. Astrophys J 725(1):658–669. https://doi.org/10.1088/0004-637X/725/1/658. arXiv:1010.0487 [astro-ph.SR]

Unno W, Osaki Y, Ando H et al (1989) Nonradial oscillations of stars

van Ballegooijen AA (1986) On the surface response of solar giant cells. Astrophys J 304:828. https://doi.org/10.1086/164219

Vasil GM, Lecoanet D, Brown BP et al (2013) Energy conservation and gravity waves in sound-proof treatments of stellar interiors. II. Lagrangian constrained analysis. Astrophys J 773(2):169. https://doi.org/10.1088/0004-637X/773/2/169

Vasil GM, Julien K, Featherstone NA (2021) Rotation suppresses giant-scale solar convection. Proc Natl Acad Sci 118(31):e2022518118. https://doi.org/10.1073/pnas.2022518118

Viviani M, Käpylä MJ, Warnecke J et al (2019) Stellar dynamos in the transition regime: multiple dynamo modes and antisolar differential rotation. Astrophys J 886(1):21. https://doi.org/10.3847/1538-4357/ab3e07

Warnecke J (2018) Dynamo cycles in global convection simulations of solar-like stars. Astron Astrophys 616:A72. https://doi.org/10.1051/0004-6361/201732413

Witzke V, Duehnen HB, Shapiro AI et al (2022) Small-scale dynamo in cool main sequence stars. II. The effect of metallicity. arXiv:2211.02722

Yadav RK, Christensen UR, Morin J et al (2015) Explaining the coexistence of large-scale and small-scale magnetic fields in fully convective stars. Astrophys J 813(2):L31. https://doi.org/10.1088/2041-8205/813/2/L31

Zhao J, Nagashima K, Bogart RS et al (2012) Systematic center-to-limb variation in measured helioseismic travel times and its effect on inferences of solar interior meridional flows. Astrophys J 749(1):L5. https://doi.org/10.1088/2041-8205/749/1/L5. arXiv:1203.1904 [astro-ph.SR]

Zhao J, Bogart RS, Kosovichev AG et al (2013) Detection of equatorward meridional flow and evidence of double-cell meridional circulation inside the Sun. Astrophys J 774(2):L29. https://doi.org/10.1088/2041-8205/774/2/L29. arXiv:1307.8422 [astro-ph.SR]

Authors and Affiliations

Hideyuki Hotta[1] · Yuto Bekki[2] · Laurent Gizon[2,3] · Quentin Noraz[4,5] · Mark Rast[6]

✉ L. Gizon
gizon@mps.mpg.de

H. Hotta
hotta.h@isee.nagoya-u.ac.jp

Y. Bekki
bekki@mps.mpg.de

Q. Noraz
quentin.noraz@astro.uio.no

M. Rast
mark.rast@colorado.edu

[1] Institute for Space-Earth Environmental Research, Nagoya University, Chikusa-ku, Nagoya, Aichi, 464-8601, Japan

[2] Max-Planck-Institut für Sonnensystemforschung, Justus-von-Liebig-Weg 3, Göttingen, 37077, Germany

[3] Institut für Astrophysik, Georg-August-Universtät Göttingen, Friedrich-Hund-Platz 1, Göttingen, 37077, Germany

[4] Rosseland Centre for Solar Physics, University of Oslo, P.O. Box 1029 Blindern, Oslo, NO-0315, Norway

[5] Institute of Theoretical Astrophysics, University of Oslo, P.O. Box 1029 Blindern, Oslo, NO-0315, Norway

[6] Department of Astrophysical and Planetary Sciences, Laboratory for Atmospheric and Space Physics, University of Colorado, Boulder, CO 80309, USA

Space Science Reviews (2023) 219:87
https://doi.org/10.1007/s11214-023-01027-0

Dynamics of the Tachocline

Antoine Strugarek[1] · Bernadett Belucz[2,3,4] · Allan Sacha Brun[1] · Mausumi Dikpati[5] ·
Gustavo Guerrero[6,7]

Received: 27 April 2023 / Accepted: 20 November 2023 / Published online: 14 December 2023
© The Author(s) 2023

Abstract

The solar tachocline is an internal region of the Sun possessing strong radial and latitudinal
shears straddling the base of the convective envelope. Based on helioseismic inversions, the
tachocline is known to be thin (less than 5% of the solar radius). Since the first theory of
the solar tachocline in 1992, this thinness has not ceased to puzzle solar physicists. In this
review, we lay out the grounds of our understanding of this fascinating region of the solar
interior. We detail the various physical mechanisms at stake in the solar tachocline, and
put a particular focus on the mechanisms that have been proposed to explain its thinness.
We also examine the full range of MHD processes including waves and instabilities that
are likely to occur in the tachocline, as well as their possible connection with active region
patterns observed at the surface. We reflect on the most recent findings for each of them, and
highlight the physical understanding that is still missing and that would allow the research
community to understand, in a generic sense, how the solar tachocline and stellar tachocline
are formed, are sustained, and evolve on secular timescales.

Keywords Solar tachocline · Magnetohydrodynamics of stars · Shear and magnetic
instabilities · Solar dynamo

1 The Solar Tachocline

1.1 What Is Known About the Solar Tachocline?

After several decades of observations, one of the great achievements of helioseismology
remains the inversion of the internal rotational profile of the Sun (Brown et al. 1989). The
surface differential rotation prevails through the whole convective zone, as seen in Fig. 1
(Thompson et al. 2003). In the radiative interior, the rotation rate is uniform and matches
the rotation rate of the convective envelope at mid-latitudes (about 430 nHz near latitude
$\pm 35°$). The transition from a differential to uniform rotation occurs in a transition layer
known as the tachocline (Hughes et al. 2007). This region lies just beneath the convection
zone and likely has a prolate form, i.e., it is located at $\sim 0.693 R_\odot$ close the equator, and at
$\sim 0.717 R_\odot$ at higher latitudes (Charbonneau et al. 1999; Basu and Antia 2003). Its thickness

Solar and Stellar Dynamos: A New Era
Edited by Manfred Schüssler, Robert H. Cameron, Paul Charbonneau, Mausumi Dikpati, Hideyuki
Hotta and Leonid Kitchatinov

Extended author information available on the last page of the article

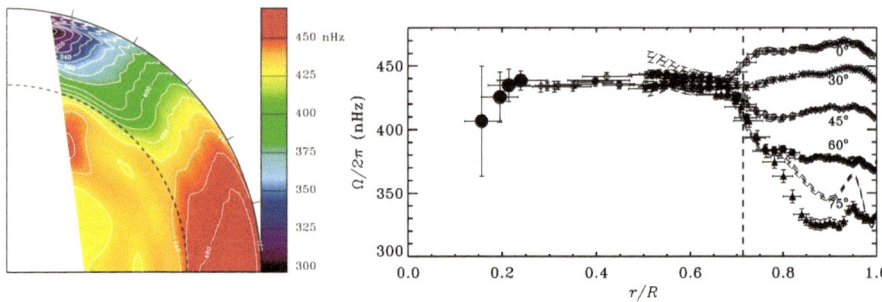

Fig. 1 *Left*: Solar differential rotation, as deduced from helioseismology (Schou et al. 1998; Thompson et al. 2003). The surface exhibits a shear layer (often dubbed *near-surface shear layer*). The bulk of the convective envelope shows a latitudinal differential rotation with mostly conical iso-contours, except perhaps at the equator where contours are more aligned with the rotation axis. At the base of the convection zone, the rotation becomes progressively solid, creating a sheared layer called the tachocline. *Right*: Radial cuts of the solar differential rotation profile at latitudes $0°$, $30°$, $45°$, $60°$ and $75°$ (the dashed line shows the results with a different inversion technique using regularized least-squares, see Thompson et al. (2003) and references therein for more details)

slightly varies according to the definition considered in the helioseismic forward modeling determinations (see, for instance Kosovichev 1996; Corbard et al. 1999; Elliott and Gough 1999; Basu and Antia 2001), yet it is certainly very thin (less than 5% of the solar radius). Recent results considering ~ 20 years of observations indicate that neither the position or the thickness of the tachocline significantly change with the solar cycle (Basu and Antia 2019). The tachocline is subject to strong latitudinal and radial shears. The radial shear in the mean azimuthal flow is the strongest (by almost an order of magnitude) and changes sign at latitude $\pm 35°$. Both the radial and the latitudinal shears exhibit significant changes in their amplitudes with the solar cycle. Interestingly, these changes have been found to be different between the cycles 23 and 24 (Basu and Antia 2019).

This strong large-scale shear drew the attention to the tachocline, and led it to be considered a prominent player in many dynamo models of the Sun known as 'interface dynamos' (Parker 1993). Indeed, such a strong large-scale shear is a very efficient converter of poloidal fields into toroidal field in mean-field models (see, in this collection, Brandenburg et al. 2023; Charbonneau and Sokoloff 2023; Käpylä et al. 2023). Nevertheless, the real role played by the tachocline in solar and stellar dynamos is still controversial. On one hand, ZDI observations suggest that the magnetic topology changes from a simple dipolar structure for young TTauri, fully-convective, objects to a complex multipolar topology for objects that have already developed a radiative core (Gregory et al. 2012). On the other hand, following earlier works from Mohanty and Basri (2003), Reiners et al. (2012), Wright and Drake (2016) unambiguously showed that the X-ray luminosity of fully-convective stars followed the same empirical trend than slightly hotter stars with a radiative core, showing no strong difference in the case a tachocline could be present. In addition, Route (2016) discovered magnetic activity cycles in ultra-cool dwarfs which are fully-convective, again showing seemingly no strong impact of the existence or non-existence of a tachocline. In the theoretical field, global numerical simulations (e.g. Strugarek et al. 2017) also showed that a tachocline was actually not necessary to obtain Sun-like cyclic dynamos, and Bice and Toomre (2020) showed that a tachocline was not necessary to generate strong wreaths of toroidal field within the convective core of M-dwarfs. It could nevertheless influence how the dynamo operates. Indeed, numerical experiments show that instabilities in the tachocline could affect the magnetic cycle period (Lawson et al. 2015), by modulating (due to insta-bilities) the Poynting flux permeating the base of the convective envelope (more on this in Sect. 3). More recently Guerrero et al. (2019) also reported a similar instability in their numerical endeavors and Brun et al. (2022) found hints in their simulations that taking the tachocline into account could affect the cycle period (see also Beaudoin et al. 2018).

1.2 Why Is It so Thin?

This simple question about the solar tachocline has puzzled and fascinated solar physicists since its discovery. Indeed, its extreme thinness remains one of the key mysteries of our star. Right after its discovery, Spiegel and Zahn (1992) produced the first hydrodynamic theory of the solar tachocline. They analyzed the long-term equilibrium of this thin shear layer, and showed that it should burrow on secular time-scales due to the phenomenon of *radiative spreading*. In a nutshell, the latitudinal shear with conical iso-contours requires an associated temperature gradient to break the Taylor-Proudman constraint. A meridional circulation then follows, spreading the latitudinal shear downwards on an Eddington-Sweet timescale, defined as

$$t_{ES} = \frac{N^2}{\Omega_\odot^2} \frac{R^2}{\kappa} ,$$ (1)

where the square of the Brunt-Väisälä frequency is defined as $N^2 = g \left(\frac{1}{\gamma} \frac{d \ln P}{dr} - \frac{d \ln \rho}{dr} \right)$, $\Omega_\odot$ is the mean rotation of the Sun, R the radius of the tachocline and κ the thermal diffusion coefficient. The Eddington-Sweet timescale based on molecular thermal diffusion reaches about 10^{20} years for the solar tachocline (Brun and Zahn 2006).

The radiative spreading (or burrowing) has been revisited multiple times ever since (Garaud and Brummell 2008; Garaud and Acevedo Arreguin 2009; Wood and Brummell 2012, 2018), confirming essentially the robustness of the mechanism highlighted by Spiegel and Zahn (1992). A back-of-the-envelope calculation quickly shows that according to this theory, the solar tachocline should have spread by at least $0.3 R_\odot$ into the radiative zone at the solar age. Hence, additional mechanisms must be at play to confine the solar tachocline to its extreme thinness smaller than 5% $R_\odot$.

Multiple physical scenarios have been proposed to confine the solar tachocline. They can be decomposed into two families: the fast tachocline scenarios, relying on physical mechanisms on timescales of months to years; and the slow tachocline scenarios, relying on physical mechanisms acting on timescales longer than a millennium. The large community debate about the solar tachocline confinement can therefore be largely attributed to the difficulty to adequately model it, because of the multiple time-scales involved in its dynamics and evolution.

The multiple physical mechanism behind confinement scenarios proposed to explain the thinness of the solar tachocline are illustrated in Fig. 2 and ordered as a function of their physical timescale (the differentiation between slow and first processes in the tachocline was first described by Gilman 2000, see e.g. the Table 1 there). The various processes can be summarized as follows (we will dive deeper into the last three in Sect. 2 and Sect. 3).

- *Anisotropic viscosity and turbulent transport*. It was originally proposed by Spiegel and Zahn (1992) that turbulent viscosity in the upper tachocline was likely to be anisotropic due to the strong stratification encountered by convective plumes penetrating from above. They modelled it by simply considering a dominant horizontal viscosity, and in this case

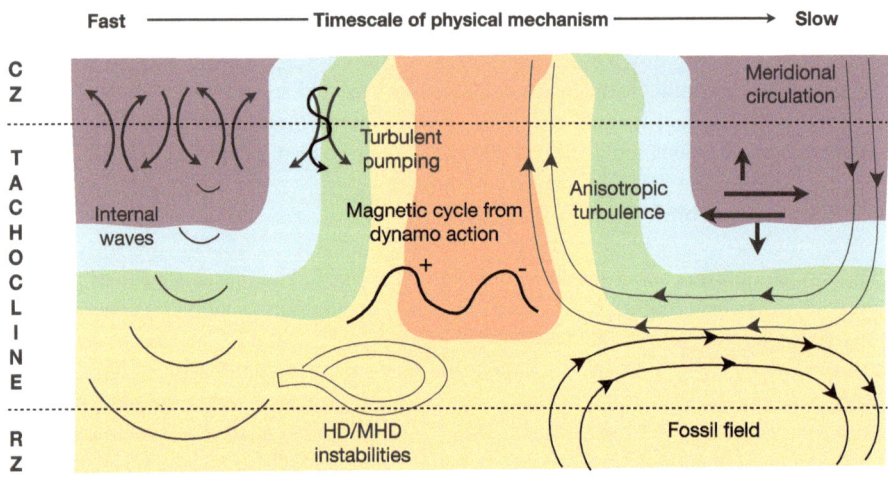

Fig. 2 Schematic of the physical processes at play in the various proposed scenarios to confine the tachocline. The typical timescale of the physical process increases from left to right. The solar differential rotation in the tachocline is illustrated by the background colored zones (see also Fig. 1)

the tachocline could be confined to its present thickness, provided the effective (turbulent) Prandtl number was large enough. This scenario was opposed by Gough and McIntyre (1998), who argued that turbulence was unlikely to transport angular momentum in a diffusive manner on such a quasi-2D layer, basing their argument on what occurs in the Earth atmosphere. In fact, this debate is not settled today. In a series of analytical papers (Kim 2005; Leprovost and Kim 2006; Kim and Leprovost 2007; Leprovost and Kim 2009), Kim and Leprovost showed quite clearly that depending on the considered characteristics of the solar tachocline (horizontal shear, radial shear, stratification, magnetism, waves) turbulent transport could actually be either diffusive or anti-diffusive. These different behaviours were confirmed to exist with numerical simulations by Miesch (2003). Indeed, the consideration of low-Pr physics changes the situation when compared to the Earth atmosphere, making the direction and amplitude of the turbulent transport in the tachocline not so obvious to estimate (Garaud 2020). Furthermore, Tobias et al. (2007) showed that in the presence of large-scale magnetic fields, turbulence could sometimes even not transport angular momentum efficiently at all. This debate was further pushed forward recently by Chen and Diamond (2020), who developed a more complete view of the problem and proposed that the tachocline could be confined by the combined effect of the relaxation of potential vorticity gradient along a resisto-elastic drag originating from tangled magnetic fields in the solar tachocline. Unsurprisingly, turbulent transport therefore still remains one of the most promising and debated origin for the solar tachocline confinement.

- *Magnetic confinement with a fossil field*. Following Spiegel and Zahn (1992), Gough and McIntyre (1998) wrote a very influential paper on the confinement of the solar tachocline, after a seminal idea by Rudiger and Kitchatinov (1997) who proposed a confinement of the tachocline due to the existence of a fossil magnetic field. Arguing that turbulent transport could not behave like a diffusion (see the previous point for a more critical view on this argument), they looked for another possible source for the tachocline confinement. They naturally turned to magnetism. In their seminal paper, they proposed that a magnetic field of fossil origin confined inside the solar radiation zone could balance out the inward radiative spreading of the tachocline through Lorentz forces. Nevertheless, magnetic fields diffuse on Ohmic, secular timescales and an additional mechanism was also required to confine the magnetic field itself, leading to a *double-confinement* scenario. The large-scale meridional flow accompanying *a priori* the burrowing of the tachocline (Wood and Brummell 2012) was in turn invoked to play this confining role, leading to an overall balance of the solar tachocline. The Gough and McIntyre (1998) scenario possesses an appealing advantage: it offers a truly large-scale theory for the confinement of the solar tachocline, that does not depend on the detailed behavior of complex, turbulent MHD flows at the top of the solar tachocline. Nevertheless, one potential problem with this scenario lies in the possibility for the Sun to possess of a magnetic-field zone between the top of the convection and the confinement layer where the fossil field Lorentz force is at play. This complex scenario was then questioned with various series of numerical simulations (e.g. Strugarek et al. 2011, Acevedo-Arreguin et al. 2013). We will come back to this active debate in Sect. 2.2, along with the most recent numerical endeavors on this topic.

- *Magnetic confinement with a dynamo field*. The dynamo magnetic field of the Sun visible at its photosphere certainly permeates deep inside the solar interior. At the very least, down to the solar tachocline along with the differential rotation and the turbulent convective motions. Many dynamo theories have historically given the tachocline an important role (see e.g. Käpylä et al. 2023). If the cyclic magnetic field of the Sun permeates the upper tachocline, could it contribute to confining it to its observed thinness? This interesting question was first addressed by Forgács-Dajka and Petrovay (2001) and further developed in a series of papers by the same authors. This scenario historically deeply interested Jean-Paul Zahn, who developed an extension of the initial development by Forgács-Dajka and Petrovay (2001). This work was later refined and published in Barnabé et al. (2017) (more on the details in Sect. 2.1).

- *Magneto-hydrodynamic instabilities*. As a sheared, stratified, magnetized thin layer, one may naturally question the global stability of the solar tachocline, and whether any MHD instabilities could actually help to explain its thinness (e.g. Watson 1981, Gilman and Fox 1997, Charbonneau et al. 1999). A general introduction to this topic can be found in Sect. 8.2 of Miesch (2005).

We now first turn to the magnetic tachocline confinement problem (Sect. 2) and move on to the detailed description of the magneto-hydrodynamic instabilities in the tachocline and their impact on our understanding of the tachocline of the solar magnetism in Sect. 3.

2 Magnetic Confinement Scenarios of the Tachocline

We now give a detailed description of two mechanisms that have been proposed to confine the tachocline to its present thickness, based on the existence of internal magnetic fields either generated by dynamo processes (Sect. 2.1) or of fossil origin (Sect. 2.2).

2.1 Confinement with a Dynamo Field

The solar magnetic cycle of 11 years, visible at the photosphere, sustains strong magnetic fields inside our star. But how strong? It is known for a fact that at the surface of the Sun, the dipolar magnetic field has a typical value of a few Gauss over the solar surface (DeRosa et al. 2012). Moreover, magnetic flux concentrations in the form of sunspots can reach magnetic field intensities of hundreds of kG. Deeper down in the solar convective envelope, where the overall dynamo is supposed to be seated, we do not have any precise observational

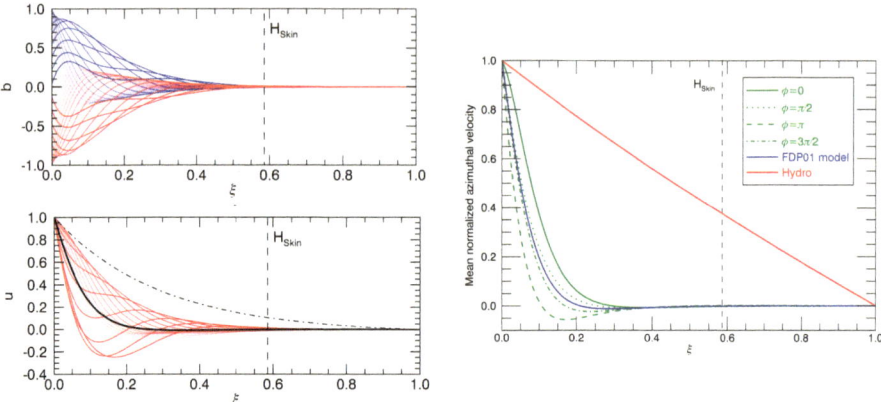

Fig. 3 *Left panels*: Original confinement scenario of Forgács-Dajka and Petrovay (2001) as revisted by Barnabé et al. (2017). The upper panel shows the toroidal field b, and the lower panel the differential rotation u, as a function of depth ξ from left to right. The skin depth $H_{skin} = \sqrt{2\eta\,P_{cyc}}$ is shown by the vertical dashed line. The color and transparency of the line label different phases in one magnetic cycle. *Right panel*: Study of the influence of the phasing between the two components a and b of the magnetic field. The angle ϕ corresponds to this dephasing. The blue curve is the reference model of Forgács-Dajka and Petrovay (2001) that differs from the green models in the boundary condition imposed on a. A reference hydrodynamic model is shown by the red curve. All models considering a dynamo cycle (red and green curve) manage to confine the tachocline regardless of the assumed relative phase ϕ

constraints on the field amplitude. Antia et al. (2000) have tried to derive an upper limit for the magnetic field amplitude through helioseismic inversion and found that it should be below 30 T – a somewhat large upper value. Models of flux emergence from the base of the convection zone leading to realistic flux concentration at the photosphere yield a toroidal field of the order of $1 - 10$ T (Fan 2009; Jouve et al. 2018).

The possibility of confining the solar tachocline with the oscillating dynamo field penetrating from the upper convective envelope was first suggested by Forgács-Dajka and Petrovay (2001), and further explored by Barnabé et al. (2017). A minimal 1D mean-field model for this scenario, which can be mathematically written as

$$\partial_\tau a = \partial_{\xi\xi}^2 a\,, \tag{2}$$

$$\partial_\tau b = \partial_{\xi\xi}^2 b - (k\delta)^2\,b - C_A au\,, \tag{3}$$

$$\partial_\tau u = \mathrm{Pm}\left(\partial_{\xi\xi}^2 u - (k\delta)^2\,u\right) + C_L ab\,, \tag{4}$$

where ξ is the depth below the bottom of the convection zone. This model traces the evolution of the azimuthally-averaged poloidal magnetic field a, toroidal magnetic field b, and differential rotation u, over a domain of depth δ below the convection zone. For the sake of simplicity, a is assumed to be constant in latitude, while the latitudinal shape of the toroidal field and of the differential rotation is characterized by a unique wave number k. The fields are coupled through the induction term $C_A au$ in Eq. (3), and the Lorentz force $C_L ab$ in Eq. (4). This simplified model allows assessing whether the Lorentz force can confine the tachocline by preventing the downward spread of the differential rotation into the underlying radiative interior. In this work, C_A, C_L and $\mathrm{Pm} = \nu/\eta$ (the magnetic Prandtl number) were varied to assess in which parameter regime a confinement was realized.

Fig. 4 Magnetic field confinement due to a convective dynamo generated in the global simulations of Matilsky et al. (2022). Panel (a) shows viscous tachocline spreading in a hydrodynamic simulation. Panel (b) shows the formation of a tachocline and a solid body rotating radiative interior in a MHD simulation

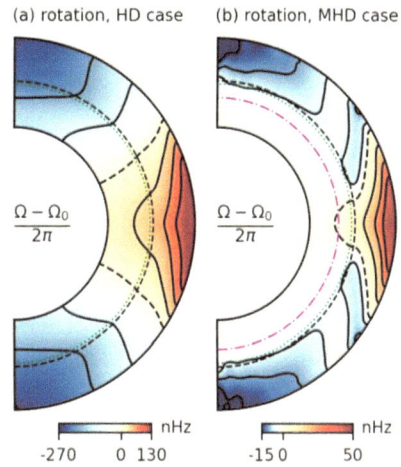

Let's consider first the original scenario proposed by Forgács-Dajka and Petrovay (2001) where an oscillating magnetic field is imposed at the top of the domain with a period $P_{cyc} = 22$ years, with the poloidal and toroidal fields being out of phase by $\pi/2$. The two left panels of Fig. 3 show the evolution of the toroidal field (upper panel) and of the differential rotation (lower panel). As expected, the toroidal field oscillates due to the boundary forcing and penetrates down to the electro-magnetic skin-depth $H_{skin} = \sqrt{2\eta P_{cyc}}$. The differential rotation u penetrates over about the same depth, as seen in the lower panel (the average over P_{cyc} is shown by the thick black line). In this case, the tachocline is indeed confined, as the initial profile shown by the dash-dotted line was actually penetrating deeper down than the converged rotation profile.

The first criticism of the work of Forgács-Dajka and Petrovay (2001) was the assumed phase between the toroidal and poloidal magnetic field. Indeed, this phasing can play a critical role to ensure that the Lorentz force is actually strong enough to confine the tachocline. This hypothesis was alleviated in Barnabé et al. (2017) (see Fig. 3), who found that the phasing of the two components of the magnetic field actually marginally affect the net differential rotation. In all cases (shown in green), the tachocline remains confined, with a somewhat different radial profile. The conclusion of this work is that this confinement scenario is actually quite robust with respect to the relative phase between the poloidal and toroidal components of the dynamo field.

The second and more important criticism to this scenario lies in its simplicity. Indeed, the original scenario did not consider that the transport in tachocline could be either diffusive or anti-diffusive, or that it could be subject to a radiative spreading (see Sect. 1.2). Again, this limitation was also alleviated in the work of Barnabé et al. (2017) who identified the region in parameter space where the confinement could be strong enough to oppose such spreading mechanisms. Interestingly, these results show that the dynamo-based confinement scenario seems robust for a large variety of turbulent transport scenarios.

The confinement due to a dynamo generated magnetic field has furthermore been recently confirmed by global simulations (Matilsky et al. 2022). Their model considers a fraction of a stable layer and a convection zone with a thermodynamic stratification that resembles the solar interior up to $0.947 R_\odot$ and rotates 3 times faster than the Sun which allows to get fast equator and slower poles. If the simulation does not include magnetic field, the angular momentum spreads from the convection zone to the stable interior because of the viscosity.

If the simulation includes magnetic field, a dynamo develops generating non-axisymmetric wreaths of toroidal field around the equator (compare left and right panels of Fig. 4). This field has a small phase difference with the poloidal magnetic field. This phase difference is instrumental in generating axisymmetric large-scale Maxwell stresses which balances the viscous stresses preventing the inwards transport of angular momentum. There is a skin-depth of penetration of the magnetic field in the stable layers such as in the models of Forgács-Dajka and Petrovay (2001), Barnabé et al. (2017).

In such a confinement scenario, the differential rotation exhibits a variability on the magnetic cycle timescale, over the electro-magnetic skin-depth. Such strong variability was not systematically detected in the Sun (Howe et al. 2011). The spatial resolution limit of helioseismic inversions is about 5% of the solar radius (Howe 2009). If the tachocline is confined by the oscillating dynamo field, this yields an upper limit of about 5×10^{10} cm^2/s for the Ohmic diffusivity (Barnabé et al. 2017).

One can summarize the present understanding of the tachocline confinement scenario by an oscillating dynamo magnetic field as follow:

- The confinement scenario based on the cyclic dynamo field of the Sun is a robust scenario with respect to the various hypothetical physical ingredients participating in the angular momentum transport
- For realistic solar parameters, this confinement scenario works only if the magnetic field is able to permeate sufficiently deep. In other words, if the turbulent Ohmic diffusivity in the upper tachocline exceeds 10^8 cm^2/s, it is very likely that the cyclic, large scale field of the Sun is able to confine the tachocline to a thickness smaller than 5% of the solar radius
- The plausibility of this confinement scenario has been confirmed through global MHD simulations of convective dynamos (Matilsky et al. 2022). Although the parameter regime of these simulations is still far from realistic, it is an encouraging alternative to explore in future simulations.

2.2 Confinement with a Fossil Field

The confinement of the solar tachocline by a fossil field was proposed by Gough and McIntyre (1998). This scenario was based on equilibrium arguments, but lacked the geometrical realism of a 3D modelling. In particular, it relies on the existence of a meridional flow in the upper tachocline which prevents the magnetic field to permeate above. Conveniently, the radiative spreading theory of Spiegel and Zahn (1992) predicts that such a meridional circulation necessarily co-exists as the differential rotation burrows down in the radiative core on secular timescales (Wood and Brummell 2012). Nevertheless, this scenario has one major difficulty (setting aside the aforementioned inner zone of tachocline assumed to be magnetic field-free). Indeed, such a meridional flow has to go down at some latitude and up at some others to conserve mass. In the upwelling part, the magnetic field is likely to be entrained into the convection zone and therefore does not remain confined. This is what was initially found in numerical simulation of this confinement scenario in Brun and Zahn (2006), Strugarek et al. (2011) using the ASH code. 3D visualizations of this phenomenon are shown in Fig. 5. Initially (left panel), a fossil magnetic field is assumed to be confined inside the radiative core of the Sun. The model is then evolved, and the magnetic field is twisted in the tachocline (as seen in the middle panel). It eventually finds a way to connect to the convective envelope (third panel) as was just explained. This connection path can prevent the fossil field scenario (Gough and McIntyre 1998) to establish if the angular momentum exchanged by magnetic torques between the convective and the radiative zones is strong enough. Indeed, the first consequence is that the radiative core is not in solid body

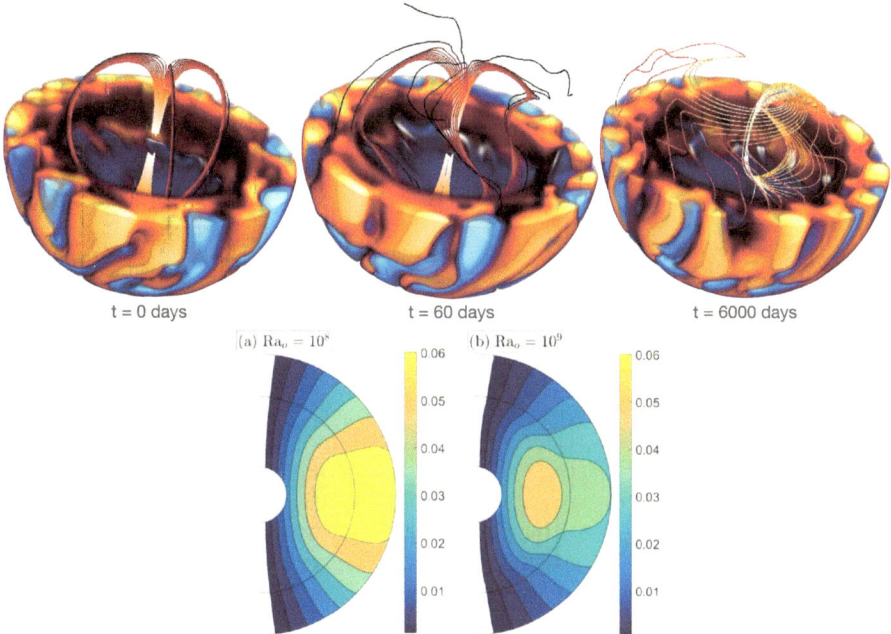

Fig. 5 Top: 3D visualizations of the tachocline confinement scenario with a fossil field, adapted from Strugarek et al. (2011). The convective motions are shown in the lower cut-out of the sphere, with orange-red volumes denoting upward motions and blue volumes downward motions. A few selected field lines are traced from the middle of the radiative zone. The initial confined dipole is shown on the left panel. In the middle panel, the field starts to be sheared in the lower tachocline. In the right panel, the field has permeated through the tachocline and is now connected to the surface. Bottom: confinement of a fossil field through convective turbulent pumping at low (left) and high (right) Rayleigh numbers, adapted from Korre et al. (2021). The black contours show the dipole field lines with values of the stream function depicted by the colormap

rotation anymore, which is of course in disagreement with the observational constraints at our disposal (see Fig. 1).

The modelling efforts of Strugarek et al. (2011) were revisited by Acevedo-Arreguin et al. (2013) because their parameter regime was preventing a true radiative burrowing of Spiegel and Zahn (1992) to occur, by instead favoring viscous burrowing. Indeed, the meridional flow in their case was solely driven by the differential rotation inside the convection zone, and was consequently less energetic and less extended than a meridional circulation accompanying a radiative burrowing (Wood and Brummell 2018). In this latter work, they unambiguously showed that the differential rotation can spread downward in various ways depending on the adimensional parameter

$$\sigma = \sqrt{\mathrm{Pr}}\frac{|N|}{2\Omega_\odot} < 1,\tag{5}$$

where $\mathrm{Pr} = \nu/\kappa$ is the Prandtl number. If $\sigma > 1$, the differential rotation spreads viscously. If $\sigma < 1$, the spread is controlled by thermal dissipation and a non-negligible meridional circulation develops as the differential rotation burrows. This regime is particularly challenging to model in the solar tachocline, because the Prandtl number is expected to be much less than 10^{-3} there. New 3D global simulations of this problem, carried out in the correct σ-parameter regime, indicate that the meridional flow can indeed halt efficiently the outward magnetic spreading in the equatorial region compared to the results of Strugarek et al. (2011). Nevertheless, even in this regime the regions where the meridional flow turns upwards still allow the magnetic field to permeate into the overlying convective envelope and magnetic torques lead to the internal radiative zone to depart from solid-body rotation (Strugarek & Brun, in prep.).

Concurrently, another mechanism may contribute to the confinement of the magnetic field in the radiative interior. The turbulent magnetic pumping is a mean-field effect that advects a large-scale field away from the regions where turbulence is strongest (Tobias et al. 1998). According to mean-field theory, the amplitude of the advection is proportional to minus the gradient of the turbulent magnetic diffusivity, $\gamma = -\frac{1}{2}\nabla\eta_T$ (Kitchatinov and Rüdiger 2008). Thus, it is expected that the field is transported from the convection zone downward into the radiative interior. Alternatively, the upwards diffusion of the fossil field could be balanced by its turbulent advection. In mean field models, this effect was verified by Kitchatinov and Rüdiger (2008). They found that total confinement can be found for $\nabla\eta_T \gtrsim 10^5$. The recent global simulations of Korre et al. (2021) explore this effect in the Boussinesq approximation and in the non-rotating case. They can progressively increase $\nabla\eta_T$ by increasing the Rayleigh (Ra) number in the convection zone, i.e., by making more vigorous convection. In their highest resolution case, with $Ra = 10^9$, their simulations evince confinement of the fossil field (see bottom panels in Fig. 5). Yet, a much higher Ra is still needed to achieve the advection values predicted by Kitchatinov and Rüdiger (2008). So far, it is also unclear how rotation may affect these results. In addition, this mechanism relies on the strength of convection in the deep convection zone, which is debated today based on various helioseismic inversions and numerical simulations results (Hanasoge et al. 2015; Hotta et al. 2023).

To conclude, the present understanding of the tachocline confinement scenario by a fossil magnetic field can be summarized as follows:

- In the correct burrowing parameter regime of the solar differential rotation (i.e. the radiative spreading), the meridional flow accompanying the burrowing can efficiently halt the outward spread of the hypothetical fossil magnetic field where the flow and the magnetic field are horizontal.
- This spread cannot be easily halted in upflow regions of the meridional circulation, and it remains in question to know how the solar convection zone can reach the level of turbulent magnetic pumping needed to compensate the loss of confinement of the fossil field in these regions.
- If the spread of the fossil field cannot be halted at some specific latitudes, a magnetic connection between the convective and radiative zone establishes. The associated angular momentum transport leads to a significant departure of the inner radiative zone from solid-body rotation, that is incompatible with present constraints from helioseismology.

3 Global (M)HD of the Solar Tachocline

After having reviewed the confinement mechanisms of the solar tachocline, we now turn to the more generic MHD phenomenon that could be taking place in this particular region of our Sun, and that can affect the large-scale flows and the magnetism of our star.

Differential rotation (DR) inferred from observations in the tachocline (see Fig. 1) can be sustained by the angular momentum transport in latitude and radius, due to the influence of rotation on convection. This differential rotation can itself be unstable to global hydrodynamic disturbances of low longitudinal wave number m. If the tachocline differential rotation is perturbed by some random disturbances or by disturbances that include some longitudinal structures in the form of normal modes, these disturbances either can grow by extracting energy from the differential rotation, or decay away by transferring energy to the differential rotation (Gilman and Cally 2007; Dikpati 2012). In the nonlinear regime they can quasiperiodically exchange energies with the differential rotation, and form patterns that have properties of Rossby waves. In particular, the disturbance patterns propagate in the longitudinal direction and have a tilt in latitude. Because of their latitudinal tilts they are able to either extract energy from the DR or can give energy back to the DR.

In addition, a large amount of toroidal magnetic field is expected to exist at the solar tachocline due to its subadiabatic stratification and the strong radial and latitudinal shear. Therefore, toroidal fields in the tachocline are likely closely coupled by MHD processes to the differential rotation and the aforementioned hydrodynamic disturbances, leading to global MHD instabilities. Their behavior depends on the relative energy present in the differential rotation compared to that of the toroidal field (Mestel and Moss 1983; Charbonneau and MacGregor 1993; Charbonneau 2004; Spruit 2004). The disturbances of toroidal fields, coexisting with differential rotation, can again produce modes which retain some properties of Rossby waves, particularly their phase speed in longitude and latitudinal tilt structures (Gilman and Fox 1997). But both differential rotation and toroidal fields change these speeds significantly compared to the well known retrograde propagating Rossby waves for a uniformly rotating thin spherical shell of constant thickness, first found by Haurwitz (1940).

Moreover, the presence of magnetic field within the stably stratified radiative interior of stars (the tachocline being subadiabatic qualifies it as such a region) further leads to a large range of MHD instabilities (Spruit 1999; Jouve et al. 2015). Several MHD instabilities present in stellar radiative interiors have hence been studied in details, for instance that of a large scale poloidal field with closed field lines (Markey and Tayler 1973; Wright 1973; Braithwaite 2007), that of strong toroidal magnetic field (akin to the tachocline situation) (Tayler 1973; Pitts and Tayler 1985; Guerrero et al. 2019) and the MRI (Menou and Le Mer 2006; Masada 2011; Jouve et al. 2020) to cite only a few emblematic ones. It was for instance shown that the ratio between the poloidal and toroidal magnetic field components or energies is key to maintaining or not the magnetic structure into a stable configuration on long timescales (Wright 1973; Tayler 1980; Braithwaite and Spruit 2004; Braithwaite 2008, 2009; Duez and Mathis 2010; Monteiro et al. 2023; Fuller and Mathis 2023). Special attention has been given to the interaction between a large scale field with a large scale mean flow (meridional circulation or differential rotation, Mestel and Moss 1977, Mestel and Weiss 1987). In the situation where a large scale shear is present and maintained, it was shown that a dynamo loop could exist, using the shear as Ω-effect and the generation of non axisymmetric fields and flows through MHD instabilities as α-effect (Spruit 2002; Cline et al. 2003; Braithwaite 2006; Zahn et al. 2007; Zaire et al. 2022; Petitdemange et al. 2023). However, such dynamo loop in the radiative interior, with the large scale shear being maintained by external processes such as the convective envelope above or the angular momentum extraction due to stellar wind torque, are not easy to realize as demonstrated in Zahn et al. (2007). Indeed, they required an effective α effect originating from the correlations of the field and flow perturbations, i.e. corresponding to strong electromotive force $\epsilon = \langle v' \times b' \rangle$. Such correlation can be provided by the MHD instabilities but are not easy to achieve or maintain on long timescales because they can easily be damped when the large scale field aligns with the large scale shear flow or vice and versa, a phenomenon known as Ferraro's law of iso-rotation. Nevertheless, it is remarkable that the joint presence of magnetic fields (either poloidal, toroidal, or both) and large scale shears yields a rich realm of physical mechanisms that are likely acting within stars such as our Sun and in their tachoclines. Since both HD and MHD global processes are likely to operate in the tachocline, we discuss them in more details in the following subsections.

3.1 Global HD Instabilities

In strictly 2D spherical shells, a DR profile of the solar type[1] is unstable to low longitudinal wave numbers only if the equator-to-pole difference in rotation is nearly 30% (Watson 1981). Nevertheless, the instability can occur for smaller equator-to-pole contrast if (i) Ω is expressed using both the $\sin^2(\theta)$ and $\sin^4(\theta)$ terms, such as $\Omega = \Omega_0 - s_1 \sin^2 \theta - s_2 \sin^4 \theta$, and/or (ii) 3D effects are considered (e.g. the threshold for the DR instability drops substantially in a shallow-water model when moving from purely 2D to quasi-3D, see Dziembowski and Kosovichev 1987, Charbonneau et al. 1999, Garaud 2001) In the case (i), the tachocline differential rotation can be still unstable in 2D. In the case (ii), particularly for a layer like the convective overshoot layer of the tachocline that is only slightly subadiabatic on average (thus allowing more vertical motion), differential rotation can be unstable for much smaller amplitude than 30% equator-to-pole differential rotation (Dikpati and Gilman 2001a). In this latter case, there also appear bulges and depressions in the top surface of the tachocline (corresponding to the tachocline–convection zone interface). Because Coriolis and pressure gradient forces almost balance (as in a geostrophic balance), the perturbation flow becomes clockwise on the bulges, and counterclockwise in the depressions. These also are properties of Rossby waves patterns, whether stable or unstable.

Vertical motions are also correlated with the bulges and depressions, leading to non-zero kinetic helicity. This in turn provides an additional source of alpha effect for solar dynamo models (Dikpati and Gilman 2001b) that could play a role in parity selection in solar dynamo models (Bonanno et al. 2002; Belucz et al. 2015).

3.2 Rossby Waves

In the Earth's atmosphere, where Rossby waves were first discovered, the forcing primarily comes from differential heating from the Sun; this is thermal forcing. Coriolis forces are always present in a rotating body, but they do not provide a source of energy to drive the Rossby waves. They only alter the direction of fluid motion in the wave. In the case of the Sun, unlike the Earth's atmosphere, the energy is coming from the mechanical forcing of differential rotation at the top of the tachocline. However, like in the Earth's atmosphere, the Coriolis forces again can alter the direction of plasma flow, but can't provide energy.

Very much like the Earth's Rossby waves, solar Rossby waves were also first predicted theoretically before they were directly observed (Dziembowski and Kosovichev 1987; Gilman and Fox 1997; Zaqarashvili et al. 2010). These works found that the disturbances that perturb the tachocline differential rotation form the patterns that have Rossby waves-like properties such as the ones described in Sect. 3.1. Later on, various observations have indicated the existence of Rossby waves in the Sun (McIntosh et al. 2017; Löptien et al. 2018; Harris et al. 2022).

In fact, even before the role of Rossby waves as perturbation to the global flows and magnetic fields, the idea that the Sun might have an MHD dynamo driven by Rossby waves was explored by Gilman (1968, 1969a,b). That model assumed the existence of a 'thermal

---

[1] such as expressed by $\Omega = \Omega_0 - s \sin^2 \theta$ form, in which Ω_0 denotes the core-rotation rate (or, in other words, the rotation rate at $32°$ latitude) and s the amplitude of equator-to-pole differential rotation.

wind' inside the convection zone, with latitudinal and radial rotation gradients as well as an associated latitudinal specific entropy gradient. Gilman found that Rossby waves and DR can together drive a cyclic dynamo with a period within an order of magnitude of the observed sunspot cycle. The solar tachocline, for which the model assumptions were plausible, was not discovered until almost thirty years later, and observational evidence of Rossby waves in the Sun has been found only over the past five years.

So-called thermal Rossby waves have been discovered theoretically for both the convection zones and radiative interiors (Glatzmaier and Gilman 1981; Hindman and Jain 2022). Those may be more like Earth's Rossby waves, in that they contain thermodynamic effects and rely for some of their properties on the outward decline in plasma density, as well as on the spherical geometry of a thick layer as opposed to thin layer that is the tachocline. Thermal Rossby waves are yet to be detected in the Sun.

It is worth noting that in the presence of a latitudinal DR that is stable to perturbations, Rossby waves can theoretically still exist. In that case, their patterns do not have tilts in latitude, and so they cannot extract energy from the reference state. Their phase velocity differs from the uniform rotation case by an amount close to that of the local rotation speed at the latitude where the relative amplitude of the Rossby wave peaks. In an unstable differential rotation, Rossby waves behave similarly, but they have tilts in latitude and a phase speed in longitude close to the rotation speed of the latitude where the unstable mode peaks. These properties are similar for purely 2D waves, and the quasi-3D waves characteristic of hydrodynamic shallow water models (Dikpati and Gilman 2001a).

With the addition of a toroidal field, the picture changes significantly, depending on the latitudinal profile and strength of the field. We first note that Rossby waves are still nearly hydrodynamic (with retrograde propagation). In addition, a very slow prograde wave starts to appear, and vanishes in the limit of zero toroidal field (Dikpati et al. 2020; Hori et al. 2020). For much stronger fields, magnetic effects overpower Coriolis forces, leading to Alfvén waves propagating with nearly equal speeds in both prograde and retrograde directions. Unstable MHD waves for even moderately strong but latitudinally compact toroidal fields propagate at the rotational speed of the latitude where the toroidal field peaks. Therefore, these waves are likely to be retrograde at high latitudes (poleward of sunspot zones), but could be prograde for a certain range in low latitudes (see, e.g. Dikpati et al. 2018).

It is to be noted that all Rossby waves are inertial waves, however all inertial waves are not Rossby waves. Recently, solar inertial oscillations were studied by Triana et al. (2022) in a 3D spherical shell model of the convection zone containing a homogeneous, incompressible, viscous fluid, using an eigen-system formalism. Retrograde as well as prograde Rossby modes were found, along with other inertial modes, which are distinct from Rossby modes and have radial velocities comparable to horizontal velocities. Rossby modes found by Triana et al. (2022) are similar to those found by Bekki et al. (2022) using a similar model but in a fully compressible regime.

3.3 Global MHD Instabilities

The addition of a toroidal field to 2D, quasi-3D shallow-water type, and 3D thin-shell type hydrodynamical models has a destabilizing effect, as we noted in the previous section. In that case, instabilities for low longitudinal wave numbers can occur down to very small differential rotations. Which longitudinal wave number is most unstable depends on the strength and profile of the toroidal field. Narrower toroidal bands render higher longitudinal wavenumbers unstable. But very narrow bands of about 2° latitudinal width are not unstable because they cease to sense the difference in rotation rate across the band. So there is an optimum bandwidth for instability in the context of the solar DR profile (Gilman and Dikpati 2000).

The existence of unstable toroidal bands at particular latitudes with low longitudinal wave number could provide templates for the longitude and latitude distribution of active regions on the solar surface. The bulges in the tachocline that happen to contain strong toroidal field may indeed be the source for emerging magnetic flux that produces new active regions (Dikpati et al. 2017).

Global HD and MHD instabilities in the tachocline evolve non-linearly due to the interactions among differential rotation, magnetic fields and Rossby waves, which are essentially the disturbance patterns. Nonlinear evolution of these instabilities can produce many interesting features which are discussed in the following two subsections, for pure 2D models and quasi-3D thin-shell shallow-water type models.

3.3.1 Nonlinear Evolution of 2D HD and MHD Instabilities

In a pure 2D tachocline, nonlinear evolution of the HD instabilities can produce high-latitude jets. Due to the action of Reynolds stresses, the angular momentum can be extracted from the differential rotation specifically from the mid-latitudes, and can be deposited on the poleward sides, creating prograde jets there. Traces of such tachocline jets were observationally detected by helioseismic analysis (Christensen-Dalsgaard et al. 2005).

For a 2D magnetized tachocline, nonlinear evolution of MHD instabilities lead to two interesting features. In the case of broad toroidal fields open into clam-shell pattern (Cally 2001). Conversely, a narrow toroidal band produces tipping or deformation patterns depending on the field strength. If the toroidal band harbours strong fields, it behaves like a 'steel' ring, and it tips to produce an $m = 1$ longitudinal pattern (Cally et al. 2003). If the field is weak, the ring deforms and produces $m > 1$ patterns in longitude.

3.3.2 Nonlinear Evolution of Quasi-3D Instabilities

In quasi-3D thin-shell and shallow-water type models, the formation of high-latitude jets, as well as the clamshell opening of broad toroidal fields and tipping of narrow bands occur too. Nevertheless, additional important nonlinear effects have been found in quasi-3D model simulations, which are not present in 2D models. Due to the presence of potential energy, along with kinetic and magnetic energies in a shallow-water model, nonlinear interactions among differential rotation, toroidal field and Rossby waves can produce the 'Tachocline Nonlinear Oscillations' (TNOs), which arise as a consequence of quasi-periodic exchange of energies among these three reservoirs. These oscillations are produced in both HD and MHD cases, very much like 'nonlinear Orr mechanism' in fluid mechanics. The period of oscillation is typically 6-18 months in the context of the solar interior. This oscillation period interestingly coincides with the observations of Rieger-type periodicity, or 'seasons' of solar activity, characterized by intervals of more intense activity alternating with more quiet periods. This suggests that the origin of such 'seasons' of could be the TNOs (Dikpati et al. 2017).

Figure 6 describes the physics of TNOs in an ideal hydrodynamic system, in which they occur due to nonlinear exchange between kinetic energy of differential rotation and perturbation energy of Rossby waves. We note that the TNO physics in the MHD case is a bit more complex in the MHD case, and defer the interested reader to Fig. 6 of Dikpati et al. (2018) for their extended description. An interesting property of TNOs is that they do not damp out in several years' runs in an ideal MHD simulation. Evidence for their existence has been

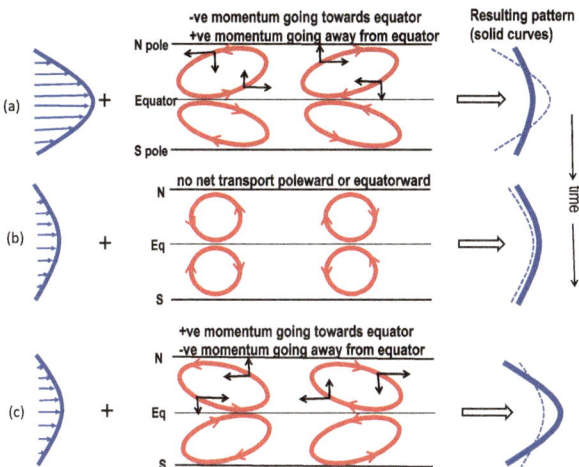

Fig. 6 Panels a, b, c show schematically how Rossby waves and tachocline latitudinal differential rotation nonlinearly interact. Time goes from top panel to bottom. Energy from unperturbed differential rotation (thick blue curve in left frame of panel(a)) is extracted by perturbation flow-patterns (eastward-oriented red ellipses in the north hemisphere, plotted in middle frame of panel (a)), because their eastward tilts transport angular momentum away from the equator; consequently the polar region spins up and the equatorial region spins down, resulting into a decrease in pole-to-equator differential rotation (thick blue curve in right frame of panel a), compared to the original (thin dashed curve). Panel (b) shows that tilts become neutral when no more energy can be extracted out of differential rotation. However, the perturbation flow-patterns overshoot from neutral tilt to acquire westward tilts, as shown in panel (c) (middle frame); hence flow-patterns can transport positive angular momentum back from high to low latitudes (see the black arrows on the flow-patterns of the bottom panel), spinning up the equatorial region again. Consequently, differential rotation is restored back, and becomes the source of energy to be extracted by the perturbation flow-patterns. In a nearly dissipationless system this process repeats, very much like nonlinear Orr mechanism in fluid dynamics, and leads to an oscillatory exchange of energies between Rossby waves and tachocline differential rotation (Dikpati et al. 2017)

furthermore obtained recently in a fully nonlinear direct numerical simulation of tachocline instabilities using generalized quasilinear approximation by Plummer et al. (2019), where the inclusion of magnetic diffusivity is shown to create an abrupt transition of short-term to long-term oscillations between kinetic and magnetic energies.

Another notable nonlinear effect arises from the action of the magnetic stress (often called the 'mixed stress'), which arises due to the cross-correlation between perturbation magnetic and velocities fields in unstable global MHD modes. This stress can cause extraction of energy from the center of the band, and deposition of energy on the both shoulders of the band. In the nonlinear evolution, this process can continue, and at a point the energy at the band-center can be depleted so much as to cause the band to split into two. The poleward part of the split-ring slips fast towards the pole due to curvature stress, which is responsible for causing the famous "poleward-slip" instability of a toroidal ring (Spruit and van Ballegooijen 1982). A 6° toroidal ring splitting into two can be seen in this ring-split movie. The toroidal ring is displayed in white arrow-vectors on the colormap, which represents deformation of thin fluid-layer's top-surface (red/orange denotes bulges and blue/dark-blue depressions). The movie shows that the ring is just split into two at t = 6.7023 (24.5 days), and the split part of the ring is seen to move poleward at t = 7.6556 (30 days). The snapshots are presented in dimensionless time (dimensional time can be obtained by multiplying by 3.65 in the unit of days).

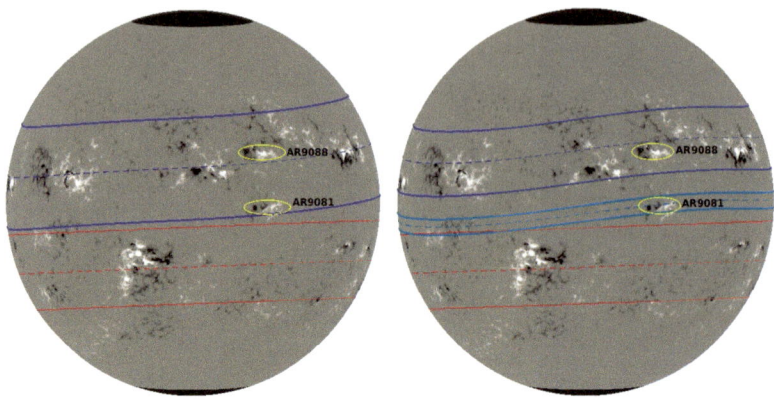

Fig. 7 Left panel displays active regions' toroid-pattern for July 19, 2000, if all the active regions emerged from one wide toroidal magnetic band from the base of the convection zone; the right panel displays two split toroid-patterns if active regions emerged from spot-producing toroidal rings that have undergone dynamical splitting. North and South toroids are displayed in blue and red respectively (Dikpati et al. 2021)

The equatorward part can move slowly to the equator because of the energy getting deposited from the band-center to the equatorward shoulder of the band. This provides a possible physical mechanism for a pair of spots, often emerging at the same longitude but separated in latitude by more than 20-degrees (Dikpati et al. 2021). The Fig. 7 shows an observational evidence of a pair of spots that could have emerged from split-ring.

MHD Rossby waves could have an effect on the solar magnetic activity cycle (Raphaldini and Raupp 2015) and on space weather (Dikpati and McIntosh 2020). Here for the general readers, we display a diagram to show how magnetically modified solar Rossby waves can create spatio-temporal patterns on dynamo-generated magnetic fields at the tachocline, very much like the "jet streams" produced by the interaction of Earths' atmospheric Rossby waves with the mean zonal flow. As the "jet streams" steer the terrestrial weather by causing cold wave or draught, respectively transporting cold from high to mid-latitudes or heat from low to high latitudes, similarly spatio-temporal magnetic patterns created by the interaction of solar MHD Rossby waves could determine the plausible latitude-longitude locations of flux emergence on the solar surface (see, e.g. Fig. 8). We point the reader to the comprehensive review by Zaqarashvili et al. (2021) for more details about Rossby waves and their roles in producing various astrophysical phenomena.

3.4 Magnetic Buoyancy Instability

Concomitantly to global HD and MHD instabilities, perturbations to a tube or layer of magnetic field stored at the tachocline may trigger the buoyancy instability. Over the years, it has been assumed that this instability is responsible for the emergence of magnetic flux from the tachocline to the solar surface, leading perhaps for the formation of sunspots. Furthermore, the turbulence generated by this process may in principle generate a net electromotive force, and therefore an α-effect.

The most relevant aspect of the buoyancy instability is that the conditions for its excitation are easily fulfilled for the magnetic fields developed by the strong radial shear at tachocline levels. The general idea was proposed by Parker (1955) considering a magnetized tube of plasma in pressure equilibrium with its surroundings. If the temperature of the plasma inside the tube is the same as in the exterior, the density inside will be smaller and

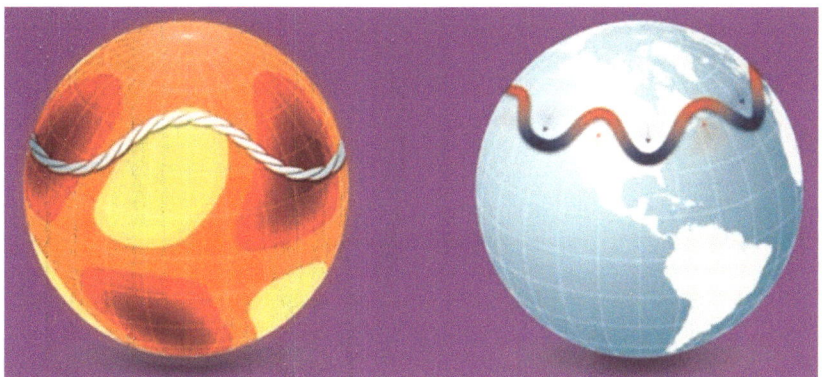

Fig. 8 Large-scale Rossby waves propagating around the Sun may play a role in triggering solar storms (yellow and deep red regions indicate where sunspot activity is unlikely and more likely, respectively). These waves are akin to Rossby waves in Earth's atmosphere, which influence weather, including potential heat waves (red arrows) and cold snaps (blue arrows) (schematic figure adapted from SCIENCE NEWS BY AGU, Eos.org 51, vol. 101, no. 8, August 2020)

the tube will become a bubble. This out of equilibrium situation was ever since extensively studied formally as an instability. For instance, Newcomb (1961) used the energy principle of Bernstein et al. (1958) to find the stability condition for interchange modes where the wavenumber along the field tube tends to zero (i.e., there is no deformation of the tube along the direction of the magnetic field). The instability turned out to depend on the vertical gradient of the magnetic field (Thomas and Nye 1975). The use of the energy principle for an unstable parcel of gas was also performed for non-ideal plasmas (Acheson and Gibbons 1978). Yet this approach becomes insufficient for more realistic models including rotation and/or shear (Tobias and Hughes 2004). These aspects, however, were studied through local instability analysis for interchange and, three-dimensional, undular modes (e.g., Gilman 1970; Acheson and Gibbons 1978; Hughes 1985a,b). In general, small magnetic diffusivity and large thermal conductivity have a destabilizing effect (since the instability may develop faster than the magnetic field diffusion, and the positive entropy gradient of a stable layer may decrease, respectively). Rotation was found to have a stabilizing effect, yet the instability depends strongly on the vertical profile of the magnetic field, and rotation-induced shear may therefore contribute to make axisymmetric the easily excited non-axisymmetric buoyant modes. A complete review can be found in (Hughes 2007). More recently, Gilman (2018b,a) extended the linear stability analysis to the spherical domain, including rotation and vertical shear, and exploring together rotational and buoyancy instabilities. His results indicate that the subadiabatic tachocline is unstable to negative vertical shear, such as it is observed at high latitudes in the Sun. Therefore, the magnetic field diffuses rapidly there and the buoyancy instability becomes inefficient. On the other hand, positive shear, as in the solar lower latitudes, is a stable configuration to rotational instabilities, and the buoyancy instability develops for magnetic fields above ~ 9 kG. Above this threshold, plenty of unstable modes may become unstable, depending on the amplitude of the magnetic field.

The non-linear evolution of buoyant magnetic fields has been extensively studied through MHD numerical simulations. Initial studies were performed in two-dimensional domains and explored the evolution of single flux tubes (Schuessler 1979; Moreno-Insertis and Emonet 1996; Fan et al. 1998), or the development of interchange modes in a magnetic layer (Cattaneo and Hughes 1988). These studies found the development of the mushroom or umbrella shape in the emerging structures. These structures develop strong vorticity at their edges which ends destroying their coherence. Cattaneo et al. (1990) found in 2D simulations that the addition of twist to the initial magnetic field prevents the development of vorticity and facilitates the emergence of the field. It turned out, however, that in the three-dimensional case the axisymmetric perturbations are unstable only during the first stages of evolution (Matthews et al. 1995; Wissink et al. 2000; Fan 2001), or remain stable (Fan 2001), depending on the ambient stratification. The undular 3D modes are dominant, and the bending of the magnetic field lines helps the magnetic structures to remain cohesive during the emergence (Abbett et al. 2001).

Concurrently, the buoyancy instability has been studied under the assumption that below the convection zone the magnetic field is organized in the form of thin flux tubes with radius much smaller than the local pressure scale height. The properties of such tubes change along them but not over their cross section. Thus, the evolution equations may be reduced to one dimension (Spruit 1981). The simulations commonly consider the full longitudinal domain, and impose non-axisymmetric perturbations, which, after the instability develops, forms arch structures anchored at the base of the tube in the solar convection zone (e.g., Choudhuri 1989; Ferriz-Mas and Schüssler 1993) or the subadiabatic tachocline (e.g., Caligari et al. 1995). Because computing the evolution of thin flux tubes is generally inexpensive, a large amount of work has been performed under this approximation, exploring the role of rotation or the conditions for reproducing the tilt angle observed in active regions (e.g., Choudhuri and Gilman 1987; D'Silva and Choudhuri 1993). In order to overcome the effect of the Coriolis force, and to emerge with the tilt angles compatible with observations, the initial field strength of the thin flux tubes must be $\geq 10^5$ G. A complete description of results and references can be found in Fan (2021). The most recent work on this topic (Weber et al. 2011, 2013; Weber and Fan 2015) explored the evolution of thin flux tubes in the presence of turbulent convection provided by a global simulation (Miesch et al. 2006). The results show an interesting interplay between buoyancy and advection, with initial flux tubes with strength between $40 - 50$ kG being the ones that better resemble observational tilt angles, and with rapid rising times. It is worth noting that the thin flux tube approximation is not except of criticism. Different works have shown that the internal dynamics of the tube, neglected on the averaged formulation of the thin flux tube approximation, may play a relevant role during the emerging process (Hughes et al. 1998, Hughes and Falle 1998, or more recently Martínez-Sykora et al. 2015).

In the same spirit, Cartesian three-dimensional MHD simulations have explored the evolution of imposed magnetic tubes in the presence of convection (of course this not exactly happening at the stable tachocline but above it, see Fan et al. 2003), the self generation of a magnetic layer by a vertical shear (Vasil and Brummell 2008, 2009), or even the vertical shear beneath a convectively unstable layer (Guerrero and Käpylä 2011). Convection strongly affects the evolution of a magnetic structure and creates arched tubes anchored by the downflow lanes with a fat magnetic flux configuration frozen-in with the convectively rising plasma. Under shear conditions more extreme that in the solar tachocline, undular magnetic structures develop and may become buoyancy unstable (Fig. 9(a) and (b)). If the shear is sustained by a mechanical forcing then a self-sustained dynamo can exist because an incoherent, yet different from zero, α-effect can persist (Guerrero and Käpylä 2011). An oscillatory behavior can also be found if the field is replenished by the boundary condition (Kersalé et al. 2007).

Three dimensional simulations have also been performed in spherical coordinates considering the emergence of magnetic flux tubes in adiabatic (Jouve and Brun 2007; Fan 2008; Fournier et al. 2017, e.g.,) and superadiabatic layers (Jouve and Brun 2009). Even

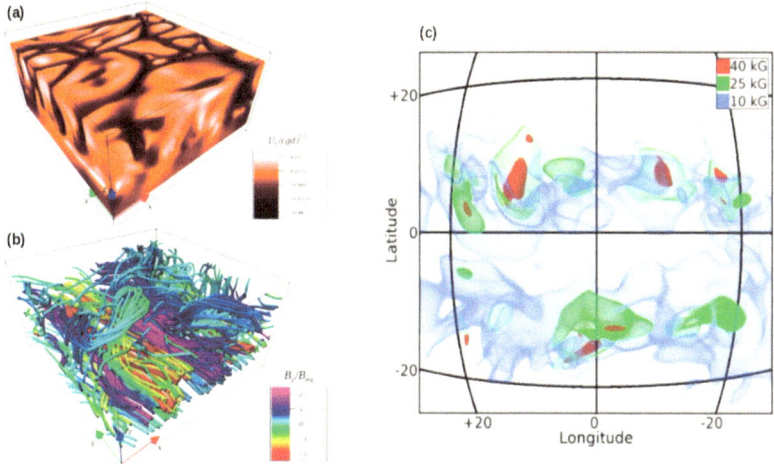

Fig. 9 Panels (a) and (b): convection pattern and magnetic field lines resulting from a numerical simulation with forced radial shear beneath a convectively unstable layer. If the shear is sufficiently strong, the magnetic elements become buoyant forming Ω-loops. This magnetic field modifies the pattern of convection (adapted from Guerrero and Käpylä 2011). Panel (c): volume rendering of the buoyant magnetic elements developed from a global dynamo simulation. The iso-surface in blue is in equipartition with the kinetic energy whereas the regions in red have four times more energy (adapted from Nelson et al. 2013)

though these works preclude the development of the buoyancy instability in the subadiabatic tachocline, they clearly expose how complex it is for a buoyant structures to cross a strongly stratified layer, specially in the presence of turbulent convection and its associated large scale mean flows (Jouve and Brun 2009) or of background dynamo magnetic fields (Pinto and Brun 2013). Reproducing the observed features of active regions, therefore, requires a fine tuning of parameters as the magnitude of the magnetic field, twist, and wave-number of the instability. Note that the origin and magnitude of the twist considered in these models do not have proper justification. Most recently Manek et al. (2018), Manek and Brummell (2021) explored the evolution of twisted magnetic tubes embedded in an adiabatic layer with a background magnetic field. They found that there is a lower limit of the background field that allows the tube emergence, and that magnetic structures where the twist and the background field are aligned rise more easily than tubes with the opposite configuration. Finally, magnetic buoyant elements have been observed in convective global dynamo simulations where the magnetic field is not imposed but develops self-consistently (Nelson et al. 2011, 2014; Fan and Fang 2014). These works showed that convection may not be a hindrance to magnetic flux emergence but may contribute to the rise of magnetic loops (see Fig. 9(c)). It is not yet clear, however, whether these buoyant magnetic elements may reach the surface levels and form bipolar regions. Another kind of simulations have been performed to explore the conditions of the surface emergence phenomena (e.g., see the review of Cheung and Isobe 2014, and references therein). These models are far from the scope of this review.

3.5 Double Band and the Extended Solar Cycle

The examination of torsional oscillations (Guerrero et al. 2016), the coronal green line emission (Tappin and Altrock 2013) and the studies of ephemeral active regions (Martin 2018) provide hints of the existence of extended solar cycles. In these observables, the solar activity cycles also manifest at higher latitudes (around 60-70 degrees) than the classical butterfly

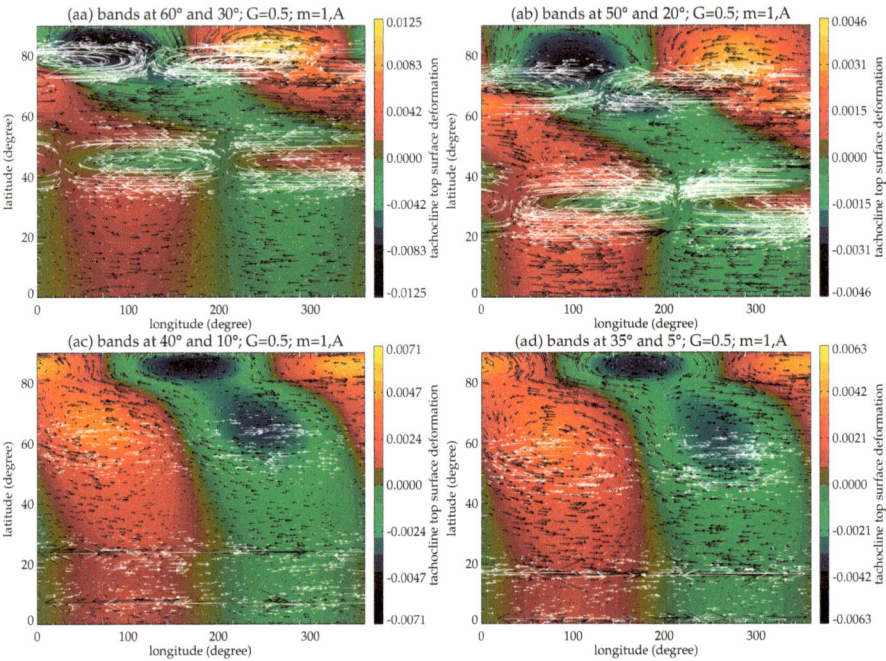

Fig. 10 Snapshots of velocity and magnetic disturbance patterns are shown in the latitude-longitude plane for the asymmetric $m = 1$ modes. Black arrows represent magnetic field vector, white arrows flow vectors and color map the tachocline top surface deformation (red-yellow denote the bulging of top surface, and green-blue the depression). This figure shows the eigenfunctions for the overshoot part of tachocline ($G = 0.5$) and the bands at $60° - 30°$ (aa), $50° - 20°$ (ab), $40° - 10°$ (ac) and $35° - 5°$ (ad) (Belucz et al. 2023)

diagram for the sunspot cycle. It is generally observed at these latitudes ahead of the start of the new sunspot cycle. This high-latitude component of the solar cycle gradually moves towards the equator, maintaining a separation of about 30° (see, e.g., the Fig. 1 of Belucz et al. 2023) with the low-latitude active cycle's branch. When the active cycle's branch reaches near the equator to be ready to annihilate with its opposite-hemisphere counterparts, the high-latitude branch reaches about 30-35 degrees, and could then start producing sunspots of the next cycle. Motivated by these observations, one may consider the possibility that there are double, subsurface magnetic bands with opposite directions in each hemisphere, with a low-latitude toroidal strip and a weak high-latitude strip. Belucz et al. (2023) studied the global HD/MHD instabilities of such a double structure to understand the global MHD processes and their contribution to properties of solar activity and the solar cycle. The thin solar tachocline can be modelled in a shallow-water formalism as a 3D thin fluid shell with a rigid bottom and deformable top surface. The flows and fields are much larger in the latitude-longitude than in the radial direction. This condition is validated in the tachocline because of its subadiabatic stratification.

Motivated by observations of extended solar cycle, Belucz et al. (2023) considered a double-band system located within the solar tachocline and consisting of two oppositely-directed toroidal bands in each hemisphere, and study their MHD instability in a quasi-3D MHD shallow-water model. As we make the two oppositely-directed bands in each hemisphere migrate equatorward in the model (Belucz et al. 2023), it is found that the high- and low-latitude bands interact in the same hemisphere when they are at latitude 60° and 30°

(Fig. 10). Here, band-interaction occurs by teleconnection mechanism – we briefly describe the physics below, before elaborating on the MHD of the double-band in the solar tachocline.

Here, band-interaction occurs by a 'teleconnection' mechanism. While teleconnection has long been known in meteorology (Lorenz 1951) and atmospheric circulation (Blackmon et al. 1984), this concept is relatively new for the Sun (see, e.g., Leamon et al. 2021). Teleconnection is essentially a contemporaneous correlation among various global fluid properties and parameters at remotely separated regions and their influence on each other. Teleconnection was demonstrated in a pioneering paper by Wallace and Gutzler (1981) to be created by planetary waves in the North Atlantic and North Pacific Oscillations. In the Sun, Rossby waves act as perturbations to the unperturbed DR and magnetic fields, nonlinearly exchanging energy among the perturbations and reference states of the system. This long range communication can happen in latitude (even across the equator) as well as in longitude, from one side of the Sun to the other. In the case of dynamo-generated magnetic bands in a solar interior, the basic physics of the global connections was described in Fig. 4 of Cally et al. (2003).

The main conclusion from the MHD of double-band (Belucz et al. 2023) is that, if an extended solar cycle is represented by the two magnetic bands in the interior, generated by dynamo, namely a magnetic band at sunspot latitudes to represent an active-cycle and an oppositely-directed magnetic band at high-latitude band at 60-degree latitude to represent the extended part of the cycle, the model shows that the high- and low-latitude bands interact and telecommunicate among themselves, but not with their opposite-hemisphere counterparts, until this double-band-system has migrated enough towards the equator. The cross-equatorial communication starts when the low-latitude bands in the north and south hemispheres are closer than 15-degree from the equator, because then the low-latitude bands reach the optimal distance between them to start the teleconnection. This is approximately the phase of the solar cycle when the cycle activity reaches the peak. The low-latitude bands in the North and South start the teleconnection across the equator to indicate the start of the declining phase.

4 Conclusions and Perspectives

The basic physics of the solar tachocline began to be revealed to solar physicists with the seminal paper of Spiegel and Zahn (1992). Since then, the physics of this intriguing layer has led to many debates in the research community. In particular, its extreme thinness continues to be a source of questioning. The tachocline is a turbulent, doubly-sheared, highly stratified, low-Pr, and magnetized thin layer. It is a promising subject of research for any fluid-dynamics enthusiast. We have seen in the preceding sections that advanced, global 3D models could be developed to address in a major way the dynamics of the tachocline. On the other hands, such models are not resolved enough to model adequately all the small-scale turbulent transport in the solar tachocline. In that respect, several routes can be highlighted to continue deciphering the mysteries of the tachocline:

- In the past two years, the study of turbulent transport in the solar tachocline has been again revisited. On one hand, Garaud (2020) has developed in a clear manner the argument that the low-Pr regime of the solar tachocline makes the estimate of the net turbulent transport non-trivial. On the other hand, the effect of magnetic field on the transport is still not fully understood. Following the important work of Tobias et al. (2007), Chen and Diamond (2020) have developed a theoretical argument for a complex tachocline confinement leveraging dynamo-generated tangled magnetic fields. We are today in the position of simulating the highly turbulent solar tachocline in a more and more realistic regime thanks to the massive development of 3D global numerical codes. Dedicated, high-resolution numerical experiment could be built to address frontally this problem, e.g. taking into account the convective penetration in the upper tachocline and the associated magnetic field sustained by dynamo action above.

- The role of magnetoshear instabilities in the dynamics of the tachocline and of the solar dynamo is progressing actively recently. These have been found in some global models of the solar interior (Lawson et al. 2015; Guerrero et al. 2019; Jouve et al. 2020), and were interpreted to have different effects on the global dynamo. While the solar dynamo is responsible for generating the spot-producing magnetic fields and their migration in latitude, the global MHD of the tachocline provides a mechanism to create the longitudinal distribution of active regions. As discussed in Sect. 3.3.2, there exists a major opportunity to link via data assimilation the observed spatio-temporal distribution of active regions with the simulated "imprints" of their latitude-longitude locations created by the global tachocline instabilities of the spot-producing magnetic fields (see, e.g. Fig. 12 of Dikpati and McIntosh 2020).

- The tachocline coincides at least in part with the convective-radiative transition in Sun-like stars. Angular momentum can be transported from one region to the other, especially when the star changes its structure or when the external convection zone spins down due to magnetic torques from its wind. In the theory of cool-stars' rotation, the coupling between these two regions plays an important role in the pre-main sequence and early main sequence phases. For instance, Gallet and Bouvier (2013) parameterized such angular momentum exchange timescale to reproduce the rotational distribution observed in open clusters (see also Benbakoura et al. 2019, Ahuir et al. 2020). An assessment of this coupling timescale in realistic models of the tachocline (see e.g. Brun et al. 2011) would therefore be extremely useful today to improve our understanding of the secular evolution of the Sun and solar-type stars.

Funding AS and ASB acknowledge partial financial support by DIM-ACAV+ ANAIS2 project, ERC Whole Sun Synergy grant #810218 and STARS2 starting grant #207430, ANR Toupies, ANR STORMGENESIS #ANR-22-CE31-0013-01, INSU/PNST, Solar Orbiter and PLATO CNES funds, and GENCI via project 1623. MD acknowledges the support for this work from the National Center for Atmospheric Research, which is a major facility sponsored by the National Science Foundation under cooperative agreement 1852977, and also acknowledges support from several NASA grants, namely NASA-LWS award number 80NSSC20K0355, NASA-HSR award number 80NSSC21K1676, subaward from JHU/APL's NASA-HSR grant with award number 80NSSC21K1678 and subaward from Stanford's COFFIES Phase II NASA-DRIVE Center with award number 80NSSC22M0162. Bernadett Belucz's work is supported by Newton International Fellowship of The Royal Society, program number NIF-R1-192417. GG acknowledges support from NASA grants NNX14AB70G, 80NSSC20K0602, and 80NSSC20K1320.

Declarations

Competing Interests The authors declare they have no conflicts of interest.

References

Abbett WP, Fisher GH, Fan Y (2001) The effects of rotation on the evolution of rising omega loops in a stratified model convection zone. Astrophys J 546(2):1194–1203. https://doi.org/10.1086/318320. arXiv: astro-ph/0008501 [astro-ph]

Acevedo-Arreguin LA, Garaud P, Wood TS (2013) Dynamics of the solar tachocline – III. Numerical solutions of the Gough and McIntyre model. Mon Not R Astron Soc 434(1):720–741. https://doi.org/10.1093/mnras/stt1065. arXiv:1304.3167 [astro-ph.SR]

Acheson DJ, Gibbons MP (1978) Magnetic instabilities of a rotating gas. J Fluid Mech 85:743–757. https://doi.org/10.1017/S0022112078000907

Ahuir J, Brun AS, Strugarek A (2020) From stellar coronae to gyrochronology: a theoretical and observational exploration. A&A 635:A170. https://doi.org/10.1051/0004-6361/201936974. arXiv:2002.00696 [astro-ph.SR]

Antia HM, Chitre SM, Thompson MJ (2000) The Sun's acoustic asphericity and magnetic fields in the solar convection zone. A&A 360:335–344. arXiv:astro-ph/0005587 [astro-ph]

Barnabé R, Strugarek A, Charbonneau P, Brun AS, Zahn JP (2017) Confinement of the solar tachocline by a cyclic dynamo magnetic field. A&A 601:A47. https://doi.org/10.1051/0004-6361/201630178. arXiv: 1703.02374 [astro-ph.SR]

Basu S, Antia HM (2001) A study of possible temporal and latitudinal variations in the properties of the solar tachocline. Mon Not R Astron Soc 324(2):498–508. https://doi.org/10.1046/j.1365-8711.2001.04364.x. arXiv:astro-ph/0101314 [astro-ph]

Basu S, Antia HM (2003) Changes in Solar Dynamics from 1995 to 2002. Astrophys J 585(1):553–565. https://doi.org/10.1086/346020. arXiv:astro-ph/0211548 [astro-ph]

Basu S, Antia HM (2019) Changes in solar rotation over two solar cycles. Astrophys J 883(1):93. https://doi.org/10.3847/1538-4357/ab3b57. arXiv:1908.05282 [astro-ph.SR]

Beaudoin P, Strugarek A, Charbonneau P (2018) Differential rotation in solar-like convective envelopes: influence of overshoot and magnetism. Astrophys J 859(1):61. https://doi.org/10.3847/1538-4357/aabfef

Bekki Y, Cameron RH, Gizon L (2022) Theory of solar oscillations in the inertial frequency range: linear modes of the convection zone. A&A 662:A16. https://doi.org/10.1051/0004-6361/202243164. arXiv: 2203.04442 [astro-ph.SR]

Belucz B, Dikpati M, Forgács-Dajka E (2015) A babcock-leighton solar dynamo model with multi-cellular meridional circulation in advection- and diffusion-dominated regimes. Astrophys J 806(2):169. https://doi.org/10.1088/0004-637X/806/2/169. arXiv:1504.00420 [astro-ph.SR]

Belucz B, Dikpati M, McIntosh SW, Leamon RJ, Erdélyi R (2023) Magnetohydrodynamic instabilities of double magnetic bands in a shallow-water tachocline model. I. Cross-equatorial interactions of bands. Astrophys J 945(1):32. https://doi.org/10.3847/1538-4357/acb43b

Benbakoura M, Réville V, Brun AS, Le Poncin-Lafitte C, Mathis S (2019) Evolution of star-planet systems under magnetic braking and tidal interaction. A&A 621:A124. https://doi.org/10.1051/0004-6361/201833314. arXiv:1811.06354 [astro-ph.SR]

Bernstein IB, Frieman EA, Kruskal MD, Kulsrud RM (1958) An energy principle for hydromagnetic stability problems. Proc R Soc Lond Ser A 244(1236):17–40. https://doi.org/10.1098/rspa.1958.0023

Bice CP, Toomre J (2020) Probing the influence of a tachocline in simulated M-dwarf dynamos. Astrophys J 893(2):107. https://doi.org/10.3847/1538-4357/ab8190. arXiv:2001.05555 [astro-ph.SR]

Blackmon ML, Lee YH, Wallace JM (1984) Horizontal structure of 500 mb height fluctuations with long, intermediate and short time scales. J Atmos Sci 41(6):961–980. https://doi.org/10.1175/1520-0469(1984)041<0961:HSOMHF>2.0.CO;2.

Bonanno A, Elstner D, Rüdiger G, Belvedere G (2002) Parity properties of an advection-dominated solar alpha^2 omega-dynamo. A&A 390:673–680. https://doi.org/10.1051/0004-6361:20020590. arXiv:astro-ph/0204308 [astro-ph]

Braithwaite J (2006) A differential rotation driven dynamo in a stably stratified star. A&A 449(2):451–460. https://doi.org/10.1051/0004-6361:20054241. arXiv:astro-ph/0509693 [astro-ph]

Braithwaite J (2007) The stability of poloidal magnetic fields in rotating stars. A&A 469(1):275–284. https://doi.org/10.1051/0004-6361:20065903. arXiv:0705.0185 [astro-ph]

Braithwaite J (2008) On non-axisymmetric magnetic equilibria in stars. Mon Not R Astron Soc 386(4):1947–1958. https://doi.org/10.1111/j.1365-2966.2008.13218.x. arXiv:0803.1661 [astro-ph]

Braithwaite J (2009) Axisymmetric magnetic fields in stars: relative strengths of poloidal and toroidal components. Mon Not R Astron Soc 397(2):763–774. https://doi.org/10.1111/j.1365-2966.2008.14034.x. arXiv:0810.1049 [astro-ph]

Braithwaite J, Spruit HC (2004) A fossil origin for the magnetic field in A stars and white dwarfs. Nature 431(7010):819–821. https://doi.org/10.1038/nature02934. arXiv:astro-ph/0502043 [astro-ph]

Brandenburg A, Elstner D, Masada Y, Pipin V (2023) Turbulent processes and mean-field dynamo. Space Sci Rev 219:55. https://doi.org/10.1007/s11214-023-00999-3

Brown TM, Christensen-Dalsgaard J, Dziembowski WA, Goode P, Gough DO, Morrow CA (1989) Inferring the Sun's internal angular velocity from observed p-mode frequency splittings. Astrophys J 343:526. https://doi.org/10.1086/167727

Brun AS, Zahn JP (2006) Magnetic confinement of the solar tachocline. A&A 457(2):665–674. https://doi.org/10.1051/0004-6361:20053908. arXiv:astro-ph/0610069 [astro-ph]

Brun AS, Miesch MS, Toomre J (2011) Modeling the dynamical coupling of solar convection with the radiative interior. Astrophys J 742(2):79. https://doi.org/10.1088/0004-637X/742/2/79

Brun AS, Strugarek A, Noraz Q, Perri B, Varela J, Augustson K, Charbonneau P, Toomre J (2022) Powering stellar magnetism: energy transfers in cyclic dynamos of Sun-like stars. Astrophys J 926(1):21. https://doi.org/10.3847/1538-4357/ac469b. arXiv:2201.13218 [astro-ph.SR]

Caligari P, Moreno-Insertis F, Schussler M (1995) Emerging flux tubes in the solar convection zone. I. Asymmetry, tilt, and emergence latitude. Astrophys J 441:886. https://doi.org/10.1086/175410

Cally PS (2001) Nonlinear evolution of 2d tachocline instabilities. Sol Phys 199(2):231–249. https://doi.org/10.1023/A:1010390814663

Cally PS, Dikpati M, Gilman PA (2003) Clamshell and tipping instabilities in a two-dimensional magnetohydrodynamic tachocline. Astrophys J 582(2):1190–1205. https://doi.org/10.1086/344746

Cattaneo F, Hughes DW (1988) The nonlinear breakup of a magnetic layer – instability to interchange modes. J Fluid Mech 196:323–344. https://doi.org/10.1017/S0022112088002721

Cattaneo F, Chiueh T, Hughes DW (1990) A new twist on the solar cycle. Mon Not R Astron Soc 247:6P–9P

Charbonneau P (2004) Three single stars (see how they spin) (invited review). In: Maeder A, Eenens P (eds) Stellar rotation, vol 215, p 366

Charbonneau P, MacGregor KB (1993) Angular momentum transport in magnetized stellar radiative zones. II. The solar spin-down. Astrophys J 417:762. https://doi.org/10.1086/173357

Charbonneau P, Sokoloff D (2023) Evolution of solar and stellar dynamo theory. Space Sci Rev 219:35. https://doi.org/10.1007/s11214-023-00980-0

Charbonneau P, Christensen-Dalsgaard J, Henning R, Larsen RM, Schou J, Thompson MJ, Tomczyk S (1999) Helioseismic constraints on the structure of the solar tachocline. Astrophys J 527(1):445–460. https://doi.org/10.1086/308050

Charbonneau P, Dikpati M, Gilman PA (1999) Stability of the solar latitudinal differential rotation inferred from helioseismic data. Astrophys J 526(1):523–537. https://doi.org/10.1086/307989

Chen CC, Diamond PH (2020) Potential vorticity mixing in a tangled magnetic field. Astrophys J 892(1):24. https://doi.org/10.3847/1538-4357/ab774f. arXiv:2003.04944 [astro-ph.SR]

Cheung MCM, Isobe H (2014) Flux emergence (theory). Living Rev Sol Phys 11(1):3. https://doi.org/10.12942/lrsp-2014-3

Choudhuri AR (1989) The evolution of loop structures in flux rings within the solar convection zone. Sol Phys 123(2):217–239. https://doi.org/10.1007/BF00149104

Choudhuri AR, Gilman PA (1987) The influence of the Coriolis force on flux tubes rising through the solar convection zone. Astrophys J 316:788. https://doi.org/10.1086/165243

Christensen-Dalsgaard J, Corbard T, Dikpati M, Gilman PA, Thompson MJ (2005) Jets in the solar tachocline as diagnostics of global MHD processes. In: Sankarasubramanian K, Penn M, Pevtsov A (eds) Large-scale structures and their role in solar activity. Astronomical society of the Pacific conference series, vol 346, p 115

Cline KS, Brummell NH, Cattaneo F (2003) Dynamo action driven by shear and magnetic buoyancy. Astrophys J 599(2):1449–1468. https://doi.org/10.1086/379366

Corbard T, Blanc-Féraud L, Berthomieu G, Provost J (1999) Non linear regularization for helioseismic inversions. Application for the study of the solar tachocline. A&A 344:696–708. https://doi.org/10.48550/arXiv.astro-ph/9901112. arXiv:astro-ph/9901112 [astro-ph]

DeRosa ML, Brun AS, Hoeksema JT (2012) Solar magnetic field reversals and the role of dynamo families. Astrophys J 757(1):96. https://doi.org/10.1088/0004-637X/757/1/96. arXiv:1208.1768 [astro-ph.SR]

Dikpati M (2012) Nonlinear evolution of global hydrodynamic shallow-water instability in the solar tachocline. Astrophys J 745(2):128. https://doi.org/10.1088/0004-637X/745/2/128. arXiv:1110.2100 [astro-ph.SR]

Dikpati M, Gilman PA (2001a) Analysis of hydrodynamic stability of solar tachocline latitudinal differential rotation using a shallow-water model. Astrophys J 551(1):536–564. https://doi.org/10.1086/320080

Dikpati M, Gilman PA (2001b) Flux-transport dynamos with α-effect from global instability of tachocline differential rotation: a solution for magnetic parity selection in the Sun. Astrophys J 559(1):428–442. https://doi.org/10.1086/322410

Dikpati M, McIntosh SW (2020) Space weather challenge and forecasting implications of Rossby waves. Space Weather 18(3):e02109. https://doi.org/10.1029/2019SW002109

Dikpati M, Cally PS, McIntosh SW, Heifetz E (2017) The origin of the "seasons" in space weather. Sci Rep 7:14750. https://doi.org/10.1038/s41598-017-14957-x

Dikpati M, Belucz B, Gilman PA, McIntosh SW (2018) Phase speed of magnetized Rossby waves that cause solar seasons. Astrophys J 862(2):159. https://doi.org/10.3847/1538-4357/aacefa

Dikpati M, McIntosh SW, Bothun G, Cally PS, Ghosh SS, Gilman PA, Umurhan OM (2018) Role of interaction between magnetic Rossby waves and tachocline differential rotation in producing solar seasons. Astrophys J 853(2):144. https://doi.org/10.3847/1538-4357/aaa70d

Dikpati M, Gilman PA, Chatterjee S, McIntosh SW, Zaqarashvili TV (2020) Physics of magnetohydrodynamic Rossby waves in the Sun. Astrophys J 896(2):141. https://doi.org/10.3847/1538-4357/ab8b63

Dikpati M, Norton AA, McIntosh SW, Gilman PA (2021) Dynamical splitting of spot-producing magnetic rings in a nonlinear shallow-water model. Astrophys J 922(1):46. https://doi.org/10.3847/1538-4357/ac1359

D'Silva S, Choudhuri AR (1993) A theoretical model for tilts of bipolar magnetic regions. A&A 272:621

Duez V, Mathis S (2010) Relaxed equilibrium configurations to model fossil fields. I. A first family. A&A 517:A58. https://doi.org/10.1051/0004-6361/200913496

Dziembowski W, Kosovichev A (1987) Low frequency oscillations in slowly rotating stars. I. General properties. Acta Astron 37:313–330

Elliott JR, Gough DO (1999) Calibration of the thickness of the solar tachocline. Astrophys J 516(1):475–481. https://doi.org/10.1086/307092

Fan Y (2001) Nonlinear growth of the three-dimensional undular instability of a horizontal magnetic layer and the formation of arching flux tubes. Astrophys J 546(1):509–527. https://doi.org/10.1086/318222

Fan Y (2008) The three-dimensional evolution of buoyant magnetic flux tubes in a model solar convective envelope. Astrophys J 676(1):680–697. https://doi.org/10.1086/527317

Fan Y (2009) Magnetic fields in the solar convection zone. Living Rev Sol Phys 6(1):4. https://doi.org/10.12942/lrsp-2009-4

Fan Y (2021) Magnetic fields in the solar convection zone. Living Rev Sol Phys 18(1):5. https://doi.org/10.1007/s41116-021-00031-2

Fan Y, Fang F (2014) A simulation of convective dynamo in the solar convective envelope: maintenance of the solar-like differential rotation and emerging flux. Astrophys J 789(1):35. https://doi.org/10.1088/0004-637X/789/1/35. arXiv:1405.3926 [astro-ph.SR]

Fan Y, Zweibel EG, Lantz SR (1998) Two-dimensional simulations of buoyantly rising, interacting magnetic flux tubes. Astrophys J 493(1):480–493. https://doi.org/10.1086/305122

Fan Y, Abbett WP, Fisher GH (2003) The dynamic evolution of twisted magnetic flux tubes in a three-dimensional convecting flow. I. Uniformly buoyant horizontal tubes. Astrophys J 582(2):1206–1219. https://doi.org/10.1086/344798

Ferriz-Mas A, Schüssler M (1993) Instabilities of magnetic flux tubes in a stellar convection zone I. Equatorial flux rings in differentially rotating stars. Geophys Astrophys Fluid Dyn 72(1):209–247. https://doi.org/10.1080/03091929308203613

Forgács-Dajka E, Petrovay K (2001) Tachocline confinement by an oscillatory magnetic field. Sol Phys 203(2):195–210. https://doi.org/10.1023/A:1013389631585. arXiv:astro-ph/0106133 [astro-ph]

Fournier Y, Arlt R, Ziegler U, Strassmeier KG (2017) 3D simulations of rising magnetic flux tubes in a compressible rotating interior: the effect of magnetic tension. A&A 607:A1. https://doi.org/10.1051/0004-6361/201629989. arXiv:1707.06781 [astro-ph.SR]

Fuller J, Mathis S (2023) Linking the interiors and surfaces of magnetic stars. Mon Not R Astron Soc 520(4):5573–5585. https://doi.org/10.1093/mnras/stad475. arXiv:2301.11914 [astro-ph.SR]

Gallet F, Bouvier J (2013) Improved angular momentum evolution model for solar-like stars. A&A 556:A36. https://doi.org/10.1051/0004-6361/201321302. arXiv:1306.2130 [astro-ph.SR]

Garaud P (2001) Latitudinal shear instability in the solar tachocline. Mon Not R Astron Soc 324(1):68–76. https://doi.org/10.1046/j.1365-8711.2001.04245.x

Garaud P (2020) Horizontal shear instabilities at low Prandtl number. Astrophys J 901(2):146. https://doi.org/10.3847/1538-4357/ab9c99. arXiv:2006.07436 [astro-ph.SR]

Garaud P, Acevedo Arreguin L (2009) On the penetration of meridional circulation below the solar convection zone. II. Models with convection zone, the Taylor-proudman constraint, and applications to other stars. Astrophys J 704(1):1–16. https://doi.org/10.1088/0004-637X/704/1/1. arXiv:0906.1756 [astro-ph.SR]

Garaud P, Brummell NH (2008) On the penetration of meridional circulation below the solar convection zone. Astrophys J 674(1):498–510. https://doi.org/10.1086/524837. arXiv:0708.0258 [astro-ph]

Gilman PA (1968) The general circulation of the solar atmosphere: large thermally driven Rossby waves. Astron J Suppl 73:61

Gilman PA (1969a) A Rossby-wave dynamo for the Sun, I. Sol Phys 8(2):316–330. https://doi.org/10.1007/BF00155379

Gilman PA (1969b) A Rossby-wave dynamo for the Sun, II. Sol Phys 9(1):3–18. https://doi.org/10.1007/BF00145722

Gilman PA (1970) Instability of magnetohydrostatic stellar interiors from magnetic buoyancy. I. Astrophys J 162:1019. https://doi.org/10.1086/150733

Gilman PA (2000) Fluid dynamics and MHD of the solar convection zone and tachocline: current understanding and unsolved problems – (invited review). Sol Phys 192:27–48. https://doi.org/10.1023/A:1005280502744

Gilman PA (2018a) Magnetic buoyancy and magnetorotational instabilities in stellar tachoclines for solar- and antisolar-type differential rotation. Astrophys J 867(1):45. https://doi.org/10.3847/1538-4357/aae08e

Gilman PA (2018b) Magnetic buoyancy and rotational instabilities in the tachocline. Astrophys J 853(1):65. https://doi.org/10.3847/1538-4357/aaa4f4

Gilman PA, Cally PS (2007) Global MHD instabilities of the tachocline. In: Hughes DW, Rosner R, Weiss NO (eds) The solar tachocline, p 243

Gilman PA, Dikpati M (2000) Joint instability of latitudinal differential rotation and concentrated toroidal fields below the solar convection zone. II. Instability of narrow bands at all latitudes. Astrophys J 528(1):552–572. https://doi.org/10.1086/308146

Gilman PA, Fox PA (1997) Joint instability of latitudinal differential rotation and toroidal magnetic fields below the solar convection zone. Astrophys J 484(1):439–454. https://doi.org/10.1086/304330

Glatzmaier GA, Gilman PA (1981) Compressible convection in a rotating spherical shell. III. Analytic model for compressible vorticity waves. Astrophys J Suppl Ser 45:381. https://doi.org/10.1086/190716

Gough DO, McIntyre ME (1998) Inevitability of a magnetic field in the Sun's radiative interior. Nature 394(6695):755–757. https://doi.org/10.1038/29472

Gregory SG, Donati JF, Morin J, Hussain GAJ, Mayne NJ, Hillenbrand LA, Jardine M (2012) Can we predict the global magnetic topology of a pre-main-sequence star from its position in the Hertzsprung-Russell diagram? Astrophys J 755(2):97. https://doi.org/10.1088/0004-637X/755/2/97. arXiv:1206.5238 [astro-ph.SR]

Guerrero G, Käpylä PJ (2011) Dynamo action and magnetic buoyancy in convection simulations with vertical shear. A&A 533:A40. https://doi.org/10.1051/0004-6361/201116749. arXiv:1102.3598 [astro-ph.SR]

Guerrero G, Smolarkiewicz PK, de Gouveia Dal Pino EM, Kosovichev AG, Mansour NN (2016) Understanding solar torsional oscillations from global dynamo models. Astrophys J Lett 828(3):16. https://doi.org/10.3847/2041-8205/828/1/L3. arXiv:1608.02278 [astro-ph.SR]

Guerrero G, Del Sordo F, Bonanno A, Smolarkiewicz PK (2019) Global simulations of Tayler instability in stellar interiors: the stabilizing effect of gravity. Mon Not R Astron Soc 490(3):4281–4291. https://doi.org/10.1093/mnras/stz2849. arXiv:1909.02897 [astro-ph.SR]

Guerrero G, Zaire B, Smolarkiewicz PK, de Gouveia Dal Pino EM, Kosovichev AG, Mansour NN (2019) What sets the magnetic field strength and cycle period in solar-type stars? Astrophys J 880(1):6. https://doi.org/10.3847/1538-4357/ab224a. arXiv:1810.07978 [astro-ph.SR]

Hanasoge S, Miesch MS, Roth M, Schou J, Schüssler M, Thompson MJ (2015) Solar dynamics, rotation, convection and overshoot. Space Sci Rev 196(1–4):79–99. https://doi.org/10.1007/s11214-015-0144-0. arXiv:1503.08539 [astro-ph.SR]

Harris J, Dikpati M, Hewins IM, Gibson SE, McIntosh SW, Chatterjee S, Kuchar TA (2022) Tracking movement of long-lived equatorial coronal holes from analysis of long-term McIntosh archive data. Astrophys J 931(1):54. https://doi.org/10.3847/1538-4357/ac67f2

Haurwitz B (1940) Atmospheric disturbances on the rotating Earth. Trans Am Geophys Union 21(2):262–264. https://doi.org/10.1029/TR021i002p00262

Hindman BW, Jain R (2022) Radial trapping of thermal Rossby waves within the convection zones of low-mass stars. Astrophys J 932(1):68. https://doi.org/10.3847/1538-4357/ac6d64. arXiv:2205.02346 [astro-ph.SR]

Hori K, Tobias SM, Jones CA (2020) Solitary magnetostrophic Rossby waves in spherical shells. J Fluid Mech 904:R3. https://doi.org/10.1017/jfm.2020.743. arXiv:2007.10741 [physics.flu-dyn]

Hotta H, Bekki Y, Gizon L, Noraz Q, Rast MP (2023) Dynamics of large-scale solar flows. Space Sci Rev 219:77. https://doi.org/10.1007/s11214-023-01021-6

Howe R (2009) Solar interior rotation and its variation. Living Rev Sol Phys 6(1):1. https://doi.org/10.12942/lrsp-2009-1. arXiv:0902.2406 [astro-ph.SR]

Howe R, Komm R, Hill F, Christensen-Dalsgaard J, Larson TP, Schou J, Thompson MJ, Toomre J (2011) Rotation-rate variations at the tachocline: an update. In: GONG-SoHO 24: a new era of seismology of the Sun and solar-like stars. J Phys Conf Ser 271:012075. https://doi.org/10.1088/1742-6596/271/1/012075

Hughes DW (1985a) Magnetic buoyancy instabilities for a static plane layer. Geophys Astrophys Fluid Dyn 32(3):273–316. https://doi.org/10.1080/03091928508208787

Hughes DW (1985b) Magnetic buoyancy instabilities incorporating rotation. Geophys Astrophys Fluid Dyn 34:99–142. https://doi.org/10.1080/03091928508245440

Hughes DW (2007) Magnetic buoyancy instabilities in the tachocline. In: Hughes DW, Rosner R, Weiss NO (eds) The solar tachocline, p 275

Hughes DW, Falle SAEG (1998) The rise of twisted magnetic flux tubes: a high Reynolds NumberAdaptive grid calculation. Astrophys J Lett 509(1):L57–L60. https://doi.org/10.1086/311762

Hughes DW, Falle SAEG, Joarder P (1998) The rise of twisted magnetic flux tubes. Mon Not R Astron Soc 298(2):433–444. https://doi.org/10.1046/j.1365-8711.1998.01622.x

Hughes DW, Rosner R, Weiss NO (2007) The solar tachocline

Jouve L, Brun AS (2007) 3-D non-linear evolution of a magnetic flux tube in a spherical shell: the isentropic case. Astron Nachr 328(10):1104. https://doi.org/10.1002/asna.200710887. arXiv:0712.3408 [astro-ph]

Jouve L, Brun AS (2009) Three-dimensional nonlinear evolution of a magnetic flux tube in a spherical shell: influence of turbulent convection and associated mean flows. Astrophys J 701(2):1300–1322. https://doi.org/10.1088/0004-637X/701/2/1300. arXiv:0907.2131 [astro-ph.SR]

Jouve L, Gastine T, Lignières F (2015) Three-dimensional evolution of magnetic fields in a differentially rotating stellar radiative zone. A&A 575:A106. https://doi.org/10.1051/0004-6361/201425240. arXiv:1412.2900 [astro-ph.SR]

Jouve L, Brun AS, Aulanier G (2018) Interactions of twisted Ω-loops in a model solar convection zone. Astrophys J 857(2):83. https://doi.org/10.3847/1538-4357/aab5b6. arXiv:1803.04709 [astro-ph.SR]

Jouve L, Lignières F, Gaurat M (2020) Interplay between magnetic fields and differential rotation in a stably stratified stellar radiative zone. A&A 641:A13. https://doi.org/10.1051/0004-6361/202037828. arXiv:2006.08230 [astro-ph.SR]

Käpylä PJ, Browning MK, Brun AS, Guerrero G, Warnecke J (2023) Simulations of solar and stellar dynamos and their theoretical interpretation. Space Sci Rev 219:58. https://doi.org/10.1007/s11214-023-01005-6

Kersalé E, Hughes DW, Tobias SM (2007) The nonlinear evolution of instabilities driven by magnetic buoyancy: a new mechanism for the formation of coherent magnetic structures. Astrophys J Lett 663(2):L113–L116. https://doi.org/10.1086/520339. arXiv:0706.4463 [astro-ph]

Kim EJ (2005) Self-consistent theory of turbulent transport in the solar tachocline. I. Anisotropic turbulence. A&A 441(2):763–772. https://doi.org/10.1051/0004-6361:20053170

Kim EJ, Leprovost N (2007) Self-consistent theory of turbulent transport in the solar tachocline. III. Gravity waves. A&A 468(3):1025–1031. https://doi.org/10.1051/0004-6361:20065971. arXiv:astro-ph/0607546 [astro-ph]

Kitchatinov LL, Rüdiger G (2008) Diamagnetic pumping near the base of a stellar convection zone. Astron Nachr 329(4):372. https://doi.org/10.1002/asna.200810971. arXiv:0802.2415 [astro-ph]

Korre L, Brummell N, Garaud P, Guervilly C (2021) On the dynamical interaction between overshooting convection and an underlying dipole magnetic field – I. The non-dynamo regime. Mon Not R Astron Soc 503(1):362–375. https://doi.org/10.1093/mnras/stab477. arXiv:2008.01857 [astro-ph.SR]

Kosovichev AG (1996) Helioseismic constraints on the gradient of angular velocity at the base of the solar convection zone. Astrophys J Lett 469:L61. https://doi.org/10.1086/310253

Lawson N, Strugarek A, Charbonneau P (2015) Evidence of active MHD instability in EULAG-MHD simulations of solar convection. Astrophys J 813(2):95. https://doi.org/10.1088/0004-637X/813/2/95. arXiv:1509.07447 [astro-ph.SR]

Leamon RJ, McIntosh SW, Marsh DR (2021) Termination of solar cycles and correlated tropospheric variability. Earth Space Sci 8(4):e01223. https://doi.org/10.1029/2020EA001223

Leprovost N, Kim EJ (2006) Self-consistent theory of turbulent transport in the solar tachocline. II. Tachocline confinement. A&A 456(2):617–621. https://doi.org/10.1051/0004-6361:20065265. arXiv:astro-ph/0605124 [astro-ph]

Leprovost N, Kim EJ (2009) Turbulent transport and dynamo in sheared magnetohydrodynamics turbulence with a nonuniform magnetic field. Phys Rev E 80(2):026302. https://doi.org/10.1103/PhysRevE.80.026302. arXiv:0903.3352 [physics.flu-dyn]

Löptien B, Gizon L, Birch AC, Schou J, Proxauf B, Duvall TL, Bogart RS, Christensen UR (2018) Global-scale equatorial Rossby waves as an essential component of solar internal dynamics. Nat Astron 2:568–573. https://doi.org/10.1038/s41550-018-0460-x. arXiv:1805.07244 [astro-ph.SR]

Lorenz EN (1951) Seasonal and irregular variations of the northern hemisphere sea-level pressure profile. J Atmos Sci 8(1):52–59. https://doi.org/10.1175/1520-0469(1951)008<0052:SAIVOT>2.0.CO;2.

Manek B, Brummell N (2021) On the origin of solar hemispherical helicity rules: simulations of the rise of magnetic flux concentrations in a background field. Astrophys J 909(1):72. https://doi.org/10.3847/1538-4357/abd859. arXiv:2101.03472 [astro-ph.SR]

Manek B, Brummell N, Lee D (2018) The rise of a magnetic flux tube in a background field: solar helicity selection rules. Astrophys J Lett 859(2):L27. https://doi.org/10.3847/2041-8213/aac723. arXiv:1805.08806 [astro-ph.SR]

Markey P, Tayler RJ (1973) The adiabatic stability of stars containing magnetic fields. II. Poloidal fields. Mon Not R Astron Soc 163:77–91. https://doi.org/10.1093/mnras/163.1.77

Martin S (2018) Observational evidence of shallow origins for the magnetic fields of solar cycles – a review. Front Astron Space Sci 5(17). https://doi.org/10.3389/fspas.2018.00017 [astro-ph.SR]

Martínez-Sykora J, Moreno-Insertis F, Cheung MCM (2015) Multi-parametric study of rising 3D buoyant flux tubes in an adiabatic stratification using AMR. Astrophys J 814(1):2. https://doi.org/10.1088/0004-637X/814/1/2. arXiv:1507.01506 [astro-ph.SR]

Masada Y (2011) Impact of magnetohydrodynamic turbulence on thermal wind balance in the Sun. Mon Not R Astron Soc 411(1):L26–L30. https://doi.org/10.1111/j.1745-3933.2010.00987.x. arXiv:1011.2681 [astro-ph.SR]

Matilsky LI, Hindman BW, Featherstone NA, Blume CC, Toomre J (2022) Confinement of the solar tachocline by dynamo action in the radiative interior. Astrophys J Lett 940(2):L50. https://doi.org/10.3847/2041-8213/ac93ef. arXiv:2206.12920 [astro-ph.SR]

Matthews PC, Hughes DW, Proctor MRE (1995) Magnetic buoyancy, vorticity, and three-dimensional flux-tube formation. Astrophys J 448:938. https://doi.org/10.1086/176022

McIntosh SW, Cramer WJ, Pichardo Marcano M, Leamon RJ (2017) The detection of Rossby-like waves on the Sun. Nat Astron 1:0086. https://doi.org/10.1038/s41550-017-0086

Menou K, Le Mer J (2006) Magnetorotational transport in the early Sun. Astrophys J 650(2):1208–1216. https://doi.org/10.1086/507022. arXiv:astro-ph/0606358 [astro-ph]

Mestel L, Moss DL (1977) Models for rotating magnetic stars. Mon Not R Astron Soc 178:27–49. https://doi.org/10.1093/mnras/178.1.27

Mestel L, Moss DL (1983) On the decay of the toroidal component of a large-scale stellar magnetic field. Mon Not R Astron Soc 204:575–581. https://doi.org/10.1093/mnras/204.2.575

Mestel L, Weiss NO (1987) Magnetic fields and non-uniform rotation in stellar radiatives zones. Mon Not R Astron Soc 226:123–135. https://doi.org/10.1093/mnras/226.1.123

Miesch MS (2003) Numerical modeling of the solar tachocline. II. Forced turbulence with imposed shear. Astrophys J 586(1):663–684. https://doi.org/10.1086/367616

Miesch MS (2005) Large-scale dynamics of the convection zone and tachocline. Living Rev Sol Phys 2(1):1. https://doi.org/10.12942/lrsp-2005-1

Miesch MS, Brun AS, Toomre J (2006) Solar differential rotation influenced by latitudinal entropy variations in the tachocline. Astrophys J 641(1):618–625. https://doi.org/10.1086/499621

Mohanty S, Basri G (2003) Rotation and activity in mid-M to L field dwarfs. Astrophys J 583(1):451–472. https://doi.org/10.1086/345097. arXiv:astro-ph/0201455 [astro-ph]

Monteiro G, Guerrero G, Del Sordo F, Bonanno A, Smolarkiewicz PK (2023) Global simulations of Tayler instability in stellar interiors: a long-time multistage evolution of the magnetic field. Mon Not R Astron Soc 521(1):1415–1428. https://doi.org/10.1093/mnras/stad523. arXiv:2211.10536 [astro-ph.SR]

Moreno-Insertis F, Emonet T (1996) The rise of twisted magnetic tubes in a stratified medium. Astrophys J Lett 472:L53. https://doi.org/10.1086/310360

Nelson NJ, Brown BP, Brun AS, Miesch MS, Toomre J (2011) Buoyant magnetic loops in a global dynamo simulation of a young Sun. Astrophys J Lett 739(2):L38. https://doi.org/10.1088/2041-8205/739/2/L38. arXiv:1108.4697 [astro-ph.SR]

Nelson NJ, Brown BP, Brun AS, Miesch MS, Toomre J (2013) Magnetic wreaths and cycles in convective dynamos. Astrophys J 762(2):73. https://doi.org/10.1088/0004-637X/762/2/73. arXiv:1211.3129 [astro-ph.SR]

Nelson NJ, Brown BP, Sacha Brun A, Miesch MS, Toomre J (2014) Buoyant magnetic loops generated by global convective dynamo action. Sol Phys 289(2):441–458. https://doi.org/10.1007/s11207-012-0221-4. arXiv:1212.5612 [astro-ph.SR]

Newcomb WA (1961) Convective instability induced by gravity in a plasma with a frozen-in magnetic field. Phys Fluids 4(4):391–396. https://doi.org/10.1063/1.1706342

Parker EN (1955) The formation of sunspots from the solar toroidal field. Astrophys J 121:491. https://doi.org/10.1086/146010

Parker EN (1993) A solar dynamo surface wave at the interface between convection and nonuniform rotation. Astrophys J 408:707. https://doi.org/10.1086/172631

Petitdemange L, Marcotte F, Gissinger C (2023) Spin-down by dynamo action in simulated radiative stellar layers. Science 379(6629):300–303. https://doi.org/10.1126/science.abk2169. arXiv:2206.13819 [astro-ph.SR]

Pinto RF, Brun AS (2013) Flux emergence in a magnetized convection zone. Astrophys J 772(1):55. https://doi.org/10.1088/0004-637X/772/1/55. arXiv:1305.2159 [astro-ph.SR]

Pitts E, Tayler RJ (1985) The adiabatic stability of stars containing magnetic fields. IV – the influence of rotation. Mon Not R Astron Soc 216:139–154. https://doi.org/10.1093/mnras/216.2.139

Plummer A, Marston JB, Tobias SM (2019) Joint instability and abrupt nonlinear transitions in a differentially rotating plasma. J Plasma Phys 85(1):905850113. https://doi.org/10.1017/S0022377819000060. arXiv: 1809.00921 [physics.plasm-ph]

Raphaldini B, Raupp CFM (2015) Nonlinear dynamics of magnetohydrodynamic Rossby waves and the cyclic nature of solar magnetic activity. Astrophys J 799(1):78. https://doi.org/10.1088/0004-637X/799/1/78

Reiners A, Joshi N, Goldman B (2012) A catalog of rotation and activity in early-M stars. Astron J 143(4):93. https://doi.org/10.1088/0004-6256/143/4/93. arXiv:1201.5774 [astro-ph.SR]

Route M (2016) The discovery of solar-like activity cycles beyond the end of the main sequence? Astrophys J Lett 830(2):L27. https://doi.org/10.3847/2041-8205/830/2/L27. arXiv:1609.07761 [astro-ph.SR]

Rudiger G, Kitchatinov LL (1997) The slender solar tachocline: a magnetic model. Astron Nachr 318(5):273. https://doi.org/10.1002/asna.2113180504

Schou J, Antia HM, Basu S, Bogart RS, Bush RI, Chitre SM, Christensen-Dalsgaard J, Di Mauro MP, Dziembowski WA, Eff-Darwich A, Gough DO, Haber DA, Hoeksema JT, Howe R, Korzennik SG, Kosovichev AG, Larsen RM, Pijpers FP, Scherrer PH, Sekii T, Tarbell TD, Title AM, Thompson MJ, Toomre J (1998) Helioseismic studies of differential rotation in the solar envelope by the solar oscillations investigation using the Michelson Doppler imager. Astrophys J 505(1):390–417. https://doi.org/10.1086/306146

Schuessler M (1979) Magnetic buoyancy revisited: analytical and numerical results for rising flux tubes. A&A 71(1–2):79–91

Spiegel EA, Zahn JP (1992) The solar tachocline. A&A 265:106–114

Spruit HC (1981) Motion of magnetic flux tubes in the solar convection zone and chromosphere. A&A 98:155–160

Spruit HC (1999) Differential rotation and magnetic fields in stellar interiors. A&A 349:189–202. https://doi.org/10.48550/arXiv.astro-ph/9907138. arXiv:astro-ph/9907138 [astro-ph]

Spruit HC (2002) Dynamo action by differential rotation in a stably stratified stellar interior. A&A 381:923–932. https://doi.org/10.1051/0004-6361:20011465. arXiv:astro-ph/0108207 [astro-ph]

Spruit HC (2004) Angular momentum transport and mixing by magnetic fields (invited review). In: Maeder A, Eenens P (eds) Stellar rotation, vol 215, p 356

Spruit HC, van Ballegooijen AA (1982) Stability of toroidal flux tubes in stars. A&A 106(1):58–66

Strugarek A, Brun AS, Zahn JP (2011) Magnetic confinement of the solar tachocline: II. Coupling to a convection zone. A&A 532:A34. https://doi.org/10.1051/0004-6361/201116518. arXiv:1107.3665 [astro-ph.SR]

Strugarek A, Beaudoin P, Charbonneau P, Brun AS, do Nascimento JD (2017) Reconciling solar and stellar magnetic cycles with nonlinear dynamo simulations. Science 357(6347):185–187. https://doi.org/10.1126/science.aal3999. arXiv:1707.04335 [astro-ph.SR]

Tappin S, Altrock R (2013) The extended solar cycle tracked high into the Corona. Sol Phys 282(1). https://doi.org/10.1007/s11207-012-0133-3. arXiv:1209.2969 [astro-ph.SR]

Tayler RJ (1973) The adiabatic stability of stars containing magnetic fields – I. Toroidal fields. Mon Not R Astron Soc 161:365. https://doi.org/10.1093/mnras/161.4.365

Tayler RJ (1980) The adiabatic stability of stars containing magnetic fields – IV. Mixed poloidal and toroidal fields. Mon Not R Astron Soc 191:151–163. https://doi.org/10.1093/mnras/191.1.151

Thomas JH, Nye AH (1975) Convective instability in the presence of a nonuniform horizontal magnetic field. Phys Fluids 18:490. https://doi.org/10.1063/1.861158

Thompson MJ, Christensen-Dalsgaard J, Miesch MS, Toomre J (2003) The internal rotation of the Sun. Annu Rev Astron Astrophys 41:599–643. https://doi.org/10.1146/annurev.astro.41.011802.094848

Tobias SM, Hughes DW (2004) The influence of velocity shear on magnetic buoyancy instability in the solar tachocline. Astrophys J 603(2):785–802. https://doi.org/10.1086/381492

Tobias SM, Brummell NH, Clune TL, Toomre J (1998) Pumping of magnetic fields by turbulent penetrative convection. Astrophys J Lett 502(2):L177–L180. https://doi.org/10.1086/311501

Tobias SM, Diamond PH, Hughes DW (2007) β-Plane magnetohydrodynamic turbulence in the solar tachocline. Astrophys J Lett 667(1):L113–L116. https://doi.org/10.1086/521978

Triana SA, Guerrero G, Barik A, Rekier J (2022) Identification of inertial modes in the solar convection zone. Astrophys J Lett 934(1):L4. https://doi.org/10.3847/2041-8213/ac7dac. arXiv:2204.13007 [astro-ph.SR]

Vasil GM, Brummell NH (2008) Magnetic buoyancy instabilities of a shear-generated magnetic layer. Astrophys J 686(1):709–730. https://doi.org/10.1086/591144

Vasil GM, Brummell NH (2009) Constraints on the magnetic buoyancy instabilities of a shear-generated magnetic layer. Astrophys J 690(1):783–794. https://doi.org/10.1088/0004-637X/690/1/783

Wallace JM, Gutzler DS (1981) Teleconnections in the geopotential height field during the northern hemisphere winter. Mon Weather Rev 109(4):784. https://doi.org/10.1175/1520-0493(1981)109<0784:TITGHF>2.0.CO;2

Watson M (1981) Shear instability of differential rotation in stars. Geophys Astrophys Fluid Dyn 16(4):285–298

Weber MA, Fan Y (2015) Effects of radiative diffusion on thin flux tubes in turbulent solar-like convection. Sol Phys 290(5):1295–1321. https://doi.org/10.1007/s11207-015-0674-3. arXiv:1503.08034 [astro-ph.SR]

Weber MA, Fan Y, Miesch MS (2011) The rise of active region flux tubes in the turbulent solar convective envelope. Astrophys J 741(1):11. https://doi.org/10.1088/0004-637X/741/1/11. arXiv:1109.0240 [astro-ph.SR]

Weber MA, Fan Y, Miesch MS (2013) Comparing simulations of rising flux tubes through the solar convection zone with observations of solar active regions: constraining the dynamo field strength. Sol Phys 287(1–2):239–263. https://doi.org/10.1007/s11207-012-0093-7. arXiv:1208.1292 [astro-ph.SR]

Wissink JG, Hughes DW, Matthews PC, Proctor MRE (2000) The three-dimensional breakup of a magnetic layer. Mon Not R Astron Soc 318(2):501–510. https://doi.org/10.1046/j.1365-8711.2000.03785.x

Wood TS, Brummell NH (2012) Transport by meridional circulations in solar-type stars. Astrophys J 755(2):99. https://doi.org/10.1088/0004-637X/755/2/99. arXiv:1501.05161 [astro-ph.SR]

Wood TS, Brummell NH (2018) A self-consistent model of the solar tachocline. Astrophys J 853(2):97. https://doi.org/10.3847/1538-4357/aaa6d5. arXiv:1801.02565 [astro-ph.SR]

Wright GAE (1973) Pinch instabilities in magnetic stars. Mon Not R Astron Soc 162:339–358. https://doi.org/10.1093/mnras/162.4.339

Wright NJ, Drake JJ (2016) Solar-type dynamo behaviour in fully convective stars without a tachocline. Nature 535(7613):526–528. https://doi.org/10.1038/nature18638. arXiv:1607.07870 [astro-ph.SR]

Zahn JP, Brun AS, Mathis S (2007) On magnetic instabilities and dynamo action in stellar radiation zones. A&A 474(1):145–154. https://doi.org/10.1051/0004-6361:20077653. arXiv:0707.3287 [astro-ph]

Zaire B, Jouve L, Gastine T, Donati JF, Morin J, Landin N, Folsom CP (2022) Transition from multipolar to dipolar dynamos in stratified systems. Mon Not R Astron Soc 517(3):3392–3406. https://doi.org/10.1093/mnras/stac2769. arXiv:2209.11652 [astro-ph.SR]

Zaqarashvili TV, Carbonell M, Oliver R, Ballester JL (2010) Quasi-biennial oscillations in the solar tachocline caused by magnetic Rossby wave instabilities. Astrophys J Lett 724(1):L95–L98. https://doi.org/10.1088/2041-8205/724/1/L95. arXiv:1011.1361 [astro-ph.SR]

Zaqarashvili TV, Albekioni M, Ballester JL, Bekki Y, Biancofiore L, Birch AC, Dikpati M, Gizon L, Gurgenashvili E, Heifetz E, Lanza AF, McIntosh SW, Ofman L, Oliver R, Proxauf B, Umurhan OM, Yellin-Bergovoy R (2021) Rossby waves in astrophysics. Space Sci Rev 217(1):15. https://doi.org/10.1007/s11214-021-00790-2

Publisher's Note Springer Nature remains neutral with regard to jurisdictional claims in published maps and institutional affiliations.

Authors and Affiliations

Antoine Strugarek[1] · Bernadett Belucz[2,3,4] · Allan Sacha Brun[1] · Mausumi Dikpati[5] · Gustavo Guerrero[6,7]

✉ B. Belucz
b.belucz@sheffield.ac.uk

A. Strugarek
antoine.strugarek@cea.fr

A.S. Brun
sacha.brun@cea.fr

M. Dikpati
dikpati@ucar.edu

G. Guerrero
guerrero@fisica.ufmg.br

[1] Université Paris-Saclay, Université Paris Cité, CEA, CNRS, AIM, 91191, Gif-sur-Yvette, France

2   Solar Physics and Space Plasma Research Center, School of Mathematics and Statistics, University of Sheffield, Sheffield, S3 7RH, UK

3   Hungarian Solar Physics Foundation, Gyula, Hungary

4   Department of Astronomy, Institute of Geography and Earth Sciences, Eötvös University, Budapest, Hungary

5   High Altitude Observatory, NCAR, 3080 Center Green Drive, Boulder, 80301, CO, USA

6   Universidade Federal de Minas Gerais, Av. Pres. Antônio Carlos, 6627, Belo Horizonte, MG, 31270-901, Brazil

7   New Jersey Institute of Technology, Newark, NJ 07103, USA

Space Science Reviews (2023) 219:63
https://doi.org/10.1007/s11214-023-01006-5

Understanding Active Region Origins and Emergence on the Sun and Other Cool Stars

Maria A. Weber[1] · Hannah Schunker[2] · Laurène Jouve[3] · Emre Işık[4,5]

Received: 27 May 2023 / Accepted: 15 September 2023 / Published online: 17 October 2023
© The Author(s) 2023

Abstract

The emergence of active regions on the Sun is an integral feature of the solar dynamo mechanism. However, details about the generation of active-region-scale magnetism and the journey of this magnetic flux from the interior to the photosphere are still in question. Shifting paradigms are now developing for the source depth of the Sun's large-scale magnetism, the organization of this magnetism into fibril flux tubes, and the role of convection in shaping active-region observables. Here we review the landscape of flux emergence theories and simulations, highlight the role flux emergence plays in the global dynamo process, and make connections between flux emergence on the Sun and other cool stars. As longer-term and higher fidelity observations of both solar active regions and their associated flows are amassed, it is now possible to place new constraints on models of emerging flux. We discuss the outcomes of statistical studies which provide observational evidence that flux emergence may be a more passive process (at least in the upper convection zone); dominated to a greater extent by the influence of convection and to a lesser extent by buoyancy and the Coriolis force acting on rising magnetic flux tubes than previously thought. We also discuss how the relationship between stellar rotation, fractional convection zone depth, and magnetic activity on other stars can help us better understand the flux emergence processes. Looking forward, we identify open questions regarding magnetic flux emergence that we anticipate can be addressed in the next decade with further observations and simulations.

Keywords Sun · Solar · Sunspot · Magnetic field · Flux emergence

1 Introduction

The Sun is a magnetically active star showing activity on a wide range of spatial scales and field strengths. An active region is defined by the appearance of a dark feature at the surface of the Sun in continuum white light observations. These features are associated with concentrations of strong magnetic fields, and often develop into fully formed, stable sunspots. Typical active regions consist of opposite polarity pairs that are predominantly east-west aligned and have sizes ranging on the order of 10s to 100s of microhemispheres with lifetimes ranging from about two days up to many weeks.

Solar and Stellar Dynamos: A New Era
Edited by Manfred Schüssler, Robert H. Cameron, Paul Charbonneau, Mausumi Dikpati, Hideyuki Hotta and Leonid Kitchatinov

Extended author information available on the last page of the article

The Sun's coherent surface flux elements such as sunspots and active regions emerge from the solar interior. However, how they arrive at the surface and their specific depth of origin is not clear. Helioseismology has placed upper bounds on the amplitude and speed of the flows at and below the surface prior to emergence (e.g. Birch et al. 2013), however any unambiguous detection of flows above the background noise remains a challenge. Therefore, numerical simulations of flux emergence through the surface of the Sun are critical to reconciling the observations with the physics of the formation of active regions.

Originally, the paradigm of an idealized, magnetically isolated flux tube was invoked to model magnetism giving rise to active regions. Here, it is assumed that the dynamo has already managed to create magnetism at the base of the convection zone a priori. Studies employing this paradigm were first conducted using the thin flux tube approximation, which treats physical quantities as averages over the tube's cross-section (e.g. Spruit 1981; Fan et al. 1993; Caligari et al. 1995). This was followed by 2D (e.g. Moreno-Insertis and Emonet 1996; Fan et al. 1998) and 3D magnetohydrodynamic (MHD) (e.g. Abbett et al. 2001; Fan 2008) approaches to resolve the flux tube cross-section and twist of magnetic field lines. As a body of work, these simulations suggest that magnetic buoyancy, the Coriolis force, and the twist of magnetic field lines in a tube play roles in the flux emergence process and are responsible for many observed characteristics of active regions. Addition of a convection velocity field further demonstrated that turbulent interior flows modulate flux emergence, provided the magnetic field strength of the flux tube is not substantially super-equipartition (e.g. Fan et al. 2003; Jouve and Brun 2009; Weber et al. 2011). However, this paradigm of idealized flux tubes built by a deep-seated dynamo mechanism has been challenged by results from 3D global convective dynamo simulations. Some demonstrate that toroidal wreaths of magnetism can be formed within the bulk of a stellar convection zone (e.g. Brown et al. 2011; Nelson et al. 2011; Augustson et al. 2015; Matilsky and Toomre 2020). Either from these toroidal wreaths (Nelson et al. 2011, 2013) or more localized regions of strong toroidal field (Fan and Fang 2014) within the magneto-convection, buoyantly rising magnetic structures – possible starspot progenitors – are spawned. Results from both idealized flux tube simulations and the buoyant magnetic structures built self-consistently by dynamo action show similarities to active region observables, but there are also many discrepancies. Further modeling work with direct comparison to active region observations is critical to elucidate the true origin of active-region-scale magnetism.

The paradigm of the idealized, isolated flux tube mechanism for producing active regions is thus now changing towards a more complete picture. A large part of the recent paradigm shift was brought about from a statistical analysis of the flows associated with emerging active regions (e.g. Schunker et al. 2016; Birch et al. 2019), emphasising the importance of solar monitoring missions. Prior to instruments such as Helioseismic and Magnetic Imager (HMI) onboard NASA's Solar Dynamics Observatory (SDO) (Scherrer et al. 2012), with high duty-cycle observations of the magnetic field and Doppler velocity at a cadence sufficiently shorter than the time it takes an active region to emerge, it was not possible to gain any statistical understanding of the emergence process in such detail. Similarly, monitoring campaigns for stars e.g., Mt Wilson, *Kepler*, BCool, LAMOST, and TESS have increased both the sample size and the time range of data, such that magnetic variability has been measured over multiple cycle periods on other stars. Although the level of precision and sampling rate in such measurements are insufficient to amass emergence statistics for other stars like we have for the Sun, they help to shape our view over general trends of active-region formation in longitude and latitude, as well as the lifetimes of surface magnetic structures.

The longest record of the eleven-year activity cycle of the Sun is defined by the number of sunspots, or cool, dark regions, visible on the Sun. Active regions are defined from this visible darkening when they are assigned an active region number by the National Oceanic and Atmospheric Administration. At the beginning of the solar cycle, sunspots appear at latitudes around 30°, and closer to the equator towards the end of the cycle, creating the observed butterfly diagram. Besides simply defining the solar cycle, active regions are found to have characteristics which correlate with the next solar cycle, suggesting that they are an integral part of the dynamo process. For example, the average tilt angle of sunspot groups over a solar cycle is anti-correlated with the amplitude of the next solar cycle (Dasi-Espuig et al. 2010; Jiao et al. 2021), and large active regions that emerge across the equator (e.g. Nagy et al. 2017) have a significant effect on the amplitude and duration of the subsequent solar cycle. Thus, to fully understand the dynamo process, it is critical to understand how active regions form.

Presumably the distribution of active regions on the surface of the Sun reflects the distribution of the global toroidal field in the interior (Işık et al. 2011), and can provide a strong constraint for their origin and the solar dynamo (e.g. Cameron et al. 2018). However, it cannot be excluded that the dynamo also produces strong field at latitudes which do not become unstable and rise to the surface. For other cool stars, the combined effects of rotation rate and fractional depth of the convection zone can lead to a possible mismatch between active regions on the surface and distributions of magnetic flux in the deeper interior due to latitudinal deflection as bundles of magnetism rise. As a result, any one-to-one association of observed surface field and the underlying dynamo in active cool stars is not necessarily straightforward (Işık et al. 2011). While it is not currently possible to directly observe the emergence of a starspot, it is possible to make proxy observations (e.g. from chromospheric indices, spectropolarimetry, Zeeman–Doppler imaging and asteroseismology; Berdyugina 2005; García et al. 2010; See et al. 2016) to infer the distribution, size, lifetime and magnetic field strength.

In this paper, we attempt to paint a comprehensive picture of the possible flux emergence process from generation of the active-region-scale magnetism in the deep interior to its appearance on the photosphere. We begin in Sect. 2, where we describe observations of the formation of active regions on the solar surface. These observations serve as inspiration and constraints for models of the generation and rise of emerging flux, which we review in Sect. 3. New observations are highlighted in Sect. 4, which support a more passive process for active region emergence than was previously understood based on flux emergence models. We then briefly review the role flux emergence plays in the solar dynamo process in Sect. 5, and discuss flux emergence leading to starspots on other cool stars in Sect. 6. In Sect. 7, we conclude with some recommendations as we move toward solving the active-region-scale flux emergence puzzle.

2 Formation of Active Regions at the Surface of the Sun

Active regions are defined by the appearance of dark spots on the visible disk of the Sun in white light, caused by strong, concentrated magnetic fields. The presence of this magnetism renders the spots cooler, and therefore darker, than the surrounding photosphere. Active region magnetic fields consist of roughly east-west aligned opposite polarity pairs, ranging from 10 up to 3000 micro-hemispheres in size, and 10^{20} to 10^{22} Mx of magnetic flux. Known as Hale's Law, bipolar active regions typically emerge with the same sign of the leading magnetic polarity in the same hemisphere, and the sign of the polarities are flipped in the opposite hemisphere (Hale et al. 1919). At the end of each 11-year sunspot cycle, the polarity orientation reverses for each hemisphere. In both hemispheres, active regions are roughly

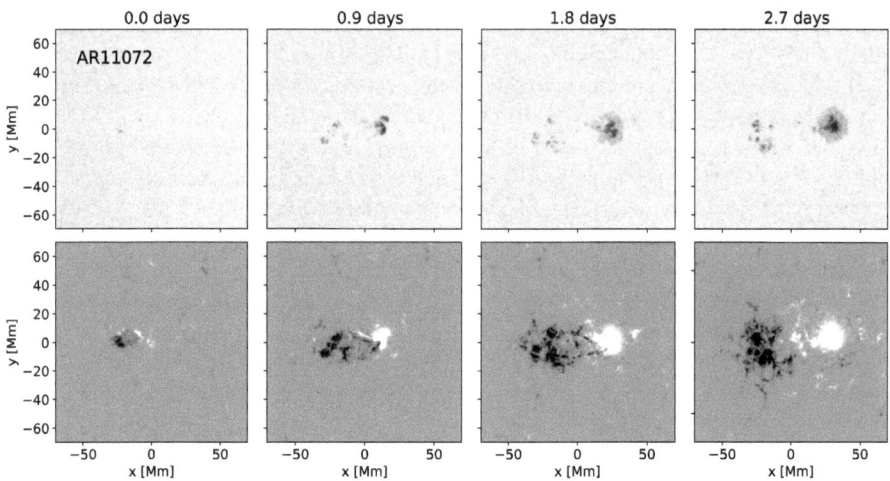

Fig. 1 Example of a typical active region, NOAA AR11072, emerging onto the surface of the Sun as observed by SDO/HMI. The top row shows Postel projected maps of the continuum intensity, and the bottom row shows maps of the line-of-sight magnetic field ±500 G. In this instance, 0 days corresponds to the emergence time 2010.05.20_17:12:00_TAI, and the maps are centred at Carrington longitude 316.43° and latitude −15.13°. The east-west direction is x and the north-south is y. Hale's Law, the formation of a sunspot in the leading polarity, and Joy's Law are evident

confined to toroidal bands of latitudes up to about 30°. They appear at higher latitudes at the beginning of each cycle and progressively closer to the equator over the duration of the roughly eleven year cycle.

The leading polarity of an active region (in the prograde direction) also tends to be closer to the equator than the following polarity (in the retrograde direction). This statistical feature is known as Joy's Law (Hale et al. 1919). Joy's Law is often quantified by the 'tilt angle' of the line drawn between the centers of leading and following polarity regions with respect to the east-west direction. Figure 1 shows the bipolar nature of a typical active region and illustrates Joy's Law, as the leading polarity of this southern-hemisphere active region is tilted closer to the equator.

An active region develops from a small magnetic bipole and grows in size as more and more magnetic flux emerges (e.g. Fig. 1). The flux-weighted centres of the polarities move further apart, predominantly in the east-west direction during the emergence process. The line-of-sight magnetic field observations show that magnetic field typically emerges as small scale features near the flux-weighted centre of the active region, which then stream towards the main polarities. Active regions have lifetimes on the order of days to weeks, where large, high-flux active regions live longer than small, low-flux regions (e.g. Schrijver and Zwaan 2000). Within the active regions, sunspots can form with peak magnetic field strengths from 2000 to 4000 G (e.g. Solanki 2003). Generally, the leading-polarity spot of the bipolar pair is larger and more coherent than the trailing-polarity region (see also Fig. 1). Active regions also have a preferred hemispheric sense of magnetic helicity, as obtained from vector magnetograms. The observations favor a left-handed (negative) twist of the field lines in the northern hemisphere, and a right-handed (positive) twist in the southern hemisphere (e.g. Pevtsov et al. 2001, 2003, 2014; Prabhu et al. 2020).

Having said how active regions *typically* form, there is a wide variation in characteristics. When two or more polarity pairs emerge in the vicinity of one another, the polarities can

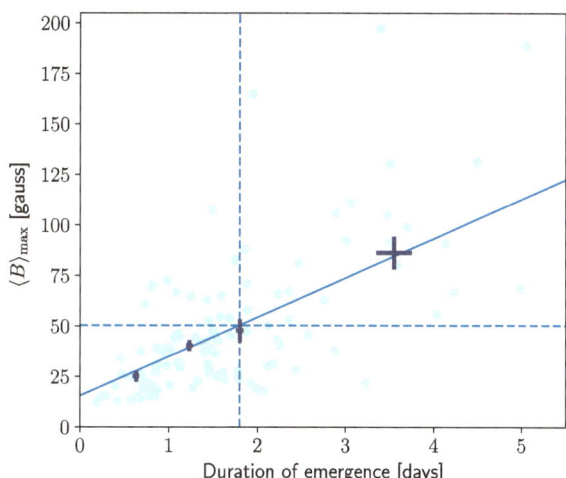

Fig. 2 Duration of the emergence process as a function of the maximum mean unsigned magnetic field $\langle B \rangle_{max}$ for each emerging active region (light blue points) and the blue diagonal line is a linear fit with slope 19.5 ± 2.2 G/day. The blue horizontal dashed line is the mean of $\langle B \rangle_{max}$, 48.6 ± 2.9 G, and the vertical dashed line is the mean duration of emergence time 1.7 ± 0.1 days. The mean and uncertainty values of $\langle B \rangle_{max}$ bins with equal number of points are shown in dark blue. See Appendix 1 for details on how the emergence time was calculated

morph into the traditional bipolar structure during the emergence process, usually leading to a more complex, multi-spot active region (e.g. AR 11158 in Schunker et al. 2019, Fig. 1). It is also common to find active regions emerging into sites of existing magnetic field from previous active regions, so-called 'nests' of activity (Castenmiller et al. 1986), where the emerging magnetic field interacts via cancellation and superposition with the existing magnetic field.

Figure 2 shows that the duration of the emergence process until magnetic flux has stopped increasing is, on average, linearly proportional to the maximum mean unsigned magnetic field $\langle B \rangle_{max}$ (see Appendix 1 for details on how the emergence time was calculated). Given that the maximum flux of an active region is known to directly correlate with the lifetime (e.g. Schrijver and Zwaan 2000), our results are consistent with Harvey (1993) (Chap. 3, Table III). Those results show that the "rise time" of active regions with a smaller maximum area is 1-2 days and increases to 3-4 days for active regions with larger maximum area (Harvey 1993). Here, we specifically avoid the term "rise time" since it implies a physical rising. What we are actually measuring is the time it takes magnetic flux to stop increasing in an active region at the surface. Figure 2 shows that the relationship is, though with considerable scatter, remarkably linear, with a slope of 19.5 ± 2.2 G per day.

As active regions are key observables for solar activity, any model of activity generation must reproduce the observed characteristics of the active regions.

3 Models of Emerging Flux

If the active-region-scale magnetism described in Sect. 2 is generated by the underlying dynamo, then it must somehow make its way from the subsurface large-scale magnetic field to the surface. The appearance of active regions evokes the idea of rising ropes of magnetism. We see arches of magnetic bundles extended above the Sun's surface. At the footpoint of these are sunspots. Within the Sun's interior is where we think these bundles of magnetic flux are born, which then rise and intersect with the photosphere to form sunspots. In this section, we briefly review models and their outcomes which describe the formation of active-region-scale magnetic structures and their rise to the photosphere (also see the reviews by Fan (2021) and by Cheung and Isobe (2014)).

3.1 Formation and Destabilization of Active-Region-Scale Magnetic Structures

The magnetism responsible for active regions is formed in the solar interior, however, the exact physical location of magnetic field generation is not known with certainty. There is a large body of work assuming that the magnetism giving rise to active regions is generated and stored at the base of the convection zone in the weakly subadiabatic overshoot region (e.g. Parker 1975; van Ballegooijen 1982; Moreno-Insertis et al. 1992; Rempel 2003). The tachocline is the name given to the region of radial and latitudinal shear at the interface between the solidly rotating radiative interior and the differentially rotating convection zone. Here it is thought that shear from differential rotation at the tachocline transforms poloidal field into toroidal field, which is amplified until it is strong enough to become buoyantly unstable. Then the magnetism subsequently rises through the convection zone to the photosphere. Beyond this shearing and storage mechanism, many studies of flux emergence, assuming the magnetism is formed as 'flux tubes' in the overshoot layer or at the very bottom of the convection zone, reproduce many properties of solar active regions (see 3.2).

Magnetic buoyancy instabilities have been considered as a means to initiate the rise of magnetic flux bundles from the overshoot region (e.g. Spruit and van Ballegooijen 1982; Spruit and van Vallegooijen 1982; Ferriz-Mas and Schüssler 1995; Caligari et al. 1995). Magnetic buoyancy is the result of a buoyant force due to the presence of a concentration of magnetism. Imagining this magnetism as a bundle or 'tube' of magnetic flux, there is a pressure balance between the gas pressure outside the tube (P_e) and the sum of the gas pressure (P_i) and magnetic pressure (P_b) inside. The gas density of the tube can be reduced if there is a condition of temperature equilibrium, allowing the tube to buoyantly rise (Parker 1955). Even if the tube is located in a subadiabatically stratified region and is in neutral density with its surroundings, a perturbation (e.g., in the transversal direction) could result in an undular instability that lifts part of the tube upward, creating an Ω-shaped loop, allowing mass to locally drain down the legs of the rising loop apex and initiating a buoyant rise. When considering thin flux tubes in mechanical equilibrium, their stability is primarily determined by their magnetic field strength and the subadiabaticity of the overshoot region (e.g. Caligari et al. 1995). It is found that the field strength of the flux tube must exceed the equipartition value of $\sim 10^4$ G by about an order of magnitude in order to develop unstable modes at sunspot latitudes in less than ~ 1 year.

Instead of considering isolated magnetic flux tubes built in the tachocline region, many studies using multi-dimensional MHD simulations have focused on the formation of buoyant instabilities within layers of uniform, horizontal magnetic field. (e.g. Cattaneo and Hughes 1988; Matthews et al. 1995; Fan 2001; Vasil and Brummell 2008, 2009). Indeed, it has been shown that regions of velocity shear can generate tube-like magnetic structures or magnetic layers (e.g. Cline et al. 2003)). Vasil and Brummell (2008) find that a velocity shear representing a tachocline-like shear can generate a strong layer of horizontal magnetic field. From this self-consistently generated magnetic layer, buoyant structures resembling undulating 'tubes' arise due to magnetic buoyancy instabilities within the magnetic layer. However, the shear required to develop the magnetic buoyancy instabilities of the magnetic layer is much stronger and the magnetic Prandtl number is much larger than what is expected in the solar tachocline (Vasil and Brummell 2009). In order to generate a twist of the magnetic field within such rising magnetic 'flux tubes', as is found in active region observations, Favier et al. (2012) showed that it was sufficient to add a weak inclined uniform field on top of the unstable horizontal magnetic layer. Indeed, in this case, the unstable undulating tubes interact with the overarching inclined field as they buoyantly rise and the field lines start to wind around the tube axis, creating an effective twist in the magnetic structure.

There is now a shifting paradigm regarding the location of active-region-scale magnetic field generation. Recent global 3D magnetohydrodynamic (MHD) dynamo simulations have compelling outcomes which suggest that active-region-scale magnetism need not be formed at the base of the convection zone. In some, toroidal wreaths of magnetism exhibiting polarity cycles are built amid the magneto-convection without the need for a tachocline (e.g. Brown et al. 2011; Augustson et al. 2015). Taking similar simulations but reducing subgrid-scale turbulent diffusion, Nelson et al. (2011, 2013, 2014) capture the generation of buoyant magnetic structures arising from magnetic wreaths – possible starspot progenitors. While typical azimuthal field strengths are a few kilogauss, buoyant loops are only spawned in regions with super-equipartition localized fields. The global dynamo simulations of Fan and Fang (2014) also exhibit super-equipartition flux bundles that rise toward the simulation upper domain. A common trait of these dynamo-generated buoyant loops is that they are continually amplified by shear and differential rotation as they rise. Unlike flux tube simulations (see 3.2), these are not isolated magnetic structures. Yet recently, Bice and Toomre (2022) found self-consistently generated flux ropes in a global 3D-MHD dynamo simulation representative of an early M-dwarf with a tachocline. The majority of the ropes remain embedded in the tachocline, while buoyant portions are lifted upward by longitudinally localized regions, or 'nests', of radial convection.

Taken together, these models and simulations of the formation and instability of buoyant magnetic structures, possible starspot progenitors, ask us to reconsider the paradigm of isolated magnetic flux tubes arising from the deep convection zone. However, as is the case with all simulations, it is important to note that all the simulations discussed here are far removed from the regime of real stars. For instance, thin flux tube models are very idealised and cannot treat the near-surface layers, while the 3D MHD models covering the entire convection zone do not yet reach magnetic Reynolds and Prandtl numbers of the solar convection zone. Yet, these simulations reproduce the observed properties of active regions remarkably well and give us a glimpse into the complex interplay of forces and mechanisms at work in stellar interiors that conspire to generate magnetic structures and facilitate their journey toward the surface.

3.2 The Flux Tube Paradigm

Isolated magnetic flux tubes in the convection zone have a long history of study because they are convenient both analytically and computationally, and had until recently been able to sufficiently explain the observed properties of active regions. In most studies, they are given an 'a priori' magnetic field strength and flux – it is taken for granted that the dynamo, via global or local processes, has somehow managed to create them - and usually assume they have been formed at the bottom of the convection zone. There are two primary types of flux tube simulations – the thin flux tube approximation (e.g. Spruit 1981; Fan et al. 1993; Caligari et al. 1995; Weber et al. 2011) and global 2D/3D MHD simulations (e.g. Emonet and Moreno-Insertis 1998; Fan et al. 2003; Fan 2008; Jouve and Brun 2009). The thin flux tube approximation takes the flux tube as so thin that there is an instantaneous balance between the pressure outside the flux tube and the gas pressure plus magnetic pressure inside the flux tube. All physical quantities are taken as averages over the tube cross-section, and the tube is essentially a 1D string of mass elements, free to be accelerated in three dimensions by bulk forces in ideal MHD, including buoyancy, magnetic tension, the Coriolis force (in the co-rotating frame), and aerodynamic drag. In order to resolve the flux tube cross section, 2D or 3D MHD simulations are used. These simultaneously solve the full set of MHD equations, and in some cases, convection. They often use the anelastic approximation, where a background density variation is allowed in radius but in which sound waves are filtered out from the system. This approximation is perfectly valid within the solar interior. But to meet the grid resolution typical of these models, the flux tubes often have a flux too large for most active regions (see Fan 2021).

Flux tube simulations have sought to explain the appearance of solar active regions, such as their latitude of emergence, tilting action in accordance with Joy's Law, and the general trend of a more coherent, less fragmented morphology for the leading polarity of an active region as depicted in Fig. 1. For all of these examples, flux tube simulations have pointed toward the Coriolis force as the driver of the phenomenon. Consider three primary forces acting on a flux tube cross-section in the frame of reference co-rotating with the medium surrounding the tube: a magnetic tension force directed toward the Sun's rotation axis, a buoyancy force directed radially outward, and the Coriolis force acting on toroidal flows within the tube. As the tube traverses the convection zone, conservation of angular momentum drives a retrograde flow within the flux tube, resulting in a Coriolis force (as mentioned above) directed inward toward the rotation axis.[1] When the magnetic field strength of the flux tube is strong (i.e. super-equipartition), the buoyancy force dominates and the flux tubes rise radially from their original latitude at the base of the convection zone. As the initial field strength of the flux tube is decreased, the outward component of buoyancy diminishes compared to the inward component of the Coriolis force, and the resulting trajectory turns more poleward, such that flux avoids emerging at lower latitudes (e.g. Choudhuri and Gilman 1987; Caligari et al. 1995). A fourth force acting on the flux tube, the drag force, is stronger for flux tubes of lower magnetic flux. As a result, flux tubes with lower initial values of magnetic flux around 10^{20}–10^{21} Mx are able to rise more radially than those of 10^{22} Mx (e.g. Choudhuri and Gilman 1987; D'Silva and Choudhuri 1993; Fan et al. 1993).

If portions of the flux tube remain anchored deeper down in the convection zone, it is found within thin flux tube simulations that the material near the apex of a rising loop will both expand and diverge (although still with net retrograde motion), leading to a Coriolis force induced tilting of the loop toward the equator (D'Silva and Choudhuri 1993). Following the Joy's Law trend, these simulations also show an increasing tilt of the flux tube legs with increasing latitude of emergence (e.g. D'Silva and Choudhuri 1993; Caligari et al. 1995). This is expected if the Coriolis force is responsible for the tilting action, as the Coriolis force is proportional to the sine of the latitude. Additionally, the tilt angle is found to increase with increasing magnetic flux at fixed field strength (Fan et al. 1994). Within thin flux tube simulations, the retrograde plasma motion near the flux tube apex contributes to a stronger magnetic field strength in the leading leg (in the direction of solar rotation) compared to the following leg (e.g. Fan et al. 1993, 1994). It is noted that plasma is evacuated out of the leading flux tube leg into the following leg. Owing to the condition of pressure balance between the flux tube and its surroundings ($P_i + P_b = P_e$), this results in a stronger magnetic field strength for the leading side of the loop compared to the following (Fan et al. 1993; Caligari et al. 1995). Here it is important to highlight that idealized flux tube simulations of all varieties are very efficient at conserving angular momentum (e.g. Fan 2008; Jouve and Brun 2009; Weber et al. 2011), yet studies utilizing local helioseismology rules out the presence of retrograde flows on the order of 100 m/s, in favor of flows not exceeding $\sim$15 m/s (Birch et al. 2013). In comparison, the buoyantly rising magnetic structures within the 3D convective dynamo simulations of Nelson et al. (2014) are weakly retrograde and are

---

[1]This effect is thought to be responsible for the predominance of high-latitude spots on rapidly rotating active stars (see Sect. 6).

actually prograde within the simulations of Fan and Fang (2014). Within these 3D convective dynamo simulations, and perhaps within the Sun itself, flux emergence processes may deviate more from the 'idealized' flux tube paradigm than originally thought.

The twist of flux tube magnetic field lines in 2D and 3D MHD simulations has been found to be important for the coherent rise of the tube and consistent tilt angles. This body of work shows that if the magnetic field is not twisted enough along the flux tube axis, the flux tube tends to break apart and lose coherence as it rises (see review by Fan 2021). However, a radius of curvature in the rising Ω-shaped flux tube can partially mitigate this (e.g. Martínez-Sykora et al. 2015). Essentially, a minimum magnetic field twist rate (i.e. angular rotation of the magnetic field lines along the flux tube axis) is needed to counteract vorticity generation in the surrounding plasma caused by the buoyancy gradient across the flux tube's cross section. As discussed in Sect. 4.3, active region magnetic fields have been observed to have a preferred helical twist that is left-handed in the Northern hemisphere and right-handed in the Southern hemisphere (Pevtsov et al. 2003). But, Fan (2008) finds the tilt of the rising flux tube ends up in the wrong direction if the twist is of the observed preferred hemispheric sign and strong enough to maintain coherence of the flux tube. If the twist of the field lines is reversed in handedness, the tilt angle is of the correct sign. Reducing the magnetic field twist per unit length also solves the hemisphere tilt problem, but then the tube becomes less coherent and looses more flux as it rises.

The flux tube simulations mentioned previously in this section (3.2) do not consider the impact of convection on the evolution of rising magnetism. However, it is absolutely clear that convection modulates flux emergence when it is included, provided that the magnetic field is not substantially super-equipartition (e.g. Fan et al. 2003; Jouve and Brun 2009; Jouve et al. 2013; Weber et al. 2011, 2013b). Convective motions and magnetic buoyancy work in concert to promote flux emergence. Convection destabilizes the tube at the base of the convection zone, forcing parts to rise. As the tube bends, mass drains down the tube legs, making the apex less dense than portions deeper down, and that part of the tube also rises buoyantly. This, in combination with convective upflows, helps the flux tube to rise toward the surface, while convective downflows can pin parts of the flux tube in deeper layers. By embedding thin flux tubes in a rotating spherical shell of solar-like convection, Weber et al. (2013b) investigated how convection statistically impacts flux tube properties that can be compared to solar active regions. Taking all their results into consideration, they attempt to constrain the as-of-yet unknown dynamo-generated magnetic field strength of active-region scale flux tubes. They find that tubes with initial field strength ≥ 40 kG are good candidates for the progenitors of large (10^{21}–10^{22}) Mx solar active regions.

In particular, Weber et al. (2013b) find that convection tends to positively contribute to the Joy's Law trend, especially for mid-field-strength flux tubes of 40–50 kG. These flux tubes also take the longest time to rise due to the competing interplay of buoyancy and drag from surrounding turbulent flows. By 'increasing the Joy's Law trend', the authors refer to a systematic effect that the addition of solar-like giant cell convection tends to boost the tilt angle at the same emergence latitude compared to simulations not subject to a convective velocity field. This is attributed, in part, to the associated kinetic helicity within the upflows. Taking all of the simulations together for tubes initiated $\pm 40°$ degrees around the equator with a magnetic flux of 10^{20}–10^{22} Mx and initial field strengths of 15–100 kG, the distribution of tilt angles peaks around $10°$ degrees. This is in good agreement with the active region observations of Howard (1996) and Stenflo and Kosovichev (2012). Furthermore, similarly peaked tilt angle distributions are found for the buoyantly rising, dynamo-generated loops from the 3D convective dynamo simulations of Nelson et al. (2014) and Fan and Fang (2014). Perhaps this is indicative of similar processes at work in both these convective dynamo simulations and the thin flux tube simulations of Weber et al. (2013b) – it is the turbulent, helical motion of convective upflows and the dynamics of the rising flux bundles themselves that contribute to the tilt angles extracted here.

3.3 Beyond Idealized Flux Tubes

In Sect. 3.2, we introduced the idealized flux tube paradigm to describe the transport of magnetism from the deep interior toward the surface. In Sect. 3.1, we noted that buoyantly rising magnetic structures have been found to arise from simulations of extended magnetic layers and form within wreaths of magnetism generated by dynamo action. In these latter two examples of MHD simulations, the buoyantly rising magnetism is *not* in the form of idealized, magnetically isolated magnetic flux tubes. While simulations of idealized flux tubes are able to reproduce some properties of solar active region observables (see Sect. 3.2), it may be unlikely that flux bundles rise within the convection zone entirely isolated from other nearby magnetic flux structures or a background field. Here we review studies of flux emergence that go beyond idealized flux tubes.

It is recognized that the presence of a background magnetic field and reconnection occurring between various magnetic flux structures have implications for the flux tube's evolution and the complexity of active regions. For example, Pinto and Brun (2013) introduce a twisted flux tube in a 3D spherical convection zone with an evolving background dynamo. In comparison to the purely hydrodynamic case of Jouve and Brun (2009), the presence of the background magnetic field introduces a 'drag' on the tube as it rises, which is dependent on the orientation of the flux tube's magnetic field with respect to the background field. In particular, flux tubes with one sign of twist seem to rise faster than the ones possessing the opposite sign. The favored handedness then depends on the preferred magnetic helicity sign of the dynamo field. By embedding a twisted toroidal flux tube in an effectively poloidal background magnetic field, Manek et al. (2018), Manek and Brummell (2021), Manek et al. (2022) show that a particular sign of twist increases the likelihood of a flux tube's rise and aligns with solar hemispheric helicity rules of active regions. Indeed, as mentioned in Sect. 2, observations show a tendency for active regions to possess a negative helicity in the Northern hemisphere and a positive one in the Southern hemisphere, although this is not a strict rule and only obeyed by only about 60% of active regions (Pevtsov et al. 2014).

Beyond the interactions between buoyant concentrations of magnetic field and the dynamo-generated smaller-scale fields, it has also been argued that the reconnections between multiple buoyantly rising structures could have strong consequences on emerging regions. In particular, these reconnections can be at the origin of complex active regions, with strongly sheared polarity inversion lines and patches of positive and negative magnetic helicity, indicating a high potential for flaring activity. Simulations of such processes were conducted initially by Linton et al. (2001) in a Cartesian geometry and then by Jouve et al. (2018) in a spherical shell including convection. In the latter, it was found that flux tubes with the same sign of axial field and same twist could merge to produce a single active region with a complicated structure and non-neutralized radial currents which could make these regions more likely to produce flares. Fully compressible calculations by Toriumi et al. (2014) were also performed to explore the possibility that the intense flare-productive active region NOAA 11158 could be the product of interacting buoyant magnetic structures.

The global models which simulate the interactions between convective motions, large-scale flows and more-or-less idealized, isolated magnetic flux tubes do not treat the uppermost layers of the convection zone and thus do not model the photospheric emergence. Firstly, the thin flux-tube approximation loses its validity above $\sim 0.98\ R_\odot$, owing to the expansion of the tube apex to maintain pressure balance, to the extent that the tube radius becomes comparable with the local pressure scale height. Secondly, the anelastic approximation also breaks down close to the photosphere where Mach number becomes of order unity. At this point, as a caution to the reader, we have to remember that the outcomes from these computational simulations serve only as touchstones for comparison to active regions. Direct comparisons between the properties of observed active regions and the results of thin flux tube simulations, as well as the magnetic bipolar structures produced at the top of the computational domain of 3D anelastic simulations, may be misleading.

Compressible simulations including radiative transfer more closely approach the physics occurring at the top of the convection zone. These simulations aim at understanding how buoyant magnetic structures would make their way through the huge gradients of density and temperature in this region. This work first started with Cheung et al. (2010) who used the MURaM code (Vögler et al. 2005) to simulate the photospheric emergence of a highly twisted semi-torus placed at the base of the computational domain (around 7 Mm below), which was then gently brought towards the surface by an imposed radial velocity field of 1 km/s. This work was then extended to investigate the effects of other geometric structures of magnetic fields introduced at the bottom of the domain. In particular, Chen et al. (2017) used the flux concentrations produced by the convective dynamo simulations of Fan and Fang (2014) as an input, with a significant rescaling of the magnetic flux contained in these concentrations to have values at the photosphere compatible with typical active regions. Subsequently, an active-region-like structure was formed.

Another example employing the MURaM code are the near-surface simulations of Birch et al. (2016), where a half torus of magnetic field, without twist, is introduced through the bottom boundary with varying speeds, up to the $\sim$500 m/s predicted rise speed of a thin flux tube (Fan 2008). They found that the strong diverging flows at the surface for structures rising quickly are incompatible with observations, which do not show a significant diverging flow when they emerge.

Using the STAGGER code to model compressible, radiative MHD of the near-surface Stein and Nordlund (2012) introduced an even less structured magnetic concentration by only imposing a relatively weak untwisted uniform horizontal field of 1 kG at the bottom boundary (at 20 Mm blow the surface). This field then rises towards the photosphere at the convective upflow speed and self-organizes into a bipolar region with one coherent polarity and one more dispersed polarity. Several observational aspects like the rise speed, the absence of a strong retrograde flow, and the asymmetry between the polarities are reproduced in these simulations.

Other types of simulations of highly stratified turbulence also spontaneously produced magnetic flux concentrations resembling active regions, albeit not at the spatial or flux scales of real active regions, without the need to advect a well-defined magnetic structure at the bottom of the domain. It is the case for example of simulations by Brandenburg et al. (2013) and then Käpylä et al. (2016) where the Negative Magnetic Pressure Instability (NEMPI) mechanism is invoked to explain the spontaneous clumping of magnetic fields into a coherent structure. The most important ingredients in such simulations seem to be the strong density stratification and the large degree of turbulence. The formation of active regions following such a mechanism would then imply that they are produced in the subsurface layers of the Sun where both strong stratification and turbulence exist. If it turns out that such mechanisms are indeed at work in the Sun, this would completely revise our understanding of the flux emergence process and its origin. However, it is yet unclear if the NEMPI would still occur in more realistic conditions (Käpylä et al. 2016) or how observed active region properties such as Joy's Law might be reproduced via this mechanism.

The flux emergence process spans many orders of magnitude of density scale heights. Owing in part to this, it is difficult to get one singular simulation that tracks flux emergence from its generation by the dynamo to its interaction with the photosphere. As described above, some work has been done to 'couple' flux emergence simulations of the deeper interior with those of a photosphere-like region. Hotta and Iijima (2020) performed the first radiative MHD simulation of a rising flux tube in a deep convection zone, although without rotation, up to the photosphere. A 10 kG flux tube is introduced at 35 Mm below the top domain. Convection then modulates the flux tube, resulting in magnetic 'roots' anchored in two downflows as deep as 80 Mm with a bipolar sunspot-like region forming at the apex of the now Ω-shaped flux bundle. More realistic simulations like these, incorporating rotational effects, will make it increasingly straightforward to directly compare to, and interpret, solar observations.

4 Observational Constraints Supporting a Passive Active Region Emergence

The formation of each active region is unique. Simulations of active region emergence, especially those in 3D with appropriate active-region-scale magnetic flux, are currently too computationally expensive to build a statistically significant sample of flux emergence scenarios (although Weber et al. 2011, 2013b, have circumvented this somewhat by performing simulations of thin flux tubes embedded within a time-varying 3D convective velocity field; see Sect. 3.2). This limiting factor makes it especially important to have a comprehensive sample of observed emerging active regions for comparison. Understanding the common properties of the emergence process is the only avenue to constrain the common physics behind flux emergence. There have been a number of statistical observational studies (e.g. Komm et al. 2008; Kosovichev and Stenflo 2008; Leka et al. 2013; Birch et al. 2013; Barnes et al. 2014; McClintock and Norton 2016) on the formation of active regions, but in this paper we will focus on the observed characteristics that can place direct constraints on the models. We refer to an 'active' emergence as being guided by the magnetic field, and 'passive' as being guided by the convection.

4.1 Geometry of the Flux Tube

There is an apparent asymmetry in the east-west proper motions of the two main active region polarities as they emerge, with the leading polarity moving prograde faster than the following polarity moves retrograde (e.g. Gilman and Howard 1985; Chou and Wang 1987). Simulations have explained this as consistent with a geometrical asymmetry in the legs of an emerging flux tube, where the leg in the prograde direction is more tangentially oriented than the following leg which is more radial (see for example Fig. 5 of Jouve et al. 2013). As this flux tube rises through the surface, the leading polarity moves more rapidly in the prograde direction than the following polarity in the retrograde direction. Modeled within the thin flux tube approach in particular, this asymmetry is due to the Coriolis force driving a counter-rotating motion of the tube plasma so that the summit of the loop moves retrograde relative to the legs (e.g. Moreno-Insertis et al. 1994; Caligari et al. 1995).

However, Schunker et al. (2016) showed that while there is an apparent asymmetry of the leading and following polarity motion of the active region with respect to the Carrington rotation rate, this east-west motion is actually *symmetric* with respect to the local rotation speed following the differential rotation profile as described by Snodgrass and Ulrich (1990).

Here, in Fig. 3 we show the separation velocities for 117 active regions (more than the sample in Schunker et al. (2016)) further supporting the initial results. The average motion of the leading and following polarities in the first day after emergence is asymmetric with respect to the Carrington rotation rate (Fig. 3, left), consistent with e.g. Chou and Wang (1987), where the mean east-west velocity of the leading polarity in the first day after emergence is 127 ± 14 ms^{-1} and the trailing polarity is -61 ± 10 ms^{-1}. However, we emphasise that *the east-west motion of the polarities about the local plasma rotation speed is symmetric* (Fig. 3, right).

While not related to the east-west motion of the individual polarities, Weber et al. (2013b) showed that by embedding thin flux tubes within solar-like convection, the average rotation rate of the center between the flux tube legs can approach the surface plasma rotation rate. However, this only occurs for strong flux tubes with initial field strengths of ≥ 60 kG, which exceeds the magnetic field strength in equipartition with convection (~ 10 kG) near the base of the convection zone (see e.g. Weber et al. 2011). Furthermore, due to strong conservation of angular momentum within the rising loop, the plasma flow at the apex of the loop is substantially retrograde (see also Sect. 3.2), in contrast to observations (Birch et al. 2013).

Based upon these outcomes from observations and simulations, we suggest that any constraints placed on models of emerging flux tubes with geometrically asymmetric legs should be carefully reconsidered. Care should also be taken when choosing the particular reference frame to study the motion of the polarities.

4.2 Rise Speed of the Flux Tube

In the absence of convection, idealized thin flux tube simulations show an upward rise speed of about 500 m/s at about 20–30 Mm below the surface (e.g. Caligari et al. 1995). To understand how this would manifest at the surface of the Sun, Birch et al. (2016) inserted a half torus of magnetic flux with a rise speed of 500 m/s through the bottom boundary of a three-dimensional, fully convective, near-surface simulation. This simulation produced a strong horizontal outflow at the surface (about 400 m/s) as it emerged. Observations of the surface flows during an emergence on the Sun do not show such a strong outflow signature, but rather flow velocities that are consistent with a rise-speed less than ~ 100 m/s, typical of convective upflows in the near-surface layers. In agreement with the observations, a flux tube that emerges naturally from a depth of 50 Mm within the radiative MHD simulations of (Hotta and Iijima 2020) does not produce any significant outflow at the surface. However, the rise speed of the flux tube is 250 m/s. This calls into question the traditional, idealized flux tube picture and suggests that the convection has an influence on the near-surface emergence process, but does not exclude thin flux tubes which may rise from the base of the convection zone with a slower speed.

Hotta and Iijima (2020) suggest that the reason their flux tube forms such a convincing active region structure is because it is initially placed across two coherent downflow regions. The downflows effectively pin the ends of the flux tube down, so that the centre emerges as a loop, implying that the influence of convection extends down to where flux tubes lie, and even instigate the emergence process, supporting some of the global models. Such correlations between rising (sinking) parts of flux tubes and upflows (downflows) were already observed in models of emergence in the bulk of the convection zone (e.g. Fan et al. 2003; Weber et al. 2011, 2013b) and in rising flux bundles generated within 3D convective dynamo simulations (Nelson et al. 2011, 2013; Fan and Fang 2014). However, the work of Hotta and Iijima (2020) shows that this interplay could also happen near the photosphere and highlights the potential importance of convective motions to bring the observed magnetic structures to the surface.

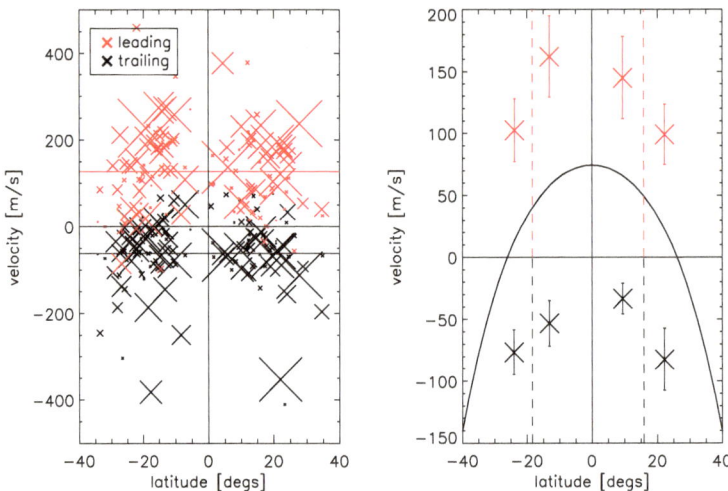

Fig. 3 Left: The mean east-west velocity relative to the Carrington rotation rate of the leading (red crosses) and trailing (black crosses) polarities over the first day after emergence for 117 active regions selected from the Solar Dynamics Observatory Helioseismic Emerging Active Regions Survey (SDO/HEARS; see Table A.1 in each of Schunker et al. (2016, 2019) for a full list of active regions in SDO/HEARS and Appendix 1 for a list of active regions that were excluded). The size of the symbols represents the size of the active region (AR 11158 is the largest). The scatter is large; this emphasises the unique nature of each active region emergence. The mean velocity of the leading polarity in the first day after emergence is $127 \pm 14 \text{ ms}^{-1}$ and the trailing polarity is $-61 \pm 10 \text{ ms}^{-1}$. Right: The average velocities in bins of polewards and equatorwards latitudes divided by the median latitude (dashed vertical lines) of the EARs. The black curve shows the differential velocity of the surface plasma relative to the Carrington rotation rate. The uncertainties are given by the rms of the velocities in each bin, divided by the square root of the number of EARs in the bin. This figure is an updated version of Fig. 11 in Schunker et al. (2016) where the average speed of the leading polarities was $121 \pm 22 \text{ ms}^{-1}$ and for the trailing polarities was $-70 \pm 13 \text{ ms}^{-1}$. Full details of the method to measure the east-west polarity speeds are described in Sect. 7 of Schunker et al. (2016)

4.3 Onset of Joy's Law

Joy's Law is the observed tendency of the leading polarity in predominantly east-west aligned active regions to be slightly closer to the equator than the following polarity. The angle these polarities make relative to the east-west direction is called the tilt angle, and it increases with the latitude of the active region, strongly suggesting that Joy's Law has its origins in the Coriolis force. In some mean-field dynamo models, Joy's Law is an important characteristic where the tilt angle acts as a non-linear feedback mechanism (e.g. Cameron et al. 2010).

Within the idealized flux tube paradigm, plasma near the rising flux tube apex will expand and diverge. This results in a Coriolis force-induced tilt of the tube axis in the sense of Joy's Law that increases with latitude (see Sect. 3.2). In this picture, the tilt angle should be present at the time of emergence. The tilt angle also depends on the flux and field strength of the magnetism (e.g. Fan et al. 1994). A larger magnetic flux, Φ, increases the buoyancy of the tube, and therefore the rise speed and effect from the Coriolis force, such that the tilt angle, α, increases for fixed magnetic field strength B ($\uparrow \Phi_{B=\text{fixed}} \Rightarrow \uparrow \alpha$, or, the thicker the tube, the larger the tilt angle). But larger magnetic field strength (for the same flux) increases the magnetic tension of the flux tube, which decreases the tilt angle due to the domination of tension over the Coriolis force ($\uparrow B_{\Phi=\text{fixed}} \Rightarrow \downarrow \alpha$, for fixed cross-sectional radius, see also Işık 2015).

Weber et al. (2013b) show that incorporating the effects of time-varying giant cell convection systematically increases the tilt angles of rising flux tubes compared to the case without convection, but does not necessarily reproduce the tilt angle trends as found in Fan et al. (1994) (also see Sect. 3.2). However, they do note that there is a larger spread in tilt angles at lower magnetic flux, as reported in some observations (e.g. Wang and Sheeley 1989; Stenflo and Kosovichev 2012). Taken together, these simulation results show that the interplay of time-varying convection, the Coriolis force, magnetic tension, and buoyancy are all factors that could influence the tilt angle of active regions on the Sun.

Schunker et al. (2020) measured the tilt angle of over 100 active region magnetic polarities throughout the emergence process, and found that on average, the polarities tend to emerge east-west aligned (i.e., with zero tilt), albeit with a large scatter, and the tilt angle develops during the emergence process. Moreover, Schunker et al. (2020) found that the latitudinal dependence of the tilt angle arises only from the north-south motion of the polarities, and the east-west motion is only dependent on the amount of flux that has emerged. They also found that there was not a statistically significant dependence of the tilt angle on the eventual maximum magnetic flux of the active regions. Schunker et al. (2020) conclude that the observed Joy's Law trend is inconsistent with a rising flux tube that has an established, latitudinally dependent tilt angle as it rises to intersect with the photosphere. We note that idealized thin flux tube models do not extend all the way to the surface (typically $\approx 0.98\ R_\odot$) where convection becomes important, however in simulations of coherent magnetic structures rising through the near-surface convection, the surface signature still reflects the orientation of the subsurface footpoints (e.g. Chen et al. 2017).

Another possibility to explain Joy's Law is the conservation of magnetic helicity in a flux tube as it rises through the surface (e.g. Berger and Field 1984; Longcope and Klapper 1997). The magnetic helicity is composed of the writhe, which measures the deformation of the flux tube axis, and the twist of the magnetic field lines about the axis. Ideally, the magnetic helicity is a conserved quantity, and changing one component necessarily requires a change in the other. In some simulations, the twist of the magnetic field about the axis of the flux tube is vital to it remaining coherent as it rises (e.g. Fan et al. 1998; Fan 2008). Within the thin flux tube context, it is shown that the writhe developed by the evolving flux tube can generate local magnetic field twist (Longcope and Klapper 1997; Fan and Gong 2000), a phenomenon which Longcope et al. (1998) term the Σ effect (see also Cheung et al. 2017). However, this effect alone is not enough to account for the observed twist of active regions (see Fan 2021). Indeed, this relationship between the twist of the magnetic field and the writhe of the flux tube (related to the tilt of the active region) has been posited as a means to explore the link between 'kink unstable' flux tubes and complex sunspot groups that have polarity orientations opposite to Hale's Law (e.g. Fuentes et al. 2003; Fan 2021). While there have been multiple studies of the helicity and twist of the surface magnetic field in active regions (e.g. Pevtsov et al. 2014, and references therein), the relationship between the twist and writhe is still ambiguous. This is probably because observations do not have access to the full three-dimensional structure of the magnetic field above the surface; only proxies for the twist (e.g. Baumgartner et al. 2022) and estimates of the helicity can be measured.

An interesting proxy for the global twist and writhe in active regions is the presence of so-called magnetic tongues (see Fig. 4, left). These structures are due to the fact that the polarities of active regions appear elongated in line-of-sight magnetograms during their emergence (López Fuentes et al. 2000). The elongation is thought to be produced by the line-of-sight projection of the azimuthal magnetic field at the peak of a twisted emerging flux-tube as it emerges through the surface. Thus, it is a proxy for the net twist of the active

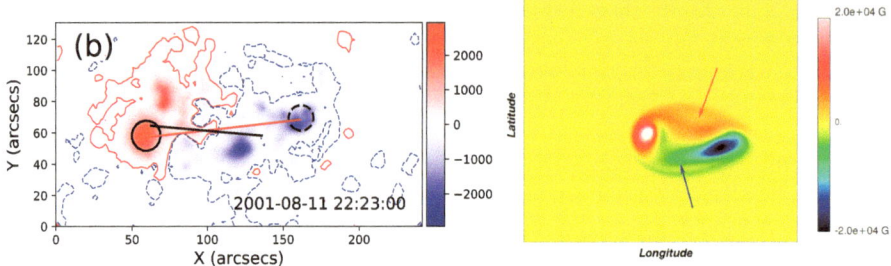

Fig. 4 Left: Example of magnetic tongues observed by SOHO/MDI on the active region 9574. The blue and red shaded areas correspond to negative and positive values of the line-of-sight magnetic field (in units of gauss) and the black circles indicate the positions of the core flux of each polarity. Right: Example of magnetic tongues simulated in a global 3D MHD model of a twisted Ω-shaped loop magnetic structure emerging close the top of the spherical shell (here $r = 0.9\ R_\odot$). Red and blue colours correspond to positive and negative radial magnetic fields and the arrows indicate the tongues of each polarity. ©AAS. Reproduced by permission from Poisson et al. (2020) (left panel) and Jouve et al. (2013) (right panel)

region flux tube, and coupled to the orientation of the polarities (as a proxy for writhe), gives a constraint on the magnetic helicity brought to the photosphere by the emergence process (Luoni et al. 2011). As the emergence proceeds, the tongues will vanish as the peak of the flux tube passes the surface and the legs of the flux tube remain. A less-biased measurement of the tilt angle will then be accessible. Magnetic tongues have also been reproduced in 3D MHD simulations where a twisted flux tube emerges through the deeper solar interior (e.g. Jouve et al. 2013, and Fig. 4, right) and closer to the photosphere (e.g. Archontis and Hood 2010; Cheung et al. 2010). Again, a clear relationship can be established between the direction of elongated tongues and the sign of the global active region twist, similarly to what is found in observations (Poisson et al. 2022).

The cycle-averaged tilt angles of sunspot groups show anti-correlation with the amplitude of the cycle (Dasi-Espuig et al. 2010; Jiao et al. 2021), also known as tilt quenching (see also Jha et al. 2020). A surface mechanism that explains this phenomenon is driven by inflows in the north-south direction towards active belts around the equator, effectively pushing the latitudinal separation of active regions polarities closer together (Jiang et al. 2010; Cameron and Schüssler 2012). In the idealized thin flux tube picture, the same effect is consistent with enhanced cooling near the base of the convection zone, where the strong toroidal flux is thought to be stored (Rempel 2003; Işık 2015): this cooling stabilizes the overshoot layer, shifting the onset of the magnetic buoyancy instability to higher field strengths, which increases the magnetic tension in the rising flux tube, which in turn quenches the tilt angle at the emergence (Işık 2015). The magnitude of local cooling is proportional to the amount of magnetic flux in the overshoot region. Global helioseismic estimation of the sound-speed change from minimum to maximum of an activity cycle (Baldner and Basu 2008) implies a degree of cooling at the base of the convection zone that is consistent with the rates required to yield the observed cycle-to-cycle changes of the mean tilt angle.

An important task for future numerical simulations then is to decide to which extent Joy's law arises from (1) the latitude-dependent Coriolis force induced by diverging flows near the tube apex below the surface (ie, angular momentum conservation of horizontally diverging flows on a rotating fluid), (2) the interplay among twist, writhe and magnetic tension, and (3) the convective flows, which impose the tilt on flux bundles.

5 Flux Emergence and the Solar Dynamo

5.1 Crucial Role of Active Regions in Global Field Reversals

The magnetic flux that emerges through the photosphere in the form of bipolar magnetic regions is likely to play a key role in recycling the global magnetic field of the Sun. The idea that flux emergence as tilted bipolar active regions could play an active part in the dynamo process dates back to the 1960s with the seminal works of Babcock (1961) and Leighton (1969). Indeed, in the so-called Babcock–Leighton (BL) model, the large-scale poloidal field owes its origin to the decay of sunspots at the photosphere. The leading polarity, closer to the equator, partially cancels with the opposite polarity in the other hemisphere, leaving a net flux to diffuse towards the pole to reverse the polar field of opposite sign. An important ingredient has then been added to this model – the large-scale meridional flow observed in the uppermost part of the convection zone (Gizon et al. 2020). These models including this flow are known as flux-transport dynamo models and are reviewed in another article of this collection (Hazra et al. 2023).

Recently, Cameron and Schüssler (2015) applied Stokes' theorem on the meridional plane of the Sun encompassing the convection zone to show that the net toroidal flux generated by differential rotation must arise solely from the magnetic flux emerging at the surface. That surface flux mainly stems from the dipole moment contribution to the poloidal field of the Sun, which the tilted active regions eventually produce in the course of an activity cycle (Cameron et al. 2018). This theoretical finding highlighted the importance of flux emergence in solar and stellar dynamo processes. Indeed, similar analysis has been conducted by Jeffers et al. (2022) on two active K-dwarf stars followed by spectropolarimetry (ϵ-Eridani and 51 Cygni A) where, similarly to the Sun, a balance is found between the generation of toroidal flux associated with the poloidal field threading through the stellar surfaces and the loss of magnetic flux associated with flux emergence.

The latitudinal distribution and the tilt angle of emerging active regions thus seems to be of utmost importance in determining the global axial dipole of the Sun. As discussed in Sect. 4.3, cycle-averaged tilt angle of sunspot groups are reported to show anti-correlation with the cycle strength (Dasi-Espuig et al. 2010; Jiao et al. 2021). This tendency has been interpreted as a manifestation of nonlinear saturation of the solar cycle. Accordingly, the effect works so as to limit further growth of the toroidal flux of the subsequent cycle. It does so by quenching the surface source for the global axial dipole moment through a lower average tilt angle of active regions. To account for the systematic effect, two physical mechanisms have been suggested: convergent flows towards emerged active regions, with the velocity depending on cycle strength (Jiang et al. 2010; Cameron et al. 2010), or a deep-seated stabilisation of flux tubes by cooling, the extent of which depends on the toroidal magnetic flux (Işık 2015, see Sect. 4.3).

Although most global MHD dynamo simulations producing large-scale magnetic cycles (e.g. Ghizaru et al. 2010; Nelson et al. 2013; Käpylä et al. 2012; Augustson et al. 2015; Hotta et al. 2016; Strugarek et al. 2017) do not extend to the photosphere, the top magnetic field boundary generally assumes a radial field or connects to a potential magnetic field. Thus, magnetic flux that reaches the top boundary is allowed to thread through the surface, producing the necessary radial magnetic field distribution that is crucial for the generation of the net toroidal flux (as described in Cameron and Schüssler 2015). However, what is still missing in those simulations is the formation of explicit starspots at the photosphere. Global dynamo simulations with more realistic near-surface layers and treatment of active-region-scale flux emergence would be ideal to better understand the physics of the emergence process.

5.2 Incorporating Flux Emergence in Babcock–Leighton Dynamo Models

Following the idea that flux emergence could play a key role in the dynamo process and that full 3D MHD global models do not yet capture all the characteristics of flux emergence, some works have been devoted to take prescriptions coming from 3D models of flux emergence and incorporate them into 2D mean-field Babcock–Leighton model. This was done, for example, in Jouve et al. (2010). Here, the idea was to take into account the fact that flux tubes do not rise instantaneously to the surface (contrary to what is assumed in the standard BL model) and that the rise speed is a non-linear function of the magnetic field strength. They found that this small (but non-linear) delay in the rise time of flux tubes could produce long-term modulation of the cycle amplitude and phase.

Recently, the idea of combining the outcomes of 3D flux emergence simulations and 2D BL models has been used to produce new 3D flux-transport BL dynamo models, where active regions would be formed according to the toroidal field self-consistently created by the shearing of the poloidal field at the base of the convection zone (Yeates and Muñoz-Jaramillo 2013; Miesch and Dikpati 2014; Kumar et al. 2019; Pipin 2022; Bekki and Cameron 2023). These new models are particularly promising to study the role of active regions in the reversal of the polar magnetic field in the Sun and possibly other cool stars. Indeed, one of their advantages is that they are less prone to the caveat of 2D models of producing too much polar flux compared to observations. Moreover, in the last two references cited above, the non-linear feedback of the Lorentz force on the large-scale flows is taken into account and the impact of flux emergence on differential rotation and meridional flows can then be assessed. As further proof on the importance of active region tilt angles on the reversal of the Sun's poloidal field, Karak and Miesch (2017) find that introducing a tilt angle scatter around the Joy's Law trend in a 3D BL dynamo induces variability in the magnetic cycle, promoting grand maxima and minima. Many improvements still need to be implemented in these models, by incorporating statistics of flux emergence and characteristics of mean flows even closer to observations for example and possibly by implementing data assimilation techniques to construct predictive models for future solar activity (see recent review by Nandy 2021, on this subject). Another improvement could also be to adapt these models to other stars with various emergence characteristics. Nonetheless, simplified 3D BL models are already very valuable tools to be used before full 3D MHD models of spot-producing dynamos can be constructed.

6 Flux Emergence on Other Cool Stars

6.1 Some Clues from Observations

Most stars with outer convection zones are capable of generating strong magnetism leading to starspots (e.g. Berdyugina 2005; Strassmeier 2009). The emergence of magnetic regions on other stars is not directly observable, however the strength and distribution of magnetic flux on the surface of stars can be inferred from observations such as light-curve variability, (Zeeman-)Doppler imaging, and interferometry (see also van Saders et al. in this volume). This is only possible for stars significantly more active than the Sun, and it would not be possible to measure these properties treating the Sun as a star. The general trend for Sun-like stars is that for a given effective temperature, the unsigned surface magnetic flux increases with rotation rate until reaching a saturation point for faster rotators (e.g. Reiners et al. 2022). There is also a preference for faster rotators to exhibit higher-latitude spots (e.g.

Berdyugina 2005), but some rapidly rotating stars and fully convective M dwarfs can exhibit spots simultaneously at high and low latitudes (e.g. Barnes et al. 1998; Jeffers et al. 2002; Barnes et al. 2015; Davenport et al. 2015).

It is then natural to wonder whether the observed trends of magnetic flux result from a link between the generation of the large-scale toroidal magnetic field and the bulk rotation rate. Stellar rotation and effective temperature also affects the amplitude, vorticity, and turn-over time of the convection; in turn impacting the star's differential rotation (i.e. shear) profile (see e.g. Brun and Browning 2017, and references therein). Some mean-field dynamos of the Sun incorporating a solar differential rotation profile find toroidal magnetic field generation with equatorward propagation near the tachocline (e.g. Charbonneau and MacGregor 1997; Dikpati and Charbonneau 1999; Dikpati and Gilman 2001). These simulations emphasized that the tachocline is a key physical component in the solar dynamo mechanism. Yet, it is observed that even fully convective M dwarfs without tachoclines exhibit starspots and the so-called magnetic 'activity-rotation correlation' (e.g. Reiners et al. 2014; Wright and Drake 2016; Reiners et al. 2022). Further, some 3D convective dynamo models demonstrate that buoyantly rising magnetic flux structures can be generated within the bulk of the convection zone (see Sect. 3.1). With the recent emphasis on the role that convection plays (both local and mean flows) in active region emergence on the Sun (see also Sect. 4), stars without tachoclines can offer some additional insights into how active-region-scale magnetism is manifested.

The observed diversity in the 3D geometry (i.e. vector components) of stellar photospheric magnetic fields poses another problem for numerical simulations of flux emergence. Zeeman-Doppler imaging of cool stars indicate that the magnetic energy in the toroidal component increases with the poloidal field for more active stars (See et al. 2015). Though with large scatter, the observational relation is steeper than one-to-one scaling for stars with masses above 0.5 $M_\odot$, with a power index of 1.25 ± 0.06. The existence of a large amount of toroidal flux on active stars provides valuable constraints for the theory of magnetic flux emergence. Further analysis and interpretation by numerical simulations are needed to understand how such magnetic landscapes occur.

6.2 Modelling the Distribution of Activity on Stars: Hints from Simulations

6.2.1 Active Nests and Longitudes

We noted in Sect. 6.1 that the unsigned surface magnetic flux increases with the rotation rate, for a given effective temperature. Whether this is due to increasing emergence frequency of active regions or larger sizes of individual active regions is unclear. These two scenarios do not exclude each other. An increased tendency for active regions to emerge near existing sites of emergence, known as active nests, is another possibility (Işık et al. 2020, see also van Saders et al. in this volume).

Observations indicate that manifestations of solar activity, including sunspots, coronal flares, and coronal streamers, are distributed inhomogeneously in longitude (e.g. Jetsu et al. 1997; Berdyugina and Usoskin 2003; Li 2011). Some other cool stars and young rapid rotators also exhibit these so-called 'active longitudes' (e.g. Järvinen et al. 2005; García-Alvarez et al. 2011; Luo et al. 2022). The cause of active longitudes is still unknown, but a few theories have been put forward. One simple suggestion is that a long-lived localization of toroidal, amplified magnetic field at the base of the convection zone could spawn the onset of a magnetic buoyancy instability, promoting a series of rising flux loops (e.g. Ruzmaikin 1998). Similarly, the convective dynamo simulations of Nelson et al. (2011, 2013) generate wreaths of magnetism within the convection zone that spawn buoyant bundles of flux when localized regions exceed a threshold field strength. Although, this effect is perhaps more closely related to the 'active nest' phenomenon described above.

Instead of relying on the localized enhancement of magnetic fields at particular longitudes, Dikpati and Gilman (2005) show that MHD instabilities within a shallow water model of the tachocline can produce simultaneous variations in the tachocline thickness and tipping instabilities of the toroidal magnetic field there. A correlation between a 'bulge' in the tachocline and a tipped toroidal band can force the magnetic field into the convection zone where it will rise buoyantly. Weber et al. (2013a) present yet another alternate theory utilizing their thin flux tube simulations embedded in solar-like convection (Weber et al. 2011, 2013b), which shows that active longitudes might also arise from the presence of rotationally aligned giant-cell convection. The simulations exhibit a pattern of flux emergence with longitudinal modes of low order and low-latitude alignment across the equator. Essentially, the extent of giant-cell upflows and the strong downflow boundaries form windows within which rising flux tubes can emerge. Although, Weber et al. (2013a) use 'active longitudes' to refer to a longitudinal alignment of flux emergence rather than repeated flux emergence at specific longitudes for multiple rotations. In reality, it is likely that both the amplification of localized magnetic fields and the effects of convective flows (which can also amplify localized fields) play a role in the active longitude and active nest phenomena.

Active longitudes have also been observed on stars in close binary systems (e.g. Berdyugina and Tuominen 1998; Berdyugina 2005). In this case tidal forcing was shown to affect the flux emergence patterns, leading to active longitudes on opposite sides of the star (Holzwarth and Schüssler 2003). An exploration of the surface distribution of flux emergence for increasing stellar activity level has now become a necessity for physics-based numerical simulations, to better understand how stellar activity patterns scale with the activity level and the rotation rate.

6.2.2 Emergence Latitudes and Tilts

Although highly simplified, simulations employing the thin flux tube approximation have been used as tools to explore the distribution of magnetic activity on stars with varying rotation rates (Schüssler et al. 1996) and spectral types (Granzer et al. 2000). These models once again point toward the importance of the rotationally-driven Coriolis force on flux tube dynamics (see also Sect. 3.2). The existence of high-latitude and polar spots on stars with more rapid rotation and/or deeper convective envelopes can be explained by angular momentum conservation of a rising flux loop, leading to an internal retrograde flow. In the co-rotating frame, this effect would be experienced as an inward directed Coriolis force component towards the rotation axis, with a magnitude that increasingly dominates the radially outward buoyancy with more rapid rotation (Schüssler and Solanki 1992). The general trend is that beyond four times the solar rotation rate, a zone of avoidance forms around the equator, where no flux emergence occur (Işık et al. 2011, 2018). In these simulations, the initial field strengths of toroidal flux tubes are assumed to be close to the analytical prediction of the onset of magnetic buoyancy instability. This limits the initial field strengths to the range 80-110 kG for the solar model with the initial tube location at the middle of the overshoot region below the convection zone. In a solar-type star rotating eight times faster, the range is in 150-350 kG, so that rotation stabilizes the tubes at a fixed field strength, owing mainly to angular momentum conservation (Işık et al. 2018, see Fig. 2). For rapidly rotating early K dwarfs and subgiants, the equatorial band of avoidance is somewhat widened in latitude, owing simply to the geometry of the convection zone boundaries: when the fractional depth

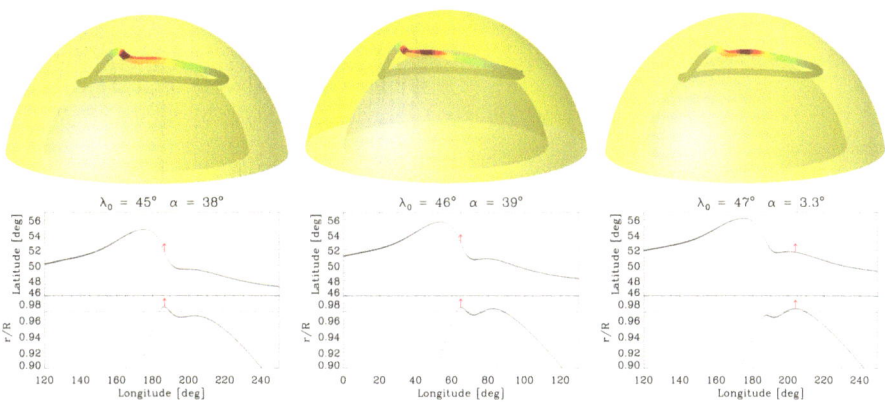

Fig. 5 Geometry of three emerging flux loops with initial latitudes λ_0 at the base of the convection zone and emergence tilt angles α. Upper panels: The parts of the tube that are beneath the outer sphere (0.97 $R_\odot$) are shaded in grey, whereas the emerged parts are brighter. The colours denote the cross-sectional tube radius (the redder the thicker). Lower panels: latitudinal and radial projections of the tubes. The horizontal line on the radial profile corresponds to the location of the outer sphere (0.97 $R_\odot$), where α is measured from footpoint locations. The red arrows denote the apex of each tube. Işık et al., A&A, 620, A177 (2018), reproduced with permission © ESO

of the convection zone increases (ie, towards cooler stars), the poleward-deflected tube apex can emerge at even higher latitudes (Granzer et al. 2000; Işık et al. 2011).

The aforementioned simulations were based on the assumption that active-region producing flux tubes were formed near the base of the convection zone in the overshoot region, in the same way as for the idealized flux tubes in the Sun. It should be noted that in these studies (Işık et al. 2011, 2018), thin flux tubes rise in the presence of a differential rotation profile $\Delta\Omega$, which is kept constant with increasing stellar rotation rate. Taking notes from the simulations of Weber et al. (2011, 2013b), it is likely that incorporating turbulent, time-varying convective flows could modify these trends.

Thin flux tube simulations have also shown that the tilt angles near emergence generally increase with the rotation rate (Işık et al. 2018). This is consistent with the Coriolis acceleration along the tube apex being proportional to the local rotation rate. An increase of the tilt angle limits flux cancellation within the emerged bipolar regions and supports stronger fields to accumulate at the rotational poles (see also Işık et al. 2007).[2] The tilt angles are not only larger in average than solar ones, but their variance is also larger, showing jumps at some emergence latitudes. Such a jump is demonstrated in Fig. 5, which shows the detailed geometry of the flux tube apex starting close to 45° latitude and emerging around 52°. When the initial latitude λ_0 is above 46°, the prograde part of the apex is intruded by a more east-west oriented and broader peak, leading to a tilt angle of about 3°. For $\lambda_0 < 46°$, the large-tilt loop (38°) emerges before the small-tilt loop. Possibly, such multiple-peaked adjacent loops emerge on active stars at certain latitudes, leading to complex active-region topologies with enhanced free energy deposits for the upper atmosphere.

M-dwarfs with masses $\leq 0.35\,M_\odot$ are fully convective, and so lack a tachocline. Yet, in at least some ways, this magnetism is similar to that observed in Sun-like stars (see Sect. 6.1). Weber and Browning (2016) embed the thin flux tube model within simulations of time-

---

[2]However, with the latitudinal distribution of emerging flux being confined to high latitudes, the stellar dynamo might not be dominated by the dipolar mode.

varying giant-cell convection to explore flux emergence trends in fully convective M dwarfs. Since there is no tachocline layer of shear, they introduce flux tubes at depths of 0.5 $R_\star$ and 0.75 $R_\star$ to sample the differing mean and local time-varying flows at each depth. A range of initial flux tube field strengths of 30-200 kG are chosen. On the lower end (30 kG), this encompasses magnetic fields that would not be too susceptible to suppression of their rise due to turbulent downflows. On the upper end (200 kG), this excludes field strengths above which the flux tubes would rise faster than they could plausibly be generated by large-scale convective eddies (Browning et al. 2016). Convection modulates the flux tubes as they rise, both promoting localized rising loops while suppressing the global rise of flux tubes (akin to magnetic pumping) for those initiated in the deeper interior at lower latitudes (see also Weber et al. 2017). Within these simulations, a robust result is a tendency for flux tubes to rise parallel to the rotation axis (see Sect. 3.2 and the first paragraph in this section), leading to a preference of mid-to-high latitude flux emergence. However, low latitude flux emergence is found in special cases where the flux tubes are initiated closer to the surface and are of strong magnetic fields, or of weaker fields and rise through regions of prograde differential rotation near the equator.

7 Moving Forward

Active regions define the solar cycle, and in some models are an integral part of the transformation of the toroidal field to the poloidal field (Sect. 5). Understanding their origins, formation and distribution will place tight constraints on their role in the solar dynamo and provide insights into these same processes in other cool stars. Typically, active-region-scale magnetism has been modeled as buoyantly rising, fibril tubes originating in the deep interior (Sect. 3.2). New observations and simulations now suggest a shifting paradigm away from these idealized, isolated flux tubes toward a paradigm with a more complex, yet realistic, interplay between rising bundles of magnetism and their surroundings.

Observations of surface magnetism demonstrate that active region flux emergence is a more 'passive' process than was originally thought (Sect. 4). The upward rise of the magnetism as detected near the surface is typical of convective upflows, placing much less of an emphasis on buoyancy in this region. However, it is not yet possible to say whether this influence of convection over buoyancy is confined only to the near-surface regions, or if it extends to the deeper origin of the magnetic structure.

In idealized flux tube simulations, the Coriolis force leads to a geometrical asymmetry in the rising loop legs and a tilting action of these legs toward the equator. The former has been used as an explanation for why the leading active region polarity moves prograde faster than the following polarity moves retrograde. However, it is shown that the east-west motion of active regions is actually symmetric with respect to the local rotation speed which varies with latitude (as described by Snodgrass and Ulrich 1990, see Sect. 4.1). Further, the observed Joy's Law trend may not be consistent with the latitudinally-dependent tilt that the legs of a flux tube acquire as it rises through the convection zone (Sect. 4.3). The examples here and in the previous paragraph are observational constraints that are inconsistent with thin flux-tube models.

No global convective dynamo models have yet been able to produce starspots, partly because they do not include a realistic surface layer. Yet, we know that the surface distribution of emerging flux and the timing of their appearance is a key ingredient in Babcock–Leighton flux-transport dynamo models. Indeed, these incorporate ingredients self-consistently generated in global convective dynamo models such as differential rotation and meridional circulation. Also, they often assume that the primary region of magnetic field generation is at the tachocline. However, some convective dynamo simulations show that rising bundles of magnetism can be built within the bulk of the convection zone. At present, the exact generation region of active-region-scale magnetism and its strength is unknown. Learning more about the distribution of starspots across stellar photospheres for both Sun-like and fully convective stars may help to better constrain the origin of coherent magnetic structures in the interior. Knowing how the patterns of flux emergence vary as a function of stellar rotation and inferred surface differential rotation will also play a role in disentangling the imprints of rotation, mean flows, and shearing regions on the flux emergence pattern.

To fully understand the extent to which flux emergence is a passive process, more stringent constraints from observations are still needed. But to understand what is happening below the surface, more simulations are critical. In particular, we suggest a strong emphasis to be placed on developing simulations that connect near-surface simulations with deeper flux emergence and dynamo models with fidelity. Further statistical analysis of solar active region emergence properties dependent on, for example, the extremes of magnetic flux and latitude, are also needed.

We conclude with some open questions regarding magnetic flux emergence in the Sun and other cool stars, raised by the observational and theoretical understanding presented, that we anticipate can be addressed in the next decade:

- What properties of active region formation are driven primarily by the influence of convection?
- To what extent do the Coriolis force, convective flows, and tension, twist, and writhe of the magnetic field contribute to the observed Joy's law trend?
- What is the important physics that must be faithfully simulated to capture the observed statistical properties of emerging active regions?
- In our Sun, where is the primary region of generation for active-region-scale magnetism - the tachocline, near-surface, bulk of the convection zone, or some combination of these?
- Can signatures of the underlying dynamo be found in patterns of magnetic activity (as reviewed here) on the photospheres of the Sun and other cool stars?

Appendix 1: Defining the Duration of Active Region Emergence

Here we describe how we computed the duration of the active region emergence process as shown in Fig. 2.

Figure 6 shows the evolution of the mean unsigned magnetic field, $\langle B \rangle$ within the central region of 49 Mm radius the map. The emergence time is defined as being 10% of the total magnetic flux 36 hours after the active region was officially named (see Schunker et al. 2016, for more details). The maps are centred on the flux-weighted centre of the active region about the time of emergence (for a detailed definition see Schunker et al. 2016; Birch et al. 2016) for four example active regions from the Solar Dynamics Observatory Helioseismic Emerging Active Regions survey (SDO/HEARS) (Schunker et al. 2016) which contains a total of 180 active regions.

The emergence of an active region ends when all of the magnetic field has appeared at the surface. We identified the time when the mean line-of-sight unsigned magnetic field $\langle B \rangle$ with a 5.3 hour cadence was a maximum. If the maximum occurred within three time intervals (≈ 16 hours) of the end of the time series it is difficult to assess whether it is still emerging (e.g. AR11103 in Fig. 6), and so we exclude these regions (35 in total). Otherwise, we fit a quadratic to the $\langle B \rangle$ values between 8 hours (two time intervals) before and 13 hours (three time intervals) after the time when the maximum mean magnetic field $\langle B \rangle_{\mathrm{max}}$

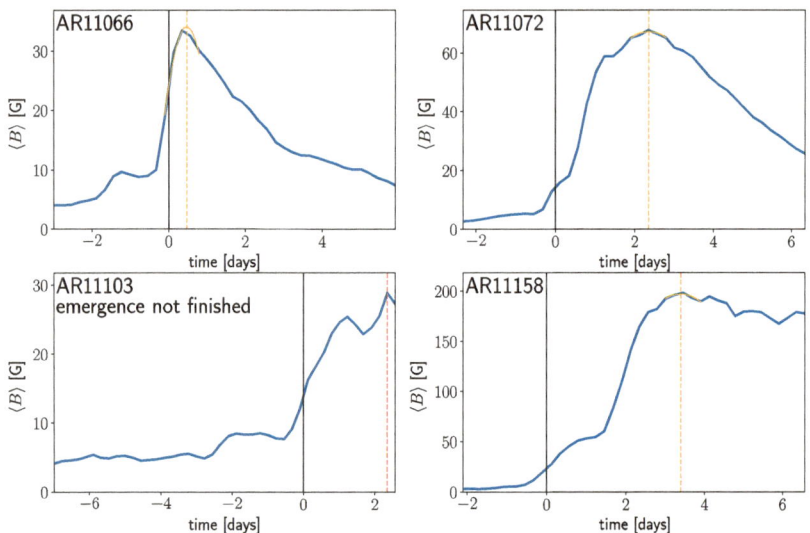

Fig. 6 Examples of the evolution of the averaged unsigned magnetic field within the central 49 Mm radius, $\langle B \rangle$, for four emerging active regions relative to the emergence time (blue). For AR11066, AR11072 and AR11158 the orange curve shows the least-squares fit of a quadratic to the peak of the curve and the dashed vertical line shows the time of the maximum value of the quadratic. Since the maximum value of $\langle B \rangle$ for AR11103 occurs within the last 16 hours of the time series, we exclude it from our sample

occurred, and we defined the time of the maximum of the quadratic function as the end time of the emergence process. There is no physical basis for fitting a quadratic, only that we found it fit the peak reasonably well (see Fig. 6). The duration of the emergence process is the difference between end of the emergence process and the emergence time.

List of 120 NOAA active region numbers included in Fig. 3: 11066, 11070, 11072, 11075, 11076, 11079, 11081, 11086, 11088, 11103, 11105, 11114, 11122, 11132, 11136, 11137, 11138, 11141, 11142, 11145, 11148, 11154, 11158, 11159, 11167, 11198, 11199, 11200, 11206, 11209, 11211, 11214, 11223, 11239, 11273, 11288, 11290, 11297, 11300, 11304, 11310, 11322, 11327, 11331, 11381, 11397, 11400, 11404, 11406, 11414, 11416, 11431, 11437, 11446, 11450, 11456, 11472, 11497, 11500, 11510, 11511, 11523, 11531, 11547, 11549, 11551, 11554, 11565, 11570, 11574, 11597, 11603, 11607, 11624, 11626, 11627, 11631, 11640, 11645, 11686, 11696, 11699, 11703, 11707, 11712, 11718, 11736, 11750, 11780, 11781, 11784, 11786, 11789, 11807, 11813, 11821, 11824, 11833, 11843, 11855, 11867, 11874, 11878, 11886, 11894, 11915, 11924, 11946, 11962, 11969, 11978, 11992, 12003, 12011, 12039, 12064, 12078, 12099, 12118, 12119.

Acknowledgements HS is the recipient of an Australian Research Council Future Fellowship Award (project number FT220100330) funded by the Australian Government and her contribution is partially funded by this grant. LJ acknowledges funding by the Institut Universitaire de France. The authors thank the reviewers for their comments, which improved the clarity of this manuscript.

Author Contribution All authors contributed equally to this work.

Funding Open Access funding enabled and organized by Projekt DEAL.

Declarations

Competing Interests The authors have no conflicts of interest to declare that are relevant to the content of this article.

References

Abbett WP, Fisher GH, Fan Y (2001) The effects of rotation on the evolution of rising omega loops in a stratified model convection zone. Astrophys J 546(2):1194–1203. https://doi.org/10.1086/318320. arXiv: astro-ph/0008501 [astro-ph]

Archontis V, Hood AW (2010) Flux emergence and coronal eruption. Astron Astrophys 514:A56. https://doi.org/10.1051/0004-6361/200913502. arXiv:1003.2333 [astro-ph.SR]

Augustson K, Brun AS, Miesch M et al (2015) Grand minima and equatorward propagation in a cycling stellar convective dynamo. Astrophys J 809(2):149. https://doi.org/10.1088/0004-637X/809/2/149. arXiv: 1410.6547 [astro-ph.SR]

Babcock HW (1961) The topology of the Sun's magnetic field and the 22-year cycle. Astrophys J 133:572. https://doi.org/10.1086/147060

Baldner CS, Basu S (2008) Solar cycle related changes at the base of the convection zone. Astrophys J 686(2):1349–1361. https://doi.org/10.1086/591514. arXiv:0807.0442 [astro-ph]

Barnes JR, Collier Cameron A, Unruh YC et al (1998) Latitude distributions and lifetimes of star-spots on G dwarfs in the α Persei cluster. Mon Not R Astron Soc 299(3):904–920. https://doi.org/10.1046/j.1365-8711.1998.01805.x

Barnes G, Birch AC, Leka KD et al (2014) Helioseismology of pre-emerging active regions. III. Statistical analysis. Astrophys J 786:19. https://doi.org/10.1088/0004-637X/786/1/19. arXiv:1307.1938 [astro-ph.SR]

Barnes JR, Jeffers SV, Jones HRA et al (2015) Starspot distributions on fully convective M dwarfs: implications for radial velocity planet searches. Astrophys J 812(1):42. https://doi.org/10.1088/0004-637X/812/1/42. arXiv:1509.05284 [astro-ph.SR]

Baumgartner C, Birch AC, Schunker H et al (2022) Impact of spatially correlated fluctuations in sunspots on metrics related to magnetic twist. Astron Astrophys 664:A183. https://doi.org/10.1051/0004-6361/202243357. arXiv:2207.02135 [astro-ph.SR]

Bekki Y, Cameron RH (2023) Three-dimensional non-kinematic simulation of the post-emergence evolution of bipolar magnetic regions and the Babcock-Leighton dynamo of the Sun. Astron Astrophys 670:A101. https://doi.org/10.1051/0004-6361/202244990. arXiv:2209.08178 [astro-ph.SR]

Berdyugina SV (2005) Starspots: a key to the stellar dynamo. Living Rev Sol Phys 2(1):8. https://doi.org/10.12942/lrsp-2005-8

Berdyugina SV, Tuominen I (1998) Permanent active longitudes and activity cycles on RS CVn stars. Astron Astrophys 336:L25–L28

Berdyugina SV, Usoskin IG (2003) Active longitudes in sunspot activity: century scale persistence. Astron Astrophys 405:1121–1128. https://doi.org/10.1051/0004-6361:20030748

Berger MA, Field GB (1984) The topological properties of magnetic helicity. J Fluid Mech 147:133–148. https://doi.org/10.1017/S0022112084002019

Bice CP, Toomre J (2022) Longitudinally modulated dynamo action in simulated M-dwarf stars. Astrophys J 928(1):51. https://doi.org/10.3847/1538-4357/ac4be0. arXiv:2202.02869 [astro-ph.SR]

Birch AC, Braun DC, Leka KD et al (2013) Helioseismology of pre-emerging active regions. II. Average emergence properties. Astrophys J 762(2):131. https://doi.org/10.1088/0004-637X/762/2/131. arXiv: 1303.1391 [astro-ph.SR]

Birch AC, Schunker H, Braun DC et al (2016) A low upper limit on the subsurface rise speed of solar active regions. Sci Adv 2(7):e1600557. https://doi.org/10.1126/sciadv.1600557. arXiv:1607.05250 [astro-ph.SR]

Birch AC, Schunker H, Braun DC et al (2019) Average surface flows before the formation of solar active regions and their relationship to the supergranulation pattern. Astron Astrophys 628:A37. https://doi.org/10.1051/0004-6361/201935591

Brandenburg A, Kleeorin N, Rogachevskii I (2013) Self-assembly of shallow magnetic spots through strongly stratified turbulence. Astrophys J Lett 776(2):L23. https://doi.org/10.1088/2041-8205/776/2/L23. arXiv:1306.4915 [astro-ph.SR]

Brown BP, Miesch MS, Browning MK et al (2011) Magnetic cycles in a convective dynamo simulation of a young solar-type star. Astrophys J 731(1):69. https://doi.org/10.1088/0004-637X/731/1/69. arXiv:1102.1993 [astro-ph.SR]

Browning MK, Weber MA, Chabrier G et al (2016) Theoretical limits on magnetic field strengths in low-mass stars. Astrophys J 818(2):189. https://doi.org/10.3847/0004-637X/818/2/189. arXiv:1512.05692 [astro-ph.SR]

Brun AS, Browning MK (2017) Magnetism, dynamo action and the solar-stellar connection. Living Rev Sol Phys 14(1):4. https://doi.org/10.1007/s41116-017-0007-8

Caligari P, Moreno-Insertis F, Schussler M (1995) Emerging flux tubes in the solar convection zone. I. Asymmetry, tilt, and emergence latitude. Astrophys J 441:886. https://doi.org/10.1086/175410

Cameron RH, Schüssler M (2012) Are the strengths of solar cycles determined by converging flows towards the activity belts? Astron Astrophys 548:A57. https://doi.org/10.1051/0004-6361/201219914. arXiv:1210.7644 [astro-ph.SR]

Cameron R, Schüssler M (2015) The crucial role of surface magnetic fields for the solar dynamo. Science 347(6228):1333–1335. https://doi.org/10.1126/science.1261470. arXiv:1503.08469 [astro-ph.SR]

Cameron RH, Jiang J, Schmitt D et al (2010) Surface flux transport modeling for solar cycles 15-21: effects of cycle-dependent tilt angles of sunspot groups. Astrophys J 719(1):264–270. https://doi.org/10.1088/0004-637X/719/1/264. arXiv:1006.3061 [astro-ph.SR]

Cameron RH, Duvall TL, Schüssler M et al (2018) Observing and modeling the poloidal and toroidal fields of the solar dynamo. Astron Astrophys 609:A56. https://doi.org/10.1051/0004-6361/201731481. arXiv:1710.07126 [astro-ph.SR]

Castenmiller MJM, Zwaan C, van der Zalm EBJ (1986) Sunspot nests - manifestations of sequences in magnetic activity. Sol Phys 105(2):237–255. https://doi.org/10.1007/BF00172045

Cattaneo F, Hughes DW (1988) The nonlinear breakup of a magnetic layer - instability to interchange modes. J Fluid Mech 196:323–344. https://doi.org/10.1017/S0022112088002721

Charbonneau P, MacGregor KB (1997) Solar interface dynamos. II. Linear, kinematic models in spherical geometry. Astrophys J 486(1):502–520. https://doi.org/10.1086/304485

Chen F, Rempel M, Fan Y (2017) Emergence of magnetic flux generated in a solar convective dynamo. I. The formation of sunspots and active regions, and the origin of their asymmetries. Astrophys J 846(2):149. https://doi.org/10.3847/1538-4357/aa85a0. arXiv:1704.05999 [astro-ph.SR]

Cheung MCM, Isobe H (2014) Flux emergence (theory). Living Rev Sol Phys 11(1):3. https://doi.org/10.12942/lrsp-2014-3

Cheung MCM, Rempel M, Title AM et al (2010) Simulation of the formation of a solar active region. Astrophys J 720(1):233–244. https://doi.org/10.1088/0004-637X/720/1/233. arXiv:1006.4117 [astro-ph.SR]

Cheung MCM, van Driel-Gesztelyi L, Martínez Pillet V et al (2017) The life cycle of active region magnetic fields. Space Sci Rev 210(1–4):317–349. https://doi.org/10.1007/s11214-016-0259-y

Chou DY, Wang H (1987) The separation velocity of emerging magnetic flux. Sol Phys 110(1):81–99. https://doi.org/10.1007/BF00148204

Choudhuri AR, Gilman PA (1987) The influence of the Coriolis force on flux tubes rising through the solar convection zone. Astrophys J 316:788. https://doi.org/10.1086/165243

Cline KS, Brummell NH, Cattaneo F (2003) Dynamo action driven by shear and magnetic buoyancy. Astrophys J 599(2):1449–1468. https://doi.org/10.1086/379366

Dasi-Espuig M, Solanki SK, Krivova NA et al (2010) Sunspot group tilt angles and the strength of the solar cycle. Astron Astrophys 518:A7. https://doi.org/10.1051/0004-6361/201014301. arXiv:1005.1774 [astro-ph.SR]

Davenport JRA, Hebb L, Hawley SL (2015) Detecting differential rotation and starspot evolution on the M dwarf GJ 1243 with Kepler. Astrophys J 806(2):212. https://doi.org/10.1088/0004-637X/806/2/212. arXiv:1505.01524 [astro-ph.SR]

Dikpati M, Charbonneau P (1999) A Babcock-Leighton flux transport dynamo with solar-like differential rotation. Astrophys J 518(1):508–520. https://doi.org/10.1086/307269

Dikpati M, Gilman PA (2001) Flux-transport dynamos with α-effect from global instability of tachocline differential rotation: a solution for magnetic parity selection in the Sun. Astrophys J 559(1):428–442. https://doi.org/10.1086/322410

Dikpati M, Gilman PA (2005) A shallow-water theory for the Sun's active longitudes. Astrophys J Lett 635(2):L193–L196. https://doi.org/10.1086/499626

D'Silva S, Choudhuri AR (1993) A theoretical model for tilts of bipolar magnetic regions. Astron Astrophys 272:621

Emonet T, Moreno-Insertis F (1998) The physics of twisted magnetic tubes rising in a stratified medium: two-dimensional results. Astrophys J 492(2):804–821. https://doi.org/10.1086/305074. arXiv:astro-ph/9711043 [astro-ph]

Fan Y (2001) Nonlinear growth of the three-dimensional undular instability of a horizontal magnetic layer and the formation of arching flux tubes. Astrophys J 546(1):509–527. https://doi.org/10.1086/318222

Fan Y (2008) The three-dimensional evolution of buoyant magnetic flux tubes in a model solar convective envelope. Astrophys J 676(1):680–697. https://doi.org/10.1086/527317

Fan Y (2021) Magnetic fields in the solar convection zone. Living Rev Sol Phys 18(1):5. https://doi.org/10.1007/s41116-021-00031-2

Fan Y, Fang F (2014) A simulation of convective dynamo in the solar convective envelope: maintenance of the solar-like differential rotation and emerging flux. Astrophys J 789(1):35. https://doi.org/10.1088/0004-637X/789/1/35. arXiv:1405.3926 [astro-ph.SR]

Fan Y, Gong D (2000) On the twist of emerging flux loops in the solar convection zone. Sol Phys 192:141–157. https://doi.org/10.1023/A:1005260207672

Fan Y, Fisher GH, Deluca EE (1993) The origin of morphological asymmetries in bipolar active regions. Astrophys J 405:390. https://doi.org/10.1086/172370

Fan Y, Fisher GH, McClymont AN (1994) Dynamics of emerging active region flux loops. Astrophys J 436:907. https://doi.org/10.1086/174967

Fan Y, Zweibel EG, Lantz SR (1998) Two-dimensional simulations of buoyantly rising, interacting magnetic flux tubes. Astrophys J 493(1):480–493. https://doi.org/10.1086/305122

Fan Y, Abbett WP, Fisher GH (2003) The dynamic evolution of twisted magnetic flux tubes in a three-dimensional convecting flow. I. Uniformly buoyant horizontal tubes. Astrophys J 582(2):1206–1219. https://doi.org/10.1086/344798

Favier B, Jouve L, Edmunds W et al (2012) How can large-scale twisted magnetic structures naturally emerge from buoyancy instabilities? Mon Not R Astron Soc 426(4):3349–3359. https://doi.org/10.1111/j.1365-2966.2012.21920.x. arXiv:1208.4787 [astro-ph.SR]

Ferriz-Mas A, Schüssler M (1995) Instabilities of magnetic flux tubes in a stellar convection zone II. Flux rings outside the equatorial plane. Geophys Astrophys Fluid Dyn 81(3):233–265. https://doi.org/10.1080/03091929508229066

Fuentes ML, Démoulin P, Mandrini CH et al (2003) Magnetic twist and writhe of active regions. On the origin of deformed flux tubes. Astron Astrophys 397:305–318. https://doi.org/10.1051/0004-6361:20021487. arXiv:1411.5626 [astro-ph.SR]

García RA, Mathur S, Salabert D et al (2010) CoRoT reveals a magnetic activity cycle in a Sun-like star. Science 329(5995):1032. https://doi.org/10.1126/science.1191064. arXiv:1008.4399 [astro-ph.SR]

García-Alvarez D, Lanza AF, Messina S et al (2011) Starspots on the fastest rotators in the β Pictoris moving group. Astron Astrophys 533:A30. https://doi.org/10.1051/0004-6361/201116646. arXiv:1107.5688 [astro-ph.SR]

Ghizaru M, Charbonneau P, Smolarkiewicz PK (2010) Magnetic cycles in global large-eddy simulations of solar convection. Astrophys J Lett 715(2):L133–L137. https://doi.org/10.1088/2041-8205/715/2/L133

Gilman PA, Howard R (1985) Rotation rates of leader and follower sunspots. Astrophys J 295:233–240. https://doi.org/10.1086/163368

Gizon L, Cameron RH, Pourabdian M et al (2020) Meridional flow in the Sun's convection zone is a single cell in each hemisphere. Science 368(6498):1469–1472. https://doi.org/10.1126/science.aaz7119

Granzer T, Schüssler M, Caligari P et al (2000) Distribution of starspots on cool stars. II. Pre-main-sequence and ZAMS stars between 0.4 $M_\odot$ and 1.7 $M_\odot$. Astron Astrophys 355:1087–1097

Hale GE, Ellerman F, Nicholson SB et al (1919) The magnetic polarity of Sun-spots. Astrophys J 49:153. https://doi.org/10.1086/142452

Harvey KL (1993) Magnetic bipoles on the Sun. PhD thesis. Utrecht University

Hazra G, Nandy D, Kitchatinov L et al (2023) Mean field models of flux transport dynamo and meridional circulation in the Sun and stars. Space Sci Rev 219:39. https://doi.org/10.1007/s11214-023-00982-y arXiv:2302.09390 [astro-ph.SR]

Holzwarth V, Schüssler M (2003) Dynamics of magnetic flux tubes in close binary stars. II. Nonlinear evolution and surface distributions. Astron Astrophys 405:303–311. https://doi.org/10.1051/0004-6361:20030584. arXiv:astro-ph/0304498 [astro-ph]

Hotta H, Iijima H (2020) On rising magnetic flux tube and formation of sunspots in a deep domain. Mon Not R Astron Soc 494(2):2523–2537. https://doi.org/10.1093/mnras/staa844. arXiv:2003.10583 [astro-ph.SR]

Hotta H, Rempel M, Yokoyama T (2016) Large-scale magnetic fields at high Reynolds numbers in magnetohydrodynamic simulations. Science 351(6280):1427–1430. https://doi.org/10.1126/science.aad1893

Howard RF (1996) Axial tilt angles of active regions. Sol Phys 169(2):293–301. https://doi.org/10.1007/BF00190606

Işık E (2015) A mechanism for the dependence of sunspot group tilt angles on cycle strength. Astrophys J Lett 813(1):L13. https://doi.org/10.1088/2041-8205/813/1/L13. arXiv:1510.04323 [astro-ph.SR]

Işık E, Schüssler M, Solanki SK (2007) Magnetic flux transport on active cool stars and starspot lifetimes. Astron Astrophys 464(3):1049–1057. https://doi.org/10.1051/0004-6361:20066623. arXiv:astro-ph/0612399 [astro-ph]

Işık E, Schmitt D, Schüssler M (2011) Magnetic flux generation and transport in cool stars. Astron Astrophys 528:A135. https://doi.org/10.1051/0004-6361/201014501. arXiv:1102.0569 [astro-ph.SR]

Işık E, Solanki SK, Krivova NA et al (2018) Forward modelling of brightness variations in Sun-like stars. I. Emergence and surface transport of magnetic flux. Astron Astrophys 620:A177. https://doi.org/10.1051/0004-6361/201833393. arXiv:1810.06728 [astro-ph.SR]

Işık E, Shapiro AI, Solanki SK et al (2020) Amplification of brightness variability by active-region nesting in solar-like stars. Astrophys J Lett 901(1):L12. https://doi.org/10.3847/2041-8213/abb409. arXiv:2009.00692 [astro-ph.SR]

Järvinen SP, Berdyugina SV, Tuominen I et al (2005) Magnetic activity in the young solar analog AB Dor. Active longitudes and cycles from long-term photometry. Astron Astrophys 432(2):657–664. https://doi.org/10.1051/0004-6361:20041998

Jeffers SV, Barnes JR, Collier Cameron A (2002) The latitude distribution of star-spots on He 699. Mon Not R Astron Soc 331(3):666–672. https://doi.org/10.1046/j.1365-8711.2002.05143.x

Jeffers SV, Cameron RH, Marsden SC et al (2022) The crucial role of surface magnetic fields for stellar dynamos: ϵ Eridani, 61 Cygni A, and the Sun. Astron Astrophys 661:A152. https://doi.org/10.1051/0004-6361/202142202. arXiv:2201.07530 [astro-ph.SR]

Jetsu L, Pohjolainen S, Pelt J et al (1997) Is the longitudinal distribution of solar flares nonuniform? Astron Astrophys 318:293–307

Jha BK, Karak BB, Mandal S et al (2020) Magnetic field dependence of bipolar magnetic region tilts on the Sun: indication of tilt quenching. Astrophys J Lett 889(1):L19. https://doi.org/10.3847/2041-8213/ab665c. arXiv:1912.13223 [astro-ph.SR]

Jiang J, Işık E, Cameron RH et al (2010) The effect of activity-related meridional flow modulation on the strength of the solar polar magnetic field. Astrophys J 717(1):597–602. https://doi.org/10.1088/0004-637X/717/1/597. arXiv:1005.5317 [astro-ph.SR]

Jiao Q, Jiang J, Wang ZF (2021) Sunspot tilt angles revisited: dependence on the solar cycle strength. Astron Astrophys 653:A27. https://doi.org/10.1051/0004-6361/202141215. arXiv:2106.11615 [astro-ph.SR]

Jouve L, Brun AS (2009) Three-dimensional nonlinear evolution of a magnetic flux tube in a spherical shell: influence of turbulent convection and associated mean flows. Astrophys J 701(2):1300–1322. https://doi.org/10.1088/0004-637X/701/2/1300. arXiv:0907.2131 [astro-ph.SR]

Jouve L, Proctor MRE, Lesur G (2010) Buoyancy-induced time delays in Babcock-Leighton flux-transport dynamo models. Astron Astrophys 519:A68. https://doi.org/10.1051/0004-6361/201014455. arXiv:1005.2283 [astro-ph.SR]

Jouve L, Brun AS, Aulanier G (2013) Global dynamics of subsurface solar active regions. Astrophys J 762(1):4. https://doi.org/10.1088/0004-637X/762/1/4. arXiv:1211.7251 [astro-ph.SR]

Jouve L, Brun AS, Aulanier G (2018) Interactions of twisted Ω-loops in a model solar convection zone. Astrophys J 857(2):83. https://doi.org/10.3847/1538-4357/aab5b6. arXiv:1803.04709 [astro-ph.SR]

Käpylä PJ, Mantere MJ, Brandenburg A (2012) Cyclic magnetic activity due to turbulent convection in spherical wedge geometry. Astrophys J Lett 755(1):L22. https://doi.org/10.1088/2041-8205/755/1/L22. arXiv:1205.4719 [astro-ph.SR]

Käpylä PJ, Brandenburg A, Kleeorin N et al (2016) Magnetic flux concentrations from turbulent stratified convection. Astron Astrophys 588:A150. https://doi.org/10.1051/0004-6361/201527731. arXiv:1511.03718 [astro-ph.SR]

Karak BB, Miesch M (2017) Solar cycle variability induced by tilt angle scatter in a Babcock-Leighton solar dynamo model. Astrophys J 847(1):69. https://doi.org/10.3847/1538-4357/aa8636. arXiv:1706.08933 [astro-ph.SR]

Komm R, Morita S, Howe R et al (2008) Emerging active regions studied with ring-diagram analysis. Astrophys J 672:1254–1265. https://doi.org/10.1086/523998

Kosovichev AG, Stenflo JO (2008) Tilt of emerging bipolar magnetic regions on the Sun. Astrophys J Lett 688:L115. https://doi.org/10.1086/595619

Kumar R, Jouve L, Nandy D (2019) A 3D kinematic Babcock Leighton solar dynamo model sustained by dynamic magnetic buoyancy and flux transport processes. Astron Astrophys 623:A54. https://doi.org/10.1051/0004-6361/201834705. arXiv:1901.04251 [astro-ph.SR]

Leighton RB (1969) A magneto-kinematic model of the solar cycle. Astrophys J 156:1. https://doi.org/10.1086/149943

Leka KD, Barnes G, Birch AC et al (2013) Helioseismology of pre-emerging active regions. I. Overview, data, and target selection criteria. Astrophys J 762:130. https://doi.org/10.1088/0004-637X/762/2/130. arXiv:1303.1433 [astro-ph.SR]

Li J (2011) Active longitudes revealed by large-scale and long-lived coronal streamers. Astrophys J 735(2):130. https://doi.org/10.1088/0004-637X/735/2/130. arXiv:1104.5537 [astro-ph.SR]

Linton MG, Dahlburg RB, Antiochos SK (2001) Reconnection of twisted flux tubes as a function of contact angle. Astrophys J 553(2):905–921. https://doi.org/10.1086/320974

Longcope DW, Klapper I (1997) Dynamics of a thin twisted flux tube. Astrophys J 488(1):443–453. https://doi.org/10.1086/304680

Longcope DW, Fisher GH, Pevtsov AA (1998) Flux-tube twist resulting from helical turbulence: the Σ-effect. Astrophys J 507(1):417–432. https://doi.org/10.1086/306312

López Fuentes MC, Demoulin P, Mandrini CH et al (2000) The counterkink rotation of a non-hale active region. Astrophys J 544(1):540–549. https://doi.org/10.1086/317180. arXiv:1412.1456 [astro-ph.SR]

Luo X, Gu S, Xiang Y et al (2022) Active longitudes and starspot evolution of the young rapidly rotating star USNO-B1.0 1388-0463685 discovered in the Yunnan-Hong Kong survey. Mon Not R Astron Soc 514(1):1511–1521. https://doi.org/10.1093/mnras/stac1406

Luoni ML, Démoulin P, Mandrini CH et al (2011) Twisted flux tube emergence evidenced in longitudinal magnetograms: magnetic tongues. Sol Phys 270(1):45. https://doi.org/10.1007/s11207-011-9731-8

Manek B, Brummell N (2021) On the origin of solar hemispherical helicity rules: simulations of the rise of magnetic flux concentrations in a background field. Astrophys J 909(1):72. https://doi.org/10.3847/1538-4357/abd859. arXiv:2101.03472 [astro-ph.SR]

Manek B, Brummell N, Lee D (2018) The rise of a magnetic flux tube in a background field: solar helicity selection rules. Astrophys J Lett 859(2):L27. https://doi.org/10.3847/2041-8213/aac723. arXiv:1805.08806 [astro-ph.SR]

Manek B, Pontin C, Brummell N (2022) The rise of buoyant magnetic structures through convection with a background magnetic field. Astrophys J 929(2):162. https://doi.org/10.3847/1538-4357/ac5828. arXiv:2204.13078 [astro-ph.SR]

Martínez-Sykora J, Moreno-Insertis F, Cheung MCM (2015) Multi-parametric study of rising 3D buoyant flux tubes in an adiabatic stratification using AMR. Astrophys J 814(1):2. https://doi.org/10.1088/0004-637X/814/1/2. arXiv:1507.01506 [astro-ph.SR]

Matilsky LI, Toomre J (2020) Exploring bistability in the cycles of the solar dynamo through global simulations. Astrophys J 892(2):106. https://doi.org/10.3847/1538-4357/ab791c. arXiv:1912.08158 [astro-ph.SR]

Matthews PC, Hughes DW, Proctor MRE (1995) Magnetic buoyancy, vorticity, and three-dimensional flux-tube formation. Astrophys J 448:938. https://doi.org/10.1086/176022

McClintock BH, Norton AA (2016) Tilt angle and footpoint separation of small and large bipolar sunspot regions observed with HMI. Astrophys J 818:7. https://doi.org/10.3847/0004-637X/818/1/7

Miesch MS, Dikpati M (2014) A three-dimensional Babcock-Leighton solar dynamo model. Astrophys J Lett 785(1):L8. https://doi.org/10.1088/2041-8205/785/1/L8. arXiv:1401.6557 [astro-ph.SR]

Moreno-Insertis F, Emonet T (1996) The rise of twisted magnetic tubes in a stratified medium. Astrophys J Lett 472:L53. https://doi.org/10.1086/310360

Moreno-Insertis F, Schüssler M, Ferriz-Mas A (1992) Storage of magnetic flux tubes in a convective overshoot region. Astron Astrophys 264(2):686–700

Moreno-Insertis F, Caligari P, Schüssler M (1994) Active region asymmetry as a result of the rise of magnetic flux tubes. Sol Phys 153(1–2):449–452. https://doi.org/10.1007/BF00712518

Nagy M, Lemerle A, Labonville F et al (2017) The effect of "rogue" active regions on the solar cycle. Sol Phys 292(11):167. https://doi.org/10.1007/s11207-017-1194-0. arXiv:1712.02185 [astro-ph.SR]

Nandy D (2021) Progress in solar cycle predictions: sunspot cycles 24-25 in perspective. Sol Phys 296(3):54. https://doi.org/10.1007/s11207-021-01797-2. arXiv:2009.01908 [astro-ph.SR]

Nelson NJ, Brown BP, Brun AS et al (2011) Buoyant magnetic loops in a global dynamo simulation of a young Sun. Astrophys J Lett 739(2):L38. https://doi.org/10.1088/2041-8205/739/2/L38. arXiv:1108.4697 [astro-ph.SR]

Nelson NJ, Brown BP, Brun AS et al (2013) Magnetic wreaths and cycles in convective dynamos. Astrophys J 762(2):73. https://doi.org/10.1088/0004-637X/762/2/73. arXiv:1211.3129 [astro-ph.SR]

Nelson NJ, Brown BP, Sacha Brun A et al (2014) Buoyant magnetic loops generated by global convective dynamo action. Sol Phys 289(2):441–458. https://doi.org/10.1007/s11207-012-0221-4. arXiv:1212.5612 [astro-ph.SR]

Parker EN (1955) The formation of sunspots from the solar toroidal field. Astrophys J 121:491. https://doi.org/10.1086/146010

Parker EN (1975) The generation of magnetic fields in astrophysical bodies. X. Magnetic buoyancy and the solar dynamo. Astrophys J 198:205–209. https://doi.org/10.1086/153593

Pevtsov AA, Canfield RC, Latushko SM (2001) Hemispheric helicity trend for solar cycle 23. Astrophys J Lett 549(2):L261–L263. https://doi.org/10.1086/319179

Pevtsov AA, Maleev VM, Longcope DW (2003) Helicity evolution in emerging active regions. Astrophys J 593(2):1217–1225. https://doi.org/10.1086/376733

Pevtsov AA, Berger MA, Nindos A et al (2014) Magnetic helicity, tilt, and twist. Space Sci Rev 186(1–4):285–324. https://doi.org/10.1007/s11214-014-0082-2

Pinto RF, Brun AS (2013) Flux emergence in a magnetized convection zone. Astrophys J 772(1):55. https://doi.org/10.1088/0004-637X/772/1/55. arXiv:1305.2159 [astro-ph.SR]

Pipin VV (2022) On the effect of surface bipolar magnetic regions on the convection zone dynamo. Mon Not R Astron Soc 514(1):1522–1534. https://doi.org/10.1093/mnras/stac1434. arXiv:2112.09460 [astro-ph.SR]

Poisson M, Démoulin P, Mandrini CH et al (2020) Active-region tilt angles from white-light images and magnetograms: the role of magnetic tongues. Astrophys J 894(2):131. https://doi.org/10.3847/1538-4357/ab8944. arXiv:2004.07345 [astro-ph.SR]

Poisson M, Grings F, Mandrini CH et al (2022) Bayesian approach for modeling global magnetic parameters for the solar active region. Astron Astrophys 665:A101. https://doi.org/10.1051/0004-6361/202244058

Prabhu A, Brandenburg A, Käpylä MJ et al (2020) Helicity proxies from linear polarisation of solar active regions. Astron Astrophys 641:A46. https://doi.org/10.1051/0004-6361/202037614. arXiv:2001.10884 [astro-ph.SR]

Reiners A, Schüssler M, Passegger VM (2014) Generalized investigation of the rotation-activity relation: favoring rotation period instead of Rossby number. Astrophys J 794(2):144. https://doi.org/10.1088/0004-637X/794/2/144. arXiv:1408.6175 [astro-ph.SR]

Reiners A, Shulyak D, Käpylä PJ et al (2022) Magnetism, rotation, and nonthermal emission in cool stars. Average magnetic field measurements in 292 M dwarfs. Astron Astrophys 662:A41. https://doi.org/10.1051/0004-6361/202243251. arXiv:2204.00342 [astro-ph.SR]

Rempel M (2003) Thermal properties of magnetic flux tubes. II. Storage of flux in the solar overshoot region. Astron Astrophys 397:1097–1107. https://doi.org/10.1051/0004-6361:20021594

Ruzmaikin A (1998) Clustering of emerging magnetic flux. Sol Phys 181(1):1–12. https://doi.org/10.1023/A:1016563632058

Scherrer PH, Schou J, Bush RI et al (2012) The Helioseismic and Magnetic Imager (HMI) investigation for the Solar Dynamics Observatory (SDO). Sol Phys 275(1–2):207–227. https://doi.org/10.1007/s11207-011-9834-2

Schrijver CJ, Zwaan C (2000) Solar and stellar magnetic activity. Cambridge University Press, Cambridge. https://doi.org/10.1017/CBO9780511546037

Schunker H, Braun DC, Birch AC et al (2016) SDO/HMI survey of emerging active regions for helioseismology. Astron Astrophys 595:A107. https://doi.org/10.1051/0004-6361/201628388. arXiv:1608.08005 [astro-ph.SR]

Schunker H, Birch AC, Cameron RH et al (2019) Average motion of emerging solar active region polarities. I. Two phases of emergence. Astron Astrophys 625:A53. https://doi.org/10.1051/0004-6361/201834627. arXiv:1903.11839 [astro-ph.SR]

Schunker H, Baumgartner C, Birch AC et al (2020) Average motion of emerging solar active region polarities. II. Joy's law. Astron Astrophys 640:A116. https://doi.org/10.1051/0004-6361/201937322. arXiv:2006.05565 [astro-ph.SR]

Schüssler M, Solanki SK (1992) Why rapid rotators have polar spots. Astron Astrophys 264:L13–L16

Schüssler M, Caligari P, Ferriz-Mas A et al (1996) Distribution of starspots on cool stars. I. Young and main sequence stars of 1 M$_\odot$. Astron Astrophys 314:503–512

See V, Jardine M, Vidotto AA et al (2015) The energy budget of stellar magnetic fields. Mon Not R Astron Soc 453(4):4301–4310. https://doi.org/10.1093/mnras/stv1925. arXiv:1508.01403 [astro-ph.SR]

See V, Jardine M, Vidotto AA et al (2016) The connection between stellar activity cycles and magnetic field topology. Mon Not R Astron Soc 462(4):4442–4450. https://doi.org/10.1093/mnras/stw2010. arXiv:1610.03737 [astro-ph.SR]

Snodgrass HB, Ulrich RK (1990) Rotation of Doppler features in the solar photosphere. Astrophys J 351:309. https://doi.org/10.1086/168467

Solanki SK (2003) Sunspots: an overview. Astron Astrophys Rev 11(2–3):153–286. https://doi.org/10.1007/s00159-003-0018-4

Spruit HC (1981) Motion of magnetic flux tubes in the solar convection zone and chromosphere. Astron Astrophys 98:155–160

Spruit HC, van Ballegooijen AA (1982) Stability of toroidal flux tubes in stars. Astron Astrophys 106(1):58–66

Spruit HC, van Vallegooijen AA (1982) Erratum - stability of toroidal flux tubes in stars. Astron Astrophys 113:350

Stein RF, Nordlund Å (2012) On the formation of active regions. Astrophys J Lett 753(1):L13. https://doi.org/10.1088/2041-8205/753/1/L13. arXiv:1207.4248 [astro-ph.SR]

Stenflo JO, Kosovichev AG (2012) Bipolar magnetic regions on the Sun: global analysis of the SOHO/MDI data set. Astrophys J 745:129. https://doi.org/10.1088/0004-637X/745/2/129. arXiv:1112.5226 [astro-ph.SR]

Strassmeier KG (2009) Starspots. Astron Astrophys Rev 17(3):251–308. https://doi.org/10.1007/s00159-009-0020-6

Strugarek A, Beaudoin P, Charbonneau P et al (2017) Reconciling solar and stellar magnetic cycles with nonlinear dynamo simulations. Science 357(6347):185–187. https://doi.org/10.1126/science.aal3999. arXiv:1707.04335 [astro-ph.SR]

Toriumi S, Iida Y, Kusano K et al (2014) Formation of a flare-productive active region: observation and numerical simulation of NOAA AR 11158. Sol Phys 289(9):3351–3369. https://doi.org/10.1007/s11207-014-0502-1. arXiv:1403.4029 [astro-ph.SR]

van Ballegooijen AA (1982) The overshoot layer at the base of the solar convective zone and the problem of magnetic flux storage. Astron Astrophys 113:99–112

Vasil GM, Brummell NH (2008) Magnetic buoyancy instabilities of a shear-generated magnetic layer. Astrophys J 686(1):709–730. https://doi.org/10.1086/591144

Vasil GM, Brummell NH (2009) Constraints on the magnetic buoyancy instabilities of a shear-generated magnetic layer. Astrophys J 690(1):783–794. https://doi.org/10.1088/0004-637X/690/1/783

Vögler A, Shelyag S, Schüssler M et al (2005) Simulations of magneto-convection in the solar photosphere. Equations, methods, and results of the MURaM code. Astron Astrophys 429:335–351. https://doi.org/10.1051/0004-6361:20041507

Wang YM, Sheeley NR Jr (1989) Average properties of bipolar magnetic regions during sunspot cycle 21. Sol Phys 124(1):81–100. https://doi.org/10.1007/BF00146521

Weber MA, Browning MK (2016) Modeling the rise of fibril magnetic fields in fully convective stars. Astrophys J 827(2):95. https://doi.org/10.3847/0004-637X/827/2/95. arXiv:1606.00380 [astro-ph.SR]

Weber MA, Fan Y, Miesch MS (2011) The rise of active region flux tubes in the turbulent solar convective envelope. Astrophys J 741(1):11. https://doi.org/10.1088/0004-637X/741/1/11. arXiv:1109.0240 [astro-ph.SR]

Weber MA, Fan Y, Miesch MS (2013a) A theory on the convective origins of active longitudes on solar-like stars. Astrophys J 770(2):149. https://doi.org/10.1088/0004-637X/770/2/149. arXiv:1305.1904 [astro-ph.SR]

Weber MA, Fan Y, Miesch MS (2013b) Comparing simulations of rising flux tubes through the solar convection zone with observations of solar active regions: constraining the dynamo field strength. Sol Phys 287(1–2):239–263. https://doi.org/10.1007/s11207-012-0093-7. arXiv:1208.1292 [astro-ph.SR]

Weber MA, Browning MK, Boardman S et al (2017) The suppression and promotion of magnetic flux emergence in fully convective stars. In: Nandy D, Valio A, Petit P (eds) Living around active stars. IAU Symposium, vol 328. Cambridge University Press, pp 85–92. https://doi.org/10.1017/S1743921317003830. arXiv:1703.04982

Wright NJ, Drake JJ (2016) Solar-type dynamo behaviour in fully convective stars without a tachocline. Nature 535(7613):526–528. https://doi.org/10.1038/nature18638. arXiv:1607.07870 [astro-ph.SR]

Yeates AR, Muñoz-Jaramillo A (2013) Kinematic active region formation in a three-dimensional solar dynamo model. Mon Not R Astron Soc 436(4):3366–3379. https://doi.org/10.1093/mnras/stt1818. arXiv:1309.6342 [astro-ph.SR]

Publisher's Note Springer Nature remains neutral with regard to jurisdictional claims in published maps and institutional affiliations.

Authors and Affiliations

Maria A. Weber[1] · Hannah Schunker[2] · Laurène Jouve[3] · Emre Işık[4,5]

✉ M.A. Weber
mweber@deltastate.edu

✉ E. Işık
isik@mps.mpg.de

H. Schunker
hannah.schunker@newcastle.edu.au

L. Jouve
ljouve@irap.omp.eu

1   Division of Mathematics and Sciences, Delta State University, 1003 W Sunflower Rd, Cleveland, 38733, MS, United States

2   School of Information and Physical Sciences, University of Newcastle, University Drive, Callaghan, 2308, NSW, Australia

3   IRAP/OMP/CNRS, Université Toulouse 3, 14 Avenue Edouard Belin, Toulouse, 31400, France

4   Max-Planck-Institut für Sonnensystemforschung, Justus-von-Liebig-Weg 3, 37077, Göttingen, Germany

5   Department of Computer Science, Turkish-German University, Şahinkaya Cd. 94, Beykoz, 34800, Istanbul, Turkey

Space Science Reviews (2023) 219:31
https://doi.org/10.1007/s11214-023-00978-8

Surface Flux Transport on the Sun

Anthony R. Yeates[1] · Mark C.M. Cheung[2] · Jie Jiang[3] · Kristof Petrovay[4] · Yi-Ming Wang[5]

Received: 31 January 2023 / Accepted: 9 May 2023 / Published online: 19 May 2023
© The Author(s) 2023, corrected publication 2023

Abstract

We review the surface flux transport model for the evolution of magnetic flux patterns on the Sun's surface. Our underlying motivation is to understand the model's prediction of the polar field (or axial dipole) strength at the end of the solar cycle. The main focus is on the "classical" model: namely, steady axisymmetric profiles for differential rotation and meridional flow, and uniform supergranular diffusion. Nevertheless, the review concentrates on recent advances, notably in understanding the roles of transport parameters and – in particular – the source term. We also discuss the physical justification for the surface flux transport model, along with efforts to incorporate radial diffusion, and conclude by summarizing the main directions where researchers have moved beyond the classical model.

Keywords Sun · Solar magnetic field · Solar photosphere · Solar activity

Solar and Stellar Dynamos: A New Era
Edited by Manfred Schüssler, Robert H. Cameron, Paul Charbonneau, Mausumi Dikpati, Hideyuki Hotta and Leonid Kitchatinov

✉ A.R. Yeates
   anthony.yeates@durham.ac.uk

   M.C.M. Cheung
   mark.cheung@csiro.au

   J. Jiang
   jiejiang@buaa.edu.cn

   K. Petrovay
   k.petrovay@astro.elte.hu

   Y.-M. Wang
   yi.wang@nrl.navy.mil

1   Department of Mathematical Sciences, Durham University, Durham, UK

2   CSIRO, Space & Astronomy, Marsfield, NSW, Australia

3   School of Space and Environment, Beihang University, Beijing, People's Republic of China

4   Department of Astronomy, Eötvös Loránd University, Budapest, Hungary

5   Space Science Division, Naval Research Laboratory, Washington, DC, USA

1 Introduction

The surface flux transport (hereafter SFT) model is based on an elegant and simple idea, originally formulated by Leighton (1964): radial magnetic flux on the solar surface behaves like a passive scalar field. In other words, flux is carried around by horizontal plasma flows but with no back reaction on these flows.

Despite its simplicity, the SFT model has proven remarkably successful at replicating the magnetic flux patterns on the real solar surface (photosphere). Figure 1 shows an SFT simulation for Solar Cycle 24, where new active regions have been inserted based on magnetograph observations. With appropriate parameters, the time-latitude "magnetic butterfly diagram" in the SFT model (Fig. 1a) is a good match for the observed time-latitude plot (Fig. 1c) at all latitudes. In general, the success of the SFT model has led to important applications both as (i) an inner boundary condition for extrapolations of the magnetic field in the solar atmosphere, and (ii) an outer boundary constraint on models for the solar interior dynamo.

In this review, our focus is on understanding the model itself: both its key ingredients and fundamental behaviour when applied in the solar regime. Details about applications, particularly to the solar atmosphere, may be found in previous review articles (Sheeley 2005; Mackay and Yeates 2012; Wang 2017). In the solar dynamo context, the SFT model has been used to constrain theories and models of the magnetic field in the solar interior (e.g., Cameron et al. 2012; Cameron and Schüssler 2015; Jiang et al. 2014b; Lemerle and Charbonneau 2017; Whitbread et al. 2019; Hazra 2021). But it is also a valuable practical tool for solar cycle prediction, enabling predictions to be made of the polar field at the end of the

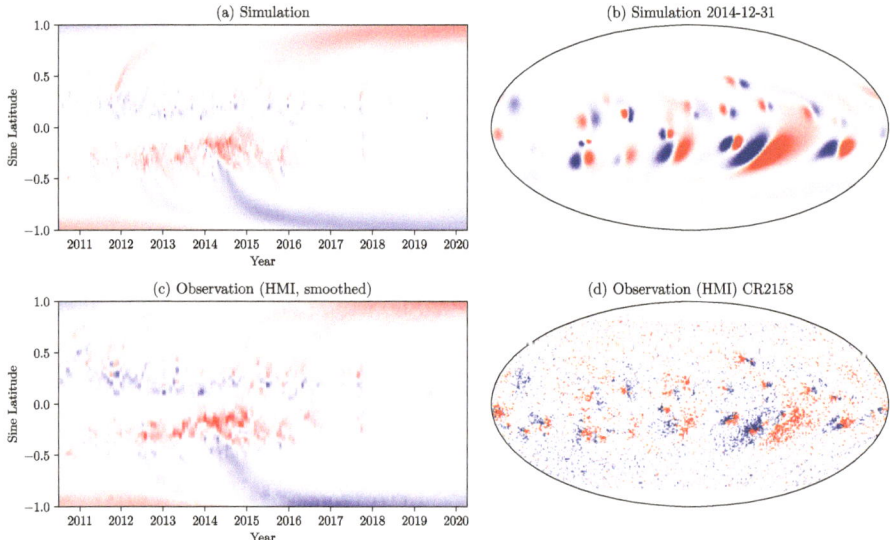

Fig. 1 An SFT model for Solar Cycle 24 with emerging regions derived from SDO/HMI SHARPs data (following the method of Yeates 2020). Panel (a) shows the longitude-averaged field $\langle B_r \rangle$ in the simulation, and (b) shows a snapshot of the two-dimensional field B_r on 31 December 2014. For comparison, (c) shows a magnetic butterfly diagram (or super-synoptic map) constructed from SDO/HMI pole-corrected synoptic maps (Sun 2018), smoothed to a comparable resolution to the simulation. The individual, unsmoothed synoptic map for Carrington rotation CR2158 is shown in (d). Red/blue denote positive/negative values, capped at $\pm 10\,\mathrm{G}$ in (a,c) and $\pm 50\,\mathrm{G}$ in (b,d)

current solar cycle, and hence – through well-established correlations – the amplitude of the following solar activity cycle (e.g., Cameron et al. 2016; Iijima et al. 2017; Jiang et al. 2018; Upton and Hathaway 2018; Bhowmik and Nandy 2018; Jiang et al. 2023). Understanding the origin and limitations of such polar field predictions requires an understanding of the SFT model itself, which is what we seek to provide here.

The review is organised as follows. In Sect. 2, we present the basic equations of the "classical" SFT model. Section 3 discusses the imposed flows in the model, including the importance of including meridional flow and recent work on constraining the flow parameters. Section 4 discusses the source term representing new flux emergence, which is fundamental to the flux patterns that the model predicts. Section 5 examines the important question of whether the SFT model – usually seen as purely phenomenological – can be derived from physical principles. We conclude in Sect. 6 with an overview of model features beyond our "classical" version.

2 Fundamentals of the Classical Model

Denoting the radial magnetic field distribution by $B_r(\theta, \phi, t)$, the equation for a passive scalar field is

$$\frac{\partial B_r}{\partial t} + \nabla_h \cdot \left(\mathbf{u}_h B_r\right) = \eta \nabla_h^2 B_r + S, \tag{1}$$

where $\mathbf{u}_h$ is the imposed advection velocity, and η is the diffusivity. In the classical model, B_r represents the large-scale mean field; the model does not resolve the smaller-scale motions of supergranular convection, but rather models these with the turbulent diffusivity η. This was introduced by Leighton (1964) to parameterise the "random walk" of individual magnetic flux elements due to the changing pattern of supergranular flows. For SFT it is necessary to include also a prescribed source term $S(\theta, \phi, t)$ that describes the emergence of new magnetic flux, typically in the form of active regions. In a more complete physical model, S would arise self-consistently through Faraday's induction equation (to be discussed in Sects. 5 and 6), but in the classical SFT model it is a prescribed model input. Throughout we will use subscript h to denote the "horizontal" components of a vector, meaning those tangential to the solar surface.

In the classical SFT model, the diffusivity η is uniform and constant, and most authors assume a steady, axisymmetric imposed velocity of the form

$$\mathbf{u}_h(\theta) = R_\odot \sin \theta \, \Omega(\theta) \mathbf{e}_\phi + u_\theta(\theta) \mathbf{e}_\theta. \tag{2}$$

Thus $\Omega(\theta)$ represents the angular velocity of solar differential rotation, and $u_\theta(\theta)$ represents the meridional circulation. The choice of these flows is important and will be discussed further in Sect. 3. Relaxing the classical assumptions is considered in Sect. 6 (except for the addition of an exponential decay term which is discussed in Sect. 5).

2.1 Dimensionless Form

Ignoring S, we can consider non-dimensionalization of equation (1) by defining dimensionless variables $\mathbf{u}_h' = \mathbf{u}_h / U_0$, $\nabla_h' = R_\odot \nabla_h$ and $t' = t U_0 / R_\odot$, where U_0 is a typical flow speed. Then (1) becomes

$$\frac{\partial B_r}{\partial t'} + \nabla_h' \cdot \left(\mathbf{u}_h' B_r\right) = \frac{1}{\mathrm{Rm}} \nabla_h'^2 B_r, \tag{3}$$

suggesting that the behaviour (in the absence of new emergence) is controlled by the dimensionless magnetic Reynolds number

$$\mathrm{Rm} = \frac{R_\odot U_0}{\eta}. \tag{4}$$

In effect, it is only the relative speed of advective to diffusive transport that matters.

2.2 Explicit Form

Writing out (1) explicitly in spherical coordinates, and assuming (2), gives the standard SFT equation

$$\frac{\partial B_r}{\partial t} + \frac{1}{R_\odot \sin\theta} \frac{\partial}{\partial\theta} \left(\sin\theta \, u_\theta \, B_r \right) + \Omega(\theta) \frac{\partial B_r}{\partial\phi} =$$

$$\frac{\eta}{R_\odot^2 \sin\theta} \frac{\partial}{\partial\theta} \left(\sin\theta \frac{\partial B_r}{\partial\theta} \right) + \frac{\eta}{R_\odot^2 \sin^2\theta} \frac{\partial^2 B_r}{\partial\phi^2} + S. \tag{5}$$

In some applications it suffices to consider the longitude-averaged field,

$$\langle B_r \rangle (\theta, t) = \frac{1}{2\pi} \int_0^{2\pi} B_r(\theta, \phi, t) \, d\phi. \tag{6}$$

Integrating (5), we find that $\langle B_r \rangle$ obeys the one-dimensional equation

$$\frac{\partial \langle B_r \rangle}{\partial t} + \frac{1}{R_\odot \sin\theta} \frac{\partial}{\partial\theta} \left(\sin\theta \, u_\theta \langle B_r \rangle \right) = \frac{\eta}{R_\odot^2 \sin\theta} \frac{\partial}{\partial\theta} \left(\sin\theta \frac{\partial \langle B_r \rangle}{\partial\theta} \right) + \langle S \rangle, \tag{7}$$

showing in particular that differential rotation has no effect on the evolution of $\langle B_r \rangle$ (Leighton 1964). On the other hand, the differential rotation – being the fastest flow – plays an important role in determining the two-dimensional flux patterns seen on the solar surface. By increasing the length of the polarity inversion lines in and between active regions, it also speeds up the diffusive cancellation of non-axisymmetric components of B_r (Sheeley and DeVore 1986).

2.3 Implementation

Although some analytical analysis is possible (see Sheeley and DeVore 1986; DeVore 1987, and also Sect. 4 below), for most applications it is usual to solve (5) or (7) with numerical methods. This dates right back to the original paper of Leighton (1964). The most natural numerical approach would be a spectral method based on spherical harmonics, as implemented for example by Mackay et al. (2002) or Baumann et al. (2004) (see Baumann 2005, for more details). However, care is needed in treating the source term S, since newly-emerging active regions are typically highly localized in space and usually require filtering in spectral space to avoid the Gibb's phenomenon ("ringing"). A more straightforward approach is to use a simple explicit finite-volume method, provided that care is taken in both the discretization and the source term to conserve magnetic flux (i.e., preserve $\int_0^{2\pi} \int_0^\pi B_r(\theta, \phi, t) \sin\theta \, d\theta d\phi = 0$). The resulting time-step restriction is typically not a severe problem on modern machines, given the two-dimensional nature and modest resolutions typically used (for example, a 360×180 mesh). Much higher resolutions would not be consistent with the mean-field assumption of the classical model (alternatives are discussed in Sect. 6).

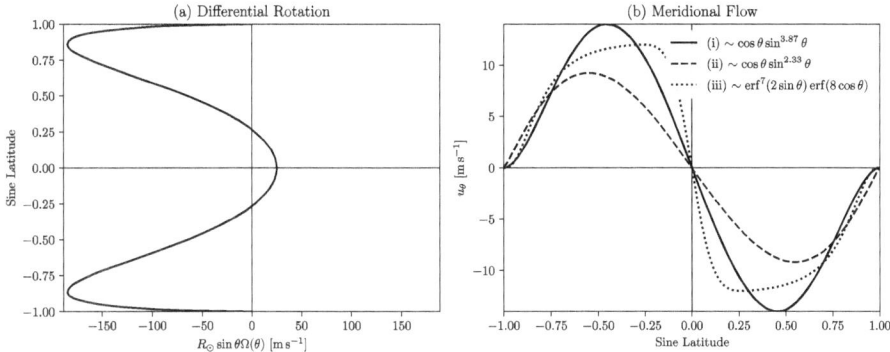

Fig. 2 Velocity profile of differential rotation in the Carrington frame (a), and some example optimized profiles of meridional flow velocity (b), including (i) the simulation shown in Fig. 1; (ii) the Cycle 21 simulation of Whitbread et al. (2017); and (iii) the Cycle 21 simulation of Lemerle et al. (2015). The corresponding values of Δ_u are (i) $0.7 \times 10^{-7}\,\mathrm{s}^{-1}$, (ii) $0.4 \times 10^{-7}\,\mathrm{s}^{-1}$, and (iii) $1.6 \times 10^{-7}\,\mathrm{s}^{-1}$

3 Flows

3.1 Differential Rotation

The solar surface differential rotation is well constrained observationally (see, e.g. Beck 2000) and usually treated as a fixed constraint. Typically, SFT models use a steady axisymmetric angular velocity profile such as

$$\Omega(\theta) = 0.18 - 2.396\cos^2\theta - 1.787\cos^4\theta \quad \left[^\circ\,\mathrm{day}^{-1}\right] \tag{8}$$

as determined by Snodgrass and Ulrich (1990). The constant term here is written in the Carrington frame that is usually adopted for SFT simulations. The resulting velocity profile is shown in Fig. 2(a). As mentioned above, the differential rotation affects only the non-axisymmetric component of B_r, not the axisymmetric component $\langle B_r \rangle$, and will not be discussed further.

3.2 Meridional Flow

Although the only large-scale flow included by Leighton (1964) was the differential rotation, it became clear from subsequent investigation of the SFT model that adding a meridional flow gives more realistic magnetic flux distributions (DeVore et al. 1984). In particular, a poleward flow is needed in order to concentrate the magnetic field into polar caps at the end of the solar cycle – compare Figs. 3(a) and (b). Otherwise, once B_r has become approximately axisymmetric it will tend to the slowest decaying ($\ell = 1$) eigenmode of the diffusion operator, which is the dipole $B_r \sim \cos\theta$. (A pure dipole is not seen in Fig. 3c because it requires a few more years: the decay time for the next higher mode, $l = 2$, is $R_\odot^2 / [\eta l(l+1)] \approx 6\,\mathrm{yr}$.)

Observational evidence now clearly supports the existence of a surface meridional flow (Hanasoge 2022) although it is much slower than the differential rotation and potentially more variable. As such, different modellers have used different flow profiles. Typical examples have a single peak in each hemisphere, but vary in their latitudinal profiles. Figure 2(b)

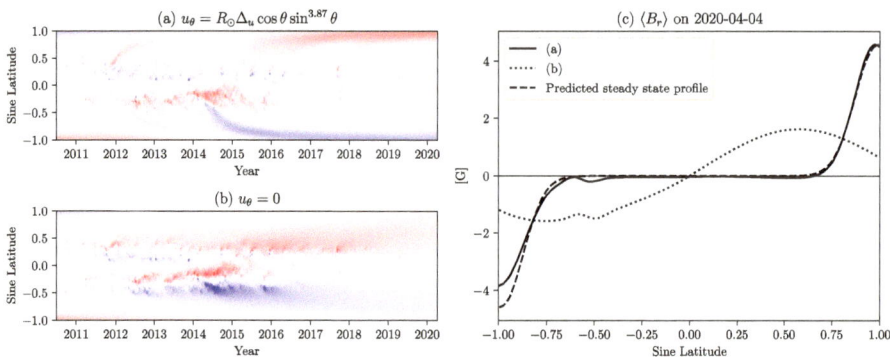

Fig. 3 Effect of meridional flow in the simulation from Fig. 1, showing latitude-time plots of $\langle B_r \rangle$ when the flow is included (a) or omitted (b). Panel (c) shows the latitudinal profiles of $\langle B_r \rangle$ at the end of the simulation. The dashed curve shows the (near) steady-state profile (11) for the case with flow. (After Fig. 3 of Sheeley 2005)

illustrates three profiles: (i) and (ii) come from the simple two-parameter family

$$u_\theta(\theta) = -R_\odot \Delta_u \cos\theta \sin^p \theta, \tag{9}$$

where Δ_u is the flow divergence at the equator, and larger values of p lead to flows more concentrated near the equator (the speed peaks at $\cos\theta = \pm(1+p)^{-1/2}$). Profile (iii) in Fig. 2(b) has the more complex form

$$u_\theta(\theta) = -\frac{\sqrt{\pi} R_\odot \Delta_u}{2w \, \text{erf}^q(v)} \, \text{erf}^q(v\sin\theta) \, \text{erf}(w\cos\theta), \tag{10}$$

which allows the gradient to be concentrated nearer to the equator (see Wang 2017).

It is non-trivial to determine the precise eigenmodes of equation (7) when meridional flow is included (DeVore 1987), even with a simple flow profile such as (9). However, one can determine a useful approximation by seeking a perfectly axisymmetric steady state $B_r(\theta)$ that balances the poleward advection with diffusion. For example, for the flow profile (9), equation (7) can be solved in an individual hemisphere to give the steady state solution

$$B_r(\theta) = B_r(0) \exp\left[-\frac{\text{Rm}_0 \sin^{1+p}\theta}{(1+p)} \right]. \tag{11}$$

Here $\text{Rm}_0 = R_\odot^2 \Delta_u / \eta$, which is the magnetic Reynolds number Rm from (4) with the specific choice $U_0 = R_\odot \Delta_u$, highlighting explicitly the dependence of the solution on the magnetic Reynolds number. The amplitude $B_r(0)$ will depend on the initial condition and source term S and cannot be determined directly. The solution (11) can only be an approximation to the slowest-decaying eigenfunction because it is necessarily non-zero at the equator, and will therefore generate a discontinuity at the equator when applied in both hemispheres with opposite sign. However, this discontinuity is small for typical values of Rm_0 and will lead to diffusive cancellation only on a timescale much longer than the solar cycle (cf. Cameron et al. 2010). Indeed, Fig. 3(c) shows that (11) gives an excellent approximation to the latitudinal B_r profile at the end of the example simulation in Fig. 3(a), particularly in the Northern hemisphere. (In the Southern hemisphere there is a remnant active region at low latitude that modifies the profile.) This simulation used $\eta = 425\,\text{km}^2\text{s}^{-1}$, $p = 3.87$, $\Delta_u = 6.9 \times 10^{-8}\,\text{s}^{-1}$, and consequently $\text{Rm}_0 \approx 79$.

3.3 Parameter Optimization

The primary flow parameters to choose are the meridional flow profile $u_\theta(\theta)$ and the diffusivity coefficient η. The basic effects of varying these parameters were investigated in the 1980s (DeVore et al. 1984; Wang et al. 1989). A more systematic parameter study was published by Baumann et al. (2004), who explored the results of varying both η and the meridional flow amplitude (in addition to properties of the source term), albeit varying only one parameter at a time and not the shape of the meridional flow profile.

More recent studies have explored the parameter space more widely, and have also attempted to optimize the parameters directly against synoptic magnetogram observations. The two most general studies are Lemerle et al. (2015) and Whitbread et al. (2017), who both allow the strength and shape of $u_\theta(\theta)$ to vary, in addition to η. For u_θ, Whitbread et al. (2017) allowed for profiles of the form (9), whereas Lemerle et al. (2015) allow for the more general (but still single-peaked) form (10). The optimal profiles from both studies for data from Cycle 21 are shown in Fig. 2. At present, it is not possible to select confidently between these solutions using observations, though helioseismic measurements of the plasma flow suggest equatorial slopes Δ_u in the range $[0.6 - 1.2] \times 10^{-7}\,\mathrm{s}^{-1}$ – somewhere between profiles (i) and (iii) in Fig. 2. Measurements based on magnetic feature tracking give lower equatorial slopes more like that of profile (ii), but it has been suggested that these are contaminated by supergranular diffusion (Dikpati et al. 2010; Wang 2017). A recent list of observations is given in Jiang et al. (2023).

Both Lemerle et al. (2015) and Whitbread et al. (2017) used the same genetic optimization algorithm, PIKAIA (Charbonneau and Knapp 1995). These two studies differed in their chosen goodness-of-fit functions, although both were ultimately derived from comparing to observed $B_r(\theta, \phi)$ maps. Whitbread et al. (2017) gave more weight to lower latitudes (where magnetogram observations are more reliable), whereas Lemerle et al. (2015) gave additional weight to the mid-latitude "transport regions" (because they represent the result of the model evolution rather than only the active region emergence) and to the axial dipole strength. At the other extreme, a further parameter study by Petrovay and Talafha (2019) focused only on optimizing the high latitude (polar) field, albeit in the 1D model. This study used a synthetic (averaged) source term and fitted to average cycle properties from Wilcox Observatory polar field measurements, such as reversal time or width of the polar cap.

A robust finding in these optimization studies is a degeneracy between η and the amplitude of u_θ. This is illustrated by Fig. 4, which shows that there is a long ridge of near-optimal solutions in parameter space. Increasing both parameters together tends to lead to a equally (or nearly equally) well-matched solution, perhaps explaining why different groups have been able to use quite different values of η – for example, Cameron et al. (2010) use $\eta = 250\,\mathrm{km}^2\mathrm{s}^{-1}$ as their standard value whereas the simulation in Fig. 1 used $\eta = 425\,\mathrm{km}^2\,\mathrm{s}^{-1}$. This degeneracy makes sense given the appearance of the magnetic Reynolds number Rm in equation (3), which is essentially the ratio of η to $|u_\theta|$. It means that SFT simulations can not be used to constrain both the meridional flow and diffusion from magnetogram observations alone.

When optimizing the model individually for different solar cycles, Whitbread et al. (2017) found some cycle-to-cycle variation in the optimal speeds and diffusivities. This is understandable given the phenomenological nature of the model (to be discussed further in Sect. 5). Indeed, when simulating multiple cycles, Wang et al. (2002) had previously varied the meridional flow speed from cycle to cycle so as to avoid unrealistic drift of the polar field over time. On the other hand, other authors have avoided this problem by varying instead the tilts of emerging active regions (Cameron et al. 2010, see also Sect. 4.3), or adding an addition decay term (to be discussed in Sect. 5). In reality it is likely that the effective mean-field

Fig. 4 Fitness function χ^{-2} as a function of meridional flow amplitude $u_0 = \max_\theta |u_\theta|$ (horizontal axis) and diffusivity $\eta_R \equiv \eta$ (vertical axis), from the optimization study of Lemerle et al. (2015). Black lines show the optimum value and blue lines the limit of the acceptable region ($\chi^{-2} \geq 93\% \chi_{\max}^{-2}$). (© AAS. Reproduced with permission.)

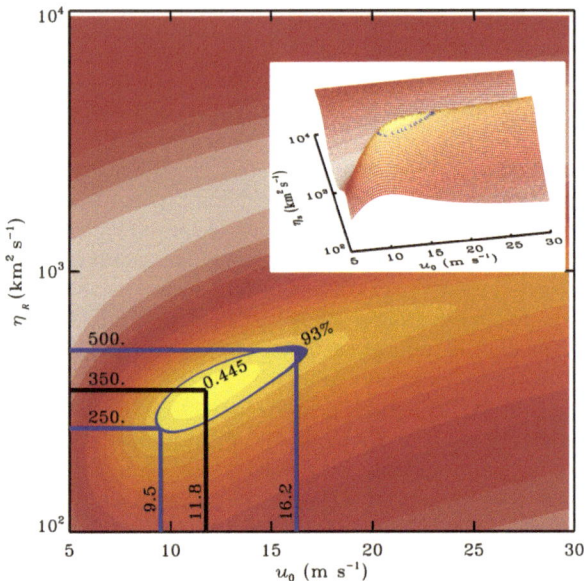

meridional flow varies even over the course of a single Solar Cycle (see Sect. 6). Interestingly, Hung et al. (2017) have shown – in the context of a flux-transport (interior) dynamo model – that a time-dependent meridional flow may be recovered from surface magnetic data through variational data-assimilation, and in future this approach could also be applied to SFT.

4 The Source Term

The magnetic flux patterns in the SFT model are determined in large part by the source term $S(\theta, \phi, t)$, which – in the classical mean-field model – represents the emergence of new macroscopic active regions on the solar surface. Since the classical SFT equation (1) is linear in B_r, the solution is a superposition of solutions for each individual active region, so it is insightful to consider the evolution of one of these regions in isolation. Since most SFT simulations follow the evolution for periods of years, it is usual to emerge each active region instantaneously in time, so that

$$S(\theta, \phi, t) = \sum_i B_r^{(i)}(\theta, \phi)\delta(t - t^{(i)}), \qquad (12)$$

where $B_r^{(i)}(\theta, \phi)$ is the magnetic field of an individual active region emerging at $t = t^{(i)}$.

Traditionally, SFT models treat each active region as a bipolar magnetic region (BMR). Figure 5 shows the shape used by Van Ballegooijen et al. (1998), with circular flux patches centred on the poles (θ_-, ϕ_-) and (θ_+, ϕ_+) and having the form

$$B_r(\theta, \phi) = B_0 \left\{ \exp\left[-\frac{2(1 - \cos\beta_+)}{(b\rho_0)^2}\right] - \exp\left[-\frac{2(1 - \cos\beta_-)}{(b\rho_0)^2}\right] \right\}, \qquad (13)$$

Fig. 5 Positive and negative contours of B_r for a BMR of the Van Ballegooijen et al. (1998) form (13). The size is exaggerated ($\rho_0 = 25°$) compared to a real active region. This example follows Joy's Law in that the leading (rightmost) polarity is closest to the equator

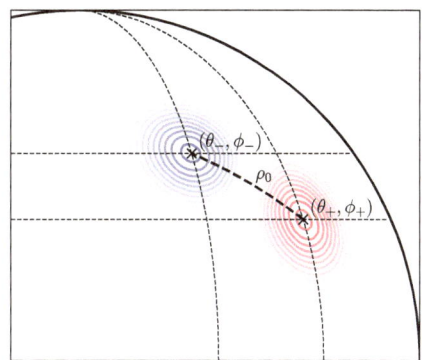

where

$$\cos \beta_\pm = \cos \theta_\pm \cos \theta + \sin \theta_\pm \sin \theta \cos(\phi - \phi_\pm), \tag{14}$$

$$\cos \rho_0 = \cos \theta_+ \cos \theta_- + \sin \theta_+ \sin \theta_- \cos(\phi_+ - \phi_-). \tag{15}$$

Thus $\beta_\pm(\theta, \phi)$ denote the heliocentric angles from each pole, and ρ_0 the heliocentric angle between them. Van Ballegooijen et al. (1998) took $b = 0.4$. For some purposes, one can approximate (13) with a pair of Dirac-delta sources,

$$B_r(\theta, \phi) = \frac{\Phi_0}{R_\odot^2 \sin \theta} \Big[\delta(\theta - \theta_+)\delta(\phi - \phi_+) - \delta(\theta - \theta_-)\delta(\phi - \phi_-) \Big], \tag{16}$$

where Φ_0 gives the flux of each polarity, defined assuming flux balance as

$$\Phi_0 = \frac{R_\odot^2}{2} \int_S |B_r| \sin \theta \, d\theta \, d\phi. \tag{17}$$

Although the precise chosen shape for BMRs varies between implementations (see Yeates 2020, for another variation), the key properties are the magnetic flux, Φ_0, and pole locations, (θ_-, ϕ_-) and (θ_+, ϕ_+). The latter may equivalently be specified by giving the co-ordinates of the BMR centre (θ_0, ϕ_0) along with the separation ρ_0 as in (15) and tilt angle γ_0, typically defined by

$$\tan \gamma_0 = \frac{\theta_+ - \theta_-}{\sin \theta_0 (\phi_+ - \phi_-)}. \tag{18}$$

Together these BMR properties determine both the short-term and long-term evolution of the region.

After a new region emerges in the model, much of its magnetic flux cancels by supergranular diffusion. This models the observed process of flux cancellation at the polarity inversion line (PIL) between the positive and negative polarities. This cancellation rate is enhanced as the region is sheared by differential rotation and the PIL lengthened. On short-timescales (days) it is possible to approximate the solar surface as a Cartesian plane. Assuming a linear shear flow profile for the differential rotation, Lagrangian variables can be used to solve the Cartesian form of (1) for the exact evolution $B_r(\theta, \phi, t)$ of a tilted BMR (we will see an example in Sect. 4.1). On longer timescales, it is necessary to follow the evolution numerically.

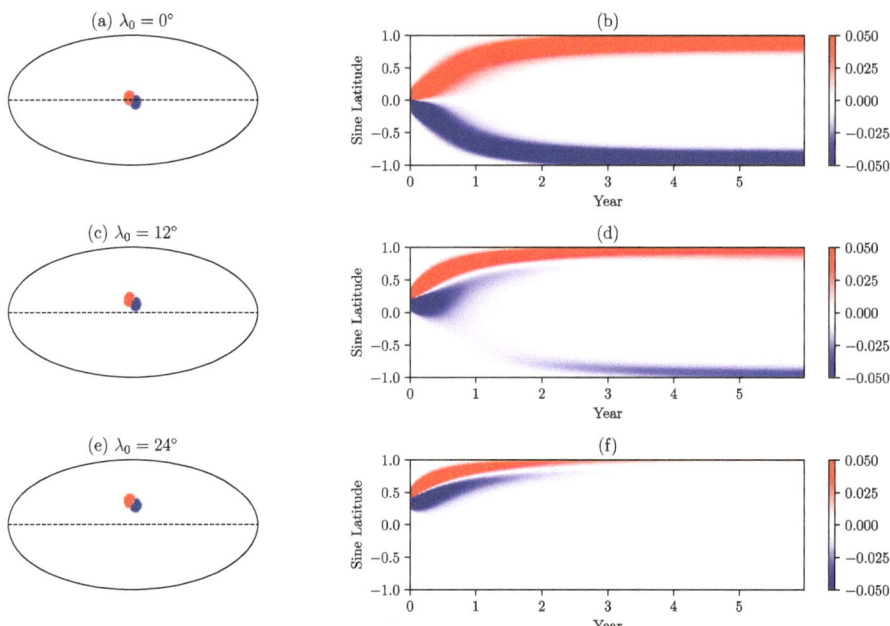

Fig. 6 Long term evolution of three identical BMRs emerged at different latitudes ($\lambda_0 = \pi/2 - \theta_0$) in the SFT model. Left column shows the initial BMRs and right column the time-latitude plot of $\langle B_r \rangle$ in each case

It takes approximately 2 years for the non-axisymmetric component of B_r to cancel completely (Wang and Sheeley 1991). Whether or not any axisymmetric B_r remains on a longer timescale depends on how much flux of one polarity escapes across the equator so that the two polarities are pushed to opposite poles by the meridional flow. An untilted region will send both polarities equally to each pole and so leave no asymptotic contribution at the end of the solar cycle. In a similar way, a (tilted) region that is nearer to the equator will produce a greater asymptotic contribution, because more flux escapes across the equator before being cancelled. This important effect is illustrated in Fig. 6, where the same BMR is inserted at three different latitudes.

4.1 Dipole Amplification Factor of a BMR

A common way to measure the end-of-cycle contribution of an individual BMR is through its axial dipole strength, which is the axisymmetric spherical harmonic coefficient of B_r with lowest degree,

$$b_{1,0}(t) = \frac{3}{4\pi} \int_0^{2\pi} \int_0^\pi B_r \cos\theta \sin\theta \, d\theta \, d\phi = \frac{3}{2} \int_0^\pi \langle B_r \rangle \cos\theta \sin\theta \, d\theta. \qquad (19)$$

By linearity of the classical SFT model, the total axial dipole strength will be the sum of the individual contributions from all of the active regions.

At the time of emergence, a BMR with the simple form (16) has

$$b_{1,0}(t_{\text{em}}) = \frac{3\Phi_0}{4\pi R_\odot^2} \int_0^\pi \left[\delta(\theta - \theta_+) - \delta(\theta - \theta_-) \right] \cos\theta \, d\theta \qquad (20)$$

$$= \frac{3\Phi_0}{4\pi R_\odot^2} \left(\cos\theta_+ - \cos\theta_- \right) \tag{21}$$

$$= \frac{3\Phi_0}{2\pi R_\odot^2} \sin\left(\frac{\theta_- - \theta_+}{2} \right) \sin\left(\frac{\theta_+ + \theta_-}{2} \right) \tag{22}$$

$$\approx -\frac{3\Phi_0}{4\pi R_\odot^2} \rho_0 \sin\gamma_0 \sin\theta_0. \tag{23}$$

Here we have defined the central colatitude $\theta_0 = (\theta_+ + \theta_-)/2$ and recognized that for tilt angle γ_0 and heliocentric angle ρ_0 between the poles, their latitudinal separation is $(\theta_+ - \theta_-) = \rho_0 \sin\gamma_0$ (assuming $\theta_+ > \theta_-$). Thus, as noted by Wang and Sheeley (1991), the axial dipole strength of a newly-emerged BMR depends on its flux, its latitudinal pole separation, and the cosine of its emergence latitude.

Importantly, the axial dipole strength of a BMR can change under the ensuing SFT evolution: it will be amplified if the BMR emerged near the equator, or will decay if the BMR emerged far from the equator. It was first recognized by Jiang et al. (2014a) that the "dipole amplification factor"

$$f_\infty = \lim_{t \to \infty} \frac{b_{1,0}(t)}{b_{1,0}(t_{em})} \tag{24}$$

is well approximated by a Gaussian function of latitude, of the form

$$f_\infty(\lambda_0) = A \exp\left(-\frac{\lambda_0^2}{2\lambda_R^2} \right), \tag{25}$$

where $\lambda_0 = \pi/2 - \theta_0$ is the central latitude of the BMR. (It is convenient to work in terms of latitude $\lambda = \pi/2 - \theta$ rather than colatitude θ.) Fig. 7 shows the functional form measured in several different numerical SFT models, where we note that both the amplitude A and width λ_R depend on the model. Once these parameters are known, equation (25) – coupled with the linearity of the SFT evolution equation (5) or (7) – allows the net axial dipole strength at the end of a solar cycle to be determined algebraically just by adding up the contributions of the individual BMRs, without the need to solve the evolution equation.

Fig. 7 Latitude dependence of the dipole amplification factor for BMRs in different published SFT models. The solid lines show Gaussian fits. Reproduced from Petrovay et al. (2020)

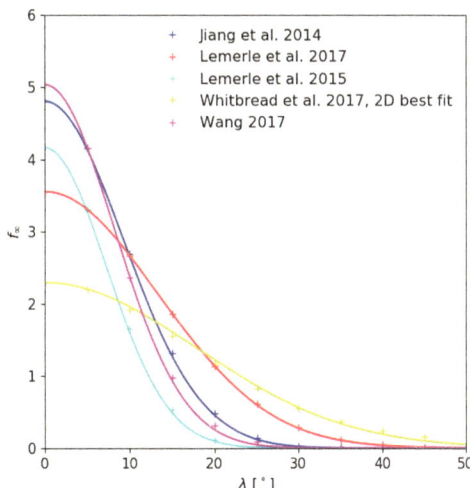

The interpretation of Fig. 7 is that only BMRs that emerge with latitude $|\lambda_0| < \lambda_R$ will contribute to the global dipole moment at the end of the solar cycle. Petrovay et al. (2020) call λ_R the "dynamo effectivity range", and give the following simple physical derivation. To give a lasting contribution, a BMR must be close enough to the equator that some of its leading-polarity flux is able to cross the equator by diffusion, in opposition to the meridional flow. The timescale for advective separation at the equator is Δ_u^{-1}, where $\Delta_u = R_\odot^{-1} u_\theta'(\pi/2)$ is the equatorial divergence of $u_\theta(\theta)$. Equating this to the diffusion timescale $(\lambda R_\odot)^2/\eta$ from latitude λ to the equator suggests that

$$\lambda_R \approx \sqrt{\frac{\eta}{R_\odot^2 \Delta_u}} = \mathrm{Rm}_0^{-1/2}. \tag{26}$$

Note the reappearance of the magnetic Reynolds number from Sect. 3.2. Petrovay et al. (2020) computed $f_\infty(\lambda_0)$ for numerical solutions with several different u_θ profiles, and in most cases found that the Gaussian width λ_R was indeed well approximated by $\mathrm{Rm}_0^{-1/2}$, the exception being a flow where u_θ peaks at a very low latitude compared to observations.

Petrovay et al. (2020) went further and derived (25) analytically. (Readers not interested in the details may skip to Sect. 4.2.) The trick is to recognize that the final dipole moment – once the B_r distribution has become (near) axisymmetric – will be proportional to the remaining net magnetic flux in each hemisphere. (There will also be a coefficient depending on the latitudinal profile of the near-steady state as in (11).) Because it is determined purely by flux crossing the equator, the evolution of the net hemispheric flux can be quite well approximated by a Cartesian SFT model near the equator, which has the advantage of being analytically tractable. Thus Petrovay et al. (2020) consider the "low-latitude limit" of (7),

$$\frac{\partial \langle B_r \rangle}{\partial t} + \frac{1}{R_\odot} \frac{\partial}{\partial \lambda} \left(u_\lambda \langle B_r \rangle \right) = \frac{\eta}{R_\odot^2} \frac{\partial^2 \langle B_r \rangle}{\partial \lambda^2}. \tag{27}$$

By choosing the linearised meridional flow $u_\lambda = R_\odot \Delta_u \lambda$, we can define the Lagrangian coordinate $\ell = e^{-\Delta_u t} \lambda$ and new time variable $\tau = \left(1 - e^{-2\Delta_u t}\right)/(2\mathrm{Rm}_0)$ to reduce Equation (27) to a standard diffusion equation

$$\frac{\mathrm{D}}{\mathrm{D}\tau} \left(e^{\Delta_u t} \langle B_r \rangle \right) = \frac{\partial^2}{\partial \ell^2} \left(e^{\Delta_u t} \langle B_r \rangle \right), \tag{28}$$

where $\mathrm{D}/\mathrm{D}\tau$ denotes the partial derivative with ℓ kept constant rather than λ. Equation (28) may be solved for a variety of initial conditions using standard techniques.

If the initial condition consists of a single (monopole) point source,

$$\langle B_r \rangle (\lambda, 0) = \frac{\Phi_0}{2\pi R_\odot^2} \delta(\lambda - \lambda_0), \tag{29}$$

then solving (28) gives

$$e^{\Delta_u t} \langle B_r \rangle = \frac{\Phi_0}{2\pi R_\odot^2 \sqrt{4\pi\tau}} \exp\left(-\frac{(\ell - \lambda_0)^2}{4\tau}\right), \tag{30}$$

which for large t is approximately

$$\langle B_r \rangle (\lambda, t) \sim \frac{\Phi_0 \sqrt{\mathrm{Rm}_0} e^{-\Delta_u t}}{2\pi R_\odot^2 \sqrt{2\pi}} \exp\left(\frac{-\mathrm{Rm}_0 \left(e^{-\Delta_u t} \lambda - \lambda_0\right)^2}{2}\right). \tag{31}$$

In the approximation (31), the flux difference between the hemispheres is

$$\Phi_N - \Phi_S = 2\pi R_\odot^2 \left(\int_0^\infty \langle B_r \rangle \, d\lambda - \int_{-\infty}^0 \langle B_r \rangle \, d\lambda \right) = \Phi_0 \, \mathrm{erf} \left(\sqrt{\frac{\mathrm{Rm}_0}{2}} \lambda_0 \right), \qquad (32)$$

valid for either sign of λ_0.

For a BMR we must combine two point sources as in (16), each contributing half of the flux Φ_0, so

$$\Phi_N - \Phi_S = \frac{\Phi_0}{2} \left[\mathrm{erf} \left(\sqrt{\frac{\mathrm{Rm}_0}{2}} \lambda_+ \right) - \mathrm{erf} \left(\sqrt{\frac{\mathrm{Rm}_0}{2}} \lambda_- \right) \right] \qquad (33)$$

$$\approx \frac{\Phi_0 \sqrt{\mathrm{Rm}_0} (\lambda_+ - \lambda_-)}{\sqrt{2\pi}} \exp \left(-\frac{\mathrm{Rm}_0 \lambda_0^2}{2} \right), \qquad (34)$$

where we recognize the finite difference as an approximation of the derivative at $\lambda_0 = (\lambda_- + \lambda_+)/2$. We therefore expect that, to a good approximation, $b_{1,0}(t) \to a(\Phi_N - \Phi_S)/R_\odot^2$ as $t \to \infty$, for some constant a that depends on the (normalized) shape of the steady B_r profile (thus only on u_θ and D). At the initial time, Equation (23) gives $b_{1,0}(0) = 3\Phi_0(\lambda_+ - \lambda_-) \cos\lambda_0/(4\pi R_\odot^2)$. Approximating $\cos\lambda_0 \approx 1$, the ratio is therefore

$$f_\infty \approx \frac{a\sqrt{8\pi \mathrm{Rm}_0}}{3} \exp \left(-\frac{\mathrm{Rm}_0 \lambda_0^2}{2} \right). \qquad (35)$$

Thus we recover (25) with $\lambda_R = \mathrm{Rm}_0^{-1/2}$ as claimed. Moreover, for a known asymptotic profile of $B_r(\theta)$, we can determine a and hence also predict the amplitude A.

4.2 Non-bipolar Source Regions

Real solar active regions cannot always be represented as simple, symmetric BMRs. Even a region with two polarities will be effectively "multipolar" if the polarities are asymmetric in shape, and this will modify the evolution of $b_{1,0}$ compared to a symmetric BMR. This was investigated by Iijima et al. (2019), who ran SFT simulations with Gaussian BMRs of the form (13), but where the leading polarity has a narrower width than the following polarity (controlled by the b parameter in (13)). When calibrated to the observed level of sunspot area asymmetry, their SFT simulation gave a more realistic evolution of both $b_{1,0}$ and the magnetic butterfly diagram, as compared to a reference simulation with equally-sized polarities. In particular, they noted that a wider following polarity leads to more following polarity flux crossing the equator, cancelling some of the trans-equatorial leading polarity flux and weakening the asymptotic contribution of the region. Similarly, Wang et al. (2021) found for asymmetric BMRs with more diffuse following polarity, that f_∞ is systematically reduced (see their Fig. 4). As an illustration, Fig. 8 shows an example of the SFT evolution for an asymmetric region inserted directly with its observed shape; in this case, the effect is sufficiently extreme to reverse the sign of $b_{1,0}$ altogether compared to a symmetric BMR.

Jiang et al. (2019) considered the SFT evolution of a more complex "δ-type" flux distribution. They showed that $b_{1,0}$ changed sign during the SFT evolution, ending up with a completely different end-of-cycle contribution than would be expected for a BMR emerging at the same latitude with the same flux and same initial $b_{1,0}$. Wang et al. (2021) showed further that the dipole amplification f_∞ is no longer a simple function of emergence latitude

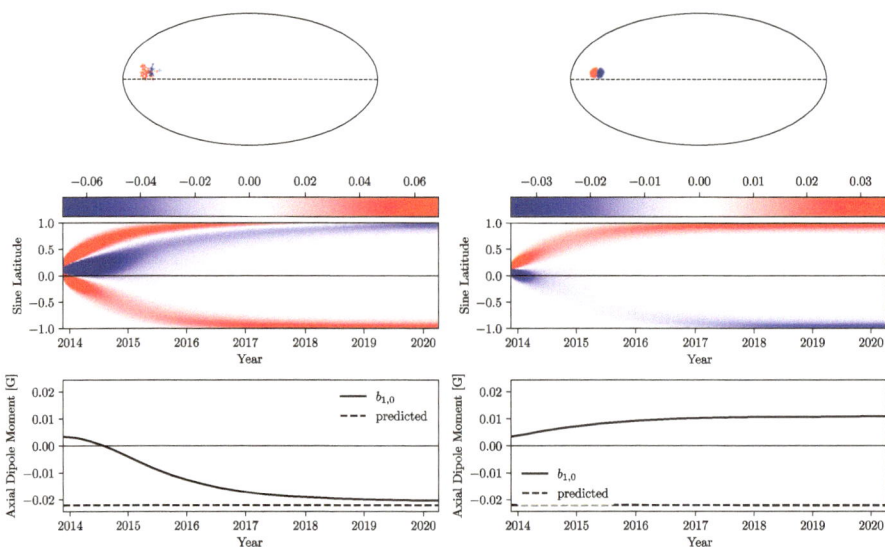

Fig. 8 Evolution of an active region with asymmetric bipolar shape, taken from the simulation in Fig. 1. The left column shows the region with its observed shape, with second row showing $\langle B_r \rangle$ and third row the axial dipole strength $b_{1,0}$. The right column shows the evolution of an "equivalent" symmetric BMR having the same initial flux and $b_{1,0}$. The dashed line shows the final $b_{1,0}$ predicted by equation (36) using the observed magnetogram

for such complex regions. However, the net effect of all of the real complex and asymmetric regions seems to be a reduction in the net end-of-cycle dipole, at least for Cycle 24. Evidence for this comes from Yeates (2020), who compared an SFT simulation of that cycle where all active regions emerged with their observed flux distributions to a simulation where they were all approximated by symmetric BMRs with the same flux and initial $b_{1,0}$. The net $b_{1,0}$ at the end of the cycle was overestimated by 24% when the regions were modelled with BMRs.

For predicting the dipole contributions of more complex regions, Wang et al. (2021) showed that (35) can be generalized to regions with non-bipolar shapes, by treating them as a superposition of point sources. In particular, for an active region with initial flux distribution $B_r(\theta, \phi, 0)$, combining the hemispheric flux differences (32) predicts that the axial dipole strength at the end of the cycle would be

$$\lim_{t \to \infty} b_{1,0}(t) \approx \frac{a}{R_\odot^2} \int_S B_r(\theta, \phi, 0) \, \mathrm{erf}\left[\sqrt{\frac{\mathrm{Rm}_0}{2}} \left(\frac{\pi}{2} - \theta\right)\right] \sin\theta \, \mathrm{d}\theta \mathrm{d}\phi, \qquad (36)$$

where a is the coefficient in the relation $b_{1,0} \approx a(\Phi_N - \Phi_S)/R_\odot^2$. Wang et al. (2021) verified this prediction against SFT simulations for 84 regions during Cycle 24. It gives an accurate prediction for the region in Fig. 8.

4.3 Modelling the Source Term

It is not always viable to use observations of real individual active regions to construct the source term. This situation arises when working with historical data, when running SFT models into the future for forecasting purposes, or just in conceptual simulations studying

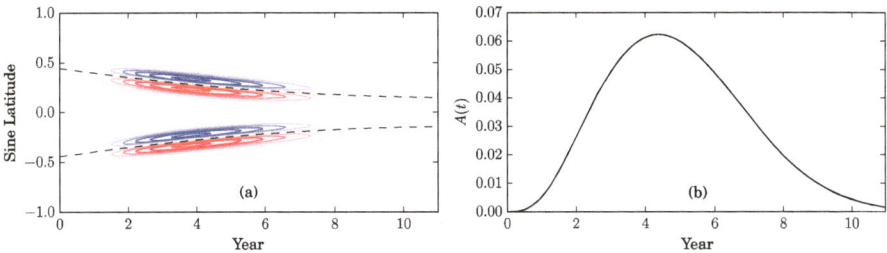

Fig. 9 The smooth source term $\langle S \rangle (\lambda, t)$ used by Petrovay and Talafha (2019). In (a), red/blue contours show $\langle S \rangle$, and dashed lines indicate $\pm \sin[\lambda_0(t)]$ from (39). Panel (b) shows the overall cycle shape $A(t)$ from (38) with $a = 0.00185$, $b = 4.058$, $c = 0.71$

the underlying physics. In such cases the source term $S(\theta, \phi, t)$ needs to be modelled, either as a smooth function (*e.g.*, Cameron and Schüssler 2007; Petrovay and Talafha 2019) or as random realizations of active regions drawn from a statistical distribution (*e.g.*, Schrijver 2001; Mackay and Lockwood 2002; Baumann et al. 2004; Jiang et al. 2018; Wang and Lean 2021).

The smooth function approach has primarily been used in the 1D SFT model, (7), for example by Petrovay and Talafha (2019), who use a pair of flux rings in each hemisphere, shown in Fig. 9(a) and given by

$$
\langle S \rangle (\lambda, t) = (-1)^n A(t) \left\{ \exp \left(-\frac{[\lambda - \lambda_+(t)]^2}{2\delta_\lambda^2} \right) - \exp \left(-\frac{[\lambda - \lambda_-(t)]^2}{2\delta_\lambda^2} \right) \right.
$$
$$
\left. + \exp \left(-\frac{[\lambda + \lambda_+(t)]^2}{2\delta_\lambda^2} \right) - \exp \left(-\frac{[\lambda + \lambda_-(t)]^2}{2\delta_\lambda^2} \right) \right\}. \tag{37}
$$

This model incorporates a number of observed solar cycle features:

(i) All polarities alternate according to the solar cycle number, n.
(ii) The cycle has an asymmetrical shape in time, shown in Fig. 9(b) and given by the Hathaway et al. (1994) observed fit

$$
A(t) = a(t - t_{\min}) \left(\exp \left[\frac{(t - t_{\min})^2}{b^2} \right] - c \right)^{-1}, \tag{38}
$$

where $t_{\min}$ is the start of the cycle.

(iii) The centres $\pm\lambda_0$ of each pair of flux rings, *i.e.* $\lambda_0 = (\lambda_+ + \lambda_-)/2$ shown by dashed lines in Fig. 9(a), migrate equatorward at the rate

$$
\lambda_0(t) = 26.4 - 34.2 \left(\frac{t}{T} \right) + 16.1 \left(\frac{t}{T} \right)^2 \quad [^\circ] \tag{39}
$$

fitted empirically by Jiang et al. (2011), where T is the cycle length (11 years).

(iv) The separation $\Delta_\lambda = \lambda_- - \lambda_+$ decreases as λ_0 approaches the equator, according to

$$
\Delta_\lambda(t) = 0.5 \frac{\sin \lambda_0(t)}{\sin 20^\circ} \quad [^\circ]. \tag{40}
$$

This models the longitude-averaged effect of the well-established Joy's Law (van Driel-Gesztelyi and Green 2015), whereby BMRs emerging at lower latitude have (on average) smaller tilt angle $|\gamma_0|$, defined in (18).

The statistical BMR approach is similar, except the functions above are treated as overall distributions from which discrete BMRs are chosen at random. For the longest historical simulations, which date back to 1700 (Jiang et al. 2018; Wang et al. 2021), the only observational input is the sunspot number time series – equivalent to emergence rate, $A(t)$. For 20th Century simulations, data on the areas and locations of individual sunspot groups can be used (*e.g.*, Cameron et al. 2010). However, even here the magnetic flux and tilt angle (equivalently axial dipole strength) must be chosen at random as they are not available observationally before the onset of routine magnetograms in the 1970s.

The tilt angle is problematic as Joy's Law, as modelled in (40), holds only for the mean, and there is known to be very significant scatter (*e.g.*, Wang and Sheeley 1989; Yeates 2020). Recent studies have shown that individual "rogue" active regions – defined as those with dipole moments significantly different from Joy's Law expectation at their latitude – can have a significant effect on the overall polar field at the end of the cycle (Jiang et al. 2015; Nagy et al. 2017). In light of (35), such rogue regions must typically emerge near to the equator, although their relative contribution depends on Rm_0 and would be reduced if Rm_0 were large. Nevertheless, simulations based on statistical source terms without individual dipole moment data should be treated with caution, particularly for prediction.

The widely-accepted $\alpha\Omega$ paradigm for the solar dynamo suggests a "self-consistent" way to build a fully synthetic SFT model: set the amount flux emerging through the source term in cycle n proportional to the axial dipole strength at the end of cycle $n - 1$. Talafha et al. (2022) modified the one-dimensional model of Petrovay and Talafha (2019) to use such an approach. They used this model to systematically study the impact of two possible nonlinearities in the source term: tilt quenching (where BMRs are less tilted in strong cycles) and latitude quenching (where BMRs emerge at higher latitudes in strong cycles). SFT simulations show that both effects act to reduce the axial dipole produced in strong cycles (Cameron et al. 2010; Jiang 2020). They are both therefore possible saturation mechanisms to explain why the solar dynamo doesn't exhibit runaway exponential growth. Talafha et al. (2022) showed that the relative impact of tilt versus latitude quenching on the end-of-cycle axial dipole depends primarily on the dynamo effectivity range λ_R in equation (26). In particular, for small λ_R, latitude quenching reduces the end-of-cycle dipole more than tilt quenching, and *vice versa* for large λ_R. However, the amount of tilt and/or latitude quenching present on the real Sun remains under debate.

5 Physical Justification

As introduced by Leighton (1964), the SFT model is purely phenomenological. But can equation (1) be derived from known physical laws? The relevant law governing the evolution of the large-scale magnetic field is the mean-field MHD (magnetohydrodynamic) induction equation,

$$\frac{\partial B_r}{\partial t} = \mathbf{e}_r \cdot \nabla \times \left(\mathbf{u} \times \mathbf{B} - \eta \nabla \times \mathbf{B} \right), \tag{41}$$

where $\mathbf{u}$ is the plasma velocity and – anticipating the form of (1) – we have made a simple approximation for the turbulent electromotive force of the form $-\eta\nabla \times \mathbf{B}$ (cf. McCloughan and Durrant 2002). Thus η represents turbulent diffusivity, not ohmic resistivity (which is negligible in the highly conducting photosphere). This assumption of a turbulent diffusivity is discussed further in Sect. 5.3 below.

Consider the first term in (41). Decomposing $\mathbf{u} = \mathbf{u}_h + u_r \mathbf{e}_r$, where $\mathbf{e}_r \cdot \mathbf{u}_h = 0$, and similarly $\mathbf{B} = \mathbf{B}_h + B_r \mathbf{e}_r$, we can write

$$\mathbf{e}_r \cdot \nabla \times (\mathbf{u} \times \mathbf{B}) = \nabla \cdot (u_r \mathbf{B}_h) - \nabla \cdot (\mathbf{u}_h B_r). \tag{42}$$

The last term is precisely the advection term in the SFT equation (1), while the term $\nabla \cdot (u_r \mathbf{B}_h)$ represents flux emergence, so corresponds to the source term S in (1). Thus the SFT model is incorporating the correct advection terms.

Now consider the diffusion term in (41). For simplicity, we will assume that $\eta = \eta(r)$ only, in which case

$$-\mathbf{e}_r \cdot \nabla \times (\eta \nabla \times \mathbf{B}) = \eta \nabla_h^2 B_r + R_\eta. \tag{43}$$

This has the diffusion term from (1) plus an additional remainder term

$$R_\eta = -\frac{\eta}{R_\odot} \nabla_h \cdot \mathbf{B} - \frac{\eta}{R_\odot \sin\theta} \frac{\partial}{\partial\theta} \left(\sin\theta \frac{\partial B_\theta}{\partial r} \right) - \frac{\eta}{R_\odot \sin\theta} \frac{\partial}{\partial\phi} \left(\frac{\partial B_\phi}{\partial r} \right). \tag{44}$$

Using $\nabla \cdot \mathbf{B} = 0$, this may be rewritten entirely in terms of B_r, simplifying to

$$R_\eta = \frac{\eta}{R_\odot^2} \frac{\partial^2}{\partial r^2} (r^2 B_r). \tag{45}$$

Thus in mean-field MHD there is an additional term representing the radial diffusion of magnetic flux that is missing from the original SFT equation (1). Physically, this incorporates the fact that the surface magnetic field is connected to the interior; for example, the decay of active regions can be slowed if they remain connected to deeper layers of the convection zone where the diffusivity is lower (Wilson et al. 1990; Whitbread et al. 2019).

One way to justify the classical SFT model is to assume that $B_\theta, B_\phi \approx 0$ in the near-surface region of the solar convection zone. It then follows from (44) that $R_\eta = 0$. To some extent this is justified by vector magnetogram observations at the photosphere (for a recent discussion, see Virtanen et al. 2019; for a theoretical argument, see van Ballegooijen and Mackay 2007). If this radial-field approximation is not made, then self-consistent computation of the radial diffusion term R_η would require simulation of the three-dimensional magnetic field in the solar convection zone. However, two approaches have been used to parametrize (45) in SFT models without the need for three-dimensional simulations, and these will be considered next.

5.1 Exponential Decay Term

The most common parametrization for the radial diffusion term (45) is to assume that $R_\eta \approx -B_r/\tau$, so that (1) becomes

$$\frac{\partial B_r}{\partial t} + \nabla_h \cdot (\mathbf{u}_h B_r) = \eta \nabla_h^2 B_r - \frac{B_r}{\tau} + S. \tag{46}$$

Multiplying by $e^{t/\tau}$ shows that

$$\frac{\partial}{\partial t} (e^{t/\tau} B_r) + \nabla_h \cdot (\mathbf{u}_h e^{t/\tau} B_r) = \eta \nabla_h^2 (e^{t/\tau} B_r) + e^{t/\tau} S. \tag{47}$$

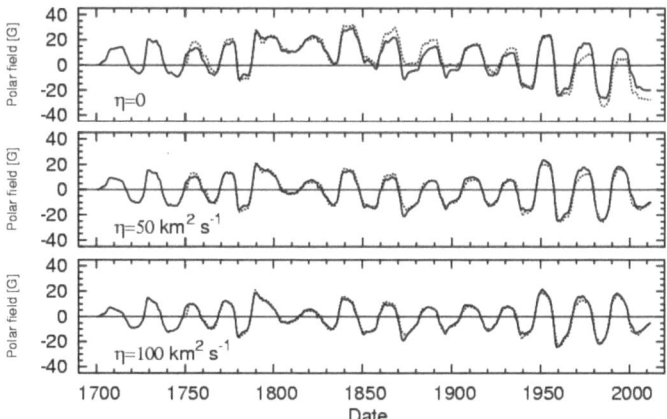

Fig. 10 Application of the radial diffusion term to reduce spurious cycle-to-cycle memory in the SFT model, from Baumann et al. (2006). The top row shows the north polar field (above 75° latitude) in a simulation with no radial diffusion, while the middle and bottom rows show the same simulation with $\eta_0 = 50\,\mathrm{km^2 s^{-1}}$ and $100\,\mathrm{km^2 s^{-1}}$ according to the prescription in Sect. 5.2. The simulation uses random emerging BMRs proportional to the observed sunspot numbers. The dashed line shows a simulation started in 1750, illustrating how the memory of the initial conditions persists. (© ESO. Reproduced with permission.)

Thus if B_r^∞ denotes the solution to the original equation (1), corresponding to $\tau \to \infty$, then the solution with finite τ but all other parameters the same is $B_r = e^{-t/\tau} B_r^\infty$. In other words, the solution decays exponentially at uniform rate τ^{-1}. For example, the dipole amplification factor (35) for a BMR would become

$$f_\infty \approx \frac{\sqrt{8\pi \,\mathrm{Rm_0}}}{3} \exp\left(-\frac{\mathrm{Rm_0}\lambda_0^2}{2}\right) \exp\left(-\frac{t}{\tau}\right), \qquad (48)$$

reflecting continuing decay of the magnetic field due to the new term.

The first application of such a decay term was by Schrijver et al. (2002), who motivated it not by consideration of radial diffusion but purely as a necessary addition to reduce the "memory" of the polar field (equivalently $b_{1,0}$) over multiple solar cycles. Without it, the varying amount of polar field production caused by the differing sunspot numbers in different cycles led to an unrealistic drift in the polar field over time, rather than the regular reversals that are observed. This drift is illustrated (for another SFT model) in the top panel of Fig. 10.

The optimization studies discussed in Sect. 3 have also looked for the optimum τ in shorter simulations where long-term memory is not an issue. With their simplified source term, Petrovay and Talafha (2019) found that a decay term (with τ in the range 5–10 yr) was essential, otherwise $b_{1,0}$ reversed too late for all of the flow profiles and parameters tried. And in simulations of Cycle 23 driven by idealised BMRs, Whitbread et al. (2017) found that a decay term with $\tau < 5$ yr helped to reduce unrealistically high values of $b_{1,0}$. However, they found that emerging active regions with observed shapes reduced $b_{1,0}$ in itself (as did Yeates 2020, for Cycle 24), and the optimization did not strongly select for a particular τ. Moreover, the fit of the optimum model did not improve significantly when the decay term was included in the model compared to when it was not. Lemerle et al. (2015) also found that τ was not strongly constrained by the optimization process, with acceptable solutions found for suitable parameter combinations with τ in the range from 7–32 yr. In summary, the presence of a decay term as required by Schrijver et al. (2002) does not seem to be ruled out by observations.

It should be noted that, in principle, an additional decay term is not the only way to reduce the cycle-to-cycle memory of $b_{1,0}$ in the model. Alternatives that have been adopted include imposed cycle-to-cycle variations in either the meridional flow speed (Wang et al. 2002) or the tilt angles of emerging BMRs (Cameron et al. 2010). It is difficult to choose definitively between these options with only about four solar cycles of full magnetogram observations.

5.2 Diffusive Interior Model

An improved parametrization for (45) was suggested by Baumann et al. (2006). They observed that if one assumes a purely diffusive evolution with uniform diffusivity $\eta = \eta_0$ throughout the convection zone, then the term R_η may be approximated using only B_r on the solar surface.

Specifically, Baumann et al. (2006) consider a purely poloidal field $\mathbf{B} = \nabla \times \nabla \times (\mathbf{r}P)$ inside the convection zone $R_b < r < R_\odot$, with boundary conditions $B_r(R_b, \theta, \phi) = 0$ and $B_\theta(R_\odot, \theta, \phi) = B_\phi(R_\odot, \theta, \phi) = 0$. Under a purely diffusive decay

$$\frac{\partial \mathbf{B}}{\partial t} = -\eta_0 \nabla \times (\nabla \times \mathbf{B}) \tag{49}$$

with η_0 constant, and a suitable gauge choice for P, this reduces to the scalar problem

$$\frac{\partial P}{\partial t} = \eta_0 \nabla^2 P, \quad \frac{\partial}{\partial r}(rP)\bigg|_{r=R_\odot} = P\bigg|_{r=R_b} = 0. \tag{50}$$

The solution, omitting the monopole term, may be written as an expansion

$$P(r, \theta, \phi, t) = \sum_{n=0}^{\infty} \sum_{l=1}^{\infty} \sum_{m=-l}^{l} \left[a_{l,n} j_l(k_{l,n}r) + c_{l,n} y_l(k_{l,n}r) \right] Y_l^m(\theta, \phi) e^{-\eta_0 k_{l,n}^2 t}, \tag{51}$$

where Y_l^m are spherical harmonics and j_l, y_l are spherical Bessel functions of the first and second kinds. Linearity of (50) allows Baumann et al. (2006) to set $a_{l,n} = 1$ without loss of generality, so the inner boundary condition fixes the other coefficient

$$c_{l,n} = -\frac{j_l(k_{l,n}R_b)}{y_l(k_{l,n}R_b)}. \tag{52}$$

The upper boundary condition then gives

$$l\left[j_l(k_{l,n}R_\odot) y_l(k_{l,n}R_b) - y_l(k_{l,n}R_\odot) j_l(k_{l,n}R_b) \right] =$$
$$k_{l,n}R_\odot \left[j_{l-1}(k_{l,n}R_\odot) y_l(k_{l,n}R_b) - y_{l-1}(k_{l,n}R_\odot) j_l(k_{l,n}R_b) \right]. \tag{53}$$

This equation must be solved numerically for each l and n to determine the eigenvalues $k_{l,n}$, which give the decay times $\tau_{l,n} = (\eta_0 k_{l,n}^2)^{-1}$ for each component, where l is the spherical harmonic degree and n is the radial mode number. Since the SFT model does not give the subsurface radial structure, Baumann et al. (2006) propose to keep only the modes with

$n = 0$, which are the slowest decaying modes for each l. They modify the SFT equation (1) to

$$\frac{\partial B_r}{\partial t} + \nabla_h \cdot \left(\mathbf{u}_h B_r\right) = \eta \nabla_h^2 B_r - \sum_{l=1}^{\infty} \sum_{m=-l}^{l} \frac{b_{l,m}(t)}{\tau_{l,0}} Y_l^m(\theta, \phi) + S, \tag{54}$$

where $b_{l,m}(t)$ are the spherical harmonic coefficients in the expansion of B_r,

$$B_r(\theta, \phi, t) = \sum_{l=1}^{\infty} \sum_{m=-l}^{l} b_{l,m}(t) Y_l^m(\theta, \phi). \tag{55}$$

The interior diffusivity η_0 that determines $\tau_{l,0}$ is taken to be different from the coefficient η of the classical diffusion term.

Note that, since radial modes with $n > 0$ are neglected, the effect on $b_{1,0}$ is identical to the simple exponential decay term, with $\tau = \tau_{1,0} = (\eta_0 k_{1,0}^2)^{-1}$. Accordingly, Baumann et al. (2006) showed that their alternative form of the decay term can also reduce the spurious long-term memory of the SFT model, as illustrated in the middle and bottom rows of Fig. 10. They found that diffusivity values in the range $\eta_0 = 50 - 100 \, \mathrm{km^2 s^{-1}}$ gave polar field evolutions consistent with recent observations. For $R_b = 0.7 R_\odot$, and since $k_{1,0} \approx 5.46$, this corresponds to decay times for $b_{1,0}$ in the range $\tau_{1,0} \approx 5 - 10 \, \mathrm{yr}$. In their model driven by idealized BMRs, Whitbread et al. (2017) found an optimum $\eta_0 = 190 \, \mathrm{km^2 s^{-1}}$, giving a decay time $\tau_{1,0} = 2.7 \, \mathrm{yr}$, in agreement with the τ found by optimizing the simple exponential decay term. Virtanen et al. (2017) also adopted the Baumann et al. (2006) model, but in a simulation where active regions had observed shapes; they found a value $\eta_0 = 100 \, \mathrm{km^2 s^{-1}}$ to give reasonable results.

It is worth remarking that these implementations of (54) have used different diffusivities for η (the classical horizontal diffusion) and η_0 (which determines $\tau_{l,0}$). Moreover, the extra term in (54) includes both radial and horizontal diffusion due to the interior diffusivity η_0. If one evaluates the radial diffusion term (45) for a single mode of the interior solution (51), one obtains

$$\frac{\eta_0}{R_\odot^2} \frac{\partial^2}{\partial r^2} \left(r^2 B_r\right) = -\eta_0 \left(k_{l,n}^2 - \frac{l(l+1)}{R_\odot^2}\right) B_r, \tag{56}$$

giving a decay time $\tau'_{l,n} = \eta_0^{-1} [k_{l,n}^2 - l(l+1)/R_\odot^2]^{-1}$ for radial diffusion alone. However, for small l the difference from $\tau_{l,n}$ is negligible.

5.3 Other Turbulent Transport Effects

If we drop the simple assumption of a turbulent diffusion in the mean-field induction equation (41), then there are a wealth of possible transport effects that could be explored in SFT models. One such effect – expected to be present from numerical convection simulations – is turbulent pumping (Petrovay 1994), which adds $-\boldsymbol{\gamma} \times \mathbf{B}$ to the turbulent electromotive force (mathematically equivalent to $\mathbf{u}$). Downward pumping ($\gamma_r < 0$) in a region near the surface could reduce the aforementioned diffusive link of active regions to deeper layers (Cameron et al. 2012; Karak and Cameron 2016). This is because it will tend to make the magnetic field lines radial, and – as noted earlier – if $B_\theta, B_\phi \approx 0$ in some region near the surface then it follows from (44) that $R_\eta = 0$, so no additional radial diffusion term should be included in the SFT model. Latitudinal pumping ($\gamma_\theta \neq 0$) is also found to be very strong in convection simulations. However, this relies on a significant influence of rotation on the turbulence, which is weaker nearer the surface than in deeper layers.

6 Beyond the Classical Model

Several have sought to improve on the classical SFT model described in the previous sections. We therefore conclude this review by outlining some of these developments.

6.1 Improved Small-Scale Flows

The approximation of small-scale flows by a uniform supergranular diffusivity, D, is perhaps the greatest simplification in the classical model. Three main approaches for improving the fidelity of the small-scale flow model have been applied.

Computationally cheapest is the method of Worden and Harvey (2000), whose primary aim was to improve the unobserved or poorly observed regions of synoptic maps. For this application, the classical diffusion model is not ideal because it does not reproduce the "clumping" of magnetic flux on supergranular network boundaries that is clearly evident in observed portions of the map. To better reproduce this, Worden and Harvey (2000) replaced the diffusion with a "random attractor" term added to each pixel in the map (without increasing the resolution compared to the classical SFT model). This is shown in Fig. 11(a). They also added a random emergence term to each pixel to sustain the small-scale background field. This background field was found not to affect the diffusion of large-scale flux patterns, but it gives a more accurate net flux in quiet regions (Fig. 11b). The technique was successful in improving the appearance of simulated maps, and continues to be used in the Air Force Data-Assimilative Photospheric flux Transport model (ADAPT; Arge et al. 2010; Hickmann et al. 2015).

A second approach is to dispense completely with parametrization of the small-scale flows, and model them directly through the advection term. This requires higher spatial and

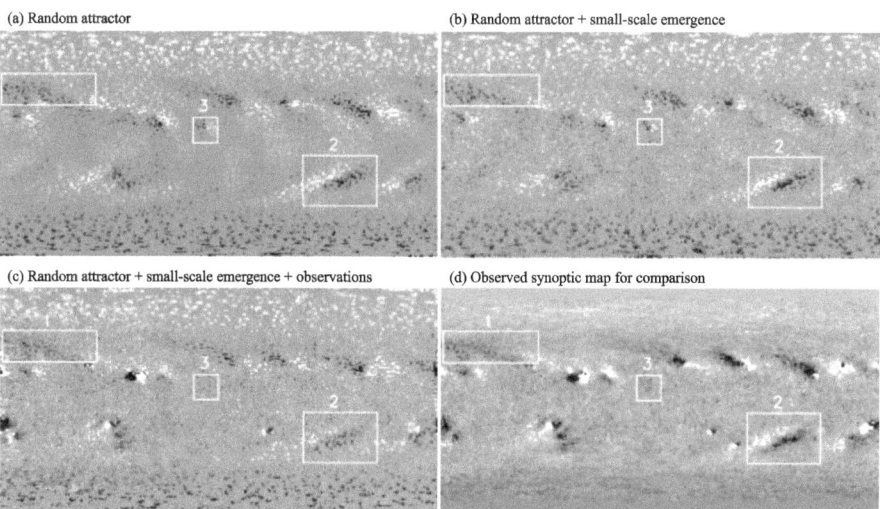

Fig. 11 Illustration of the "random attractor" model for flux dispersal, taken from several figures of Worden and Harvey (2000). Panels (a)-(c) show simulated maps after evolving for 27 days, all starting from a synoptic map for CR1928 but with successively more model components included. (Differential rotation and meridional flow were included in all three cases.) Panel (d) shows the "ground truth": an observed synoptic map for CR1929. In all cases B_r is shown in greyscale (white positive, black negative). (© Springer. Reproduced with permission.)

temporal resolution so as to resolve individual convective cells on the computational grid. Nevertheless, it has been applied successfully in the Advective Flux Transport (AFT) model (Upton and Hathaway 2014b, 2018). In this model, the small-scale flows are randomly imposed, based on a vector spherical harmonic decomposition of the form

$$
u_\theta(\theta, \phi) = \sum_{l=1}^{l_{max}} \sum_{m=0}^{l} \left(S_l^m \frac{\partial Y_l^m(\theta, \phi)}{\partial \theta} + T_l^m \frac{1}{\sin\theta} \frac{\partial Y_l^m(\theta, \phi)}{\partial \phi} \right),
\tag{57}
$$

$$
u_\phi(\theta, \phi) = \sum_{l=1}^{l_{max}} \sum_{m=0}^{l} \left(S_l^m \frac{1}{\sin\theta} \frac{\partial Y_l^m(\theta, \phi)}{\partial \phi} - T_l^m \frac{\partial Y_l^m(\theta, \phi)}{\partial \theta} \right),
\tag{58}
$$

where the complex amplitudes S_l^m and T_l^m determine the curl-free and divergence-free components of $\mathbf{u}_h$ and are chosen to match the spectrum to observations. Hathaway et al. (2000) found that observed Doppler flows could be well matched by a two-component spectrum, comprising a supergranular component centred on $l = 110$ and a granular component centered on $l = 4000$.

The third approach is to dispense with a computational grid altogether and model the magnetic flux by a discrete ensemble of individual flux "concentrations". This was implemented by Schrijver (2001) whose main aim was to simulate cool stars other than the Sun, and who therefore wanted to include the mixed-polarity network of small-scale magnetic flux because of its contribution to chromospheric emission. The discrete model of Schrijver (2001) includes (i) emergence of both active regions and ephemeral regions as BMRs, (ii) a large-scale random walk dispersal as well as differential rotation and meridional flow, (iii) a model for fragmentation and coalescence of flux concentrations, and (iv) cancellation of flux between opposite polarity fragments. The model has been successfully applied over all latitudes (Schrijver and Title 2001) and over a full 11-year cycle (Schrijver and Liu 2008). A similar model in Cartesian geometry was applied by Martin-Belda and Cameron (2016) to study the dispersion of a single active region.

One notable new feature that all three of these models have in common is nonlinearity: the rate of magnetic flux dispersal is chosen to depend on the local magnetic field strength, $|B_r|$. In particular, dispersal is suppressed in strong-field regions, compared to the classical diffusion model. This better represents real active regions which suppress shedding of the magnetic flux by supergranulation (Schrijver 1989). The effect is particularly important for more active stars (Schrijver 2001) but is still clearly observed on the Sun.

6.2 Fluctuating Large-Scale Flows

The classical model neglects fluctuations in the meridional flow and differential rotation, keeping them steady for periods of a solar cycle or longer. However, observations do suggest variations over the course of the cycle, particularly in the meridional flow. For example, Hathaway and Rightmire (2010) estimated the flow from cross-correlating latitudinal strips in magnetograms over Solar Cycle 23, and found that the dominant Legendre component, $P_2^1 \sim \sin(2\theta)$, reduced in amplitude from $11.5 - 13 \, \mathrm{ms}^{-1}$ at cycle minimum to only $8.5 \, \mathrm{ms}^{-1}$ at cycle maximum.

A plausible cause of meridional flow variations is the observed inflow toward active regions determined by helioseismology (Gizon et al. 2001). In SFT simulations, Jiang et al. (2010) showed that an axisymmetric meridional inflow toward the activity belts leads to a significant decrease of the polar field, suggesting that such meridional flow variations could be a significant ingredient in the SFT model. And Cameron et al. (2010) pointed out that the variations in P_2^1 found by Hathaway and Rightmire (2010) could be explained by this inflow, without the need for an overall modulation of meridional flow speed.

Other studies have accounted for the observed dependence of inflow speed on the active region magnetic flux, through applying a nonlinear velocity that depends on $|B_r|$. Whilst more detailed models for magnetic back-reaction on flows and transport coefficients have been introduced in dynamo models (Rempel 2006), SFT studies have so far been limited to simple parametrizations. De Rosa and Schrijver (2006) added a velocity of the form

$$\delta \mathbf{u}(\theta, \phi, t) = \alpha \nabla |\overline{B_r}|^\beta \tag{59}$$

to the discrete SFT model – where $\overline{B_r}$ denotes a Gaussian smoothing of the original B_r with width $15°$ – but found that the observed flow speeds ($50\,\mathrm{ms}^{-1}$) prevented altogether the dispersal of active regions. However, Martin-Belda and Cameron (2016) did not find this problem and proposed that the original calculations of De Rosa and Schrijver (2006) were underestimating the flux dispersal because they continued to apply the nonlinear damping of dispersal within the active region, while the inflows alone could themselves account for the damping effect. Cameron and Schüssler (2012) proposed an axisymmetric parametrization

$$\delta u_\theta(\theta, t) = c_0 \int_0^\pi \frac{\sin(\theta')}{\sin(30°)} \frac{\mathrm{d}\langle |B_r| \rangle}{\mathrm{d}\theta'} e^{-(\theta-\theta')^2/\sigma} \, \mathrm{d}\theta', \tag{60}$$

which corresponds to a Gaussian smoothing of the derivative in latitude (with σ chosen to give width $20°$). The $\sin(\theta')$ factor suppresses unrealistically strong fluctuations at high latitudes, and an amplitude $c_0 = 9.2\,\mathrm{m\,s}^{-1}\mathrm{G}^{-1}$ gives comparable inflow speeds to Gizon et al. (2001). Again, the presence of inflows reduces the axial dipole at the end of the solar cycle, by about 30% in a moderate cycle (Martin-Belda and Cameron 2017), with about a 9% variation between cycles suggesting that this nonlinearity could conceivably help to saturate the Babcock-Leighton dynamo. Nagy et al. (2020) coupled an SFT model with flux-dependent inflows to such a dynamo model. They confirmed that inflows do indeed tend to have a stabilizing effect on cycle amplitudes, although they also greatly increase the probability of the dynamo entering a grand minimum of reduced activity – a nonlinear effect which is not apparent from SFT alone. On the other hand, Yeates (2014) found that the inflows in a BMR-driven SFT model for Cycle 23 gave poorer matches to the observed butterfly diagram and dipole reversal time.

A more pragmatic approach is to impose the observed flow variations directly, as in the AFT model (Upton and Hathaway 2014b), where the best-fit Legendre coefficients are extracted from 27-day averaged velocity fields derived from magnetogram cross-correlation. These then determine $\mathbf{u}_h(\theta, t)$ in the model, allowing variations in both meridional flow and differential rotation. Using data from Solar Cycle 23, Upton and Hathaway (2014a) found that the fluctuating meridional flow in the AFT model actually increased the axial dipole strength by 20% compared to a simulation where the meridional flow was fixed to a steady latitudinal profile. Thus it is possible that meridional flow variations can increase the axial dipole as well as reduce it.

6.3 Observational Data Assimilation

In applications where the aim is to recreate as accurately as possible the real Sun at an observed time, it makes sense to construct magnetic maps that combine SFT model results with real observations. The role of the SFT model is then to fill in unobserved (or poorly observed) parts of the solar surface, such as high latitudes or the far side of the Sun. This approach is central to the model of Worden and Harvey (2000), as illustrated in Fig. 11(c) which shows the result of combining daily magnetogram observations with the simulation. The observations are weighted more highly near disk-centre and also eastward of Central Meridian (where the time since previous observation is greatest). Similar assimilation of observed magnetograms has been applied in the discrete SFT model (Schrijver and DeRosa 2003) and in the AFT model of Upton and Hathaway (2014b).

A more sophisticated approach to data assimilation has been implemented in the ADAPT model, which includes several different sequential data-assimilation methods such as ensemble Kalman filtering (Hickmann et al. 2015). The concept is to perform an ensemble of model runs. Each is adjusted at intervals using the observed magnetogram data, with observations being given greater weight in areas where the model runs disagree with one another.

Unfortunately, difficulties arise in driving time-dependent coronal magnetic field simulations from SFT models with data assimilation. In such simulations, the required photospheric boundary condition is the tangential electric field $\mathbf{E}_h$, not simply B_r. In the classical SFT model, the natural electric field would be

$$\mathbf{E}_h = -\mathbf{u} \times \mathbf{B} + \eta \nabla \times \mathbf{B} + \mathbf{E}_S, \tag{61}$$

where $\mathbf{E}_S$ accounts for the source term (i.e., $-\mathbf{e}_r \cdot \nabla \times \mathbf{E}_S = S$). When S comprises individual active regions that have no net magnetic flux, a well-behaved electric field can be determined (e.g., Yeates and Bhowmik 2022). But if the magnetic flux is unbalanced over a larger region then it is impossible to find a localized $\mathbf{E}_S$ as would be expected from Ohm's Law (Yeates 2017). This can be a problem when observed magnetograms are incorporated directly, particularly when active regions straddle the edge of the assimilation region so that only one polarity is included. If the flux imbalance is corrected by spreading it over the full Sun, the resulting spurious electric fields lead to generation of significant spurious electric currents in time-dependent coronal simulations (Weinzierl et al. 2016). Of course, this problem is not restricted to data assimilation, but could arise from the use of any unbalanced source term.

In practice the simplest way to ensure flux balance is to rephrase the right-hand side of equation (5) as $-\mathbf{e}_r \cdot \nabla \times \mathbf{E}_h$, then apply a "constrained transport" discretization with a staggered mesh (Yee 1966). Here E_θ and E_ϕ are defined at cell edges, and B_r at cell centres. Such a numerical scheme is used, for example, by Yeates (2014). When assimilating magnetograms into the SFT model in this framework, one would estimate $\mathbf{E}_h$ from the observed front-side evolution. In the case of a flux imbalance, this would automatically create a balancing polarity just outside the observed region, minimizing disruption to the global topology of the coronal magnetic field. However, it remains the case that systematic errors in observed magnetograms, especially centre-to-limb variations of the errors, are not well understood. A better understanding of these errors will require forward modelling with radiative MHD and Stokes polarimetric inversions.

A final remark is that the simplified decay term B_r/τ from Sect. 5.1 may also be written as the curl of an electric field. In particular, we would need $\mathbf{e}_r \cdot \nabla \times \mathbf{E} = B_r/\tau$. For example, writing $\mathbf{E} = -\nabla \times (\Psi \mathbf{e}_r)$, we could determine Ψ and hence $\mathbf{E}$ by solving the Poisson equation

$$\nabla_h^2 \Psi = \frac{B_r}{\tau}, \tag{62}$$

which has a unique solution on the sphere since $\int_S B_r \, dS = 0$. Of course, this does not mean that this approximation is a good representation of the real radial diffusion term (45); for example, this particular $\mathbf{E}$ will not be localized to the active region itself.

Acknowledgements We thank the International Space Science Institute for supporting the workshop where this review originated. The collaboration of the authors was also facilitated by support from the International Space Science Institute through ISSI Team 474. The SDO data used in Figs. 1, 3 and 8 are courtesy of NASA and the SDO/HMI science team.

We thank the two anonymous reviewers for improving the article.

Funding ARY was supported by STFC (UK) consortium grant ST/W00108X/1. JJ was supported by the National Natural Science Foundation of China (grant Nos. 12173005 and 11873023). KP acknowledges support by the European Union's Horizon 2020 research and innovation programme under grant agreement No. 955620.

Declarations

Competing Interests The authors have no competing interests to declare that are relevant to the content of this article.

References

Arge CN, Henney CJ, Koller J et al (2010) Air force data assimilative photospheric flux transport (ADAPT) model. In: Maksimovic M, Issautier K, Meyer-Vernet N et al (eds) Twelfth international solar wind conference, pp 343–346. https://doi.org/10.1063/1.3395870

Baumann I (2005) Magnetic flux transport on the Sun. PhD thesis, Göttingen. https://www.sidc.be/users/evarob/Literature/PhDs/Baumann_magnetic%20flux%20transport%20on%20the%20sun.pdf

Baumann I, Schmitt D, Schüssler M et al (2004) Evolution of the large-scale magnetic field on the solar surface: a parameter study. Astron Astrophys 426:1075–1091

Baumann I, Schmitt D, Schüssler M (2006) A necessary extension of the surface flux transport model. Astron Astrophys 446:307–314

Beck JG (2000) A comparison of differential rotation measurements - (invited review). Sol Phys 191:47–70

Bhowmik P, Nandy D (2018) Prediction of the strength and timing of sunspot cycle 25 reveal decadal-scale space environmental conditions. Nat Commun 9:5209

Cameron RH, Schüssler M (2007) Solar cycle prediction using precursors and flux transport models. Astrophys J 659:801–811

Cameron RH, Schüssler M (2012) Are the strengths of solar cycles determined by converging flows towards the activity belts? Astron Astrophys 548:A57

Cameron RH, Schüssler M (2015) The crucial role of surface magnetic fields for the solar dynamo. Science 347:1333–1335

Cameron RH, Jiang J, Schmitt D et al (2010) Surface flux transport modeling for solar cycles 15–21: effects of cycle-dependent tilt angles of sunspot groups. Astrophys J 719:264–270

Cameron RH, Schmitt D, Jiang J et al (2012) Surface flux evolution constraints for flux transport dynamos. Astron Astrophys 542:A127

Cameron RH, Jiang J, Schüssler M (2016) Solar cycle 25: another moderate cycle? Astrophys J Lett 823:L22

Charbonneau P, Knapp B (1995) Genetic algorithms in astronomy and astrophysics. Astrophys J Suppl Ser 101:309

De Rosa ML, Schrijver CJ (2006) Consequences of large-scale flows around active regions on the dispersal of magnetic field across the solar surface. In: Fletcher K, Thompson M (eds) Proceedings of SOHO 18/GONG 2006/HELAS I. Beyond the spherical Sun, p 12

DeVore CR (1987) The decay of the large-scale solar magnetic field. Sol Phys 112:17–35

DeVore CR, Sheeley NR Jr, Boris JP (1984) The concentration of the large-scale solar magnetic field by a meridional surface flow. Sol Phys 92:1–14

Dikpati M, Gilman PA, Ulrich RK (2010) Physical origin of differences among various measures of solar meridional circulation. Astrophys J 722:774–778

Gizon L, Duvall JTL, Larsen RM (2001) Probing surface flows and magnetic activity with time-distance helioseismology. In: Brekke P, Fleck B, Gurman JB (eds) Recent insights into the physics of the Sun and heliosphere: highlights from SOHO and other space missions, IAU Symposium, vol 203. p 189

Hanasoge SM (2022) Surface and interior meridional circulation in the Sun. Living Rev Sol Phys 19:3

Hathaway DH, Rightmire L (2010) Variations in the Sun's meridional flow over a solar cycle. Science 327:1350–1352

Hathaway DH, Wilson RM, Reichmann EJ (1994) The shape of the sunspot cycle. Sol Phys 151:177–190

Hathaway DH, Beck JG, Bogart RS et al (2000) The photospheric convection spectrum. Sol Phys 193:299

Hazra G (2021) Recent advances in the 3D kinematic Babcock-Leighton solar dynamo modeling. J Astrophys Astron 42:22

Hickmann KS, Godinez HC, Henney CJ et al (2015) Data assimilation in the adapt photospheric flux transport model. Sol Phys 290:1105–1118

Hung CP, Brun AS, Fournier A et al (2017) Variational estimation of the large-scale time-dependent meridional circulation in the Sun: proofs of concept with a solar mean field dynamo model. Astrophys J 849:160

Iijima H, Hotta H, Imada S et al (2017) Improvement of solar-cycle prediction: plateau of solar axial dipole moment. Astron Astrophys 607:L2

Iijima H, Hotta H, Imada S (2019) Effect of morphological asymmetry between leading and following sunspots on the prediction of solar cycle activity. Astrophys J 883:24

Jiang J (2020) Nonlinear mechanisms that regulate the solar cycle amplitude. Astrophys J 900:19

Jiang J, Işik E, Cameron RH et al (2010) The effect of activity-related meridional flow modulation on the strength of the solar polar magnetic field. Astrophys J 717:597–602

Jiang J, Cameron RH et al (2011) The solar magnetic field since 1700. I. Characteristics of sunspot group emergence and reconstruction of the butterfly diagram. Astron Astrophys 528:A82

Jiang J, Cameron RH, Schüssler M (2014a) Effects of the scatter in sunspot group tilt angles on the large-scale magnetic field at the solar surface. Astrophys J 791:5

Jiang J, Hathaway DH, Cameron RH et al (2014b) Magnetic flux transport at the solar surface. Space Sci Rev 186:491–523

Jiang J, Cameron RH, Schüssler M (2015) The cause of the weak solar cycle 24. Astrophys J Lett 808:L28

Jiang J, Wang JX, Jiao QR et al (2018) Predictability of the solar cycle over one cycle. Astrophys J 863:159

Jiang J, Song Q, Wang JX et al (2019) Different contributions to space weather and space climate from different big solar active regions 871: 16

Jiang J, Zhang Z, Petrovay K (2023) Comparison of physics-based prediction models of solar cycle 25. J Atmos Sol-Terr Phys 243:106,018

Karak BB, Cameron R (2016) Babcock–Leighton solar dynamo: the role of downward pumping and the equatorward propagation of activity. Astrophys J 832:94

Leighton RB (1964) Transport of magnetic fields on the Sun. Astrophys J 140:1547–1562

Lemerle A, Charbonneau P (2017) A coupled 2 × 2D Babcock–Leighton solar dynamo model. II. Reference dynamo solutions. Astrophys J 834:133

Lemerle A, Charbonneau P, Carignan-Dugas A (2015) A coupled 2 × 2D Babcock–Leighton solar dynamo model. I. Surface magnetic flux evolution. Astrophys J 810:78

Mackay DH, Lockwood M (2002) The evolution of the Sun's open magnetic flux. II. Full solar cycle simulations. Sol Phys 209:287–309

Mackay DH, Yeates AR (2012) The Sun's global photospheric and coronal magnetic fields: observations and models. Living Rev Sol Phys 9:6

Mackay DH, Priest ER, Lockwood M (2002) The evolution of the Sun's open magnetic flux. I. A single bipole. Sol Phys 207:291–308

Martin-Belda D, Cameron RH (2016) Surface flux transport simulations: effect of inflows toward active regions and random velocities on the evolution of the Sun's large-scale magnetic field. Astron Astrophys 586:A73

Martin-Belda D, Cameron RH (2017) Inflows towards active regions and the modulation of the solar cycle: a parameter study. Astron Astrophys 597:A21

McCloughan J, Durrant CJ (2002) A method of evolving synoptic maps of the solar magnetic field. Sol Phys 211:53–76

Nagy M, Lemerle A, Labonville F et al (2017) The effect of "rogue" active regions on the solar cycle. Sol Phys 292:167

Nagy M, Lemerle A, Charbonneau P (2020) Impact of nonlinear surface inflows into activity belts on the solar dynamo. J Space Weather Space Clim 10:62

Petrovay K (1994) Theory of passive magnetic field transport. In: Rutten RJ, Schrijver CJ (eds) Solar surface magnetism. NATO ASI Series C, vol 433. Springer, Dordrecht p 415

Petrovay K, Talafha M (2019) Optimization of surface flux transport models for the solar polar magnetic field. Astron Astrophys 632:A87

Petrovay K, Nagy M, Yeates AR (2020) Towards an algebraic method of solar cycle prediction. I. Calculating the ultimate dipole contributions of individual active regions. J Space Weather Space Clim 10:50

Rempel M (2006) Flux-transport dynamos with Lorentz force feedback on differential rotation and meridional flow: saturation mechanism and torsional oscillations. Astrophys J 647:662–675

Schrijver CJ (1989) The effect of an interaction of magnetic flux and supergranulation on the decay of magnetic plages. Sol Phys 122:193–208

Schrijver CJ (2001) Simulations of the photospheric magnetic activity and outer atmospheric radiative losses of cool stars based on characteristics of the solar magnetic field. Astrophys J 547:475–490

Schrijver CJ, DeRosa ML (2003) Photospheric and heliospheric magnetic fields. Sol Phys 212:165–200

Schrijver CJ, Liu Y (2008) The global solar magnetic field through a full sunspot cycle: observations and model results. Sol Phys 252:19–31

Schrijver CJ, Title AM (2001) On the formation of polar spots in Sun-like stars. Astrophys J 551:1099–1106

Schrijver CJ, DeRosa ML, Title AM (2002) What is missing from our understanding of long-term solar and heliospheric activity? Astrophys J 577:1006–1012

Sheeley NR Jr (2005) Surface evolution of the Sun's magnetic field: a historical review of the flux-transport mechanism. Living Rev Sol Phys 2:5

Sheeley NR Jr, DeVore CR (1986) The decay of the mean solar magnetic field. Sol Phys 103:203–224

Snodgrass HB, Ulrich RK (1990) Rotation of Doppler features in the solar photosphere. Astrophys J 351:309

Sun X (2018) Polar field correction for HMI line-of-sight synoptic data. arXiv e-prints https://arxiv.org/abs/1801.04265

Talafha M, Nagy M, Lemerle A et al (2022) Role of observable nonlinearities in solar cycle modulation. Astron Astrophys 660:A92

Upton LA, Hathaway DH (2014a) Effects of meridional flow variations on solar cycles 23 and 24. Astrophys J 792:142

Upton LA, Hathaway DH (2014b) Predicting the Sun's polar magnetic fields with a surface flux transport model. Astrophys J 780:5

Upton LA, Hathaway DH (2018) An updated solar cycle 25 prediction with adt: the modern minimum. Geophys Res Lett 45:8091–8095

van Ballegooijen AA, Mackay DH (2007) Model for the coupled evolution of subsurface and coronal magnetic fields in solar active regions. Astrophys J 659:1713–1725

Van Ballegooijen AA, Cartledge NP Priest ER (1998) Magnetic flux transport and the formation of filament channels on the Sun. Astrophys J 501:866–881

van Driel-Gesztelyi L, Green LM (2015) Evolution of active regions. Living Rev Sol Phys 12:1

Virtanen IOI, Virtanen II, Pevtsov AA et al (2017) Reconstructing solar magnetic fields from historical observations. II. Testing the surface flux transport model. Astron Astrophys 604:A8

Virtanen II, Pevtsov AA, Mursula K (2019) Structure and evolution of the photospheric magnetic field in 2010-2017: comparison of SOLIS/VSM vector field and B_{LOS} potential field. Astron Astrophys 624:A73

Wang YM (2017) Surface flux transport and the evolution of the Sun's polar fields. Space Sci Rev 210:351–365

Wang YM, Lean JL (2021) A new reconstruction of the Sun's magnetic field and total irradiance since 1700. Astrophys J 920:100

Wang YM, Nash AG, Sheeley NR Jr (1989) Magnetic flux transport on the Sun. Science 245:712–718

Wang YM, Sheeley NR Jr (1989) Average properties of bipolar magnetic regions during sunspot cycle 21. Sol Phys 124:81–100

Wang YM, Sheeley NR Jr (1991) Magnetic flux transport and the Sun's dipole moment: new twists to the Babcock–Leighton model. Astrophys J 375:761–770

Wang YM, Lean J, Sheeley NR Jr (2002) Role of a variable meridional flow in the secular evolution of the Sun's polar fields and open flux. Astrophys J 577:L53–L57

Wang ZF, Jiang J, Wang JX (2021) Algebraic quantification of an active region contribution to the solar cycle. Astron Astrophys 650:A87

Weinzierl M, Yeates AR, Mackay DH et al (2016) A new technique for the photospheric driving of non-potential solar coronal magnetic field simulations. Astrophys J 823:55

Whitbread T, Yeates AR, Muñoz Jaramillo A et al (2017) Parameter optimization for surface flux transport models. Astron Astrophys 607:A76

Whitbread T, Yeates AR, Muñoz Jaramillo A (2019) The need for active region disconnection in 3d kinematic dynamo simulations. Astron Astrophys 627:A168

Wilson PR, McIntosh P, Snodgrass HB (1990) The reversal of the solar polar magnetic fields. I. The surface transport of magnetic flux. Sol Phys 127:1–9

Worden J, Harvey J (2000) An evolving synoptic magnetic flux map and implications for the distribution of photospheric magnetic flux. Sol Phys 195:247–268

Yeates AR (2014) Coronal magnetic field evolution from 1996 to 2012: continuous non-potential simulations. Sol Phys 289:631–648

Yeates AR (2017) Sparse reconstruction of electric fields from radial magnetic data. Astrophys J 836:131

Yeates AR (2020) How good is the bipolar approximation of active regions for surface flux transport? Sol Phys 295:119

Yeates AR, Bhowmik P (2022) Automated driving for global nonpotential simulations of the solar corona. Astrophys J 935:13

Yee K (1966) Numerical solution of inital boundary value problems involving Maxwell's equations in isotropic media. IEEE Trans Antennas Propag 14:302–307

Space Science Reviews (2023) 219:58
https://doi.org/10.1007/s11214-023-01005-6

Simulations of Solar and Stellar Dynamos and Their Theoretical Interpretation

Petri J. Käpylä[1,2] · Matthew K. Browning[3] · Allan Sacha Brun[4] · Gustavo Guerrero[5,6] · Jörn Warnecke[7]

Received: 26 May 2023 / Accepted: 15 September 2023 / Published online: 11 October 2023
© The Author(s) 2023

Abstract

We review the state of the art of three dimensional numerical simulations of solar and stellar dynamos. We summarize fundamental constraints of numerical modelling and the techniques to alleviate these restrictions. Brief summary of the relevant observations that the simulations seek to capture is given. We survey the current progress of simulations of solar convection and the resulting large-scale dynamo. We continue to studies that model the Sun at different ages and to studies of stars of different masses and evolutionary stages. Both simulations and observations indicate that rotation, measured by the Rossby number which is the ratio of rotation period and convective turnover time, is a key ingredient in setting the overall level and characteristics of magnetic activity. Finally, efforts to understand global 3D simulations in terms of mean-field dynamo theory are discussed.

Keywords Dynamo · Magnetohydrodynamics · Simulation · Turbulence

1 Introduction

The intriguing coherence of the solar magnetic cycle has fascinated researchers for more than a century starting from Hale's discovery of magnetic field in sunspots (Hale 1908; Hale et al. 1919), and early attempts to build simple models (Larmor 1919; Cowling 1933). The first successful models of the solar cycle made use of mean-field approximations yielding equations where only the large-scale contributions were explicitly computed, whereas the small scales were characterised by physically plausible parameterizations (Parker 1955; Steenbeck and Krause 1969). Mean-field models opened up new avenues in studying solar and stellar magnetism but their Achilles' heel is the parameterizations of the small scales which are in general untractable analytically in parameter regimes relevant to stars. This is due to the closure problem of turbulence rendering such models susceptible to fine-tuning. A review of modern mean-field theory is presented elsewhere in this collection (Brandenburg et al. 2023).

Rapidly increasing computing power allowed for the first direct solutions of the equations of (magneto)hydrodynamics in spherical shells in the late 1970s and early 1980s (Gilman

Solar and Stellar Dynamos: A New Era
Edited by Manfred Schüssler, Robert H. Cameron, Paul Charbonneau, Mausumi Dikpati, Hideyuki Hotta and Leonid Kitchatinov

Extended author information available on the last page of the article

1977; Gilman and Miller 1981; Gilman 1983; Glatzmaier 1985). Prior to the these simulations and the discovery of the internal rotation profile of the Sun, the angular velocity was generally assumed to be constant in cylindrical surfaces and to decrease as a function of radius, in which case the propagation of the dynamo wave from a mean-field $\alpha\Omega$ dynamo is predicted to be equatorward given typical assumptions regarding the influence of the Coriolis force on convective eddies (Parker 1955; Steenbeck and Krause 1969; Yoshimura 1975); see, however, Roberts and Stix (1972). This changed definitively when helioseismology revealed that the angular velocity is actually increasing with radius in the bulk of the solar convection zone (e.g. Duvall et al. 1984; Schou et al. 1998) which lead to the "dynamo dilemma" (Parker 1987). This dilemma was also captured by the early 3D simulations where solar-like differential rotation with fast equator and slow poles was qualitatively reproduced, but where the dynamo waves propagated toward the poles, contrary to the Sun (Gilman 1983; Glatzmaier 1985).

After this, the interest in 3D simulations of solar and stellar dynamos waned and was not rekindled until the early 2000s, starting with the development of the ASH (Anelastic Spherical Harmonic) code (e.g. Miesch et al. 2000; Elliott et al. 2000; Brun and Toomre 2002). While many of the early studies concentrated on the Sun (e.g. Brun et al. 2004; Browning et al. 2006; Miesch et al. 2008), a proliferation of models from various groups using different codes occurred in the 2010s when simulations of more rapidly rotating Suns started to yield cycles and equatorward migration more or less routinely (e.g. Ghizaru et al. 2010; Käpylä et al. 2010; Brown et al. 2011; Käpylä et al. 2012; Nelson et al. 2013; Augustson et al. 2015; Mabuchi et al. 2015; Simitev et al. 2015). Furthermore, simulations of main-sequence stars other than the Sun also started to appear covering the mass range from fully convective M dwarfs (e.g. Dobler et al. 2006; Browning 2008; Yadav et al. 2015; Bice and Toomre 2020; Käpylä 2021) to F stars with thin surface convection zones (Augustson et al. 2013; Breton et al. 2022), as well as core convection, dynamos, and interaction with fossil fields in more massive A, B, and O stars (e.g. Featherstone et al. 2009; Augustson et al. 2016). Models exploring stellar magnetism outside of the main sequence have also started to appear, including pre-main sequence stars (e.g. Emeriau-Viard and Brun 2017), red giants (e.g. Dorch 2004; Brun and Palacios 2009), and newly born neutron stars (e.g. Raynaud et al. 2020; Masada et al. 2022).

Parallel to the developments in simulations, observational data and knowledge regarding stellar magnetism has also experienced explosive growth. We now have dozens of stars with observed cycles from long-term observing campaigns monitoring chromospheric emission (e.g. Baliunas et al. 1995). However, the systematics of these cycles as a function of stellar rotation are still under debate (e.g. Brandenburg et al. 2017; Boro Saikia et al. 2018b; Olspert et al. 2018; Bonanno and Corsaro 2022). Zeeman-Doppler imaging has also revealed polarity reversals (e.g. Kochukhov et al. 2013; Boro Saikia et al. 2018), as well as large-scale non-axisymmetric and dipole-dominated magnetic fields in rapidly rotating late-type stars (e.g. Kochukhov 2021). Finally, magnetic activity saturates when the stellar Rossby number $Ro = P_{rot}/\tau_{conv}$, which is the ratio of the rotation period and the convective turnover time, is less than about 0.1, such that for lower Ro the activity and magnetic field strength is roughly constant (e.g. Wright et al. 2018; Reiners et al. 2022). These basic observations are crucial constraints for the numerical simulations. Nevertheless, the Sun still poses the stringest constraints to simulations due its proximity and access to its interior structure through helioseismology. Somewhat surprisingly, the current 3D simulations struggle to reproduce not only the dynamo, but also the convective amplitudes and the differential rotation of the Sun, often yielding anti-solar (slow equator, fast poles) solutions with nominally solar luminosity and rotation rate (e.g. Matt et al. 2011; Käpylä et al. 2014; Gastine et al. 2014; Hotta et al.

2015; Brun et al. 2017). This issue has been dubbed the convective conundrum (O'Mara et al. 2016) and poses arguably the greatest challenge in the field of stellar dynamo simulations today. It has also been suggested that the Sun is close to a transition where its dynamo efficiency diminishes (e.g. van Saders et al. 2016), possibly due to a shift from solar-like to anti-solar differential rotation, making it difficult to capture by simulations (e.g. Käpylä et al. 2014; Brun et al. 2022). Our aim in the following is to review the current successes and shortcomings of current simulations in capturing the relevant observations.

The remainder of the review is organised as follows: the basic equations and physics are discussed in Sect. 2 and the limitations of the numerical approach are reviewed in Sect. 3. The relevant observations and the main results of current 3D simulations of various types of stars are reviewed in Sect. 4 and Sect. 5, respectively. Section 6 gives an overview of the comparisons between mean-field models and global simulations. Finally, we conclude in Sect. 7 with an overview of the state of the field, current challenges, and possible future directions.

2 Relevant Physics and Equations

Stellar convection zones are described by the equations of magnetohydrodynamics (MHD), describing the time evolution of the magnetic field and conservation of mass, momentum, and energy:

$$\frac{\partial \boldsymbol{B}}{\partial t} = \nabla \times (\boldsymbol{u} \times \boldsymbol{B} - \eta \mu_0 \boldsymbol{J}), \tag{1}$$

$$\frac{\partial \rho}{\partial t} = -\nabla \cdot (\rho \boldsymbol{u}), \tag{2}$$

$$\rho \frac{\partial \boldsymbol{u}}{\partial t} = -\nabla \cdot (\rho \boldsymbol{u} \boldsymbol{u}) + \rho \boldsymbol{g} - \nabla p - 2\rho \boldsymbol{\Omega}_0 \times \boldsymbol{U} + \boldsymbol{J} \times \boldsymbol{B} + \nabla \cdot \boldsymbol{F}^{\mathrm{visc}}, \tag{3}$$

$$\rho T \frac{\partial s}{\partial t} = -\nabla \cdot (\rho s \boldsymbol{u}) + \nabla \cdot \boldsymbol{\mathcal{F}} + \mathcal{H} + 2\nu\rho \mathbf{S}^2 + \eta\mu_0 \boldsymbol{J}^2, \tag{4}$$

where $\boldsymbol{B}$ is the magnetic field, $\boldsymbol{u}$ is the velocity, η is the magnetic diffusivity, μ_0 is the permeability of vacuum, $\boldsymbol{J} = \mu_0^{-1}\nabla \times \boldsymbol{B}$ is the current density, ρ is the fluid density, $\boldsymbol{g} = -\nabla\phi$ is the acceleration due to gravity, where ϕ is the gravitational potential, p is the gas pressure, $\boldsymbol{\Omega}_0$ is the rotation rate of the star, $\boldsymbol{F}^{\mathrm{visc}}$ is the viscous force, s is the specific entropy, and $\boldsymbol{\mathcal{F}} = \boldsymbol{\mathcal{F}}^{\mathrm{rad}} + \boldsymbol{\mathcal{F}}^{\mathrm{SGS}}$ describes radiative and any subgrid-scale (SGS) fluxes that are present. $\mathcal{H}$ describes additional cooling and heating that is sometimes used instead of, or in addition to, the radiative flux (e.g. Ghizaru et al. 2010; Guerrero et al. 2019; Matilsky et al. 2019), or to take into account heating due to nuclear reactions in the core of the star (e.g. Dobler et al. 2006; Käpylä 2021; Brun et al. 2022). Most often the gas is assumed to be fully ionised and to obey the ideal gas equation $p = \mathcal{R}\rho T$, where $\mathcal{R} = c_{\mathrm{P}} - c_{\mathrm{V}}$ is the gas constant and c_{P} and c_{V} are the specific heat capacities in constant pressure and volume, respectively (see, however, e.g. Hotta et al. 2015; Strugarek et al. 2016, for other approaches).

The viscous force is given by

$$\nabla \cdot \boldsymbol{F}^{\mathrm{visc}} = \nabla \cdot (2\nu\rho \mathbf{S}) \tag{5}$$

where ν is the kinematic viscosity and

$$\mathsf{S}_{ij} = \tfrac{1}{2}(U_{i;j} + U_{j;i}) - \tfrac{1}{3}\delta_{ij}\nabla \cdot \boldsymbol{U}, \tag{6}$$

is the traceless rate of strain tensor where the semicolons denote covariant derivatives (cf. Mitra et al. 2009). In principle ν is a function of density and temperature according to Spitzer (1962), but in practice the current simulations adapt various physically or numerically motivated formulations that are geared toward minimizing diffusion and maximising numerical stability on a given grid resolution (for a review, see e.g. Miesch et al. 2015).

Due to the short mean-free path of photons in stellar interiors, radiation is typically modeled via the diffusion approximation with

$$\mathcal{F}^{\text{rad}} = -K\nabla T, \tag{7}$$

where K is the radiative conductivity which is related to the opacity κ of the matter via

$$K = \frac{16\sigma_{\text{SB}}T^3}{3\kappa\rho}, \tag{8}$$

where σ_{SB} is the Stefan–Boltzmann constant. The radiative conductivity is often taken to be a fixed function of radius resulting either from a stellar evolution model (e.g. Brun et al. 2011; Hotta et al. 2022), or a simpler fixed analytic prescription producing a qualitatively similar behavior (e.g. Käpylä et al. 2013; Warnecke 2018). Alternatively, K can also be taken to be dependent on the ambient thermodynamic state in solar-like stars via the Kramers opacity law (e.g. Käpylä et al. 2020; Viviani and Käpylä 2021) with

$$\kappa \propto \rho T^{-7/2}, \tag{9}$$

which allows a non-linear back-reaction of, for example, rotation and magnetic fields (e.g. Käpylä et al. 2019). Radiative cooling and heating can also be included via the heating/cooling term,

$$\mathcal{H} = -\nabla \cdot \mathcal{F}^{\text{rad}}, \tag{10}$$

as is often done in the simulations with the RAYLEIGH code (e.g. Featherstone and Hindman 2016; Bice and Toomre 2022). Yet another approach is to relax the thermodynamics toward a fixed reference state using a Newtonian cooling term as is done in the EULAG simulations (e.g. Ghizaru et al. 2010; Passos and Charbonneau 2014; Strugarek et al. 2018; Guerrero et al. 2019).

Typical numerical methods need to include some form of subgrid-scale (SGS) diffusion in the entropy equation to ensure numerical stability. In some methods, such as those used in the ASH, RAYLEIGH, and PENCIL CODE, this is explicitly included in a term that is proportional to the entropy gradient (e.g. Brun et al. 2004; Käpylä et al. 2013; Matilsky and Toomre 2020)

$$\mathcal{F}^{\text{SGS}} = -\chi_{\text{SGS}}\rho T\nabla s, \tag{11}$$

where χ_{SGS} is the SGS thermal diffusivity that is responsible for turbulent diffusion at unresolved scales. This definition implicitly assumes that $ds/dr < 0$, that is, that the turbulent diffusion is due to unresolved Schwarzschild unstable convection, and the SGS term contributes to a positive (outward) energy flux. Often it is advantageous to decouple the SGS diffusion from the mean stratification such that the SGS diffusion is applied not on the total entropy s but, for example, to deviations from the spherically symmetric mean state $s' = s - \langle s\rangle$, where the overbar denotes suitable averaging, typically over the horizontal directions. This leads to a vanishing mean SGS flux, $\langle\mathcal{F}^{\text{SGS}}\rangle \approx 0$. This is advantageous if part of the convection zone is weakly stably stratified, or a stably stratified radiative layer is taken into account below the convection zone (e.g. Brun et al. 2011; Käpylä et al. 2020). An alternative approach is to include SGS effects implicitly such that the effective diffusion at small scales is determined by the numerical scheme itself. This is done in, for example, the EULAG (e.g. Ghizaru et al. 2010) and R2D2 codes (e.g. Hotta et al. 2014).

In practice, all of the diffusion coefficients in the simulations are much larger than their counterparts in stars, e.g. such that $\nu \gg \nu_\star$, $\eta \gg \eta_\star$, where the subscript $\star$ refers to stellar values. Furthermore, the radiative diffusivity $\chi = K/(c_P\rho)$ is also practically always much smaller than χ_{SGS} (see Appendix A of Käpylä et al. 2017). Therefore all of the current simulations need to be understood as large-eddy simulations (LES), where the small unresolved scales typically modeled as enhanced diffusion coefficients. This is to be contrasted with direct numerical simulations (DNS) where $\nu = \nu_\star$, $\eta = \eta_\star$, and $\chi_{SGS} = 0$. Furthermore, some models (e.g. Strugarek et al. 2016; Hotta et al. 2022) dispense with the explicit physical diffusion terms completely in order to minimize the diffusion on resolved scales while exerting adaptive diffusion at scales near the grid scale. These models are referred to as implicit LES (iLES) because the diffusion is built in into the numerical scheme without reference to physical diffusion terms.

2.1 Dimensionless Parameters and Diagnostics

A number of non-dimensional parameters arise in the analysis of the MHD equations and which define the simulations. These include the Rayleigh number which describes the efficiency of convection

$$Ra = \frac{gd^4}{\nu\chi}\left(-\frac{1}{c_P}\frac{ds}{dr}\right),$$ (12)

where d is a length scale; typically taken to be the shell thickness, the thermal and magnetic Prandtl numbers describe the relative importance of various diffusion terms:

$$Pr = \frac{\nu}{\chi}, \quad Pr_M = \frac{\nu}{\eta},$$ (13)

and the Taylor number

$$Ta = \frac{4\Omega_0^2 d^4}{\nu^2},$$ (14)

which measures the strength of rotation. The latter is related to the Ekman number via $Ek = 2Ta^{-1/2}$. Most often the relevant thermal Prandtl number is based on the SGS diffusion

$$Pr_{SGS} = \frac{\nu}{\chi_{SGS}},$$ (15)

because $\chi_{SGS} \gg \chi$. For completeness, the viscosity ν used in simulations is also an effective or SGS viscosity because it is always much larger than the real physical value. However, it has the same functional form as the physical viscosity whereas a term corresponding to the SGS entropy diffusion does not appear in the original equations. Additionally the geometry and the resulting density stratification are input parameters of the models, along with the boundary conditions applied to the various quantities.

Table 1 Orders of magnitude of some dimensionless parameters in the main sequence phase of the Sun in the bulk of convection zone and in a core convection zone of a 20 $M_\odot$ O9 star. Typical values from current global 3D simulations of stellar convection and dynamos are listed in the last column

Parameter	Sun ($M_\odot$)	O9 (20 $M_\odot$)	Simulations
Ra	10^{20}	10^{24}	10^9
Pr	10^{-6}	10^{-5}	$10^{-1} \ldots 10$
Pr$_M$	10^{-3}	10^3	$10^{-1} \ldots 10$
Re	10^{13}	10^{11}	10^4
Pe	10^7	10^6	10^4
Re$_M$	10^{10}	10^{14}	10^4
$\Delta\rho$	10^6	3	10^2
Ro	$0.1 \ldots 1$	1^1	$10^{-2} \ldots 10^3$

Note: Solar values are from Ossendrijver (2003) and Schumacher and Sreenivasan (2020) whereas the values for the O9 star are from Jermyn et al. (2022). $\Delta\rho$ is the ratio of the fluid density between the bottom and top of the convection zone. We note that in Augustson et al. (2019) the Prandtl numbers for the O9 star are somewhat lower, i.e., Pr $= 10^{-6}$ and Pr$_M = 10$; see their Fig. 2.

[1]Estimated using Ro $= P_{\mathrm{rot}}/\tau_{\mathrm{conv}}$ where τ_{conv} was taken from Fig. 74 of Jermyn et al. (2022) and the solar rotation period $P_\odot = 27$ days was used as a reference value for P_{rot}.

The most common diagnostic parameters used to describe the simulations include the Reynolds and Péclet numbers

$$\mathrm{Re} = \frac{u_{\mathrm{rms}}\ell}{\nu}, \quad \mathrm{Re_M} = \frac{u_{\mathrm{rms}}\ell}{\eta} = \mathrm{Pr_M Re}, \quad \mathrm{Pe} = \frac{u_{\mathrm{rms}}\ell}{\chi} = \mathrm{Pr\,Re}, \tag{16}$$

where u_{rms} is the rms-velocity and ℓ is a length scale, both of which are outcomes of the simulations. The latter is often defined as the integral scale; see e.g., Yadav et al. (2015b). The magnetic Reynolds number is of particular interest for dynamo simulations due to the bifurcations related to the excitation of large-scale and small-scale dynamos (SSD) (e.g. Brandenburg and Subramanian 2005; Rempel et al. 2023). The rotational effect on the flow is measured by the fluid Rossby (inverse Coriolis) number

$$\mathrm{Ro_f} = \frac{u_{\mathrm{rms}}}{2\Omega_0\ell} \propto \mathrm{Co}^{-1}. \tag{17}$$

An alternative way to define the Rossby number, which automatically takes the changing length scale into account is

$$\mathrm{Ro}_\omega = \frac{\omega_{\mathrm{rms}}}{2\Omega_0} \propto \mathrm{Co}_\omega^{-1}, \tag{18}$$

where ω_{rms} is the rms-vorticity with $\boldsymbol{\omega} = \nabla \times \boldsymbol{u}$. Order of magnitude estimates for some of these parameters in the deep parts of the solar convection zone and in a core convection zone of a 20 $M_\odot$ O star in comparison to current simulations are listed in Table 1.

Our discussion above has assumed that the *dimensional* MHD equations are being solved, in which case these non-dimensional parameters are diagnostic outputs of the simulations. An alternative approach, also employed by many authors (see, e.g. Gastine et al. 2016; Brown et al. 2020) is to non-dimensionalize the governing equations at the beginning; in this case the various non-dimensional parameters discussed here appear directly in the equations, and serve as input parameters that specify the problem. To illustrate the procedure, suppose we choose to measure lengths in units of a characteristic length ℓ_c, times in units of some time τ_c, velocities in units of u_c, and temperatures in units of T_c. That is, we assume $x = \ell_c x_{nd}$, $t = \tau_c t_{nd}$, and so forth, where the "nd" subscript denotes non-dimensional variables. Then, to take a simple example, the dimensional Boussinesq momentum equation in the absence of rotation or magnetism in a plane layer,

$$\frac{\partial \boldsymbol{u}}{\partial t} + \boldsymbol{u} \cdot \nabla \boldsymbol{u} = -\nabla \varpi + \nu \nabla^2 \boldsymbol{u} + \alpha g T \hat{z}, \tag{19}$$

where $\varpi \sim P/\rho$ is a reduced pressure and other symbols take their usual meanings, would be rewritten as

$$\frac{u_c}{\tau_c} \frac{\partial \boldsymbol{u}_{nd}}{\partial t_{nd}} + \frac{u_c^2}{\ell_c} \boldsymbol{u}_{nd} \cdot \nabla_{nd} \boldsymbol{u}_{nd} = -\frac{\varpi_c}{\ell_c} \nabla_{nd} \varpi_{nd} + \frac{u_c}{\ell_c^2} \nu \nabla_{nd}^2 \boldsymbol{u}_{nd} + \alpha g T_c T_{nd} \hat{z}, \tag{20}$$

where we have retained nd subscripts on all non-dimensional quantities (including the spatial and temporal derivatives), and where $\hat{z}$ is the vertical unit vector. Many choices of the characteristic scales τ_c, ℓ_c, etc., are possible, and in general these will each yield slightly different forms of the non-dimensional equations. A common choice is to measure lengths in units of the convection zone thickness ($\equiv L$), times in units of a thermal diffusion time across that length ($\tau_c = L^2/\chi$), and to take $u_c = L/\tau_c$ for consistency; upon substitution (and simplification) we then find

$$\frac{\partial \boldsymbol{u}_{nd}}{\partial t_{nd}} + \boldsymbol{u}_{nd} \cdot \nabla_{nd} \boldsymbol{u}_{nd} = -\frac{1}{Ma^2} \nabla_{nd} \varpi_{nd} + Pr \nabla_{nd}^2 \boldsymbol{u}_{nd} + RaPr T_{nd} \hat{z}, \tag{21}$$

now involving $Ma^2 = (P_c/\rho_c)/u_c^2$, $Pr = \nu/\chi$, and $Ra = g\alpha T_c L^3/(\nu\chi)$ (versions of the Mach, Prandtl, and Rayleigh numbers) as *input* parameters.

An advantage of this approach is that it is easier to avoid inadvertently "running the same simulation twice" – that is, conducting calculations with different luminosities, rotation rates, etc., that are nonetheless functionally equivalent (because they have the same governing non-dimensional parameters). On the other hand, it is sometimes difficult to "re-dimensionalize" such calculations and so make contact with any given astrophysical object; for illustrations of the procedure and its ambiguities, see discussions in Jones et al. (2011) and Yadav et al. (2016). We generally adopt the "dimensional" view throughout the remainder of this review.

2.2 Relevant Time and Length Scales in Stars

The structure of a star is determined by its mass M, luminosity L, chemical composition μ, and rotation rate Ω_0, the latter corresponding to its age. This information, along with material properties such as viscosity, opacity, equation of state, and nuclear energy production rate is in principle enough to construct a time-dependent model of the evolution of the star (e.g. Kippenhahn et al. 2012). However, in practice the evolution of main-sequence stars occurs over the nuclear timescale τ_n which is of the order of $\tau_n \approx 10^{10}$ yr for the Sun, which is much longer than what can be covered in any 3D dynamo simulation. Chemical evolution due to nuclear reactions occurs also in this timescale and therefore the solar and stellar dynamo simulations assume that the stellar structure is given and fixed in the course of the simulations (see, however Emeriau-Viard and Brun 2017). By the same token, the gravitational potential is assumed to be fixed and spherically symmetric. Rotational evolution of stars also happens on timescales of 10^8 to 10^9 years (e.g. Skumanich 1972; Barnes 2003; Gallet and Bouvier 2013) such that in 3D simulations the rotation rate of the star is assumed to be fixed. There is an ongoing debate based observational results suggesting magnetic braking slows down around the solar age which might be due to a transition to anti-solar differential rotation and a corresponding change in the dynamo (e.g. van Saders et al. 2016).

In general, the thermal evolution of the star in 3D simulations still occurs on a Kelvin–Helmholtz timescale

$$\tau_{\mathrm{KH}} = \frac{GM^2}{2RL}, \tag{22}$$

where G is the gravitational constant. The Kelvin-Helmholtz time for the solar convection zone is of the order of 10^5 years which is still about two orders of magnitude longer than the longest global 3D simulations to date (Passos and Charbonneau 2014; Käpylä et al. 2016). Various ways to overcome or circumvent this issue are discussed below in Sect. 3.1. The final timescale related to stellar structure is the free-fall or acoustic timescale $\tau_{\mathrm{ac}} = \sqrt{R^3/GM}$, which is of the order of 30 minutes for the Sun. The limitations the timescales impose on simulations are discussed further in Sect. 3.2.

In terms of global dynamos the most important timescales are the rotation period P_{rot} and the convective turnover time,

$$\tau_{\mathrm{conv}} = \frac{\ell}{u_{\mathrm{conv}}}, \tag{23}$$

where ℓ is the convective length scale and u_{conv} a suitably averaged convective velocity. The convective turnover time τ_{conv} can be estimated from solar surface observation where granules overturn on a timescale of a few minutes. Knowledge of τ_{conv} in deeper layers relies heavily on theoretical estimates, for example, from mixing length models (e.g. Böhm-Vitense 1958). These assume that the length scale is proportional to the pressure scale height. At the same time, convective velocities decrease such that τ_{conv} is of the order of a month near the base of the solar convection zone (e.g. Stix 2002). Stellar observations indicate that dynamo efficiency of stars is related to the Rossby number (see, Sect. 5.2)

$$\mathrm{Ro} = \frac{P_{\mathrm{rot}}}{\tau_{\mathrm{conv}}}. \tag{24}$$

The Rossby number is the only non-dimensional diagnostic that the simulations can capture relatively accurately; see Table 1 and Sect. 3.1. Equations (24) and (17) are related via $\mathrm{Ro} = 4\pi \mathrm{Ro_f}$.

Another timescale that the simulations need to capture is the activity cycle period τ_{cyc} which is 22 years for the Sun, and which varies between years to decades for stars other than the Sun (e.g. Baliunas et al. 1995; Hall et al. 2009; Olspert et al. 2018). It is, however, practically always necessary to run simulations considerably longer because establishing the global dynamo and reaching the final saturated dynamo mode takes typically significantly longer (e.g. Käpylä et al. 2013; Matilsky and Toomre 2020). Taking the thermal relaxation also into account, the integration times are typically at least an order of magnitude longer than the cycles established. The necessity to run such long times is one of the major limiting factors in the quest to reach astrophysically relevant parameter regimes.

In principle the relevant length scales vary between the depth of the convection zone ΔR (or the radius of the star R for fully convective stars), and the Kolmogorov length scale ℓ_ν where kinetic energy is dissipated into heat due to viscosity (or to the magnetic dissipation

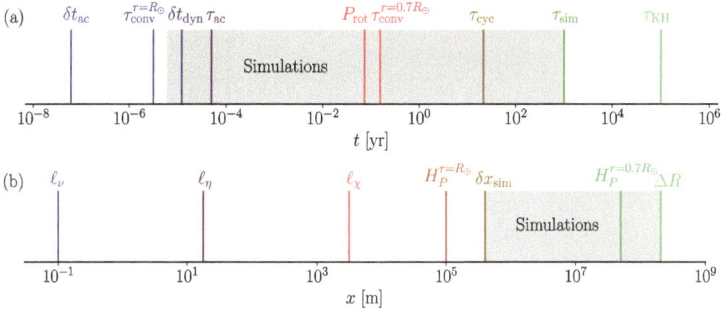

Fig. 1 Orders of magnitudes of time (a) and length (b) scales in the deep solar convection zone. The gray shaded area indicates regions accessible to a typical current global simulation. δ_{ac} and δ_{dyn} are hypothetical acoustic and dynamical timesteps in a direct simulation of the solar convection zone; see Section 3.2

scale ℓ_η in cases where $Pr_M > 1$). According to the Kolmogorov turbulence phenomenology (e.g. Frisch 1995), ℓ_ν can be estimated from the Reynolds numbers at the integral (L) and Kolmogorov scales,

$$\ell_\nu = L \left(\frac{Re_L}{Re_{\ell_\nu}} \right)^{-3/4}. \tag{25}$$

For the Sun, $L = \Delta R = 2 \cdot 10^8$ m and $Re_L \gtrsim 10^{12}$ (e.g. Ossendrijver 2003; Jermyn et al. 2022) and $Re_{\ell_\nu} = 1$ by definition. These estimates yield an upper limit of order of magnitude $\ell_\nu \approx 0.1$ m near the base of the solar convection zone. More detailed calculations yield values between 0.01 m (Kupka and Muthsam 2017) and 0.06 m (Schumacher and Sreenivasan 2020). Furthermore, the dissipation scales of magnetic fields and temperature fluctuations can be estimated from $\ell_\eta = \ell_\nu Pr_M^{-3/4}$, and $\ell_\chi = \ell_\nu Pr^{-3/4}$, respectively. Even though Pr_M, $Pr \ll 1$, these scales are also very small in comparison to the depth of the convection zone or the radius of the star. As will be discussed below, all of the scales where physical diffusion occurs are several orders of magnitude smaller than what can be achieved in any current or foreseeable simulations (see also Kupka and Muthsam 2017).

Another length scale that plays an important role is the pressure scale height $H_P = -dr/d \ln p$ which is related to the vertical scale of convection cells. Near the surface H_P is of the order of 100 km in the photosphere of the Sun. On the other hand, at the base of the convection zone $H_P \approx 5 \cdot 10^4$ km. This reflects the fact that near the surface the pressure and density decrease very rapidly and that capturing both the deep and photospheric convection in a single model is therefore very challenging. In total the solar convection zone encompasses more than 20 pressure scale heights. This translates to a density difference of 10^6 between the photosphere and the base of the convection zone. Estimates of the relevant temporal and length scales in the deep parts of the solar convection zone are summarized in Fig. 1; see also related discussion in Käpylä et al. (2013) and their Fig. 1.

3 Numerical Approach to Stellar Dynamos

3.1 Simulation Strategy

The main difficulty in solar and stellar dynamo simulations is that it is not feasible to match the dimensionless parameters with those of real stars as seen from the comparison of stellar and simulation parameters in Table 1. The only general exception to this is the Rossby number but even there we face the situation that only some of the convective scales in stars are strongly affected by rotation. For example, in the Sun the near-surface layers are practically unaffected by rotation whereas at the base of the convection zone the fluid Rossby number based on mixing length estimates of ℓ and u_{rms} is of the order of 0.1 (e.g. Ossendrijver 2003). This is to be contrasted with the Earth's dynamo where the magnetic Reynolds number is $\mathrm{Re_M} \approx 10^3$, which is within reach of current simulations (e.g. Aubert and Gillet 2021) and where all convective scales are strongly rotationally constrained ($\mathrm{Ro} \approx 10^{-6}$); see, e.g., Roberts and King (2013).

A path that stellar dynamo simulations often follow to approach physically relevant regimes is to assume a fixed convective Rossby number (Gilman 1977), given by

$$\mathrm{Ro_c} = \left(\frac{\mathrm{Ra}}{\mathrm{PrTa}} \right)^{1/2}. \tag{26}$$

Here, the stellar luminosity fixes the level of driving through the Rayleigh number, and the stellar rotation rate is fixed by observations. Using typical estimates for Ra, Pr, and Ta for the Sun (Ossendrijver 2003, see also Table 1) we arrive at $\mathrm{Ro_c} \approx 0.1 \ldots 1$. In simulations, the (SGS) Prandtl number $\mathrm{Pr_{SGS}} = \nu/\chi_{\mathrm{SGS}}$ is often fixed and changing the diffusivities ν and χ_{SGS} leads to $\mathrm{Ra} \propto \nu^{-2}$, $\mathrm{Ta} \propto \nu^{-2}$ and $\mathrm{Ro_c} = \mathrm{const}$. An obvious limitation is that the Prandtl number in simulations is typically close to unity whereas in stars $\mathrm{Pr} \ll 1$ (e.g. Augustson et al. 2019; Schumacher and Sreenivasan 2020; Jermyn et al. 2022). A similar argument applies to $\mathrm{Pr_M}$ in late-type stars whereas in the core convection zones of massive O and B stars $\mathrm{Pr_M} \gg 1$ (e.g. Augustson et al. 2016). Furthermore, in iLES models the values of the dimensionless parameters are typically unknown and not precisely controllable, although it is often possible to determine these *a posteriori* (e.g. Strugarek et al. 2016; Hotta et al. 2022). Nevertheless, the strategy in both LES and iLES models is to try to capture the stellar Rossby number with unrealistic Prandtl numbers and to resolve enough scales in an effort to reach sufficiently high Re, $\mathrm{Re_M}$, and Pe such that the large scale results are no longer affected. However, it is still questionable whether such a regime has been reached even in the highest resolution simulations to date (e.g. Hotta et al. 2022; Guerrero et al. 2022).

3.2 Limitations of Current Numerical Simulations

The challenge of doing direct numerical simulations (DNS) of stars is illustrated by considering the solar convection zone where the fluid Reynolds number is of the order of at least 10^{12} (e.g. Ossendrijver 2003; Jermyn et al. 2022). With this estimate and Eq. (25), the ratio of the system scale L, here taken to be the depth of the solar convection zone or 200 Mm, to the Kolmogorov scale ℓ_ν, is

$$\frac{L}{\ell_\nu} = \left(\frac{\mathrm{Re}_L}{\mathrm{Re}_{\ell_\nu}} \right)^{3/4}. \tag{27}$$

With $\mathrm{Re}_{\ell_\nu} = 1$, we obtain $L/\ell_\nu = 10^9$. This ratio gives the order of magnitude of grid points that is required to capture all of the physically relevant scales in the solar convection zone. Thus a direct 3D simulation requires 10^{27} grid points. A somewhat lower, but still unattainable, number was reported in Chan and Sofia (1986).

Current state-of-the-art global simulations have of the order of 10^{10} grid points and are run on a few times 10^4 CPU cores. Assuming ideal weak scaling, where the computation time remains constant when the number of CPUs is increased proportional to the grid size, a DNS of the solar convection zone requires 10^{21} CPU cores. Using a current 96-core AMD Epyc™ 9654 CPU with 360 W thermal design power as a reference,[1] gives a total power consumption of $3.8 \cdot 10^{21}$ W, corresponding roughly to a M9V main-sequence red dwarf. It is clear that such power is neither available nor meaningful to be spent. Although reaching an asymptotic regime where the large-scale dynamics are unaffected by the addition of further small scales is very likely possible at a significantly lower resolution, it is clear that even the highest resolution current simulations are not there yet (e.g. Käpylä et al. 2017; Hotta et al. 2022).

Furthermore, the timestep in such hypothetical DNS of the solar convection zone is of the order of $\delta t \approx \ell_\nu / \max(c_{\mathrm{sig}}^{\max})$, where $c_{\mathrm{sig}}^{\max}$ is the maximum signal propagation speed. In anelastic models this is set by the maximum flow velocity which is of the order of 1 km s^{-1}, whereas in the fully compressible case this is the sound speed c_{s}, which at the base of the convection zone is around 200 km s^{-1}. This gives $\delta t = \delta t_{\mathrm{dyn}} \approx 2 \times 10^{-4}$ s for anelastic and $\delta t = \delta t_{\mathrm{ac}} \approx 10^{-6}$ s for fully compressible models. In practice, the resolution is much lower and corresponding estimates for a high-resolution global simulation with 500 uniformly spaced grid points in radius gives $\delta t_{\mathrm{ac}} \approx 2$ s and $\delta t_{\mathrm{dyn}} \approx 10$ minutes. The latter is still longer than the convective turnover time near the surface of the Sun, where $\tau_{\mathrm{conv}}^{r=R_\odot} \approx 1$ minute. The surface of the Sun is extremely challenging to be taken into account in a global model due to a combination of very small length scales and short time scales and the Mach number approaching unity. Therefore a full Sun simulation requires a numerical scheme capable of dealing with practically all Mach numbers and multiscale convection. Furthermore, the boundary region where radiative cooling takes place near the surface is extremely thin, around 10 km, in comparison to the depth of the convection zone (Kupka and Muthsam 2017). Typically simulations either do not reach all the way to the photosphere, or consider a shell reaching to $R = R_\odot$ but with a much lower density stratification than in the Sun, and the boundary layer near the surface is made artificially thicker to resolve it numerically.

Another constraint arises due to a widening discrepancy of the timescales involved when resolution is increased: as was discussed earlier, a simulation of the Sun needs to cover at least a solar cycle or preferably several cycles to be considered viable such that the simulated time $\tau_{\mathrm{sim}} \gtrsim \tau_{\mathrm{cyc}}$. For the sake of argument, an acceptable maximum wall-clock time that a simulation is permitted to run to is taken to be a year. This requires that the star in the simulation has to evolve at least 22 times faster than in real time. However, when the grid resolution is increased, the timestep in explicit time-stepping methods decreases in proportion with the grid spacing δx, and the computational cost of simulation increases with δx^4. This corresponds to Re^3, making it very difficult to reach high Re. Even if the numerical scheme has ideal weak scaling, the time to solution doubles every time the resolution is doubled, which typically cannot be avoided. This poses stringent constraints on either the length, or the grid resolution, of simulations targeting global stellar dynamos. The timestep constraints can, to a certain degree, be alleviated by the use of implicit time stepping methods (e.g. Viallet et al. 2011) or by the use of local subgrids and timesteps (e.g. Popovas et al. 2022).

A further complication arises due to the thermal relaxation. In anelastic models, where the real stellar luminosity is often used, the Kelvin-Helmholtz time is much longer than the integration times of simulations. However, this is a worst-case scenario because deep stellar convection zones are nearly adiabatic which is exploited in the simulation setups. Thermal relaxation can still take a prohibitively long time if a stably stratified radiative layer

---

[1]https://www.amd.com/en/products/cpu/amd-epyc-9654.

is retained below the convection zone. This issue is sometimes alleviated by adjusting the radiative conductivity in the overshoot layer below the convection zone (e.g. Brun et al. 2017). However, this can lead to over- or underestimation of convective overshooting depending when and how such adjustments are made (Käpylä 2019). Another possibility is to adjust the thermodynamic state and fluctuations recursively toward an equilibrium solution (Anders et al. 2018, 2020), although this method has yet to gain widespread adoption in compressible or global 3D simulations.

In fully compressible simulations the timestep would be very short because it is determined by the sound speed at the base of the convection zone. This has been circumvented by the reduced sound speed technique (RSST) where the sound speed is artificially lowered such that the timestep issue is alleviated (e.g. Hotta et al. 2014). Another approach is the enhanced luminosity method (ELM) where a luminosity that is much higher than in real stars is used (e.g. Käpylä et al. 2013, 2020), leading to a higher Mach number and therefore a diminished gap between the acoustic and dynamical timescales, as well as a correspondingly shorter Kelvin-Helmholtz time. Given that the luminosity enhancement is sufficiently large, it is possible to resolve the Kelvin-Helmholtz timescale using fully compressible MHD equations (e.g. Käpylä 2023). The cost of this method is that in addition to higher flow velocities, also the thermodynamic fluctuations are enhanced, and a direct comparison with observations requires the use of scaling relations. Furthermore, to achieve the same Rossby number as in a real star with realistic luminosity, the rotation rate has to be increased in proportion to the Mach number, which would lead to unrealistically large centrifugal force (e.g. Navarrete et al. 2022).

3.3 Numerical Methods and Codes

There are a variety of codes solving the MHD equations in spherical shells and targeting solar and stellar global dynamos. The first and still popular approach is to adopt the anelastic approximation where the sound waves are filtered out by neglecting the time derivative in continuity equation. Then it is convenient to use spherical harmonics to solve for the horizontal dynamics whereas the vertical discretisation is often done with finite differences or Chebychev polynomials. Codes using this approach include ASH (e.g. Clune et al. 1999; Brun et al. 2004; Jones et al. 2011; Brun et al. 2022), RAYLEIGH (Featherstone et al. 2022), and MAGIC (e.g. Gastine and Wicht 2012). EULAG is another anelastic code but instead of spherical harmonics, it relies on a second-order accurate multidimensional positive-definite advection transport algorithm (MPDATA) and implicit time stepping (e.g. Smolarkiewicz and Charbonneau 2013).

Another popular technique is to use the fully compressible formulation and using some flavour of finite difference methods. This typically leads to coordinate singularities at the axis and at the centre of the star, which are circumvented by either omitting regions near the axis (spherical wedge, cf. Käpylä et al. 2012; Mabuchi et al. 2015), using partially overlapping grids (yin-yang grid cf. Hotta et al. 2015), or embedding a spherical star into a Cartesian cube (star-in-a-box model, cf. Dobler et al. 2006; Käpylä 2021). The Mach number issue of fully compressible simulations is dealt with by RSST and ELM methods that were discussed above. Codes using fully compressible formulation include the PENCIL CODE (Pencil Code Collaboration et al. 2021), R2D2 (Hotta et al. 2015), and the code used in Mabuchi et al. (2015). Further methods include the DEDALUS framework which uses spectral methods and is capable of solving incompressible, anelastic, and fully compressible equations in varying geometries (Brown et al. 2020; Burns et al. 2020; Anders et al. 2022a), and the DISPATCH framework where various solvers for compressible flows are possible and which uses local subdomains and timesteps (Nordlund et al. 2018; Popovas et al. 2022).

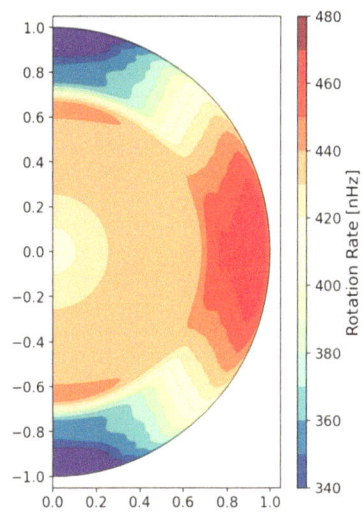

Fig. 2 Solar differential rotation profile $\Omega/2\pi$ as a function of radius and latitude as inverted from helioseismology using the first 6 years of HMI 72-day analysis. Reddish colors indicating fast rotation and blueish tones slow rotation (adapted from Larson and Schou (2018) using the data archived at http://jsoc.stanford.edu/HMI/Global_products.html)

4 Relevant Solar and Stellar Observations

The dynamo simulations discussed in this review aim, ultimately, to capture the flows and magnetism occurring in real stars. Here, we briefly describe some of the most pertinent observational constraints on these processes, which also serve to motivate and guide our work. Perhaps the most obvious characteristics that any dynamo simulation would hope to match are the Sun's observed differential rotation and its periodic cycle of magnetic activity.

In Fig. 2, we sample the solar interior rotation profile as revealed by helioseismology (e.g. Schou et al. 1998). The Sun's surface differential rotation – with a fast equator and a slow pole – imprints through the convection zone, with nearly solid-body rotation in the portions of the radiative zone below that are accessible to these global-scale inversions. There are two prominent shear layers – the tachocline near the base of the convection zone and the near-surface shear layer (NSSL) in the upper portions of the convection zone. The apparent width of the tachocline in this representation reflects the width of the inversion kernels used; its true width is thought to be narrower. Note, too, that although the shear within the convection zone has not spread through the radiative interior, the *average* rotation rate of the interior is commensurate with the average rate of the envelope; since the Sun is continuously losing angular momentum via its magnetized wind, and so spun more rapidly in the past, this observation implies some level of coupling between the two regions (Gilman et al. 1989; MacGregor and Brenner 1991; Spiegel and Zahn 1992; Gough and McIntyre 1998; Brun et al. 2011; Matt et al. 2015).

Ideally, a simulation would self-consistently capture at least a few key attributes of this profile: e.g., the overall pole-to-equator shear within the convection zone; the fact that iso-contours of Ω are more nearly aligned with radius than they are with the rotation axis, in evident tension with the Taylor-Proudman theorem (e.g. Miesch et al. 2006); and the existence and properties of both the NSSL and the tachocline. In practice each of these remain a challenge, though as discussed later in this review (Sect. 5.1), the latest 3D MHD global simulations of solar interior dynamics and angular momentum transport do manage to capture many of these elements without undue tinkering.

The most striking observational constraints on the magnetism involve its systematic evolution with space and time, as sampled in the famous "butterfly diagram." An example is

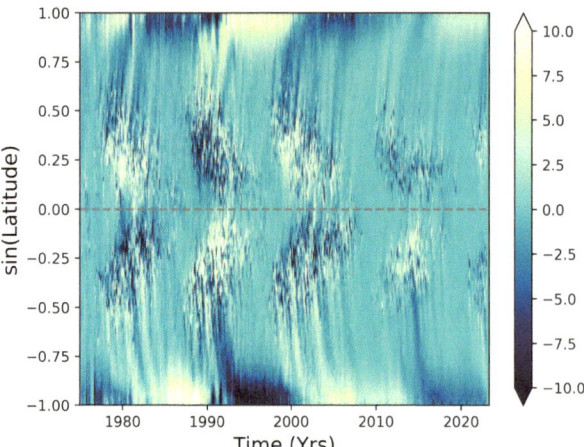

Fig. 3 Solar butterfly diagram of the line of sight surface magnetic field up to Carrington rotation 2265 (Brun et al. 2020)

provided in Fig. 3, which shows the longitudinally-averaged line of sight component of the magnetic field for every Carrington rotation since 1975 (based on Wilcox, GONG and Solis synoptic map data; Brun et al. 2020). Strong fields emerge at mid-latitudes and then, over the course of roughly 11 years, appear progressively nearer the equator; the polarity of these emergent fields is the same for most of the low-latitude events in the Northern hemisphere, and opposite to that in the Southern; the overall polarity of the field flips at the end of each 11-year period. There is also a prominent polar branch of activity, which is at its strongest when the equatorial branch is at its ebb; the polarity of this polar branch matches that of the *following* equatorward branch. The polarity of the poloidal field thus reverses when active region emergence is at its maximum. The overall number of sunspots visible over the solar surface rises and falls over the course of the cycle, in line with the surface distribution of the strongest fields.

On the whole, the Sun's ordered field exhibits dipole parity throughout most of the cycle (it is antisymmetric about the equator), but there are periods near cycle maximum during which the parity is mostly quadrupolar. The relative amplitudes of the dipolar and quadrupolar modes are shown in Fig. 4 over the past few cycles (Brun et al. 2020). These relations constitute another powerful constraint on dynamo models. For example, there is evidence that around the pronounced period of low surface activity known as the Maunder minimum, the Sun's observed surface activity was predominantly confined to one hemisphere, indicating different parity relations; the implications of this finding for dynamos generally (and grand minima in particular) have been considered by, e.g., Sokoloff and Nesme-Ribes (1994) and in many subsequent papers.

Very recently, new constraints have begun to emerge from the study of inertial and Rossby wave modes that propagate within the convection zone. These toroidal modes have recently been observed in helioseismic maps of near-surface horizontal flows obtained by HMI aboard SDO (Gizon et al. 2021); see also Hanson et al. (2022) for another recent detection of solar inertial modes. Though modeling of these modes is still in its infancy (Bekki et al. 2022; Triana et al. 2022) they appear to hold great promise for constraining aspects of the convection that would be difficult or impossible to estimate by other means. As a first application, Gizon et al. (2021) illustrate that these modes constrain the superadiabaticity and turbulent diffusivity of the deep solar convection zone.

Finally, we turn briefly to observations of other stars. Many aspects of observational stellar magnetism are treated in other reviews (e.g. Reiners 2012; Brun and Browning 2017;

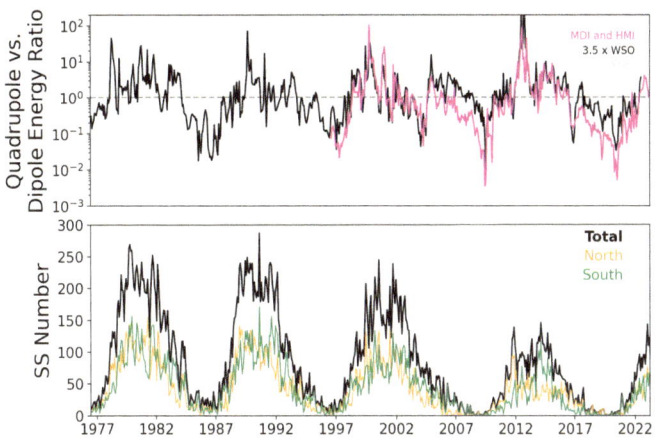

Fig. 4 Top: Ratio of the solar magnetic dipole and quadrupole energies over the last few cycles (MDI and HMI Data). We note that the quadrupole modes dominate during the maximum cycle phase and this has already started for cycle 25 in 2022. This strong quadrupolar component also explains the time lag between the northern and southern hemispheres that can reach up to 18 months. Bottom: Sunspot numbers from all data and separately from northern and southern hemispheres. Adapted from Brun et al. (2020)

Jeffers et al. 2023) so we note only a few key constraints that must eventually be matched by simulations. The most celebrated of such constraint is the "rotation-activity correlation," sampled in Figs. 5 and 6. In stars with convective envelopes, surface measurements of magnetic activity first increase with rotation rate, then plateau ("saturate") at a certain point. Here we show examples in which activity is measured by coronal emission (Wright et al. 2018), here including both fully convective and partially-convective stars in Fig. 5(a); by chromospheric Hα emission (as a fraction of the bolometric luminosity) in a large sample of M dwarfs (Newton et al. 2017) in Fig. 5(b); via Zeeman Doppler Imaging (See et al. 2019), here providing an estimate of the large-scale dipole field observed at the surface (Fig. 6(a)); and via measurements of the surface magnetic field strength as revealed by the Zeeman broadening of spectral lines (Reiners et al. 2022); see Fig. 6(b). Typically in these studies the influence of rotation is characterized via a simple estimate of the Rossby number where the convective turnover time is typically based on simple empirical relations that work well for main-sequence stars (e.g. Noyes et al. 1984). When viewed in this way, many different types of stars – including those with and without a stable radiative region – appear to exhibit the same basic relationships. Comparisons of young and evolved main-sequence stars also suggest a similar level of activity as function of Ro in terms of Ca II H&K emission, provided that the convective turnover time is an outcome of 1D stellar models (Lehtinen et al. 2020).

The rotational velocity of a solar-like star changes systematically over time, as it loses angular momentum through a magnetised wind, so over the course of its life it will trace a variety of positions on this rotation-activity correlation. Indeed, measurements of rotation rate are used in "gyrochronology" as proxies for age, because many stars are found to exhibit a common, tight relation between spin rate and time – the so-called Skumanich law $\Omega \propto t^{-1/2}$; see discussions in Skumanich (1972), Soderblom (1983), and Barnes (2003). Lately there have been some indications that this relationship may break down at late ages (e.g. van Saders et al. 2016), which may provide additional constraints on the dynamo for old stars (e.g. Metcalfe and van Saders 2017). In a related vein, there is some evidence for

Fig. 5 Relations between rotation rate (quantified by the Rossby number) and proxies of magnetic activity. (*a*) X-ray emission (normalized to the bolometric luminosity) for a sample of both fully and partially convective stars (Wright et al. 2018); (*b*) a measure of chromospheric Hα emission, normalised to the bolometric luminosity and relative to that in inactive stars, in a sample of M dwarfs (Newton et al. 2017) (©AAS. Reproduced with permission)

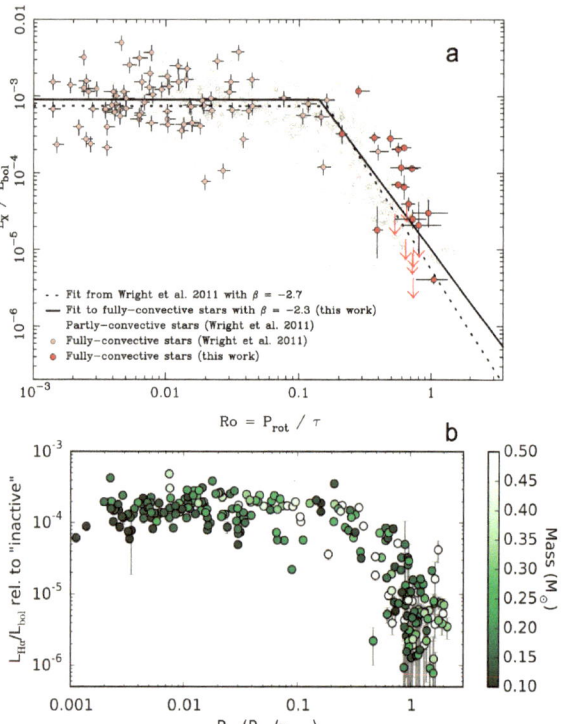

enhanced stellar activity in a subset of slowly-rotating stars, which some authors have suggested may be linked to the presence of strong anti-solar differential rotation (Brandenburg and Giampapa 2018). Global dynamo simulations seem to capture this effect; see Karak et al. (2015), Warnecke and Käpylä (2020), Brun et al. (2022), and Noraz et al. (2023).

There has been a rising interest in efforts to detect anti-solar differential rotation observationally partly due to it being a robustly appearing feature in simulations. This is a challenging task because anti-solar differential rotation is expected to occur in slowly rotating stars where even the detection of the rotation period is difficult. Nevertheless, detections have been made from giant stars using Doppler imaging (e.g. Weber et al. 2005) by tracking the drift of spots on the surfaces of stars. Furthermore, Reinhold and Arlt (2015) developed a method to distinguish between solar-like and anti-solar differential rotation from long-term photometry. In their sample of 50 Kepler stars, they found that 10-20 per cent of these stars are likely to have anti-solar differential rotation. Noraz et al. (2022) identified 22 solar-type Kepler stars as anti-solar differential rotation candidates by inferring the Rossby numbers these stars are likely to have based on their rotation rate, interior structure, and metallicity. Finally, Benomar et al. (2018) used asteroseismology to estimate the latitudinal differential rotation in a sample of 40 Kepler stars. While none of the stars in their sample unambiguously showed anti-solar differential rotation, the methodology could in principle be used to detect it.

Together, these observations of the Sun and other stars constitute powerful constraints that models would hope to satisfy. In the following sections, we will examine the extent to which they actually do so.

Fig. 6 Similar to Fig. 5 but showing measures of magnetic field as a function of Ro. (*a*) Zeeman Doppler imaging estimates of dipole component of surface magnetic field (See et al. 2019) (©AAS. Reproduced with permission). (*b*) Average surface magnetic field as measured by Zeeman broadening of spectral lines (Reiners et al. 2022) (©ESO. Reproduced with permission)

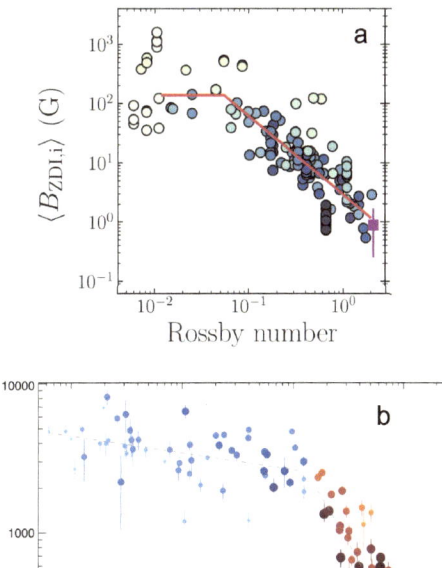

5 Simulations of Solar and Stellar Dynamos

5.1 Convection and Dynamo in the Current Sun

In addition to the fundamental numerical restrictions discussed above, simulations of the current Sun are challenging due to the possible proximity of the transition from solar-like to anti-solar differential rotation. This transition occurs around Ro ≈ 1 (e.g. Käpylä et al. 2011a; Gastine et al. 2014; Brun et al. 2017) and current simulations of the Sun appear to lie close to this in parameter space. Simulations with the nominal solar luminosity and rotation rate land predominantly in the anti-solar regime (e.g. Käpylä et al. 2014), which is one of the manifestations of the convective conundrum (O'Mara et al. 2016), or the lower than expected velocity amplitudes and Rossby number in the Sun in comparison to theoretical estimates and simulations, which is discussed in more detail elsewhere (e.g. Hanasoge et al. 2016). Therefore simulations targeting the Sun often resort to lowering the Rossby number artificially to obtain a solar-like differential rotation profile. This is most often done by suppressing the convective velocity by enhancing radiative diffusion (e.g. Käpylä et al. 2014; Fan and Fang 2014; Hotta et al. 2016; Noraz 2022), lowering the luminosity (e.g. Hotta et al. 2015; Guerrero et al. 2022), or by increasing the rotation rate.

The importance of reproducing the solar interior rotation lies in the fact that the dynamo process crucially relies on flows of various scales to maintain the observed magnetic field. At the very least, the large-scale flows are employed by all of the currently predominant solar dynamo models (Charbonneau 2020). Furthermore, it is also likely that turbulent effects, such as an α effect due to helical convection-driven turbulence, are important in the dynamo process. This highlights the importance of accurate modelling of convection, essentially necessitating that the solar velocity field has to be sufficiently well reproduced first before one should expect success in reproducing the dynamo. An important step in this is the identification of the relevant force balances that need to be reproduced in simulations. Following such an approach, a path in parameter space may be found that leads to solar-like results with feasible numerical cost. Such "path approach" is quite commonly used now in geodynamo modelling (Aubert et al. 2017; Aubert and Gillet 2021).

The convective conundrum is arguably the greatest obstacle in achieving the goal of simulating the solar dynamo successfully. Several ideas have recently been invoked to alleviate the discrepancy between models and reality. One of these ideas is that this is a manifestation of rotationally constrained convection in the interior of the Sun. In this scenario the maximum scale of convection in the deep parts of the solar convection zone is not giant cells, as is expected from mixing length models, but it matches instead that of the supergranulation, which is also detected from surface observations (Featherstone and Hindman 2016; Vasil et al. 2021). However, some such studies assume from the outset that convection is rotationally constrained in the Sun although it is unclear if this really is the case. For example, Featherstone and Hindman (2016) match a rapidly rotating simulation with the Sun based on the fact that the velocity power spectrum peaks at supergranular scale. No further independent check, for example by means of a Rossby number depending only on stellar parameters and not on estimates of convective velocities or scales (e.g. Käpylä 2023), is made that the simulation matches the relevant solar parameters. There is currently no simulation of solar convection that unambiguosly reaches a rotationally constrained state. Another caveat is that while simulations of rotationally constrained convection do produce smaller convective scales that become smaller as rotation becomes more rapid (e.g. Viviani et al. 2018), the main contribution to differential rotation in such models is still due to giant cell convection (e.g. Käpylä 2023) or thermal Rossby waves that have not yet been detected in the Sun.

Another idea that has gained popularity recently is that the deep parts of the solar convection zone can be weakly stably stratified. This is thought to result from strong driving of convection in the near-surface layers, whence plumes of cool low entropy material plough through the whole convection zone and deep into the stably stratified layers below. Such idea of *cool entropy rain* was put forward by Spruit (1997) and later incorporated into a modified mixing length model by Brandenburg (2016). In the extreme versions of these models only a very thin layer (down to a few Mm) near the surface of the convection zone is Schwarzschild unstable and the rest of the convection zone is weakly subadiabatic and mixed by the entropy rain. Such effects were explored in 3D simulations by Nelson et al. (2018) by means of a boundary condition consisting of localised cooling patches. Although non-rotating simulations often find relatively deep subadiabatic layers (e.g. Roxburgh and Simmons 1993; Tremblay et al. 2015; Hotta 2017; Käpylä et al. 2017), their effect in global simulations appears to be weak (Käpylä et al. 2019; Viviani and Käpylä 2021). This could also be due to the modest resolutions and supercriticality of convection in those studies.

Furthermore, the influence of the thermal Prandtl number has also recently been studied. In particular, several studies have concentrated on cases where the effective Prandtl number is greater than unity (e.g. O'Mara et al. 2016; Bekki et al. 2017; Karak et al. 2018). This was motivated by the observation that the overall velocities are decreased in high-Pr convection. However, this also coincides with more effective downward flux of angular momentum, exacerbating the problems related to anti-solar differential rotation (Karak et al. 2018). Another recent study (Käpylä 2023) confirmed these ideas and showed that the Prandtl number dependence is relevant in the regime $\mathrm{Pr} \gtrsim 1$, whereas for $\mathrm{Pr} \lesssim 1$, the parameter regime relevant for the Sun, no statistically significant dependence was detected.

Finally, the role of magnetism in shaping the solar rotation profile is also a viable option to explain the convective conundrum. Whereas early forays into this field yielded somewhat contradictory results with some studies finding essentially no dependence on magnetic fields (Karak et al. 2015), others reported a flip from anti-solar to solar-like differential rotation (Fan and Fang 2014; Simitev et al. 2015). All of these simulations were made at relatively modest magnetic Reynolds numbers, and it is likely that a SSD was not excited in these models. The recent high-resolution implicit large-eddy simulations (iLES) (Hotta and Kusano 2021; Hotta et al. 2022), have reached a regime where small-scale magnetic fields are generated throughout the convection zone and turn a hydrodynamically anti-solar run to a solar-like solution at the highest resolution. Somewhat worryingly, these simulations have yet to show convergence as a function of resolution such that the flows at large scales changes significantly even between the two highest resolutions. Recently, Käpylä (2023) reported that it is easier to excite solar-like differential rotation for higher Re_M from simulations with explicit diffusivities where Re_M exceeded the threshold for SSD. However, this effect is much less drastic than in the iLES simulations.

The radial shear in the solar convection zone occurs predominantly in the boundary layers which are difficult to incorporate in global simulations. The near-surface shear layer is thought to be generated in the weakly rotationally constrained small-scale convection in the outermost parts of the solar convection zone (e.g. Kitchatinov 2016) or due to gyroscopic pumping effects (e.g. Miesch and Hindman 2011). Capturing this in global simulations is challenging because a very high resolution is required to capture the near-surface small-scale convection resulting from a steep decrease of fluid density. First such simulations were presented by Hotta et al. (2015), who were able to capture some aspects of the NSSL. However, these simulations were hydrodynamic and no corresponding dynamo solutions have been presented so far. The NSSL has also been suggested to shape the global solar dynamo (Brandenburg 2005), but no direct evidence supporting or refuting this theory is currently available.

The other boundary layer at the interface of the convective and radiative layers, the tachocline, is perhaps even more challenging to capture in simulations. The main challenge is that the solar tachocline is very thin (certainly less than five per cent of solar radius, but likely much less), and it has been confined now for five billion years. Estimates of radiative spreading for the Sun suggest that the tachocline should be much thicker at the current age of the Sun so there has to be a mechanims preventing this. Several magnetic scenarios have been invoked to explain this, including a dipolar fossil field in the radiative core or a cyclic dynamo in the convection zone. Some current simulations do exhibit tachocline-like features, but they are spreading into the radiative core at rates that are much higher than expected for the Sun (e.g. Brun et al. 2011, 2017). Furthermore, iLES models also produce tachoclines at relatively low resolutions, although their confinement mechanism is yet to be understood (Guerrero et al. 2013, 2016, 2022). In all of the aforementioned simulations the diffusivities in the radiative interior were either explicitly or implicitly greatly reduced. In apparent contradiction to these models, Matilsky et al. (2022) were able to obtain a relatively thin tachocline and essentially rigidly rotating radiative core in a simulation where the diffusivities were not decreased but which housed a cycling non-axisymmetric dynamo in the convection zone. Furthermore, this model has also strong horizontal flows in the radiative interior. The actual process of tachocline confinement is still unclear in this case, although it does share some characteristics with the cyclic dynamo confinement process suggested by Forgács-dajka and Petrovay (2001); see also Barnabé et al. (2017).

Given the difficulties in reproducing the solar flows, it is then hardly surprising that dynamo simulations have had a hard time reproducing the solar large-scale magnetism. The

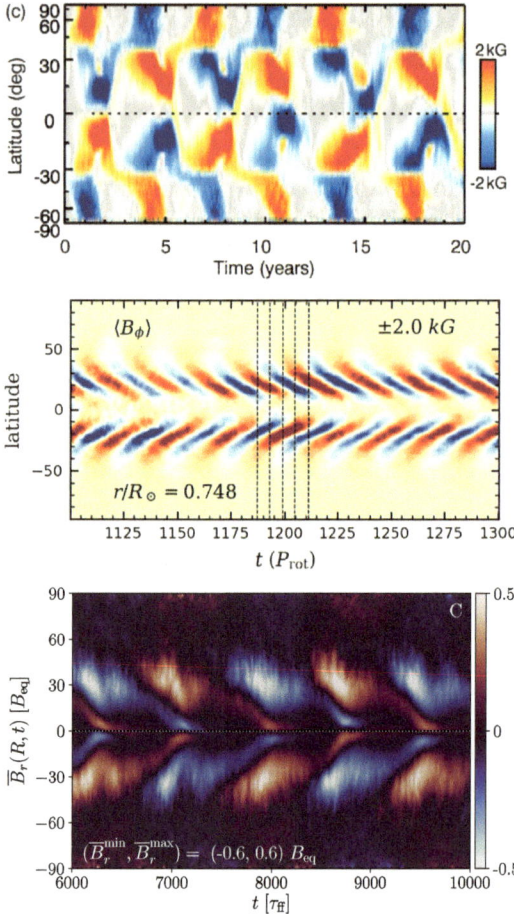

Fig. 7 Butterfly diagrams from Augustson et al. (2015) (top), Matilsky and Toomre (2020) (middle; ©AAS. Reproduced with permission) and Käpylä (2022) (bottom)

most severe issue is the difficulty in obtaining solar-like equatorward migration of activity belts in simulations with solar-like differential rotation. Nevertheless, several simulations have appeared showing equatorward migration and which capture many aspects of solar observations. For example, Käpylä et al. (2012) reported equatorward migration from spherical wedge simulations that were later shown to be in accordance with a Parker–Yoshimura dynamo wave resulting from a mid-latitude minimum of angular velocity which is not present in the Sun (Warnecke et al. 2014, 2018). A similar mid-latitude dip is seen also in the equatorward migrating solutions of Augustson et al. (2015). Further examples of equatorward propagating solutions have been reported by Duarte et al. (2016), Matilsky and Toomre (2020), Strugarek et al. (2017, 2018), and Brun et al. (2022); see also Fig. 7. The latter authors argued that a non-linear interplay between the magnetic fields and differential rotation can lead to solar-like long period cyclic dynamos. Another recent example shows equatorward migration near the surface but poleward migration at depth in a star-in-a-box model (Käpylä 2022), where a spherical star is embedded into a Cartesian cube. In such models the boundary of the star is immersed into the domain and, in theory, allows for a more realistic magnetic boundary condition. This was shown to be important in that if the exterior was made a poor conductor, the global dynamo solution changed from oscillatory to quasi-static. This confirms earlier results of Warnecke et al. (2013, 2016), where the influence of a simplified coronal layer as upper boundary on the flow and magnetic field evolution was studied.

These simulation results and their interpretation add to the ongoing debate regarding the location and dominant physical mechanisms of the solar global dynamo. The current state of affairs is particularly clearly manifested by the wide variety of mean-field models that have been put forward to explain the solar cycle. A popular class of models include flux-transport (e.g. Dikpati and Charbonneau 1999) and Babcock-Leighton (e.g. Cameron and Schüssler 2017) dynamos where a minimal set of physically plausible ingredients, such as differential rotation and meridional circulation and the decay of active regions near the solar surface are taken into account. These models typically rely on buoyantly rising flux ropes (e.g. Caligari et al. 1995) that are the result of strong shear in the tachocline, and the turbulent flows in the convection zone play only the role of turbulent diffusion. On the other hand, distributed turbulent dynamos take into acount a variety of effects that arise in mean-field electrodynamics and assume magnetic field generation throughout the convection zone (e.g. Brandenburg et al. 1992; Käpylä et al. 2006; Pipin and Kosovichev 2013). The obvious drawback of mean-field models is that the individual effects can be adjusted which leads to a great temptation to fine-tune the models. The main advantage of 3D simulations is that this freedom is greatly reduced (although by no means completely eliminated). A major part of the debate regarding the solar dynamo revolves around the relevance of the tachocline.

Unfortunately the advent of 3D simulations with and without tachoclines has not provided conclusive evidence one way or the other. For example, Guerrero et al. (2016) studied both cases with EULAG simulations and found that dynamos operating solely in the convection zone had shorter cycles and intermittent turns-off of activity. On the other hand, dynamos operating at tachocline levels result in long, coherent, magnetic cycles. Whereas the dynamos operating in the convection zone are understood as distributed $\alpha\Omega$ dynamos, those operating at the tachocline may be of $\alpha^2\Omega$ type, with the α effect generated by instabilities that extract energy from the magnetic field (e.g., Tayler or buoyancy instabilities; see Guerrero et al. 2019). Furthermore, simulations with (Käpylä 2022) and without (Käpylä et al. 2012; Warnecke 2018) tachoclines from other models produce cyclic dynamos that share some characteristics of the solar cycle. A general conclusion is that long cycles appear to be generated at depth while shorter ones have their origin near the surface (e.g. Käpylä et al. 2016).

Apart from the non-linear dynamo invoked by Strugarek et al. (2017) and Brun et al. (2022), another physical mechanism proposed to be responsible for the equatorward migration include helicity inversion in the deep parts of the convection zone. This is typically encountered in overshoot regions below the convection zone (e.g. Ossendrijver et al. 2001). However, a much deeper helicity inversion was obtained in simulations by Duarte et al. (2016), and which resulted in a change of the propagation direction of the dynamo wave in accordance with the Parker-Yoshimura rule. In these simulations convection was, however, inefficient in the bulk of the convection zone which is not the situation in the solar convection zone. Such reversed helicity configuration can perhaps arise if much of the convection zone is stably stratified as in the proposed entropy rain scenario, but there is currently no simulation that has produced this.

A yet further possibility is that the solar dynamo is driven predominantly by the kinetic helicity as in classical α^2 dynamos that can lead to equatorward migration if the α effect has a sign change at the equator (Baryshnikova and Shukurov 1987; Rädler and Bräuer 1987). Such helicity profile is expected due to symmetry arguments theoretically, and is a standard outcome in convection simulations (e.g. Brun et al. 2004). The equatorward migrating dynamo wave has been demonstrated by three-dimensional forced turbulence simulations (e.g. Mitra et al. 2010; Warnecke et al. 2011), but no definitive evidence from convection exists.

There have also been attempts to capture the long term modulations in dynamo cycles from simulations. These studies have extended the simulations to cover several tens of cycles corresponding to up to a millenium in solar time (e.g. Passos and Charbonneau 2014; Augustson et al. 2015; Käpylä et al. 2016). These simulations revealed modulation of activity and occasional periods of low activity reminiscent of grand minima (Augustson et al. 2015; Käpylä et al. 2016). Such grand minima states can arise due to an interplay of symmetric and anti-symmetric dynamo modes (e.g. Tobias 1997) or stochastic fluctuations in the buoyancy driving or in the conventional α effect (e.g. Ossendrijver 2000; Brandenburg and Spiegel 2008). However, the modulations and minima in simulations are clearly weaker compared to the solar observations. This is perhaps not too surprising because to run these simulations sufficiently long they need to be done at modest resolutions and cannot therefore be highly supercritical.

5.2 The Sun at Different Ages

During its 4.5 billion years of evolution, the Sun has experienced various changes in its internal constitution, and therefore, the extent of its convection zones and resulting large-scale flows and magnetic fields. In this section we describe numerical simulations of the Sun, or Sun-like stars, corresponding to these evolutionary stages from the formation to the current age.

5.2.1 The Pre-Main Sequence Phase

Significant structural changes occurred early in the solar evolution during the pre-main sequence (PMS) stage, as the newly formed object is still contracting. Objects at this stage, with masses similar to the solar mass, are called TTauri stars. While the temperature at the center of the protostar is still increasing, the opacity of the gas is high and the transport of energy occurs entirely due to convection. The actual rotation rate of the Sun in the TTauri phase is unknown, but models can be constructed to characterise it (e.g. Ahuir et al. 2020). Moreover, observations of open clusters have found distributions of rotational periods between roughly 1 and 10 days for solar-like stars with ages around 3 Myrs (see e.g., Gallet and Bouvier 2013).

The large-scale magnetic fields of TTauri stars are predominantly dipolar with field strengths of the order of kG (e.g. Johns-Krull 2007). Fields with a similar topology are also often observed in low mass, fully convective and rapidly rotating M dwarfs (e.g. Kochukhov 2021). Therefore, despite the difference in mass, simulations of TTauri stars and low-mass M dwarfs are, to some extent, comparable. However, the latter will be discussed in detail in Sect. 5.3.3. Following the evolution further, the protoplanetary disc disappears after about $10^6 - 10^7$ years since the beginning of the collapse. The protostar continues to contract, and therefore its angular velocity increases. Simultaneously, the star starts to develop a radiative core. Both, the angular velocity and the radiative zone increase before the star reaches the zero age main sequence (ZAMS) after about 5×10^7 years. As mentioned above, during the TTauri phase the magnetic field of a solar-like star is mainly dipolar. Observations suggest increasing complexity of the magnetic topology once the star develops a radiative zone (Gregory et al. 2012).

There are currently only a few simulation studies that specifically target dynamos in the TTauri and PMS phases of stellar evolution. One such example is the study of Zaire et al.

(2017) who considered models in the fully and partially convective phases of PMS evolution. While the differential rotation was more pronounced in the latter evolutionary phase, the resulting quasi-steady predominantly quadrupolar magnetic field configurations were quite similar. A more complete study was performed by Emeriau-Viard and Brun (2017), where five epochs between the TTauri stage and the ZAMS were studied for a 1 $M_\odot$ star. Each epoch is characterised by a diffrent internal structure and rotation period. The sequence of simulations shows a decreasing dipole contribution to the magnetic field as a function of age. However, even in the early fully convective phase, the dipole constitutes only about ten per cent of the total magnetic energy. In this case the azimuthally averaged large-scale magnetic field is cyclic with poleward migration, reminiscent of the simulations of fully convective M dwarfs (e.g. Brown et al. 2020; Käpylä 2021). The transition between dipole-dominated and multipolar dynamos is discussed in more detail from the perspective of simulations in Sect. 5.3.3.

5.2.2 Main Sequence Sun-Like Stars: Rotational Evolution of Differential Rotation and Dynamos

On the main sequence, the rotation rate of stars decreases following the observational Sku-manich (1972) law, associated to the loss of angular momentum due to magnetized stellar winds. There have been some speculative ideas about the origin of the non-saturated and saturated regimes of the rotation-activity relationship of stars (e.g. Kawaler 1988; Matt et al. 2015). For instance, Wright et al. (2011) suggested that a turbulent (interface) dynamo is at work in rapidly (slowly) rotating stars. However, the fact that fully convective stars also follow the power law for slow rotation (Wright and Drake 2016) suggests the possibility of a general dynamo theory for all main sequence stars. Nevertheless, to the date, there is no agreement about this theory. One of the main hindrances is the difficulty in observing differential rotation as a function of Ro (e.g. Reinhold and Arlt 2015; Benomar et al. 2018), but a new approach has been recently proposed by Noraz et al. (2022) using KEPLER data. Equally difficult is obtaining unambiguous measurements of dynamo cycle periods and systematics as a function of rotation as already discussed. Numerical simulations, on the other hand, can be performed at arbitrary rotation rates corresponding to different ages of the star. Below, we summarize the relevant findings for the differential rotation and dynamos for a solar mass star from its youth to the present age and beyond (see also Noraz et al. 2023).

The Rossby number dependence of large-scale mean flows has been studied in various papers (e.g. Ballot et al. 2007; Käpylä et al. 2011a,b; Guerrero et al. 2013; Gastine et al. 2014; Featherstone and Miesch 2015; Brun et al. 2017). These studies considered different rotation rates for roughly the same structural model resembling the solar interior. Irrespective of the numerical scheme, the results confirmed that the relative radial differential rotation $\Delta\Omega/\Omega$ changes from positive (solar-like differential rotation), for small Ro_f, to negative (anti-solar) for large Ro_f (see top panel of Fig. 8, adapted from Gastine et al. 2014), with the transition happening near $Ro_f = 1$. In Viviani et al. (2018), the modulus of the absolute differential rotation was also found to decrease rapidly with the rotation rate for $Ro_f \lesssim 0.1$; see bottom panel of Fig. 8 and Fig. 8 of Brun et al. (2022). This decrease, however, can be due to low supercriticality of convection at such low Ro_f. General consensus from simulations is that for sufficiently rapid rotation the differential rotation is negligibly small. This has implications on the theoretical interpretation of dynamos in the classical mean-field dynamo framework; see Sect. 6. Another characteristic is the appearance of non-axisymmetric convective modes, or active nests, in the rapidly rotating regime, $Ro_f \ll 1$ (Brown et al. 2008). Such non-axisymmetric convection has recently been suggested to be the origin of stellar ac-

Fig. 8 Top panel: Measure of the radial differential rotation $\alpha_e = \Delta\Omega/\Omega_0$ as a function of Ro_c at the equator for several studies in the literature. Adapted from Gastine et al. (2014). Bottom panel: Modulus of the absolute latitudinal differential rotation from simulations of a solar-like star Viviani et al. (2018), where $\tilde{\Omega}$ is the rotation rate normalized by the solar rotation (©ESO. Reproduced with permission). The dotted line corresponds to the transition of anti-solar to solar-like differential rotation at $\mathrm{Ro}_f \approx 0.35$ and the dashed line separates the two regimes of differential rotation dependence at $\mathrm{Ro}_f \approx 0.08$

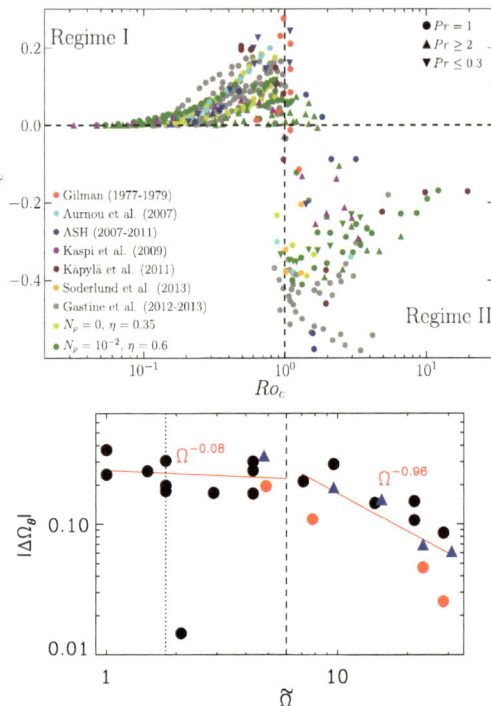

tive longitudes (Bice and Toomre 2022). Regarding the structure of convection, it is evident from all simulations that rotation breaks the broad convective cells observed in non-rotating or slowly rotating simulations. Quantitatively, Featherstone and Hindman (2016) found that the harmonic degree where the spectrum has a maximum, ℓ_{peak}, scales with the Rossby number as $\ell_{\mathrm{peak}} \sim \mathrm{Ro}_f^{-1/2}$ (see also Viviani et al. 2018). This means that the faster the rotation, the smaller the scales where most of the kinetic energy is contained. This applies to regions of the convection zone where $\mathrm{Ro}_f \lesssim 1$; in the near-surface layers $\mathrm{Ro}_f \gg 1$ even in the most rapidly rotating stars, and the size of photospheric convection cells is likely independent of stellar rotation.

As discussed in Sect. 4, the rotation of stars slows down due to magnetic braking as they age. The young Sun was therefore a much faster rotator than what it is today. Simulations of rapidly rotating ($\mathrm{Ro}_f \lesssim 0.1$) young solar-like stars have been performed by several groups using various numerical methods. Quite surprisingly, the results in this parameter regime are rather inhomogeneous: there are three distinct dynamo modes that have been reported from such studies. First, there is the large-scale dipole-dominated solutions (e.g. Gastine et al. 2012; Yadav et al. 2015b; Zaire et al. 2022) that are reminiscent of the geodynamo and those of Saturn and Jupiter. Another outcome is that the large-scale magnetic fields are dominated by a non-axisymmetric $m = 1$ mode that often propagates either in retro- or prograde fashion (e.g. Cole et al. 2014; Yadav et al. 2015b; Viviani et al. 2018; Viviani and Käpylä 2021; Navarrete et al. 2022). Finally, some simulations produce predominantly axisymmetric but multipolar large-scale fields in such rapidly rotating setting (e.g. Strugarek et al. 2018). It is unclear why there is such a variety in magnetic field topologies in this regime. A possible cause is that the dynamo is sensitive to relatively minor differences in the

Fig. 9 Top panel: Summary of the fraction of differential rotation energy from total kinetic energy as a function of Ro_ω from the simulations of Strugarek et al. (2017) and Brun et al. (2022). The colours of the symbols indicate the type, or the lack of, magnetic cycles. Adapted from Brun et al. (2022). Bottom panel: Comparison of the ratio of the rotation period and the cycle periods as a function of the Coriolis number ($\mathrm{Co} = \mathrm{Ro}_f^{-1}$) from the studies of Strugarek et al. (2018) (blue), Warnecke (2018) (red), Guerrero et al. (2019), and Käpylä (2022). Estimated location of the Sun is indicated by the symbol $\odot$. Adapted from Käpylä (2022)

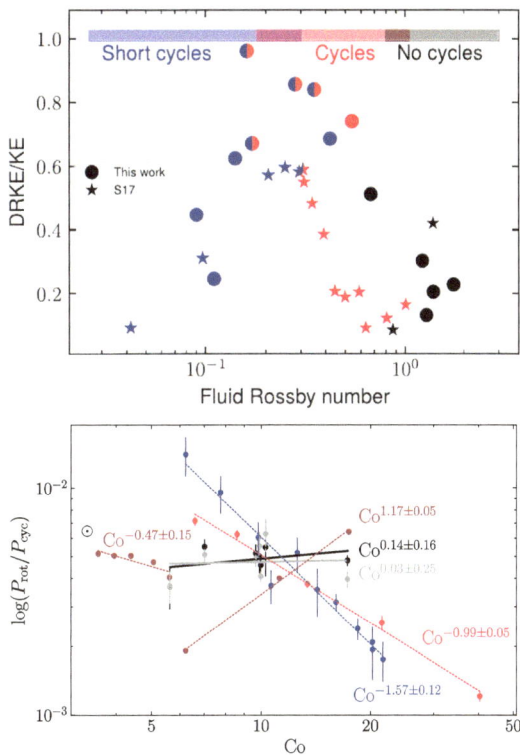

boundary conditions (see, e.g. Warnecke et al. 2016) and/or other details of the simulation setups (Orvedahl et al. 2021).

Two further regimes of dynamos can be distinguished on either side of transition between solar-like to anti-solar differential rotation. When rotation is rapid enough such that a solar-like differential rotation is produced, the large-scale fields are predominantly axisymmetric and often cyclic (e.g. Brown et al. 2010; Ghizaru et al. 2010; Käpylä et al. 2012; Nelson et al. 2013; Augustson et al. 2015; Warnecke 2018; Matilsky and Toomre 2020). As mentioned above, in some cases this behavior continues to much more rapid rotation whereas in others non-axisymmetric or dipolar dynamos modes take over. When rotation is slow enough and the differential rotation is anti-solar, the magnetic fields are predominantly axisymmetric and quasi-steady (e.g. Käpylä et al. 2017; Strugarek et al. 2018; Warnecke 2018). In the transition between the two regimes, the dynamo may excite the two modes simultaneously (Viviani et al. 2019). The appearance of cycles appears to be related to the strength of the differential rotation such that long decadal cycles, such as in the Sun, appear in a relatively narrow range of Rossby numbers where the differential rotation is solar-like and it is sufficiently strong (Guerrero et al. 2019; Warnecke and Käpylä 2020; Brun et al. 2022, see the top panel of Fig. 9).

Regardless of the commonalities between different modeling approaches, the simulated magnetic cycles seem to be sensitive to the subtleties of the models, and there is no clear agreement regarding the scaling of cycle periods as a function of rotation as can be seen in the bottom panel of Fig. 9 compiling the results of Strugarek et al. (2018), Warnecke (2018), Guerrero et al. (2019), and Käpylä (2022). Furthermore, none of the scaling laws from simulations seem to unambiguously agree with the observed cycles that are sometime grouped in activity branches with $P_{\rm rot}/P_{\rm cyc} \propto {\rm Co}^{\alpha}$, where $\alpha > 0$ (e.g. Saar and Brandenburg 1999; Brandenburg et al. 2017). It is worth mentioning, however, that the observational results may also present problems and, depending on the used analysis techniques, the distinct activity branches may not even exist (e.g. Bonanno and Corsaro 2022).

As shown in Fig. 8 (top), the amplitude of the differential rotation has maxima on either side of the solar-like to anti-solar transition. The strong shear in the anti-solar regime has been speculated to lead to enhanced magnetic activity (Brandenburg and Giampapa 2018). While this has not been systematically studied, some evidence of enhanced magnetic energy for anti-solar differential rotation has been found in simulations (e.g. Karak et al. 2015; Warnecke and Käpylä 2020; Brun et al. 2022). A much more unambiguous observational fact is that the stellar magnetic activity and magnetic fields saturate for ${\rm Ro} \lesssim 0.1$ (e.g. Reiners et al. 2022, see also Sect. 4). This does not appear to happen in simulations, where the magnetic energy typically increases with rotation even for the fastest rotation considered so far (e.g. Warnecke and Käpylä 2020). Furthermore, the ratio of magnetic to kinetic energy increases roughly proportional to ${\rm Ro}_{\rm f}^{-1}$ (Augustson et al. 2019; Warnecke and Käpylä 2020) in accordance with Magneto-Archimedean-Coriolis (MAC) balance. Brun et al. (2022) showed that the large-scale surface fields follow even a steeper increasing trend $\propto {\rm Ro}_{\rm f}^{-1.4}$ with decreasing Rossby number. However, this is consistent with stellar observations in the magnetically non-saturated regime (See et al. 2019; Brun et al. 2022). Nevertheless, it is unclear why the simulations deviate from observations in that they do not show indications of saturation when the Rossby number is decreased. However, the magnetic fields of rapidly rotating stars appear to follow a similar scaling law as the geo- and planetary dynamos as well as rapidly rotating convective dynamo simulations (Christensen et al. 2009; Yadav et al. 2013).

Another possibility is that small-scale dynamo action could also play a role in this by diverting energy to smaller scale fields that are more rapidly diffused. It is likely that most rapidly rotating stellar dynamo simulations currently do not have SSDs because of too low ${\rm Re}_{\rm M}$ and reduced integral scale of turbulence due to the rotational constraint. This is also manifested by the relative dearth of studies concentrating on the interactions between SSD and LSD. The studies of Käpylä et al. (2017) and Ortiz-Rodríguez et al. (2023) considered simulations of solar-like and fully convective stars, respectively, and where ${\rm Pr}_{\rm M}$ was varied while all of the other parameters were fixed. Both studies found that when ${\rm Pr}_{\rm M}$ (and consequently ${\rm Re}_{\rm M}$) is increased sufficiently such that an SSD is excited, the relatively regular cycles of large-scale fields at low and moderate ${\rm Re}_{\rm M}$ are disrupted and differential rotation is severely quenched. On the other hand, Hotta et al. (2016) found that also large-scale fields were amplified when an SSD was excited in comparison to a run where it was absent. The SSD and its interaction with LSD is discussed in more detail in Rempel et al. (2023).

It is also plausible that missing surface physics, such as the lack of spot formation in the simulations contributes to the issue of saturation of magnetism at small Ro. There are a handful of global simulations that achieve dynamo action and buoyantly rising flux bundles (Nelson et al. 2013, 2014; Fan and Fang 2014). These buoyant structures are much larger than active regions in the Sun or thin flux tubes that are envisaged to be essential for flux transport dynamos. Furthermore, none of the current simulations produce surface features that are comparable to sunspots. The main reason for this is that the spatial resolution of current global simulations is likely much too coarse to capture the spot formation process self-consistently, irrespective if the origin of spots is buoyant rise of thin flux tubes or some local MHD instability. Instead of trying to solve the entire problem in a single model, Chen et al. (2017) presented a hybrid solution where the suitably scaled magnetic field from a global dynamo simulation from Fan and Fang (2014) was used as a lower boundary condition in a high-resolution Cartesian surface convection simulation that has the resolution to capture spot formation. Coupled modelling efforts such as this are likely to be the best bet in the near future to capture self-consistent dynamo action and spot formation.

5.3 Stars Other than the Sun

Most stars are not like the Sun. A large majority (in our galaxy, anyway) are M dwarfs, which are less massive than the Sun, considerably less luminous (ranging down to about $10^{-3} L_\odot$), and in some cases convective throughout their interiors. At the other end of the H-R diagram, stars more massive than the Sun have convective cores and predominantly stable (radiative) envelopes, accompanied in some cases by thin near-surface convection zones. These high-mass stars can be thousands of times more luminous than the Sun; thus across the main sequence, luminosities vary by a factor of more than a million. These enormous variations in luminosity, and in the geometry and stratification as well, must influence the convection and dynamo action occurring in these stars. In this section, we therefore briefly describe our current understanding of dynamo action in main-sequence stars at lower and higher masses. Our focus here is on what has been revealed by basic theory and by simulations.

From the standpoint of dynamo theory, stars vary not just in their luminosity but also their rotation rate, their geometry, their stratification, and their microphysics. For example the core of a massive star is only weakly stratified, whereas an M dwarf (or the convective envelope of a Sun-like stars) is much more strongly stratified in density and temperature. The microscopic diffusivities vary in such a way that Pr_M is very small in many stellar convection zones but greater than unity in some (e.g. Augustson et al. 2019; Jermyn et al. 2022). Many of these variations are intertwined. For example, the influence rotation has on the convection is typically encapsulated by some version of the Rossby number $Ro \sim P_{rot}/\tau_{conv}$; because τ_{conv} depends on the flow velocity, the rotational influence can vary from star to star (and with radius within a given star) even if the rotation period is constant.

Faced with the bewildering array of possible variations in all of these parameters, would-be simulators tend to try one of two different approaches. One is to focus on a particular physical effect – stratification, for example – and try to elucidate how this affects convection, differential rotation, and magnetism, typically in an idealized setting (e.g., a Cartesian layer). Another is to attempt to model a given astrophysical object in some detail, adopting spherical coordinate systems and stratifications (of density, temperature, or entropy) that mimic those in a 1D stellar model. In neither approach is it possible for simulations to approach the actual parameter regimes attained in real stars, although some balances can be recovered that help in understanding, e.g. the type of differential rotation that occur in real stars. Before diving into a "by object" discussion, we turn first to a high-level overview of the effects of rotation, stratification, and geometry on dynamo action in general. Then we turn to summaries of how these effects play out in specific types of stars.

5.3.1 Effects of Rotation, Stratification, and Geometry in Stellar Models

All stars rotate, all have some level of density stratification, and all convect in a spherical geometry. Here we provide a very brief overview of how each of these effects likely influence the flows and magnetism. Perhaps the clearest message from the last few decades of research on convection and dynamo action is, "rotation matters a lot." It affects the convective flows, it affects their transport of heat and angular momentum, and it affects the magnetism these flows can build. Some of these effects (e.g., how rotation affects magnetic field morphology) are understood only qualitatively, whereas for others (e.g., how it affects heat transport) there is now a more quantitative picture.

To begin, consider the influence of rotation on the convective flows in the absence of magnetism. It is well-known that rotation tends to stabilise a system against convection, increasing the critical Rayleigh number for onset ($Ra_c \propto Ek^{-4/3}$, with Ek the Ekman number defined previously), while the most unstable wavenumber shifts to higher wavenumber (k) with more rapid rotation rate (lower Ek) (e.g. Chandrasekhar 1961). (Less obviously, consider a numerical simulation not too far from onset – more specifically, one for which "diffusion free" scalings do not yet apply – with, for example, a fixed heat flux or temperature contrast, and also some fixed rotation rate. In this system changing the numerical diffusivities – or in an implicit LES simulation changing the resolution – will also affect the rotational influence, via changing Ek, and so also the flows.) In a global spherical geometry, whenever rotation play a major role in the dynamics the most prominent convective modes close to onset are wave-like convective rolls, aligned with the rotation axis and arising from the conservation of potential vorticity, variously called "thermal Rossby waves," "Busse columns," or "banana cells" depending on the community (see, e.g. Busse 1970; Gilman 1977; Busse 1983; Featherstone and Hindman 2016; Bekki et al. 2022). The systematic prograde tilt of these cells, and the associated Reynolds stress, plays a major role in redistributing angular momentum in most global-scale simulations of stellar convection; the differential rotation is then also a strong function of rotation rate (Rossby number), as summarized elsewhere (e.g. Gastine et al. 2014).

The heat transport by the convection is also strongly influenced by rotation: for a given fixed heat flux, the convection tends to become "less efficient" as the rotation rate increases – that is, it requires a larger entropy gradient to carry the same flux (see, e.g., discussions in Stevenson 1979; Aurnou et al. 2020). Quantitatively, several theoretical approaches supported by numerical simulations – including appeals to a balance between Coriolis, inertial, and buoyancy forces (so-called "CIA balance") (e.g Vasil et al. 2021), "rotating mixing length" theory (Stevenson 1979; Barker et al. 2014; Currie et al. 2020) asymptotic theory at low Rossby number (Julien et al. 2012) – yield the same diffusion-free dependence of key quantities, like the temperature gradient, when rotation is rapid enough. In these models the amplitude of the convective motions goes down at higher rotation rates, the temperature gradient goes up, and the typical wavenumber of the flow increases, with important consequences for the dynamo.

The dependence of dynamo action itself on rotation is less quantitatively understood, though it is still possible to make some broad statements that are supported by theory and simulation. When rotation is dynamically significant, it is widely thought to affect the strength and morphology of dynamo-generated magnetic fields as discussed in more detail in Sect. 5.2.2. (see also, e.g., Schrinner 2013; Augustson et al. 2019; Orvedahl et al. 2021). Furthermore, because the envelope convection zones of low-mass stars are strongly stratified (in density and temperature), many authors have also sought to understand what role this stratification plays in determining various properties of the flows and magnetism. We mention here only a few. Broadly, the presence of density stratification breaks the up-down symmetry in Boussinesq systems, so it changes the convective dynamics – and also presumably the magnetism – in a variety of ways. Strongly stratified systems tend to have strong, narrow downflows and broader, weaker upflows (e.g. Hurlburt et al. 1984; Cattaneo et al. 1991); their energy dissipation budget is very different (e.g., Hewitt et al. 1975; Currie and Browning 2017); they can establish different profiles of kinetic helicity (Duarte et al. 2016) and may drive different types of zonal flows (e.g., Glatzmaier et al. 2009). If diffusion-free scalings (like mixing-length theory) apply, or simply on dimensional grounds, the velocity of the convective flows is expected to vary in amplitude across a stratified convection zone, in a way not found in Boussinesq systems. This basic fact has profound consequences for the convective dynamics in a star like the Sun, since we then expect regions near the photosphere (which are very low-density) to undergo such rapid convective motions that their Rossby number is enormous (i.e., the influence of rotation is negligible there); meanwhile the deeper flows must be influenced by rotation, at least within some finite distance of the transition to the radiative interior. The presence of density stratification may make it harder to sustain a strong global-scale dipole in some regimes (e.g., Gastine et al. 2012), but it is also clear that highly ordered fields are still realizable even when the stratification is strong (see discussions in Raynaud et al. 2015, Menu et al. 2020, Zaire et al. 2022).

Relatively little work has directly addressed the influence of the *geometry* of the convection zone, while keeping other factors constant. We briefly note only a few specific results. Goudard and Dormy (2008) showed that, in rotating Boussinesq MHD simulations with fixed temperature boundaries, changing the aspect ratio of the domain (i.e., changing the depth of the convection zone) had a strong effect on the nature of their dynamo solutions. In deep domains, they found steady "Earth-like" dipolar solutions; if the convection zone was gradually made thinner, they found a transition to "Sun-like" dynamo wave solutions. Separately, Camisassa and Featherstone (2022) have recently investigated the role of geometry in determining the Reynolds stresses – and hence the differential rotation – in anelastic simulations of rotating convective envelopes. They argue that the well-known transition from solar to anti-solar differential rotation occurred when columnar convective structures attained a diameter roughly equivalent to the shell depth. In the sections that follow, we explore how these different dynamical trends play out in models of specific stars.

5.3.2 Dynamo Action in High-Mass Stars: Convective Cores and Radiative Envelopes

Massive stars on the main sequence possess convective cores, with a predominantly stable radiative envelope above this. The high luminosities established by nuclear fusion in these stars mean that we expect the convection to be vigorous; the accompanying high temperatures mean that we expect these convection motions should generally act as magnetic dynamos, since all plausible estimates of the magnetic Reynolds number Re_M are very high; see Table 1. It may well be that dynamo action is occurring in the radiative layers as well, as discussed below. Both the flows and fields have lately been targets of intense scrutiny – partly because asteroseismology has begun to provide powerful new constraints on this topic, particularly for evolved stars (see, e.g., Stello et al. 2016, Ji et al. 2023) and also because of the implications these hold for later stages of stellar evolution (e.g., Fuller and Ma 2019). In this section, we briefly review what has been learned by simulations focusing on these types of stars. A more thorough description can be found in Brun and Browning (2017).

Early global simulations covering some aspects of the problem include Kuhlen et al. (2003), Browning et al. (2004), and Brun et al. (2005). Taken together, these papers provided the first numerical estimates of the flows and magnetism that might be generated in 3D massive star cores. They also gave some estimates of the overshooting and penetration from the convective zone into the surrounding stable envelope, the gravity wave response there, and the differential rotation arising from the interplay of convection, magnetism, and rotation. Later work has pushed towards more realistic (turbulent) flows, has examined a variety of initial states (e.g., strong magnetic "fossil" fields) and srutinized each facet of this complicated problem more systematically. We refer the interested reader to Meakin and Arnett (2007), Featherstone et al. (2009), Gilet et al. (2013), Rogers (2015), Augustson et al. (2016), Edelmann et al. (2019), Breton et al. (2022), and Baraffe et al. (2023) as a representative sample of relevant work.

There has lately also been sustained interest in the topic of magnetism generated by dynamo action in the radiative envelope, typically as a result of the interaction between shear and magnetic instabilities (e.g., Spruit 1999). Numerical work – beginning with Braithwaite (2006) and followed by many others since (e.g., Zahn et al. 2007, Duez et al. 2010, Jouve et al. 2015, Vidal et al. 2018, Petitdemange et al. 2023, Ji et al. 2023) – has examined the circumstances under which such dynamo action could occur, and the strength of the resulting fields.

Taken as a whole, these simulations are unequivocal about a few points. Very strong fields can plausibly be established by dynamo action in the cores of some massive stars, whether rotation is dynamically significant or not; for example, because the convection is rapid, even the "equipartition-scale" field (equating ρu_{conv}^2 with $B^2/(4\pi)$ and assuming MLT scalings for the convective velocity u_{conv}) would suggest $B \geq 10^6$ G in the interior of a B-type star. Some of these stars rotate very rapidly, and the saturation strength of the field in this case is less clear (see, e.g., discussions in Augustson et al. 2019), but it seems likely that strong fields are the norm rather than the exception. The differential rotation established within the core is less certain, because it is surely influenced by the strength of the magnetism – which is likewise influenced by the shear – but broadly these simulations appear to obey trends similar to those in simulations of convective envelopes. "Solar-like" differential rotation (i.e., with a prograde equator) is established when rotation is dynamically significant and the magnetism is not too strong; "anti-solar" profiles arise when the influence of rotation is weaker; magnetism reduces the shear and may yield solid-body rotation if it grows strong enough. Convective motions overshoot into the radiative envelope, though the extent of this effect is still being actively investigated (e.g., Anders et al. 2022b, Baraffe et al. 2023), and excite a substantial gravity-wave response that may be detectable even in main-sequence stars (e.g., Breton et al. 2022).

Within the radiative zone itself, the latest simulations (Petitdemange et al. 2023, Ji et al. 2023) now appear to be capturing some aspects of the long-envisioned "Tayler-Spruit" dynamo. This differs in some important respects from the picture originally envisioned by Spruit and debated in Zahn et al. (2007); for example, in the Petitdemange et al. (2023) simulations only *subcritical* dynamo action is found, alongside some other dynamo instability. The saturation amplitude of the field is still very much under debate; see discussions in Spruit (1999), Fuller et al. (2019), and Ji et al. (2023).

5.3.3 Low-Mass Stars and the Transition to Full Convection

Main-sequence stars less massive than the Sun have deeper convective envelopes (as a fraction of the total stellar radius). Below a mass of about 0.35 $M_\odot$ stars are – in standard 1D models – convective throughout their interiors. This transition occurs at a spectral type of around M3, so "M dwarfs" in general hold special interest theoretically, as probes of the various roles that rotation and stratification play in the dynamo. They are also interesting astronomically: the large majority of stars in our galaxy are M dwarfs, and they are popular targets in the quest to find and characterise exoplanets (e.g., Trifonov et al. 2018).

Here, we briefly review attempts to model these stars numerically. In terms of fundamental fluid dynamics, these stars have many similarities with pre-main sequence stars (which are also fully convective, but can have different internal heating profiles and rotational constraints), so in places our discussion parallels that in Sect. 5.2.1. In some other respects the flows in these objects resemble those in giant gaseous planets, so we also draw comparisons to the extensive literature on planetary dynamos. Finally, there are also many similarities with the flow in massive-star cores, which share the same geometry (i.e., a full sphere of convection) but are much less strongly stratified than a main-sequence M dwarf; see Table 1.

The first 3D MHD simulations that aimed specifically to model low-mass fully convective stars were reported in Dobler et al. (2006). They used the PENCIL CODE, solving the fully compressible equations to model a spherical star (established via volumetric heating and cooling terms) embedded in a Cartesian grid. These first models included only a fairly weak density stratification (with ρ at the center of the star about a factor of three greater than at its photosphere). Later, Browning (2008) conducted the first anelastic simulations (with the ASH code) that mimicked low-mass M dwarfs, including a stronger density stratification (about a factor of 100 across the deep spherical shell) and more complex flows. Subsequent work has sampled much lower diffusivities, more extreme density stratifications, and varying rotational influences. We note in particular the simulations of Gastine et al. (2012) and Yadav et al. (2015, 2016), modelling anelastic dynamos in a deep spherical shell with the MAGIC CODE; Brown et al. (2020), who considered a full spherical geometry (i.e., in spherical coordinates but with no singularity at $r = 0$) using the DEDALUS framework; Käpylä (2021), who considered "star-in-a-box" models akin to those of Dobler et al. (2006) but in a substantially different parameter regime; and Bice and Toomre (2020, 2022), modelling deep (anelastic) shells of convection with the RAYLEIGH code.

Although there are many differences between these models, some broad trends are now reasonably clear, and largely parallel those realised in other objects and geometries (as discussed elsewhere in this review). When magnetism is weak or absent, both "solar" and "anti-solar" differential rotation can be realised, depending upon the rotational influence (i.e., some version of the Rossby number). When magnetism is strong, it reduces this differential rotation and – if the field gets strong enough – can essentially eliminate it, leading to solid-body rotation. The spatial structure of the field – e.g., the fraction of the magnetism in axisymmetric components, or the strength of the dipole or quadrupole moment – is intertwined with rotation, shear, and density stratification in a complex manner.

When rotation is strong and shear is weak, the field tends to develop a large-scale dipolar component; see discussions in, e.g., Gastine et al. (2012), Schrinner et al. (2012), Yadav et al. (2015), and an array of related works modeling planetary dynamos with weak stratifications (e.g., Christensen and Aubert 2006, Schwaiger et al. 2021). In real stars, presumably any large-scale field generation is accompanied by vigorous SSD, so the overall field likely consists of a very wide range of scales. The dipolar solutions are, at least in some parameter regimes, more difficult to realise when the density stratification is strong and when the convective supercriticality is high (e.g., Gastine et al. 2012); but there are now multiple examples of highly-stratified, vigorous convection that exhibit strong dipoles (e.g., Yadav et al. 2015), including some at surprisingly modest rotational constraints (e.g., up to Ro ~ 0.4 in Zaire et al. 2022). The question of what exactly delineates these states from one another is a topic of very active investigation (see, e.g., discussions in Menu et al. 2020, Tassin et al. 2021, Zaire et al. 2022).

When shear is also present, a variety of solutions are possible, including propagating dynamo waves. Examples abound; see, for example, Yadav et al. (2016), Käpylä (2021), and Bice and Toomre (2022). Again, it seems reasonably clear that the rotational influence is the most crucial control parameter, but many details remain unclear. Other topics of active interest include the prevalence of non-axisymmetric features in the field (e.g., Bice and Toomre 2022, Käpylä 2021), or modes in which the bulk of the magnetism is confined to one hemisphere (e.g., Gallet and Pétrélis 2009, Gastine et al. 2012, Brown et al. 2020). Furthermore, Käpylä (2021) reported that the qualitative succession of dynamo modes as a function of rotation appears to be the same in simulations of fully and partially convective stars: when rotation increases, the predominantly axisymmetric steady and cyclic solutions at slow rotation give way to non-axisymmetric dynamos at rapid rotation. In these models a similar succession happens with the cycles, so that for moderate rotation the dynamo waves typically propagate in latitude; when rotation is more rapid, the large-scale magnetic structure drifts in longitude.

6 Connections to Mean-Field Dynamo Theory

Finding out which physical processes lead to the observed magnetic field evolution in 3D convective dynamo simulations is very challenging. An often-used approach is to interpret the outcome of the simulations in terms of mean-field dynamo theory. Mean-field theory provides a well-established theoretical foundation, which can be used to analyse the complex 3D simulation in a simplified way (e.g. Krause and Rädler 1980; Brandenburg and Subramanian 2005); see also Brandenburg et al. (2023). Technically this is done by computing mean-field transport coefficients from 3D simulations and using them in a corresponding mean-field model. This approach has been successfully used to pinpoint the cause of magnetic field evolution in many simulations. In mean-field theory the magnetic and velocity fields are divided into a mean or averaged part and a fluctuation, e.g. $\boldsymbol{B} = \overline{\boldsymbol{B}} + \boldsymbol{B}'$. Only the mean fields are explicitly solved for, whereas the (correlations of) fluctuations are parameterized in terms of the mean. Azimuthal averages are often used for solar and stellar dynamos such that the resulting mean field is axisymmetric. However, such an average is not well suited for rapidly rotating stars, where non-axisymmetric $m = 1, 2$ modes often dominate (Viviani et al. 2018). Applying the mean-field approach to the induction equation leads to the emergence of an additional term, the electromotive force $\overline{\boldsymbol{\mathcal{E}}} = \overline{\boldsymbol{u}' \times \boldsymbol{B}'}$. In mean-field dynamo theory this term is parameterized in terms of the $\overline{\boldsymbol{B}}$ and its gradients, assuming that the mean fields vary slowly in space and time (Krause and Rädler 1980),

$$\overline{\mathcal{E}}_i = a_{ij} \overline{B}_j + b_{ijk} \frac{\partial \overline{B}_j}{\partial x_k} + \cdots, \tag{28}$$

where the dots represent higher order derivatives that are most often neglected. The tensors $\boldsymbol{a}$ and $\boldsymbol{b}$ can then be further divided into symmetric and anti-symmetric parts (e.g. Rädler 1980), yielding an equivalent representation

$$\overline{\boldsymbol{\mathcal{E}}} = \boldsymbol{\alpha} \cdot \overline{\boldsymbol{B}} + \boldsymbol{\gamma} \times \overline{\boldsymbol{B}} - \boldsymbol{\beta} \cdot (\boldsymbol{\nabla} \times \overline{\boldsymbol{B}}) - \boldsymbol{\delta} \times (\boldsymbol{\nabla} \times \overline{\boldsymbol{B}}) - \boldsymbol{\kappa} \cdot (\boldsymbol{\nabla} \overline{\boldsymbol{B}})^{(s)}, \tag{29}$$

where $(\boldsymbol{\nabla} \overline{\boldsymbol{B}})^{(s)}$ is the symmetric part of the magnetic field gradient tensor. The coefficients $\boldsymbol{\alpha}$ and $\boldsymbol{\beta}$ are rank two tensors, $\boldsymbol{\gamma}$ and $\boldsymbol{\delta}$ are vectors, and $\boldsymbol{\kappa}$ is a rank three tensor. These coefficients can be associated with different turbulent effects important for the magnetic field evolution: the α effect (Steenbeck et al. 1966) leads to field amplification via helical flows; γ describes the turbulent pumping, which acts like mean flow (e.g. Rädler 1968; Roberts and Soward 1975); β describes turbulent diffusion; and the δ effect, also known as the Rädler effect (Rädler 1969) or the shear-current effect (e.g. Rogachevskii and Kleeorin 2003), can lead to dynamo action in non-helical turbulence in the presence of shear; and finally, the κ effect, whose physical interpretation is currently unclear.

The main challenge is to determine these coefficients from a 3D global convective dynamo simulation. The ultimate goal is to be able to reproduce the magnetic field solution with a mean-field model using the obtained coefficients.

6.1 Using Proxies Based on Flow and Magnetic Field Properties

The simplest approach to compare 3D convection simulations with mean-field theory is to use approximate proxies of turbulent transport coefficients based on the flow and magnetic field properties. Assuming isotropy and homogeneity and applying the second order correlation approximation (SOCA), the turbulent transport tensors reduce to scalars α_K and β, (e.g. Krause and Rädler 1980; Brandenburg et al. 2023)

$$\alpha_K = -\frac{1}{3}\tau\overline{\boldsymbol{\omega}' \cdot \boldsymbol{u}'}, \quad \beta = \frac{1}{3}\tau\overline{\boldsymbol{u}'^2},\tag{30}$$

where τ is the correlation time of the flow, and where $\boldsymbol{\omega}' = \nabla \times \boldsymbol{u}'$. The expressions of α_K and β are valid in the kinematic regime where the back-reaction of the magnetic field on the flow is neglected. In a more general approach, the minimal tau-approximation, this backreaction is retained and this leads to an additional magnetic contribution to the α effect (Pouquet et al. 1976; Blackman and Field 2002)

$$\alpha_M = \frac{1}{3}\tau\overline{\rho}^{-1}\overline{\boldsymbol{J}' \cdot \boldsymbol{B}'},\tag{31}$$

where $\boldsymbol{J}' = \mu_0^{-1}\nabla \times \boldsymbol{B}'$. α_M can be interpreted as a consequence of magnetic helicity conservation (see, e.g. Brandenburg et al. 2023). This approach has been used in many simulation studies to interpret the magnetic field evolution (Charbonneau 2020). For example, Warnecke et al. (2014) used the α proxy to conclude that the equatorward migration found in their, and in previous work, is due to an $\alpha\Omega$ Parker dynamo wave driven by a region of negative radial shear. Similarly, Duarte et al. (2016), explained the equatorward migration in their simulations by the inversion of α and positive radial shear. Guerrero et al. (2016, 2019) concluded that the dynamos in their simulations with tachoclines were driven by α_M below the convection zone. However, it is necessary to bear in mind the approximate nature of analyses based on proxies: often the agreement between the simulation and the behavior suggested by the proxy is qualitative at best and the 3D simulations contain a rich variety of non-linear interactions that are omitted in such analyses.

6.2 Direct Measurements of Coefficients

The alternative is to measure the coefficients in Eq. (29) directly from simulations. There are currently two commonly used methods for this. First, $\overline{\boldsymbol{B}}$ and $\overline{\mathcal{E}}$ from the 3D dynamo simulation can be used to fit for the turbulent transport coefficients in Eq. (29) using, e.g., multidimensional regression method or singular value decomposition (SVD) (Brandenburg and Sokoloff 2002; Racine et al. 2011). On the other hand, the test field (TF) method uses a sufficiently large number of linearly independent *test fields*, that do not back-react on the solution, and evolves the corresponding $\boldsymbol{B}'$ and $\overline{\mathcal{E}}$ for each. Then it is possible to unambiguously invert for the coefficients in Eq. (29) (Schrinner et al. 2005, 2007). Both of these methods are, at best, only as good as the approximate equation (29). The validity of the results needs to be tested by inserting the derived coefficients back into Eq. (29) and to mean-field models to determine how faithfully they capture the $\overline{\mathcal{E}}$ and time evolution of the mean field in the 3D simulation.

The SVD method has the issue that Eq. (29) is underdetermined: there are 27 unknown parameters and only three components of $\overline{\boldsymbol{B}}$ and $\overline{\mathcal{E}}$. This is typically overcome by considering the time dependence of $\overline{\boldsymbol{B}}$ and $\overline{\mathcal{E}}$ leading to an overdetermined system. Furthermore, if $\overline{\boldsymbol{B}}$

does not vary in time the SVD method has problems to converge. Despite these difficulties this method has been used to explain the dynamos in several simulations (e.g. Racine et al. 2011; Augustson et al. 2013, 2015). Simard et al. (2016) found that the coefficients related the gradients of $\overline{B}$ (β, δ, κ) are much less important than α and γ. Furthermore, Simard et al. (2013) could reproduce the mean-field evolution of 3D global simulation using α and γ determined with the SVD method, basically assuming it is an $\alpha^2\Omega$ dynamo. However they had to assume a higher turbulent diffusivity than what was measured. In follow-up studies the authors could explain and reproduce a dual dynamo action (Beaudoin et al. 2016) and generate Grand Minima-like events by including α quenching (Simard and Charbonneau 2020). Recently, Shimada et al. (2022) analysed the simulations of Hotta et al. (2016) with the SVD method and found that the turbulent diffusion decreases with increasing Re_M. However, it is unlikely that Eq. (29), and hence the SVD method, is valid at high Re_M where a SSD is excited.

In the TF method, a set of 9 linearly independent test fields are used to uniquely determine the 27 unknowns. First developed for the geodynamo (Schrinner et al. 2005, 2007, 2011, 2012; Schrinner 2011), it has subsequently been used for many solar and stellar dynamo simulations (Gent et al. 2017; Warnecke et al. 2018; Warnecke 2018; Viviani et al. 2019; Warnecke and Käpylä 2020). An important result is that the turbulent pumping is typically larger than the meridional circulation in global convective dynamo simulations, rendering the flux-transport dynamo scenario unlikely in those cases (Warnecke et al. 2018), because the total (meridional plus turbulent) advection generally does not have closed streamlines. The conclusion of Warnecke et al. (2014) that the equatorward migration in these kind of simulation is explained by a Parker dynamo wave was confirmed with the α effect from the TF method (Warnecke et al. 2018, 2021). This was later used to explain the cycle period dependence on rotation in 3D dynamo simulations (Warnecke 2018). Gent et al. (2017) analyzed the simulations of Käpylä et al. (2016) and found that the turbulent transport coefficients – particularly γ – vary significantly during long-term modulation of the cyclic mean magnetic field. In the work of Viviani et al. (2019), the first cyclic dynamo in the anti-solar differential rotation regime was explained to be of $\alpha^2\Omega$ type using test-field coefficients. Furthermore, Warnecke and Käpylä (2020) studied the transport coefficients as functions of rotation rate and found that $\alpha\Omega$ dynamos, appear to be possible in a relative narrow range in Ro. The trace of α agrees with α_K in pattern and amplitude in a Rossby number range spanning three orders of magnitude (Warnecke and Käpylä 2020). As a result of the test-field analysis (Warnecke and Käpylä 2020) a magnetic influence on α as described in Eq. (31) could be ruled out in their simulations. Putting all the turbulent transport coefficients into a mean-field model, the evolution of the mean magnetic field of the 3D simulation was reproduced in terms of period and pattern (Warnecke et al. 2021); see Fig. 10. Notably the full spectrum of coefficients was needed to fully reproduce the field evolution. This suggests that all of the turbulent mean-field effects play important roles in this simulation which is a good representation of current global dynamo simulations. Furthermore, the authors concluded that the assumptions of Eq. (29) are reasonably well justified in the simulations (Warnecke et al. 2021) given that the $\mathcal{E}$ is also reproduced reasonably satisfactorily (Warnecke et al. 2018; Viviani et al. 2019).

6.3 Remaining Issues

One of the important issues is that the results of the two methods, SVD and TF, do not fully agree with each other. The SVD method seems to produce satisfactory results for EULAG-MHD simulations (Simard et al. 2013; Beaudoin et al. 2016; Simard and Charbonneau 2020). However, Warnecke et al. (2018) showed that the coefficients determined

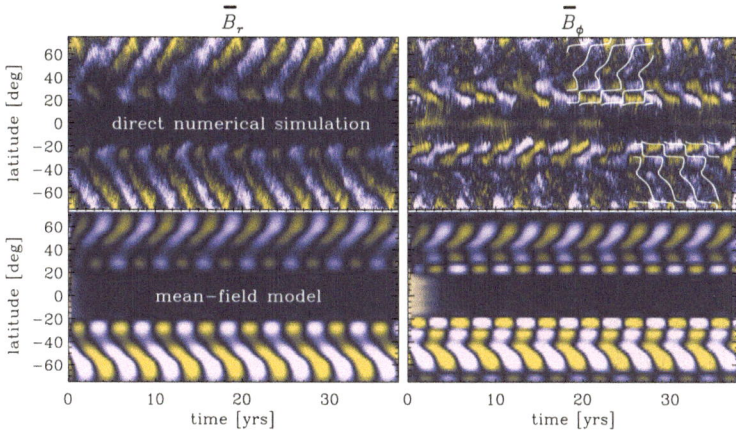

Fig. 10 Comparison of direct numerical simulations of Warnecke (2018) and Warnecke and Käpylä (2020) (top) with a corresponding mean-field model (bottom) where the turbulent transport coefficients have been obtain with the test field method. The radial (left) and toroidal (right) mean field is shown as a function of latitude and time. The white contours on the top right panel indicate the corresponding field of the mean-field model. Adapted from Warnecke et al. (2021)

with the TF and SVD methods give quite different results and that the SVD coefficients related to the derivative of $\overline{B}$ are indeed less important. This discrepancy raises questions regarding the overlap of the validity ranges and the underlying assumptions of both methods. A possible reason for the discrepancy is that non-locality (in space and/or time), which is neglected in Eq. (29), plays a role: the SVD uses the actual field whereas in TF only very large-scale gradients of test fields are retained. On the other hand, the actual magnetic fields the SVD uses are not necessarily linearly independent, leading to errors in the inversion.

SVD and TF methods reveal that turbulent transport effects play an important role in the dynamic and evolution of the large-scale magnetic field. Warnecke et al. (2021) showed that it is necessary to include practically all of the possible turbulent effects to reproduce the 3D simulation results in detail. This makes the corresponding mean-field models quite complex which to a certain extent defeats the purpose of mean-field modelling where the hope has been to capture the large-scale behavior of complicated 3D systems by a much simpler lower-dimensional model. However, the complexity of dynamos operating in these simulation also hints in the direction that the Sun and other stars are more complex than simple mean-field models including the Babcock-Leighton model can describe.

Another issue is related to the appearance of the SSD instability when global simulations reach more realistic high-$\mathrm{Re_M}$ regimes. In this case Eq. (29) is no longer valid, because a contribution to $\overline{\mathcal{E}}$ that is independent of $\overline{B}$ is possible, and non-linearity due to B' becomes important. If small-scale magnetic fields due to the SSD are dynamically important, neither the SVD nor TF method will work in their current form. Efforts to generalize the TF method to incorporate the effect of the SSD have been taken by Käpylä et al. (2022), leading to four flavours of the method in that regime. Although these flavours should in principle agree, this is not always the case, especially for high $\mathrm{Re_M}$. Hence, it is of great importance to extend the SVD and TF methods to more realistic parameter regimes to incorporate effects such as the SSD.

7 Conclusions and Future Prospects

A central challenge in simulating Solar or stellar dynamos is that, as discussed above, the interiors of real stars are characterised by extremely low diffusivities (of momentum, heat, and magnetism), and possess motion and magnetism over an extraordinarily broad range of spatial and temporal scales. No simulation, now or in the near future, can capture all these scales simultaneously. The hope of many modelers, though, is that at least *some* aspects of the dynamics, particularly on the largest scales, may become independent of the small-scale details at high enough resolution (low enough diffusivity). There is considerable debate – even amongst the authors of this review! – about the extent to which present-day simulations are nearing this diffusion-free, resolution-independent regime, and reasons for both optimism and pessimism. Here, we briefly highlight a few of these.

Current global dynamo simulations of stars routinely capture solar-like differential rotation and cyclic magnetism. Sometimes these models also reproduce equatorward migrating activity akin to the Sun. This occurs at a Rossby number regime where differential rotation is relatively strong. These results seem to be fairly robust irrespective of the numerical method or other details of the simulations. Also the theoretical understanding of the physical mechanisms driving the magnetism has developed significantly in the recent years with more advanced analysis tools such as the test-field method, full energy transfer and field production (electromotive force) analysis, and with direct comparisons to mean-field models.

Although our understanding of the physics of convection and resulting dynamo action has increased, new challenges have also been encountered. The most intriguing of these is the fact that current simulations struggle to reproduce solar convection and the resulting differential rotation at the solar luminosity and rotation rate. Given that this is a necessary requirement to get the dynamo right it is not a huge surprise that reproducing the solar dynamo remains challenging. This "convective conundrum" is the modern equivalent to the "dynamo dilemma" of the 1980s. The latter lead to a revival of old, and the conception of new, ideas about solar and stellar dynamos and a similar process is at work with respect to solar and stellar convection at the moment. The various ideas related to solving the conundrum include entropy rain and deep weakly subadiabatic convection, the influence of strong small-scale magnetism, and rotationally constrained deep convection. Research on this topic is very active and evolving rapidly, and, far from being stumped by the challenge posed by the convective conundrum, activity in the modelling of stellar convection and dynamos has instead been invigorated.

A key issue with the Sun is that even though the deep convection zone is highly likely rotationally dominated with Ro ≪ 1, there are many scales in the upper convection zone and near the surface where the rotational influence on the flow is weak. The Sun is also perhaps close to the transition to anti-solar differential rotation, which has some observational support, making it difficult to maintain a solar-like differential rotation profile if the simulations are not sufficiently near the correct parameter regime. Identifying the correct force balance prevailing in the solar convection zone is therefore key to this problem. With this information, simulations can be designed such that they follow a path leading to the correct balances and hopefully to solar-like results. This is a practice adopted in simulations of the geodynamo and perhaps a similar approach can be adopted for the Sun.

On the other hand, the issue regarding rotational influence is not as severe in stars that rotate more rapidly than the Sun and which are further away from the solar-like to anti-solar differential rotation transition. There are indications that simulations capture the characteristics of dynamos, such as non-axisymmetric large-scale fields, and dipole dominated dynamos in M dwarfs, in such stars more accurately. Although this is encouraging, we should also bear in mind that the observational data from other stars is not as accurate and detailed as the data we have from the Sun.

A common characteristic of all of the current simulations is the fact that it is not possible to model the surface layers, where the density drops vertiginously, accurately enough. This could be one of the reasons why none of the current simulations form spots that could play a role in the dynamo process via a Babcock–Leighton type effect, and their magnetic activity does not become independent of rotation at sufficiently low Ro. Self-consistent spot formation has not been reported even in local simulations to say nothing about global simulations. Therefore capturing spot formation in global simulations is perhaps as challenging, or even more challenging, than cracking the convective conundrum. Nevertheless, recent spot formation studies in more idealised simulation setups serve as a guide for the design of future global simulations that aim at achieving this.

All of these developments happen on a background where modelers have started to realize that the holy grail of stellar dynamo simulations – an asymptotic regime where results are independent of resolution or diffusivity – remains elusive, and that the computational cost of adding another data point at a higher resolution is already prohibitive. This begs the question whether it is feasible for everyone to try to beat everyone else in this very difficult task or whether it is better to combine resources for a collaborative effort where the resources of at least a large part of the field are directed in producing the "next generation" transformative simulations.

Acknowledgements The authors thank the two anonymous referees for their constructive comments on the manuscript. PJK acknowledges the financial support by the Deutsche Forschungsgemeinschaft Heisenberg programme (grant No. KA 4825/4-1). This project has received funding from the European Research Council (ERC) under the European Union's Horizon 2020 research and innovation program (Project UniSDyn, grant agreement no 818665) (JW). This work was done in collaboration with the COFFIES DRIVE Science Center. A.S.B. acknowledges support by European research council for ERC grant Stars2 (grant no. 207430) and Whole Sun (grant no. 810218), by CNRS/INSU through the Sun-Earth French national program, by French space agency CNES with Solar Orbiter grant and local fundings by Observatory OSUPS, Universities of Paris-Saclay and Paris-Cité. MB acknowledges support from the European Research Council under ERC grant agreement number 037705 (CHASM), and support from the UK Science and Technology Facilities Council (STFC). GG acknowledges support from NASA grants NNX14AB70G, 80NSSC20K0602, and 80NSSC20K1320.

Funding Open Access funding enabled and organized by Projekt DEAL.

Declarations

Competing Interests The authors declare no conflict of interest.

References

Ahuir J, Brun AS, Strugarek A (2020) From stellar coronae to gyrochronology: a theoretical and observational exploration. Astron Astrophys 635:A170. https://doi.org/10.1051/0004-6361/201936974. arXiv:2002.00696 [astro-ph.SR]

Anders EH, Brown BP, Oishi JS (2018) Accelerated evolution of convective simulations. Phys Rev Fluids 3(8):083502. https://doi.org/10.1103/PhysRevFluids.3.083502. arXiv:1807.06687 [physics.flu-dyn]

Anders EH, Vasil GM, Brown BP et al (2020) Convective dynamics with mixed temperature boundary conditions: why thermal relaxation matters and how to accelerate it. Phys Rev Fluids 5(8):083501. https://doi.org/10.1103/PhysRevFluids.5.083501. arXiv:2003.00026 [physics.flu-dyn]

Anders EH, Bauer EB, Jermyn AS et al (2022a) Moosinesq convection in the cores of moosive stars. arXiv:2204.00002 [astro-ph.SR]

Anders EH, Jermyn AS, Lecoanet D et al (2022b) Stellar convective penetration: parameterized theory and dynamical simulations. Astrophys J 926(2):169. https://doi.org/10.3847/1538-4357/ac408d. arXiv:2110.11356 [astro-ph.SR]

Aubert J, Gillet N (2021) The interplay of fast waves and slow convection in geodynamo simulations nearing Earth's core conditions. Geophys J Int 225(3):1854–1873. https://doi.org/10.1093/gji/ggab054. arXiv:2102.06552 [physics.geo-ph]

Aubert J, Gastine T, Fournier A (2017) Spherical convective dynamos in the rapidly rotating asymptotic regime. J Fluid Mech 813:558–593. https://doi.org/10.1017/jfm.2016.789. arXiv:1611.04776 [physics.geo-ph]

Augustson KC, Brun AS, Toomre J (2013) Dynamo action and magnetic cycles in F-type stars. Astrophys J 777:153. https://doi.org/10.1088/0004-637X/777/2/153

Augustson K, Brun AS, Miesch M et al (2015) Grand minima and equatorward propagation in a cycling stellar convective dynamo. Astrophys J 809:149. https://doi.org/10.1088/0004-637X/809/2/149. arXiv:1410.6547 [astro-ph.SR]

Augustson KC, Brun AS, Toomre J (2016) The magnetic furnace: intense core dynamos in B stars. Astrophys J 829(2):92. https://doi.org/10.3847/0004-637X/829/2/92. arXiv:1603.03659 [astro-ph.SR]

Augustson KC, Brun AS, Toomre J (2019) Rossby and magnetic Prandtl number scaling of stellar dynamos. Astrophys J 876(1):83. https://doi.org/10.3847/1538-4357/ab14ea

Aurnou JM, Horn S, Julien K (2020) Connections between nonrotating, slowly rotating, and rapidly rotating turbulent convection transport scalings. Phys Rev Res 2(4):043115. https://doi.org/10.1103/PhysRevResearch.2.043115. arXiv:2009.03447 [physics.flu-dyn]

Baliunas SL, Donahue RA, Soon WH et al (1995) Chromospheric variations in main-sequence stars. II. Astrophys J 438:269. https://doi.org/10.1086/175072

Ballot J, Brun AS, Turck-Chièze S (2007) Simulations of turbulent convection in rotating young solarlike stars: differential rotation and meridional circulation. Astrophys J 669:1190–1208. https://doi.org/10.1086/521617. arXiv:0707.3943

Baraffe I, Clarke J, Morison A et al (2023) A study of convective core overshooting as a function of stellar mass based on two-dimensional hydrodynamical simulations. Mon Not R Astron Soc 519(4):5333–5344. https://doi.org/10.1093/mnras/stad009. arXiv:2301.02604 [astro-ph.SR]

Barker AJ, Dempsey AM, Lithwick Y (2014) Theory and simulations of rotating convection. Astrophys J 791(1):13. https://doi.org/10.1088/0004-637X/791/1/13. arXiv:1403.7207 [astro-ph.SR]

Barnabé R, Strugarek A, Charbonneau P et al (2017) Confinement of the solar tachocline by a cyclic dynamo magnetic field. Astron Astrophys 601:A47. https://doi.org/10.1051/0004-6361/201630178. arXiv:1703.02374 [astro-ph.SR]

Barnes SA (2003) On the rotational evolution of solar- and late-type stars, its magnetic origins, and the possibility of stellar gyrochronology. Astrophys J 586(1):464–479. https://doi.org/10.1086/367639. arXiv:astro-ph/0303631 [astro-ph]

Baryshnikova I, Shukurov A (1987) Oscillatory α^2-dynamo: numerical investigation. Astron Nachr 308:89–100

Beaudoin P, Simard C, Cossette JF et al (2016) Double dynamo signatures in a global MHD simulation and mean-field dynamos. Astrophys J 826(2):138. https://doi.org/10.3847/0004-637X/826/2/138

Bekki Y, Hotta H, Yokoyama T (2017) Convective velocity suppression via the enhancement of the subadiabatic layer: role of the effective Prandtl number. Astrophys J 851:74. https://doi.org/10.3847/1538-4357/aa9b7f. arXiv:1711.05960 [astro-ph.SR]

Bekki Y, Cameron RH, Gizon L (2022) Theory of solar oscillations in the inertial frequency range: amplitudes of equatorial modes from a nonlinear rotating convection simulation. Astron Astrophys 666:A135. https://doi.org/10.1051/0004-6361/202244150. arXiv:2208.11081 [astro-ph.SR]

Benomar O, Bazot M, Nielsen MB et al (2018) Asteroseismic detection of latitudinal differential rotation in 13 Sun-like stars. Science 361(6408):1231–1234. https://doi.org/10.1126/science.aao6571. arXiv:1809.07938 [astro-ph.SR]

Bice CP, Toomre J (2020) Probing the influence of a tachocline in simulated M-dwarf dynamos. Astrophys J 893(2):107. https://doi.org/10.3847/1538-4357/ab8190. arXiv:2001.05555 [astro-ph.SR]

Bice CP, Toomre J (2022) Longitudinally modulated dynamo action in simulated M-dwarf stars. Astrophys J 928(1):51. https://doi.org/10.3847/1538-4357/ac4be0. arXiv:2202.02869 [astro-ph.SR]

Blackman EG, Field GB (2002) New dynamical mean-field dynamo theory and closure approach. Phys Rev Lett 89(26):265007. https://doi.org/10.1103/PhysRevLett.89.265007. astro-ph/0207435

Böhm-Vitense E (1958) Über die Wasserstoffkonvektionszone in Sternen verschiedener Effektivtemperaturen und Leuchtkräfte. Z Astrophys 46:108

Bonanno A, Corsaro E (2022) On the origin of the dichotomy of stellar activity cycles. Astrophys J Lett 939(2):L26. https://doi.org/10.3847/2041-8213/ac9c05. arXiv:2210.11305 [astro-ph.SR]

Boro Saikia S, Lueftinger T, Jeffers SV et al (2018) Direct evidence of a full dipole flip during the magnetic cycle of a sun-like star. Astron Astrophys 620:L11. https://doi.org/10.1051/0004-6361/201834347. arXiv:1811.11671 [astro-ph.SR]

Boro Saikia S, Marvin CJ, Jeffers SV et al (2018b) Chromospheric activity catalogue of 4454 cool stars. Questioning the active branch of stellar activity cycles. Astron Astrophys 616:A108. https://doi.org/10.1051/0004-6361/201629518. arXiv:1803.11123 [astro-ph.SR]

Braithwaite J (2006) A differential rotation driven dynamo in a stably stratified star. Astron Astrophys 449(2):451–460. https://doi.org/10.1051/0004-6361:20054241. arXiv:astro-ph/0509693 [astro-ph]

Brandenburg A (2005) The case for a distributed solar dynamo shaped by near-surface shear. Astrophys J 625:539–547. https://doi.org/10.1086/429584. astro-ph/0502275

Brandenburg A (2016) Stellar mixing length theory with entropy rain. Astrophys J 832:6. https://doi.org/10.3847/0004-637X/832/1/6. arXiv:1504.03189 [astro-ph.SR]

Brandenburg A, Giampapa MS (2018) Enhanced stellar activity for slow antisolar differential rotation? Astrophys J Lett 855(2):L22. https://doi.org/10.3847/2041-8213/aab20a. arXiv:1802.08689 [astro-ph.SR]

Brandenburg A, Sokoloff D (2002) Local and nonlocal magnetic diffusion and alpha-effect tensors in shear flow turbulence. Geophys Astrophys Fluid Dyn 96:319–344. https://doi.org/10.1080/03091920290032974. astro-ph/0111563

Brandenburg A, Spiegel EA (2008) Modeling a Maunder minimum. Astron Nachr 329(4):351. https://doi.org/10.1002/asna.200810973. arXiv:0801.2156 [astro-ph

Brandenburg A, Subramanian K (2005) Astrophysical magnetic fields and nonlinear dynamo theory. Phys Rep 417:1–209. https://doi.org/10.1016/j.physrep.2005.06.005. astro-ph/0405052

Brandenburg A, Moss D, Tuominen I (1992) Stratification and thermodynamics in mean-field dynamos. Astron Astrophys 265:328–344

Brandenburg A, Mathur S, Metcalfe TS (2017) Evolution of co-existing long and short period stellar activity cycles. Astrophys J 845(1):79. https://doi.org/10.3847/1538-4357/aa7cfa. arXiv:1704.09009 [astro-ph.SR]

Pencil Code Collaboration, Brandenburg A, Johansen A et al (2021) The Pencil Code, a modular MPI code for partial differential equations and particles: multipurpose and multiuser-maintained. J Open Sour Softw 6(58):2807. https://doi.org/10.21105/joss.02807

Brandenburg A, Elstner D, Masada Y et al (2023) Turbulent processes and mean-field dynamo. Space Sci Rev 219:55. https://doi.org/10.1007/s11214-023-00999-3. arXiv:2303.12425 [astro-ph.SR]

Breton SN, Brun AS, García RA (2022) Stochastic excitation of internal gravity waves in rotating late F-type stars: a 3D simulation approach. Astron Astrophys 667:A43. https://doi.org/10.1051/0004-6361/202244247. arXiv:2208.14759 [astro-ph.SR]

Brown BP, Browning MK, Brun AS et al (2008) Rapidly rotating suns and active nests of convection. Astrophys J 689:1354–1372. https://doi.org/10.1086/592397. arXiv:0808.1716

Brown BP, Browning MK, Brun AS et al (2010) Persistent magnetic wreaths in a rapidly rotating Sun. Astrophys J 711:424–438. https://doi.org/10.1088/0004-637X/711/1/424. arXiv:1011.2831 [astro-ph.SR]

Brown BP, Miesch MS, Browning MK et al (2011) Magnetic cycles in a convective dynamo simulation of a young solar-type star. Astrophys J 731:69. https://doi.org/10.1088/0004-637X/731/1/69. arXiv:1102.1993 [astro-ph.SR]

Brown BP, Oishi JS, Vasil GM et al (2020) Single-hemisphere dynamos in M-dwarf stars. Astrophys J Lett 902(1):L3. https://doi.org/10.3847/2041-8213/abb9a4. arXiv:2008.02362 [astro-ph.SR]

Browning MK (2008) Simulations of dynamo action in fully convective stars. Astrophys J 676:1262–1280. https://doi.org/10.1086/527432. arXiv:0712.1603

Browning MK, Brun AS, Toomre J (2004) Simulations of core convection in rotating A-type stars: differential rotation and overshooting. Astrophys J 601:512–529. https://doi.org/10.1086/380198. astro-ph/0310003

Browning MK, Miesch MS, Brun AS et al (2006) Dynamo action in the solar convection zone and tachocline: pumping and organization of toroidal fields. Astrophys J Lett 648:L157–L160. https://doi.org/10.1086/507869. arXiv:astro-ph/0609153

Brun AS, Browning MK (2017) Magnetism, dynamo action and the solar-stellar connection. Living Rev Sol Phys 14:4. https://doi.org/10.1007/s41116-017-0007-8

Brun AS, Palacios A (2009) Numerical simulations of a rotating red giant star. I. Three-dimensional models of turbulent convection and associated mean flows. Astrophys J 702:1078–1097. https://doi.org/10.1088/0004-637X/702/2/1078

Brun AS, Toomre J (2002) Turbulent convection under the influence of rotation: sustaining a strong differential rotation. Astrophys J 570:865–885. https://doi.org/10.1086/339228. astro-ph/0206196

Brun AS, Miesch MS, Toomre J (2004) Global-scale turbulent convection and magnetic dynamo action in the solar envelope. Astrophys J 614:1073–1098. https://doi.org/10.1086/423835. arXiv:astro-ph/0610073

Brun AS, Browning MK, Toomre J (2005) Simulations of core convection in rotating A-type stars: magnetic dynamo action. Astrophys J 629:461–481. https://doi.org/10.1086/430430. arXiv:astro-ph/0610072

Brun AS, Miesch MS, Toomre J (2011) Modeling the dynamical coupling of solar convection with the radiative interior. Astrophys J 742:79. https://doi.org/10.1088/0004-637X/742/2/79

Brun AS, Strugarek A, Varela J et al (2017) On differential rotation and overshooting in solar-like stars. Astrophys J 836:192. https://doi.org/10.3847/1538-4357/aa5c40. arXiv:1702.06598 [astro-ph.SR]

Brun AS, Pui Hung C, Fournier A et al (2020) A solar cycle 25 prediction based on 4D-var data assimilation approach. In: Kosovichev A, Strassmeier S, Jardine M (eds) Solar and stellar magnetic fields: origins and manifestations, Cambridge University Press, Cambridge, pp 138–146. https://doi.org/10.1017/S1743921320003993

Brun AS, Strugarek A, Noraz Q et al (2022) Powering stellar magnetism: energy transfers in cyclic dynamos of Sun-like stars. Astrophys J 926(1):21. https://doi.org/10.3847/1538-4357/ac469b. arXiv:2201.13218 [astro-ph.SR]

Burns KJ, Vasil GM, Oishi JS et al (2020) Dedalus: a flexible framework for numerical simulations with spectral methods. Phys Rev Res 2(2):023068. https://doi.org/10.1103/PhysRevResearch.2.023068. arXiv: 1905.10388 [astro-ph.IM]

Busse FH (1970) Thermal instabilities in rapidly rotating systems. J Fluid Mech 44:441–460. https://doi.org/10.1017/S0022112070001921

Busse FH (1983) Generation of mean flows by thermal convection. Physica D Nonlinear Phenom 9(3):287–299. https://doi.org/10.1016/0167-2789(83)90273-7

Caligari P, Moreno-Insertis F, Schussler M (1995) Emerging flux tubes in the solar convection zone. I. Asymmetry, tilt, and emergence latitude. Astrophys J 441:886. https://doi.org/10.1086/175410

Cameron RH, Schüssler M (2017) An update of Leighton's solar dynamo model. Astron Astrophys 599:A52. https://doi.org/10.1051/0004-6361/201629746. arXiv:1611.09111 [astro-ph.SR]

Camisassa ME, Featherstone NA (2022) Solar-like to antisolar differential rotation: a geometric interpretation. Astrophys J 938(1):65. https://doi.org/10.3847/1538-4357/ac879f. arXiv:2208.05591 [astro-ph.SR]

Cattaneo F, Brummell NH, Toomre J et al (1991) Turbulent compressible convection. Astrophys J 370:282–294. https://doi.org/10.1086/169814

Chan KL, Sofia S (1986) Turbulent compressible convection in a deep atmosphere. III - Tests on the validity and limitation of the numerical approach. Astrophys J 307:222–241. https://doi.org/10.1086/164409

Chandrasekhar S (1961) Hydrodynamic and hydromagnetic stability

Charbonneau P (2020) Dynamo models of the solar cycle. Living Rev Sol Phys 17(1):4. https://doi.org/10.1007/s41116-020-00025-6

Chen F, Rempel M, Fan Y (2017) Emergence of magnetic flux generated in a solar convective dynamo. I. The formation of sunspots and active regions, and the origin of their asymmetries. Astrophys J 846(2):149. https://doi.org/10.3847/1538-4357/aa85a0. arXiv:1704.05999 [astro-ph.SR]

Christensen UR, Aubert J (2006) Scaling properties of convection-driven dynamos in rotating spherical shells and application to planetary magnetic fields. Geophys J Int 166:97–114. https://doi.org/10.1111/j.1365-246X.2006.03009.x

Christensen UR, Holzwarth V, Reiners A (2009) Energy flux determines magnetic field strength of planets and stars. Nature 457(7226):167–169. https://doi.org/10.1038/nature07626

Clune TC, Elliot JR, Miesch MS et al (1999) Computational aspects of a code to study rotating turbulent convection in spherical shells. Parallel Comput 25:361–380. https://doi.org/10.1016/S0167-8191(99)00009-5

Cole E, Käpylä PJ, Mantere MJ et al (2014) Azimuthal dynamo wave in spherical shell convection. Astrophys J Lett 780:L22. arXiv:1309.6802 [astro-ph.SR]

Cowling TG (1933) The magnetic field of sunspots. Mon Not R Astron Soc 94:39–48. https://doi.org/10.1093/mnras/94.1.39

Currie LK, Browning MK (2017) The magnitude of viscous dissipation in strongly stratified two-dimensional convection. Astrophys J Lett 845(2):L17. https://doi.org/10.3847/2041-8213/aa8301. arXiv:1707.08858 [astro-ph.SR]

Currie LK, Barker AJ, Lithwick Y et al (2020) Convection with misaligned gravity and rotation: simulations and rotating mixing length theory. Mon Not R Astron Soc 493(4):5233–5256. https://doi.org/10.1093/mnras/staa372. arXiv:2002.02461 [astro-ph.SR]

Dikpati M, Charbonneau P (1999) A Babcock-Leighton flux transport dynamo with solar-like differential rotation. Astrophys J 518:508–520. https://doi.org/10.1086/307269

Dobler W, Stix M, Brandenburg A (2006) Magnetic field generation in fully convective rotating spheres. Astrophys J 638:336–347. https://doi.org/10.1086/498634. arXiv:astro-ph/0410645

Dorch SBF (2004) Magnetic activity in late-type giant stars: numerical MHD simulations of non-linear dynamo action in Betelgeuse. Astron Astrophys 423:1101–1107. https://doi.org/10.1051/0004-6361:20040435. arXiv:astro-ph/0403321 [astro-ph]

Duarte LDV, Wicht J, Browning MK et al (2016) Helicity inversion in spherical convection as a means for equatorward dynamo wave propagation. Mon Not R Astron Soc 456:1708–1722. https://doi.org/10.1093/mnras/stv2726. arXiv:1511.05813 [astro-ph.SR]

Duez V, Braithwaite J, Mathis S (2010) On the stability of non-force-free magnetic equilibria in stars. Astrophys J Lett 724(1):L34–L38. https://doi.org/10.1088/2041-8205/724/1/L34. arXiv:1009.5384 [astro-ph.SR]

Duvall TL Jr, Dziembowski WA, Goode PR et al (1984) Internal rotation of the Sun. Nature 310(5972):22–25. https://doi.org/10.1038/310022a0

Edelmann PVF, Ratnasingam RP, Pedersen MG et al (2019) Three-dimensional simulations of massive stars. I. Wave generation and propagation. Astrophys J 876(1):4. https://doi.org/10.3847/1538-4357/ab12df. arXiv:1903.09392 [astro-ph.SR]

Elliott JR, Miesch MS, Toomre J (2000) Turbulent solar convection and its coupling with rotation: the effect of Prandtl number and thermal boundary conditions on the resulting differential rotation. Astrophys J 533(1):546–556. https://doi.org/10.1086/308643

Emeriau-Viard C, Brun AS (2017) Origin and evolution of magnetic field in PMS stars: influence of rotation and structural changes. Astrophys J 846(1):8. https://doi.org/10.3847/1538-4357/aa7b33. arXiv:1709.04667 [astro-ph.SR]

Fan Y, Fang F (2014) A simulation of convective dynamo in the solar convective envelope: maintenance of the solar-like differential rotation and emerging flux. Astrophys J 789:35. https://doi.org/10.1088/0004-637X/789/1/35. arXiv:1405.3926 [astro-ph.SR]

Featherstone NA, Hindman BW (2016) The emergence of solar supergranulation as a natural consequence of rotationally constrained interior convection. Astrophys J Lett 830:L15. https://doi.org/10.3847/2041-8205/830/1/L15. arXiv:1609.05153 [astro-ph.SR]

Featherstone NA, Hindman BW (2016) The spectral amplitude of stellar convection and its scaling in the high-Rayleigh-number regime. Astrophys J 818:32. https://doi.org/10.3847/0004-637X/818/1/32. arXiv:1511.02396 [astro-ph.SR]

Featherstone NA, Miesch MS (2015) Meridional circulation in solar and stellar convection zones. Astrophys J 804:67. https://doi.org/10.1088/0004-637X/804/1/67. arXiv:1501.06501 [astro-ph.SR]

Featherstone NA, Browning MK, Brun AS et al (2009) Effects of fossil magnetic fields on convective core dynamos in A-type stars. Astrophys J 705(1):1000–1018. https://doi.org/10.1088/0004-637X/705/1/1000

Featherstone NA, Edelmann PVF, Gassmoeller R et al (2022) Rayleigh 1.1.0. https://doi.org/10.5281/zenodo.6522806

Forgács-dajka E, Petrovay K (2001) Tachocline confinement by an oscillatory magnetic field. Sol Phys 203(2):195–210. https://doi.org/10.1023/A:1013389631585. arXiv:astro-ph/0106133 [astro-ph]

Frisch U (1995) Turbulence. Cambridge University Press, Cambridge

Fuller J, Ma L (2019) Most black holes are born very slowly rotating. Astrophys J Lett 881(1):L1. https://doi.org/10.3847/2041-8213/ab339b. arXiv:1907.03714 [astro-ph.SR]

Fuller J, Piro AL, Jermyn AS (2019) Slowing the spins of stellar cores. Mon Not R Astron Soc 485(3):3661–3680. https://doi.org/10.1093/mnras/stz514. arXiv:1902.08227 [astro-ph.SR]

Gallet F, Bouvier J (2013) Improved angular momentum evolution model for solar-like stars. Astron Astrophys 556:A36. https://doi.org/10.1051/0004-6361/201321302. arXiv:1306.2130 [astro-ph.SR]

Gallet B, Pétrélis F (2009) From reversing to hemispherical dynamos. Phys Rev E 80(3):035302. https://doi.org/10.1103/PhysRevE.80.035302. arXiv:0907.4428 [astro-ph.EP]

Gastine T, Wicht J (2012) Effects of compressibility on driving zonal flow in gas giants. Icarus 219:428–442. https://doi.org/10.1016/j.icarus.2012.03.018. arXiv:1203.4145 [astro-ph.EP]

Gastine T, Duarte L, Wicht J (2012) Dipolar versus multipolar dynamos: the influence of the background density stratification. Astron Astrophys 546:A19. https://doi.org/10.1051/0004-6361/201219799. arXiv:1208.6093 [astro-ph.SR]

Gastine T, Yadav RK, Morin J et al (2014) From solar-like to antisolar differential rotation in cool stars. Mon Not R Astron Soc 438:L76–L80. https://doi.org/10.1093/mnrasl/slt162. arXiv:1311.3047 [astro-ph.SR]

Gastine T, Wicht J, Aubert J (2016) Scaling regimes in spherical shell rotating convection. J Fluid Mech 808:690–732. https://doi.org/10.1017/jfm.2016.659. arXiv:1609.02372 [physics.flu-dyn]

Gent FA, Käpylä MJ, Warnecke J (2017) Long-term variations of turbulent transport coefficients in a solarlike convective dynamo simulation. Astron Nachr 338:885–895. https://doi.org/10.1002/asna.201713406. arXiv:1709.00390 [astro-ph.SR]

Ghizaru M, Charbonneau P, Smolarkiewicz PK (2010) Magnetic cycles in global large-eddy simulations of solar convection. Astrophys J Lett 715:L133–L137. https://doi.org/10.1088/2041-8205/715/2/L133

Gilet C, Almgren AS, Bell JB et al (2013) Low Mach number modeling of core convection in massive stars. Astrophys J 773:137. https://doi.org/10.1088/0004-637X/773/2/137

Gilman PA (1977) Nonlinear dynamics of Boussinesq convection in a deep rotating spherical shell. I. Geophys Astrophys Fluid Dyn 8:93–135. https://doi.org/10.1080/03091927708240373

Gilman PA (1983) Dynamically consistent nonlinear dynamos driven by convection in a rotating spherical shell. II - Dynamos with cycles and strong feedbacks. Astrophys J Suppl Ser 53:243–268. https://doi.org/10.1086/190891

Gilman PA, Miller J (1981) Dynamically consistent nonlinear dynamos driven by convection in a rotating spherical shell. Astrophys J Suppl Ser 46:211–238. https://doi.org/10.1086/190743

Gilman PA, Morrow CA, Deluca EE (1989) Angular momentum transport and dynamo action in the Sun: implications of recent oscillation measurements. Astrophys J 338:528. https://doi.org/10.1086/167215

Gizon L, Cameron RH, Bekki Y et al (2021) Solar inertial modes: observations, identification, and diagnostic promise. Astron Astrophys 652:L6. https://doi.org/10.1051/0004-6361/202141462. arXiv:2107.09499 [astro-ph.SR]

Glatzmaier GA (1985) Numerical simulations of stellar convective dynamos. II - Field propagation in the convection zone. Astrophys J 291:300–307. https://doi.org/10.1086/163069

Glatzmaier G, Evonuk M, Rogers T (2009) Differential rotation in giant planets maintained by density-stratified turbulent convection. Geophys Astrophys Fluid Dyn 103(1):31–51. https://doi.org/10.1080/03091920802221245. arXiv:0806.2002 [astro-ph]

Goudard L, Dormy E (2008) Relations between the dynamo region geometry and the magnetic behavior of stars and planets. Europhys Lett 83:59001. https://doi.org/10.1209/0295-5075/83/59001. arXiv:0901.0828 [astro-ph.EP]

Gough DO, McIntyre ME (1998) Inevitability of a magnetic field in the Sun's radiative interior. Nature 394(6695):755–757. https://doi.org/10.1038/29472

Gregory SG, Donati JF, Morin J et al (2012) Can we predict the global magnetic topology of a pre-main-sequence star from its position in the Hertzsprung-Russell diagram? Astrophys J 755(2):97. https://doi.org/10.1088/0004-637X/755/2/97. arXiv:1206.5238 [astro-ph.SR]

Guerrero G, Smolarkiewicz PK, Kosovichev AG et al (2013) Differential rotation in solar-like stars from global simulations. Astrophys J 779:176. https://doi.org/10.1088/0004-637X/779/2/176. arXiv:1310.8178 [astro-ph.SR]

Guerrero G, Smolarkiewicz PK, de Gouveia Dal Pino EM et al (2016) On the role of tachoclines in solar and stellar dynamos. Astrophys J 819:104. https://doi.org/10.3847/0004-637X/819/2/104. arXiv:1507.04434 [astro-ph.SR]

Guerrero G, Zaire B, Smolarkiewicz PK et al (2019) What sets the magnetic field strength and cycle period in solar-type stars? Astrophys J 880(1):6. https://doi.org/10.3847/1538-4357/ab224a. arXiv:1810.07978 [astro-ph.SR]

Guerrero G, Stejko AM, Kosovichev AG et al (2022) Implicit large-eddy simulations of global solar convection: effects of numerical resolution in nonrotating and rotating cases. Astrophys J 940(2):151. https://doi.org/10.3847/1538-4357/ac9af3. arXiv:2208.05738 [astro-ph.SR]

Hale GE (1908) On the probable existence of a magnetic field in Sun-spots. Astrophys J 28:315. https://doi.org/10.1086/141602

Hale GE, Ellerman F, Nicholson SB et al (1919) The magnetic polarity of Sun-spots. Astrophys J 49:153. https://doi.org/10.1086/142452

Hall JC, Henry GW, Lockwood GW et al (2009) The activity and variability of the Sun and Sun-like stars. II. Contemporaneous photometry and spectroscopy of bright solar analogs. Astron J 138(1):312–322. https://doi.org/10.1088/0004-6256/138/1/312

Hanasoge S, Gizon L, Sreenivasan KR (2016) Seismic sounding of convection in the Sun. Annu Rev Fluid Mech 48:191–217. https://doi.org/10.1146/annurev-fluid-122414-034534. arXiv:1503.07961 [astro-ph.SR]

Hanson CS, Hanasoge S, Sreenivasan KR (2022) Discovery of high-frequency retrograde vorticity waves in the Sun. Nat Astron 6:708–714. https://doi.org/10.1038/s41550-022-01632-z

Hewitt JM, McKenzie DP, Weiss NO (1975) Dissipative heating in convective flows. J Fluid Mech 68:721–738. https://doi.org/10.1017/S002211207500119X

Hotta H (2017) Solar overshoot region and small-scale dynamo with realistic energy flux. Astrophys J 843:52. https://doi.org/10.3847/1538-4357/aa784b. arXiv:1706.06413 [astro-ph.SR]

Hotta H, Kusano K (2021) Solar differential rotation reproduced with high-resolution simulation. Nat Astron 5:1100–1102. https://doi.org/10.1038/s41550-021-01459-0. arXiv:2109.06280 [astro-ph.SR]

Hotta H, Rempel M, Yokoyama T (2014) High-resolution calculations of the solar global convection with the reduced speed of sound technique. I. The structure of the convection and the magnetic field without the rotation. Astrophys J 786:24. https://doi.org/10.1088/0004-637X/786/1/24. arXiv:1402.5008 [astro-ph.SR]

Hotta H, Rempel M, Yokoyama T (2015) High-resolution calculation of the solar global convection with the reduced speed of sound technique. II. Near surface shear layer with the rotation. Astrophys J 798:51. https://doi.org/10.1088/0004-637X/798/1/51. arXiv:1410.7093 [astro-ph.SR]

Hotta H, Rempel M, Yokoyama T (2016) Large-scale magnetic fields at high Reynolds numbers in magneto-hydrodynamic simulations. Science 351(6280):1427–1430. https://doi.org/10.1126/science.aad1893.

Hotta H, Kusano K, Shimada R (2022) Generation of solar-like differential rotation. Astrophys J 933(2):199. https://doi.org/10.3847/1538-4357/ac7395. arXiv:2202.04183 [astro-ph.SR]

Hurlburt NE, Toomre J, Massaguer JM (1984) Two-dimensional compressible convection extending over multiple scale heights. Astrophys J 282:557–573. https://doi.org/10.1086/162235

Jeffers SV, Kiefer R, Metcalfe TS et al (2023) Stellar activity cycles. Space Sci Rev 219:A54. https://doi.org/10.1007/s11214-023-01000-x. arXiv:2309.14138 [astro-ph.SR]

Jermyn AS, Anders EH, Lecoanet D et al (2022) An atlas of convection in main-sequence stars. Astrophys J Suppl Ser 262(1):19. https://doi.org/10.3847/1538-4365/ac7cee. arXiv:2206.00011 [astro-ph.SR]

Ji S, Fuller J, Lecoanet D (2023) Magnetohydrodynamic simulations of the Tayler instability in rotating stellar interiors. Mon Not R Astron Soc 521(4):5372–5383. https://doi.org/10.1093/mnras/stad910. arXiv:2209.08104 [astro-ph.SR]

Johns-Krull CM (2007) The magnetic fields of classical T Tauri stars. Astrophys J 664(2):975–985. https://doi.org/10.1086/519017. arXiv:0704.2923 [astro-ph]

Jones CA, Boronski P, Brun AS et al (2011) Anelastic convection-driven dynamo benchmarks. Icarus 216(1):120–135. https://doi.org/10.1016/j.icarus.2011.08.014

Jouve L, Gastine T, Lignières F (2015) Three-dimensional evolution of magnetic fields in a differentially rotating stellar radiative zone. Astron Astrophys 575:A106. https://doi.org/10.1051/0004-6361/201425240. arXiv:1412.2900 [astro-ph.SR]

Julien K, Rubio AM, Grooms I et al (2012) Statistical and physical balances in low Rossby number Rayleigh-Bénard convection. Geophys Astrophys Fluid Dyn 106(4–5):392–428. https://doi.org/10.1080/03091929.2012.696109

Käpylä PJ (2019) Overshooting in simulations of compressible convection. Astron Astrophys 631:A122. https://doi.org/10.1051/0004-6361/201834921

Käpylä PJ (2021) Star-in-a-box simulations of fully convective stars. Astron Astrophys 651:A66. https://doi.org/10.1051/0004-6361/202040049. arXiv:2012.01259 [astro-ph.SR]

Käpylä PJ (2022) Solar-like dynamos and rotational scaling of cycles from star-in-a-box simulations. Astrophys J 931:L17. https://doi.org/10.3847/2041-8213/ac6e6b. arXiv:2202.04329 [astro-ph.SR]

Käpylä PJ (2023) Transition from anti-solar to solar-like differential rotation: dependence on Prandtl number. Astron Astrophys 669:A98. https://doi.org/10.1051/0004-6361/202244395. arXiv:2207.00302 [astro-ph.SR]

Käpylä PJ, Korpi MJ, Tuominen I (2006) Solar dynamo models with α-effect and turbulent pumping from local 3D convection calculations. Astron Nachr 327:884. https://doi.org/10.1002/asna.200610636. arXiv:astro-ph/0606089

Käpylä PJ, Korpi MJ, Brandenburg A et al (2010) Convective dynamos in spherical wedge geometry. Astron Nachr 331:73. https://doi.org/10.1002/asna.200911252. arXiv:0909.1330 [astro-ph.SR]

Käpylä PJ, Mantere MJ, Brandenburg A (2011a) Effects of stratification in spherical shell convection. Astron Nachr 332:883. https://doi.org/10.1002/asna.201111619. arXiv:1109.4625 [astro-ph.SR]

Käpylä PJ, Mantere MJ, Guerrero G et al (2011b) Reynolds stress and heat flux in spherical shell convection. Astron Astrophys 531:A162. https://doi.org/10.1051/0004-6361/201015884. arXiv:1010.1250 [astro-ph.SR]

Käpylä PJ, Mantere MJ, Brandenburg A (2012) Cyclic magnetic activity due to turbulent convection in spherical wedge geometry. Astrophys J Lett 755:L22. https://doi.org/10.1088/2041-8205/755/1/L22. arXiv:1205.4719 [astro-ph.SR]

Käpylä PJ, Mantere MJ, Cole E et al (2013) Effects of enhanced stratification on equatorward dynamo wave propagation. Astrophys J 778:41. https://doi.org/10.1088/0004-637X/778/1/41. arXiv:1301.2595 [astro-ph.SR]

Käpylä PJ, Käpylä MJ, Brandenburg A (2014) Confirmation of bistable stellar differential rotation profiles. Astron Astrophys 570:A43. arXiv:1401.2981 [astro-ph.SR]

Käpylä MJ, Käpylä PJ, Olspert N et al (2016) Multiple dynamo modes as a mechanism for long-term solar activity variations. Astron Astrophys 589:A56. https://doi.org/10.1051/0004-6361/201527002. arXiv:1507.05417 [astro-ph.SR]

Käpylä PJ, Käpylä MJ, Olspert N et al (2017) Convection-driven spherical shell dynamos at varying Prandtl numbers. Astron Astrophys 599:A4. https://doi.org/10.1051/0004-6361/201628973. arXiv:1605.05885 [astro-ph.SR]

Käpylä PJ, Rheinhardt M, Brandenburg A et al (2017) Extended subadiabatic layer in simulations of overshooting convection. Astrophys J Lett 845:L23. https://doi.org/10.3847/2041-8213/aa83ab. arXiv:1703.06845 [astro-ph.SR]

Käpylä PJ, Viviani M, Käpylä MJ et al (2019) Effects of a subadiabatic layer on convection and dynamos in spherical wedge simulations. Geophys Astrophys Fluid Dyn 113:149–183. https://doi.org/10.1080/03091929.2019.1571584. arXiv:1803.05898 [astro-ph.SR]

Käpylä PJ, Gent FA, Olspert N et al (2020) Sensitivity to luminosity, centrifugal force, and boundary conditions in spherical shell convection. Geophys Astrophys Fluid Dyn 114(1–2):8–34. https://doi.org/10.1080/03091929.2019.1571586

Käpylä MJ, Rheinhardt M, Brandenburg A (2022) Compressible test-field method and its application to shear dynamos. Astrophys J 932(1):8. https://doi.org/10.3847/1538-4357/ac5b78. arXiv:2106.01107 [physics.flu-dyn]

Karak BB, Käpylä PJ, Käpylä MJ et al (2015) Magnetically controlled stellar differential rotation near the transition from solar to anti-solar profiles. Astron Astrophys 576:A26. https://doi.org/10.1051/0004-6361/201424521. arXiv:1407.0984 [astro-ph.SR]

Karak BB, Miesch M, Bekki Y (2018) Consequences of high effective Prandtl number on solar differential rotation and convective velocity. Phys Fluids 30(4):046602. https://doi.org/10.1063/1.5022034. arXiv:1801.00560 [astro-ph.SR]

Kawaler SD (1988) Angular momentum loss in low-mass stars. Astrophys J 333:236. https://doi.org/10.1086/166740

Kippenhahn R, Weigert A, Weiss A (2012) Stellar structure and evolution. Springer, Berlin. https://doi.org/10.1007/978-3-642-30304-3

Kitchatinov LL (2016) Rotational shear near the solar surface as a probe for subphotospheric magnetic fields. Astron Lett 42:339–345. https://doi.org/10.1134/S1063773716050054. arXiv:1601.04855 [astro-ph.SR]

Kochukhov O (2021) Magnetic fields of M dwarfs. Astron Astrophys Rev 29(1):1. https://doi.org/10.1007/s00159-020-00130-3. arXiv:2011.01781 [astro-ph.SR]

Kochukhov O, Mantere MJ, Hackman T et al (2013) Magnetic field topology of the RS CVn star II Pegasi. Astron Astrophys 550:A84. https://doi.org/10.1051/0004-6361/201220432. arXiv:1301.1680 [astro-ph.SR]

Krause F, Rädler KH (1980) Mean-field magnetohydrodynamics and dynamo theory. Pergamon Press, Oxford

Kuhlen M, Woosley WE, Glatzmaier GA (2003) 3D anelastic simulations of convection in massive stars. In: Turcotte S, Keller SC, Cavallo RM (eds) 3D stellar evolution, ASP Conference Series, vol 293. p 147. arXiv:astro-ph/0210557

Kupka F, Muthsam HJ (2017) Modelling of stellar convection. Living Rev Comput Astrophys 3:1. https://doi.org/10.1007/s41115-017-0001-9

Larmor J (1919) How could a rotating body such as the sun become a magnet. In: Report of the British association for the advancement of science, pp 159–160. https://www.biodiversitylibrary.org/item/96028

Larson TP, Schou J (2018) Global-mode analysis of full-disk data from the Michelson Doppler Imager and the Helioseismic and Magnetic Imager. Sol Phys 293(2):29. https://doi.org/10.1007/s11207-017-1201-5

Lehtinen JJ, Spada F, Käpylä MJ et al (2020) Common dynamo scaling in slowly rotating young and evolved stars. Nat Astron 4:658–662. https://doi.org/10.1038/s41550-020-1039-x. arXiv:2003.08997 [astro-ph.SR]

Mabuchi J, Masada Y, Kageyama A (2015) Differential rotation in magnetized and non-magnetized stars. Astrophys J 806:10. https://doi.org/10.1088/0004-637X/806/1/10. arXiv:1504.01129 [astro-ph.SR]

MacGregor KB, Brenner M (1991) Rotational evolution of solar-type stars. I. Main-sequence evolution. Astrophys J 376:204. https://doi.org/10.1086/170269

Masada Y, Takiwaki T, Kotake K (2022) Convection and dynamo in newly born neutron stars. Astrophys J 924(2):75. https://doi.org/10.3847/1538-4357/ac34f6. arXiv:2001.08452 [astro-ph.HE]

Matilsky LI, Toomre J (2020) Exploring bistability in the cycles of the solar dynamo through global simulations. Astrophys J 892(2):106. https://doi.org/10.3847/1538-4357/ab791c. arXiv:1912.08158 [astro-ph.SR]

Matilsky LI, Hindman BW, Toomre J (2019) The role of downflows in establishing solar near-surface shear. Astrophys J 871:217. https://doi.org/10.3847/1538-4357/aaf647. arXiv:1810.00115 [astro-ph.SR]

Matilsky LI, Hindman BW, Featherstone NA et al (2022) Confinement of the solar tachocline by dynamo action in the radiative interior. Astrophys J Lett 940(2):L50. https://doi.org/10.3847/2041-8213/ac93ef. arXiv:2206.12920 [astro-ph.SR]

Matt SP, Do Cao O, Brown BP et al (2011) Convection and differential rotation properties of G and K stars computed with the ASH code. Astron Nachr 332:897. https://doi.org/10.1002/asna.201111624. arXiv:1111.5585 [astro-ph.SR]

Matt SP, Brun AS, Baraffe I et al (2015) The mass-dependence of angular momentum evolution in Sun-like stars. Astrophys J Lett 799(2):L23. https://doi.org/10.1088/2041-8205/799/2/L23. arXiv:1412.4786 [astro-ph.SR]

Meakin CA, Arnett D (2007) Anelastic and compressible simulations of stellar oxygen burning. Astrophys J 665:690–697. https://doi.org/10.1086/519372. astro-ph/0611317

Menu MD, Petitdemange L, Galtier S (2020) Magnetic effects on fields morphologies and reversals in geo-dynamo simulations. Phys Earth Planet Inter 307:106542. https://doi.org/10.1016/j.pepi.2020.106542. arXiv:2007.05530 [physics.flu-dyn]

Metcalfe TS, van Saders J (2017) Magnetic evolution and the disappearance of Sun-like activity cycles. Sol Phys 292(9):126. https://doi.org/10.1007/s11207-017-1157-5. arXiv:1705.09668 [astro-ph.SR]

Miesch MS, Hindman BW (2011) Gyroscopic pumping in the solar near-surface shear layer. Astrophys J 743:79. https://doi.org/10.1088/0004-637X/743/1/79. arXiv:1106.4107 [astro-ph.SR]

Miesch MS, Elliott JR, Toomre J et al (2000) Three-dimensional spherical simulations of solar convection. I. Differential rotation and pattern evolution achieved with laminar and turbulent states. Astrophys J 532:593–615. https://doi.org/10.1086/308555

Miesch MS, Brun AS, Toomre J (2006) Solar differential rotation influenced by latitudinal entropy variations in the tachocline. Astrophys J 641:618–625. https://doi.org/10.1086/499621

Miesch MS, Brun AS, DeRosa ML et al (2008) Structure and evolution of giant cells in global models of solar convection. Astrophys J 673(1):557–575. https://doi.org/10.1086/523838. arXiv:0707.1460 [astro-ph]

Miesch M, Matthaeus W, Brandenburg A et al (2015) Large-eddy simulations of magnetohydrodynamic turbulence in heliophysics and astrophysics. Space Sci Rev 194:97–137. https://doi.org/10.1007/s11214-015-0190-7. arXiv:1505.01808 [astro-ph.SR]

Mitra D, Tavakol R, Brandenburg A et al (2009) Turbulent dynamos in spherical shell segments of varying geometrical extent. Astrophys J 697:923–933. https://doi.org/10.1088/0004-637X/697/1/923. arXiv:0812.3106

Mitra D, Tavakol R, Käpylä PJ et al (2010) Oscillatory migrating magnetic fields in helical turbulence in spherical domains. Astrophys J Lett 719:L1–L4. https://doi.org/10.1088/2041-8205/719/1/L1. arXiv:0901.2364 [astro-ph.SR]

Navarrete FH, Schleicher DRG, Käpylä PJ et al (2022) Origin of eclipsing time variations in post-common-envelope binaries: role of the centrifugal force. Astron Astrophys 667:A164. https://doi.org/10.1051/0004-6361/202243917. arXiv:2205.03163 [astro-ph.SR]

Nelson NJ, Brown BP, Brun AS et al (2013) Magnetic wreaths and cycles in convective dynamos. Astrophys J 762:73. https://doi.org/10.1088/0004-637X/762/2/73. arXiv:1211.3129 [astro-ph.SR]

Nelson NJ, Brown BP, Sacha Brun A et al (2014) Buoyant magnetic loops generated by global convective dynamo action. Sol Phys 289:441–458. https://doi.org/10.1007/s11207-012-0221-4. arXiv:1212.5612 [astro-ph.SR]

Nelson NJ, Featherstone NA, Miesch MS et al (2018) Driving solar giant cells through the self-organization of near-surface plumes. Astrophys J 859:117. https://doi.org/10.3847/1538-4357/aabc07. arXiv:1804.01166 [astro-ph.SR]

Newton ER, Irwin J, Charbonneau D et al (2017) The Hα emission of nearby M dwarfs and its relation to stellar rotation. Astrophys J 834(1):85. https://doi.org/10.3847/1538-4357/834/1/85. arXiv:1611.03509 [astro-ph.SR]

Noraz Q (2022) PhD thesis, University of Paris-Cité. https://www.theses.fr/2022UNIP7018

Noraz Q, Breton SN, Brun AS et al (2022) Hunting for anti-solar differentially rotating stars using the Rossby number. An application to the Kepler field. Astron Astrophys 667:A50. https://doi.org/10.1051/0004-6361/202243890. arXiv:2208.12297 [astro-ph.SR]

Noraz Q, Brun AS, Strugarek A (2023) Magnetochronology of solar-type star dynamos. Submitted to Astron Astrophys

Nordlund Å, Ramsey JP, Popovas A et al (2018) DISPATCH: a numerical simulation framework for the exa-scale era - I. Fundamentals. Mon Not R Astron Soc 477(1):624–638. https://doi.org/10.1093/mnras/sty599. arXiv:1705.10774 [astro-ph.IM]

Noyes RW, Weiss NO, Vaughan AH (1984) The relation between stellar rotation rate and activity cycle periods. Astrophys J 287:769–773. https://doi.org/10.1086/162735

Olspert N, Lehtinen JJ, Käpylä MJ et al (2018) Estimating activity cycles with probabilistic methods. II. The Mount Wilson Ca H&K data. Astron Astrophys 619:A6. https://doi.org/10.1051/0004-6361/201732525. arXiv:1712.08240 [astro-ph.SR]

O'Mara B, Miesch MS, Featherstone NA et al (2016) Velocity amplitudes in global convection simulations: the role of the Prandtl number and near-surface driving. Adv Space Res 58:1475–1489. https://doi.org/10.1016/j.asr.2016.03.038. arXiv:1603.06107 [astro-ph.SR]

Ortiz-Rodríguez CA, Käpylä PJ, Navarrete FH et al (2023) Simulations of dynamo action in slowly rotating M dwarfs: Dependence on dimensionless parameters. Astron Astrophys. https://doi.org/10.1051/0004-6361/202244666. arXiv:2305.16447 [astro-ph.SR]

Orvedahl RJ, Featherstone NA, Calkins MA (2021) Large-scale magnetic field saturation and the Elsasser number in rotating spherical dynamo models. Mon Not R Astron Soc 507(1):L67–L71. https://doi.org/10.1093/mnrasl/slab097

Ossendrijver MAJH (2000) Grand minima in a buoyancy-driven solar dynamo. Astron Astrophys 359:364–372

Ossendrijver M (2003) The solar dynamo. Astron Astrophys Rev 11:287–367. https://doi.org/10.1007/s00159-003-0019-3

Ossendrijver M, Stix M, Brandenburg A (2001) Magnetoconvection and dynamo coefficients: dependence of the α effect on rotation and magnetic field. Astron Astrophys 376:713–726. https://doi.org/10.1051/0004-6361:20011041. astro-ph/0108274

Parker EN (1955) Hydromagnetic dynamo models. Astrophys J 122:293. https://doi.org/10.1086/146087

Parker EN (1987) The dynamo dilemma. Sol Phys 110:11–21. https://doi.org/10.1007/BF00148198

Passos D, Charbonneau P (2014) Characteristics of magnetic solar-like cycles in a 3D MHD simulation of solar convection. Astron Astrophys 568:A113. https://doi.org/10.1051/0004-6361/201423700

Petitdemange L, Marcotte F, Gissinger C (2023) Spin-down by dynamo action in simulated radiative stellar layers. Science 379(6629):300–303. https://doi.org/10.1126/science.abk2169. arXiv:2206.13819 [astro-ph.SR]

Pipin VV, Kosovichev AG (2013) The mean-field solar dynamo with a double cell meridional circulation pattern. Astrophys J 776:36. https://doi.org/10.1088/0004-637X/776/1/36. arXiv:1302.0943 [astro-ph.SR]

Popovas A, Nordlund Å, Szydlarski M (2022) Global MHD simulations of the solar convective zone using a volleyball mesh decomposition. I. Pilot. arXiv:2211.09564 [astro-ph.SR]

Pouquet A, Frisch U, Léorat J (1976). J Fluid Mech 77:321–354. https://doi.org/10.1017/S0022112076002140

Racine É, Charbonneau P, Ghizaru M et al (2011) On the mode of dynamo action in a global large-eddy simulation of solar convection. Astrophys J 735:46. https://doi.org/10.1088/0004-637X/735/1/46

Rädler KH (1968) On the electrodynamics of conducting fluids in turbulent motion. II. Turbulent conductivity and turbulent permeability. Z Naturforsch Teil A 23:1851–1860. https://doi.org/10.1515/zna-1968-1124

Rädler KH (1969) On some electromagnetic phenomena in electrically conducting turbulently moving matter, especially in the presence of Coriolis forces. Geod Geophys Veröff 13:131–135

Rädler KH (1980) Mean-field approach to spherical dynamo models. Astron Nachr 301(3):101–129. https://doi.org/10.1002/asna.2103010302

Rädler KH, Bräuer HJ (1987) On the oscillatory behaviour of kinematic mean-field dynamos. Astron Nachr 308:101–109

Raynaud R, Petitdemange L, Dormy E (2015) Dipolar dynamos in stratified systems. Mon Not R Astron Soc 448(3):2055–2065. https://doi.org/10.1093/mnras/stv122. arXiv:1503.00165 [astro-ph.SR]

Raynaud R, Guilet J, Janka HT et al (2020) Magnetar formation through a convective dynamo in protoneutron stars. Sci Adv 6(11):eaay2732. https://doi.org/10.1126/sciadv.aay2732. arXiv:2003.06662 [astro-ph.HE]

Reiners A (2012) Observations of cool-star magnetic fields. Living Rev Sol Phys 9(1):1. https://doi.org/10.12942/lrsp-2012-1. arXiv:1203.0241 [astro-ph.SR]

Reiners A, Shulyak D, Käpylä PJ et al (2022) Magnetism, rotation, and nonthermal emission in cool stars. Average magnetic field measurements in 292 M dwarfs. Astron Astrophys 662:A41. https://doi.org/10.1051/0004-6361/202243251. arXiv:2204.00342 [astro-ph.SR]

Reinhold T, Arlt R (2015) Discriminating solar and antisolar differential rotation in high-precision light curves. Astron Astrophys 576:A15. https://doi.org/10.1051/0004-6361/201425337. arXiv:1501.07817 [astro-ph.SR]

Rempel M, Bhatia T, Bellot Rubio L et al (2023) Small-scale dynamos: from idealized models to solar and stellar applications. Space Sci Rev 219(5):36. https://doi.org/10.1007/s11214-023-00981-z. arXiv:2305.02787 [astro-ph.SR]

Roberts PH, King EM (2013) On the genesis of the Earth's magnetism. Rep Prog Phys 76(9):096801. https://doi.org/10.1088/0034-4885/76/9/096801

Roberts PH, Soward AM (1975) A unified approach to mean field electrodynamics. Astron Nachr 296(2):49–64. https://doi.org/10.1002/asna.19752960202

Roberts PH, Stix M (1972) Ac-effect dynamos, by the Bullard-Gellman formalism. Astron Astrophys 18:453

Rogachevskii I, Kleeorin N (2003) Electromotive force and large-scale magnetic dynamo in a turbulent flow with a mean shear. Phys Rev E 68(3):036301. https://doi.org/10.1103/PhysRevE.68.036301. astro-ph/0209309

Rogers TM (2015) On the differential rotation of massive main-sequence stars. Astrophys J Lett 815(2):L30. https://doi.org/10.1088/2041-8205/815/2/L30. arXiv:1511.03809 [astro-ph.SR]

Roxburgh LW, Simmons J (1993) Numerical studies of convective penetration in plane parallel layers and the integral constraint. Astron Astrophys 277:93

Saar SH, Brandenburg A (1999) Time evolution of the magnetic activity cycle period. II. Results for an expanded stellar sample. Astrophys J 524:295–310. https://doi.org/10.1086/307794

Schou J, Antia HM, Basu S et al (1998) Helioseismic studies of differential rotation in the solar envelope by the solar oscillations investigation using the Michelson Doppler imager. Astrophys J 505:390–417. https://doi.org/10.1086/306146

Schrinner M (2011) Global dynamo models from direct numerical simulations and their mean-field counterparts. Astron Astrophys 533:A108. https://doi.org/10.1051/0004-6361/201116642. arXiv:1105.2912 [astro-ph.SR]

Schrinner M (2013) Rotational threshold in global numerical dynamo simulations. Mon Not R Astron Soc 431:L78–L82. https://doi.org/10.1093/mnrasl/slt012. arXiv:1212.6910 [astro-ph.SR]

Schrinner M, Rädler KH, Schmitt D et al (2005) Mean-field view on rotating magnetoconvection and a geodynamo model. Astron Nachr 326:245–249. https://doi.org/10.1002/asna.200410384

Schrinner M, Rädler KH, Schmitt D et al (2007) Mean-field concept and direct numerical simulations of rotating magnetoconvection and the geodynamo. Geophys Astrophys Fluid Dyn 101:81–116. https://doi.org/10.1080/03091920701345707. astro-ph/0609752

Schrinner M, Petitdemange L, Dormy E (2011) Oscillatory dynamos and their induction mechanisms. Astron Astrophys 530:A140. https://doi.org/10.1051/0004-6361/201016372. arXiv:1101.1837 [astro-ph.SR]

Schrinner M, Petitdemange L, Dormy E (2012) Dipole collapse and dynamo waves in global direct numerical simulations. Astrophys J 752:121. https://doi.org/10.1088/0004-637X/752/2/121. arXiv:1202.4666 [astro-ph.SR]

Schumacher J, Sreenivasan KR (2020) Colloquium: unusual dynamics of convection in the Sun. Rev Mod Phys 92(4):041001. https://doi.org/10.1103/RevModPhys.92.041001

Schwaiger T, Gastine T, Aubert J (2021) Relating force balances and flow length scales in geodynamo simulations. Geophys J Int 224(3):1890–1904. https://doi.org/10.1093/gji/ggaa545. arXiv:2011.14701 [physics.geo-ph]

See V, Matt SP, Finley AJ et al (2019) Do non-dipolar magnetic fields contribute to spin-down torques? Astrophys J 886(2):120. https://doi.org/10.3847/1538-4357/ab46b2. arXiv:1910.02129 [astro-ph.SR]

Shimada R, Hotta H, Yokoyama T (2022) Mean-field analysis on large-scale magnetic fields at high Reynolds numbers. Astrophys J 935(1):55. https://doi.org/10.3847/1538-4357/ac7e43. arXiv:2207.01639 [astro-ph.SR]

Simard C, Charbonneau P (2020) Grand minima in a spherical non-kinematic $\alpha^2\Omega$ mean-field dynamo model. J Space Weather Space Clim 10:9. https://doi.org/10.1051/swsc/2020006

Simard C, Charbonneau P, Bouchat A (2013) Magnetohydrodynamic simulation-driven kinematic mean field model of the solar cycle. Astrophys J 768:16. https://doi.org/10.1088/0004-637X/768/1/16

Simard C, Charbonneau P, Dubé C (2016) Characterisation of the turbulent electromotive force and its magnetically-mediated quenching in a global EULAG-MHD simulation of solar convection. Adv Space Res 58:1522–1537. https://doi.org/10.1016/j.asr.2016.03.041. arXiv:1604.01533 [astro-ph.SR]

Simitev RD, Kosovichev AG, Busse FH (2015) Dynamo effects near the transition from solar to anti-solar differential rotation. Astrophys J 810:80. https://doi.org/10.1088/0004-637X/810/1/80. arXiv:1504.07835 [astro-ph.SR]

Skumanich A (1972) Time scales for Ca II emission decay, rotational braking, and lithium depletion. Astrophys J 171:565. https://doi.org/10.1086/151310

Smolarkiewicz PK, Charbonneau P (2013) EULAG, a computational model for multiscale flows: an MHD extension. J Comp Physiol 236:608–623. https://doi.org/10.1016/j.jcp.2012.11.008

Soderblom DR (1983) Rotational studies of late-type stars. II. Ages of solar-type stars and the rotational history of the Sun. Astrophys J Suppl Ser 53:1–15. https://doi.org/10.1086/190880

Sokoloff D, Nesme-Ribes E (1994) The Maunder minimum: a mixed-parity dynamo mode? Astron Astrophys 288:293–298

Spiegel EA, Zahn JP (1992) The solar tachocline. Astron Astrophys 265:106–114

Spitzer L (1962) Physics of fully ionized gases. Interscience, New York

Spruit H (1997) Convection in stellar envelopes: a changing paradigm. Mem Soc Astron Ital 68:397. astro-ph/9605020

Spruit HC (1999) Differential rotation and magnetic fields in stellar interiors. Astron Astrophys 349:189

Steenbeck M, Krause F (1969) On the dynamo theory of stellar and planetary magnetic fields. I. AC dynamos of solar type. Astron Nachr 291:49–84. https://doi.org/10.1002/asna.19692910201

Steenbeck M, Krause F, Rädler KH (1966) Berechnung der mittleren Lorentz-Feldstärke $\overline{v \times B}$ für ein elektrisch leitendes Medium in turbulenter, durch Coriolis-Kräfte beeinflußter Bewegung. Z Naturforsch Teil A 21:369. https://doi.org/10.1515/zna-1966-0401

Stello D, Cantiello M, Fuller J et al (2016) A prevalence of dynamo-generated magnetic fields in the cores of intermediate-mass stars. Nature 529(7586):364–367. https://doi.org/10.1038/nature16171. arXiv:1601.00004 [astro-ph.SR]

Stevenson DJ (1979) Turbulent thermal convection in the presence of rotation and a magnetic field: a heuristic theory. Geophys Astrophys Fluid Dyn 12(1):139–169. https://doi.org/10.1080/03091927908242681

Stix M (2002) The Sun: an introduction. Springer, Berlin

Strugarek A, Beaudoin P, Brun AS et al (2016) Modeling turbulent stellar convection zones: sub-grid scales effects. Adv Space Res 58:1538–1553. https://doi.org/10.1016/j.asr.2016.05.043. arXiv:1605.08685 [astro-ph.SR]

Strugarek A, Beaudoin P, Charbonneau P et al (2017) Reconciling solar and stellar magnetic cycles with nonlinear dynamo simulations. Science 357:185–187. https://doi.org/10.1126/science.aal3999. arXiv:1707.04335 [astro-ph.SR]

Strugarek A, Beaudoin P, Charbonneau P et al (2018) On the sensitivity of magnetic cycles in global simulations of solar-like stars. Astrophys J 863:35. https://doi.org/10.3847/1538-4357/aacf9e. arXiv:1806.09484 [astro-ph.SR]

Tassin T, Gastine T, Fournier A (2021) Geomagnetic semblance and dipolar-multipolar transition in top-heavy double-diffusive geodynamo models. Geophys J Int 226(3):1897–1919. https://doi.org/10.1093/gji/ggab161. arXiv:2101.03879 [physics.geo-ph]

Tobias SM (1997) The solar cycle: parity interactions and amplitude modulation. Astron Astrophys 322:1007–1017

Tremblay PE, Ludwig HG, Freytag B et al (2015) Calibration of the mixing-length theory for convective white dwarf envelopes. Astrophys J 799:142. https://doi.org/10.1088/0004-637X/799/2/142. arXiv:1412.1789 [astro-ph.SR]

Triana SA, Guerrero G, Barik A et al (2022) Identification of inertial modes in the solar convection zone. Astrophys J Lett 934(1):L4. https://doi.org/10.3847/2041-8213/ac7dac. arXiv:2204.13007 [astro-ph.SR]

Trifonov T, Kürster M, Zechmeister M et al (2018) The CARMENES search for exoplanets around M dwarfs. First visual-channel radial-velocity measurements and orbital parameter updates of seven M-dwarf planetary systems. Astron Astrophys 609:A117. https://doi.org/10.1051/0004-6361/201731442. arXiv:1710.01595 [astro-ph.EP]

van Saders JL, Ceillier T, Metcalfe TS et al (2016) Weakened magnetic braking as the origin of anomalously rapid rotation in old field stars. Nature 529:181–184. https://doi.org/10.1038/nature16168. arXiv:1601.02631 [astro-ph.SR]

Vasil GM, Julien K, Featherstone NA (2021) Rotation suppresses giant-scale solar convection. Proc Natl Acad Sci 118(31):e2022518118. https://doi.org/10.1073/pnas.2022518118

Viallet M, Baraffe I, Walder R (2011) Towards a new generation of multi-dimensional stellar evolution models: development of an implicit hydrodynamic code. Astron Astrophys 531:A86. https://doi.org/10.1051/0004-6361/201016374. arXiv:1103.1524 [astro-ph.IM]

Vidal J, Cébron D, Schaeffer N et al (2018) Magnetic fields driven by tidal mixing in radiative stars. Mon Not R Astron Soc 475(4):4579–4594. https://doi.org/10.1093/mnras/sty080. arXiv:1711.09612 [astro-ph.SR]

Viviani M, Käpylä MJ (2021) Physically motivated heat-conduction treatment in simulations of solar-like stars: effects on dynamo transitions. Astron Astrophys 645:A141. https://doi.org/10.1051/0004-6361/202038603. arXiv:2006.04426 [astro-ph.SR]

Viviani M, Warnecke J, Käpylä MJ et al (2018) Transition from axi- to nonaxisymmetric dynamo modes in spherical convection models of solar-like stars. Astron Astrophys 616:A160. https://doi.org/10.1051/0004-6361/201732191. arXiv:1710.10222 [astro-ph.SR]

Viviani M, Käpylä MJ, Warnecke J et al (2019) Stellar dynamos in the transition regime: multiple dynamo modes and antisolar differential rotation. Astrophys J 886(1):21. https://doi.org/10.3847/1538-4357/ab3e07. arXiv:1902.04019 [astro-ph.SR]

Warnecke J (2018) Dynamo cycles in global convection simulations of solar-like stars. Astron Astrophys 616:A72. https://doi.org/10.1051/0004-6361/201732413. arXiv:1712.01248 [astro-ph.SR]

Warnecke J, Käpylä MJ (2020) Rotational dependence of turbulent transport coefficients in global convective dynamo simulations of solar-like stars. Astron Astrophys 642:A66. https://doi.org/10.1051/0004-6361/201936922. arXiv:1910.06776 [astro-ph.SR]

Warnecke J, Brandenburg A, Mitra D (2011) Dynamo-driven plasmoid ejections above a spherical surface. Astron Astrophys 534:A11. https://doi.org/10.1051/0004-6361/201117023. arXiv:1104.0664 [astro-ph.SR]

Warnecke J, Käpylä PJ, Mantere MJ et al (2013) Spoke-like differential rotation in a convective dynamo with a coronal envelope. Astrophys J 778:141. https://doi.org/10.1088/0004-637X/778/2/141. arXiv:1301.2248 [astro-ph.SR]

Warnecke J, Käpylä PJ, Käpylä MJ et al (2014) On the cause of solar-like equatorward migration in global convective dynamo simulations. Astrophys J Lett 796:L12. https://doi.org/10.1088/2041-8205/796/1/L12. arXiv:1409.3213 [astro-ph.SR]

Warnecke J, Käpylä PJ, Käpylä MJ et al (2016) Influence of a coronal envelope as a free boundary to global convective dynamo simulations. Astron Astrophys 596:A115. https://doi.org/10.1051/0004-6361/201526131. arXiv:1503.05251 [astro-ph.SR]

Warnecke J, Rheinhardt M, Tuomisto S et al (2018) Turbulent transport coefficients in spherical wedge dynamo simulations of solar-like stars. Astron Astrophys 609:A51. https://doi.org/10.1051/0004-6361/201628136. arXiv:1601.03730 [astro-ph.SR]

Warnecke J, Rheinhardt M, Viviani M et al (2021) Investigating global convective dynamos with mean-field models: full spectrum of turbulent effects required. Astrophys J Lett 919(2):L13. https://doi.org/10.3847/2041-8213/ac1db5. arXiv:2105.07708 [astro-ph.SR]

Weber M, Strassmeier KG, Washuettl A (2005) Indications for anti-solar differential rotation of giant stars. Astron Nachr 326:287–291. https://doi.org/10.1002/asna.200410391

Wright NJ, Drake JJ (2016) Solar-type dynamo behaviour in fully convective stars without a tachocline. Nature 535:526–528. https://doi.org/10.1038/nature18638. arXiv:1607.07870 [astro-ph.SR]

Wright NJ, Drake JJ, Mamajek EE et al (2011) The stellar-activity-rotation relationship and the evolution of stellar dynamos. Astrophys J 743:48. https://doi.org/10.1088/0004-637X/743/1/48. arXiv:1109.4634 [astro-ph.SR]

Wright NJ, Newton ER, Williams PKG et al (2018) The stellar rotation-activity relationship in fully convective M dwarfs. Mon Not R Astron Soc 479(2):2351–2360. https://doi.org/10.1093/mnras/sty1670. arXiv:1807.03304 [astro-ph.SR]

Yadav RK, Gastine T, Christensen UR et al (2013) Consistent scaling laws in anelastic spherical shell dynamos. Astrophys J 774(1):6. https://doi.org/10.1088/0004-637X/774/1/6. arXiv:1304.6163 [astro-ph.SR]

Yadav RK, Christensen UR, Morin J et al (2015) Explaining the coexistence of large-scale and small-scale magnetic fields in fully convective stars. Astrophys J Lett 813:L31. https://doi.org/10.1088/2041-8205/813/2/L31. arXiv:1510.05541 [astro-ph.SR]

Yadav RK, Gastine T, Christensen UR et al (2015b) Formation of starspots in self-consistent global dynamo models: polar spots on cool stars. Astron Astrophys 573:A68. https://doi.org/10.1051/0004-6361/201424589. arXiv:1407.3187 [astro-ph.SR]

Yadav RK, Christensen UR, Wolk SJ et al (2016) Magnetic cycles in a dynamo simulation of fully convective M-star Proxima Centauri. Astrophys J Lett 833:L28. https://doi.org/10.3847/2041-8213/833/2/L28. arXiv:1610.02721 [astro-ph.SR]

Yoshimura H (1975) Solar-cycle dynamo wave propagation. Astrophys J 201:740–748. https://doi.org/10.1086/153940

Zahn JP, Brun AS, Mathis S (2007) On magnetic instabilities and dynamo action in stellar radiation zones. Astron Astrophys 474(1):145–154. https://doi.org/10.1051/0004-6361:20077653. arXiv:0707.3287 [astro-ph]

Zaire B, Guerrero G, Kosovichev AG et al (2017) Magnetic field generation in PMS stars with and without radiative core. In: Nandy D, Valio A, Petit P (eds) Living around active stars. Cambridge University Press, Cambridge, pp 30–37. https://doi.org/10.1017/S1743921317003970. arXiv:1711.02057

Zaire B, Jouve L, Gastine T et al (2022) Transition from multipolar to dipolar dynamos in stratified systems. Mon Not R Astron Soc 517(3):3392–3406. https://doi.org/10.1093/mnras/stac2769. arXiv:2209.11652 [astro-ph.SR]

Authors and Affiliations

Petri J. Käpylä[1,2]  **· Matthew K. Browning[3] · Allan Sacha Brun[4] · Gustavo Guerrero[5,6] · Jörn Warnecke[7]**

✉ P.J. Käpylä
  pkapyla@leibniz-kis.de

  M.K. Browning
  M.K.M.Browning@exeter.ac.uk

  A.S. Brun
  sacha.brun@cea.fr

G. Guerrero
guerrero@fisica.ufmg.br

J. Warnecke
warnecke@mps.mpg.de

[1]   Institute for Astrophysics and Geophysics, University of Göttingen, Friedrich-Hund-Platz 1, Göttingen, 37077, Germany

[2]   Leibniz Institute for Solar Physics (KIS), Schöneckstraße 6, Freiburg, 79104, Germany

[3]   Department of Physics & Astronomy, University of Exeter, Stocker Road, Exeter, EX4 4QL, UK

[4]   Département d'Astrophysique/AIM, Univ. Paris-Saclay and Univ. de Paris Cité, CEA, CNRS, Gif-sur-Yvette, 91191, France

[5]   Physics Department, Universidade Federal de Minas Gerais, Av. Antonio Carlos 6627, Belo Horizonte, MG 31270-901, Brazil

[6]   Physics Department, New Jersey Institute of Technology, 323 Dr Martin Luther King Jr Blvd, Newark, NJ 07103, USA

[7]   Max Planck Institute for Solar System Research, Justus-von-Liebig-Weg 3, Göttingen, 37077, Germany

Space Science Reviews (2023) 219:36
https://doi.org/10.1007/s11214-023-00981-z

Small-Scale Dynamos: From Idealized Models to Solar and Stellar Applications

Matthias Rempel[1] · Tanayveer Bhatia[2] · Luis Bellot Rubio[3] ·
Maarit J. Korpi-Lagg[2,4,5]

Received: 17 March 2023 / Accepted: 7 June 2023 / Published online: 4 July 2023
© The Author(s) 2023

Abstract

In this article we review small-scale dynamo processes that are responsible for magnetic field generation on scales comparable to and smaller than the energy carrying scales of turbulence. We provide a review of critical observation of quiet Sun magnetism, which have provided strong support for the operation of a small-scale dynamo in the solar photosphere and convection zone. After a review of basic concepts we focus on numerical studies of kinematic growth and non-linear saturation in idealized setups, with special emphasis on the role of the magnetic Prandtl number for dynamo onset and saturation. Moving towards astrophysical applications we review convective dynamo setups that focus on the deep convection zone and the photospheres of solar-like stars. We review the critical ingredients for stellar convection setups and discuss their application to the Sun and solar-like stars including comparison against available observations.

Keywords Small-scale dynamo · Stellar magnetism · Quiet Sun · Cool stars · Convection

1 Introduction

The Sun is the only star that can be scrutinized in detail at high spatial and temporal resolution. Observations show that magnetic fields are ubiquitous in the quiet Sun—the areas of the solar surface away from active regions and the enhanced network. They cover the whole solar surface, all the time, irrespective of the phase of the solar cycle. Quiet Sun (QS) fields can be classified into network and internetwork (IN) fields. The former are relatively strong and vertical, and occupy the outer boundaries of supergranular cells. The latter are weaker and highly inclined, and can be found in the cell interiors. In high-resolution observations, both fields are detected as magnetic flux concentrations of opposite signs organized on sub-arcsecond scales. Observations of quiet Sun magnetism over the past few decades have lent growing support for a dynamo process that operates in the IN independently from the large-scale dynamo (LSD) responsible for the solar cycle. We provide a detailed review of solar observations in Sect. 2.

The theoretical concept of a small-scale dynamo (SSD) instability, independent of the presence of symmetry-breaking effects, such as helicity, dates back to Kazantsev (1968).

Solar and Stellar Dynamos: A New Era
Edited by Manfred Schüssler, Robert H. Cameron, Paul Charbonneau, Mausumi Dikpati,
Hideyuki Hotta and Leonid Kitchatinov

Extended author information available on the last page of the article

As identified early on, the magnetic Prandtl number, $\mathrm{Pm} = \nu/\eta$, where ν is the kinematic viscosity and η the magnetic resistivity, plays a critical role for SSD action and we provide a more detailed account of the theoretical concepts in Sect. 3. It was first suggested by Petrovay and Szakaly (1993) that this dynamo instability may be the origin of the weak IN fields.

Numerical simulations of SSD started with convective setups and solar surface simulations with 3D radiative transfer at Pm around one, and have recently been extended towards the numerically more challenging lower Pm-regime. The evolution of numerical models is described in Sect. 4. We focus on investigations of the role of Pm during kinematic and saturated phases (Sect. 4.1), deep convection setups (Sect. 4.2), surface convection setups (Sect. 4.3) and the possibility of an SSD in the radiative interior of solar-like stars (Sect. 4.4). Section 5 provides an overview of recent applications to solar-like stars. Here we focus on the effects of small-scale magnetism on stellar structure, the basal chromospheric flux and contributions to short-term stellar variability that has to be considered for exoplanet detection. We conclude the review with an outlook in Sect. 6.

2 Solar Observations

Because of their abundance, quiet Sun fields are important contributors to the flux budget of the solar photosphere. The total unsigned longitudinal magnetic flux of the quiet Sun has been estimated to be 8×10^{23} Mx at any time (Gošić 2015), similar to the total flux carried by active regions at solar maximum ($\sim 6 \times 10^{23}$ Mx; Jin et al. 2011). Network fields contribute about 80–85% of this flux, while the IN supplies the remaining 20–15% (Wang et al. 1995; Gošić 2015). However, the IN is extremely dynamic and evolves very rapidly, with flux appearance rates from 120 to 1100 Mx cm^{-2} day^{-1} (e.g., Gošić et al. 2016; Smitha et al. 2017) that surpass those of active regions by orders of magnitude (0.1 Mx cm^{-2} day^{-1}; Schrijver and Harvey 1994). IN fields are also the main contributors to the total energy budget of the solar photosphere (Trujillo Bueno et al. 2004; Rempel 2014).

Pushed and dragged by granular and supergranular convective flows, quiet Sun magnetic fields undergo frequent interactions between them and with other pre-existing fields, particularly in the IN. Such interactions are believed to trigger magnetic reconnection events at different heights in the atmosphere, releasing energy and contributing to the heating of the chromosphere and corona locally and globally. This role has been recognized only recently, as high resolution observations became available both from ground and from space. Quiet Sun fields also contribute to atmospheric heating by channeling waves from the photosphere to higher atmospheric layers (Jess et al. 2023).

To understand these fields we must determine their properties and origin. From an observational point of view, they turn out to be very different from active region fields. For example, IN fields are weaker, more dynamic, and do not show significant changes with the solar cycle (Hinode Review Team et al. 2019). Therefore, it is likely that their origin is also different from the LSD-related active regions and magnetic fluctuations originating from this component through tangling by turbulence. Cascading from large to small scales by active region decay seems not to be viable because of the low flux emergence rate of active regions and the lack of solar cycle variations. An SSD was proposed by Petrovay and Szakaly (1993) as the origin of the weak IN fields. SSD simulations are indeed able to reproduce many statistical properties of IN fields, but important questions remain open, as for example whether or not they can explain the spatial distribution of the flux, the field strength and field inclination distributions, or the flux emergence processes observed in the IN. On the other hand, the quiet Sun network seems to show some variation with the solar cycle

(Korpi-Lagg et al. 2022). The variation is certainly smaller than that of active regions, but it suggests that the network gets contributions from both the LSD and the SSD.

By characterizing the properties, dynamics and temporal evolution of quiet Sun fields observationally, it should be possible to constrain SSD models and help set the most appropriate physical ingredients such as boundary conditions and Pm regimes. For example, a good determination of the magnetic field strength at the surface can provide information on the dynamics of the upper convection zone (CZ), where the bottom boundary condition is usually placed (see Sect. 4.3.1). Also, a detailed characterization of the flux emergence process in granules and intergranular lanes based on high-resolution observations can help validate the mechanisms suggested by numerical simulations. Finally, by studying the formation and evolution of the quiet network, one may get information about the possible interplay and feedback between SSD and LSD.

2.1 Diagnostic Techniques

Our current understanding of quiet Sun fields has been gained through the interpretation of polarimetric measurements of spectral lines formed in the solar photosphere. When the polarization signals are very weak (near the noise level), quantitative analyses are usually not possible and only general properties of the observed signals can be derived, such as their amplitudes, line asymmetries, lifetimes or spatial distribution. However, when the signals are strong it is possible to interpret them by inverting the radiative transfer equation, which gives precise inferences of the vector magnetic field and other atmospheric parameters (see reviews by del Toro Iniesta and Ruiz Cobo 2016; Trujillo Bueno and del Pino Alemán 2022).

The physical mechanisms that leave imprints of the magnetic field on the polarization profiles of spectral lines are the Zeeman and Hanle effects. Both can be used to diagnose the photospheric field, depending on the strength and position of the magnetic features on the solar disk (for details see Bellot Rubio and Orozco Suárez 2019). Zeeman measurements provide spatially resolved observations, but are affected by possible cancellation of the mixed-polarity fields that may coexist in the resolution element. Hanle measurements do not suffer from these cancellation effects, but in turn they are spatially unresolved.

The main problem faced by spectropolarimetric observations of the quiet Sun is that the signals are extremely weak (of order 10^{-3} of the continuum intensity or smaller). To properly detect them, long integrations are needed. This worsens the effective spatial resolution and the cadence of the observations, but also lowers the noise level, which is important to cope with the different sensitivity of the linear and circular polarization signals to weak magnetic fields in the Zeeman regime. Because of such a sensitivity difference, the same amount of noise in linear and circular polarization results in much larger transverse fields than longitudinal fields if interpreted as a real signal. This intrinsic bias is known to affect the determination of the magnetic field components, particularly the field inclination, which is usually inferred to be more horizontal than the true one (e.g., Borrero and Kobel 2011). Indeed, noise is often the main reason for the discrepancies between analyses, hence the need to reduce it as much as possible in the first place and then minimize its effects with appropriate techniques.

In what follows we summarize the main results derived from the observations. Aspects that are particularly relevant for comparison with numerical models include the spatial distribution of the fields on small scales, the mean longitudinal flux density, the magnetic field strength and inclination distributions, their stratification with height, the temporal evolution of the fields (in particular the flux emergence modes), solar cycle variations (total unsigned flux, polarity imbalance, latitudinal distribution), and the impact of quiet Sun fields on the chromosphere.

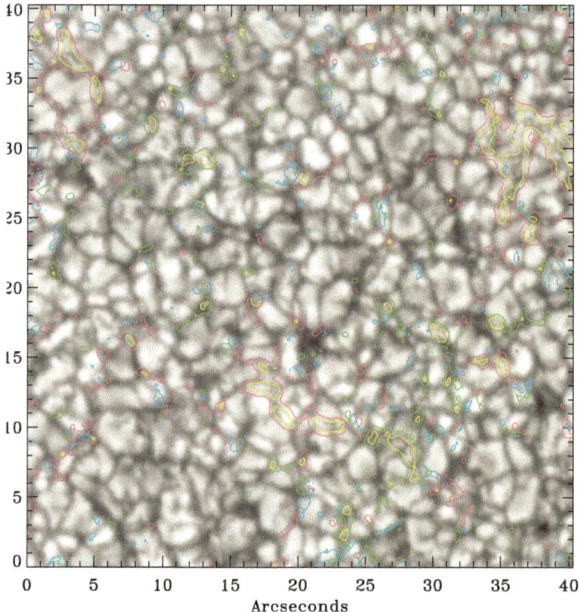

Fig. 1 Spatial distribution of quiet Sun magnetic fields as observed by the Hinode spectropolarimeter (SP) at disk center. Red and green contours show positive and negative apparent longitudinal flux densities larger than 24 Mx cm^{-2} (10 times above the noise level), while the yellow contours show strong longitudinal flux densities of more than 100 Mx cm^{-2}. Blue contours represent apparent horizontal flux densities in excess of 122 Mx cm^{-2} (three times the corresponding noise level). The longitudinal signals are preferentially located in intergranular lanes. The strong horizontal signals are spatially separated from the vertical signals, and are mostly seen above or at the edges of granules. Reproduced with permission from Lites et al. (2008), copyright by AAS

2.2 Properties of Quiet Sun Magnetic Fields

2.2.1 Spatial Distribution on Small Scales

For a long time, quiet Sun magnetic fields were thought to occupy only the intergranular space. However, with sufficient spatial resolution and polarimetric sensitivity, granules also show clear polarization signals. The spatial distribution of the field can be seen in Fig. 1. The fields in intergranular lanes tend to be stronger, more vertical, and more concentrated than in granular cells, which explains their higher visibility. They carry most of the magnetic flux (Beck and Rezaei 2009). However, flux emergence is observed to take place preferentially in granules in the form of small-scale magnetic loops (Centeno et al. 2007; Martínez González and Bellot Rubio 2009; Fischer et al. 2020) and sheets (Orozco Suárez et al. 2008; Fischer et al. 2019), putting constraints on the mechanisms that bring the field to the surface (see Sect. 4.3.1 for details). Once at the surface, the field is dragged by the horizontal granular motions to the intergranular lanes, where it can be further amplified.

2.2.2 Level of Quiet Sun Magnetism

Traditionally, the magnetization of the quiet Sun has been quantified in terms of the longitudinal flux density. This parameter is defined as $\phi = f\, B_{\mathrm{LOS}}$, with f the fraction of the

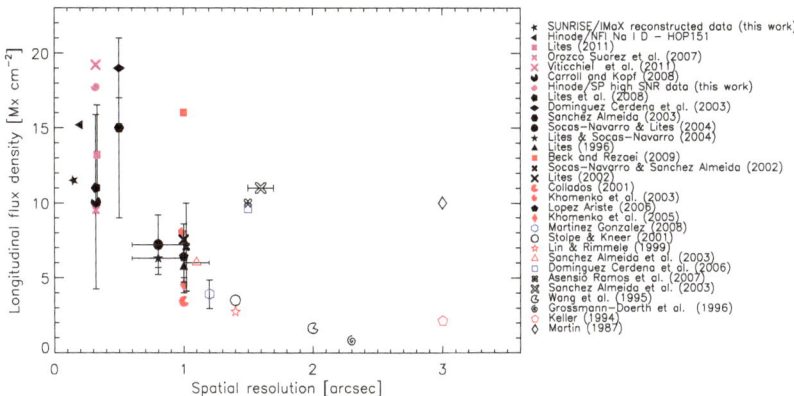

Fig. 2 Compilation of unsigned longitudinal flux densities in the quiet Sun, as a function of the spatial resolution of the observations. From Bellot Rubio and Orozco Suárez (2019)

resolution element covered by the magnetic field (assumed to be homogeneous) and B_{LOS} the longitudinal component of the field. The longitudinal flux density can be derived relatively easily from measurements of the circular polarization profile at a single wavelength or through the inversion of full Stokes profiles, which explains its popularity. In the weak field regime, ϕ is proportional to the circular polarization signal.

Unfortunately, this parameter is very dependent on the spatial resolution and polarimetric sensitivity of the observations. At low spatial resolution, the magnetic filling factor f tends to be small, decreasing the flux density values. Also, the amount of Zeeman cancellation may be significant, especially for magnetograph observations that are not based on full spectral line profiles. This leads to a further reduction of ϕ. The polarimetric sensitivity, on the other hand, affects the estimates through the noise: the larger the noise is, the higher the mean flux density values will be, unless provision is made to exclude pixels without clear polarization signal from the analysis. Therefore, high spatial resolution and high sensitivity are essential for reliable estimations of the flux density in the quiet Sun.

With increasing spatial resolution, the mean unsigned longitudinal flux density derived from Zeeman-sensitive spectral lines increases until approximately 0.5 arcsec, where it levels off and seems to remain constant at about $10–20 \, \mathrm{Mx \, cm}^{-2}$ (Fig. 2). The apparent longitudinal flux densities of $7–11 \, \mathrm{Mx \, cm}^{-2}$ reported by Danilovic et al. (2010b) from Hinode/SP measurements using the method of Lites et al. (2008) are consistent with these values.

However, despite the apparent agreement between the estimates reported at high spatial resolution, it is important to remember that the longitudinal flux density only provides a lower limit to the intrinsic longitudinal field, as it also depends on the actual filling factor of the observations, which is unknown but certainly different from 1. This dependency makes it very difficult to compare the observed flux densities with simulations, where the magnetic filling factor is always unity. The magnetic field strength is less problematic and should be preferred for quantitative analyses, especially now that powerful techniques are available to retrieve it from the observations.

2.2.3 Field Strength Distribution

Stokes inversions of varying degrees of sophistication have been used to determine the magnetic field strength, field inclination and magnetic filling factor on a pixel by pixel basis

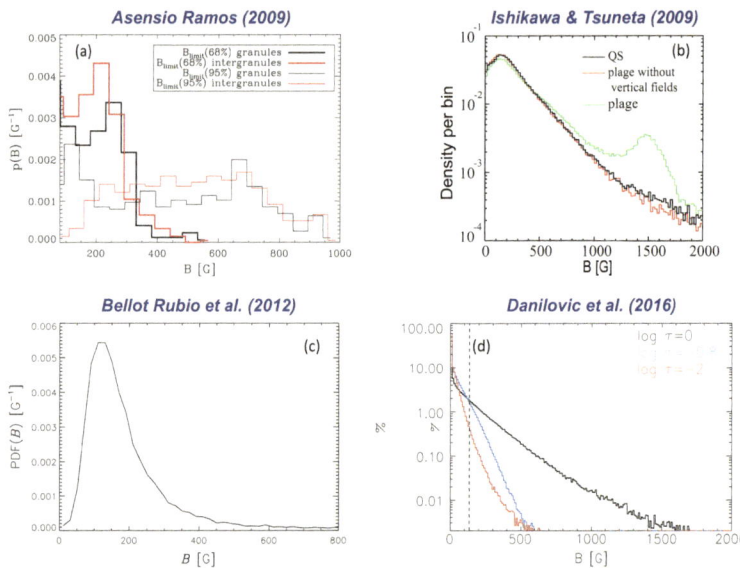

Fig. 3 Field strength distributions in the IN determined from four different inversions of Hinode/SP measurements. Clockwise from top left: Asensio Ramos (2009), Ishikawa and Tsuneta (2009), Bellot Rubio and Orozco Suárez (2012), and Danilovic et al. (2016b). Reproduced with permission, copyright by AAS (panels a, c) and by ESO (panels b, d)

from spectropolarimetric observations of Zeeman sensitive lines. Stokes inversions allow to disentangle the actual contribution of each of these parameters to the longitudinal magnetic flux, providing much richer information. Thus, they are more appropriate for comparison with simulations.

According to the inversions, IN fields are weak for the most part (Fig. 3). Although the details vary between analyses, this conclusion seems to be robust. The field strength distribution shows a preponderance of weak fields in the hG range and a long tail toward kG fields. Network fields tend to be stronger than IN fields, with a hump at 1–1.5 kG. For a summary of these results, see Sect. 4.5 of Bellot Rubio and Orozco Suárez (2019).

From the inferred field strength distribution it is possible to compute the mean field strength over the observed field of view. The resulting values are influenced by the noise of the Stokes profiles, the diagnostic technique employed, and even the way the averaging is done (in particular whether pixels with noisy signals are included or excluded from the analysis). Also the inclusion or exclusion of the network influences the results. But they tend to be much larger than the average longitudinal flux densities shown in Fig. 2. Lites et al. (2008) reported a value of $\langle B \rangle = 185$ G, while Orozco Suárez and Bellot Rubio (2012) found $\langle B \rangle = 220$ G with $\langle |B_z| \rangle = 64$ G, and Danilovic et al. (2016b) derived $\langle B \rangle = 130$ G at $\tau = 1$ with $\langle |B_z| \rangle = 84$ G. These values represent upper limits to the true mean field strengths, as in the first two cases only pixels with clear signals were considered (hence biasing the mean toward the stronger fields) and in the last case all pixels were included (hence adding some contribution from photon noise in pixels with no polarization signal). In general, the field strength is found to decrease with height in the photosphere (Danilovic et al. 2016b). Hanle-effect inversions of molecular lines also show a rapid drop of the field strength with height, from 95 G at $z = 200$ km to 5 G at 400 km (Milić and Faurobert 2012).

These results are compatible with the average field strengths determined from spatially unresolved scattering polarization measurements of the Sr I 460.7 nm line using the Hanle effect. The observed center-to-limb variation of the fractional linear polarization of the Sr I line can be reproduced by a volume-filling magnetic field with an isotropic distribution of orientations and a homogeneous strength of 60 G or, alternatively, an exponential distribution of field strengths with $\langle B \rangle = 130$ G between 200 and 400 km above $\tau = 1$ (Trujillo Bueno et al. 2004). It was concluded that most of the fields contributing to the Hanle depolarization of this line are located in intergranular space and have strengths between 2 and 300 G, well within the Hanle saturation regime. Fields above granules are much weaker and do not seem to contribute significantly to the observed depolarization. By including this small contribution in the fit to the center-to-limb variation of the Sr I linear polarization, an exponential distribution with $\langle B \rangle = 15$ G was inferred in granular cells.

Recent results from multi-line inversions of intensity profiles aimed to avoid the problems of noise in the Stokes polarization spectra also confirm that granules harbor weaker fields (Trelles Arjona et al. 2021). According to these inversions, the average field strength in granules and intergranular lanes is 16 G and 76 G, respectively, in the optical depth range from 1 to 0.1. The average field strength across the field of view is 46 G.

2.2.4 Field Inclination Distribution

The existence of inclined fields in the quiet Sun has been known since the mid 1990s, when small patches of Horizontal Internetwork Fields were discovered and characterized using full Stokes spectropolarimetric measurements at a resolution of about 1 arcsec (Lites et al. 1996). Subsequent analyses at similar resolutions but based on different spectral lines and inversion codes have confirmed them (e.g., Khomenko et al. 2003; Martínez González et al. 2007; Beck and Rezaei 2009).

With the significantly better spatial resolution of 0.3 arcsec provided by the Hinode spectropolarimeter, the transverse apparent flux density in quiet Sun areas at the disk center was found to be about 5 times larger than the longitudinal apparent flux density, suggesting that most of the IN fields are actually very inclined (Lites et al. 2008). Similar results were obtained also from ground-based observations (Beck and Rezaei 2009).

Nearly all the inversions performed to date, using different model atmospheres, codes and assumptions, result in a field inclination distribution at the disk center dominated by highly inclined fields (e.g., Orozco Suárez et al. 2007; Lites et al. 2008; Beck and Rezaei 2009; Orozco Suárez and Bellot Rubio 2012; Asensio Ramos and Martínez González 2014; Danilovic et al. 2016b; Martínez González et al. 2016). Some examples are given in Fig. 4. The distribution usually shows a maximum at 90°, representing horizontal fields, and has tails decreasing toward 0° and 180° (vertical fields). While some contamination by noise cannot be completely ruled out (Borrero and Kobel 2011), it is unlikely that all the inclined fields inferred in the solar IN are a consequence of noise in the linear polarization measurements. Still, the exact shape of the inclination distribution remains a matter of debate, particularly the amplitude of the peak at 90°, which exhibits significant differences between analyses.

Also, there is an ongoing discussion on whether the inclination distribution is isotropic, quasi-isotropic, dominantly horizontal, or dominantly vertical (Bellot Rubio and Orozco Suárez 2019). This is an important question that can shed light on the origin of the fields or the way they appear on the solar surface. Observations outside of the disk center may hold the key to answering it. If the field is isotropic, the inclination distribution should not change with the heliocentric angle. Unfortunately, studies of the center-to-limb behavior of

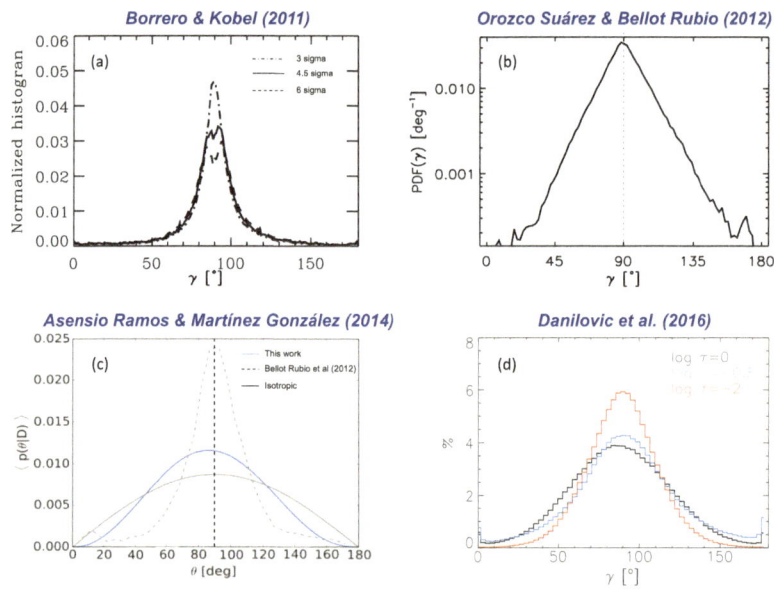

Fig. 4 Field inclination distribution in the solar IN determined from four different inversions of Hinode/SP measurements at disk center. Inclinations are measured with respect to the local vertical, with 90° corresponding to horizontal fields and 0°/180° representing fields pointing to/away from the observer. Clockwise from top left: Borrero and Kobel (2011), Orozco Suárez and Bellot Rubio (2012), Asensio Ramos and Martínez González (2014), and Danilovic et al. (2016b). Reproduced with permission, copyright by ESO (panels a, c, d) and by AAS (panel b)

the inclination distribution are very scarce (e.g., Orozco Suárez and Katsukawa 2012). The efforts have rather focused on determining the variation of the circular and linear polarization amplitudes, since they are not biased by photon noise. The results of these analyses are still controversial, but there seems to be a variation of the weakest polarization signals with heliocentric angle which would not be compatible with an isotropic distribution of field orientations (Lites et al. 2008; Borrero and Kobel 2013; Lites et al. 2014, 2017).

On the simulation side, the field turns out to be predominantly horizontal in the mid and upper photosphere, where most spectral lines used for diagnostics are formed (Sect. 4.3.3). Interestingly, this seems to be a natural outcome of both near-surface magnetoconvection and SSD action. The MHD simulations of Steiner et al. (2008) show a prevalence of inclined fields at a height of 500 km, where the ratio of horizontal to vertical field components is 2 to 5.6, in agreement with Lites et al. (2008). In these simulations, the horizontal field strength reaches a maximum in the upper photosphere because overshooting convective motions expel the horizontal field upwards to layers where vertical flows are no longer present, allowing the field to accumulate there. The same dynamic effect is observed in SSD simulations, resulting in fields that are more horizontal in the upper photosphere and ratios of horizontal to vertical field consistent with the observations (see Schüssler and Vögler 2008; Rempel 2014, and Sect. 4.3.3).

Thus, observed inclination distributions that are predominantly horizontal in the mid photosphere would be compatible with SSD action, but do not necessarily imply its existence. On the contrary, isotropic or quasi-isotropic distributions would pose a serious problem for simulations. Determining which distribution better represents the quiet Sun IN requires higher sensitivity measurements, such as those to be provided by the Daniel K. Inouye Solar

Telescope (DKIST; Rimmele et al. 2020) and the European Solar Telescope (EST; Quintero Noda et al. 2022). The new observations will also allow the height dependence of the field to be examined through inversions with vertical gradients of the parameters. The studies carried out so far suggest that the field inclination varies with height, becoming more horizontal in the mid photosphere (Danilovic et al. 2016b, see Fig. 4d). This may just reflect the existence of small-scale loop-like structures straddling a few granules all over the solar surface, as the loop tops naturally have more horizontal fields and are located higher in the atmosphere than the footpoints, but such an idea requires further observational verification.

2.3 Flux Emergence in the Quiet Sun

Magnetic flux emergence is an ubiquitous process in the quiet Sun. It happens on a wide range of spatial scales (from mesogranular to granular and subgranular scales) and on short timescales, but long duration observations are needed to characterize its statistical properties. Such observations are difficult to obtain, as they require stable conditions for hours. IN magnetic fields appear on the solar surface in two flavors: as bipolar pairs or clusters of mixed-polarity elements, and as individual unipolar features. The latter are features of given polarity without any clear associated opposite-polarity element in the surroundings (Fig. 5a). It has recently been shown that about 55% of the total IN flux appears in bipolar form, while the rest is unipolar (Gošić et al. 2022). The physical properties of these two populations turn out to be different, which suggests different origins.

Studying the modes of appearance of quiet Sun fields before they interact with photospheric convective flows is key to understanding their nature through comparisons with numerical simulations. Small-scale bipolar emergence occurs in the form of magnetic Ω-loops (Centeno et al. 2007; Martínez González and Bellot Rubio 2009) and magnetic sheets (Fischer et al. 2019). Magnetic loops emerge into the photosphere above or at the edges of granules, showing linear polarization in between two-opposite circular polarization patches (Fig. 5b). These signals represent the horizontal field of the loop top and the vertical field of the loop legs, respectively. The linear polarization signals show up prominently in high-sensitivity spectropolarimetric observations (Danilovic et al. 2010a; Ishikawa and Tsuneta 2010, 2011; Martínez González et al. 2012; Kianfar et al. 2018; Gošić et al. 2021). It has been suggested that the magnetic topology of these loops may explain the field strength and field inclination distributions observed in the quiet Sun IN (Lites et al. 2008; Bellot Rubio and Orozco Suárez 2019). Small-scale loops are common features in magnetoconvection simulations (e.g., Stein and Nordlund 2006; Abbett 2007; Moreno-Insertis et al. 2018) and SSD simulations (Schüssler and Vögler 2008; Danilovic et al. 2010b), where they are seen as individual entities or as bundles thereof.

Horizontal magnetic fields flanked by vertical fields have been observed to emerge also in granular lanes produced by horizontal vortex tubes (Fischer et al. 2020). An example is shown in Fig. 5c. The magnetic field has a loop-like structure with strengths of several hundred G and is detected at the late stage of the granular lane development, similarly to the loops described above. However, the formation mechanism seems to be different. The vortex tube grabs pre-existing horizontal fields in the adjacent intergranular lanes, located at or below the surface, and takes them to the granular interior, where they are transported to the surface and eventually back to the intergranular space by the granular upflows. In this way, vortex tubes provide a mechanism for the local recirculation of magnetic field required by the SSD to operate on the solar surface (Sect. 4.3.1).

Another type of bipolar flux emergence in the quiet Sun involves large sheets of horizontal fields covering a full granule (Fischer et al. 2019, see Fig. 5d), which has been identified

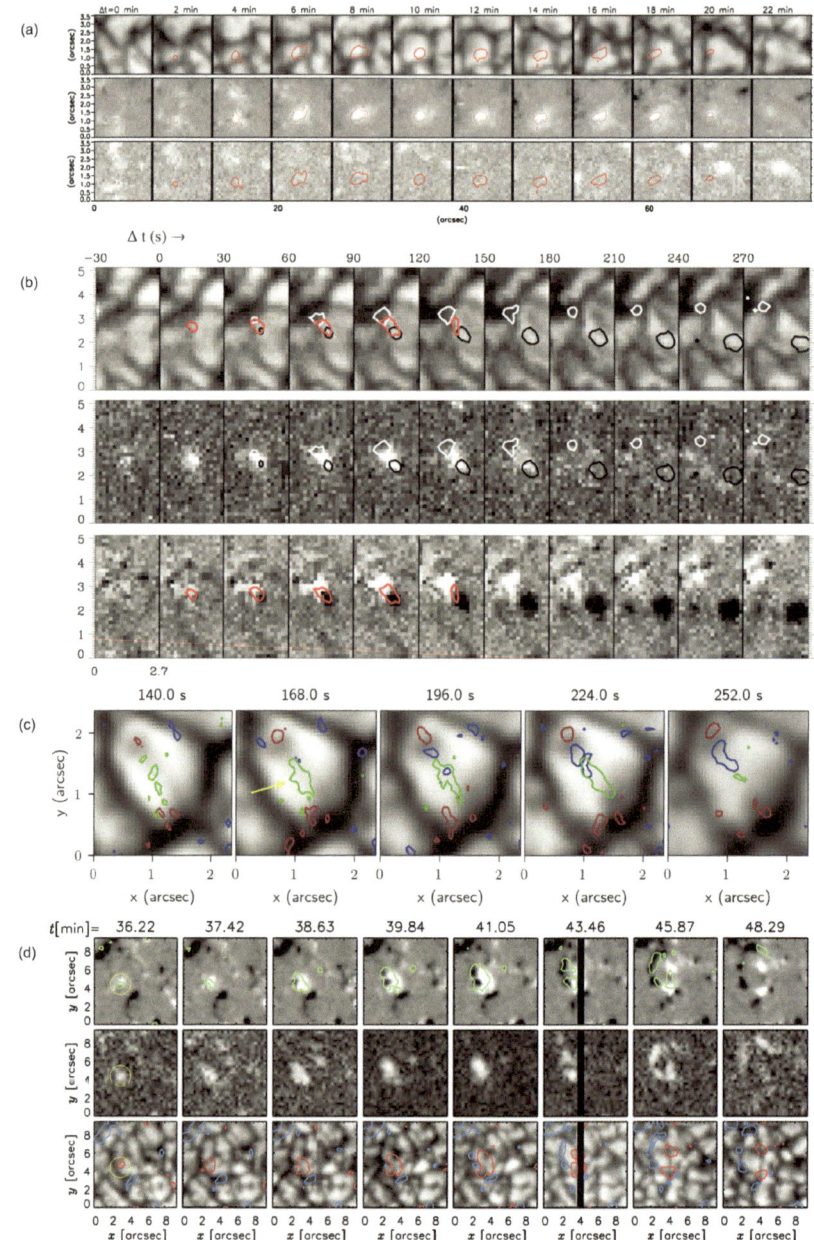

Fig. 5 Examples of magnetic flux appearance in the quiet Sun. (a) Unipolar features. (b) Bipolar magnetic loops. (c) Horizontal fields in granular lanes. (d) Sheets of horizontal fields covering a granule. The different rows show continuum intensity, circular and linear polarization in (a), continuum intensity, linear and circular polarization in (b), continuum intensity in (c), and circular, linear polarization and continuum intensity in (d). Circular polarization patches are indicated with red contours in (a), white and black contours in (b), and red and blue contours in (c) and (d). Linear polarization patches are outlined with red contours in (b) and green contours in (c) and (d). Adapted from Orozco Suárez et al. (2008), Martínez González and Bellot Rubio (2009), Fischer et al. (2020), and Fischer et al. (2019). Reproduced with permission, copyright by ESO (panels a, d) and by AAS (panels b, c)

also in simulations (Moreno-Insertis et al. 2018). The sheet fragments as it expands to the granular edges, leaving only the footpoints that can be observed as opposite-polarity patches in circular polarization maps. This form of flux emergence may explain the small clusters of mixed-polarity elements observed in longitudinal magnetograms such as those analyzed by Gošić et al. (2022), but an unambiguous confirmation is not possible until the appearance sites of the cluster members are determined.

While bipolar flux emergence should be considered the dominant form of flux appearance in the quiet Sun, unipolar appearances still provide a considerable fraction of the magnetic flux present on the solar surface. However, they pose an important problem, as the opposite polarity that must be associated with every unipolar feature seems to be missing. Clearly, this is a detection problem. Very likely, the signals are there but cannot be seen due to insufficient polarimetric sensitivity. Indeed, it has been suggested that these unipolar features do not bring new magnetic fields to the solar surface, but are the result of very weak background flux that is hidden in the noise until some mechanism concentrates it, becoming visible above the detection threshold (Lamb et al. 2008; Gošić et al. 2022). The exact mechanism is presently unknown, but it might be related to converging horizontal flows at mesogranular lanes and vertices (Yelles Chaouche et al. 2011; Requerey et al. 2017). Also the nature of the background flux is unknown, in particular whether it would be produced by SSD action or by the LSD.

Upon appearance on the solar surface, magnetic fields interact with the local granular motions and are dragged by supergranular flows towards the edges of the supergranular cells (e.g., Gošić et al. 2014). Bipolar features can be followed for some time until the footpoints cancel or merge with other magnetic features, losing their identity. Both the magnetic topology and the evolution of these features carry information on the origin of the fields, and are therefore important parameters to be compared with numerical simulations. A specific open question is the role of IN fields in the formation of the quiet network outlining the boundaries of supergranular cells. This will be briefly discussed in the following section.

2.4 Contribution of the IN to the Quiet Sun Network

The quiet Sun network is believed to be formed by active region decay, ephemeral regions and IN fields. It shows a variation with the solar cycle, which reflects its connection with the LSD (see Sect. 2.5). However, the contribution of the various components to the network flux is still under discussion.

It has been shown using Hinode/SOT data that about 40% of the total IN flux eventually ends up in the quiet Sun network (Gošić et al. 2014). According to those observations, the IN transfers magnetic flux to the network at a rate of 1.5×10^{24} Mx day^{-1} over the entire solar surface. This means that the IN supplies as much flux as is present in the network in only 9–13 hours, and could maintain it.

The results of Gošić et al. (2014) suggest that the IN is an important source of flux for the network, in agreement with the increasing evidence of a surface SSD contributing to the network field (Rempel 2014). On short-time scales, the IN flux transferred to the network may provide the seed for further amplification of the field up to kG values by convective collapse at the site of converging mesogranular and supergranular flows (Requerey et al. 2017, 2018), explaining the larger abundance of strong, long-lived flux concentrations in the network compared with the IN. This process would be consistent with the first mechanism of network formation described in Sect. 4.3.4.

To verify or refute this idea, an observational investigation of the appearance and evolution of magnetic flux in the network must be carried out, considering also the adjacent IN.

Particularly important is the site of appearance of new flux within the granulation pattern, as well as the interaction between network and IN fields, with a view to determine how the flux is eventually deposited in the network. Such an analysis has never been performed, although there exist observations with sufficient spatial resolution and sensitivity to detect weak magnetic flux appearing on subgranular scales (for example, the Hinode Solar Optical Telescope provides 0.3 arcsec and has a sensitivity of 10^{-3}, respectively).

2.5 Solar Cycle Variations of Quiet Sun Fields

Despite being plagued with difficulties, the search for possible variations of the quiet Sun magnetism with the solar cycle has been pursued vigorously in the last decades. This is because the detection of temporal and/or latitudinal variations would link the quiet Sun magnetism with the LSD responsible for the solar magnetic cycle. The main challenge is the need of very stable, homogeneous observations over periods of time spanning years. Few instruments are capable of providing such observations. Space-borne measurements are preferred, but also ground-based Hanle effect observations have been used to that end.

As summarized in Sect. 3.2 of Hinode Review Team et al. (2019), evidence for temporal variations of the polarization signals produced by quiet Sun fields is very marginal, if present at all. Most studies do not find significant changes or they are within the statistical uncertainties of the observations.

An analysis of 5.5 years of Hinode/SP data taken in quiet regions at the disk center revealed no measurable variation of either the magnetic flux or size distribution of IN patches with time (Buehler et al. 2013). Similar Hinode/SP observations were used to derive the transverse and longitudinal magnetic flux densities of very weak IN regions at different positions on the solar disk and study their variation from 2008 to 2015 (Jin and Wang 2015). No change in the flux density or the ratio of transverse to longitudinal flux was found. Following a different approach, synoptic Hinode/SP observations taken at various positions along the central meridian were used by Lites et al. (2014) to investigate the long-term evolution of the magnetic flux density in quiet IN regions. The transverse and unsigned longitudinal fluxes were found to be independent of the solar cycle at all solar latitudes, while the signed longitudinal fluxes (i.e., the polarity imbalance) showed clear changes in the polar regions and some hints of variation in the activity belt from 20 to 60 degrees latitude.

Full-disk observations in the near-infrared Fe I line at 1564.8 nm did not show notable changes in the properties of the polarization signals from IN regions for much of solar cycle 24 (Hanaoka and Sakurai 2020).

Faurobert and Ricort (2021) performed a Fourier analysis of the spatial fluctuations of the longitudinal flux density in small $10'' \times 10''$ IN regions along the central meridian, using synoptic data from the Hinode SP between 2008 and 2016. On scales smaller than $0.5''$ they did not find significant variations of the magnetic fluctuations with the solar cycle at any latitude. On granular scales, up to about $2.5''$, the power of the spatial fluctuations did not show variations at low and mid latitudes either, but a decrease was observed at high latitudes during solar maximum. The lack of changes on scales smaller than $0.5''$ indicates the presence of a time-independent magnetic field in the IN. However, the variation detected on larger scales at high latitude suggests that also the LSD contributes to the magnetism of the IN, although not homogeneously over the solar surface.

Using 12 years of SDO/HMI data, the rms longitudinal flux density in quiet $1° \times 1°$ IN regions at the central meridian was found to be nearly constant over time, with an average value of 6 Mx cm^{-2} (Korpi-Lagg et al. 2022). This was interpreted to reflect a real independence of the quiet IN magnetism on the solar activity cycle, or the inability of HMI to detect changes due to insufficient sensitivity. By contrast, the rms flux density in 15° windows did show a statistically significant correlation with the solar cycle, with the maximum of the curve lagging the sunspot number cycle by about half a year. The difference is that these 15° windows contain both network and IN fields, whereas the 1° windows avoid the network fields. This suggests that the quiet network is indeed affected by the large-scale solar dynamo, although on supergranular timescales its evolution seems to be determined by interactions with IN fields, presumably reflecting a contribution from the SSD (see Sect. 2.4). Jin and Wang (2019) also found a variation of the network with the solar cycle using full-disk SDO/HMI magnetograms, but in this case the quiet network flux showed an anti-correlation with the sunspot number, due primarily to a reduction of the network area (the magnetic flux density of quiet network patches was observed to increase by about 6% at solar maximum). Further analyses are needed to clarify the exact variation of the quiet network with the sunspot cycle and the contribution of the SSD to its formation and maintenance.

Following a different strategy, the scattering polarization measurements of Kleint et al. (2010) did not show significant changes in the amount of Hanle depolarization of selected C_2 molecular lines over two years spanning the minimum phase of solar cycle 23. The observed lines are formed almost exclusively in granules, so they sample only the weakest fields of the quiet Sun. The latest analysis, covering almost a full solar cycle, still shows no clear changes in the Hanle depolarization with time (Ramelli et al. 2019). It is important to continue this type of synoptic Hanle programs and possibly extend them to the photospheric Sr I 460.7 nm line (which is formed also in the intergranular lanes), as they provide an independent way to examine the cycle dependence of the quiet Sun magnetism.

All these results point to no or little variations of the weak IN fields with the solar cycle, which supports the view that they are generated by an SSD and not by an LSD cascading down to smaller spatial scales. By contrast, the quiet network seems to show a variation with the solar cycle, indicating some contribution from the LSD in addition to a possible one from the SSD, but its amplitude and phase are not well determined yet.

2.6 Quiet Sun Chromospheric Fields

Compared with the photosphere, little is known about quiet Sun fields in the chromosphere. This is due to the very weak polarization signals they produce, the lack of spectropolarimetric observations with sufficient sensitivity to detect them, and the challenging interpretation of the measurements, which usually requires non-local thermodynamic equilibrium analyses (see the review by de la Cruz Rodríguez and van Noort 2017). However, some spectropolarimetric observations of quiet Sun fields have been made in the chromosphere, mainly to study how the longitudinal field changes with height in different magnetic structures (e.g., Gošić et al. 2018; Morosin et al. 2020; Esteban Pozuelo et al. 2023).

SSD action is not expected to occur at chromospheric heights due to the absence of turbulent flows, but there is an interest in understanding the influence on the chromosphere of quiet Sun fields generated deeper down. Network fields—whether produced by AR decay, ephemeral regions or IN fields—are essentially vertical and can reach the chromosphere and above, fanning out with height to form large-scale magnetic canopies. Most IN fields, on the contrary, seem to be low-lying structures confined to the photosphere (for an illustration, see the data-driven magnetofrictional simulations of Gošić et al. 2022). However, in some cases they manage to reach higher layers. Approximately 25% of the small-scale magnetic loops that emerge in the photosphere are observed to rise to chromospheric layers, producing localized brightenings and polarization signals there (Martínez González and Bellot Rubio 2009). These early results have recently been confirmed using full Stokes spectropolarimetry at higher spatial resolution by Gošić et al. (2021), who reported the detection of circular polarization in the chromospheric Mg I b$_2$ 517.3 and Ca II 854.2 nm lines produced by IN magnetic features ascending through the atmosphere.

These IN features may carry magnetic flux and energy to the chromosphere and so they have attracted much interest as a potential source of chromospheric heating in the quiet Sun (Ishikawa and Tsuneta 2009; Martínez González et al. 2010). The quiet Sun is particularly difficult to explain due to its continuous presence and large area coverage, which require a ubiquitous heating mechanism. Acoustic waves contribute to the heating, but their energy flux is not sufficient to explain the observed chromospheric emission (Fossum and Carlsson 2005; Molnar et al. 2023), hence the need to resort to mechanisms involving magnetic fields. The energy deposition triggered by IN features is probably caused by magnetic reconnection of the emerging field and pre-existing chromospheric fields, or reconnection of different chromospheric flux systems perturbed by the rising photospheric fields. Recently, it has been shown that the cancellation of IN fields can lead to local temperature enhancements of up to 2000 K in the low chromosphere (from $\log \tau_5 = -4$ to -6.5), explaining the appearance of strong brightenings at the position of the cancelling features (Gošić et al. 2018). However, the energy input estimated from the flux cancellation rates observed in these high-resolution observations falls short of being able to account for the quiet Sun chromospheric radiative losses, by an order of magnitude or so. If a large fraction of flux emergence and cancellation events go unnoticed because of the noise, then IN fields could still provide an important contribution to the heating of the chromosphere, but this needs to be confirmed with more sensitive spectropolarimetric measurements.

Another potential source of chromospheric heating involves the interaction of IN features and network fields described in Sect. 2.4. This interaction is detected observationally as continuous mergings and cancellations of magnetic flux at the edges of supergranular cells (Gošić et al. 2014). The reorganization of network field lines triggered by IN fields may be the source of the solar campfires recently discovered by Solar Orbiter. Campfires are small transient brightenings observed in EUV 17.4 nm images which have a tendency to occur at the edges of the photospheric network (Berghmans et al. 2021). The EUV campfires indicate transition region/coronal temperature enhancements and have been related to the cancellation of quiet Sun features (Panesar et al. 2021; Kahil et al. 2022), although in some cases no opposite-polarity fields could be detected at the location of the campfires. The origin of these features and their contribution to chromospheric/coronal heating are still being investigated, as several magnetic topologies leading to reconnection of field line bundles seem to be able to produce campfire-like events in simulations (Chen et al. 2021). Aspects of quiet-star chromospheres are covered in Sect. 5.2.

3 Basic Theoretical Concepts

We start our discussion about the theoretical and numerical studies of SSD with a short summary of the basic theoretical concepts. We then consider more idealized numerical models, used for studying the basic properties and parameter dependencies of SSD and comparisons with theory, and finally move on to surface convection simulations, which are calibrated and can be compared against observational data.

An SSD refers to the sustained and rapid amplification of magnetic field fluctuations at spatial scales smaller than the forcing scale in a plasma system. In the case of turbulent convection, the forcing scale is the scale at which the kinetic energy spectrum peaks, which is the scale of convective cells, that is believed to vary strongly as function of depth: according to the mixing-length theory, the convective cells are small and turn over fast near the surface, and get progressively larger and slower as function of depth (Vitense 1953). The scale of convection is thought to behave this way because the convective bubbles moving up or downward are assumed to be able to travel a distance that is a fraction of the local pressure scale height, which decreases throughout the CZ, especially strongly near the surface. In the solar CZ, the plasma is turbulent enough to rapidly amplify magnetic fluctuations at or smaller than the scale of convective cells. The environment required for this dynamo instability to operate is such that the influence of rotation is weak and the flow is largely non-helical. This is in contrast to a LSD where the amplification of fields occurs at scales larger than the forcing scale, which is a consequence of symmetry breaking (due to helicity, inhomogeneities, anisotropies, etc.) at small scales, facilitating an inverse cascade of magnetic energy from small to large scales, further assisted by large-scale non-uniformities in the velocity field (see, e.g. Charbonneau 2020). This would suggest that the LSD would preferentially occur in the deeper layers of the CZ, where rotational influence on convection is strong, while SSD would operate nearly unimpeded in the surface layers, where the turbulent part of LSD would have only little chances of existing. As already hinted towards from observations, the situation is likely to be much more complex, with these two dynamo instabilities being excited together over a large fraction of the CZ, and non-linearly influencing each other.

Let us start our brief theoretical discussion by introducing the most important dimensionless control parameters of SSD-active systems. In the following, we use η for magnetic diffusivity, ν for kinematic viscosity and U as the typical velocity at the largest scale, L, of the inertial range. The latter also enters discussion of power spectra in terms of the scale of forcing/energy injections as $k_i \sim 1/L$. Then, the fluid Reynolds number is defined as $\text{Re} = UL/\nu$ and the magnetic Reynolds number as $\text{Re}_M = UL/\eta$. The magnetic Prandtl number is defined as the ratio of the two, namely $\text{Pm} = \text{Re}_M/\text{Re}$, and can hence also be written as $\text{Pm} = \nu/\eta$. The Reynolds numbers in the solar and stellar plasmas are both large, but the magnetic one is estimated, using the Spitzer formulae, to be a few orders of magnitude smaller than Re (see e.g., Brandenburg and Subramanian 2005). The solar Pm values are estimated to be in the range of 10^{-6} to 10^{-4} (see, e.g., Schumacher and Sreenivasan 2020), and the ones in even cooler stars even lower due to their increased density and lower interior temperatures.

Batchelor (1950) discussed the possibility of magnetic field amplification in a turbulent flow by drawing an analogy between the time evolution of vorticity in incompressible turbulence and the induction equation for the magnetic field. His model predicted no SSD for Pm < 1 plasmas, however. The first rigorous mathematical treatment was performed by Kazantsev (1968), where it was shown that for the simple case of Gaussian zero-mean, homogeneous and isotropic velocity field that is δ-correlated in time (Kraichnan ensemble; Kraichnan 1968), the evolution of magnetic energy (or equivalently, the magnetic correlation tensor) can be expressed in a form similar to the Schrödinger equation with an effective "mass" and a "potential" that depends on the velocity correlation tensor. A description of this tensor then completes the system. The simplest assumption is to take the scale dependence of velocity fluctuations as $\langle \delta u(l) \rangle \sim l^\alpha$, where α can range from 0 to 1, corresponding to a "rough" and "smooth" velocity field, respectively. In the case of Kolmogorov turbulence (Kolmogorov 1941), $\alpha = 1/3$. The bound-state solutions of the equation, then, describe exponentially growing modes. When η is non-zero, this potential becomes repulsive at both the smallest and largest scales, allowing dynamo action to take place only if there is sufficient scale separation between the integral and the dissipative scales (see, e.g., Sect. 3.2 and

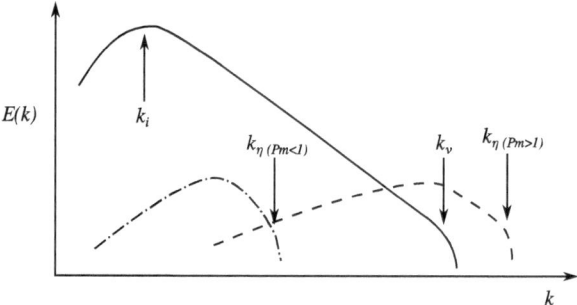

Fig. 6 Schematic of power spectrum for magnetic and kinetic energy in the low and high Pm regime. The solid line illustrates the kinetic energy for a turbulent plasma with energy injection at spatial wave number k_i and dissipation at k_ν. The dot-dashed (dashed) line show the magnetic energy power spectra in the kinematic growth phase with dissipation at scales $k_{\eta(Pm<1)}$ ($k_{\eta(Pm>1)}$) for the Pm < 1 (Pm > 1) case

Fig. 4 of Vincenzi (2001) for an illustration. In other words, there exists a critical magnetic Reynolds number Re_M^{crit} corresponding to this scale separation and dynamo action is possible only when $Re_M > Re_M^{crit}$. This quantity depends on Re through Pm, the latter of which is independent of scales and flow properties.

According to Kazantsev's model, the amplification occurs at the timescale of the turbulent eddy turnover time, which is very short in comparison to the timescales required for the amplification of the large-scale field. In the kinematic growth phase, when the magnetic fluctuations are still weak, the structures generated have the thickness around the resistive scale, but are curved up to the scale of the turbulent eddies. Hence, the peak in the magnetic power spectrum is at the resistive scale, but all scales grow with the same growth rate. At scales larger than the resistive scale, a positive power law of $k \propto 3/2$ is predicted (Kazantsev 1968), hence called the Kazantsev spectrum, while at smaller scales, the spectrum can fall off very steeply, following the so-called Macdonald function (for details, see e.g. Brandenburg and Subramanian 2005; Rincon 2019).

In low-Pm plasmas, as the solar one, the resistivity, η, is much larger than the fluid viscosity, ν, meaning that the dissipation of the fluid motions occurs at scales much smaller than the scales at which magnetic fields dissipate (for an illustration of how the spatial power spectrum of magnetic and kinetic energy look like for low and high Pm cases, see Fig. 6). In effect, the turbulent eddies can dissipate into much smaller-scale structures than the magnetic structures, due to which the magnetic fluctuations must grow within the inertial range of the turbulent spectrum. The smoother magnetic structure, therefore, sees the turbulent eddies as a rough field around it (comparable to the "rough" flow in the Kazantsev picture); these circumstances are to be contrasted with high Pm fluids, where a smooth velocity field would be acting on smaller magnetic structures. The amplification of the magnetic fluctuation is more challenging in the former case of a rough velocity field, and hence the critical Reynolds number for dynamo action, Re_M^{crit}, is elevated. At the incompressible limit or near it (weak compressibility), Re_M^{crit} ranges between 30–60 for high-Pm plasma (e.g., Brandenburg and Subramanian 2005), while for low Pm values of around 400 have been analytically computed (e.g., Kleeorin and Rogachevskii 2012).

How the SSD non-linearly saturates remains an open problem. Extending the analytical work on the Kazantsev model, it has been proposed that the SSD can grow magnetic fluctuations at the resistive scale up to and exceeding the equipartition with kinetic energy of turbulence, but that the generated field would be concentrated into resistive-scale ropes, hence not being volume filling, and therefore being energetically rather insignificant (e.g. Subramanian 1998). The non-linear regime poses a formidable problem for analytical studies, but numerical studies can be attempted.

4 Numerical Models

Reaching the extreme-low Pms of the solar and stellar CZs is impossible currently and will most likely remain so in the future, at least for explicit-diffusion codes. The simulations conducted with these codes are also referred to as direct numerical simulations (DNS), although they are not quite fulfilling this definition in a strict sense, as orders of magnitude elevated diffusivities are used than in the real object; hence, hereafter we refer to these type of models as DNS-like.

An alternative to explicit diffusivity schemes is the usage of implicit large-eddy simulations (hereafter ILES), where the diffusive terms are replaced with numerical counterparts, providing diffusion only close to the grid scale, while leaving well-resolved scales unaffected. This has the advantage to maximize the Reynolds numbers in the flow. However, the actual values of the dimensionless control parameters then become ill-defined. There are various incarnations of these techniques, ranging from hyperviscous operators (see, e.g., Stein and Nordlund 1998) to slope-limited diffusion schemes (see, e.g., Rempel 2014), to mention just a couple (for a more thorough review, see Miesch et al. 2015).

Third type of numerical schemes are the so-called explicit large-eddy simulations (hereafter ELES), where the solved equations are filtered at a spatial scale larger than the grid scale, and (ideally) physically motivated subgrid-scale (SGS) models are used to describe the terms representing the scales left out by the filtering procedure. The best known example is the Smagorinsky scheme (Smagorinsky 1963), that develops a concept of turbulent eddy viscosity as an SGS model. More involved closures for the magnetized case, where more terms than the viscosity itself need to be described by the SGS model, have also been developed (see, e.g., Grete et al. 2017), although their usage in convection modeling is still limited. In the most complicated cases the ELES should also account for the influence of small-scale turbulence on large-scale dynamics, such as the inductive action of helical small-scale turbulence on large scales, dubbed the α effect, usually referred to as backscatter in LES terminology.

ILES schemes have been immensely successful for solar surface simulations with radiation transport (e.g. Stein and Nordlund 1998; Rempel 2014), as will be described later on in this chapter. However, concrete indications that ILES schemes might not be sufficient and appropriate for convective systems with LSD dynamo and SSD together come from comparisons like that of DNS-type simulations of Käpylä et al. (2017) and ILES-type ones of Hotta et al. (2016) showing markedly different behaviour (also to be discussed later in this chapter). Nevertheless, numerical models give valuable insights to the regimes where analytical approaches completely fail, such as studying the saturation mechanisms, and the interactions between SSD and LSD.

4.1 The Quest for Finding and Studying Low-Pm Small-Scale Dynamos

Numerical studies of SSD in solar and stellar contexts are challenging due to the low Pm of these environments. As discussed earlier, the critical Re_M required for the onset of SSD mechanism is expected to be in the order of hundreds, already requiring high resolution. But there is also the added challenge of Re required being orders of magnitude larger than Re_M

to reach the physically relevant low-Pm regime. Hence, limited numerical studies exist with only moderately small Pm, with the minimal achieved values currently being approximately 0.003 by Warnecke et al. (2023). Only a fraction of studies pushing to the low-Pm regime include convection as the driver of the background turbulence, while the majority use some sort of idealized forcing function. It has been argued, however, that the SSD properties are only weakly dependent on the type of the forcing used (see, e.g., Moll et al. 2011). With the current computational resources we are, however, finally at the limit of being able to answer the first imminent question of whether the analytic predictions of Re_M^{crit} are correct, or if the threshold is lower/even higher than expected.

4.1.1 Kinematic Phase

While numerical evidence for SSD at Pm $= 1$ in incompressible, homogeneous, isotropic, and non-helical setting was obtained several decades ago by Meneguzzi et al. (1981), and the literature of high-Pm number SSD in varying setups is abundant (see, e.g., the reviews by Brandenburg and Subramanian 2005; Rincon 2019), the first numerical evidence for low-Pm dynamos dates back to only 15 years (Iskakov et al. 2007). This initial evidence was restricted to simplified, forced, setups with hyperdiffusivity. The encouraging result from these studies is that the Re_M^{crit} does not continue increasing with Pm, but appears to have a maximum (Iskakov et al. 2007). A more recent DNS-like study by Warnecke et al. (2023) shows that after plateauing, Re_M^{crit} even starts decreasing again, approaching values of the order of a hundred. This non-monotonic behavior appears now to be firmly associated with the magnetic energy peak falling into the so-called bottleneck range of the kinetic spectrum (e.g. Brandenburg 2011; Warnecke et al. 2023). The bottleneck refers to both experimentally (reported in several papers since She and Jackson 1993) and numerically (reported since Dobler et al. 2003) confirmed inefficiency of the kinetic energy cascade at wavenumbers somewhat smaller than the viscous scale. This is seen most clearly by plotting the kinetic energy power spectrum (as in Fig. 6) compensated by $E(k)k^{5/3}$, which is the expected spectral kinetic energy cascade of the flow field for incompressible and homogeneous systems (Kolmogorov 1941), called the Kolmogorov law and cascade. For ideal Kolmogorov turbulence, it should result in a flat line in the inertial scale, followed by a drop-off around the viscous scale. However experiments and simulations show a "hump" that forms near the viscous scale before the drop-off occurs. This hump, especially when its lower wavenumber side, where the deviation from Kolmogorov scale is positive, overlaps with the energy-carrying scale wavenumber of the magnetic fluctuations, has been attributed to decreased SSD efficiency (Warnecke et al. 2023).

In the kinematic regime, one is additionally interested in determining the growth rate and its dependence on the key system parameters, and the evolution of the energy spectrum of the magnetic fluctuations, for which links to the Kazantzev theory can be established. Such links can be obtained by mapping an appropriate turbulence model, such as a one from Kolmogorov theory, to the Kazantsev theory. For example, the predicted growth rate by Kazantzev theory, $\gamma \propto u_d/l_d$, can be estimated by replacing estimates of dissipation scale velocity, u_d, and length scale, l_d, from Kolmogorov theory. Similarly, estimating Re_M^{crit} for different Pm has been based on using the roughness of the turbulent flow at different scales as a mapping (see, e.g., Boldyrev and Cattaneo 2004). Also, the effects of compressibility have been taken into account by assuming a linear relation between the transverse and the longitudinal component of the velocity correlation tensor for the two extreme cases of divergence-free (Kolmogorov turbulence) and irrotational (Burgers turbulence) flow and assuming the same form for the longitudinal component (see, e.g., Schober et al. 2012). For solar-like weakly compressible flows all (semi-)analytical models indicate that a kinematic dynamo exists for the case of low Pm, and differ only in the details of the estimated Re_M^{crit} values for different flow fields.

The sparse set of numerical models so far indicate the following: for fixed Re_M, Schekochihin et al. (2007) reported growth rates monotonically decreasing with decreasing Pm (achieved through increasing Re) in the range Pm ≈ 0.1–1. For even smaller values, the growth rates were observed to tend towards a constant value. These findings led Schekochihin et al. (2007) to hypothesize that an asymptotic positive value of the growth rate would exist for high Re_M and low Pm values. However, such an asymptotic value has not been found yet, even for lower Pm studies (Warnecke et al. 2023). While, at Pm > 1 regime, the growth rates retrieved from numerical experiments are closely consistent with the $Re_M^{1/2}$ scaling expected from Kolmogorov turbulence (reviewed, e.g., by Brandenburg and Subramanian 2005), the low-Pm simulations are less consistent with it (Iskakov et al. 2007; Schekochihin et al. 2007; Warnecke et al. 2023). A better match to a logarithmic scaling law, growth rate being proportional to $\ln\left(Re_M/Re_M^{crit}\right)$, valid near the onset of the dynamo action (Kleeorin and Rogachevskii 2012), was reported by Warnecke et al. (2023). They concentrated on mapping the region near Re_M^{crit}, hence this result might not be so unexpected. However, this also does not reveal the true scaling of growth rate, for which simulations far removed from Re_M^{crit} would be required. This is an extremely challenging task numerically.

As per the expected magnetic energy spectrum from Kazantsev theory, the simplified low-Pm models do not yield a direct agreement either (Iskakov et al. 2007; Schekochihin et al. 2007; Warnecke et al. 2023). Usually, for numerical convenience, the spectrum is cut short at low wavenumbers, so that the maximum range of higher wavenumbers could be modelled. Hence, in the setups trying to minimize Pm, the expected $k^{3/2}$ scaling of magnetic energy often cannot be seen by design. Nevertheless, at higher wavenumbers between the forcing and resistive scales, the magnetic spectra develop a cascade with negative power laws of varying steepness. Schekochihin et al. (2007) report spectral indices down to $-11/3$ at low Re_M and of -1 at higher Re_M, and Warnecke et al. (2023) find a slope approximately of -3 for their simulations near Re_M^{crit}. As per as the visual appearance (see Fig. 7), the magnetic field exhibits less obvious folded structures having the width at the resistive scale, as would be expected from the Kazantsev model (Iskakov et al. 2007; Schekochihin et al. 2007; Warnecke et al. 2023). In the case of Pm $= 1$ (left column, panels (a) and (c)), the correlation of magnetic field strength with high/low turbulent speeds is not very strong, while the magnetic field tends not to be volume filling. In the small Pm case (right column, panels (b) and (d)), the magnetic field has a clearly higher filling factor, and shows even a weaker correlation with the turbulent velocity field. Less folded and thicker structures are seen in the low-Pm case in comparison to Pm of unity.

With turbulent convection, SSD was established for Pm > 1 simulations since the 1990s, the first successful convection simulation with self-sustaining magnetic fluctuations having been reported by Nordlund et al. (1992). Setups with Boussinesq approximation have been confirmed to exhibit SSD action (e.g., Cattaneo 1999), and the same applies also to stratified and compressible setups (e.g., Nordlund et al. 1992; Vögler and Schüssler 2007; Pietarila Graham et al. 2010; Hotta et al. 2015; Bekki et al. 2017).

Some attempts to go down to Pms of 0.1 have been performed in local stratified domains (Käpylä et al. 2018) in DNS-like deep convection setups, but no SSD action has been detected at the lower limit. In Käpylä et al. (2018) models, the growth and decay rates at variable Pm were found to be closely compatible with $Re_M^{1/2}$ scaling, which is different from the behavior seen in the simplified setups described above. Also, at small wavenumbers, the Kazantsev spectrum was not prominent, suggesting that the large-scale motions present in

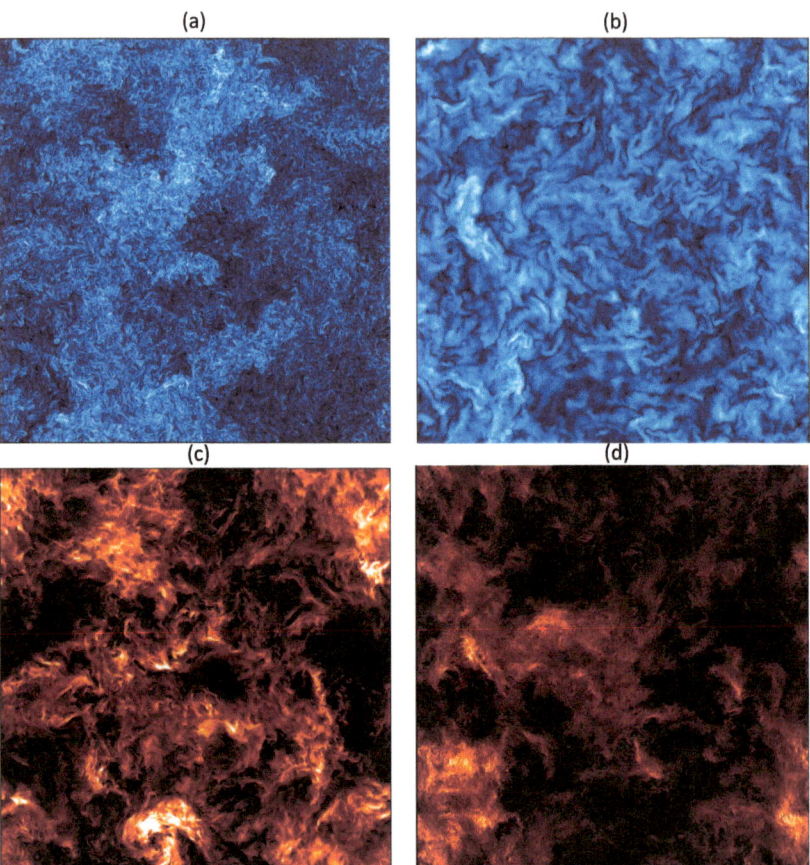

Fig. 7 Two-dimensional slices of the magnetic field strength (upper row) and speed (lower row), from Pm $= 1$ simulation (left column) and Pm $= 0.005$ (right column) simulations (reproduced from the models reported in Warnecke et al. 2023). The Pm $= 1$ models were run with the resolution of 1024^3 and Re $=$ Re$_M$ $= 4096$, while the low Pm runs with the resolution of 4096^3, Re $\approx 33{,}000$ and Re$_M$ $= 165$. In this model setup, the Reynolds numbers are defined using the forcing wavenumber, and are hence to be multiplied by 2π to match the definitions used in this paper. The simulation domain is fully periodic and has dimensions of $k_1 = 2\pi$ in all directions, while the white-in-time plane wave forcing has a mean wave number of $k_f = 2k_1$

the deep convection setups induce excess power to the magnetic energy spectrum. In models of solar surface convection with ILES schemes, however, SSD at around Pm ≈ 0.1 has been found (Rempel 2018; Brandenburg and Rempel 2019). In these works a low (numerical) Pm was realized by combining a more diffusive scheme for the induction equation with a less diffusive scheme for the momentum equation. The resulting effective (numerical) Pm was estimated from the solution based on the resulting effective diffusivities.

Global or semi-global convection simulations in spherical geometries have been reported to exhibit SSD, albeit still limited to Pm $= 1$ regime (Hotta et al. 2016; Käpylä et al. 2017; Hotta and Kusano 2021). In such setups, rotation and stratification are included by default, raising the question whether the fluctuations are genuinely produced by SSD, or rather from a tangling of field generated from a LSD (see, e.g., the discussion in Brandenburg and Subramanian 2005). The most convincing experiments address this by removing the mean field at each time step, hence allowing for the detection of SSD-generated fluctuations alone (for the method, see e.g. Käpylä et al. 2022).

In summary, at Pm < 1, the numerical challenges have not yet enabled the detection of SSD in global models of turbulent convection. In the light of the evidence obtained from the more simplified systems, however, the earlier strong doubts about the existence of SSD in turbulent convection at low-Pm have recently been alleviated. Moreover, this question will be directly addressable in the near future with codes capable of taking advantage of accelerator platforms (see, e.g., Wright et al. 2021; Pekkilä et al. 2022).

4.1.2 Non-linear Saturation

Simulating low Pm SSD is an extremely challenging task from a computational power requirement perspective. The simultaneous requirement of high Re and an $Re_M > Re_M^{crit}$ for dynamo action requires extremely high resolutions and long integration times. Hence, even the most simplified setups operate in the kinematic regime, where the generated magnetic field has negligible back reaction on the flow. The non-linear regime of the SSD has been mainly studied for Pm $\geq$ 1, while only a handful of studies have been able to address the Pm < 1 regime. Non-linear studies, however, are required to draw any conclusions on the effects of the SSD-generated fluctuations on the dynamics of systems like solar and stellar CZs, and about the interactions of the two dynamo instabilities (namely LSD and SSD).

The study of Brandenburg (2011) used DNS-like simulations to investigate SSD in forced non-helical turbulence in the non-linear regime adopting the following strategy. They ran a Pm = 1 setup up to saturation, and then kept decreasing the kinematic viscosity while keeping Re_M roughly constant, and continuing the integration from the saturated state with the new parameter values. Their study led to two important findings. Firstly, most of the energy was found out to be dissipated via Joule heating before reaching to the viscous dissipation scale, hence allowing to decrease the kinematic viscosity even further than estimated for the specific grid resolution. Secondly, the saturation strength of the SSD was only weakly dependent on Pm, reducing from roughly 40 percent of equipartition with turbulence to near 10 percent when the Pm was changed by two orders of magnitude, when computed from volume-averages. However, at all scales, the magnetic energy was sub-dominant to the kinetic energy. Interestingly, the bottleneck effect, dominating the dynamics in the kinematic regime, was suppressed in the nonlinear regime. An attempt with a similar strategy with turbulent convection was undertaken by Käpylä et al. (2018), who found that the decrease of the saturation strength of the magnetic field was somewhat, but not detrimentally, stronger with Pm.

Brandenburg (2014) and Sahoo et al. (2011) performed further idealized simulations in the non-linear regime, reporting on the kinetic and magnetic dissipation rates, and their dependence on Pm. Brandenburg (2014) studied both helically and non-helically forced cases, the former also allowing for LSD. Interestingly, the ratio of kinetic to magnetic dissipation was observed to exhibit a positive power-law behaviour with Pm in both scenarios, albeit with somewhat varying power law index in helical versus non-helical cases. This implies that in the case of low-Pm dynamos the energy being pumped into the system through the kinetic energy reservoir (be it forcing, convection, shear, …) is converted by the Lorentz force to magnetic energy more efficiently for smaller Pm, and then dissipated through resistive dissipation rather than through the viscous one.

At first sight these findings appear contradictory, how can the saturation field strength be mostly independent of Pm and the Lorentz force work vary strongly with Pm at the same time? Brandenburg and Rempel (2019) further studied the SSD saturation using both forced

DNS and convective ILES models and confirmed a similar behavior using both approaches. They further analyzed the transfer function of the Lorentz force in spectral space and found that there is a regime on small scales where the Lorentz force work can be positive, dubbed "reversed dynamo", as in the case of a normal dynamo, the flow is doing work against the Lorentz force, and a negative work contribution would be expected. The wavenumber at which this reversed dynamo regime is entered depends strongly on Pm. For a sufficiently low Pm the reversed dynamo is completely absent, but it is growing with increasing Pm until the positive energy transfer of the reversed dynamo on small scales almost completely offsets the negative transfer on large scales. As a consequence a high-Pm dynamo can have in the saturated state a strong field with little net Lorentz force work, while the transfer of energy from kinetic to magnetic energy on large scales remains strong. On small scales this energy is returned to kinetic energy and dissipated through viscosity. While the magnetic energy cascade extends to scales smaller than the kinetic energy cascade, there is very little energy left in that cascade. In the low-Pm regime the Lorentz force work is negative on all scales and most energy is dissipated through resistivity. While the kinetic energy cascade extends to much smaller scales, there is very little energy left in that cascade. As a consequence the saturated dynamo behaves in both cases like a Pm $\sim$ 1 dynamo since kinetic and magnetic energy cascades terminate at a similar scale, which is given by the larger of the viscous and resistive dissipation scales.

4.2 Models of Deep Convection

SSD action has not yet been found in DNS-like models of deep convection with low Pm. This is mostly due to lack of computing resources to properly investigate this regime. There is, however, rich literature and exciting findings at Pm ≥ 1 regime. This regime is either selected on purpose by using explicit viscosity schemes and setting $\nu \geq \eta$, or using ILES schemes, which results in effective Pms close to unity by inspecting the spectral cut-off scales.

SSD magnetic fields have been studied in local Cartesian convection setups, to maximize the fluid Reynolds numbers. In these studies Hotta et al. (2015, 2016), Bekki et al. (2017), an efficient SSD was found to operate, which also resulted in suppression of convective velocities near the base of the CZ. There, the magnetic and kinetic energies were found to be nearly in equipartition, resulting in the suppression of convective velocities by a factor of two relative to a purely hydrodynamic solution due to the Lorentz force feedback. The enthalpy flux was not, however, observed to be quenched thanks to a simultaneous suppression of horizontal mixing of entropy by the magnetic fluctuations. These results are suggestive of SSD aiding to resolve the convection conundrum by reducing the convective velocities while increasing the convective flux. The work of Karak et al. (2018), motivated by these results, studied cases of large thermal Prandtl numbers conjectured to be due to the suppression of thermal diffusion by the strong magnetic fluctuations. They could not provide support to these results, however. Therefore, as of writing of this chapter, the issue remains unresolved.

Global and semi-global simulations of solar and stellar magnetism have also recently reached parameter regimes where SSDs are obtained (Hotta et al. 2014, 2016; Käpylä et al. 2017; Hotta and Kusano 2021), but unless rotation is deliberately suppressed as in Hotta et al. (2014), LSD cannot be ruled out, and perhaps even then not completely, as anisotropies due to density stratification and inhomogeneties due to boundary conditions would still be present, giving a faint possibility of LSD to be excited.

All these models suggest that a vigorous SSD would have profound repercussions for the LSD and differential rotation, but the results appear rather divergent and dependent on the viscosity schemes used. For example, Hotta et al. (2016) report on non-monotonic behavior of the LSD in the presence of SSD—at low resolution and Reynolds numbers with explicit diffusion scheme, the LSD and cyclic behavior is obtained, while it gets irregular and sub-critical at medium resolution, when switching to an ILES scheme. Increasing the resolution further, LSD is revived again, attaining saturation strengths larger than in the lowest resolution case. Whether this behavior is due to the change of the explicit/ILES schemes remains unclear, as no such non-monotonicity is observed when explicit schemes are used throughout (Käpylä et al. 2017).

Instead, Käpylä et al. (2017) report monotonically increasing values of the mean magnetic field, although the growth clearly slows down, and perhaps tends towards an asymptotic constant value. At the same time the differential rotation is strongly reduced. This can be traced back to the growing small-scale Maxwell stresses which oppose the small-scale Reynolds ones, and hence through that route lead to weaker differential rotation when the SSD becomes more vigorous. This could explain the tendency of the mean magnetic field growth slowing down as function of Re_M, hence reflecting only the quenching of the differential rotation rather than any asymptotics.

In the global magnetoconvection model of Hotta and Kusano (2021), on the other hand, increasing the resolution and consequently the Reynolds numbers to higher values than used before in similar type of ILES calculations, again at the effective Pm $= 1$ limit, has been shown to lead to superequipartition of magnetic energy due to SSD at the small scales. This has been observed to result in more solar-like rotation profiles, meaning more radial than cylindrical isocontours of the angular velocity. In these computations, however, no LSD has yet been reported to be excited. This could be due to the difficulty of integrating long enough, albeit the simulations extend already up to 4000 days of solar evolution.

4.3 Development of Solar Surface Small-Scale Dynamo Simulations over the Past 2 Decades

Using an incompressible (Bousinesq) convective SSD simulation Cattaneo (1999) demonstrated that highly intermittent small-scale field with a saturation field strengths of about 20% equipartition (averaged over the simulation domain) could be reached in a stellar photosphere. This work suggested that substantial fraction of the quiet Sun magnetic field with strength of a few 10 G could originate from SSD action. Later Bercik et al. (2005) studied turbulent dynamos in solar like (F-M type) stars using anelastic simulations and found comparable results when applied to the Sun. Their work suggested that SSDs may explain the observed lower limits for X-ray fluxes from solar-like main sequence stars. This study is covered in more detail in Sect. 5. The first comprehensive SSD simulation of the solar photosphere (including compressibility, radiation transport, open bottom boundary conditions and an equation of state accounting for partial ionization) was presented by Vögler and Schüssler (2007). The simulation produced a mean vertical magnetic field amplitude of about 30 G in the photosphere and was subsequently compared in detail to observations through forward modeling of spectral lines using both Zeeman (Danilovic et al. 2010b) and Hanle (Shchukina and Trujillo Bueno 2011) diagnostics. It was found that these simulations fell short by a factor of 2–3 compared to Hinode observations of Zeeman polarisation in the Fe I 6302 Å lines; an even larger discrepancy by an order of magnitude was found comparing the Hanle depolarization in the Sr I 4607 Å line, suggesting that in addition to being too weak, the magnetic field was also falling off with height too rapidly. It was found by Rempel (2014) that increasing the resolution alone was insufficient to address the discrepancy. A critical component was to account for magnetic field that is transported into the photosphere from the deeper layers of the CZ (see Sect. 4.3.1 for further detail). These improved simulations were again compared to observations through forward modeling (Danilovic et al. 2016a,b; del Pino Alemán et al. 2018) and it was found that simulations with a mean vertical field strength of around 60–80 G at optical depth unity in the photosphere were in agreement with constraints from both Zeeman and Hanle diagnostics. Khomenko et al. (2017) presented SSD simulations that included the Biermann Battery term (Biermann 1950) in the induction equation. It was found that this term produces at the edge of granules continuously seed fields with a strength of around 10^{-6} G, which can be amplified by the dynamo to saturation field strength within a few hours of time. While such fields are too weak to make a difference for the saturated dynamo state, this work highlights that fundamental physical processes do provide a lower bound for the quiet Sun magnetic field that is independent from external seeds (e.g. galactic magnetic field amplified during the star formation process).

4.3.1 Deep Versus Shallow Recirculation

A general challenge of near surface dynamo simulations is the treatment of the bottom boundary. Since closed boundary conditions enforce in the usually adopted shallow domains unphysical recirculation, most photospheric convection setups use bottom boundaries that are open for convective flows and mimic the presence of a deep CZ. Such an open bottom boundary does make SSD simulations ill-posed, since, dependent on the details of the boundary condition, magnetic energy can leave or enter the simulation domain. The work of Vögler and Schüssler (2007) used boundary conditions that do not allow for a Poynting flux at the bottom boundary in inflow regions. While downflows do transport energy out of the domain (owing to resistive transport right at the boundary), inflows do not transport energy back into the domain. This setup is conservative and demonstrates dynamo action in the presence of little local recirculation and continuous loss of magnetic energy towards the deep CZ, which was surmised to be a large hurdle for dynamo action in the photosphere (Stein et al. 2003). As descibed in Sect. 4.3 these models reached a vertical field strength of 30 G at optical depth unity, which is about a factor of 2–3 lower than implied by observations. The deeper parts of the solar convection have a larger magnetic Reynolds [Prandtl] numbers than the photosphere (10^9 [10^{-3}] instead of 10^5 [10^{-5}]), which should enable SSD action over a wide range of scales. Using ILES simulations Hotta et al. (2015) found super-equipartition fields near the base of the CZ in SSD simulations of the deep CZ. In addition, the deeper CZ will host magnetic field produced by the LSD, which may modulate the quiet Sun network field in addition as found in observations (Korpi-Lagg et al. 2022).

A fraction of this field is transported back to the surface and will appear in the photosphere and boost the amplification of field in the surface layers. Rempel (2014) captured this effect by allowing for the transport of horizontal field through the bottom boundary and by considering simulations with a closed boundary and complete recirculation and found and increase of the photospheric saturation field strength by about a factor of 2. Magnetic field that reaches the photosphere from deeper layers (deep recirculation) has undergone substantial horizontal expansion and appears as a rather smooth seed field in the center of granules, while magnetic field being brought back into the photosphere as a consequence of local downflow/upflow mixing (shallow recirculation) appears as a smaller-scale turbulent field at the edge of granules (Rempel 2018).

4.3.2 Energy Transfers, Saturation and Total Power of the Dynamo

Saturation of dynamo action at large magnetic Reynolds numbers is in general not a property of the velocity field, but rather about the relation of velocity and magnetic field to each other.

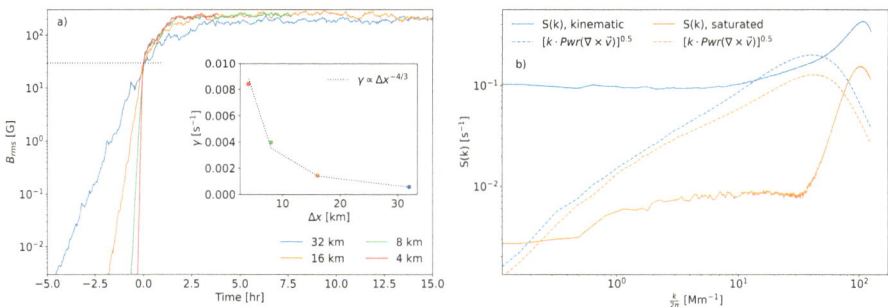

Fig. 8 Panel a): Transition from kinematic to saturated phase in ILES SSD simulations of the solar photosphere for models presented in Rempel (2014) with grid spacings from 32 down to 4 km. Presented is the time evolution of B_{rms} in the photosphere (optical depth of unity). All models start with the same seed field (around 0.001 G) and the curves are shifted such that the transition from kinematic to saturated phase lines up (B_{rms} of 30 G). The growth rate during the kinematic phase depends strongly on resolution as shown in the insert. Panel b): Saturation process of a ILES photospheric dynamo simulation. Shown are the effective shear rate (see text) and the vorticity spectrum during the kinematic and saturated phase

This was demonstrated by Tilgner and Brandenburg (2008), Cattaneo and Tobias (2009) for both small and LSD setups. They found that the velocity field of the saturated dynamo simulation remains to be an efficient growing dynamo in the kinematic regime, highlighting that saturation is not a property of the velocity field alone and requires a continuous adjustment of the velocity field to small changes in the saturated magnetic field solution.

Figure 8a) shows for various simulations from Rempel (2014) the transition from kinematic phase to saturated phase. We show here B_{rms} in the photosphere, which reaches in the saturated state values around 200–250 G, corresponding to about 30–40% of the equipartition value. While the models show large differences in the their kinematic growth rate depending on their resolution, after passing threshold of about $B_{rms} = 30$ G, the remaining slow growth is similar and requires a few hours to reach the final saturation value. The saturation process was further studied in Rempel (2014) by looking at the effective shear rate defined in wavenumber space as

$$T_{MS}(k) = \frac{1}{8\pi}\widehat{\vec{B}}(k) \cdot \left[\widehat{(\vec{B}\cdot\nabla)\vec{v}}\right]^{*}(k) + c.c. \tag{1}$$

$$E_M(k) = \frac{1}{8\pi}\widehat{\vec{B}}(k) \cdot \widehat{\vec{B}}^{*}(k) \tag{2}$$

$$S(k) = T_{MS}(k)/E_M(k) \tag{3}$$

During the kinematic growth phase $S(k)$ (Fig. 8b) has over a wide range of scales a value corresponding to the magnitude of vorticity (given by the quantity $\sqrt{k\cdot\mathrm{Pwr}(\nabla\times\vec{v})}$) on small scales. In the saturated state $S(k)$ has dropped over a wide range of scales by a factor of around 30 to values comparable to the magnitude of vorticity on the larges scales, while there is only a small reduction of the vorticity by less than a factor of 2 on the smallest scales. In the saturated state of the dynamo $S(k)$ is small due to combination a misalignment of the shear and magnetic field (reducing $(\vec{B}\cdot\nabla)\vec{v}$) and an induced field being mostly orthogonal to the existing field, minimizing the energy transfer while velocity shear remains mostly unchanged.

The growth rate of the ILES SSD shows in the explored range a very strong dependence on resolution (insert in Fig. 8a) in the form of $\gamma \propto (\Delta x)^{-4/3}$, which is significantly steeper than the scaling of vorticity amplitude $\omega \sim k \cdot v \sim k^{2/3}$ (based on Kolmogorov scaling). For the case with 4 km grid spacing the growth rate for magnetic energy (twice the rate of B_{rms}) is with $0.017\,\mathrm{s}^{-1}$ about $1/6$ of the rate given by vorticity (Fig. 8b). It is currently uncertain how this scaling will change at higher resolution.

How much energy is required to maintain the small-scale magnetic field of the Sun? We can provide an estimate based on the previous discussion. With the small value of $\mathrm{Pm} \sim 10^{-5}$ in the photosphere and values not much larger than 10^{-3} throughout the CZ, the dynamo is operating in a regime where the Lorentz force transfers energy from kinetic to magnetic energy on all scales and therefore maximises the power of the dynamo by minimising the energy lost to viscous dissipation. Rempel (2018) estimated from photospheric ILES dynamo simulations that about $150\,\mathrm{erg\,cm}^{-3}\,\mathrm{s}^{-1}$ are available in the uppermost 1.5 Mm of the CZ to power the SSD, which is integrated over the whole solar surface about 30% of the solar luminosity. The total power of the SSD integrated over the volume of the CZ is bounded by the total pressure buoyancy driving, which is on the order of a few solar luminosities based on mixing length models. We note that the power of the dynamo can exceed a solar luminosity, since it is not a sink of energy. Through Ohmic dissipation this energy is returned to internal energy. If there would be no SSD a similar amount of energy would be dissipated through viscosity instead. Since the SSD alone is already capable of consuming most of the available convective driving, a LSD can only grow at the expense of the SSD as it has been suggested by Cattaneo and Tobias (2014). This would imply that the total power of the combined small- and LSD does not change much with rotation, while the structuring of magnetic field does change. Maintaining a magnetic field clumped into sunspots and starspots requires less energy than maintaining a field structured on the smallest scales, which maximises Ohmic dissipation. With a mean vertical magnetic field strength of around 60–80 G in the photosphere, the unsigned flux of the quiet Sun corresponds to about a 100 active regions at any given time. Reorganizing that amount of magnetic flux into starspots would turn the Sun into a very active star by comparison, while dramatically reducing Ohmic dissipation.

4.3.3 Anisotropy of Magnetic Field in Photosphere

As summarized in Sect. 2, observations indicate a significant anisotropy of the magnetic field above the photosphere in the sense that the horizontal field components are stronger than the vertical ones. The dynamo simulation of Vögler and Schüssler (2007) did show a similar preference for horizontal field as reported in Schüssler and Vögler (2008). It was found that at the height the Hinode lines are sensitive to, the strength of the horizontal field component is 4–5 times stronger than the vertical field component. Rempel (2014) analyzed SSD simulation in a wider and deeper domain that had the top boundary located about 1.5 Mm above the average $\tau_c = 1$ level. In these simulations it was found that the ratio of horizontal to vertical field peaks at a height of about 450 km (see Fig. 9). While the ratio of horizontal to vertical field reached values as high as 5 during the growth-phase of the dynamo, the ratio dropped to about 2.5 when the dynamo is saturated (Fig. 9a). Figure 9b) shows the strength of the vertical and horizontal field components individually. During the kinematic growth phase both drop monotonically with height, however, the vertical field component drops more rapidly with height, which leads to a peak in their ratio at around 450 km height. During the saturated phase the horizontal field strength does show a distinct peak, while the vertical field strength continues to drop monotonically. However, the magnitude of the magnetic field anisotropy is lower compared to the kinematic phase (a ratio of about 2.5 instead of 5).

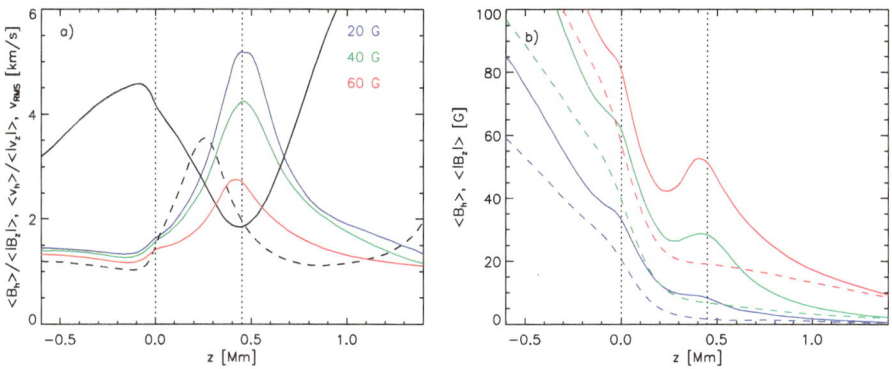

Fig. 9 Magnetic field anisotropy in upper CZ and photosphere. a): Ratio of horizontal to vertical field amplitude for 3 different field strengths (blue, green, red). The black solid line shows the convective RMS velocity, the black dashed line the velocity anisotropy. The peak of magnetic field anisotropy about 450 km above $\tau_c = 1$ coincides with the minimum of convective RMS velocity. b) Height variation of vertical (dashed) and horizontal (solid) field amplitudes. The fully saturated dynamo (red line) shows a distinct local peak of horizontal field amplitude at the location of minimum RMS velocity

What is the origin of the field anisotropy and specifically the origin of the peak in the horizontal field component? Obviously the velocity field is anisotropic above the granulation layer where overturning motions lead to preference of horizontal flows. However, the maximum flow field anisotropy is found at a height of about 250 km (black dashed line), which is 200 km lower than the height of peak anisotropy in the magnetic field. The height of peak field anisotropy does coincide with the minimum of the convective RMS velocity (black solid line). This in combination with the distinct peak in horizontal field strength may point to the diamagnetic part of turbulent pumping as the mechanism that expels horizontal magnetic field from the photosphere into lower chromosphere where it accumulates in the region with the lowest turbulence intensity. We note that this explanation is at best qualitative since SSD simulations do not have a large scale mean-field, however, the horizontal field overlying the photosphere is organized on scales larger than granules.

Alternative to the approach of using inversions to infer magnetic field anisotropy from observed Stokes profiles, this can also be achieved by analyzing the properties of the Stokes signals directly, specifically their center-to-limb variation. The simulation highlighted in Fig. 9 was compared to Hinode observations by Lites et al. (2017) and a good agreement between the CLV of synthetic and observed Stokes Q and U was found. This suggests that current photospheric SSD simulations do reproduce to a large extent the observed magnetic field anisotropy. A further test for these models would be multi-height observations that map out the amplitude of horizontal magnetic field in order to test the prediction of a peak in the horizontal magnetic field amplitude at a height of about 450 km.

4.3.4 Quiet Sun Network Field, Relevance for Coronal Heating

There are two common misconceptions: firstly, SSDs can only produce zero-mean fluctuations on the scale of granules and their downflow lanes, and secondly, that LSDs cannot produce zero-mean small-scale fluctuations through tangling of turbulent motions at the scale of convection. As a consequence both dynamos contribute to the organization of magnetic field on the observable scales from granules to super-granules. Here we focus specifically on the contribution from the SSD to network field.

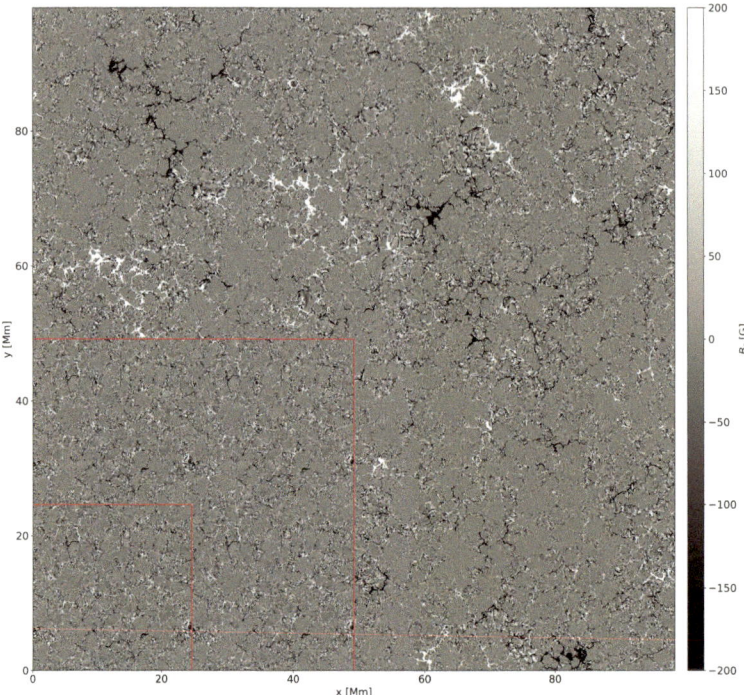

Fig. 10 Comparison of B_z at optical depth unity in two dynamo simulations with different domain sizes. The full horizontal extent shows results from a simulation in a $98.304 \times 98.304 \times 18.432$ Mm^3 domain. The small red box indicates a simulation in a $24.576 \times 24.576 \times 7.68$ Mm^3. For better visibility we periodically extended this simulation as a 2×2 tile (large red box). The SSD simulation in the wider and deeper domain does produce a mixed polarity network structure on scales larger than 10 Mm, which is absent in the smaller domain

Observations of the quiet Sun during a solar cycle minimum do show a quiet Sun mixed polarity network field (see, e.g., Fig. 13 in Lites et al. 2008), which raises the question of whether this network field is still a remnant of the LSD or if it is part of the quiet Sun and maintained by the SSD. Figure 10 compares 2 SSD simulations, one in a $98.304 \times 98.304 \times 18.432$ Mm^3 domain and one in a $24.576 \times 24.576 \times 7.68$ Mm^3 (see also Rempel 2014, 2020). Both simulations were set up with zero net flux on the scale of the simulation domain and both simulations have a mean vertical field strength of about 60 G. However, in the case of the wider and deeper simulation, we find larger scale flux imbalances that lead to the formation of network field. The larger simulation has on a scale of 24.576×24.576 Mm (extent of small simulation domain) an average flux imbalance corresponding to a mean vertical field of about 10 G. These flux imbalances are comparable to the imbalance found in the Hinode observations of Lites et al. (2008) (see Rempel (2020) for further discussion). These simulations suggest that the SSD does make at least a partial contribution to the quiet Sun network field, while at the same time there are also solar cycle dependent contributions from the LSD as found by Korpi-Lagg et al. (2022). We will discuss this further in Sect. 4.3.5 with regard to quiet Sun contributions to solar irradiance variations.

There are essentially two processes at work that lead to the emergence of a larger-scale network: (1) The dynamo is mostly saturated for the field strengths present in the solar photosphere. While the dynamo is very fast during the kinematic phase, simulations indicate that the kinematic phase ends for field strengths that are about 10% of typical quiet Sun field strengths (see Fig. 8a). For stronger field the remaining growth time-scale is on the order of several hours, which means that the magnetic field can get organized by photospheric flows on scales larger than granulation. (2) As discussed in Sect. 4.3.1 at least 50% of the small-scale field present in the photosphere originates from deeper layers. In a deep, heavily stratified domain the field that is brought up from deeper layers is organized on scales of the deep seated, larger-scale convection and this organization is imprinted on the photosphere and further organized by (1). Further experiments (Rempel, 2022, private comm.) point towards (2) as the critical effect. Setups without deep recirculation (i.e. zero field in upflow regions at the bottom boundary) do not show a network structure even if deeper domains are used.

The quiet Sun network plays a critical role for shaping the upper solar atmosphere. Flux imbalance on larger-scales leads to stronger field reaching the corona, which is in turn critical for maintaining the quiet Sun corona. The need for a small flux imbalance (corresponding to about 5 G) was identified in models by Amari et al. (2015). Small-scale dynamo simulations in sufficiently deep domains with deep recirculation naturally produce such a flux imbalance on super-granular scales and have demonstrated that they can maintain a quiet Sun corona at temperatures in the 1–1.5 million K range (Rempel 2017; Chen et al. 2022).

4.3.5 Irradiance Properties of Quiet Sun Magnetic Field

While most small-scale magnetic fields in the quiet Sun are too weak to influence the radiative properties of the photosphere, simulations predict a small amount kG strength flux concentrations ($\lesssim 1\%$) in the photosphere which can enhance radiative losses similar to flux concentrations in solar network regions (regions of photosphere with a significant magnetic flux imbalance). As long as the quiet Sun can be considered as not varying (in the global sense) this will lead to an offset in the total and spectral solar irradiance (TSI/SSI) compared to a hypothetical non-magnetic Sun. However, if the quiet Sun varies over solar-cycle or even longer time-scales the quiet Sun magnetic field can make a contribution to the observed variation of TSI and SSI. We have to distinguish here between the network and the internetwork magnetic field. It is known from observations that quiet Sun network does have some residual variation with the cycle (Korpi-Lagg et al. 2022) on the order of 6 G, and has therefore contributions from both LSD and SSD. It is an open question whether the lowest level of quiet Sun network during the cycle minimum is a representation of the SSD contribution alone and therefore the lowest activity state possible or if a further drop of activity is possible. While observational support for a cycle variation of the internetwork field is marginal (see Sect. 2.5), we cannot rule it out completely. In order to assess the contribution of quiet Sun field it is necessary to derive the TSI or SSI sensitivity to changes of the quiet Sun field strength. This was investigated by Rempel (2020) who computed the TSI and SSI for quiet Sun (zero net flux) and weak network setups and found that a 7 G change of $\langle |B_z| \rangle$ on the $\tau_c = 1$ level (about 10% variation) causes about a 0.1% change in TSI. Given that this is about the total observed TSI change over the solar cycle and most of that is explained through contributions from the active Sun, there is very little room for the quiet Sun to vary over the solar cycle (meaning in addition to the quiet Sun network fraction that results from the LSD and is already considered in irradiance models). TSI provides more stringent constraints on quiet Sun variability than direct measurements of the magnetic field (see, e.g., Lites et al. 2014; Buehler et al. 2013; Meunier 2018; Faurobert and Ricort 2021). While this does not rule out longer-term variations of the quiet Sun, they are very unlikely to happen. The models from Rempel (2020) were used by Yeo et al. (2020) to reconstruct solar TSI starting from HMI magnetograms (using radiative MHD simulations to translate HMI magnetograms into irradiance taking into account the full HMI data pipeline). It was found that 97% of the observed TSI variation can be accounted for that way, which gives a significant confidence that radiative MHD simulations capture the radiative properties of magnetic flux concentrations to a significant degree. Using this model they provided a lower bound for a grand minimum irradiance of no more than 2 W/m^2 lower than the 2019 solar cycle minimum. To this end SSD simulations were used to represent the lowest activity state of the Sun assuming that only the internetwork field is present. This is a lower bound, since as discussed in Sect. 4.3.4 a significant fraction of the network field present during a solar minimum could originate from a SSD and therefore would be part of the lowest possible activity state as well (i.e. present even during a grand minimum).

4.4 Radiative Zone

The SSD is usually understood to operate in the CZ, where the field is amplified by turbulent convection. However, no convection is possible in the stably-stratified radiative zones of stars. Does a source of turbulence still exist in these conditions, and does it have the required strength and properties to drive SSD? At the tachocline, shear instabilities, gravity waves as well as convective overshoot could, potentially, drive turbulence. Turbulence, in this case, is then no longer isotropic and comes under the realm of stably-stratified systems, an actively researched topic worth its own multiple review articles (see, e.g., Riley and Lindborg 2012; Lindborg 2006; Cheng et al. 2020, and references therein).

The essential feature of such turbulence is the existence of multiple scales over which the characteristics of turbulence change significantly. The length scales for the largest eddies in the vertical direction (with wavenumbers k_v) are much smaller than those in the horizontal direction (with wavenumber k_h), i.e., $k_v \gg k_h$. Apart from the usual Reynolds number, another dimensionless number enters into the picture: Froude number (Fr). This can be defined in the horizontal and vertical direction in terms of the Brunt-Väisälä frequency N as $\mathrm{Fr}_h = Uk_h/N$, $\mathrm{Fr}_v = Uk_v/N$. Here, Fr_h can be understood as an inverse of the degree of stratification (in stably stratified system, $\mathrm{Fr}_h < 1$) and Fr_v can be understood as the ratio of inertial and buoyancy forces. Then there exists the Kolmogorov dissipation wave number $k_v \sim (\varepsilon/\eta^3)^{1/4}$ (Kolmogorov 1941) (where the kinetic energy gets dissipated), the Dougherty-Ozmidov wavenumber $k_O \sim (N^3/\varepsilon)^{1/2}$ (Dougherty 1961; Ozmidov 1965) (where energy in buoyant motions becomes comparable to the kinetic energy), and the buoyancy wavenumber $k_b \sim N/u$ (the length scale corresponding to the adiabatic displacement of a parcel with velocity u in the vertical direction). Under certain assumptions (Lindborg 2006), these wavenumbers (or, equivalently, length scales) can be described in terms of Re and Fr_h. In addition, in the presence of magnetic fields, there also exists the resistive wavenumber k_η (where magnetic energy gets dissipated).

The existence of SSD in solar radiative zone was investigated by Skoutnev et al. (2021) recently. In their paper, they consider a single $\mathrm{Fr} = u_{\mathrm{rms}}/l_i$, where $l_i \sim 1/k_i$ is the integral length scale. Hence, we shall use the same notation to describe their results. They considered 2 situations: i) $k_b < k_v < k_O$, ii) $k_b < k_O < k_v$, along with $k_\eta \lesssim k_v$ for $\mathrm{Pm} \lesssim 1$. In the first case, the small scale motions down to k_O are suppressed by viscosity making the flow more laminar and unsuitable for SSD growth. For the second case, however, Fr comes into the scaling. As $k_v \sim \mathrm{Re}^{3/4}\varepsilon^{-1/4}$ and $k_O \sim \varepsilon^{-1/2}\mathrm{Fr}^{-3/2}$, taking their ratio gives $k_v/k_O \sim (\mathrm{ReFr}^2)^{3/4}\varepsilon^{1/4}$. For fixed values of Pm, they explored the $\mathrm{Re} - \mathrm{Fr}$ parameter space, and investigated the critical value of Re (Re^c) above which SSD action was possible. They found a scaling of $\mathrm{Re}^c \sim \mathrm{Fr}^{-2}$ for $\mathrm{Pm} \geq 1$, reducing the parameter space from $\mathrm{Re} - \mathrm{Fr} - \mathrm{Pm}$

to Rb − Pm, where Rb = ReFr² is defined as a buoyancy Reynolds number. The two cases can then be distinguished as Rb < 1 and Rb > 1, respectively. Their main result was the existence of a critical Rb of around 3 to 9 (for high and low Pm regime, respectively) above which SSD action was possible. Since SSD fields are expected to be the fastest growing fields, these fields could influence the operation of instabilities associated with the generation of large-scale field at the base of the CZ (Spruit 1999). They also calculated the critical Rb from estimated solar values to be around a 100 (an order of magnitude higher than the critical value), which would imply that such a stably-stratified SSD could, in principle, exist in the solar interior. However, the sensitive dependence of this value on poorly constrained parameters like u, l_i for the solar interior makes this statement somewhat tentative.

5 SSD on Other Cool Stars

The Sun is the only star which we can resolve well enough to study detailed properties of granulation and quiet small-scale solar magnetism. For other stars the Zeeman-Doppler imaging inversion technique can be used to study their surface magnetic field (ZDI; see, e.g., Semel 1989). Due to the lack of surface resolution, this method has the capacity to trace the large-scale magnetic structures only. To estimate the total surface magnetic field, the Zeeman broadening and intensification can be used (see, e.g., Kochukhov et al. 2020). They demonstrated that only a fraction of the total magnetic field is recovered by ZDI. By combining ZDI with Zeeman broadening and intensification measurements, it is thus possible, in principle, to estimate how much of the magnetic field is hidden in small-scale structures (Trelles Arjona et al. 2021). Hence, it is well motivated to extrapolate our understanding of these phenomena to other stars similar to the Sun. In this review we consider the possible impact of this quiet star magnetism from the perspective of observations as well as theoretical modelling.

5.1 MHD Simulations of Near-Surface Stellar Convection

The first 3D MHD studies of SSD action in other stars, to the best of our knowledge, were conducted by Bercik et al. (2005). In this study, the authors consider anelastic simulations of near surface convection. The authors find the magnetic energy in the saturated phase of SSD evolution to be around 6.7% of the total kinetic energy. The X-ray and Mg II luminosities were then estimated by fitting the unsigned magnetic flux to empirical luminosity relations. This study, however, is anelastic and without proper radiative transfer, assumptions which fail near the surface. Despite these simplifications, a reasonable fit to the floor of X-ray and Mg II flux for cool main-sequence stars ranging in spectral class from F0 to M0 (Figs. 6 and 7) is reproduced, leading credence to the fact that SSD fields could, in principle, be responsible for lower limit of X-ray and chromospheric fluxes.

In addition, studies of fully compressible 3D MHD near-surface convection with realistic treatment of radiation have been conducted with varying strengths of imposed magnetic field, for example, by Beeck et al. (2015) using the MURaM code and Steiner et al. (2014) and Salhab et al. (2018) using the CO5BOLD code (Freytag et al. 2012). Since these studies have an imposed magnetic field, they cannot be considered quiet-star SSD simulations and are more comparable to a "plage" simulation, however, the weak field cases (20 G for the former, and 50 G for the latter) should provide an idea of the effect of effect of magnetic fields expected to arise from an SSD mechanism. The main takeaways from these simulations are that the small-scale magnetic flux concentrations form in the intergranular lanes

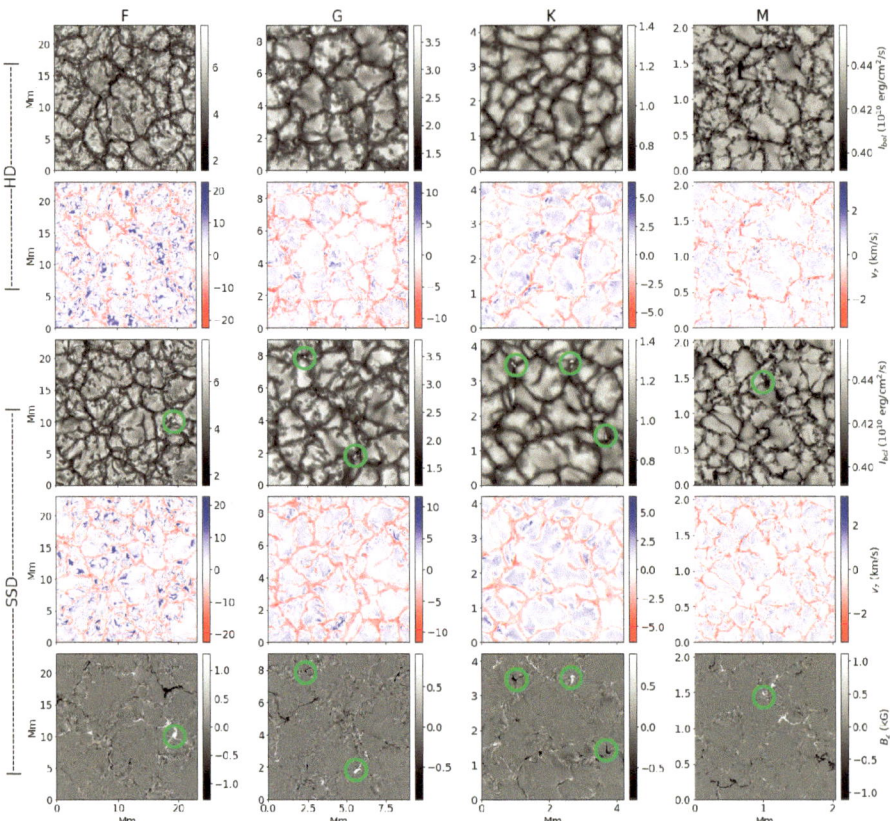

Fig. 11 Emergent intensity and surface vertical velocities in different stellar types for models with and without magnetic field. *From top to bottom:* Snapshot of the bolometric intensity and v_z at $\tau = 1$ for the hydrodynamic case (*row 1 and 2*), bolometric intensity and v_z at $\tau = 1$ for the SSD case (*row 3 and 4*) and the corresponding vertical magnetic field at $\tau = 1$ (*row 5, from left to right*) for spectral types F, G, K and M, respectively. The green circles indicate the bright points and corresponding magnetic field concentrations. Figure adapted from Bhatia et al. (2022)

with field values largely independent of stellar type at the $\tau = 1$ surface and that magnetic bright-points and filigree structure are apparent in the intergranular lanes, with highest contribution to bolometric intensity (at least for disk center) from around the G to K-type stars.

More recent simulations demonstrate that SSD action results in generation of magnetic fields of similar strength for all stars (F3V, G2V, K0V and M0V) considered (see Fig. 12, left panel; Bhatia et al. 2022) as well as reproduce the aforementioned features of magnetic field distribution (see Fig. 11 for a snapshot of said simulations). These simulations consist of two setups: one with self-consistently generated SSD fields with an open bottom boundary (same as *OSb* in Rempel 2014) and one that is purely hydrodynamic (HD). Compared to the pure HD runs, the SSD runs show small changes in the density and pressure stratification ($\sim$1%) for the F-star (Fig. 12, right panel), which is attributed to a decrease in turbulent pressure. This decrease in turbulent pressure occurs because the amplification of SSD fields takes away energy from the kinetic energy reservoir, decreasing the velocities (and, consequently, turbulent pressure). This effect is noticeable for the F-star because the kinetic energy is within an order of magnitude of the internal energy near the surface, whereas it is

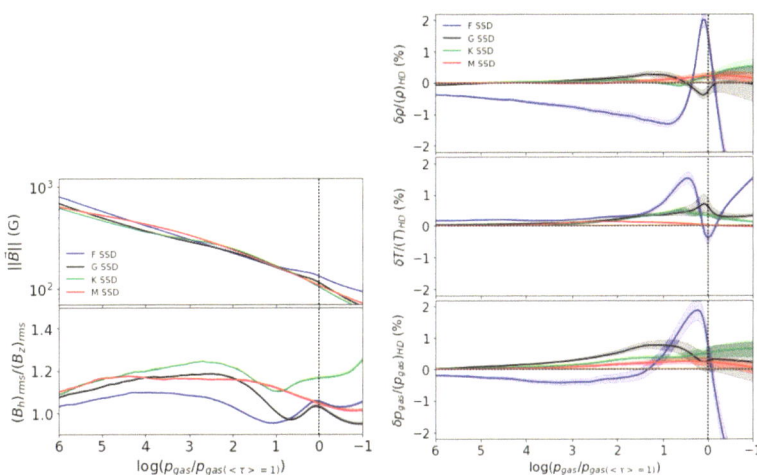

Fig. 12 *Left:* Horizontally averaged magnetic field magnitude (*top*) and the ratio of the horizontal r.m.s field strength to the vertical r.m.s field strength *bottom*. *Right:* Deviations in (*from top to bottom*) ρ, T and p_{gas} relative to hydrodynamic cases. The vertical axis gives the geometric deviations as a percentage relative to the hydrodynamic case. The horizontal axis is the number of pressure scale heights $\log_{10}(p_{gas}/p_{gas(\tau=1)})$, calculated for the HD cases, below the surface (*dotted vertical black line*). The shaded regions correspond to 1-σ standard error ($e = \sigma/\sqrt{N}$, N is the number of snapshots) computed over time averaging of snapshots. Figures adapted from Bhatia et al. (2022)

significantly smaller for other stars. At the photosphere, it is found that the effect of SSD fields is to reduce the upflow velocities for all cases, slightly decrease the average granule size for all cases and slightly increase the disk-center bolometric intensity for the G and K-star (Bhatia et al. 2023, submitted). The changes in velocity and intensity imply possible influence on high-resolution spectra. It must be noted that these simulations use gray radiative transfer (Rosseland mean opacities), so any analysis above the lower photosphere must be taken with a grain of salt. The authors are currently working on studies with 12-opacity bins, similar to what is used in Magic et al. (2015). In addition, these simulations have an effective Pm of $\sim$1.

5.2 Abundance Determination

The determination of stellar abundances is a vast field on its own, with applications ranging from planet formation and stellar evolution to cosmology (see, e.g., Jofré et al. 2019). The key ingredients to abundance determination are high resolution stellar spectra, precise atomic/molecular data and grids of model stellar atmospheres. As with other topics touched here, a complete discussion of these is well outside the scope of this article. Here, we focus on the last point regarding stellar models.

Traditionally 1D stellar models have been used to generate model spectra. However, these models rely on free parameters to model expected effects of convection (α mixing-length) and turbulence (η_{turb} microturbulent velocity). Asplund (2005) and Nissen and Gustafsson (2018) review the consequences for spectral line formation and abundance determination when effects of non-LTE and 3D convection are included in model atmospheres (see Sect. 2.3 of the first and Sect. 2.4 of the second reference, especially), and highlight the need for better models, especially for metal-poor stars. Section 2.6 of Nissen and Gustafsson (2018) also discusses the possible effect of SSD upon abundance determinations by various authors (Fabbian et al. 2012; Shchukina and Trujillo Bueno 2015), with the broad picture being that the effect of SSD fields on abundance determinations is small and field-morphology dependent (with strong and organized field concentrations having the largest effect), but potentially non-negligible for stars other than the Sun. It must be noted that the spectral syntheses in these studies were carried out in LTE, and thus are not directly comparable to observations.

5.3 Basal Chromospheric Flux

Stellar chromospheres have historically been studied by considering activity indicators based on chromospheric emission lines, like the S-index (Baliunas et al. 1995). A review of stellar chromospheric activity is outside the scope of this article, and has been covered extensively elsewhere (e.g., Hall 2008; Linsky 2017; de Grijs and Kamath 2021, and references therein). The aspect of stellar chromospheric activity that concerns us is the minimum activity level or, equivalently, the basal chromospheric flux. Here we consider the term "basal" to refer to the minimum level of chromospheric activity that is independent of the stellar magnetic cycle.

For studies of stellar activity, the contribution from basal flux is usually estimated empirically from inactive stars (Mittag et al. 2013). This basal activity is color-dependent and was initially assumed to be due to acoustic heating (Schrijver 1987). The presence of a corona for solar-like "inactive" stars (Judge et al. 2004), as well as comparisons of observed solar chromospheric intensity fluctuations to models have brought this assumption into question (Fossum and Carlsson 2005). As described in Sect. 2.6, the current understanding is that acoustic waves alone are probably not sufficient to heat the chromosphere and that the magnetic field has an important role to play in the transport and dissipation of energy (Jefferies et al. 2006; Rajaguru et al. 2019). The precise details and mechanisms involved, however, remain poorly understood.

In the absence of sufficiently high-resolution observations, numerous simulations of the solar chromosphere illustrate the importance of magnetic fields in transferring energy to the chromosphere (see Sect. 5 of Carlsson et al. 2019, and references therein for an overview). Recent simulations of solar chromosphere using non-equilibrium ionization, non local-thermodynamic-equilibrium (Martínez-Sykora et al. 2019) indicated a dynamo-like process acting in the chromosphere converting kinetic to magnetic energy. Subsequent inclusion of ambiploar diffusion seems to, at least qualitatively, reproduce the observed Mg II emission line profiles (Martínez-Sykora et al. 2023). Current "realistic" simulations of the solar chromosphere using MURaM with a chromospheric extension (Przybylski et al. 2022) show that the net vertically directed Poynting flux at the base of the IN chromosphere (with fields generated from an SSD mechanism) provides sufficient energy (Withbroe and Noyes 1977) to heat the chromosphere (Przybylski, in prep.). In addition, these simulations seem to reproduce the observed Mg II line profiles significantly better than existing models (Ondratschek, in prep.). All of this recent progress in simulations of solar chromosphere, coupled with the fact the basal chromospheric flux is expected to be quite universal for stars with an outer CZ (Schröder et al. 2012), makes the prospect of modelling stellar chromospheres an achievable goal in the near future.

5.4 Stellar Variability and Exoplanet Detection

In the last two decades, the field of exoplanet detection has exploded. Dedicated exoplanet hunting missions like Kepler/K2, TESS and, soon, PLATO, in hand with follow-up radial velocity (RV) observations from ground-based echelle spectrometers like CARMENES (Quirrenbach et al. 2016) and ESPRESSO-VLT (the latter with resolving power over 190,000 in the visible wavelength region; Pepe et al. 2021), have made it possible to detect and study rocky exoplanets (Pepe et al. 2013; Ribas et al. 2018). The instrumental precision of these modern spectrometers are within 10 cm/s, making it possible, in principle, to detect and characterize an earth-like planet orbiting around a Sun-like star (the RV contribution of the earth to the Sun is around 9 cm/s).

However, stellar variability remains the single largest source of uncertainty in current observations. The strategies commonly used to account for stellar variability at small timescales (granulation, p-mode oscillations, etc.) revolve around tweaking observation frequency and exposure times (Dumusque et al. 2011) to essentially average them out as best as possible. This can become prohibitively costly to approach noise level of less than 10 cm/s, as granulation is expected to be correlated for timescales much longer than few minutes (Meunier et al. 2015). Another approach is to model effects of stellar convection using realistic radiative-MHD simulations to model contributions of granulation to RV signal (Cegla et al. 2013). On the theoretical side, simulations of SSD in solar convection (Hotta et al. 2015), as well as near-surface convection for other cool main-sequence stars (Bhatia et al. 2022) have shown a consistent reduction in convective velocities. This reduction in convective velocities may be expected to influence the RV signal of granulation and characteristics of pressure modes.

In addition, the time-averaged stellar photospheric lines show a asymmetry due to granular motions, termed convective blueshift (Dravins 1987). This is one of the few observable signatures of stellar granulation and the degree of this blueshift for different spectral lines at different formation heights show a sort of universal scaling with effective temperature for different stellar types (Gray 2009; Liebing et al. 2021). Shporer and Brown (2011) showed using a simplified model how convective blueshift could influence the measurement of spin-orbit angles. Bhatia et al. (2023, submitted) show that in simulations of stellar photospheres with SSD fields, there is a reduction not only in convective velocities but also the scale of granulation (granules appear to be slightly smaller in presence of SSD fields), especially so for hotter spectral types.

Lastly, the center-to-limb variation of spectral intensities (used, among other things, for characterizing exoplanetary transits) is usually estimated from 1D model atmospheres. However, there have been discrepancies in comparison between true limb darkening from exoplanetary transit and model atmosphere based limb darkening (Howarth 2011), which could be accounted for by better models. Existing 3D hydrodynamic simulations already show a systematic difference in limb darkening calculated from 1D atmospheres (Magic et al. 2015). To improve precision of stellar photometry (important for determination of stellar radii as well as for transmission spectroscopy; de Wit and Seager 2013), it is necessary to account for these effects.

6 Outlook

Small-scale dynamo simulations that describe the process in solar and stellar CZs and photosphere require ingredients that go in several aspects beyond the simplified setups that are used to study the fundamentals of the dynamo processes, and their kinematic and non-linear phases, important in their own right. At a minimum, these setups require stratification and turbulence that is driven through convection, which results in the case of photospheric simulations from a volumetric cooling term based on radiative losses (typically computed from full 3D radiative transfer). Furthermore, the near surface layers of the Sun and Sun-like stars require an equation of state that accounts for the partial ionization of the most abundant elements. Simulations of the photosphere of the Sun and Sun-like stars typically use rather shallow domains that do not reach to the base of the convective envelope and just capture the uppermost 5–10 density scale-heights. In such setups it is not uncommon to use an open lower boundary condition (allowing for vertical mass and convective energy flux) that mimics the deeper CZ. Such open boundary conditions lead to a less well determined dynamo problem, since solutions and their saturation will depend on the Poynting flux crossing the lower boundary, specifically the Poynting flux in upflow regions. Such dependence is not unphysical as it describes the not-directly-observable coupling between the photosphere and the deeper layers of the CZ. Ultimately it will be required to conduct SSD and LSD simulations in domains that reach from photosphere to the base of CZ to account for the full interaction of all convective scales in the system. Simulations that include all of the above listed ingredients are often referred as "realistic" or "comprehensive" simulations, however, their realism is, as in most simulations, determined by the affordable resolution, which limits the achievable Re and Re_M and typically constrains values of $Pm = Re_M/Re$ close to unity. It is not uncommon for these models to use implicit large eddy simulations (ILES) in which diffusion terms are arising from the employed numerical scheme and are typically of a hyper-diffusive nature (higher order than Laplacian) with a non-linear dependence on the solution properties.

Although in simple cases ILES and DNS-type models do produce results that are in good agreement, they tend to disagree in situations where conditions for LSD onset are also fulfilled, that is, where rotation and its non-uniformities together with stratification are also allowed for. This seems to indicate that it does matter for the large-scale dynamics and magnetism how and at which scale the magnetic dissipation takes place, and investigating these issues further is an important future direction in deep convection models. LSD-SSD interactions cannot be ruled out as one decisive mechanism contributing to the convection conundrum and its solution.

The past couple of years have brought important verification steps of SSD excitation both in low-Pm and highly and stably stratified plasmas. While there is now nearly no doubt of the ubiquitous existence of SSD in solar and stellar convection and radiation zones, this raises the need of considering its role in various new scenarios, opening up new, exciting research avenues. However, in view of the drastic differences between $Pm \approx 1$ systems and those with low Pm, albeit known mostly only in the kinematic regime, far-reaching conclusions from $Pm = 1$ models should be avoided, and verification at lower Pm should always be pursued.

Solar observations will continue to play a critical role in constraining solar dynamo models. Of particular importance is to resolve the question of the isotropy of internetwork fields at different heights in the atmosphere and to study their temporal evolution from emergence until disappearance. The spatial distribution of the emergence sites in the granulation pattern may place additional constraints on the SSD mechanism and should be investigated further. DKIST will make it possible to tackle these questions with unprecedented sensitivity, providing precise linear polarization measurements over a much larger fraction of the solar surface than has been possible until now. DKIST will also obtain the first spatially and temporally resolved Hanle measurements ever, which will open new avenues for studying the weakest magnetic fields of the quiet solar photosphere. In order to allow for a comparison between simulations and observations we will need in the future higher resolution photospheric SSD simulations that also explore the low Pm regime present in the photosphere. At this point it is unknown if differences between the currently realized $Pm \sim 1$ (ILES) simulations and the $Pm \sim 10^{-5}$ regime of the solar photosphere will be detected or not.

Lastly, we note that SSD mechanism, as per the evidence presented in this review, is expected to be active across the HR diagram for all stars with a convective zone. The effects of such a field on stellar atmospheres are only just beginning to be understood. Better models, together with some of the most precise instruments currently available for observations are pushing boundaries of not just detection and characterization of exoplanets but also detailed understanding of stellar structure and variability.

Acknowledgements MJKL acknowledges fruitful discussions with Dr. Jörn Warnecke and Dr. Thomas Hackman, and funding from the European Research Council (ERC) under the European Union's Horizon 2020 research and innovation program (Project UniSDyn, grant agreement no 818665). LBR acknowledges financial support from the Spanish MICIN/AEI 10.13039/501100011033 through grants RTI2018-096886-B-C5, PID2021-125325OB-C5, and PCI2022-135009-2, co-funded by "ERDF A way of making Europe", and through the "Center of Excellence Severo Ochoa" grant CEX2021-001131-S awarded to Instituto de Astrofísica de Andalucía. TB is grateful to Damein Przybylski for some great discussions on the role of magnetic fields in the solar chromosphere. This material is based upon work supported by the National Center for Atmospheric Research, which is a major facility sponsored by the National Science Foundation under Cooperative Agreement No. 1852977. MR received partial funding from NASA grant NNX17AI30G.

Author Contribution All authors contributed equally to this work.

Funding Note Open Access funding enabled and organized by Projekt DEAL.

Declarations

Competing Interests The authors declare no conflict of interests.

References

Abbett WP (2007) The magnetic connection between the convection zone and corona in the quiet Sun. Astrophys J 665(2):1469–1488. https://doi.org/10.1086/519788

Amari T, Luciani JF, Aly JJ (2015) Small-scale dynamo magnetism as the driver for heating the solar atmosphere. Nature 522(7555):188–191. https://doi.org/10.1038/nature14478

Asensio Ramos A (2009) Evidence for quasi-isotropic magnetic fields from Hinode quiet-Sun observations. Astrophys J 701(2):1032–1043. https://doi.org/10.1088/0004-637X/701/2/1032. arXiv:0906.4230 [astro-ph.SR]

Asensio Ramos A, Martínez González MJ (2014) Hierarchical analysis of the quiet-Sun magnetism. Astron Astrophys 572:A98. https://doi.org/10.1051/0004-6361/201423860. arXiv:1410.5953 [astro-ph.SR]

Asplund M (2005) New light on stellar abundance analyses: departures from LTE and homogeneity. Annu Rev Astron Astrophys 43(1):481–530. https://doi.org/10.1146/annurev.astro.42.053102.134001

Baliunas SL, Donahue RA, Soon WH et al (1995) Chromospheric variations in main-sequence stars. II. Astrophys J 438:269. https://doi.org/10.1086/175072

Batchelor GK (1950) On the spontaneous magnetic field in a conducting liquid in turbulent motion. Proc R Soc Lond Ser A 201(1066):405–416. https://doi.org/10.1098/rspa.1950.0069

Beck C, Rezaei R (2009) The magnetic flux of the quiet Sun internetwork as observed with the Tenerife infrared polarimeter. Astron Astrophys 502:969–979. https://doi.org/10.1051/0004-6361/200911727. arXiv:0903.3158

Beeck B, Schüssler M, Cameron RH et al (2015) Three-dimensional simulations of near-surface convection in main-sequence stars. III. The structure of small-scale magnetic flux concentrations. Astron Astrophys 581:A42. https://doi.org/10.1051/0004-6361/201525788. arXiv:1505.04739 [astro-ph.SR]

Bekki Y, Hotta H, Yokoyama T (2017) Convective velocity suppression via the enhancement of the subadiabatic layer: role of the effective Prandtl number. Astrophys J 851(2):74. https://doi.org/10.3847/1538-4357/aa9b7f. arXiv:1711.05960 [astro-ph.SR]

Bellot Rubio L, Orozco Suárez D (2019) Quiet Sun magnetic fields: an observational view. Living Rev Sol Phys 16(1):1. https://doi.org/10.1007/s41116-018-0017-1

Bellot Rubio LR, Orozco Suárez D (2012) Pervasive linear polarization signals in the quiet Sun. Astrophys J 757(1):19. https://doi.org/10.1088/0004-637X/757/1/19. arXiv:1207.0692 [astro-ph.SR]

Bercik DJ, Fisher GH, Johns-Krull CM et al (2005) Convective dynamos and the minimum X-ray flux in main-sequence stars. Astrophys J 631(1):529–539. https://doi.org/10.1086/432407. arXiv:astro-ph/0506027 [astro-ph]

Berghmans D, Auchère F, Long DM et al (2021) Extreme-UV quiet Sun brightenings observed by the Solar Orbiter/EUI. Astron Astrophys 656:L4. https://doi.org/10.1051/0004-6361/202140380. arXiv:2104.03382 [astro-ph.SR]

Bhatia TS, Cameron RH, Solanki SK et al (2022) Small-scale dynamo in cool stars. I. Changes in stratification and near-surface convection for main-sequence spectral types. Astron Astrophys 663:A166. https://doi.org/10.1051/0004-6361/202243607. arXiv:2206.00064 [astro-ph.SR]

Biermann L (1950) Über den Ursprung der Magnetfelder auf Sternen und im interstellaren Raum (mit einem Anhang von A. Schlüter). Z Naturforsch A 5(2):65–71. https://doi.org/10.1515/zna-1950-0201

Boldyrev S, Cattaneo F (2004) Magnetic-field generation in Kolmogorov turbulence. Phys Rev Lett 92(14):144501. https://doi.org/10.1103/PhysRevLett.92.144501. arXiv:astro-ph/0310780 [astro-ph]

Borrero JM, Kobel P (2011) Inferring the magnetic field vector in the quiet Sun. I. Photon noise and selection criteria. Astron Astrophys 527:A29. https://doi.org/10.1051/0004-6361/201015634. arXiv:1011.4380 [astro-ph.SR]

Borrero JM, Kobel P (2013) Inferring the magnetic field vector in the quiet Sun. III. Disk variation of the Stokes profiles and isotropism of the magnetic field. Astron Astrophys 550:A98. https://doi.org/10.1051/0004-6361/201118239. arXiv:1212.0788 [astro-ph.SR]

Brandenburg A (2011) Nonlinear small-scale dynamos at low magnetic Prandtl numbers. Astrophys J 741:92. https://doi.org/10.1088/0004-637X/741/2/92. arXiv:1106.5777 [astro-ph.SR]

Brandenburg A (2014) Magnetic Prandtl number dependence of the kinetic-to-magnetic dissipation ratio. Astrophys J 791:12. https://doi.org/10.1088/0004-637X/791/1/12. arXiv:1404.6964 [astro-ph.SR]

Brandenburg A, Rempel M (2019) Reversed dynamo at small scales and large magnetic Prandtl number. Astrophys J 879(1):57. https://doi.org/10.3847/1538-4357/ab24bd. arXiv:1903.11869 [astro-ph.SR]

Brandenburg A, Subramanian K (2005) Astrophysical magnetic fields and nonlinear dynamo theory. Phys Rep 417:1–209. https://doi.org/10.1016/j.physrep.2005.06.005. arXiv:astro-ph/0405052

Buehler D, Lagg A, Solanki SK (2013) Quiet Sun magnetic fields observed by Hinode: support for a local dynamo. Astron Astrophys 555:A33. https://doi.org/10.1051/0004-6361/201321152. arXiv:1307.0789 [astro-ph.SR]

Carlsson M, De Pontieu B, Hansteen VH (2019) New view of the solar chromosphere. Annu Rev Astron Astrophys 57:189–226. https://doi.org/10.1146/annurev-astro-081817-052044

Cattaneo F (1999) On the origin of magnetic fields in the quiet photosphere. Astrophys J Lett 515(1):L39–L42. https://doi.org/10.1086/311962

Cattaneo F, Tobias SM (2009) Dynamo properties of the turbulent velocity field of a saturated dynamo. J Fluid Mech 621:205. https://doi.org/10.1017/S0022112008004990

Cattaneo F, Tobias SM (2014) On large-scale dynamo action at high magnetic Reynolds number. Astrophys J 789(1):70. https://doi.org/10.1088/0004-637X/789/1/70. arXiv:1405.3071 [astro-ph.SR]

Cegla HM, Shelyag S, Watson CA et al (2013) Stellar surface magneto-convection as a source of astrophysical noise. I. Multi-component parameterization of absorption line profiles. Astrophys J 763(2):95. https://doi.org/10.1088/0004-637X/763/2/95. arXiv:1212.0236 [astro-ph.SR]

Centeno R, Socas-Navarro H, Lites B et al (2007) Emergence of small-scale magnetic loops in the quiet-Sun internetwork. Astrophys J Lett 666(2):L137–L140. https://doi.org/10.1086/521726. arXiv:0708.0844 [astro-ph]

Charbonneau P (2020) Dynamo models of the solar cycle. Living Rev Sol Phys 17(1):4. https://doi.org/10.1007/s41116-020-00025-6

Chen Y, Przybylski D, Peter H et al (2021) Transient small-scale brightenings in the quiet solar corona: a model for campfires observed with Solar Orbiter. Astron Astrophys 656:L7. https://doi.org/10.1051/0004-6361/202140638. arXiv:2104.10940 [astro-ph.SR]

Chen F, Rempel M, Fan Y (2022) A comprehensive radiative magnetohydrodynamics simulation of active region scale flux emergence from the convection zone to the corona. Astrophys J 937(2):91. https://doi.org/10.3847/1538-4357/ac8f95. arXiv:2106.14055 [astro-ph.SR]

Cheng Y, Li Q, Argentini S et al (2020) A model for turbulence spectra in the equilibrium range of the stable atmospheric boundary layer. J Geophys Res, Atmos 125(5):e2019JD032191. https://doi.org/10.1029/2019JD032191

Danilovic S, Beeck B, Pietarila A et al (2010a) Transverse component of the magnetic field in the solar photosphere observed by SUNRISE. Astrophys J Lett 723(2):L149–L153. https://doi.org/10.1088/2041-8205/723/2/L149. arXiv:1008.1535 [astro-ph.SR]

Danilovic S, Schüssler M, Solanki SK (2010b) Probing quiet Sun magnetism using MURaM simulations and Hinode/SP results: support for a local dynamo. Astron Astrophys 513:A1. https://doi.org/10.1051/0004-6361/200913379. arXiv:1001.2183 [astro-ph.SR]

Danilovic S, Rempel M, van Noort M et al (2016a) Observed and simulated power spectra of kinetic and magnetic energy retrieved with 2D inversions. Astron Astrophys 594:A103. https://doi.org/10.1051/0004-6361/201527917. arXiv:1607.06242 [astro-ph.SR]

Danilovic S, van Noort M, Rempel M (2016b) Internetwork magnetic field as revealed by two-dimensional inversions. Astron Astrophys 593:A93. https://doi.org/10.1051/0004-6361/201527842. arXiv:1607.00772 [astro-ph.SR]

de Grijs R, Kamath D (2021) Stellar chromospheric variability. Universe 7(11):440. https://doi.org/10.3390/universe7110440

de la Cruz Rodríguez J, van Noort M (2017) Radiative diagnostics in the solar photosphere and chromosphere. Space Sci Rev 210(1–4):109–143. https://doi.org/10.1007/s11214-016-0294-8. arXiv:1609.08324 [astro-ph.SR]

de Wit J, Seager S (2013) Constraining exoplanet mass from transmission spectroscopy. Science 342(6165):1473–1477

del Pino Alemán T, Trujillo Bueno J, Štěpán J et al (2018) A novel investigation of the small-scale magnetic activity of the quiet Sun via the hanle effect in the Sr I 4607 Å line. Astrophys J 863(2):164. https://doi.org/10.3847/1538-4357/aaceab. arXiv:1806.07293 [astro-ph.SR]

del Toro Iniesta JC, Ruiz Cobo B (2016) Inversion of the radiative transfer equation for polarized light. Living Rev Sol Phys 13:4. https://doi.org/10.1007/s41116-016-0005-2. arXiv:1610.10039 [astro-ph.SR]

Dobler W, Haugen NE, Yousef TA et al (2003) Bottleneck effect in three-dimensional turbulence simulations. Phys Rev E 68(2):026304. https://doi.org/10.1103/PhysRevE.68.026304. arXiv:astro-ph/0303324 [astro-ph]

Dougherty JP (1961) The anisotropy of turbulence at the meteor level. J Atmos Terr Phys 21(2):210–213. https://doi.org/10.1016/0021-9169(61)90116-7

Dravins D (1987) Stellar granulation. I – the observability of stellar photospheric convection. Astron Astrophys 172(1–2):200–224

Dumusque X, Udry S, Lovis C et al (2011) Planetary detection limits taking into account stellar noise. I. Observational strategies to reduce stellar oscillation and granulation effects. Astron Astrophys 525:A140. https://doi.org/10.1051/0004-6361/201014097. arXiv:1010.2616 [astro-ph.EP]

Esteban Pozuelo S, Asensio Ramos A, de la Cruz Rodríguez J et al (2023) Estimating the longitudinal magnetic field in the chromosphere of quiet-Sun magnetic concentrations. Astron Astrophys 672:A141. https://doi.org/10.1051/0004-6361/202245267. arXiv:2302.04258 [astro-ph.SR]

Fabbian D, Moreno-Insertis F, Khomenko E et al (2012) Solar Fe abundance and magnetic fields. Towards a consistent reference metallicity. Astron Astrophys 548:A35. https://doi.org/10.1051/0004-6361/201219335. arXiv:1209.2771 [astro-ph.SR]

Faurobert M, Ricort G (2021) Magnetic flux structuring of the quiet Sun internetwork. Center-to-limb analysis of solar-cycle variations. Astron Astrophys 651:A21. https://doi.org/10.1051/0004-6361/202140705. arXiv:2105.08657 [astro-ph.SR]

Fischer CE, Borrero JM, Bello González N et al (2019) Observations of solar small-scale magnetic flux-sheet emergence. Astron Astrophys 622:L12. https://doi.org/10.1051/0004-6361/201834628. arXiv:1901.05870 [astro-ph.SR]

Fischer CE, Vigeesh G, Lindner P et al (2020) Interaction of magnetic fields with a vortex tube at solar subgranular scale. Astrophys J Lett 903(1):L10. https://doi.org/10.3847/2041-8213/abbada. arXiv:2010.05577 [astro-ph.SR]

Fossum A, Carlsson M (2005) High-frequency acoustic waves are not sufficient to heat the solar chromosphere. Nature 435(7044):919–921. https://doi.org/10.1038/nature03695

Freytag B, Steffen M, Ludwig HG et al (2012) Simulations of stellar convection with CO5BOLD. J Comput Phys 231(3):919–959. https://doi.org/10.1016/j.jcp.2011.09.026. arXiv:1110.6844 [astro-ph.SR]

Gošić M (2015) The solar internetwork. PhD thesis, Universidad de Granada, Spain

Gošić M, Bellot Rubio LR, Orozco Suárez D et al (2014) The solar internetwork. I. Contribution to the network magnetic flux. Astrophys J 797(1):49. https://doi.org/10.1088/0004-637X/797/1/49. arXiv:1408.2369 [astro-ph.SR]

Gošić M, Bellot Rubio LR, del Toro Iniesta JC, Orozco Suárez D et al (2016) The solar internetwork. II. Flux appearance and disappearance rates. Astrophys J 820:35. https://doi.org/10.3847/0004-637X/820/1/35. arXiv:1602.05892 [astro-ph.SR]

Gošić M, de la Cruz Rodríguez J, De Pontieu B et al (2018) Chromospheric heating due to cancellation of quiet Sun internetwork fields. Astrophys J 857(1):48. https://doi.org/10.3847/1538-4357/aab1f0. arXiv: 1802.07392 [astro-ph.SR]

Gošić M, De Pontieu B, Bellot Rubio LR, Sainz Dalda A et al (2021) Emergence of internetwork magnetic fields through the solar atmosphere. Astrophys J 911(1):41. https://doi.org/10.3847/1538-4357/abe7e0. arXiv:2103.02213 [astro-ph.SR]

Gošić M, Bellot Rubio LR, Cheung MCM, Orozco Suárez D et al (2022) The solar internetwork. III. Unipolar versus bipolar flux appearance. Astrophys J 925(2):188. https://doi.org/10.3847/1538-4357/ac37be. arXiv:2111.03208 [astro-ph.SR]

Gray DF (2009) The third signature of stellar granulation. Astrophys J 697(2):1032–1043. https://doi.org/10.1088/0004-637X/697/2/1032

Grete P, Vlaykov DG, Schmidt W et al (2017) Comparative statistics of selected subgrid-scale models in large-eddy simulations of decaying, supersonic magnetohydrodynamic turbulence. Phys Rev E 95(3):033206. https://doi.org/10.1103/PhysRevE.95.033206. arXiv:1703.00858 [physics.flu-dyn]

Hall JC (2008) Stellar chromospheric activity. Living Rev Sol Phys 5(1):2. https://doi.org/10.12942/lrsp-2008-2

Hanaoka Y, Sakurai T (2020) Internetwork magnetic fields seen in Fe I 1564.8 nm full-disk images. Astrophys J 904(1):63. https://doi.org/10.3847/1538-4357/abbc07. arXiv:2009.12751 [astro-ph.SR]

Hinode Review Team, Al-Janabi K Antolin P et al (2019) Achievements of Hinode in the first eleven years. Publ Astron Soc Jpn 71(5):R1. https://doi.org/10.1093/pasj/psz084

Hotta H, Kusano K (2021) Solar differential rotation reproduced with high-resolution simulation. Nat Astron 5:1100–1102. https://doi.org/10.1038/s41550-021-01459-0. arXiv:2109.06280 [astro-ph.SR]

Hotta H, Rempel M, Yokoyama T (2014) High-resolution calculations of the solar global convection with the reduced speed of sound technique. I. The structure of the convection and the magnetic field without the rotation. Astrophys J 786(1):24. https://doi.org/10.1088/0004-637X/786/1/24. arXiv:1402.5008 [astro-ph.SR]

Hotta H, Rempel M, Yokoyama T (2015) Efficient small-scale dynamo in the solar convection zone. Astrophys J 803(1):42. https://doi.org/10.1088/0004-637X/803/1/42. arXiv:1502.03846 [astro-ph.SR]

Hotta H, Rempel M, Yokoyama T (2016) Large-scale magnetic fields at high Reynolds numbers in magnetohydrodynamic simulations. Science 351(6280):1427–1430. https://doi.org/10.1126/science.aad1893. http://science.sciencemag.org/content/351/6280/1427

Howarth ID (2011) On stellar limb darkening and exoplanetary transits. Mon Not R Astron Soc 418(2):1165–1175. https://doi.org/10.1111/j.1365-2966.2011.19568.x. arXiv:1106.4659 [astro-ph.SR]

Ishikawa R, Tsuneta S (2009) Comparison of transient horizontal magnetic fields in a plage region and in the quiet Sun. Astron Astrophys 495(2):607–612. https://doi.org/10.1051/0004-6361:200810636. arXiv: 0812.1631 [astro-ph]

Ishikawa R, Tsuneta S (2010) Spatial and temporal distributions of transient horizontal magnetic fields with deep exposure. Astrophys J Lett 718(2):L171–L175. https://doi.org/10.1088/2041-8205/718/2/L171. arXiv:1103.5812 [astro-ph.SR]

Ishikawa R, Tsuneta S (2011) The relationship between vertical and horizontal magnetic fields in the quiet Sun. Astrophys J 735(2):74. https://doi.org/10.1088/0004-637X/735/2/74. arXiv:1103.5556 [astro-ph.SR]

Iskakov AB, Schekochihin AA, Cowley SC et al (2007) Numerical demonstration of fluctuation dynamo at low magnetic Prandtl numbers. Phys Rev Lett 98(20):208501. https://doi.org/10.1103/PhysRevLett.98.208501. arXiv:astro-ph/0702291 [astro-ph]

Jefferies SM, McIntosh SW, Armstrong JD et al (2006) Magnetoacoustic portals and the basal heating of the solar chromosphere. Astrophys J Lett 648(2):L151–L155. https://doi.org/10.1086/508165

Jess DB, Jafarzadeh S, Keys PH et al (2023) Waves in the lower solar atmosphere: the dawn of next-generation solar telescopes. Living Rev Sol Phys 20(1):1. https://doi.org/10.1007/s41116-022-00035-6. arXiv:2212.09788 [astro-ph.SR]

Jin CL, Wang JX (2015) Does the variation of solar intra-network horizontal field follow sunspot cycle? Astrophys J 807(1):70. https://doi.org/10.1088/0004-637X/807/1/70

Jin CL, Wang JX (2019) Magnetic flux participation in solar surface magnetism during solar cycle 24. Res Astron Astrophys 19(5):069. https://doi.org/10.1088/1674-4527/19/5/69

Jin CL, Wang JX, Song Q et al (2011) The Sun's small-scale magnetic elements in solar cycle 23. Astrophys J 731:37. https://doi.org/10.1088/0004-637X/731/1/37. arXiv:1102.3728 [astro-ph.SR]

Jofré P, Heiter U, Soubiran C (2019) Accuracy and precision of industrial stellar abundances. Annu Rev Astron Astrophys 57:571–616. https://doi.org/10.1146/annurev-astro-091918-104509. arXiv:1811.08041 [astro-ph.SR]

Judge PG, Saar SH, Carlsson M et al (2004) A comparison of the outer atmosphere of the "flat activity" star τ Ceti (G8 V) with the Sun (G2 V) and α Centauri A (G2 V). Astrophys J 609(1):392–406. https://doi.org/10.1086/421044

Kahil F, Hirzberger J, Solanki SK et al (2022) The magnetic drivers of campfires seen by the Polarimetric and Helioseismic Imager (PHI) on Solar Orbiter. Astron Astrophys 660:A143. https://doi.org/10.1051/0004-6361/202142873. arXiv:2202.13859 [astro-ph.SR]

Käpylä PJ, Käpylä MJ, Olspert N et al (2017) Convection-driven spherical shell dynamos at varying Prandtl numbers. Astron Astrophys 599:A4. https://doi.org/10.1051/0004-6361/201628973. arXiv:1605.05885 [astro-ph.SR]

Käpylä PJ, Käpylä MJ, Brandenburg A (2018) Small-scale dynamos in simulations of stratified turbulent convection. Astron Nachr 339(127):127–133. https://doi.org/10.1002/asna.201813477. arXiv:1802.09607 [astro-ph.SR]

Käpylä MJ, Rheinhardt M, Brandenburg A (2022) Compressible test-field method and its application to shear dynamos. Astrophys J 932(1):8. https://doi.org/10.3847/1538-4357/ac5b78. arXiv:2106.01107 [physics.flu-dyn]

Karak BB, Miesch M, Bekki Y (2018) Consequences of high effective Prandtl number on solar differential rotation and convective velocity. Phys Fluids 30(4):046602. https://doi.org/10.1063/1.5022034. arXiv:1801.00560 [astro-ph.SR]

Kazantsev AP (1968) Enhancement of a magnetic field by a conducting fluid. Sov Phys JETP 26:1031

Khomenko EV, Collados M, Solanki SK et al (2003) Quiet-sun inter-network magnetic fields observed in the infrared. Astron Astrophys 408:1115–1135. https://doi.org/10.1051/0004-6361:20030604

Khomenko E, Vitas N, Collados M et al (2017) Numerical simulations of quiet Sun magnetic fields seeded by the Biermann battery. Astron Astrophys 604:A66. https://doi.org/10.1051/0004-6361/201630291. arXiv:1706.06037 [astro-ph.SR]

Kianfar S, Jafarzadeh S, Mirtorabi MT et al (2018) Linear polarization features in the quiet-Sun photosphere: structure and dynamics. Sol Phys 293(8):123. https://doi.org/10.1007/s11207-018-1341-2. arXiv:1807.04633 [astro-ph.SR]

Kleeorin N, Rogachevskii I (2012) Growth rate of small-scale dynamo at low magnetic Prandtl numbers. Phys Scr 86(1):018404. https://doi.org/10.1088/0031-8949/86/01/018404. arXiv:1112.3926 [astro-ph.SR]

Kleint L, Berdyugina SV, Shapiro AI et al (2010) Solar turbulent magnetic fields: surprisingly homogeneous distribution during the solar minimum. Astron Astrophys 524:A37. https://doi.org/10.1051/0004-6361/201015285

Kochukhov O, Hackman T, Lehtinen JJ et al (2020) Hidden magnetic fields of young suns. Astron Astrophys 635:A142. https://doi.org/10.1051/0004-6361/201937185. arXiv:2002.10469 [astro-ph.SR]

Kolmogorov A (1941) The local structure of turbulence in incompressible viscous fluid for very large Reynolds' numbers. Akad Nauk SSSR Dokl 30:301–305

Korpi-Lagg MJ, Korpi-Lagg A, Olspert N et al (2022) Solar-cycle variation of quiet-Sun magnetism and surface gravity oscillation mode. Astron Astrophys 665:A141. https://doi.org/10.1051/0004-6361/202243979. arXiv:2205.04419 [astro-ph.SR]

Kraichnan RH (1968) Small-scale structure of a scalar field convected by turbulence. Phys Fluids 11(5):945–953. https://doi.org/10.1063/1.1692063

Lamb DA, DeForest CE, Hagenaar HJ et al (2008) Solar magnetic tracking. II. The apparent unipolar origin of quiet-Sun flux. Astrophys J 674(1):520–529. https://doi.org/10.1086/524372

Liebing F, Jeffers SV, Reiners A et al (2021) Convective blueshift strengths of 810 F to M solar-type stars. Astron Astrophys 654:A168. https://doi.org/10.1051/0004-6361/202039607. arXiv:2108.03859 [astro-ph.SR]

Lindborg E (2006) The energy cascade in a strongly stratified fluid. J Fluid Mech 550:207–242. https://doi.org/10.1017/S0022112005008128

Linsky JL (2017) Stellar model chromospheres and spectroscopic diagnostics. Annu Rev Astron Astrophys 55(1):159–211. https://doi.org/10.1146/annurev-astro-091916-055327

Lites BW, Leka KD, Skumanich A et al (1996) Small-scale horizontal magnetic fields in the solar photosphere. Astrophys J 460:1019. https://doi.org/10.1086/177028

Lites BW, Kubo M, Socas-Navarro H et al (2008) The horizontal magnetic flux of the quiet-sun internetwork as observed with the hinode spectro-polarimeter. Astrophys J 672:1237–1253. https://doi.org/10.1086/522922

Lites BW, Centeno R, McIntosh SW (2014) The solar cycle dependence of the weak internetwork flux. Publ Astron Soc Jpn 66:S4. https://doi.org/10.1093/pasj/psu082

Lites BW, Rempel M, Borrero JM et al (2017) Are internetwork magnetic fields in the solar photosphere horizontal or vertical? Astrophys J 835(1):14. https://doi.org/10.3847/1538-4357/835/1/14

Magic Z, Chiavassa A, Collet R et al (2015) The STAGGER-grid: a grid of 3D stellar atmosphere models. IV. Limb darkening coefficients. Astron Astrophys 573:A90. https://doi.org/10.1051/0004-6361/201423804. arXiv:1403.3487 [astro-ph.SR]

Martínez González MJ, Bellot Rubio LR (2009) Emergence of small-scale magnetic loops through the quiet solar atmosphere. Astrophys J 700:1391–1403. https://doi.org/10.1088/0004-637X/700/2/1391. arXiv:0905.2691

Martínez González MJ, Collados M, Ruiz Cobo B et al (2007) Low-lying magnetic loops in the solar internetwork. Astron Astrophys 469:L39–L42. https://doi.org/10.1051/0004-6361:20077505. arXiv:0705.1319

Martínez González MJ, Manso Sainz R, Asensio Ramos A et al (2010) Small magnetic loops connecting the quiet surface and the hot outer atmosphere of the Sun. Astrophys J Lett 714(1):L94–L97. https://doi.org/10.1088/2041-8205/714/1/L94. arXiv:1003.1255 [astro-ph.SR]

Martínez González MJ, Manso Sainz R, Asensio Ramos A et al (2012) Dead calm areas in the very quiet Sun. Astrophys J 755(2):175. https://doi.org/10.1088/0004-637X/755/2/175. arXiv:1206.4545 [astro-ph.SR]

Martínez González MJ, Pastor Yabar A, Lagg A et al (2016) Inference of magnetic fields in the very quiet Sun. Astron Astrophys 596:A5. https://doi.org/10.1051/0004-6361/201628449. arXiv:1804.10089 [astro-ph.SR]

Martínez-Sykora J, Hansteen VH, Gudiksen B et al (2019) On the origin of the magnetic energy in the quiet solar chromosphere. Astrophys J 878(1):40. https://doi.org/10.3847/1538-4357/ab1f0b

Martínez-Sykora J, de la Cruz Rodríguez J, Gošić M et al (2023) Chromospheric heating from local magnetic growth and ambipolar diffusion under nonequilibrium conditions. Astrophys J Lett 943(2):L14. https://doi.org/10.3847/2041-8213/acafe9

Meneguzzi M, Frisch U, Pouquet A (1981) Helical and nonhelical turbulent dynamos. Phys Rev Lett 47(15):1060–1064. https://doi.org/10.1103/PhysRevLett.47.1060

Meunier N (2018) Solar chromospheric emission and magnetic structures from plages to intranetwork: contribution of the very quiet Sun. Astron Astrophys 615:A87. https://doi.org/10.1051/0004-6361/201730817. arXiv:1804.00869 [astro-ph.SR]

Meunier N, Lagrange AM, Borgniet S et al (2015) Using the Sun to estimate Earth-like planet detection capabilities. VI. Simulation of granulation and supergranulation radial velocity and photometric time series. Astron Astrophys 583:A118. https://doi.org/10.1051/0004-6361/201525721

Miesch M, Matthaeus W, Brandenburg A et al (2015) Large-Eddy simulations of magnetohydrodynamic turbulence in heliophysics and astrophysics. Space Sci Rev 194:97–137. https://doi.org/10.1007/s11214-015-0190-7. arXiv:1505.01808 [astro-ph.SR]

Milić I, Faurobert M (2012) Hanle diagnostics of weak solar magnetic fields: inversion of scattering polarization in c_2 and MGH molecular lines. Astron Astrophys 547:A38. https://doi.org/10.1051/0004-6361/201219737

Mittag M, Schmitt JHMM, Schröder KP (2013) Ca II H+K fluxes from S-indices of large samples: a reliable and consistent conversion based on PHOENIX model atmospheres. Astron Astrophys 549:A117. https://doi.org/10.1051/0004-6361/201219868

Moll R, Pietarila Graham J, Pratt J et al (2011) Universality of the small-scale dynamo mechanism. Astrophys J 736(1):36. https://doi.org/10.1088/0004-637X/736/1/36. arXiv:1105.0546 [astro-ph.SR]

Molnar ME, Reardon KP, Cranmer SR et al (2023) Constraining the systematics of (acoustic) wave heating estimates in the solar chromosphere. Astrophys J 945(2):154. https://doi.org/10.3847/1538-4357/acbc75. arXiv:2302.04253 [astro-ph.SR]

Moreno-Insertis F, Martinez-Sykora J, Hansteen VH et al (2018) Small-scale magnetic flux emergence in the quiet Sun. Astrophys J Lett 859(2):L26. https://doi.org/10.3847/2041-8213/aac648. arXiv:1806.00489 [astro-ph.SR]

Morosin R, de la Cruz Rodríguez J, Vissers GJM et al (2020) Stratification of canopy magnetic fields in a plage region. Constraints from a spatially-regularized weak-field approximation method. Astron Astrophys 642:A210. https://doi.org/10.1051/0004-6361/202038754. arXiv:2006.14487 [astro-ph.SR]

Nissen PE, Gustafsson B (2018) High-precision stellar abundances of the elements: methods and applications. Astron Astrophys Rev 26(1):6. https://doi.org/10.1007/s00159-018-0111-3. arXiv:1810.06535 [astro-ph.SR]

Nordlund A, Brandenburg A, Jennings RL et al (1992) Dynamo action in stratified convection with overshoot. Astrophys J 392:647. https://doi.org/10.1086/171465

Orozco Suárez D, Bellot Rubio LR (2012) Analysis of quiet-Sun internetwork magnetic fields based on linear polarization signals. Astrophys J 751(1):2. https://doi.org/10.1088/0004-637X/751/1/2. arXiv:1203.1440 [astro-ph.SR]

Orozco Suárez D, Katsukawa Y (2012) On the distribution of quiet-Sun magnetic fields at different heliocentric angles. Astrophys J 746(2):182. https://doi.org/10.1088/0004-637X/746/2/182

Orozco Suárez D, Bellot Rubio LR, del Toro Iniesta JC, Tsuneta S et al (2007) Quiet-Sun internetwork magnetic fields from the inversion of Hinode measurements. Astrophys J Lett 670(1):L61–L64. https://doi.org/10.1086/524139. arXiv:0710.1405 [astro-ph]

Orozco Suárez D, Bellot Rubio LR, del Toro Iniesta JC et al (2008) Magnetic field emergence in quiet Sun granules. Astron Astrophys 481:L33–L36. https://doi.org/10.1051/0004-6361:20079032. arXiv:0712.2663

Ozmidov RV (1965) On the turbulent exchange in a stably stratified ocean. Izv Acad Sci USSR Atmos Ocean Phys 1:861–871

352 Springer

Panesar NK, Tiwari SK, Berghmans D et al (2021) The magnetic origin of solar campfires. Astrophys J Lett 921(1):L20. https://doi.org/10.3847/2041-8213/ac3007. arXiv:2110.06846 [astro-ph.SR]

Pekkilä J, Väisälä MS, Käpylä MJ et al (2022) Scalable communication for high-order stencil computations using cuda-aware MPI. Parallel Comput 111:102904. https://doi.org/10.1016/j.parco.2022.102904

Pepe F, Cameron AC, Latham DW et al (2013) An Earth-sized planet with an Earth-like density. Nature 503(7476):377–380. https://doi.org/10.1038/nature12768. arXiv:1310.7987 [astro-ph.EP]

Pepe F, Cristiani S, Rebolo R et al (2021) ESPRESSO at VLT. On-sky performance and first results. Astron Astrophys 645:A96. https://doi.org/10.1051/0004-6361/202038306. arXiv:2010.00316 [astro-ph.IM]

Petrovay K, Szakaly G (1993) The origin of intranetwork fields: a small-scale solar dynamo. Astron Astrophys 274:543

Pietarila Graham J, Cameron R, Schüssler M (2010) Turbulent small-scale dynamo action in solar surface simulations. Astrophys J 714(2):1606–1616. https://doi.org/10.1088/0004-637X/714/2/1606. arXiv:1002.2750 [astro-ph.SR]

Przybylski D, Cameron R, Solanki SK et al (2022) Chromospheric extension of the MURaM code. Astron Astrophys 664:A91. https://doi.org/10.1051/0004-6361/202141230. arXiv:2204.03126 [astro-ph.SR]

Quintero Noda C, Schlichenmaier R, Bellot Rubio LR, Löfdahl MG et al (2022) The European Solar Telescope. Astron Astrophys 666:A21. https://doi.org/10.1051/0004-6361/202243867. arXiv:2207.10905 [astro-ph.SR]

Quirrenbach A, Amado PJ, Caballero JA et al (2016) CARMENES: an overview six months after first light. In: Evans CJ, Simard L, Takami H (eds) Ground-based and airborne instrumentation for astronomy VI. Proc SPIE 9908, p 990812. https://doi.org/10.1117/12.2231880

Rajaguru SP, Sangeetha CR, Tripathi D (2019) Magnetic fields and the supply of low-frequency acoustic wave energy to the solar chromosphere. Astrophys J 871(2):155. https://doi.org/10.3847/1538-4357/aaf883

Ramelli R, Bianda M, Berdyugina S et al (2019) Measurement of the evolution of the magnetic field of the quiet photosphere over a solar cycle. In: Belluzzi L, Casini R, Romoli M et al (eds) Solar polariation workshop 8, p 283

Rempel M (2014) Numerical simulations of quiet Sun magnetism: on the contribution from a small-scale dynamo. Astrophys J 789(2):132. https://doi.org/10.1088/0004-637X/789/2/132. arXiv:1405.6814 [astro-ph.SR]

Rempel M (2017) Extension of the MURaM radiative MHD code for coronal simulations. Astrophys J 834(1):10. https://doi.org/10.3847/1538-4357/834/1/10. arXiv:1609.09818 [astro-ph.SR]

Rempel M (2018) Small-scale dynamo simulations: magnetic field amplification in exploding granules and the role of deep and shallow recirculation. Astrophys J 859(2):161. https://doi.org/10.3847/1538-4357/aabba0. arXiv:1805.08390 [astro-ph.SR]

Rempel M (2020) On the contribution of quiet-Sun magnetism to solar irradiance variations: constraints on quiet-Sun variability and grand-minimum scenarios. Astrophys J 894(2):140. https://doi.org/10.3847/1538-4357/ab8633. arXiv:2004.01795 [astro-ph.SR]

Requerey IS, Del Toro Iniesta JC, Bellot Rubio LR et al (2017) Convectively driven sinks and magnetic fields in the quiet-Sun. Astrophys J Suppl Ser 229(1):14. https://doi.org/10.3847/1538-4365/229/1/14. arXiv:1610.07622 [astro-ph.SR]

Requerey IS, Cobo BR, Gošić M et al (2018) Persistent magnetic vortex flow at a supergranular vertex. Astron Astrophys 610:A84. https://doi.org/10.1051/0004-6361/201731842. arXiv:1712.01510 [astro-ph.SR]

Ribas I, Tuomi M, Reiners A et al (2018) A candidate super-Earth planet orbiting near the snow line of Barnard's star. Nature 563(7731):365–368. https://doi.org/10.1038/s41586-018-0677-y. arXiv:1811.05955 [astro-ph.EP]

Riley JJ, Lindborg E (2012) Recent progress in stratified turbulence. Cambridge University Press, Cambridge. https://doi.org/10.1017/CBO9781139032810.008

Rimmele TR, Warner M, Keil SL et al (2020) The Daniel K. Inouye Solar Telescope – observatory overview. Sol Phys 295(12):172. https://doi.org/10.1007/s11207-020-01736-7

Rincon F (2019) Dynamo theories. J Plasma Phys 85(4):205850401. https://doi.org/10.1017/S0022377819000539. arXiv:1903.07829 [physics.plasm-ph]

Sahoo G, Perlekar P, Pandit R (2011) Systematics of the magnetic-Prandtl-number dependence of homogeneous, isotropic magnetohydrodynamic turbulence. New J Phys 13(1):013036. https://doi.org/10.1088/1367-2630/13/1/013036. arXiv:1006.5585 [nlin.CD]

Salhab RG, Steiner O, Berdyugina SV et al (2018) Simulation of the small-scale magnetism in main-sequence stellar atmospheres. Astron Astrophys 614:A78. https://doi.org/10.1051/0004-6361/201731945

Schekochihin AA, Iskakov AB, Cowley SC et al (2007) Fluctuation dynamo and turbulent induction at low magnetic Prandtl numbers. New J Phys 9(8):300. https://doi.org/10.1088/1367-2630/9/8/300. arXiv:0704.2002 [physics.flu-dyn]

Schober J, Schleicher D, Bovino S et al (2012) Small-scale dynamo at low magnetic Prandtl numbers. Phys Rev E 86(6):066412. https://doi.org/10.1103/PhysRevE.86.066412. arXiv:1212.5979 [astro-ph.CO]

Schrijver CJ (1987) Magnetic structure in cool stars. XI. Relations between radiative fluxes mesuring stellar activity, and evidence for two components in stellar chromospheres. Astron Astrophys 172:111–123

Schrijver CJ, Harvey KL (1994) The photospheric magnetic flux budget. Sol Phys 150:1–18. https://doi.org/10.1007/BF00712873

Schröder KP, Mittag M, Pérez Martínez MI et al (2012) Basal chromospheric flux and Maunder minimum-type stars: the quiet-Sun chromosphere as a universal phenomenon. Astron Astrophys 540:A130. https://doi.org/10.1051/0004-6361/201118363. arXiv:1202.3314 [astro-ph.SR]

Schumacher J, Sreenivasan KR (2020) Colloquium: unusual dynamics of convection in the Sun. Rev Mod Phys 92(4):041001. https://doi.org/10.1103/RevModPhys.92.041001

Schüssler M, Vögler A (2008) Strong horizontal photospheric magnetic field in a surface dynamo simulation. Astron Astrophys 481:L5–L8. https://doi.org/10.1051/0004-6361:20078998. arXiv:0801.1250

Semel M (1989) Zeeman-Doppler imaging of active stars. I – basic principles. Astron Astrophys 225:456–466

Shchukina N, Trujillo Bueno J (2011) Determining the magnetization of the quiet Sun photosphere from the Hanle effect and surface dynamo simulations. Astrophys J Lett 731(1):L21. https://doi.org/10.1088/2041-8205/731/1/L21. arXiv:1103.5652 [astro-ph.SR]

Shchukina N, Trujillo Bueno J (2015) The impact of surface dynamo magnetic fields on the solar iron abundance. Astron Astrophys 579:A112. https://doi.org/10.1051/0004-6361/201425569

She ZS, Jackson E (1993) On the universal form of energy spectra in fully developed tur bulence. Phys Fluids A 5(7):1526–1528. https://doi.org/10.1063/1.858591

Shporer A, Brown T (2011) The impact of the convective blueshift effect on spectroscopic planetary transits. Astrophys J 733(1):30. https://doi.org/10.1088/0004-637X/733/1/30. arXiv:1103.0775 [astro-ph.EP]

Skoutnev V, Squire J, Bhattacharjee A (2021) Small-scale dynamo in stably stratified turbulence. Astrophys J 906(1):61. https://doi.org/10.3847/1538-4357/abc8ee. arXiv:2008.01025 [physics.flu-dyn]

Smagorinsky J (1963) General circulation experiments with the primitive equations. Mon Weather Rev 91:99. https://doi.org/10.1175/1520-0493(1963)091<0099:GCEWTP>2.3.CO;2

Smitha HN, Anusha LS, Solanki SK et al (2017) Estimation of the magnetic flux emergence rate in the quiet Sun from sunrise data. Astrophys J Suppl Ser 229:17. https://doi.org/10.3847/1538-4365/229/1/17. arXiv:1611.06432 [astro-ph.SR]

Spruit HC (1999) Differential rotation and magnetic fields in stellar interiors. Astron Astrophys 349:189–202. arXiv:astro-ph/9907138 [astro-ph]

Stein RF, Nordlund Å (1998) Simulations of solar granulation. I. General properties. Astrophys J 499:914–933. https://doi.org/10.1086/305678

Stein RF, Nordlund Å (2006) Solar small-scale magnetoconvection. Astrophys J 642(2):1246–1255. https://doi.org/10.1086/501445

Stein RF, Bercik D, Nordlund Å (2003) Solar surface magneto-convection. In: Pevtsov AA, Uitenbroek H (eds) Current theoretical models and future high resolution solar observations: preparing for ATST, p 121. arXiv:astro-ph/0209470

Steiner O, Rezaei R, Schaffenberger W et al (2008) The horizontal internetwork magnetic field: numerical simulations in comparison to observations with Hinode. Astrophys J Lett 680(1):L85. https://doi.org/10.1086/589740. arXiv:0801.4915 [astro-ph]

Steiner O, Salhab R, Freytag B et al (2014) Properties of small-scale magnetism of stellar atmospheres. Publ Astron Soc Jpn 66:S5. https://doi.org/10.1093/pasj/psu083

Subramanian K (1998) Can the turbulent galactic dynamo generate large-scale magnetic fields? Mon Not R Astron Soc 294:718–728. https://doi.org/10.1046/j.1365-8711.1998.01284.x arXiv:astro-ph/9707280 [astro-ph]

Tilgner A, Brandenburg A (2008) A growing dynamo from a saturated Roberts flow dynamo. Mon Not R Astron Soc 391(3):1477–1481. https://doi.org/10.1111/j.1365-2966.2008.14006.x. arXiv:0808.2141 [astro-ph]

Trelles Arjona JC, Martínez González MJ Ruiz Cobo B (2021) Mapping the hidden magnetic field of the quiet Sun. Astrophys J Lett 915(1):L20. https://doi.org/10.3847/2041-8213/ac0af2. arXiv:2106.10546 [astro-ph.SR]

Trujillo Bueno J, del Pino Alemán T (2022) Magnetic field diagnostics in the solar upper atmosphere. Annu Rev Astron Astrophys 60:415–453. https://doi.org/10.1146/annurev-astro-041122-031043

Trujillo Bueno J, Shchukina N, Asensio Ramos A (2004) A substantial amount of hidden magnetic energy in the quiet sun. Nature 430:326–329. https://doi.org/10.1038/nature02669. arXiv:astro-ph/0409004

Vincenzi D (2001) The Kraichnan-Kazantsev dynamo. arXiv e-prints physics/0106090. arXiv:physics/0106090 [physics.flu-dyn]

Vitense E (1953) Die Wasserstoffkonvektionszone der Sonne. Z Astrophys 32:135

Vögler A, Schüssler M (2007) A solar surface dynamo. Astron Astrophys 465:L43–L46. https://doi.org/10.1051/0004-6361:20077253. arXiv:astro-ph/0702681

Wang J, Wang H, Tang F et al (1995) Flux distribution of solar intranetwork magnetic fields. Sol Phys 160:277–288. https://doi.org/10.1007/BF00732808

Warnecke J, Korpi-Lagg M, Gent F et al (2023) Numerical evidence for a small-scale dynamo approaching solar magnetic Prandtl numbers. Nat Astron (accepted). https://doi.org/10.21203/rs.3.rs-1819381/v1

Withbroe GL, Noyes RW (1977) Mass and energy flow in the solar chromosphere and corona. Annu Rev Astron Astrophys 15:363–387. https://doi.org/10.1146/annurev.aa.15.090177.002051

Wright E, Przybylski D, Rempel M et al (2021) Refactoring the MPS/University of Chicago Radiative MHD (MURaM) model for GPU/CPU performance portability using OpenACC directives. In: Proceedings of the platform for advanced scientific computing conference, PASC '21. Association for Computing Machinery, New York. https://doi.org/10.1145/3468267.3470576

Yelles Chaouche L, Moreno-Insertis F, Martínez Pillet V et al (2011) Mesogranulation and the solar surface magnetic field distribution. Astrophys J Lett 727(2):L30. https://doi.org/10.1088/2041-8205/727/2/L30. arXiv:1012.4481 [astro-ph.SR]

Yeo KL, Solanki SK, Krivova NA et al (2020) The dimmest state of the Sun. Geophys Res Lett 47(19):e90243. https://doi.org/10.1029/2020GL090243. arXiv:2102.09487 [astro-ph.SR]

Publisher's Note Springer Nature remains neutral with regard to jurisdictional claims in published maps and institutional affiliations.

Authors and Affiliations

Matthias Rempel[1]  · **Tanayveer Bhatia[2]**  · **Luis Bellot Rubio[3]**  · **Maarit J. Korpi-Lagg[2,4,5]**

✉ M.J. Korpi-Lagg
   maarit.korpi-lagg@aalto.fi

   M. Rempel
   rempel@ucar.edu

   T. Bhatia
   bhatia@mps.mpg.de

   L. Bellot Rubio
   lbellot@iaa.es

[1]   High Altitude Observatory, National Center for Atmospheric Research, P.O. Box 3000, Boulder, 80307, CO, USA

[2]   Department of Sun and Heliosphere, Max Planck Institute for Solar System Research, Justus-von-Liebig-Weg 3, Göttingen, 37077, Germany

[3]   Instituto de Astrofísica de Andalucía, CSIC, Glorieta de la Astronomía s/n, Granada, 18008, Spain

[4]   Department of Computer Science, Aalto University, Konemiehentie 2, Espoo, 00076, Finland

[5]   Nordic Institute for Theoretical Physics, KTH Royal Institute of Technology and Stockholm University, Hannes Alvéns väg 12, Stockholm, 10691, Sweden

Space Science Reviews (2023) 219:55
https://doi.org/10.1007/s11214-023-00999-3

Turbulent Processes and Mean-Field Dynamo

Axel Brandenburg[1,2,3,4] · Detlef Elstner[5] · Youhei Masada[6] · Valery Pipin[7]

Received: 26 March 2023 / Accepted: 1 September 2023 / Published online: 28 September 2023
© The Author(s) 2023

Abstract
Mean-field dynamo theory has important applications in solar physics and galactic magnetism. We discuss some of the many turbulence effects relevant to the generation of large-scale magnetic fields in the solar convection zone. The mean-field description is then used to illustrate the physics of the α effect, turbulent pumping, turbulent magnetic diffusivity, and other effects on a modern solar dynamo model. We also discuss how turbulence transport coefficients are derived from local simulations of convection and then used in mean-field models.

Keywords Large-scale dynamos · Turbulence · Stellar magnetism · Magnetic helicity

1 Introduction

The problem of solar and stellar dynamos is still an open one. In spite of tremendous progress over recent decades, we still do not understand with any degree of certainty the reason behind the equatorward migration of solar activity belts, the dependence of cycle frequency on rotation frequency, or the level of magnetic activity.[1] All models of solar and stellar magnetism rely on some assumptions. Even the most realistic simulations suffer from finite resolution and the compromises in the physics that are made. The crucial question is then, when and where we are allowed to make compromises and when not. Among those approximations is the second-order correlation approximation (SOCA), also known as the first-order smoothing approximation. These are nowadays either replaced by other approximations or by numerical techniques such as the test-field method (TFM), as will be explained later in this review.

The Sun's magnetic field exhibits a clear mean field with spatio-temporal order: antisymmetry of radial and toroidal fields about the equator and the 11-yr cycle. This mean field can well be described by an azimuthal average. The radial component of such an azimuthally averaged mean field has a typical strength of ± 10 G. This is not much compared with the

---

[1] The reader is referred to the review of Hazra et al. (2023) for a discussion of flux transport dynamos to explain some of the outstanding questions of large-scale dynamos in the Sun and stars. We comment on the main differences between the proposed models in Sects. 8.2 and 8.4 below.

Solar and Stellar Dynamos: A New Era
Edited by Manfred Schüssler, Robert H. Cameron, Paul Charbonneau, Mausumi Dikpati, Hideyuki Hotta and Leonid Kitchatinov

Extended author information available on the last page of the article

peak strength of ± 2 kG in sunspots, but much of this is "lost" in the process of averaging. Of course, whatever is lost corresponds to fluctuations, which actually play crucial parts and correlations between different fluctuations lead to various mean-field effects.

Mathematically, once an averaging procedure has been defined, we have the mean field $\overline{B}$, indicated by an overbar. Then, the difference between the actual and the mean field, B and $\overline{B}$, gives the fluctuating field as $b \equiv B - \overline{B}$. The same procedure also applies to all other quantities. This formal distinction between mean and fluctuating fields, which are sometimes also called large-scale and small-scale fields, is important in discussions with observers. Coronal mass ejections, for example, are superficially reported as being part of a large-scale field, but this may not be true anymore when we think of averaging over the full solar circumference. Thus, paradoxically, even if something is large by some standards, it may not qualify as large-scale under this formal definition of an azimuthal average.

Azimuthal averaging is not always a good recipe. Some stars have nonaxisymmetric magnetic fields, and even the Sun is believed to have what is known as active longitudes – a weak nonaxisymmetric magnetic field on top of a predominantly axisymmetric one. Those nonaxisymmetric fields might best be described through low-order Fourier mode filtering. This is probably completely fine, but slightly problematic at the formal level, because then the average of the product of mean and fluctuating fields is no longer vanishing, as it would be in the case of an azimuthal average. This mathematical property is one of several rules that are called the Reynolds rules. However, as alluded to above, the violation of this particular Reynolds rule is probably just a technicality that makes mean-field predictions less accurate. We refer here to the work of Zhou et al. (2018) for a detailed investigation. There are a number of other limitations in mean-field theories that will be discussed below.

The purpose of defining mean fields is twofold. On the one hand, they allow us to quantify large-scale magnetic, velocity, and other fields that are observed or that are present in a simulation. On the other hand, they allow us to develop predictive theories for these averages. In these theories, mean fields can sometimes emerge because of instabilities and/or because of suitable boundary conditions. This is possible because of certain mean-field effects, by which one usually means the relations between correlations of fluctuations and various mean fields. Discussing those effects is an important purpose of this review. The ultimate goal of mean-field dynamo theory is to understand and model the Sun and other stars. We therefore also discuss in this review the status of such attempts. For a *basic* introduction to mean-field theory, which is not the subject of this review, we refer to standard textbooks (Moffatt 1978; Krause and Rädler 1980; Zeldovich et al. 1983) and other reviews (Brandenburg and Subramanian 2005a; Kulsrud and Zweibel 2008; Miesch and Toomre 2009; Charbonneau 2014, 2020; Brandenburg 2018a; Tobias 2021; Brandenburg and Ntormousi 2023; Karak 2023).

2 The Golden Years of Dynamo Theory

The first mean-field model was constructed by Parker (1955). In his model, the toroidal magnetic field is generated from the dipole field by nonuniform rotation. To overcome the restrictions of Cowling's theorem (Cowling 1933), Parker suggested that the dipole magnetic field can be regenerated by cyclonic convective motions which transform emerging toroidal magnetic loops into poloidal magnetic field. The coalescing loops can amplify the dipole magnetic field. Studying the combined action of differential rotation and cyclonic motions, he found a solution in the form of a dynamo wave and formulated conditions for the equatorward propagation of dynamo waves. Steenbeck et al. (1966) and Steenbeck and Krause (1969) constructed the theoretical basis of mean-field theory, introduced the notion of the mean electromotive force (MEMF) of the turbulence and showed that the Parker effect of the cyclonic convective motions is equivalent to the effective MEMF along the large-scale field. The 1970s can be considered the golden years of mean-field dynamo theory. Back then, Schüssler (1983) stated: "dynamo theory reached the textbook state", mentioning the famous monographs by Moffatt (1978), Parker (1979), Krause and Rädler (1980), Vainshtein et al. (1980), and Zeldovich et al. (1983).

Indeed, intensive theoretical and observational studies led to the establishment of the basic solar dynamo scenario, identification of key dynamo parameters, and the formulation of a general paradigm of the nature of solar and stellar magnetism. Schüssler (1983) summarized that mean-field dynamo models can reproduce the "physics of solar activity to a great extent" including:

- Hale's polarity rule of sunspots groups,
- the time-latitude evolution of sunspot activity ("butterfly diagram"),
- reversals of the polar magnetic field,
- the phase relationship between the evolution of poloidal and toroidal magnetic fields and their consistence with the observed butterfly diagrams (Stix 1976),
- rigid rotation of the magnetic sector structure and coronal holes (Stix 1974, 1977),
- chaotic variations of dynamo activity either due to a random α effect or the dynamo nonlinearity from the Lorentz force (Leighton 1969; Yoshimura 1978; Ruzmaikin 1981), and
- a quantitative understanding of the solar torsional oscillations (Schüssler 1981; Yoshimura 1981).

We note that the first and second items are based on the assumption that sunspot groups are formed from the large-scale toroidal magnetic field. Already at the time it was recognized that the mean-field models need to take into account the fibril state of the magnetic field which we observed at the solar surface. We return to this point later.

Classical mean-field dynamo models utilize the $\alpha\Omega$ scenario using the differential rotation (Ω effect) as the source of the toroidal magnetic flux production and the α effect for the poloidal magnetic field generation. Since the seminal work of Pouquet et al. (1976), it started to become clear that the magnetic helicity results in an important nonlinear contribution to the α effect and turbulent magnetic field generation (Kleeorin and Ruzmaikin 1982).

3 Mean-Field Theory and Avoiding Some of Its Limitations

We can never expect a mean-field theory to produce an accurate representation of reality. One reason is the fact that the underlying turbulence has stochastic aspects, so each realization with slightly different initial conditions would result in a somewhat different outcome. However, there could be other reasons for discrepancies that we discuss next. Some of those discrepancies can nowadays be avoided.

3.1 Mean-Field Electrodynamics

In mean-field theory, one derives evolution equations for the averaged fields, namely the mean magnetic field $\overline{B}$, the mean velocity $\overline{U}$, and the mean thermodynamic variables such as mean specific entropy $\overline{S}$ and the mean density $\overline{\rho}$. Often, one neglects the evolution of $\overline{U}$, $\overline{S}$, and $\overline{\rho}$, which is then already an important limitation.

If one focuses on the evolution of the mean magnetic field only, one often talks about the mean-fields electrodynamics or quasi-kinematic mean-field theory, which can still be nonlinear if the various mean-field transport coefficients depend on the mean fields. If they are unaffected, one talks about kinematic mean-field theory, which is linear. Of course, once there is a dynamo, we have an exponentially growing solution, so the magnetic field would grow without limit, i.e., it would not saturate within kinematic mean-field theory. Obviously, a correct mean-field theory must be nonlinear, but even within the realm of linear theory, there are important lessons to be learnt. Below, we discuss the aspects of nonlocality, which were often omitted out of ignorance, but nowadays we know that this is often not possible and this restriction can easily be relaxed.

3.2 Nonlocality

The mean magnetic field is governed by the mean induction equation, which is sometimes also referred to as the mean-field dynamo equation. The most important term here is the mean electromotive force,

$$\overline{\mathcal{E}} = \overline{\boldsymbol{u} \times \boldsymbol{b}}, \tag{1}$$

i.e., the averaged cross product of velocity and magnetic fluctuations. In mean-field electrodynamics, it is often expressed as

$$\overline{\mathcal{E}}_i = \overline{\mathcal{E}}_{0i} + \alpha_{ij}\overline{B}_j + \eta_{ijk}\partial\overline{B}_j/\partial x_k + \cdots, \tag{2}$$

where the ellipsis denotes higher derivative terms, of which there should be infinitely many, and there should also be time derivatives. The term $\overline{\mathcal{E}}_{0i}$ is a contribution that can exist already in the absence of a mean field; see Brandenburg and Rädler (2013) for details and numerical experiments. Including only a finite number of derivatives in Eq. (2) and ignoring time derivatives is another important approximation. In fact, it is usually easier to express $\overline{\mathcal{E}}$ as a convolution between an integral kernel and the mean field. Furthermore, it is instructive to split the integral kernel into two pieces and write

$$\overline{\mathcal{E}}_i = \overline{\mathcal{E}}_{0i} + \hat{\alpha}_{ij} * \overline{B}_j + \hat{\eta}_{ijk} * \partial\overline{B}_j/\partial x_k, \tag{3}$$

where the asterisks mean a convolution in space and time, and the hats denote integration kernels. In principle, the spatial derivative can be absorbed as being part of the integral kernel, but separating the kernel into $\hat{\alpha}_{ij}$ and $\hat{\eta}_{ijk}$ has conceptual advantages, because they preserve the similarity to Eq. (2). Note also that, unlike Eq. (2), where we allowed for arbitrarily many derivatives, here, we have no other terms, because all even derivatives are already absorbed in $\hat{\alpha}_{ij}$ and all odd derivatives are absorbed in $\hat{\eta}_{ijk}$. Time derivatives can also be absorbed in both of them if the convolution with the kernels is also over time.

For the benefit of better interpretation, both α_{ij} and η_{ijk} (and analogously also for $\hat{\alpha}_{ij}$ and $\hat{\eta}_{ijk}$) can be broken down into further pieces. The α_{ij} tensor can be split into a symmetric and an antisymmetric tensor. The latter is characterized by a vector, $\gamma_i = -\frac{1}{2}\epsilon_{ijk}\alpha_{jk}$, which corresponds to a pumping velocity. Having in mind that the magnetic gradient tensor can also be split into symmetric and antisymmetric parts, where the latter is the mean current density, $\overline{\boldsymbol{J}}$, with $\overline{J}_i = -\frac{1}{2}\epsilon_{ijk}\partial\overline{B}_j/\partial x_k$, we can separate the rank-3 tensor, η_{ijk}, into a rank-2 tensor operating only on $\overline{\boldsymbol{J}}$ and the rest operating on the symmetric part of $\partial\overline{B}_j/\partial x_k$.

The convolution can only be replaced by a multiplication, as in Eq. (2), if the mean field is constant in time (which is normally never the case!) and if it varies at most linearly in space (which is normally also not the case). We return to this point further below.

3.3 Beyond SOCA and Scale Separation Limits

An important question concerns the calculation of the α_{ij} and η_{ijk} coefficients or kernels. A problem arises from the fact that the differential equations for these expressions are non-linear and therefore hard to solve analytically. A commonly used approximation is SOCA. It neglects triple (and higher) correlations in the evolution equations for the fluctuating velocity and magnetic fields. This closure can be applied when either the magnetic Reynolds number, Re_M, or the Strouhal number, St, are much smaller than unity. These limits are rather restrictive for astrophysical conditions. For example, the convection zones (CZs) of the Sun and stars are in a turbulent state with huge values of the fluid Reynolds number ($Re \gtrsim 10^{12}$), the magnetic Reynolds number ($Re_M \gtrsim 10^8$), Rayleigh number ($Ra \gtrsim 10^{20}$), and an extremely low Prandtl number ($Pr \sim 10^{-4}$–10^{-7}); see, e.g., Ossendrijver (2003).

Results of Schrinner et al. (2005) showed that SOCA does not work well when Re_M exceeds unity and St is not small. There are analytical approaches, e.g., different variants of the so-called τ approximation (Kleeorin et al. 1996; Field and Blackman 2002; Brandenburg and Subramanian 2005a), which can be applied in the high Reynolds number limit. The restrictions inherent to SOCA or the τ approximation no longer apply when calculating solutions of the underlying differential equations numerically. This is done in the TFM (Schrinner et al. 2005, 2007).

Using a set of mean magnetic fields, the TFM allows one to determine the turbulent transport coefficients for arbitrary velocity fields, provided they can be computed or otherwise represented on the computer. The velocity field can be determined either as a solution of the nonlinear Navier-Stokes equations for a forced turbulent flow or it can be obtained as a results of global convective dynamo (GCD) simulations. To compute $\overline{\mathcal{E}}$, the solution for the induction equation for the fluctuating magnetic field is needed as well. The original TFM of Schrinner et al. (2005, 2007) adopts the scale-separation assumption. It was shown that the TFM describes the dynamo processes for GCD simulations at moderate Reynolds numbers of around 50 rather well (Schrinner 2011; Warnecke et al. 2018; Viviani et al. 2018). The calculations within TFM have some technical restrictions and are currently unable to meet the very high astrophysical limits of $Re, Re_M > 10^6$. Nevertheless, the current applications of the TFM concern cases with $Re, Re_M \gg 1$, which is well beyond the SOCA limits. In recent developments of the TFM, Käpylä et al. (2022) took compressibility effects into account. They also studied the effects of the small-scale dynamo on the turbulent electromotive force (see also Rempel et al. 2023).

An alternative way of extracting the coefficients of the mean-electromotive force employs a multi-dimensional regression method (Brandenburg and Sokoloff 2002; Racine et al. 2011; Augustson et al. 2015; Simard et al. 2016). In this approach, instead of solving the equations for the fluctuations in the presence of different mean magnetic fields, as it is done in the TFM, the regression methods try to exploit the form of Eq. (2) for the dynamo-generated large-scale magnetic field. Detailed comparisons of the above method with TFM were done by Warnecke et al. (2018). It was found that the TFM gives a more accurate representation of the mean-field coefficients than the multidimensional regression method. We encourage the reader to consult this paper for further details. We return to the problem of extracting turbulent transport coefficients from GCDs in Sect. 7.

The limitations discussed so far are in principle all avoidable: (i) The evolution equations for $\overline{U}$, $\overline{S}$, and $\overline{\rho}$ can be (and have been) included (Brandenburg et al. 1992), in addition to that for $\overline{B}$, but in practice, even this is still an approximation in the sense that the full set of coefficients for the equations is not (or only approximately) known. (ii) The electromotive force can be (and has been) expressed as a convolution, which can most effectively be

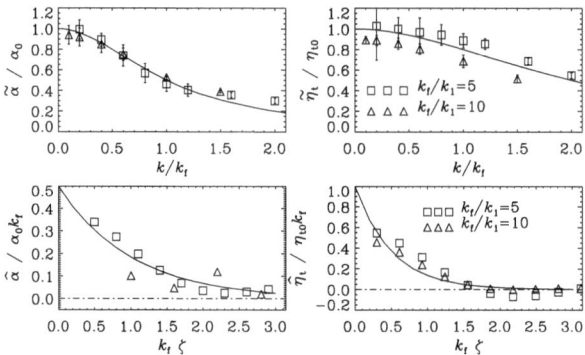

Fig. 1 Top: Dependences of the normalized $\tilde{\alpha}$ and $\tilde{\eta}_t$ on the normalized wavenumber k/k_f for isotropic turbulence forced at wavenumbers $k_f/k_1 = 5$ with $\mathrm{Re}_M = 10$ (squares) and $k_f/k_1 = 10$ with $\mathrm{Re}_M = 3.5$ (triangles), all with $\nu/\eta = 1$, using data from Brandenburg et al. (2008). The solid lines give the Lorentzian fits (4). Bottom: Normalized integral kernels $\hat{\alpha}$ and $\hat{\eta}_t$ versus $k_f\zeta$ for isotropic turbulence forced at wavenumbers $k_f/k_1 = 5$ with $\mathrm{Re}_M = 10$ (squares) and $k_f/k_1 = 10$ with $\mathrm{Re}_M = 3.5$ (triangles), all with $\nu/\eta = 1$. The solid lines are defined by (5). Adapted from Brandenburg et al. (2008)

solved by rewriting the equations as a differential equation, as will be described below. (iii) Numerical solutions can be employed to find specific values for α_{ij} and η_{ijk}; see Warnecke et al. (2018, 2021) for doing this for solar simulations using the TFM. It often turns out that analytical closure techniques are very useful as a first orientation and they are often also accurate enough for a qualitatively useful model. In special cases, when a more accurate solution is required, the answer may well be obtained numerically using the TFM. The problem is then only that numerical solutions themselves are limited in just the same way as those for a full numerical solution in the solar and stellar dynamo problems.

Figure 1 shows results for $\tilde{\alpha}(k)$ and $\tilde{\eta}_t(k)$ with $\nu/\eta = 1$. Both $\tilde{\alpha}$ and $\tilde{\eta}_t$ decrease monotonously with increasing $|k|$. The functions $\tilde{\alpha}(k)$ and $\tilde{\eta}_t(k)$ are well represented by Lorentzian fits of the form

$$\tilde{\alpha}(k) \approx \frac{\alpha_0}{1 + (k/k_f)^2}, \quad \tilde{\eta}_t(k) \approx \frac{\eta_{t0}}{1 + (k/2k_f)^2}. \tag{4}$$

Also shown in the lower part of Fig. 1 are the kernels $\hat{\alpha}(\zeta)$ and $\hat{\eta}_t(\zeta)$, obtained through Fourier transforms of the Lorentzian fits,

$$\hat{\alpha}(\zeta) \approx \tfrac{1}{2}\alpha_0 k_f \exp(-k_f|\zeta|), \quad \hat{\eta}_t(\zeta) \approx \eta_{t0} k_f \exp(-2k_f|\zeta|). \tag{5}$$

We see that the profile of $\hat{\eta}_t$ is half as wide as that of $\hat{\alpha}$, but it is not known whether this is a general property. It is important to realize that the suggested mean-field modifications employing the Lorentzian forms of the integral convolution kernels are based on empirical results. Nevertheless, they are much more accurate than the approximation of replacing the kernels by δ functions, which is done in conventional approaches.

Under suitable conditions, the accuracy of the TFM can be so high that discrepancies become apparent that are solely the result of having made unjustified approximations in the comparison. An example is the memory effect. Comparing the growth rate for a supercritical dynamo with that obtained theoretically from the coefficient obtained from the TFM can give noticeable discrepancies if the memory effect is neglected; see Fig. 1 of Hubbard and

Brandenburg (2009). The combined Fourier transformed integral kernel is of the form

$$\tilde{\alpha}(k, \omega) \approx \frac{\alpha_0}{1 + (k/k_f)^2 - i\omega\tau} , \quad \tilde{\eta}_t(k, \omega) \approx \frac{\eta_{t0}}{1 + (k/2k_f)^2 - i\omega\tau} , \tag{6}$$

where τ is well approximated by the turbulent turnover time. Even for stationary flows, the memory effect can be dramatically important (Rädler et al. 2011).

In practice, it is cumbersome to solve the integral equation in time. However, as alluded to above, it is possible to approximate this integral equation by a differential equation for $\overline{\mathcal{E}}$ with respect to space and time t of the form

$$\left(1 + \tau \frac{\partial}{\partial t} - \ell^2 \nabla^2\right) \overline{\mathcal{E}}_i = \alpha_{ij} \overline{B}_j + \eta_{ijk} \partial \overline{B}_j / \partial x_k. \tag{7}$$

This has been done in several papers (Rheinhardt and Brandenburg 2012; Rheinhardt et al. 2014; Brandenburg and Chatterjee 2018; Pipin 2023). We return to this in Sect. 5.3.

3.4 The Use of Mean-Field Theory

If mean-field theory cannot reliably be applied to a regime outside that of the direct numerical simulations (DNS), one must ask what is then the use of mean-field theory. The answer lies in the fact that mean-field theory provides us with an excellent diagnostic "tool" for approaching the problem. Particular features of a solution can usually be attributed to particular terms in the mean-field equations. This would then allow us a more informed answer by saying that the main dynamo mechanism is, for example, of $\alpha\Omega$ type, or of a specific type of a shear flow dynamo, for example. *Thus, mean-field theory may be regarded as a convenient tool for understanding what is going on rather than predicting what might be going on.*

4 The Catastrophic Quenching Problem

Since the 1990s, a problem emerged in that numerical dynamo solutions were found to depend on the value of the microphysical magnetic diffusivity. Typically, the strength of the mean-fields then decreases with increasing magnetic Reynolds number. This is unusual and does not have any correspondence with ordinary hydrodynamics where the large-scale dynamics is usually already captured at moderate fluid Reynolds numbers. In its original form, the catastrophic quenching problem refers to the finding that the volume-averaged electromotive force scales with the microphysical magnetic diffusivity, and thus goes to zero when $\eta \to 0$. To some extent, this is a problem related to the use of periodic boundary conditions. However, even for astrophysically more realistic boundary conditions, numerical simulations reveal that there is still a problem. Plasma relaxation experiments have identified the role of magnetic helicity as the main culprit in causing η-dependent large-scale dynamics and catastrophic quenching. We therefore begin by briefly reviewing the essential findings.

4.1 Lessons from Plasma Relaxation Experiments

The magnetic field is divergence-free and can therefore be written as $\boldsymbol{B} = \nabla \times \boldsymbol{A}$, where $\boldsymbol{A}$ is the magnetic vector potential. The magnetic helicity density is defined as $\boldsymbol{A} \cdot \boldsymbol{B}$. Its evolution equation follows directly from the uncurled induction equation, $\partial A/\partial t = -E - \nabla\Phi$, or, using Ohm's law, $-E = U \times B - \eta\mu_0 J$, so

$$\frac{\partial A}{\partial t} = U \times B - \eta\mu_0 J - \nabla\Phi. \tag{8}$$

It yields the evolution equation for the magnetic helicity density,

$$\frac{\partial}{\partial t}(A \cdot B) = -2\eta\mu_0 J \cdot B - \nabla \cdot (E \times A + \Phi B). \tag{9}$$

It must here be emphasized that there is an important difference to the equation for the magnetic energy density,

$$\frac{\partial}{\partial t}(B^2/2\mu_0) = -U \cdot (J \times B) - 2\eta\mu_0 J^2 - \nabla \cdot (E \times B/\mu_0). \tag{10}$$

While both Eqs. (9) and (10) have analogous terms such as dissipation $\propto J \cdot B$ versus $\propto J^2$, respectively, and flux terms $E \times A$ versus Poynting vector $E \times B/\mu_0$, respectively, there is the work against the Lorentz force, $-U \cdot (J \times B)$ in Eq. (10), which would be $U \cdot (B \times B)$ in Eq. (9), but it obviously vanishes. In statistical equilibrium, $\langle 2\eta\mu_0 J^2 \rangle$ must balance $-\langle U \cdot (J \times B)\rangle$, which implies that the current density diverges like $|J| \sim \eta^{-1/2}$. By contrast, no magnetic helicity is being produced, and also its dissipation converges to zero like $\propto \eta|J \cdot B| \to \eta^{1/2}$ as $\eta \to 0$.

Already since the 1970s, we know of the conjecture of J. B. Taylor (1974, 1986) that the magnetic field relaxes under the constraint of magnetic helicity conservation to a nearly force-free field to minimize dissipation. The approximate conservation of magnetic helicity has been verified experimentally in plasma relaxation experiments; see, e.g., Ji et al. (1995). There are obviously some differences between the solar convection zone and laboratory plasmas, for example, the role of the electron pressure in the generalized Ohm's law could play an important role in explaining why magnetic helicity changes are observed to be faster in plasma experiments than what is predicted by Ohm's law (Ji et al. 1995). Furthermore, the α effect has been identified as the main agent for converting magnetic helicity from the turbulent field to the mean field (Ji 1999).

In the context of plasma relaxation experiments, it is useful to distinguish between electromagnetic and electrostatic turbulence. This distinction refers to the curl-free and divergence-free parts of the electric field written as $E = -\nabla\Phi - \partial A/\partial t$. In plasma relaxation experiments, turbulence is mostly electrostatic. It can be affected by the electron pressure gradient $(en_e)^{-1}\nabla p_e$ in the generalized Ohm's law, where e is the unit charge, and n_e and p_e are the electron density and electron pressure, respectively; see Ji (1999) for details. This leads to a battery term $\propto \nabla n_e \times \nabla p_e$ in the equation for $\partial B/\partial t$ and to a magnetic helicity flux, which transports magnetic helicity across physical space, as opposed to wavenumber space (Ji 1999).

There is the possibility that the divergence of a magnetic helicity flux, $\overline{\mathcal{F}}_f$, itself can constitute an α effect. This corresponds to $\alpha = -\frac{1}{2}B^{-2}\nabla \cdot \overline{\mathcal{F}}_f$; see Vishniac and Cho (2001), who have derived a specific form for such a flux. The subscript 'f' indicates that the flux originates from correlations of the fluctuating magnetic field. Mean-field models of the type described below have shown that a dynamo can operate even without kinetic helicity, i.e., it is based only on shear and current helicity fluxes, provided a nondimensional scaling factor in front of the magnetic helicity flux exceeds a certain critical value (Brandenburg and Subramanian 2005c). However, there are so far no DNS that have supported this kind of behavior, nor has the proposed flux been confirmed (Hubbard and Brandenburg 2012). Nevertheless, the idea of an α effect being related to the magnetic helicity flux divergence is certainly consistent with the laboratory experiment presented in Fig. 1 of Ji (1999).

The α effect reflects the physics of the inverse cascade of magnetic helicity (Pouquet et al. 1976). In the absence of energy input, this is known to lead to a slower turbulent decay of magnetic energy $\propto t^{-2/3}$ (Hatori 1984; Biskamp and Müller 1999). In hydrodynamics, by comparison, the kinetic energy density decays like $t^{-10/7}$ or $t^{-6/5}$, depending on the initial subinertial range energy spectrum (Davidson 2000); see also Brandenburg and Larsson (2023) for a comparison with the magnetic case.

In the presence of magnetic driving by applying a voltage drop along the magnetic field, small-scale instabilities such as the tearing instability develop. This leads to a sawtooth-like time dependence in the mean toroidal magnetic flux; see Ji and Prager (2002) for a review. This can be associated with the resulting development of a mean electromotive force, $\overline{\mathcal{E}} = \overline{\boldsymbol{u} \times \boldsymbol{b}}$, along with an α effect that accomplishes the helicity transport (Ji 1999).

Unlike astrophysical dynamos, which are generally understood as self-excited ones, the plasma experiments operate in a regime where a magnetic field is always present, but it is then redistributed by the α effect. The conceptional similarities and differences have been discussed in detail by Blackman and Ji (2006). In the following, we discuss in more detail the consequences imposed by magnetic helicity evolution in astrophysical dynamos. It is important to emphasize, however, that the same physics that is used to explain the catastrophic quenching phenomenology also applies to plasma experiments such as the reversed field pinch, as was shown in corresponding mean-field simulations by Kemel et al. (2011).

4.2 Mean Fields in Periodic Domains

Under astrophysical conditions of interest, η is so small that the volume-averaged electromotive force would be negligibly small. If this result was actually astrophysically relevant, it would be a "catastrophe," i.e., it would not be possible to understand astrophysical magnetic fields as mean-field dynamos. The solution to this particular problem turned out to be that relating the volume-averaged electromotive force to the volume-averaged mean magnetic field is only of limited relevance to the problem of α effect dynamos. Any dynamo would produce a non-uniform field. Especially in a periodic domain, the mean magnetic flux through any of the faces of the periodic domain is constant in time, so if it was zero to begin with, it would always remain zero. A dynamo problem can therefore not be formulated in that case.

A proper dynamo problem should always allow for the possibility of the magnetic field to decay to zero if there is sufficient magnetic diffusivity. Simple examples of nontrivial mean fields in a periodic domain are Beltrami fields of the form

$$\overline{\boldsymbol{B}}(x) \propto \begin{pmatrix} 0 \\ \sin kx \\ \cos kx \end{pmatrix}, \quad \overline{\boldsymbol{B}}(y) \propto \begin{pmatrix} \cos ky \\ 0 \\ \sin ky \end{pmatrix}, \quad \text{or} \quad \overline{\boldsymbol{B}}(z) \propto \begin{pmatrix} \sin kz \\ \cos kz \\ 0 \end{pmatrix}, \tag{11}$$

which can be solutions of the simple α^2 dynamo problem, $\partial \overline{\boldsymbol{B}}/\partial t = \alpha \nabla \times \overline{\boldsymbol{B}} + \eta_{\mathrm{T}} \nabla^2 \overline{\boldsymbol{B}}$. Nevertheless, there is still a problem of catastrophic nature because it turned out that the time required to reach the final solution scales inversely with η. This is demonstrated in Fig. 2, where we show the evolution of one of the three planar averages. In the beginning, all three mean fields grow in a similar fashion, but at some point, only one of the three reaches a significant amplitude. Note, however, that the ultimate saturation takes a resistive time, $\tau_{\mathrm{res}} = 1/(2\eta k_1^2)$.

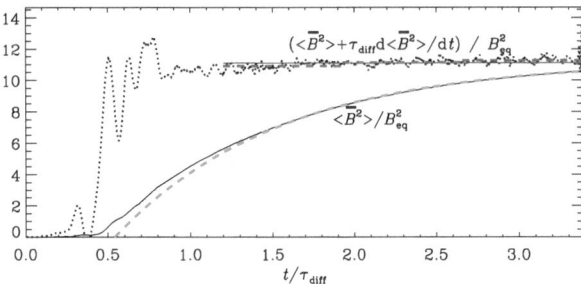

Fig. 2 Evolution of the normalized $\langle \overline{\boldsymbol{B}}^2 \rangle$ and that of $\langle \overline{\boldsymbol{B}}^2 \rangle + \tau_{\mathrm{diff}} \mathrm{d} \langle \overline{\boldsymbol{B}}^2 \rangle / \mathrm{d}t$ (dotted), compared with its average in the interval $1.2 \leq t/\tau_{\mathrm{diff}} \leq 3.5$ (horizontal blue solid line), as well as averages over three subintervals (horizontal red dashed lines). The green dashed line corresponds to Eq. (14) with $t_{\mathrm{sat}}/\tau_{\mathrm{diff}} = 0.54$. Adapted from Candelaresi and Brandenburg (2013)

4.3 Quenching Phenomenology

To understand the reason for the catastrophically slow saturation, it suffices to consider Eq. (9) for the magnetic helicity density. For periodic domains, we just have

$$\frac{\mathrm{d}}{\mathrm{d}t} \langle \boldsymbol{A} \cdot \boldsymbol{B} \rangle = -2\eta\mu_0 \langle \boldsymbol{J} \cdot \boldsymbol{B} \rangle. \tag{12}$$

This equation is gauge-independent, because the gauge transformation $\boldsymbol{A} \to \boldsymbol{A}' + \nabla\Lambda$ yields $\langle \boldsymbol{A} \cdot \boldsymbol{B} \rangle = \langle \boldsymbol{A}' \cdot \boldsymbol{B} \rangle$, with $\langle \boldsymbol{B} \cdot \nabla\Lambda \rangle = \langle \nabla \cdot (\Lambda\boldsymbol{B}) \rangle - \langle \Lambda\nabla \cdot \boldsymbol{B} \rangle = 0$, because $\nabla \cdot \boldsymbol{B} = 0$ and the domain is periodic, so the average of a divergence vanishes.

For fully helical large-scale and small-scale magnetic fields of opposite magnetic helicity, Eq. (12) becomes (Brandenburg 2001)

$$\frac{\mathrm{d}}{\mathrm{d}t} \langle \overline{\boldsymbol{B}}^2 \rangle = 2\eta k_1 k_{\mathrm{f}} B_{\mathrm{eq}}^2 - 2\eta k_1^2 \langle \overline{\boldsymbol{B}}^2 \rangle, \tag{13}$$

with the solution

$$\langle \overline{\boldsymbol{B}}^2 \rangle = B_{\mathrm{eq}}^2 \frac{k_{\mathrm{f}}}{k_1} \left[1 - e^{-2\eta k_1^2 (t - t_{\mathrm{sat}})} \right]. \tag{14}$$

This agrees with the slow saturation behavior seen first in the simulations of Brandenburg (2001); see Fig. 2. Here, t_{sat} is the time when the slow saturation phase commences; see the crossing of the green dashed line with the abscissa. Interestingly, instead of waiting until full saturation is accomplished, one can obtain the saturation value already much earlier simply by differentiating the simulation data to compute (Candelaresi and Brandenburg 2013)

$$B_{\mathrm{sat}}^2 \approx \langle \overline{\boldsymbol{B}}^2 \rangle + \tau_{\mathrm{res}} \frac{\mathrm{d}}{\mathrm{d}t} \langle \overline{\boldsymbol{B}}^2 \rangle. \tag{15}$$

Since τ_{res} involves the microphysical magnetic diffusivity, the quenching is still in that sense catastrophic.

4.4 The α Quenching Formula

A more complete description is in terms of kinetic and magnetic α effects, i.e.,

$$\alpha = \alpha_{\mathrm{K}} + \alpha_{\mathrm{M}} \approx -\frac{\tau}{3} \left(\overline{\boldsymbol{\omega} \cdot \boldsymbol{u}} - \overline{\boldsymbol{j} \cdot \boldsymbol{b}} / \overline{\rho} \right), \tag{16}$$

and observing the fact that the magnetic helicity evolution of averages and fluctuations is given by

$$\frac{d}{dt}\langle \overline{A} \cdot \overline{B}\rangle = +2\langle \overline{\mathcal{E}} \cdot \overline{B}\rangle - 2\eta\mu_0\langle \overline{J} \cdot \overline{B}\rangle, \tag{17}$$

$$\frac{d}{dt}\langle a \cdot b\rangle = -2\langle \overline{\mathcal{E}} \cdot \overline{B}\rangle - 2\eta\mu_0\langle j \cdot b\rangle. \tag{18}$$

Equation (17) allows for the possibility that magnetic helicity can be produced by the mean electromotive force, because, in general, $\overline{\mathcal{E}} \cdot \overline{B} \equiv \overline{u \times B} \cdot \overline{B} \neq 0$. (By contrast, of course, $(\overline{u} \times \overline{B}) \cdot \overline{B} = 0$.) In particular, if $\overline{\mathcal{E}} = \alpha\overline{B} - \eta_t\mu_0\overline{J}$, then, $\overline{\mathcal{E}} \cdot \overline{B} = \alpha\overline{B}^2 - \eta_t\mu_0\overline{J} \cdot \overline{B}$, which produces positive (negative) magnetic helicity of the mean field when $\alpha > 0$ ($\alpha < 0$)

Equation (18) is constructed such that the sum of Eqs. (17) and (18) yields Eq. (12). Given that $\langle a \cdot b\rangle$ is related to $\langle j \cdot b\rangle$, which, in turn, is related to a magnetic contribution to the α effect (Pouquet et al. 1976), Eq. (18) can be rewritten as an evolution equation for the total α (Brandenburg 2008),

$$\frac{d\alpha_M}{dt} = -2\eta_{t0}k_f^2\left(\frac{\alpha\overline{B}^2 - \eta_t\mu_0\overline{J} \cdot \overline{B}}{B_{eq}^2} + \frac{\alpha_M}{Re_M}\right), \tag{19}$$

which can also be expressed in the form

$$\alpha(\overline{B}) = \frac{\alpha_0 + Re_M \times \text{``extra terms''}}{1 + Re_M \overline{B}^2/B_{eq}^2} \tag{20}$$

where

$$\text{``extra terms''} = \eta_t\frac{\mu_0\overline{J} \cdot \overline{B}}{B_{eq}^2} - \frac{\nabla \cdot \overline{\mathcal{F}}_f}{2k_f^2 B_{eq}^2} - \frac{\partial\alpha/\partial t}{2k_f^2 B_{eq}^2}. \tag{21}$$

Note that the last term is here a time derivative. Equation (20) resembles the catastrophic quenching formula of Vainshtein and Cattaneo (1992), but it also shows that it needs to be extended in several important ways: when the mean field is no longer defined as a volume average, extra terms emerge that are of the same order as those in the denominator. They can therefore potentially offset the catastrophic quenching. In practice, this is only partially true, because there are also other terms, for example the aforementioned time derivative term. It is responsible for the fact that a strong field state is only reached after a resistively long time.

4.5 Magnetic Helicity Fluxes and Helicity Reversals

Magnetic helicity fluxes could in principle remove the catastrophic quenching problem, but only if preferentially small-scale magnetic helicity is being removed (Kleeorin et al. 2000). To see this, let us first consider the problem of an α^2 dynamo in insulating boundaries using the Weyl gauge, i.e.,

$$\frac{\partial}{\partial t}\overline{A} = \alpha\overline{B} - \eta_T\mu_0\overline{J}, \quad \text{with} \quad \partial_z\overline{A}_x = \partial_z\overline{A}_y = \overline{A}_z = 0. \tag{22}$$

The boundary condition implies that $\overline{B}_x = \overline{B}_y = 0$, and is therefore also referred to as the vertical field condition. In this 1-D problem, however, this boundary condition is equivalent to a proper vacuum boundary condition.

The α^2 dynamo with this boundary condition was first considered by Gruzinov and Diamond (1994), who found that the saturation field strength of such a dynamo decreases with Re_M. This was later confirmed by Brandenburg and Dobler (2001). In Fig. 3, we show the profiles of magnetic helicity, current helicity, and the magnetic helicity fluxes for Runs A of Brandenburg (2018b) with $Re_M = 180$. The computational domain is $0 \le z \le \pi/2$ with a perfect conductor boundary condition on $z = 0$ and a vertical field condition on $z = \pi/2$; see Brandenburg (2017) for the relevant mean-field models. For normalization purposes, he defined the reference values

$$C_{f0} = k_f B_{eq}^2 \quad \text{and} \quad F_{m0} = \eta_{t0} k_1^2 \int_0^{\pi/2} \overline{\boldsymbol{B}}^2 \, dz. \tag{23}$$

He emphasized that the largest contribution to the magnetic helicity density comes from the large-scale field. Near the surface ($z = \pi/2$), the (negative) magnetic helicity flux from small-scale fields is only about $0.02\, F_{m0}$, which explains why they are not efficient enough to alleviate the catastrophic dependence of the resulting mean magnetic field (Del Sordo et al. 2013; Rincon 2021).

Subsequent simulations with an outside corona indicated that the magnetic helicity changes sign at or near the outer surface (Brandenburg et al. 2009). This was just a speculation and needs to be reconsidered with the help of global models of the type considered by Warnecke et al. (2011, 2012) and Brandenburg et al. (2017a). This is shown in Fig. 4, where we present the line-of-sight averaged current helicity density, $\langle \boldsymbol{J} \cdot \boldsymbol{B} \rangle$ in the plane of the sky using a simulation of Brandenburg et al. (2017a). The quantity $\langle \boldsymbol{J} \cdot \boldsymbol{B} \rangle$ is a proxy of magnetic helicity at small scales and shows clearly the reversal of sign between the dynamo interior and the exterior.

4.6 Radial Magnetic Helicity Reversal in the Solar Wind

If the idea of alleviating catastrophic quenching by magnetic helicity fluxes is to make sense, we would expect to see signs of the expelled magnetic helicity in the solar wind. The magnetic helicity spectrum can be measured in the solar wind by determining the parity-odd contribution to the magnetic correlation tensor, which, in Fourier space, takes the form

$$\langle \tilde{B}_i(\boldsymbol{k}) \tilde{B}_j^*(\boldsymbol{k}) \rangle = \left(\delta_{ij} - \hat{k}_i \hat{k}_j \right) 2E(k) - i\hat{k}_k \epsilon_{ijk} H(k). \tag{24}$$

This would allow one to compute $H(k_z) = \text{Im}(\tilde{B}_x \tilde{B}_y^*)$ and $E(k_z) = \frac{1}{2}(|\tilde{B}_x|^2 + |\tilde{B}_y|^2)$, which also obeys the realizability condition $k_z|H(k_z)| \le 2E(k_z)$.

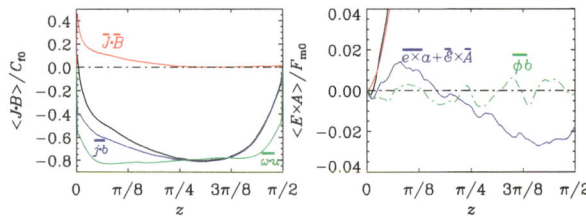

Fig. 3 Magnetic helicity, current helicity, and magnetic helicity fluxes for Run A of Brandenburg (2018b) with $Re_M = 180$. The kinetic helicity is shown in green and is found to be of similar magnitude as the current helicity of the small-scale field. The second panel shows $\overline{\boldsymbol{E} \times \boldsymbol{A}}$ near zero. The green line denotes $\overline{\phi b}$, which is seen to fluctuate around zero

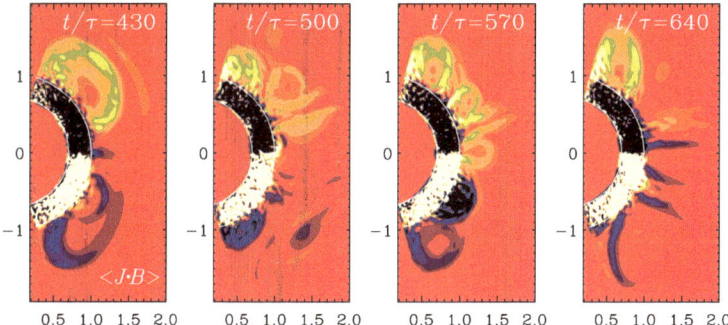

Fig. 4 Current helicity $\langle \boldsymbol{J} \cdot \boldsymbol{B} \rangle$ in the plane of the observer at four different times. Yellow and white shades denote positive values and blue and black shades denote negative values; adapted from Brandenburg et al. (2017a)

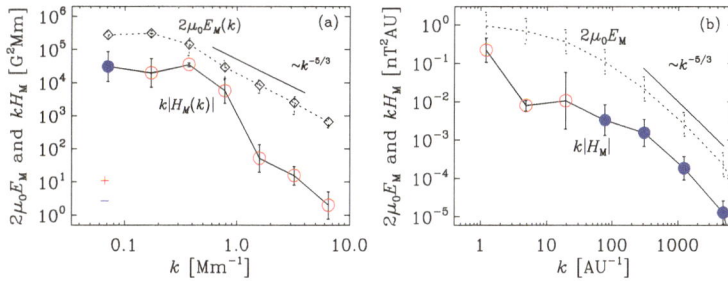

Fig. 5 Magnetic energy and magnetic helicity spectra for southern latitudes (a) at the solar surface in active region AR 11158, and (b) in the solar wind at ~ 1 AU distance (1 AU $\approx 149{,}600$ Mm). Positive (negative) signs are shown as red open (blue filled) symbols. Positive signs are the solar surface at intermediate and large k correspond to positive values in the solar wind at small k. Note that 1 G $= 10^{-4}$ T $= 10^5$ nT

The Ulysses spacecraft was the only one to cover high heliographic latitudes, where a non-vanishing sign of magnetic helicity can be expected. It turned out that $H(k)$ has, as expected from dynamo theory, different signs in the northern and southern hemispheres. It also has different signs at small and large wavenumbers. This, in itself, is also expected from an α^2 dynamo, because the α effect produces no net magnetic helicity, but it separates magnetic helicity in wavenumber space. However, the signs are opposite to what is seen at the solar surface, where the helicity in the north is negative at small length scales. In the solar wind, however, it is positive in the north and at small scales. Of course, the meaning of small is here relative and has to be with respect to larger scales, where a sign change in k has been seen. If one just assumed a linear expansion of all scales from the solar surface (radius $r = 700$ Mm, to the location of the Earth at 1 AU, we expect a corresponding expansion ratio so that a wavenumber of 1 Mm^{-1} corresponds to $1/700$ AU^{-1}. In particular, 20 Mm^{-1} corresponds to $2/70$ AU^{-1}, which is close to the wavenumber where we see a sign-change in Fig. 5. It is unexpected, however, that at the solar surface (Fig. 5b), the sign in the northern hemisphere changes from minus to plus as k increases, while in the solar wind, it changes from plus to minus. This apparent mismatch may not just be a measurement error, but it may actually be a real result and would tell us that the simpleminded picture of expelling magnetic helicity of one sign all the way to infinity may not be accurate.

When the domain is inhomogeneous, and especially when there are boundaries, magnetic helicity fluxes are possible and Eq. (18) takes the form

$$\frac{\partial}{\partial t}\overline{\boldsymbol{a} \cdot \boldsymbol{b}} = -2\overline{\boldsymbol{\mathcal{E}}} \cdot \overline{\boldsymbol{B}} - 2\eta\mu_0 \overline{\boldsymbol{j} \cdot \boldsymbol{b}} - \nabla \cdot \overline{\boldsymbol{\mathcal{F}}}_{\mathrm{f}}. \tag{25}$$

In the steady state, we have

$$\nabla \cdot \overline{\boldsymbol{\mathcal{F}}}_{\mathrm{f}} = \underbrace{-2\alpha\overline{\boldsymbol{B}}^2 + 2\eta_t\mu_0\overline{\boldsymbol{J}} \cdot \overline{\boldsymbol{B}}}_{-2\overline{\boldsymbol{\mathcal{E}}} \cdot \overline{\boldsymbol{B}}} -2\eta\mu_0\overline{\boldsymbol{j} \cdot \boldsymbol{b}}. \tag{26}$$

In the dynamo interior at the northern hemisphere, $\alpha > 0$, and, assuming $\alpha \overline{\boldsymbol{B}}^2$ to dominate the EMF, we expect $-2\overline{\boldsymbol{\mathcal{E}}} \cdot \overline{\boldsymbol{B}}$ to be negative. However, a negative flux divergence of a negative quantity would eventually make this quantity positive, which is what has been observed.

Whether or not this is really the right interpretation remains still an open question. It would clearly be useful to have an independent assessment of this interpretation.

4.7 Nonlocal Effects of $\overline{\boldsymbol{\mathcal{E}}}$ and Catastrophic Quenching

Catastrophic quenching in large-scale dynamos is a rather general property. It is a consequence of the build-up of magnetic helicity of the mean magnetic field. It has been conjectured that catastrophic quenching would be prevented if the sources of toroidal field generation are spatially separated from the sources of the poloidal field; see, e.g., Tobias and Weiss (2007). This would be the case in what is known as interface dynamos (Parker 1993). It could also be through a nonlocal α effect. Such a nonlocal α effect is an essential ingredient of the Babcock–Leighton and flux-transport dynamo models; see, Hazra et al. (2023). The studies of Brandenburg and Käpylä (2007) and Chatterjee et al. (2010) showed that a spatial separation between shear and α effects does in general not help to avoid catastrophic quenching for such types of dynamo models. It is interesting, however, that Kitchatinov and Olemskoy (2011) and later also Brandenburg et al. (2015) found that the inclusion of diamagnetic downward pumping of the toroidal magnetic field can alleviate the catastrophic quenching in the Babcock–Leighton dynamo model with a strongly nonlocal α.

The catastrophic quenching models are reasonably well reproduced by DNS when the geometries of the setups are sufficiently simple. It would therefore be worthwhile to apply DNS to conditions where turbulent pumping and a strongly nonlocal mean electromotive force can be expected. At present, however, even just the physical reality of a nonlocal α effect of Babcock–Leighton type through the decay of active regions rests mainly on the interpretation of solar observations. Turbulence simulations have so far not been able to make contact with such concepts.

5 Alternative Large-Scale Dynamo Effects

Given the difficulties encountered with α effect dynamos, there have been various attempts to construct large-scale dynamos that are not based on the α effect. A common misconception here is the idea that catastrophic quenching would not apply just because there is no α effect. This is not true, because an α_M term can always emerge regardless of whether or not there existed an original α effect. An example is the shear–current effect. It is due to the presence of shear and boundaries that a helicity can be introduced. Shear of the form

$\overline{U} = (0, Sx, 0)$ implies a finite vorticity, $\nabla \times \overline{U} = (0, 0, S)$ and boundaries would lead to a gradient vector of turbulent intensity near the boundaries. Thus, while there can be hope that catastrophic quenching may not be as strong, this may turn out not to be the case. An example of this was presented in Brandenburg and Subramanian (2005c).

5.1 Rädler and Shear–Current Effects

The Rädler effect is another large-scale dynamo effect (Rädler 1969). In the simplest representation it leads to an EMF proportional to $\mathbf{\Omega} \times \overline{J}$. It is similar to the shear–current effect. In this case it cannot change the magnetic energy of the mean field. Indeed, the energy equation for the mean field is given by

$$\frac{d}{dt} \langle \overline{B}^2/2 \rangle = \underbrace{\overline{J} \cdot (\mathbf{\Omega} \times \overline{J})}_{=0} + \underbrace{\langle \nabla \cdot [(\mathbf{\Omega} \times \overline{J}) \times \overline{B}] \rangle}_{=0 \text{ under periodicity}} . \tag{27}$$

In the general case, the generation effects due to global rotation and mean currents can be written as follows (see Krause and Rädler 1980; Kitchatinov et al. 1994a; Rädler et al. 2003; Pipin 2008):

$$\overline{\mathcal{E}}^{(\delta)} = \delta_1 \mathbf{\Omega} \times \overline{J} + \delta_2 \nabla \left(\mathbf{\Omega} \cdot \overline{B} \right) + \delta_3 \frac{\mathbf{\Omega} \left[\mathbf{\Omega} \cdot \nabla \left(\mathbf{\Omega} \cdot \overline{B} \right) \right]}{\mathbf{\Omega}^2} , \tag{28}$$

where the coefficients $\delta_{1,2,3}$ depend on the spatial profiles of the turbulent parameters such as the typical convective turnover time, the convective velocity u_{rms}, etc. The last two terms in this equation may lead to an δ^2 dynamo (Pipin and Seehafer 2009). For the solar case, the δ effect can provide an additional non-helical source of poloidal magnetic field generation. Interestingly, Pipin and Seehafer (2009) found that for the solar-type dynamos, i.e., those with equatorward propagation of the dynamo waves, the δ dynamo effect does not dominate the contributions of the α-effect. We will discuss the available scenario in the next section.

5.2 Dynamos from Negative Turbulent Magnetic Diffusivity

There are two other effects that are noteworthy, although it is not clear that either of them can play a role in stellar convection zones. One is the negative turbulent magnetic diffusivity and the other is the memory effect in conjunction with a pumping effect.

When modeling a negative turbulent magnetic diffusivity dynamo, high wavenumbers must not be destabilized at the same time. Brandenburg and Chen (2020) studied classes of dynamos with a very low critical Re_{M}. The Willis dynamo (Willis 2012) has a critical Re_{M} of 2.01, which is small compared to 6.3 for the Roberts flow and 17.9 for the ABC flow. In this dynamo, one of the two horizontally averaged field components grows exponentially, because the total magnetic diffusivity in that direction is negative (Brandenburg and Chen 2020). The other component decays and is not coupled to the former one.

As we see from Fig. 6, η_t is negative only for $k \lesssim 1.5$. The k dependence of the turbulent magnetic diffusivity can be expanded up to second order as

$$\tilde{\eta}_{yy}(k) = \tilde{\eta}_{yy}^{(0)} + \tilde{\eta}_{yy}^{(2)} k^2 + \cdots , \tag{29}$$

where the tildes indicate Fourier transformed quantities. In the proximity of $k = 1$, which corresponds to the largest scale in the computational domain of 2π, we have $\tilde{\eta}_{yy}^{(0)} \approx -0.233$

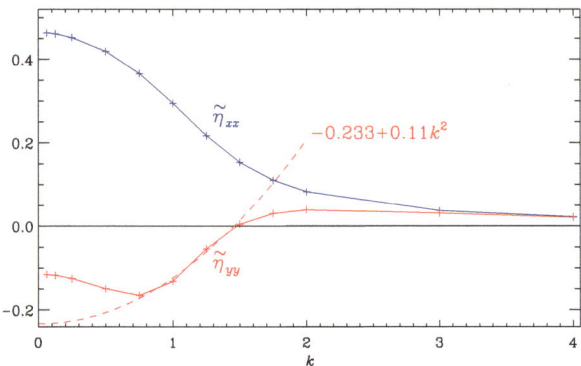

Fig. 6 Dependence of $\tilde{\eta}_{xx}$ (blue) and $\tilde{\eta}_{yy}$ (red) on k for the Willis flow in the marginally exited case with $\eta = 0.403$. The red dashed line denotes the fit $-0.233 + 0.11\,k^2$. Adapted from Brandenburg and Chen (2020)

and $\tilde{\eta}_{yy}^{(2)} \approx 0.11$. In addition, there is still the microphysical magnetic diffusivity, which is positive ($\eta = 0.403$). To a first approximation, one can just consider the equation for $\overline{A}_{yy}$, which can then be written as

$$\frac{\partial \overline{A}_{yy}}{\partial t} = \left[\eta + \tilde{\eta}_{yy}^{(0)}\right] \frac{\partial^2 \overline{A}_{yy}}{\partial z^2} - \tilde{\eta}_{yy}^{(2)} \frac{\partial^4 \overline{A}_{yy}}{\partial z^4}. \tag{30}$$

We recall that the minus sign in front of the fourth derivative corresponds to positive diffusion if $\tilde{\eta}_{yy}^{(2)}$ is positive, and so does the plus sign in front of the second derivative, unless the term in squared brackets is negative, which is the case we are considering here.

5.3 Dynamos from Pumping and Memory Effects

Pumping effects alone cannot usually lead to interesting dynamo effects, unless there is also a memory effect. This effect means that the mean electromotive force depends not just on the instantaneous mean magnetic field at that time, but also on the mean magnetic field at earlier times. It is therefore described as a convolution between a pumping kernel and the mean magnetic field. This can lead to dynamo action, as has been demonstrated by Rheinhardt et al. (2014) for the case of two of the four flow fields studied by Roberts (1972). These are flows II and III with $U_{\mathrm{II}}(x, y)$ and $U_{\mathrm{III}}(x, y)$, respectively, and are given by

$$U_{\mathrm{II}} = \begin{pmatrix} u_0 \sin Kx \cos Ky \\ -u_0 \cos Kx \sin Ky \\ u_0 \cos Kx \cos Ky \end{pmatrix}, \quad U_{\mathrm{III}} = \begin{pmatrix} u_0 \sin Kx \cos Ky \\ -u_0 \cos Kx \sin Ky \\ \frac{1}{2} u_0 (\cos 2Kx + \cos 2Ky) \end{pmatrix}, \tag{31}$$

where K is the wavenumber of the flow. Both flows have zero kinetic helicity and no α effect, but flow II is also pointwise nonhelical. A supercritical three-dimensional magnetic field with growth rate γ and wavenumber k in the z direction of the form $\boldsymbol{B} = \boldsymbol{b}_0(x, y)\, e^{\gamma t + ikz}$ is possible when $\mathrm{Re}_{\mathrm{M}} \equiv u_0/\eta K > 4.58$ and 2.9 for flows II and III, respectively; see Rheinhardt et al. (2014). Here, $\boldsymbol{b}_0(x, y)$ is the eigenfunction.

For both flows, there are xy-averaged mean fields $\overline{B}_x(z, t)$ and $\overline{B}_y(z, t)$, with waves traveling in opposite directions for flow II and in the same direction for flow III; see Figs. 6 and 8 of Rheinhardt et al. (2014), respectively. These dynamos appear to be atypical, because there is so far no other known example of a flow where pumping produces a memory effect that is strong enough to lead to dynamo action. This may well be due to the absence (until recently) of computational tools for determining the memory effect. Indeed, it was only with the development of the TFM (Schrinner et al. 2005, 2007) that the importance of the memory effect was noticed (Hubbard and Brandenburg 2009) and applied to pumping.

The dispersion relation for a problem with turbulent pumping γ and turbulent magnetic diffusion η_t is given by $\lambda = -ik\gamma - \eta_t k^2$. Since $\text{Re}\,\lambda < 0$, the solution can only decay, but it is oscillating with the frequency $\omega = \text{Im}\,\lambda = \gamma$. In the presence of a memory effect, γ is replaced by $\gamma/(1 - i\omega\tau)$, where τ is the memory time. Then, $\lambda \approx -ik\gamma (1 - i\omega\tau) - \eta_t k^2$, and $\text{Re}\,\lambda$ can be positive. This is the case for the Roberts flows II and III.

We return to nonlocality and memory effects further below in this article when we discuss concrete solar models; see Pipin (2023). One of the most obvious consequences of the memory effect is a lowering of the critical excitation conditions for the dynamo, which was already reported by Rheinhardt and Brandenburg (2012). Interestingly, for the nonlocal mean electromotive force, the lowering of the critical threshold can be accompanied by multiple instabilities of different dynamo modes that have different frequencies and spatial localization; see Pipin (2023).

5.4 Dynamos from Cross-Helicity

An alignment of velocity and magnetic field, i.e., cross helicity, plays a key role in numerous processes and phenomena of astrophysical plasmas. Krause and Rädler (1980) showed that the saturation stage of the turbulent generation is characterized by an alignment of the turbulent convective velocity and the magnetic field. This consideration does not account for the effects of cross-helicity that take place in the strongly stratified subsurface layers of the stellar convective envelope. For example, the direct numerical simulations of Matthaeus et al. (2008) showed a directional alignment of velocity and magnetic field fluctuations in the presence of gradients of either pressure or kinetic energy.

The mean electromotive force in this case is along to the mean vorticity,

$$\overline{\mathcal{E}}^{\Upsilon} = \Upsilon \nabla \times \overline{\mathbf{U}} + \cdots, \qquad (32)$$

where, $\Upsilon = \tau_c \langle \mathbf{u}^{(0)} \cdot \mathbf{b}^{(0)} \rangle$ is the cross helicity pseudoscalar, and τ_c is the turbulent turnover time. The superscripts (0) indicate quantities of the background turbulence, which exists in the absence of a mean magnetic field and a mean flow; see our comment after Eq. (2) about the $\overline{\mathcal{E}}_0$ term. In the standard mean-field framework, it is assumed that $\Upsilon = 0$; see Krause and Rädler (1980). Yoshizawa and Yokoi (1993) generalized the framework assuming $\Upsilon \neq 0$, see the comprehensive review of Yokoi (2013).

Dynamo scenarios based on cross helicity have been suggested in a number of papers (Yoshizawa and Yokoi 1993; Yoshizawa et al. 2000; Yokoi 2013). Pipin and Yokoi (2018) showed that the large-scale dynamo instability does not require the existence of a global axisymmetric mean. The mix of axisymmetric and nonaxisymmetric magnetic fields can be produced even in the case $\overline{\Upsilon = 0}$, where the overbar means azimuthal averaging. The surface magnetic field of the Sun and other similar stars tends to be organized in sunspots, plagues, ephemeral regions, super-granular magnetic network, etc. These structures tend to demonstrate the alignment of local velocity and magnetic fields (Rüdiger et al. 2011). Therefore, the cross helicity dynamo instability can contribute to dynamo generation effects that operate near the stellar surface. Stellar observations, for example those of Katsova et al. (2021), require such dynamo effects to be working in situ at the stellar surface. The solar analogs show an increase of the spottiness with an increase of the rotation rate (Berdyugina 2005). In that case, cross helicity dynamo effects can be considered as a relevant addition to the standard turbulent generation by means of convective helical motions. Rapidly rotating

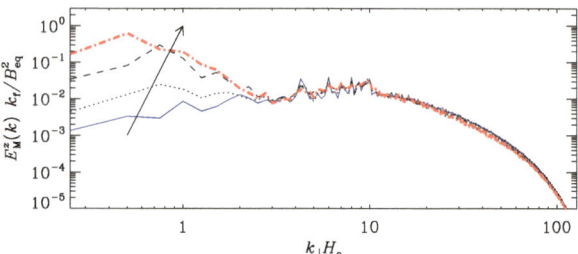

Fig. 7 Normalized spectra of B_z from a simulation of MHD turbulence with strong gravity at turbulent diffusive times $t\eta_t/H_\rho^2 \approx 0.2, 0.5, 1,$ and 2.7 with $k_f H_\rho = 10$ and $k_1 H_\rho = 0.25$. Adapted from Brandenburg et al. (2014)

M-dwarfs show the highest level of magnetic activity (Kochukhov 2021). There is a population of rapidly rotating M-dwarfs that show a rather strong dipole type magnetic field. These stars show a rather small level of differential rotation. For solid body rotation, an α^2 dynamo generates a nonaxisymmetric magnetic field (Chabrier and Küker 2006; Elstner and Rüdiger 2007). At high rotation rates, the α effect is highly anisotropic (Rüdiger and Kichatinov 1993). It cannot employ the component of the large-scale magnetic field along the rotation axis for the generation of an axial electromotive force. Results of Pipin and Yokoi (2018) show that the $\alpha^2 \Upsilon^2$ scenario can produce a strong constant dipole magnetic field. The model predicts the existence of large-scale cross helicity patterns occupying the stellar surface. We hope that this can be tested either in observations or in GCDs.

The nonlinear theory for the cross helicity effect is not yet developed. Sur and Brandenburg (2009) showed that the turbulent generation due to Υ is quenched by large-scale vorticity in a way that is similar to catastrophic quenching given by Eq. (20), i.e.,

$$\Upsilon \sim \frac{1}{1 + \mathrm{Re_M}\,\tau_c^2(\nabla \times \overline{\mathbf{U}})^2}.\tag{33}$$

One should remember that for the initialization of the cross-helicity dynamo instability we have to seed both the cross helicity and the magnetic field. The solar type model scenarios based on cross helicity require an α effect, which produces poloidal magnetic field and cross helicity at the top of the dynamo domain (Yokoi et al. 2016).

Given that cross helicity is an ideal invariant of the MHD equations, it is natural to ask whether systems with strong cross helicity exhibit inverse cascading. The answer seems to be yes; see Brandenburg et al. (2014). In Fig. 7 we demonstrate the gradual build-up of magnetic fields in the vertical direction when the system has significant cross helicity owing to the presence of a magnetic field along the direction of gravity (Rüdiger et al. 2011).

6 Mean-Field Dynamo Models

In general, the α effect, as well as any other turbulent generation effect, including the δ effect (Rädler 1969), the shear-current effect (Kleeorin et al. 2000), and the cross-helicity effect (Yokoi 2013) can generate both toroidal and poloidal magnetic fields. Therefore there can be a number of possibilities for solar-types dynamo models Krause and Rädler (1980), Yokoi et al. (2016), Pipin and Kosovichev (2018). Some of them skip the α effect altogether. For example, Seehafer and Pipin (2009) studied $\delta^\Omega\Omega$ and $\delta^W\Omega$ scenarios, where turbulent generation of the poloidal magnetic field is due to $\mathbf{\Omega} \times \overline{\mathbf{J}}$ and shear-current effect, respectively. These scenarios show oscillating solutions and the correct time-latitude diagram of the toroidal magnetic field if the meridional circulation is included. A similar possibility was mentioned earlier by Krause and Rädler (1980) for the $\delta\Omega$ scenario. However, the given scenarios result in an incorrect phase relation between activity of the toroidal and poloidal magnetic fields. The aim to search for α effect alternatives pursues double benefits. First, the nonhelical source of dynamo generations avoids the above mentioned catastrophic quenching problem. This issue is less important currently. Secondly, and it was already mentioned earlier by Köhler (1973) as well as Steenbeck and Krause (1969), the mixing length estimate of the α effect for the solar convection zone parameters results in a very strong α effect with a magnitude as strong as the convective velocity rms. Solar observations of the ratio between the typical strength of the toroidal and poloidal fields and the solar cycle period, favor an order of magnitude smaller α effect. In addition, the turbulent generation sources in the $\alpha\Omega$ scenario help reduce the given constraints. We must stress that the GCD simulations of Schrinner (2011), Schrinner et al. (2011), and Warnecke et al. (2021) showed that the mean-field models need a full spectrum of turbulent effects to describe DNS.

In the case of a solar-like star, i.e., with solar-like stratification, differential rotation, and meridional circulation profiles, the turbulent sources of the poloidal magnetic field generation due to δ, shear-current, and cross-helicity effects are likely complimentary to the α effect.

We thus arrive at the conclusion that the $\alpha^2\Omega$ dynamo is, probably, the simplest scenario for the solar dynamo. Also, this scenario seems to fit well with observations of stellar activity of young solar-type stars.

6.1 Basic Model

We discuss some results of the state-of-the-art mean-field dynamo model of a solar dynamo developed recently by Pipin and Kosovichev (2019) (hereafter PK19). The magnetic field evolution is governed by the mean-field induction equation:

$$\frac{\partial \overline{\boldsymbol{B}}}{\partial t} = \nabla \times \left(\overline{\boldsymbol{\mathcal{E}}} + \overline{\boldsymbol{U}} \times \overline{\boldsymbol{B}} - \eta\mu_0\overline{\boldsymbol{J}}\right). \tag{34}$$

The expression for the components of $\overline{\boldsymbol{\mathcal{E}}}$ reads

$$\overline{\mathcal{E}}_i = \left(\alpha_{ij} + \gamma_{ij}\right)\overline{B}_j - \eta_{ijk}\nabla_j\overline{B}_k. \tag{35}$$

Here, α_{ij} describes the turbulent generation by the α effect, γ_{ij} represents turbulent pumping, and η_{ijk} is the eddy magnetic diffusivity tensor. The α effect tensor includes the small-scale magnetic helicity density contribution, i.e., the pseudoscalar $\langle \mathbf{a} \cdot \mathbf{b}\rangle$,

$$\alpha_{ij} = C_\alpha \psi_\alpha(\beta)\alpha_{ij}^K + \alpha_{ij}^M \psi_\alpha(\beta)\frac{\langle \mathbf{a} \cdot \mathbf{b}\rangle \tau_c}{4\pi\overline{\rho}\ell_c^2}, \tag{36}$$

where C_α is the dynamo parameter characterizing the magnitude of the kinetic α effect, and α_{ij}^K and α_{ij}^M are the anisotropic versions of the kinetic and magnetic α effects, as described in PK19. Note that, unlike Eq. (16), where the two α contributions are pseudoscalars and have the same dimension, they are here tensorial where only α_{ij}^K is a pseudotensor, but α_{ij}^M is not, and they have here different dimensions.

The radial profiles of $\alpha_{ij}^{(H)}$ and $\alpha_{ij}^{(M)}$ depend on the mean density stratification, the profile of the convective velocity u_{rms}, and on the Coriolis number,

$$\mathrm{Co} = 2\Omega_0\tau_c, \tag{37}$$

where Ω_0 is the global angular velocity of the star and τ_c is the convective turnover time. The magnetic quenching function $\psi_\alpha(\beta)$ depends on the parameter $\beta = |\overline{\boldsymbol{B}}|/(\sqrt{4\pi\overline{\rho}}u_{\mathrm{rms}})$. In this model the magnetic helicity is governed by the global conservation law for the total magnetic helicity, $\langle \mathbf{A} \cdot \mathbf{B} \rangle = \langle \mathbf{a} \cdot \mathbf{b} \rangle + \overline{\boldsymbol{A}} \cdot \overline{\boldsymbol{B}}$ (see Hubbard and Brandenburg 2012; Pipin et al. 2013):

$$\left(\frac{\partial}{\partial t} + \overline{\boldsymbol{U}} \cdot \nabla \right) \langle \mathbf{A} \cdot \mathbf{B} \rangle = -\frac{\langle \mathbf{a} \cdot \mathbf{b} \rangle}{\mathrm{Re}_\mathrm{M}\,\tau_c} - 2\eta\overline{\boldsymbol{B}} \cdot \overline{\boldsymbol{J}} - \nabla \cdot \overline{\mathcal{F}}_\mathrm{f}, \tag{38}$$

where we have used $2\eta\langle \mathbf{j} \cdot \mathbf{b} \rangle = \langle \mathbf{a} \cdot \mathbf{b} \rangle / \mathrm{Re}_\mathrm{M}\,\tau_c$ (Kleeorin and Rogachevskii 1999). Also, we have introduced a diffusive flux of the small-scale magnetic helicity density, $\overline{\mathcal{F}}_\mathrm{f} = -\eta_\chi \nabla \langle \mathbf{a} \cdot \mathbf{b} \rangle$, and Re_M is the magnetic Reynolds number, for which we employ $\mathrm{Re}_\mathrm{M} = 10^6$. Following results of Mitra et al. (2010a), we put $\eta_\chi = 0.1\,\eta_T$. Here, the turbulent fluxes of the magnetic helicity are approximated by the only term which is related to the diffusive flux. Besides the diffusive helicity flux, the other turbulent fluxes of the magnetic helicity can be important for the nonlinear dynamo regimes and the catastrophic quenching problem (Kleeorin et al. 2000; Vishniac and Cho 2001; Pipin 2008; Chatterjee et al. 2011; Brandenburg and Subramanian 2005a; Kleeorin and Rogachevskii 2022; Gopalakrishnan and Subramanian 2023). The relative importance of different kinds of magnetic helicity fluxes for the dynamo should be studied further.

The above ansatz for the helicity evolution differs from that given by Eq. (18); see also papers by Kleeorin and Ruzmaikin (1982) and Kleeorin and Rogachevskii (1999). Hubbard and Brandenburg (2012) had studied the magnetic helicity evolution for shearing dynamos. They found that employing Eq. (18) in the dynamo problem can result in nonphysical fluxes of magnetic helicity over spatial scales. For the ansatz given by Eq. (18), the nonlinear dynamo models can show sharp magnetic structures inside the dynamo domain. Such structures are connected with concentrations of the magnetic helicity; see, e.g., Chatterjee et al. (2011) and Brandenburg and Chatterjee (2018). Even a strong diffusive helicity flux does not seem to correct for those irrelevant features from the numerical solution. The technical point is that the helicity fluxes, which are omitted in Eq. (18), should be consistent with the turbulent effects involved in the mean electromotive force, e.g., the rotationally induced anisotropy of the α effect, the magnetic eddy diffusivity, etc. Such calculations are currently absent. Also, we have to take into account the modulation of the magnetic helicity density by the magnetic activity. On the other hand, with the magnetic helicity evolution equation Eq. (38), Pipin et al. (2013) found that magnetic helicity density follows the large-scale dynamo wave. This alleviates the catastrophic quenching of the α effect. They showed that if we write the Eq. (38) in the form of Eq. (18), we get an additional helicity flux due to the global dynamo, Rewriting Eq. (38) in the form of Eq. (18) we get

$$\frac{\partial \langle \mathbf{a} \cdot \mathbf{b} \rangle}{\partial t} = -2\left(\overline{\boldsymbol{\mathcal{E}}} \cdot \overline{\boldsymbol{B}}\right) - \frac{\langle \mathbf{a} \cdot \mathbf{b} \rangle}{\mathrm{Re}_\mathrm{M}\,\tau_c} + \nabla \cdot \left(\eta_\chi \nabla \langle \mathbf{a} \cdot \mathbf{b} \rangle\right) - \eta\overline{\boldsymbol{B}} \cdot \overline{\boldsymbol{J}} - \nabla \cdot \left(\overline{\boldsymbol{\mathcal{E}}} \times \overline{\boldsymbol{A}}\right) + \cdots, \tag{39}$$

where the ellipsis refers to additional helicity transport terms due to the large-scale flow. The term $\overline{\boldsymbol{\mathcal{E}}} \times \overline{\boldsymbol{A}}$ consists of the counterparts of the sources of magnetic helicity, which are represented by $-2\overline{\boldsymbol{\mathcal{E}}} \cdot \overline{\boldsymbol{B}}$, and the fluxes which result from pumping of the large-scale magnetic fields. The sources of magnetic helicity in the term $-2\overline{\boldsymbol{\mathcal{E}}} \cdot \overline{\boldsymbol{B}}$ are partly compensated in Eq. (39) by the counterparts in $\overline{\boldsymbol{\mathcal{E}}} \times \overline{\boldsymbol{A}}$. This results in the spatially homogeneous quenching of the large-scale magnetic generation and alleviation of the catastrophic quenching problem. The effect of $\overline{\boldsymbol{\mathcal{E}}} \times \overline{\boldsymbol{A}}$ was not unambiguously confirmed in DNS because of limited numerical resolution; see Del Sordo et al. (2013) and Brandenburg (2018b).

The turbulent pumping is expressed by the antisymmetric tensor γ_{ij}. The tuning of γ_{ij} for the solar-type mean-field dynamo model was discussed by Pipin (2018). We define it as follows,

$$\gamma_{ij} = \gamma_{ij}^{(\Lambda\rho)} + \frac{\alpha_{\mathrm{MLT}} u_{\mathrm{rms}}}{\gamma} \mathcal{H}(\beta)\, \hat{r}_n \varepsilon_{\mathrm{inj}}, \tag{40}$$

$$\gamma_{ij}^{(\Lambda\rho)} = 3 v_T f_1^{(a)} \left\{ (\boldsymbol{\Omega} \cdot \boldsymbol{\Lambda}^{(\rho)}) \frac{\Omega_n}{\Omega^2} \varepsilon_{\mathrm{inj}} - \frac{\Omega_j}{\Omega^2} \varepsilon_{inm} \Omega_n \Lambda_m^{(\rho)} \right\}, \tag{41}$$

where $\boldsymbol{\Lambda}^{(\rho)} = \nabla \log \overline{\rho}$, $\alpha_{\mathrm{MLT}} = 1.9$ is the mixing-length theory parameter, γ is the adiabatic law constant. In Eq. (40), the first term takes into account the mean drift of large-scale field due the gradient of the mean density, and the second one does the same for the mean-field magnetic buoyancy effect. The function $\mathcal{H}(\beta)$ takes into account the effect of the magnetic tensions. It is $\mathcal{H}(\beta) \sim \beta^2$ for small β and it saturates as β^{-2} for $\beta \gg 1$; see P22.

We employ an anisotropic diffusion tensor following the formulation of Pipin (2008) (hereafter, P08):

$$\eta_{ijk} = 3\eta_T \left\{ \left(2 f_1^{(a)} - f_2^{(d)}\right) \varepsilon_{ijk} + 2 f_1^{(a)} \frac{\Omega_i \Omega_n}{\Omega^2} \varepsilon_{jnk} \right\}, \tag{42}$$

where the functions $f_{1,2}^{(a,d)}(\Omega^*)$ are determined in P08. Analytical calculations of $\overline{\mathcal{E}}$ in the above cited paper includes the effects of a small scale dynamo. In the above expressions for $\overline{\mathcal{E}}$, we assume equipartition between kinetic energy of the turbulence and magnetic fluctuations which stem from the small-scale dynamo. It was found that for the case of slow rotation (Co $\ll 1$), the part of $\overline{\mathcal{E}}$ that depends on the gradients of $\overline{B}$ consists of an isotropic eddy diffusivity and Rädler's $\boldsymbol{\Omega} \times \overline{\boldsymbol{J}}$ effect due to the small-scale dynamo (see also Rädler et al. 2003). In the case of rapid rotation, the fluctuating magnetic fields from the small-scale dynamo contribute both to isotropic and anisotropic parts of the diffusivity. The effect appears already in the terms of order Ω^2 in the global rotation rate (Rädler et al. 2003). In particular, the part of EMF which corresponds to Eq. (42) can be written as

$$\overline{\mathcal{E}}^{\eta} = -3\eta_T \left(2 f_1^{(a)} - f_2^{(d)}\right) \overline{\boldsymbol{J}} + 6\eta_T f_1^{(a)} \boldsymbol{\Omega} \frac{\boldsymbol{\Omega} \cdot \overline{\boldsymbol{J}}}{\Omega^2}. \tag{43}$$

It is noteworthy that the full expression for $\overline{\mathcal{E}}$ obtained in P08 is complicated and includes other contributions due to the effects of global rotation $\boldsymbol{\Omega}$, mean shear, mean current density $\overline{\boldsymbol{J}}$, and the magnetic deformation tensor $(\nabla \overline{\boldsymbol{B}})$. We skip them in the application to the solar dynamo model. The analytical results about the relations of the specific effects of $\overline{\mathcal{E}}$ and the global rotation rate show qualitative agreement with the DNS of Käpylä et al. (2009a) and Brandenburg et al. (2012). Yet, a more detailed comparison of the analytical results and the GCD simulations is needed; for further discussions, see Sect. 7.

We assume that the large-scale flow is axisymmetric. It is decomposed into the sum of meridional circulation and differential rotation, $\overline{\mathbf{U}} = \overline{\mathbf{U}}^m + r \sin\theta\, \Omega(r, \theta)\, \hat{\boldsymbol{\phi}}$, where r is the radial coordinate, θ is the polar angle, $\hat{\boldsymbol{\phi}}$ is the unit vector in the azimuthal direction, and $\Omega(r, \theta)$ is the angular velocity profile. The angular momentum conservation and the equation for the azimuthal component of large-scale vorticity, $\overline{\omega} = (\nabla \times \overline{\mathbf{U}}^m)_\phi$, determine the distributions of differential rotation and meridional circulation:

$$\frac{\partial}{\partial t} \overline{\rho} r^2 \sin^2\theta\, \Omega = -\nabla \cdot \left[r \sin\theta\, \overline{\rho} \left(\hat{\mathbf{T}}_\phi + r \sin\theta\, \Omega \overline{\mathbf{U}}^m \right) \right] + \nabla \cdot \left[r \sin\theta\, \frac{\overline{B B_\phi}}{4\pi} \right], \tag{44}$$

$$\frac{\partial \omega}{\partial t} = r \sin \theta \, \nabla \cdot \left[\frac{\hat{\boldsymbol{\phi}} \times \nabla \cdot \overline{\rho} \hat{\mathbf{T}}}{r \overline{\rho} \sin \theta} - \frac{\overline{\mathbf{U}}^m \omega}{r \sin \theta} \right] + r \sin \theta \frac{\partial \Omega^2}{\partial z} - \frac{g}{c_p r} \frac{\partial \overline{s}}{\partial \theta}$$

$$+ \frac{1}{4\pi \overline{\rho}} \left(\boldsymbol{B} \cdot \nabla \right) \left(\nabla \times \boldsymbol{B} \right)_\phi - \frac{1}{4\pi \overline{\rho}} \left[\left(\nabla \times \boldsymbol{B} \right) \cdot \nabla \right] \overline{B}_\phi,$$

where $\hat{\mathbf{T}}$ is the turbulent stress tensor:

$$\hat{T}_{ij} = \langle u_i u_j \rangle - \frac{1}{4\pi \overline{\rho}} \left(\langle b_i b_j \rangle - \frac{1}{2} \delta_{ij} \langle \mathbf{b}^2 \rangle \right); \tag{45}$$

see the detailed description in Pipin and Kosovichev (2018) and PK19. Also, $\overline{\rho}$ is the mean density, $\overline{s}$ is the mean entropy; $\partial/\partial z = \cos\theta \, \partial/\partial r - \sin\theta/r \cdot \partial/\partial\theta$ is the gradient along the axis of rotation. The mean heat transport equation determines the mean entropy variations from the reference state due to the generation and dissipation of the large-scale magnetic field and large-scale flows:

$$\overline{\rho} \overline{T} \left[\frac{\partial \overline{s}}{\partial t} + \left(\overline{\mathbf{U}} \cdot \nabla \right) \overline{s} \right] = -\nabla \cdot \left(\mathbf{F}^c + \mathbf{F}^r \right) - \hat{T}_{ij} \frac{\partial \overline{U}_i}{\partial r_j} - \overline{\mathcal{E}} \cdot \overline{\boldsymbol{J}}, \tag{46}$$

where $\overline{T}$ is the mean temperature, $\mathbf{F}^r$ is the radiative heat flux, $\mathbf{F}^c$ is the anisotropic convective flux; see PK19. The last two terms in Eq. (46) take into account the convective energy gain and loss caused by the generation and dissipation of large-scale magnetic fields and large-scale flows. The reference profiles of mean thermodynamic parameters, such as entropy, density, and temperature are determined from the stellar interior model MESA (Paxton et al. 2015). The radial profile of the typical convective turnover time, τ_c, is determined from the MESA code, as well. We assume that τ_c does not depend on the magnetic field and global flows. The convective rms velocity is determined from the mixing-length approximation,

$$u_c = \frac{\ell_c}{2} \sqrt{ -\frac{g}{2c_p} \frac{\partial \overline{s}}{\partial r} }, \tag{47}$$

where $\ell_c = \alpha_{\mathrm{MLT}} H_p$ is the mixing length, $\alpha_{\mathrm{MLT}} = 1.9$ is the mixing length parameter, and H_p is the pressure height scale. Equation (47) determines the reference profiles for the eddy heat conductivity χ_T, eddy viscosity ν_T, and eddy magnetic diffusivity η_T,

$$\chi_T = \frac{\ell^2}{6} \sqrt{ -\frac{g}{2c_p} \frac{\partial \overline{s}}{\partial r} }, \tag{48}$$

$$\nu_T = \mathrm{Pr}_T \, \chi_T, \tag{49}$$

$$\eta_T = \mathrm{Pm}_T \, \nu_T. \tag{50}$$

It should be noted that stellar convection might well have convection zones with slightly subadiabatic stratification in some layers. In those cases, the enthalpy flux can no longer be transported entirely by the mean entropy gradient, but there can be an extra term that is nowadays called the Deardorff term; see Deardorff (1972). Such convection can be driven through the rapid cooling in the surface layers and is therefore sometimes referred to as entropy rain Brandenburg (2016). It is useful to stress that the Deardorff term is distinct from the usual overshoot, because there the enthalpy flux points downward, while entropy rain still produces an outward enthalpy flux. It is instead more similar to semiconvection.

Boundary conditions. At the bottom of the tachocline, $r_i = 0.68\,R$, we assume solid body rotation and perfect conductor boundary conditions. Following to the MESA solar interior model, we put the bottom of the convection zone to $r_b = 0.728\,R$. At this boundary we fix the total heat flux, $F_r^{conv} + F_r^{rad} = L_\star(r_b)/4\pi r_b^2$. We introduce the decrease by $\exp(-100\,z/R)$ for all turbulent coefficients (except the eddy viscosity and eddy diffusivity), where z is the distance from the bottom of the convection zone. The decrease of the eddy viscosity and eddy diffusivity is at most one order of magnitude for numerical stability. The meridional circulation is restricted to the convection zone. Therefore, we put the azimuthal component of the large-scale vorticity to zero, i.e., we set $\overline{\omega} = 0$ at r_b. At the top, $r_t = 0.99\,R$, we employ a stress free and black body radiating boundary. Following ideas of Moss and Brandenburg (1992), we formulate the top boundary condition in a form that allows penetration of the toroidal magnetic field to the surface:

$$\delta \frac{\eta_T}{r_{top}} B \left[1 + \left(\frac{|B|}{B_{esq}} \right) \right] + (1 - \delta)\,\mathcal{E}_\theta = 0, \tag{51}$$

Free parameters. The model employs a number of free parameters, including C_α, the turbulent Prandtl numbers $\mathrm{Pr_T}$ and $\mathrm{Pr_{M,T}}$, δ, B_{esq}, and the global rotation rate Ω_0. For the solar case we use a period of rotation of the solar tachocline determined from helioseismology, $\Omega_0/2\pi = 434$ nHz (Kosovichev et al. 1997). The best agreement of the angular velocity profile with helioseismology results is found for $\mathrm{Pr_T} = 3/4$. Also, the dynamo model reproduces the solar magnetic cycle period, ~ 20 years, if $\mathrm{Pm_T} = 10$. Results of Pipin and Kosovichev (2011) showed that the parameters δ and B_{esq} affect the drift of the equatorial drift of the toroidal magnetic field in the subsurface shear layer and magnitude of the surface toroidal magnetic field. Solar observations show the magnitude of the surface toroidal field to be about 1-2 G (Vidotto et al. 2018). To reproduce it, we use $\delta = 0.99$ and $B_{esq} = 50$ G. In what follows, we present the results of the solar dynamo model for a slightly supercritical parameter C_α (10% above the threshold). Further details of the dynamo model can be found in PK19.

Figure 8 shows profiles of the basic turbulent effects and large-scale flow distributions for the nonmagnetic case. The amplitude of the meridional circulation on the surface is about

Fig. 8 (a) Streamlines of meridional circulation and the angular velocity distribution; the magnitude of circulation velocity is of 13 m/s on the surface at the latitude of 45°. (b) Radial profiles of $\eta_T + \eta_{||}$, the rotationally induced part $\eta_{||}$, as well as ν_T. (c) Radial profiles of the α tensor at 45° latitude. (d) Streamlines of effective drift velocity from magnetically affected pumping and meridional circulation. Reproduced by permission from Pipin (2022)

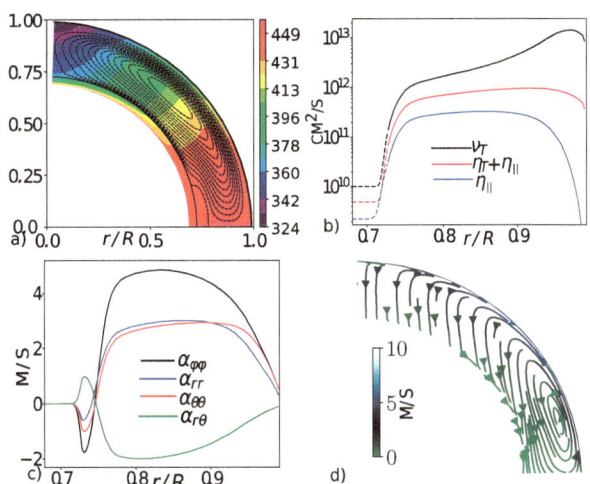

$13 \, \mathrm{m\,s^{-1}}$. In the lower part of the convection zone, the equatorward flow is about $1 \, \mathrm{m\,s^{-1}}$. The angular velocity profile is in agreement with helioseismology data.

Interestingly, the stagnation point of the meridional circulation is near the lower boundary of the subsurface shear layer, i.e., at $r = 0.9\,R$. This is in agreement with observations of Hathaway (2012) and the helioseismic inversions of Stejko et al. (2021). The structure of meridional circulation and turbulent pumping promotes an effective equatorward drift of the toroidal magnetic field below the subsurface shear layer; see Fig. 8(d).

6.2 Parker–Yoshimura Dynamo Waves and Extended Cycle

The dynamo shown in Fig. 9 demonstrates the numerical solution of the dynamo system including Eqs. (34) and (44)–(46). The time latitude diagrams of the surface radial magnetic field and the toroidal magnetic field in the upper part of the convection zone show agreement with observations of the evolution of the large-scale magnetic field of the Sun (Hathaway 2015; Vidotto et al. 2018, see also the review of Righmire in this volume). The dynamo waves propagate to the surface- and equatorward. The radial direction of propagation follows the Parker–Yoshimura rule because of the positive sign of the α effect in the main part of the convection zone and a positive latitudinal shear. It is noteworthy that at high latitude the model shows another dynamo wave family which propagates poleward along the convection zone boundary. This family follows the Parker–Yoshimura rule as well. Further we will see that the latitudinal shear plays a dominant role in this dynamo model and perhaps

Fig. 9 a) The surface radial magnetic field evolution (color image) and the toroidal magnetic field at $r = 0.9R$ (contours in range of ± 1kG); b) snapshots of the magnetic field distributions inside the convection zone for half dynamo cycle, color shows the toroidal magnetic field and contours show streamlines of the poloidal field; c) snapshots of the dynamo induced variations of zonal acceleration (color image) and streamlines of the meridional circulation variations (contours); d) variations of zonal velocity acceleration at the surface. This Figure was prepared using the data of the dynamo model of PK19

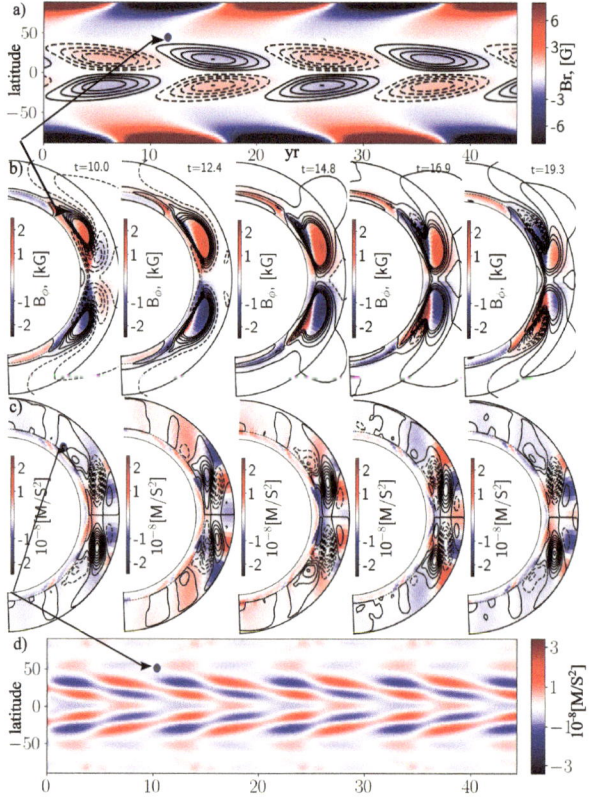

in the solar dynamo as well (see also Cameron and Schüssler 2015). The latitudinal drift of the toroidal magnetic field in this model results from turbulent pumping and meridional circulation; see Fig. 8(d). The GCD simulations of Warnecke et al. 2018, 2021 show the crucial role of turbulent pumping in the solar type dynamo model, as well. The extended mode of the dynamo cycle is another feature of their model. The toroidal magnetic dynamo wave starts at the bottom of the convection zone at around $50°$ latitude (see the marks in Fig. 9). It disappears near the solar equator after a full dynamo cycle. On the surface, the extended mode of the solar cycle is seen in the radial magnetic field evolution, in the torsional oscillations of zonal flow, and in the variations of the meridional circulation as well (Getling et al. 2021). The origin of the extended mode of the dynamo cycle is due to the distributed character of the large-scale dynamo and the interaction of the global dynamo modes, where the low order dynamo modes, e.g., dipole and octupole modes, are mainly generated in the deep part of the convection zone. The high order modes are predominantly generated in the near surface level. The phase difference between the models results in a dynamo mode of the extended length (Stenflo 1992; Obridko et al. 2021).

6.3 Torsional Oscillations

Solar zonal variations of the angular velocity ("torsional oscillations") were discovered by Howard and Labonte (1980). Since that time it was found that torsional oscillations represent a complicated wave-like pattern which consists of alternating zones of accelerated and decelerated plasma flows (Snodgrass and Howard 1985; Altrock et al. 2008; Howe et al. 2011). Ulrich (2001) found two oscillatory modes of these variations with periods of 11 and 22 years. Torsional oscillations were linked to ephemeral active regions that emerge at high latitudes during the declining phase of solar cycles, but represent the magnetic field of the subsequent cycle (Wilson et al. 1988). It is interesting that in their original paper, Howard and Labonte (1980) conjectured that the solar torsional oscillation can shear magnetic fields and induce the dynamo cycle. This idea was further elaborated upon in a number of papers. However, the idea looks unreasonable because it conflicts with Cowling's theorem. Also, the magnitude of the torsional oscillations of $3–6\,\mathrm{m\,s^{-1}}$ is too small in comparison with the magnitude of the magnetic field generated by a dynamo. The first papers by Schüssler (1981) and Yoshimura (1981) suggested that the 11-year solar torsional oscillation can be explained by the mechanical effect of the Lorentz force. The double frequency of the zonal variation results from the B^2 modulation of the large-scale flow due to the dynamo activity. On the basis of a flux-tube dynamo model, Schüssler (1981) used the simple estimate of a large-scale Lorentz force and found both the 11 and 22 year mode of the torsional oscillations. This result was further elaborated upon by Kleeorin and Ruzmaikin (1991). The further development of the mean-field theory of the solar differential rotation showed that, in addition to the large-scale Lorentz force, the dynamo induced B^2 modulation of the turbulent angular momentum fluxes is also an essential source of the torsional oscillations (Rüdiger and Kichatinov 1990; Kitchatinov et al. 1994b; Kleeorin et al. 1996; Küker et al. 1996; Rüdiger et al. 2012). Global convective dynamo simulations (e.g., Beaudoin et al. 2013; Käpylä et al. 2016; Guerrero et al. 2016) confirmed this conclusion. The strength of the solar torsional oscillations is more than two orders of magnitude less than the differential rotation. It looks like the theory of the torsional oscillations can be constructed using perturbative approximations. Models of this type (see, e.g., Tobias 1996, Covas et al. 2000, Bushby and Tobias 2007, Pipin 2015, Hazra and Choudhuri 2017) were inspired by results of Malkus and Proctor (1975). Yet, the constructed models are incomplete because they ignored the Taylor-Proudman balance, which is a key ingredient of solar differential rotation theory (see Kitchatinov 2013, and also the contribution of Hazra et al, this volume). Complete mean-field dynamo models, which take into accounts the Taylor-Proudman balance (hereafter TPB), were constructed by Brandenburg et al. (1992), Rempel (2007), and PK19. Figure 9 shows variations of the zonal acceleration for our mean-field model in following the PK19 line of work. Similar to the results of helioseismology (Howe et al. 2011; Kosovichev and Pipin 2019) and the results of Rempel (2007), snapshots of the model show that in the main part of the convection zone, the acceleration patterns are elongated along the rotation axis. This is caused by the Taylor-Proudman balance. Near the convection zone boundaries, these patterns deviate in the radial direction, which is in agreement with the above cited helioseismology results, as well. The given observation on the role of TPB shows the importance of the meridional circulation and the dynamo-induced heat transport perturbation (Spruit 2003; Rempel 2007) in the theory of torsional oscillations. This fact does not deny the importance of the large-scale Lorentz force and the magnetic modulation of the turbulent angular momentum transport. Results of Figs. 9(b) and (c) show that the positive sign of the zonal acceleration propagates from high latitudes at the bottom of the convection zone toward the equator, sticking to the equatorial edge of the dynamo wave. The torsional oscillation wave is accompanied by corresponding variations of the meridional circulation. These variations are induced by magnetic perturbations of the heat transport (see details in PK19). We emphasize that the given dynamo models also show overlapping magnetic cycles; see Fig. 9(b), similarly to what was originally proposed by Schüssler 1981]. In this case, the B^2 effect of the dynamo on the heat transport and the TPB results in about 4 to 5 meridional circulation cells along latitude. This tracks the zonal variations of angular velocity, which are caused by the mechanical action of the large-scale Lorentz force and magnetic quenching of the turbulent stresses, from polar regions to the equator. PK19 found that the induced zonal acceleration is $\sim (2\text{--}4) \times 10^{-8}\,\mathrm{m\,s^{-2}}$, which is in agreement with the observational results of Kosovichev and Pipin (2019). However, the individual forces in the angular momentum balance such that the large-scale Lorentz force, the variations of the angular momentum transport due to meridional circulation, the inertial forces, and others are by more than an order of magnitude stronger than their combined action and can reach a magnitude of $\sim 10^{-6}\,\mathrm{m\,s^{-2}}$. Therefore, the resulting pattern of the torsional oscillations forms in nonlinear balance, which includes the forces driving the angular momentum transport, the TPB, and heat perturbations due to magnetic activity in the convection zone (see details in PK19).

7 Mean-Field Models Based on the EMF Obtained from DNS

Here we provide an example of how the mean-field theory is utilized as *a tool for understanding what is going on* (see Sect. 3.4). We discuss recent studies of mean-field dynamo models constructed based on the electromotive force (EMF) obtained from direct numerical simulation (DNS) of rotating stratified convection, especially focusing on "semi-global" models. The properties of solar and stellar convection, and the various methods for extracting the information of the EMF from DNS are also summarized.

7.1 Properties of Solar and Stellar Convection

A quantitative physical description of solar and stellar dynamos, which should be the result of the nonlinear interaction of turbulent flows and magnetic fields, is a great challenge for us and constitutes a significant milestone on the long way to a full understanding of turbulence.

Even with state-of-the-art supercomputers, it is impossible to numerically simulate solar and stellar convection and its interaction with the magnetic field and to observe/analyze numerical data in detail with realistic parameters. Therefore, to say with confidence that one has fully understood the solar and stellar dynamo problem, it should be necessary to find a universal law of magneto-hydrodynamic (MHD) turbulence, build a reliable sub-grid scale (SGS) turbulence model, and then reproduce the magnetic activities of the Sun and stars quantitatively in an integrated framework by numerical models with incorporating the SGS model. This is because fluid quantities that may be verified in future observations should include the meridional distributions of fluid velocity, vorticity, kinetic helicity, and thus the turbulence model constructed on the basis of these profiles (e.g., Hanasoge et al. 2016). Only when the correctness of the turbulence model is observationally validated should our understanding of the solar and stellar dynamos as a consequence of the turbulent dynamo process be completed. In the near future, a very exciting time may come when we will be able to test and verify various turbulence models under extreme conditions inside the solar and stellar interiors.

What physical characteristics should be taken into account when constructing a turbulence model of thermal convection in the Sun and stars? Let us summarize some essential features:

1. **Extremely low dissipation**: turbulent state with $\mathrm{Re} \gtrsim 10^{12}$, $\mathrm{Re_M} \gtrsim 10^8$, and a large Pèclet number, $\mathrm{Pe} \sim 10^6$–10^9 (where $\mathrm{Pe} = \mathrm{Re} \cdot \mathrm{Pr}$).
2. **Huge separation of dissipation scales**: $\mathrm{Pr} \sim 10^{-4}$–10^{-7}, $\mathrm{Pr_M} \sim 10^4$
3. **Compressibility**: high Mach number $\mathcal{O}(1)$ in the upper convection zone makes the convective motion compressible.
4. **Anisotropy**: spin of stars (i.e., Coriolis force in a rotating system) makes fluid motions anisotropic.
5. **Inhomogeneity**: density contrast of 10^6 between top and bottom CZs results in multiscale properties of fluid motion.
6. **Non-locality**: Radiative energy loss at the CZ surface (open system), allowing the growth of cooling-driven downflow.

In view of these features, it can be seen that the characteristics of thermal convection operating inside the Sun and stars are quite different from those of isotropic turbulence. Those can be considered to some extent in DNS even with the current computing performance, as listed under 3–6, while the others, (items 1 and 2) are unreachable with current grid-based simulations. It should be emphasized, however, that higher resolution simulations using state-of-the-art supercomputers is a classical way forward in turbulence research, and the knowledge obtained from such studies in unexplored low-dissipation regimes will greatly expand the horizon of our understanding of turbulence (e.g., Kaneda et al. 2003; Hotta and Kusano 2021). Moreover, if sufficient scale separation between the turbulent and mean fields is ensured and the inertial range of the turbulent cascade is captured appropriately, there is the possibility that the evolution of mean-field components, such as large-scale flow and large-scale magnetic field, can be approximately reproduced even by simulations with enhanced dissipation compared to the actual solar and stellar values (e.g., Ossendrijver 2003). It should be remembered, however, that in spite of the rapid increase in computing power, some rather basic questions about the solar dynamo still remain, for example the equatorward migration of the sunspot belts and the formation of sunspots themselves.

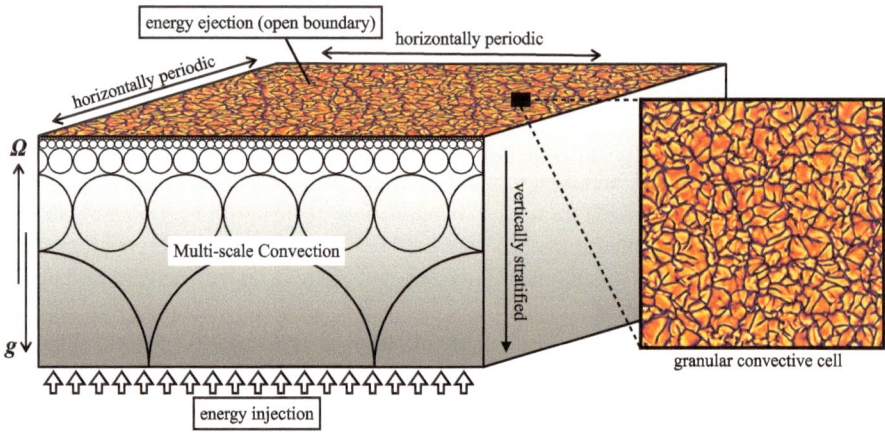

Fig. 10 Numerical setup typical for semi-global simulation of rotating stratified convection. Since the CZs of the Sun and stars are strongly stratified, there is a large separation of time scales from minutes (upper CZs) to months (bottom CZs)

7.2 Semi-Global Simulation of Rotating Stratified Convection

On our way toward a reliable SGS turbulence model for solar and stellar interiors, numerical models of convection and its dynamo should be studied, while keeping the characteristic features of solar and stellar convection, as listed under items 3–6 above, in mind. It should be noted that the underlying necessity for numerical modeling is an important component of earlier studies that applied mixing-length type concepts to the dynamo theory, which never successfully explained the magnetic activities of the Sun and stars (e.g., Brandenburg and Tuominen 1988).

In recent years, significant progress has been made in GCD simulations (e.g., Browning et al. 2006; Ghizaru et al. 2010; Käpylä et al. 2012; Masada et al. 2013; Fan and Fang 2014; Augustson et al. 2015; Hotta et al. 2016; Warnecke 2018), there is also a growing effort to extract the information of turbulent transport processes from so-called "semi-global" (or local model) MHD convection simulations with the aim of quantifying the dynamo effect of rotational stratified convection (e.g., Brandenburg et al. 1990, 1996; Nordlund et al. 1992; Brummell et al. 1998, 2002; Ossendrijver et al. 2001; Käpylä et al. 2006a, 2009b; Masada and Sano 2014b,a, 2016; Bushby et al. 2018; Masada and Sano 2022). A typical numerical setup of the semi-global model is shown in Fig. 10 schematically. In this setting, the gas is gravitationally stratified in the vertical direction, while periodicity is assumed in the horizontal directions. The governing equations (mostly compressible MHD equations) are solved in a rotating Cartesian frame, and the rotation axis is usually set to be parallel or anti-parallel to the gravity vector. Several studies have simulated the model with the tilt of the rotation axis with respect to the gravity vector, and the latitudinal dependence of the convection has been investigated (e.g., Ossendrijver et al. 2001; Käpylä et al. 2004, 2006a).

7.3 Extraction of Information of Dynamo Effects

In the semi-global studies, four-types of approaches have been used typically to extract the information of dynamo effects veiled in the convective motion. The starting point of all the four methods is common, the decomposition of the flow field (U) and magnetic field

(B) into a spatially large-scale, slowly-varying mean-component, and a small-scale, rapidly varying fluctuating component, as introduced in § 1, i.e., $U = \overline{U} + u$ and $B = \overline{B} + b$, where the lower-case represent the fluctuating component and the overbars denote the mean component. In the case of a semi-global model, a temporal and horizontal average is often used for deriving the mean component. Then, the equation of mean-field electrodynamics can be derived

$$\frac{\partial \overline{B}}{\partial t} = \nabla \times (\overline{U} \times \overline{B} + \overline{\mathcal{E}} - \eta \nabla \times \overline{B}) , \tag{52}$$

where $\overline{\mathcal{E}} = \overline{u' \times b'}$ is the mean electromotive force (EMF) due to the fluctuation of the flow and the magnetic field. The mean EMF can be described as a power series about the large-scale magnetic component and its derivatives as

$$\overline{\mathcal{E}} = \overline{u \times b} = \alpha \cdot \overline{B} + \gamma \times \overline{B} - \beta \cdot (\nabla \times \overline{B}) + \cdots , \tag{53}$$

where α represents (tonsorial form of) the α-effect, γ is the turbulent pumping, and β denotes the turbulent diffusion.

To obtain the information of the dynamo coefficients, such as α, γ, and η_T, from the MHD convection simulation, there are the following four methods (see general discussion in Sect. 3.3):

(i) Methods based on results of analytical theories, e.g., SOCA (or first-order smoothing approximation, FOSA) expressions
(ii) Imposed-field method
(iii) Test-field method
(iv) The multi-dimensional regression method based on the dynamo generated magnetic field

Method (i) involves the estimation of dynamo coefficients based on results of analytical theory. It also exploit the mixing-length approximation in final results. There, the distributions of, for example, the fluctuating components of the convection velocity (u), vorticity ($\omega = \nabla \times u$), and the resulting kinetic helicity ($\mathcal{H} = \omega \cdot u$), are directly extracted from the simulation results and used to reconstruct the turbulent α and β via their analytic forms, derived under SOCA, such as Eq. (16) and $\beta = (\tau/3)\overline{u^2}$, where τ is the correlation time of the turbulence and is often replaced by the convective turnover time, τ_c. Note that anisotropy effects are often neglected in the expressions above, but see Brandenburg and Subramanian (2007), who included them.

Method (ii) is mainly used in the analysis of the numerical results without self-sustained magnetic field. In this method, a uniform external magnetic field is imposed as the mean component to the computational domain, artificially. The α effect and turbulent pumping are then inferred from $\overline{\mathcal{E}} = \overline{u \times b}$ by directly calculating from simulation data

$$\alpha = \overline{\mathcal{E}} \cdot \overline{B}/\overline{B}^2 , \quad \gamma = -\overline{\mathcal{E}} \times \overline{B}/\overline{B}^2 . \tag{54}$$

Furthermore, one might be tempted to compute $\eta_t = \overline{\mathcal{E}} \cdot \overline{J}/(\mu_0 \overline{J}^2)$, but this would assume that $\overline{J} \cdot \overline{B}$ is vanishing, which is generally not the case for α-effect dynamos; see Hubbard et al. (2009) for details.

Method (iii) utilizes a so-called test field, as introduced by Schrinner et al. (2005, 2007) for the spherical case and Brandenburg et al. (2008) for the Cartesian case, allowing for scale dependence. In this method, the evolution equation of b'_T, the fluctuating component of the test field $\boldsymbol{B}_\mathrm{T}$, which are passive to the velocity field taken from the simulation, is solved additionally to the basic (MHD) equations. From the linear evolution of the test-field, the mean EMF is evaluated and then the full set of turbulent transport coefficients can be obtained. For example, in the case without the large-scale flow, the test-field equation is, for $\boldsymbol{b}_\mathrm{T}$,

$$\frac{\partial \boldsymbol{b}_\mathrm{T}}{\partial t} = \nabla \times \left(\boldsymbol{u} \times \boldsymbol{B}_\mathrm{T} + \boldsymbol{u} \times \boldsymbol{b}_\mathrm{T} - \overline{\boldsymbol{u} \times \boldsymbol{b}_\mathrm{T}} - \eta \nabla \times \boldsymbol{b}_\mathrm{T} \right), \tag{55}$$

with a chosen test field $\boldsymbol{B}_\mathrm{T}$ while taking $\boldsymbol{u}$ from the MHD simulation. The original method does not work in cases when magnetic fluctuations are driven through an artificially added electromotive force. In such cases, one needs to use a more general nonlinear method explored by Rheinhardt and Brandenburg (2010) and Käpylä et al. (2022). Theses magnetically driven systems are also examples where the imposed-field method gives reliable results in two dimensions with volume averaged mean fields, where no turbulent diffusion can act.

Method (iv) can be used only in the analysis of the numerical results with self-sustained magnetic fields. Since the fluctuating and mean components are all known quantities in such simulations, the mean *emf*, $\overline{\mathcal{E}} = \overline{\boldsymbol{u} \times \boldsymbol{b}}$, and the mean magnetic component, $\overline{\boldsymbol{B}}$, can be directly calculated from the simulation data. Then, the mean profiles of dynamo coefficients are inferred based on a fitting procedure via the relationship,

$$\mathcal{E}_i = \alpha_{ij} \overline{B}_j + \epsilon_{ijk} \gamma_j \overline{B}_k + \text{higher derivative terms}. \tag{56}$$

Given $\mathcal{E}_i$ and $\overline{B}_i$ which are calculated from simulation data, and then find α_{ij} and γ_i such that the residual of Eq. (56) is minimized. In the equation above, the contributions from the derivatives of the mean magnetic component to the mean *emf* are neglected (see, e.g., Racine et al. 2011; Simard et al. 2013, 2016; Shimada et al. 2022, for the fitting based analysis of the dynamo coefficient with including the contribution from the first-order derivative of the mean magnetic component). As we have mentioned in Sect. 3.3, the results of Warnecke et al. (2018) show that the TFM is more accurate compared to the regression method in obeying Eq. (56); see the detailed comparison in above cited paper.

In all cases, however, the first (and often higher) derivative terms are of the same order as the first term and can therefore not be neglected. This was already done in the work of Brandenburg and Sokoloff (2002), who typically found small diffusion coefficients in the cross-stream direction. This, however, turned out to be a shortcoming of the method and has not been borne out by more advanced measurements (Karak et al. 2014).

7.4 Transport Coefficients from Semi-Global Turbulence Simulations

Here, we briefly review the results of previous semi-global simulations, with a particular focus on the studies that have been dedicated for extracting information about dynamo coefficients.

Brandenburg et al. (1990), hereafter B90, performed turbulent 3-D magneto-convection simulations under the influence of the rotation for the semi-global model whose depth is equivalent to about one pressure scale height. They found that, due to the effect of the rotation, a systematic separation of positive and negative values of the kinetic helicity was developed in the vertical direction of the CZ, i.e., in the upper CZ, negative (positive) helicity in the northern (southern) hemisphere, while positive (negative) helicity in the northern (southern) hemisphere. Using the imposed field method, they evaluated the magnitude of the turbulent α-effect with anisotropic properties as $\alpha_V/(\tau\mathcal{H}) \sim \mathcal{O}(0.1)$ and $\alpha_H/(\tau\mathcal{H}) \sim \mathcal{O}(0.01)$, where $\mathcal{H} = \boldsymbol{\omega} \cdot \boldsymbol{u}$ and $\overline{\boldsymbol{\mathcal{E}}} = \alpha_{\mathrm{H}} \overline{\boldsymbol{B}}_{\mathrm{H}} + \alpha_{\mathrm{V}} \overline{\boldsymbol{B}}_{\mathrm{V}}$. It is interesting to note that these values are about one to two orders of magnitude smaller than $\alpha \sim \Omega d$, which is the estimation based on the mixing-length theory. Additionally, it was also suggested that the magnetic helicity showed a similar depth variation, but the sign was opposite to that of the kinetic helicity.

While α_{H} had the expected sign (opposite to that of the kinetic helicity), α_{V} was found to have the 'wrong' sign (same as that of the kinetic helicity). Such a result was subsequently also obtained by Ferriere (1993). The theoretical possibilities for such effects should be studied further. For example, Rüdiger and Pipin (2000) found that large-scale shear can affect both the sign of the α effect and kinetic helicity in magnetically driven compressible turbulence in such a way that they have the same sign, e.g., for Keplerian accretion disks. These ideas were also applied to understanding the finding of a negative α effect in stratified accretion disk simulations (Brandenburg 1998).

Ossendrijver et al. (2001) also performed the semi-global simulation with a similar model as B90. They showed that, even in the regime where the condition justifying the FOSA (or SOCA) is not satisfied, i.e., in the situation where St $= u_{\mathrm{rms}} \tau / d \gtrsim 1$ and $Re > 1$, the kinetic helicity was clearly separated into positive and negative values at the lower and upper CZs when taking temporal average of the convective motion over sufficiently long time. Using the imposed field method, they also measured the magnitude of the turbulent α-effect and obtained similar values to B90 in terms of α_H and α_V. The rotational dependence of the α-effect was also investigated in this work for the first time. They showed that, in the larger Co regime, the α_V underwent a rotational quenching, while the α_H was saturated, where Co is the Coriolis number [see Eq. (37)]. The turnover time was defined, in this work, as $\tau = d / u_{\mathrm{rms}}$. While the depth-dependence or rotational dependence of the α, which was obtained from the simulation, agreed, to some extent, with a theoretical model based on the mixing-length theory (Rüdiger and Kichatinov 1993), their amplitudes were one to two orders of magnitude smaller than those predicted from the theoretical model. Noteworthy, the critical threshold of the α effect parameter in mean-field dynamo models (see Sect. 6.1) is about same magnitude less than the mixing-length models of the solar convection zone predicts; see Sect. 6.

In Käpylä et al. (2004, 2006a), additionally to the rotational dependence, the latitudinal dependence of the turbulent α-effect was studied in the semi-global convection simulations with varying the inclination of the rotation axis with respect to the gravity vector. With the imposed field method, they found that, for slow and moderate rotation with Co < 4, the latitudinal dependence of the α followed $\cos \theta$ profile with a peak at the pole (see also, Egorov et al. 2004), while, in the rapid rotation regime with Co ≈ 10, it rather peaked much closer to the equator at $\theta \simeq 30°$. Additionally, the vertical profile of the α directly evaluated from simulation was found to be qualitatively consistent with analytic expression derived under the FOSA even when changing the latitude. A practical application of these results was the development of a kinematic mean-field solar dynamo model in Käpylä et al. (2006b). In it, the rotation profile deduced from the helioseismic observation and the meridional profiles of the α-effect and turbulent pumping obtained with the semi-global simulation of Käpylä et al. (2006a) are integrated into the framework of the α–Ω dynamo, and then the solar dynamo mean-field model was constructed. It is interesting that their kinematic dynamo model correctly reproduced many of the general features of the solar magnetic activity, for example realistic migration patterns and correct phase relation.

The existence of large-scale dynamo, i.e., self-excitation of the mean magnetic component, in rigidly-rotating convection was demonstrated for the first time in the semi-global simulation by Käpylä et al. (2009b). By changing the angular velocity, they showed that the large-scale dynamo could be excited only when the rotation is rapid enough, i.e., Co $\gtrsim 60$, with Eq. (37) as the definition of Co which is same as that used in Ossendrijver et al. (2001) and Käpylä et al. (2006a); see, e.g., Tobias et al. (2008) and Cattaneo and Hughes (2006), and Favier and Bushby (2013), for unsuccessful large-scale dynamo in rigidly-rotating convection probably due to slow rotation, and/or short integration time. From the measurements of the turbulent α-effect and the turbulent diffusivity by the TFM, they also suggested that while the magnitude of the α-effect stayed approximately constant as a function of rotation, the turbulent diffusivity decreased monotonically with increasing the angular velocity, resulting in the excitation of the large-scale dynamo in the higher Co. The reliability of the dynamo coefficients extracted with the test-field method from the simulation was validated with the one-dimensional mean-field dynamo model in which the test-field results for α and β were used as input parameters by studying the excitation of the large-scale magnetic field at the linear stage. Note that the oscillatory properties of the large-scale dynamo in rigidly-rotating convection and its possible relationship with α^2 dynamo mode with inhomogeneous α profile were also found in Käpylä et al. (2013); see, e.g., Baryshnikova and Shukurov (1987) and Mitra et al. (2010b) for the oscillatory α^2 dynamo.

7.5 Mean-Field Dynamo Models Linked with DNSs

7.5.1 Weakly-Stratified Model

Below we review recent mean-field dynamo models linked with semi-global MHD convection simulations, where the large-scale dynamo is successfully operated; see Masada and Sano (2014a,b, 2016, 2022) for a series of numerical studies.

While Käpylä et al. (2009b, 2013) were the first to demonstrate that rigidly-rotating convection can excite the large-scale dynamo as reviewed above, their simulation model was a so-called "three-layer polytrope" consisting of top and bottom stably-stratified layers and the CZ in between them. Therefore, it was suspected for a while that the essential factor for the successful large-scale dynamo observed there might be the presence of the stably-stratified layer assumed in their model rather than the rapid rotation (e.g., Favier and Bushby 2013). To pin down the key requirement for the large-scale dynamo, the impact of the stably-stratified layers on the large-scale dynamo was studied in Masada and Sano (2014b), hereafter MS14a, in which two-types of semi-global models with and without stably-stratified layers are compared with the same control parameters and the same grid spacing. It was found in this study that a large-scale dynamo was successfully operated even in the model without the stably-stratified layer, and confirmed that the key requirement for it should be a rapid rotation if we evolved the simulation for a sufficiently long time than the ohmic diffusion time. Note that a relatively weak density stratification (the density contrast between the top and bottom CZs is about 10) was assumed in the simulation model employed in this study as well as Käpylä et al. (2009b, 2013).

With these results, Masada and Sano (2014a), hereafter MS14b, explored the mechanism of the large-scale dynamo operated in the rigidly-rotating stratified convection by linking the mean-field (MF) dynamo model with the DNS. In this study, the FOSA based approach was adopted in the MF modeling. The mean vertical profiles of the kinetic helicity and root-mean square velocity were directly extracted from the simulation data and then the vertical profiles of the turbulent α, turbulent pumping (γ) and turbulent diffusivity (β) were reconstructed according to the analytic expressions of

$$\alpha(z) = -\tau_c \overline{(u_z \partial_x u_y - u_z \partial_y u_x)}, \quad \gamma(z) = -\tau_c \partial_z \overline{(u_z)^2}, \quad \beta(z) = \tau_c \overline{(u_z)^2}, \quad (57)$$

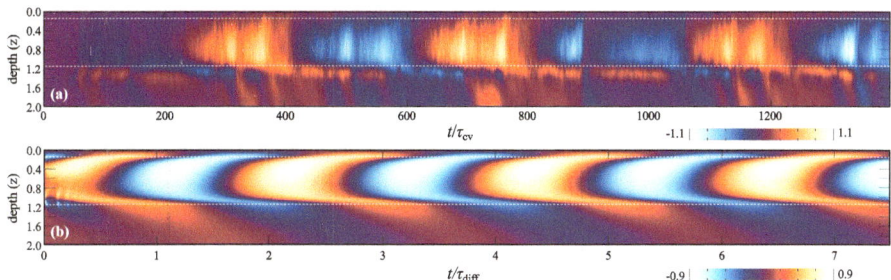

Fig. 11 Time-depth diagram $\overline{B}_x(z, t)$ for the MF model (panel (b)) and its DNS counterpart (panel (a)). For DNS result, the horizontal average of the magnetic field is shown. The orange and blue tones represent positive and negative $\overline{B}_x$ in units of $B_{cv} \equiv \overline{(\rho u^2)}^{1/2}$. Time is normalized by τ_c. Note that $\overline{B}_y$ shows a similar cyclic behavior with $\overline{B}_x$ yet with a phase delay of $\pi/2$; see MS14a,b for details

in anisotropic forms of dynamo coefficients under the FOSA (e.g., Käpylä et al. 2006a). Although recent numerical studies indicate that the small-scale current helicity, i.e., $\mathbf{j} \cdot \mathbf{b}$, is important for the α-effect when the magnetic field is dynamically important (Pouquet et al. 1976; Brandenburg and Subramanian 2005b), its contribution was ignored in this study. As the correlation time τ_c, the convective turnover time defined by $\tau = H_\rho(z)/u_{rms}$ was chosen there (H_ρ is the density scale-height as a function of the depth). By solving one-dimensional MF α^2 dynamo equation in which these profiles were used as input parameters, i.e.,

$$\frac{\partial \overline{\mathbf{B}}_h}{\partial t} = \nabla \times (\overline{\mathcal{E}} - \eta \nabla \times \overline{\mathbf{B}}_h) , \tag{58}$$

with

$$\overline{\mathcal{E}} = \alpha(z)\overline{\mathbf{B}}_h + \gamma(z)\mathbf{e}_z \times \overline{\mathbf{B}}_h - \beta(z)\nabla \times \overline{\mathbf{B}}_h , \tag{59}$$

the time-depth diagram for the mean (horizontal) magnetic component ($\overline{\mathbf{B}}_h$) was obtained. In Fig. 11, we show $\overline{B}_x(z, t)$ for the MF model (panels (b)) and its DNS counterpart (panels (a)). Note that, for ensuring the saturation of the magnetic field growth, the quenching effect was also taken into account. Since the DNS results were quantitatively reproduced by the MF α^2 dynamo, MS14b concluded that the large-scale magnetic field organized in the rigidly-rotating turbulent convection was a consequence of the oscillatory α^2 dynamo.

Reproducing the DNS results with mean-field models using coefficients from the original DNS is an important verification of the whole approach. This has been done on many occasions in the past; see, for example, the work by Gressel (2010) and Warnecke et al. (2021).

7.5.2 A Strongly Stratified Model

In MS14a,b, a weakly-stratified model, in which the density contrast between top and bottom CZs is about 10, was adopted. However, the actual Sun has very strong stratification with a density contrast of $\sim 10^6$ between top and bottom CZs, resulting in a large segregation of time scales from minutes to months. Aiming at the application to solar and stellar interiors, Masada and Sano (2016), hereafter MS16, performed a convective dynamo simulation in a strongly stratified atmosphere with a density contrast of 700 in a semi-global

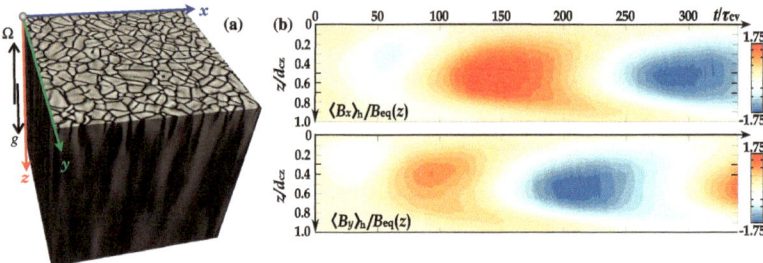

Fig. 12 (a) 3-D view of the strongly stratified convection (for the progenitor run without rotation). The black (gray) tone denotes downflows (upflows). (b) Time-depth diagrams for $\overline{B}_x$ and $\overline{B}_y$. The normalization is the equipartition field strength, $B_{eq} \equiv (\overline{\rho u^2})^{1/2}$. In MS16, a one-layer polytrope with a super-adiabaticity of $\delta \equiv \nabla - \nabla_{ad} = 1.6 \times 10^{-3}$ was used; see MS16 and MS22 for details

Fig. 13 Snapshots for (a) the horizontal distribution of B_z at the CZ surface, and (b) its Fourier filtered image. In panel (b), the small-scale structures with $k/k_c \gtrsim 8$ are eliminated for casting light on the large-scale pattern (k is the wavenumber and $k_c = 2\pi/L_h$ with the horizontal box size L_h)

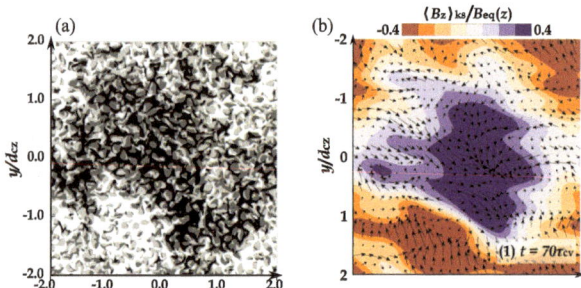

setup. Due to the strong solar- and stellar-like density stratification, multi-scale convection with a strong up-down asymmetry, i.e., slower and broader upflow regions surrounded by a network of faster and narrower downflow lanes, was developed in this simulation, as shown in Fig. 12(a). Even in such a situation, a large-scale dynamo was found to operate. As shown in Fig. 12(b), the mean magnetic field components observed there showed a time–depth evolution similar to that seen in the weakly-stratified model (MS14a,b), suggesting that an oscillatory α^2 dynamo is responsible for it. It was intriguing that, in addition to the mean horizontal component, the large-scale structures of the vertical magnetic field were spontaneously organized at the CZ surface in the case of the strongly stratified atmosphere, as shown in Fig. 13.

A possible physical origin of such surface magnetic structure formation is the negative magnetic pressure instability (NEMPI; see § 8 for details). NEMPI is a mean-field process in the momentum equation, where the Reynolds and Maxwell stresses attain a component proportional to the square of the mean magnetic field, which acts effectively like a negative pressure by suppressing the turbulent pressure. Since its growth rate becomes larger for stronger density stratification (e.g., Jabbari et al. 2014), one can imagine that it may play an important role in organizing sunspot-like large-scale magnetic field structures in the upper part of the solar CZ. Although its presence has been confirmed numerically in forced MHD turbulence (e.g., Brandenburg et al. 2011; Warnecke et al. 2013), it does not play a significant role in organizing the surface magnetic structure seen in MS16 because of their relatively rapid rotation; Ro = 0.02 was assumed there, while, according to Losada et al. (2012), Ro $\gtrsim$ 5 is required to excite the NEMPI.

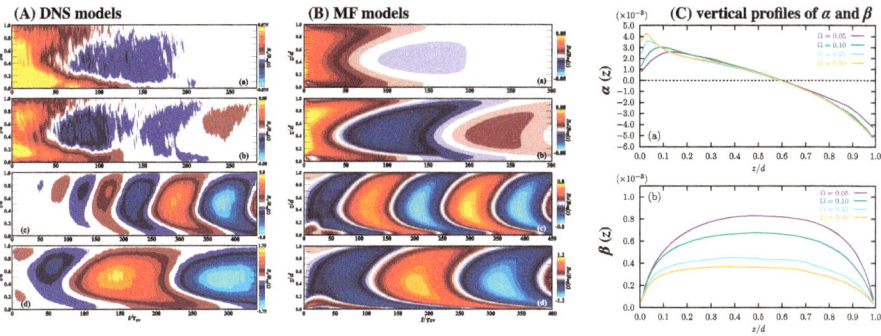

Fig. 14 Time-depth diagrams of $\overline{B}_x$ for (A) DNS models and (B) MF models. (C) Vertical profiles of the turbulent α-effect (top) and turbulent diffusivity β (bottom) which are reconstructed with the analytic expressions of Eq. (57) from the information, such as kinetic helicity and rms velocity, directly extracted from DNSs

The large-scale structure of the vertical magnetic field observed in MS16 is similar to that observed in the large-scale dynamo by forced turbulence in a strongly stratified atmosphere (Mitra et al. 2014; Jabbari et al. 2016). This implies that there would be an as-yet-unknown mechanism for the self-organization of large-scale magnetic structures, which would be inherent in a strongly stratified atmosphere.

In Masada and Sano (2022), hereafter MS22, with varying angular velocity as a control parameter, the rotational dependence of the large-scale dynamo was explored in a series of DNSs of rigidly-rotating convection. They linked its cause through MF dynamo models with DNSs where a strongly stratified polytrope was adopted as a model of the convective atmosphere, as in MS16.

In Fig. 14(a), DNS results are shown where a time-depth diagram of $\overline{B}_x$ is depicted for models with different values of Co. While in the slowly rotating model with low Co, the large-scale magnetic component fails to grow, it is found to be spontaneously organized in the rapidly-rotating models with high Co. It was found from DNS that the large-scale dynamo was excited when Co $\gtrsim$ Co$_{\mathrm{crit}}$, where Co$_{\mathrm{crit}}$ is the critical Coriolis number in the range $25 \lesssim$ Co$_{\mathrm{crit}} \lesssim 80$, with Eq. (37) as the definition of the Coriolis number. It is remarkable that Co$_{\mathrm{crit}}$, which determines the success or failure of the large-scale dynamo, was almost the same – independent of the density contrast (see Käpylä et al. 2009b) or the geometry of the simulation model (see, e.g., Käpylä et al. 2012; Warnecke 2018, for Co$_{\mathrm{crit}}$ in the global simulations); see MS22 for the quantitative comparison between models.

To explore the underlying mechanism of the rotational dependence of the large-scale dynamo, the influence of the rotation on the turbulent transport coefficients was also studied in MS22 with the FOSA-based approach similar to that adopted in MS14b. In Fig. 14(c), the vertical profiles of the turbulent α effect and turbulent diffusivity β reconstructed with the analytic expressions of Eq. (57) were shown. It can be found that, as the spin rate increases, the turbulent diffusivity decreases significantly, but the profile of the α effect remains almost unchanged. This result suggested that the rotational dependence of the large- scale dynamo observed in MS22 may be primarily due to a change in the magnitude of the turbulent diffusion. In fact, this insight was confirmed by the evidence that the MF dynamo model with incorporating the dynamo coefficients shown in Fig. 14 reproduced quantitatively the result of the DNS; see Fig. 14(b) for the time-depth diagram of $\overline{B}_x$ obtained in the MF models with using different dynamo coefficients extracted from the corresponding DNSs.

MS22 concluded that the independence (dependence) of the turbulent α effect (turbulent diffusivity) on the rotation was the essence of the rotational dependence of the dynamo. This is not only the same as the conclusion obtained by Käpylä et al. (2009b) from weakly-stratified convective dynamo simulations using the TFM, but also the same as that obtained by Shimada et al. (2022) from global solar dynamo simulation with using the "self-sustained field method". Although we don't know whether the independence (dependence) of the α-effect (turbulent diffusion) on the rotation, seen in these studies, is universal or not, it may give an important suggestion not only on the turbulence modeling but on the solar dynamo modeling.

8 Solar-Stellar Connections and Questions Beyond Standard Mean-Field Theory

The title of our review is "Turbulent processes and mean-field dynamo" Obviously, there are mean-field effects in turbulence that are not just of dynamo type, and the mean-field dynamo is not just related to the solar dynamo problem, but its relevance goes much beyond. Here, we highlight, just some effects, but we refer to the reviews of Käpylä et al. (2023) for additional examples.

8.1 Origin of Sunspots and Active Regions

An important goal in solar dynamo theory is to compute synthetic butterfly diagrams. The question then emerges from which depth to take the mean toroidal field, for example. The usual argument here is to invoke Parker's theory of sunspot formation and to postulate that the field at some depth translates directly to one at the surface. This is critical because the final result depends on the assumed depth.

It is possible that sunspots are not deeply rooted, but are actually a surface phenomenon. No successful and self-consistent model of shallow formation of active regions or sunspots exists as yet. Noteworthy in this context is NEMPI, which is a mean-field theory of the Reynolds and Maxwell stresses. This theory is extremely successful in that its results agree remarkably well with direct numerical simulations (DNS).[2] The problem is only that the effect is not strong enough to make real sunspots or active regions. Because of this remarkable agreement between theory and simulations, and because it is an important mean-field process, we shall discuss here a bit more detail.

The essence of the effect is the contribution of the turbulent hydromagnetic pressure to the horizontal force balance. The turbulent pressure is a small-scale effect, but it reacts to the large-scale magnetic field. As the magnetic field increases, it suppresses the turbulence locally, disturbing therefore the horizontal force balance. Although this large-scale magnetic field itself contributes with its own magnetic pressure to the horizontal force balance, the effect from the suppression of the turbulence is often stronger, so the net effect is a negative one. This is why the mean-field effect from a large-scale magnetic field is a negative effective

---

[2]DNS means that viscous and diffusive operators are assumed to be the physical ones, but with coefficients that are enhanced relative to the physical ones, but as small as possible. The ordering of these coefficients is often preserved so as to access the relevant regimes with small magnetic and thermal Prandtl numbers. Large eddy simulations (LES) or implicit LES, by contrast, use just numerical schemes to keep the code stable. Such schemes are often too complicated to state them as an explicit term in the equations, as if they are negligible, but they never are.

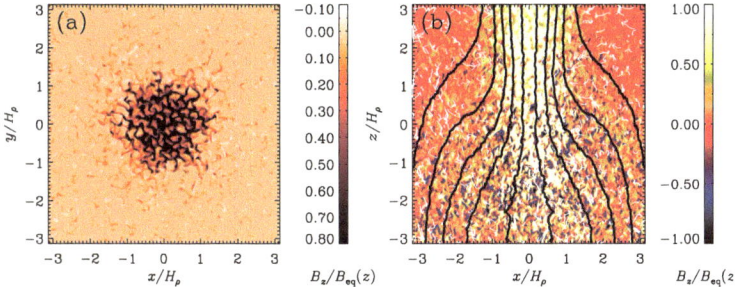

Fig. 15 Cuts of $B_z/B_{eq}(z)$ in the xy plane at the top boundary ($z/H_\rho = \pi$) and the xz plane through the middle of the spot at $y = 0$. In the xz cut, we also show magnetic field lines and flow vectors obtained by numerically averaging in azimuth around the spot axis. Adapted from Brandenburg et al. (2013)

magnetic pressure. This idea goes back to early work of Kleeorin et al. (1989, 1996), who developed the mean-field theory for this effect.

In the beginning, it was not clear what kind of numerical experiments one could try to test the a negative effective magnetic pressure effect. The first mean-field simulations were done with a uniform horizontal magnetic field (Brandenburg et al. 2010). This led to the development of magnetic flux concentrations near the surface, but those began to sink downward as time went on. A similar effect was soon also seen in DNS (Brandenburg et al. 2011). The sinking of such structures was explained by the *negative* effective magnetic pressure: a positive magnetic pressure would lead to the rise of structures (Parker 1967) while a negative one leads to a sinking. The sinking of magnetic structures had the side effect that the structures disappeared from the surface and became less prominent.

Subsequent experiments with a vertical field had a more dramatic effect on the general appearance of structures. Because the ambient field was vertical, the downflow had little effect on the magnetic flux concentrations themselves (Brandenburg et al. 2013). Figure 15 shows the spontaneous development of a magnetic spot. Brandenburg et al. (2013) found that NEMPI saturates slightly below equipartition value.

Most of the numerical experiments where done with forced turbulence, where one had explicit control over the degree of scale separation. For the results presented here $Re_M \sim 19$ and $Pr_M = 0.5$. The value of $Re_M = u_{rms}/\eta k_f$ is very small because it is based on the forcing wavenumber k_f, which is chosen to be large. Furthermore, Pr_M is chosen to be less than unity, because NEMPI is not expected to work for $Pr_M \geq 14$. This is rather different in actual stellar convection, where the development of magnetic structures takes different shapes (Stein and Nordlund 2012; Masada and Sano 2016; Käpylä et al. 2016).

8.2 Dynamo Flux Budget and Impact of the Surface Activity on the Deep Dynamo

Following Cameron and Schüssler (2015) (hereafter, CS15, also see Cameron and Schüssler 2023, this collection) we now estimate the budget of the toroidal magnetic flux in the dynamo region. Using the Stokes theorem and the induction equation Eq. (34), we define the derivative of the toroidal magnetic field flux in the northern hemisphere of the Sun as

$$\frac{\partial \Phi^N_{tor}}{\partial t} = \oint_{\delta\Sigma} \left(\overline{U} \times \overline{B} + \mathcal{E} \right) \cdot dl, \qquad (60)$$

where $\Phi^N_{tor} = \int_\Sigma \overline{B}_\phi dS$, Σ is the meridional cut of the northern hemisphere of the solar convection zone, $\delta\Sigma$ stands for the contour confining the cut and the differential dl is the

Fig. 16 Estimation of contributions of the budget equation; see Eq. (61), for the time of cycle minimum

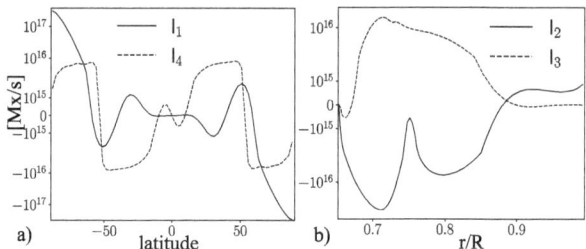

line element of $\delta\Sigma$. The same can be written for the southern hemisphere flux Φ_{tor}^S. Similarly to CS15, we use the boundary conditions, and we estimate the RHS of the Eq. (60) in the coordinate system which is co-rotating with angular velocity of the solar equator, $\overline{U}_{0\phi} = R\sin\theta\,\Omega_0$, and Ω_0 the surface angular velocity at the equator,

$$\frac{\partial\Phi_{\text{tor}}^N}{\partial t} = \int_0^{\pi/2}\overbrace{\left(\overline{U}_\phi - \overline{U}_{0\phi}\right)\overline{B}_r r_t\,d\theta}^{I_1} + \int_{r_i}^{r_t}\overbrace{\left(\overline{U}_\phi^{(\frac{\pi}{2})} - \overline{U}_{0\phi}\right)\overline{B}_\theta^{(\frac{\pi}{2})}\,dr}^{I_2} \tag{61}$$

$$+ \int_{r_i}^{r_t}\overbrace{\left(\mathcal{E}_r^{(0)} - \mathcal{E}_r^{(\frac{\pi}{2})}\right)\,dr}^{I_3} + \int_0^{\pi/2}\overbrace{\left(\mathcal{E}_\theta^{(t)}r_t - \mathcal{E}_\theta^{(i)}r_i\right)d\theta}^{I_4}$$

here, $r_t = 0.99\,R$, $r_i = 0.67\,R$, are the radial boundaries of the dynamo region. In compare to CS15 we have additional contributions in the budget equation. Figure 16 shows profiles of the kernels I_{1-4} for the period of the magnetic cycle minimum. The estimations are based on results and parameters of the mean-field model presented above. Noteworthy, the south hemisphere should show the profiles of the opposite sign (see CS15). The results for $I_{1,4}$ qualitatively similar to CS15. This is because the mean-field model show the qualitative agreement with solar observations for the time latitude evolution of the surface radial magnetic field. The diffusive decay of the toroidal magnetic flux is captured as well because of the phase shift between evolution of the poloidal and toroidal magnetic field in dynamo model and presumably in the solar dynamo as well. The model show the sharp poleward increase of I_1. This effect produces the winding of the toroidal magnetic field from poloidal component by the latitudinal shear. The effect of the radial shear, I_2, has maximum near the bottom of the convection zone, where its magnitude is less than the I_1.

Figure 17(a) shows the time evolution of the RHS contributions of Eq. (61). In our model the We see that I_2 is about factor 2 less than I_1. Winding of the toroidal field by the latitudinal shear seems to be the main generation effect in our model and, perhaps, in the solar dynamo, as well. The radial shear is less efficient because it is small in the main part of the convection zone. Also, it has the opposite sign near the convective zone boundaries. This justifies applications of simple 1-D Babcock–Leighton dynamos to the solar observations as argued by CS15. Together with the fact of the poleward increase of I_1 it explain the relative success of correlation of the polar magnetic field strength and the magnitude of the subsequent magnetic cycle for the solar cycle prediction (Choudhuri et al. 2007).

Figure 17(b) shows the budget of the toroidal flux generation rate $(I_1 + I_2)$ and loss rate $(I_3 + I_4)$ for our dynamo model. The parameters of the budget are larger than those deduced by CS15 from solar observations. The difference is because of additional generation and loss terms. The budget which includes only the surface activity contributions (green line in Fig. 17b) is less than the full case. Also, the magnitude of the generation rate by

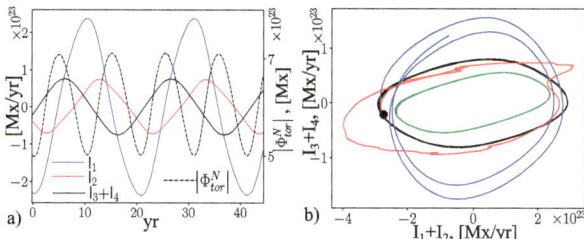

Fig. 17 (a) Time evolution of the RHS contributions of Eq. (61); (b) the dynamo models budget, black line show the standard mean-field model, green line – the budget which includes only the surface contributions ($I_{1,4}$), blue line – the run where the radial subsurface shear (region r = 0.9-0.99R) is neglected and the red line shows the model with accounts of surface spot-like activity effects. Starting point is marked by black circle

the latitudinal shear can be larger than in the solar observations because of difference in the latitudinal profiles of the surface radial magnetic field. We guess that in the dynamo model the radial magnetic field increase poleward steeper than in observations. This issue have to be studied further. Figure 17b shows the budget for another two dynamo models. In one case, we neglect the generation effect of the radial subsurface shear in region r = 0.9-0.99R. In compare to the standard case, this model shows the reduction of the generation rate, the amplitude of the generated toroidal flux, and increase of the loss rate. Therefore we conclude the importance of the subsurface shear layer for our dynamo model.

The above analysis shows the importance of surface activity for the dynamo model and perhaps for the solar dynamo as well. Sunspot activity in the form of magnetic bipolar regions is one of the most important aspects of magnetic surface activity. A consistent approach to include it in dynamo models is at present absent. Also, the origin of sunspots and their bipolar magnetic field is not well known; see Sect. 8.1. The Babcock–Leighton type and flux-transport dynamo models use a phenomenological approach. It is also applicable to mean-field models. Pipin (2022) studied the effect of surface activity on convection zone dynamos. Here, we briefly discuss some results of the paper. The emergence of bipolar magnetic regions (BMRs) is modeled using the mean electromotive force which is represented by the α and magnetic buoyancy effects acting on the unstable part of the axisymmetric magnetic field as follows:

$$\mathcal{E}_i^{(BMR)} = \alpha_\beta \delta_{i\phi} \langle B \rangle_\phi + V_\beta \left(\hat{r} \times \langle \mathbf{B} \rangle \right)_i , \qquad (62)$$

where the first term takes into account the BMR's tilt and the second term models the surface magnetic region in the bipolar form. To produce the bipolar like regions we have to restrict spatially V_β in Eq. (62) to the small scales that are typical for the solar BMR; see details in the above cited paper. Position and emergence time are chosen to be random and modulated by the large-scale magnetic activity. The BMR's α-effect parameters are random as well; see details in (Pipin 2022; Pipin et al. 2023). The given approach could be refined further using the 3D hydrodynamics, effects of the Coriolis force and the theory of the Joy's law developed recently by Kleeorin et al. (2020). Figure 18 illustrates the formation of BMR simulated in the dynamo model. It was found that the BMR effects on the dynamo are restricted to the shallow layer below the surface. At the surface, the effect of the BMR on the magnetic field generation is dominant. Compare to the standard axisymmetric mean-field model discussed in the subsections above, the nonaxisymmetric dynamo, which includes the emergence of tilted BMR, can result in additional dynamo generation of the large-scale poloidal magnetic

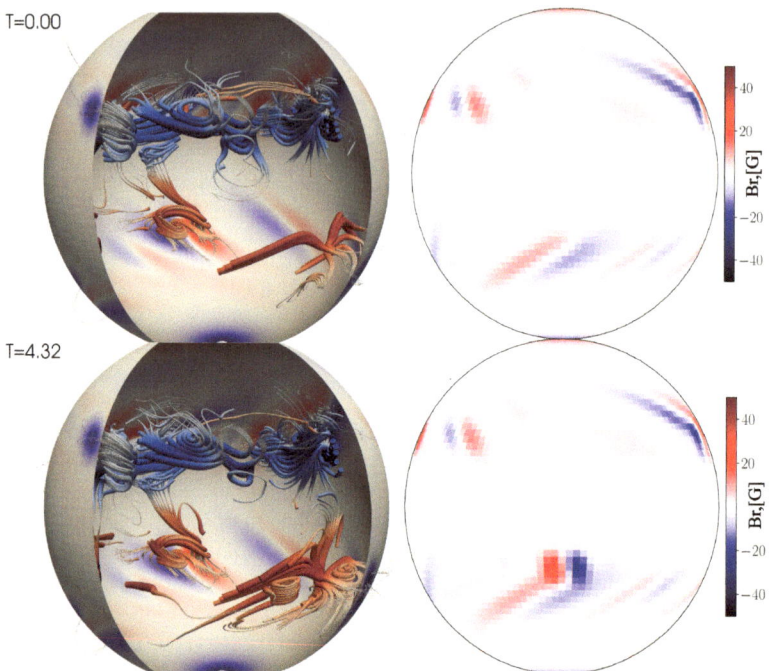

Fig. 18 Snapshots of magnetic regions in the south hemisphere in ascending phase of the magnetic cycle. The left column shows the nonaxisymmetric magnetic field lines, time is shown in days. The right column shows the radial magnetic field on the top boundary. Reproduced by permission from Pipin et al. (2023)

field and to an increase of the polar magnetic field. The red line in Fig. 17(b) shows the budget for this nonaxisymmetric dynamo model. We see an increase of the toroidal flux generation rate in the nonaxisymmetric model because of the surface BMR activity. Similar to Cameron and Schüssler (2015), we can conclude that sunspot surface activity seems to play an important part in the solar large-scale dynamo.

8.3 Effect of Corona on the Dynamo

Usually, dynamo models are limited to the star embedded in a vacuum, which is described by boundary conditions on the stellar surface. However, the boundary conditions have a determining influence on the global solutions, such as the symmetry about the equator. With the assumption of an external vacuum, all induction effects in the corona are neglected. Since the solar surface rotates differentially, the highly conductive plasma in the corona also causes induction effects through shear. Observations of coronal rotation are very scarce. There is evidence from extended coronal holes of rigid rotation in latitude (Timothy et al. 1975; Bagashvili et al. 2017). Kinematic dynamo models involving the corona with various assumptions on its rotation and conductivity give a wide range of solutions (Elstner et al. 2020). A notable influence of the corona on the dynamo in the convection zone was also observed in DNS by Warnecke et al. (2016). A too weak density contrast and too strong viscous coupling of the corona to the star in their model probably underestimates the effect of the Lorentz force in the corona. Considering a dynamical situation with dominant Lorentz force in the corona, the solution in the Sun corresponds to that with vacuum boundary condition independent of rotation and conductivity in the corona. The magnetic field in the corona varies in time to a nearly force-free solution. Further investigations of the star-corona coupling are needed to clarify the exchange of magnetic energy and helicity.

8.4 Stellar Cycle Periods

Noyes et al. (1984) developed an early understanding of the observed stellar cycle periods, P_{cyc}. In those years, there where just six stars with measured rotation and cycle periods. Remarkably, for the stars of Noyes et al. (1984), those values have not changed much with the more accurate data of Baliunas et al. (1995); see Table 1 for a comparison of their cycle periods and the more recent data sets (Olspert et al. 2018; Boro Saikia et al. 2018; Bonanno and Corsaro 2022). However, for the many stars of Baliunas et al. (1995), the modern analysis of Olspert et al. (2018) and Boro Saikia et al. (2018) showed considerable changes in the results for stellar cycle periodicities. In particular, many of the double periodicities of Baliunas et al. (1995) vanished when extended data became available, and different methods were used.

In recent work of Bonanno and Corsaro (2022), new cycle data were collected for altogether 67 stars. Their new sample includes stars with less accurate data points, so the existence of different branches was no longer a pronounced feature. In addition, many of the new data points are different from the earlier ones of Brandenburg et al. (2017b); see Table 2. As in their paper, we denote G and F dwarfs by the same blue italic symbols and K dwarfs by the same red roman symbols.

To see how strong this revision of the data is, we plot in Fig. 19 the ratios $P_{\mathrm{rot}}/P_{\mathrm{cyc}}$ versus $\log\langle R'_{\mathrm{HK}}\rangle$ for all stars of Bonanno and Corsaro (2022) and highlight with lowercase and uppercase letters the stars that were also included in the sample of Brandenburg et al. (2017b). We see that the new data are remarkably consistent with the old ones. Out of the eight stars on the branch of active stars, five where listed by Brandenburg et al. (2017b) as having two periods. Of the 16 inactive stars, three were listed with two periods, but the case of the Sun was classified by Brandenburg et al. (2017b) as somewhat different, because the 80 years Gleissberg cycle does not fit well on the active branch and, unlike all the other stars with two cycle periods, which are all younger than 3.3 Gyr, the Sun is relatively old.

The early data of Noyes et al. (1984) did already suggest

$$\omega_{\mathrm{cyc}} \propto \Omega^{1.25} \tag{63}$$

Table 1 Comparison of cycle periods P_{cyc} (in years) from Noyes et al. (1984) (NWV84), Baliunas et al. (1995) (Bal+95), and Bonanno and Corsaro (2022) (BC22). The last two columns compare the seismic age given by BC22 and the gyrochronological age as listed by Brandenburg et al. (2017b) (BMM17). The latter differ significantly, but the determined cycle periods were remarkably stable over the decades

HD	P_{cyc} [yr]			age [Gyr]	
	NWV84	Bal+95	BC22	BC22	BMM17
3651	10	13.8	14.70	—	7.2
4628	8.5	8.37	8.47	3.33	5.3
16,160	11.5	13.2	12.68	—	6.9
160,346	7	7.00	7.19	—	4.4
201,091	7	7.3	7.11	6.10	3.3
201,092	11	11.7	—	—	3.2

Table 2 Comparison of stellar cycle properties from the samples of Bonanno and Corsaro (2022) and Brandenburg et al. (2017b) (indicated as "old"). The blue italics and red roman letters refer to the stars discussed in Brandenburg et al. (2017b) and are also indicated in Fig. 19

HD	Sym	$\log\langle R'_{HK}\rangle$	$\log\langle R'^{old}_{HK}\rangle$	P_{rot} [d]	P^{old}_{rot} [d]	P_{cyc} [yr]	P^{I}_{cyc} [yr]	P^{A}_{cyc} [yr]
100,180	h	−4.83	−4.92	14.06	14.00	3.60	3.60	12.90
103,095	i	−4.90	−4.90	32.51	31.00	7.07	7.30	—
10,476	c	−4.97	−4.91	35.40	35.20	10.45	9.60	—
146,233	l	−4.95	−4.93	22.66	22.70	11.59	7.10	—
160,346	m	−4.86	−4.79	34.20	36.40	7.19	7.00	—
16,160	d	−4.94	−4.96	48.29	48.00	12.68	13.20	—
165,341	n	−4.61	−4.55	19.51	19.00	5.09	5.10	15.50
166,620	o	−5.00	−4.96	42.25	42.40	16.81	15.80	—
219,834	s	−4.93	−4.94	38.89	43.00	9.48	10.00	—
26,965	f	−4.96	−4.87	40.83	43.00	10.24	10.10	—
3651	a	−5.06	−4.99	40.50	44.00	14.70	13.80	—
4628	b	−4.95	−4.85	37.82	38.50	8.47	8.60	—
81,809	g	−4.89	−4.92	40.93	40.20	8.05	8.20	—
219,834	r	−5.10	−5.07	43.40	42.00	16.29	21.00	—
201,091	p	−4.56	−4.76	35.62	35.40	7.11	7.30	—
Sun	a	−4.94	−4.90	25.55	25.40	10.70	11.00	80.00
149,661	K	−4.61	−4.58	20.92	21.10	12.38	4.00	17.40
152,391	M	−4.46	−4.45	11.01	11.40	11.94	—	10.90
156,026	L	−4.56	−4.66	18.85	21.00	19.31	—	21.00
190,406	N	−4.76	−4.80	14.01	13.90	18.61	2.60	16.90
76,151	F	−4.68	−4.66	14.70	15.00	16.34	2.50	—
78,366	G	−4.57	−4.61	9.60	9.70	14.26	5.90	12.20
114,710	J	−4.74	−4.75	11.99	12.30	14.12	9.60	16.60
22,049	E	−4.46	−4.46	11.09	11.10	11.00	2.90	12.70

for the cycle frequency $\omega_{cyc} = 2\pi/P_{cyc}$ versus angular rotation rate Ω. This dependence is reproduced by considering free dynamo waves and assuming axisymmetric mean fields, $\boldsymbol{B} = b\hat{\boldsymbol{\phi}} + \nabla \times a\hat{\boldsymbol{\phi}}$, with $(a, b) \propto e^{i(ky-\omega t)}$ and writing $-i\omega = -i\omega_{cyc} + \lambda$, where both ω_{cyc} and λ are assumed to be real. The mean-field dynamo equations result in traveling wave solutions with a dispersion relation of the form

$$\omega_{cyc} = \sqrt{\alpha\Omega' k L/2}, \quad \lambda = \omega_{cyc} - \eta_T k^2. \tag{64}$$

At least up to moderate rotation rates, it is reasonable to assume that α and Ω' are proportional to Ω. The crucial assumption in arriving at an approximation that matches Eq. (63) is to assume that the relevant wavenumber k_y is selected not by the condition of marginal excitation, but by the assumption that $\lambda = \lambda(k)$ is maximized. Thus, k has to obey $d\lambda/dk = 0$, which yields $\omega_{cyc} \propto (\alpha\Omega')^{2/3} \propto \Omega^{4/3}$. By contrast, if the dynamo is quenched to the being marginally excited, then $\omega_{cyc} \approx \eta_T/L^2$, which would be either independent of Ω, or perhaps even decreasing with Ω, if η_T decreases with increasing Ω due to quenching.

Of course, nonlinear dynamos must always be quenched to reach a steady state. This led Brandenburg et al. (1998) to suggest that Eq. (63) could be obeyed if both α and η_T are *antiquenched* in such a way that η_T is antiquenched more slowly than α, so that ω_{cyc} would

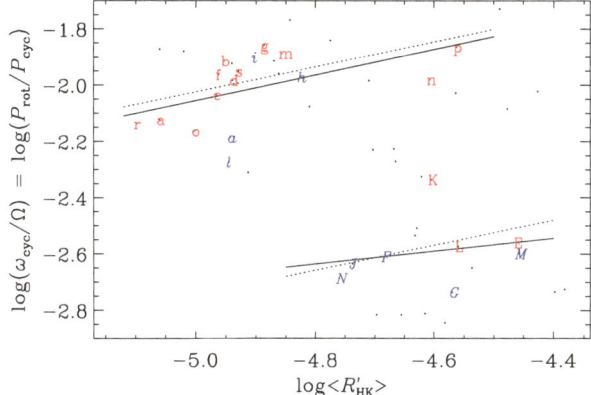

Fig. 19 $P_{\rm rot}/P_{\rm cyc}$ versus $\log\langle R'_{\rm HK}\rangle$ for all stars of Bonanno and Corsaro (2022) (small black symbols). Lowercase (uppercase) letters denote data points of Bonanno and Corsaro (2022) that were also included in the sample of Brandenburg et al. (2017b). There are refer to the data of stars given in the Table 2. The dotted lines denote the fits determined by Brandenburg et al. (2017b) while the upper (lower) solid lines denote fits to the stars of Bonanno and Corsaro (2022) with lowercase (uppercase) letters

Fig. 20 Dependence of cycle period on stellar rotation rate. Red and black crosses show the results of Brandenburg et al. (2017b), green crosses those of Lehtinen et al. (2016), orange squares the models of Warnecke (2018), and the asterisks are from the models of Pipin (2021); act/inact marks the active and inactive branches of activity, respectively; while kin/nkin stand for kinematic and nonkinematic models, respectively (adapted by permission from Pipin 2021)

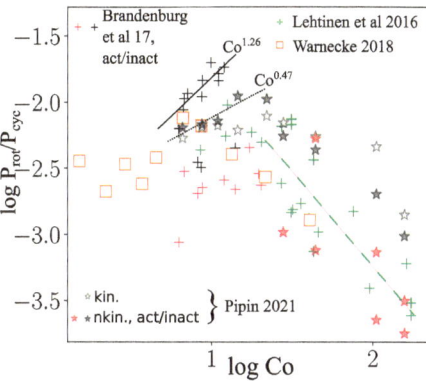

increase with increasing magnetic field strength, and yet the dynamo would still saturate. Whether this is the only viable solution to this puzzle remains still unclear.

Recently, a number of numerical dynamo models were applied to investigate the relation of the cycle period on the stellar rotation rate in solar analogs (Pipin 2015; Strugarek et al. 2017; Warnecke 2018; Viviani et al. 2018; Hazra et al. 2019; Noraz et al. 2022). Figure 20 shows some of these results including the results of observations of Brandenburg et al. (2017b) and the survey of Lehtinen et al. (2016). It is interesting that the saturation branch of stellar activity for the young solar analogs with rotation periods of less than 10 days is well reproduced in the very different solar-like dynamo models including various GCD simulations (Strugarek et al. 2017; Warnecke 2018; Viviani et al. 2018), flux transport model of Hazra et al. (2019) and mean-field model of Pipin (2021). In Fig. 20, this branch is marked by the green line.

The mean cycle period in this branch is almost independent of stellar rotation rate. The nonkinematic nonlinear model of Pipin (2021) shows multiple periods along this line. Pipin (2021) found that saturation of dynamo activity is accompanied by a depression of latitudinal shear, a concentration of magnetic activity to the surface, and changes in the meridional circulations from one cell to multiple cells per hemisphere structure. According to the conclusions of this paper, it is clear that in the saturated state, dynamo waves do not follow the Parker–Yoshimura rule. Their cycle period is determined by turbulent diffusion and meridional circulation. This is why the predictions of flux transport and nonkinematic mean-field dynamo models coincide. The independence of the cycle period on the rotation rate can be typical for the dynamo solutions which show concentration of the magnetic activity toward the boundaries of the dynamo (see Pipin 2015; Pipin and Kosovichev 2016).

The inactive branch of the nonkinematic mean-field dynamo models shows a fairly strong positive inclination (see Fig. 20), which is absent in the kinematic models. We see that the dynamo model can reproduce a power law $\sim Co^{0.5}$, avoiding the antiquenching concept of Brandenburg et al. (1998). In fact, the nonkinematic dynamo models show a frequency doubling phenomenon for models in the range from 10 to 15 days rotation period (see Figs. 3 and 8 of Pipin 2021). The frequency doubling or the second harmonic generation is known from nonlinear optics. This is typical for wave propagation in nonlinear media. In dynamo waves, second harmonics are generated because of B^2 effects such as the magnetic feedback on the large-scale flow, magnetic helicity conservation, and magnetic buoyancy effects. The second harmonics can be found in the solar activity as well (Sokoloff et al. 2020). For the solar case, they are subdominant. However, they can become dominant for fast rotating stars. This makes the interpretation of magnetic activity cycles difficult (Stepanov et al. 2020).

It is important to note that the GCD simulations of Viviani et al. (2018) show an increased level of nonaxisymmetry with an increase of rotation rate and a transition to preferred nonaxisymmetric dynamos for solar-type stars with a rotation period of less than 14 days. In simulations, this transition happens when the rotation profile changes from an antisolar to a solar-like profile with a faster equator and slow poles. Stellar Zeeman Doppler Imaging observations of Donati and Landstreet (2009) and See et al. (2016), for example, show that the magnetic topology depends on stellar mass and rotation rate. In a certain interval of rotation periods (below 14 d) and stellar mass, the results of Viviani et al. (2018) are compatible with the observational findings mentioned above.

Summarizing, we suggest that the Parker–Yoshimura dynamo regime can work for solar analogs rotating with periods above 14–15 days. In an interval of stellar rotation periods between 10 and 15 days, frequency doubling and a transition to the nonaxisymmetric dynamos occurs. For lower rotation periods, the dynamo transits to a saturation stage. It can be characterized by high magnetic activity and multiple dynamo periods which are independent of stellar rotation rate. This new picture should be further improved by including possible dynamo effects of surface activity in the form of bipolar magnetic regions and star spots.

Another interesting observation is that different types of stars, including partially convective main sequence solar-type stars, fully convective M-dwarfs, and evolved post-main sequence giant stars show similar scaling with the Rossby number for the unsaturated regime (see, e.g., Wright and Drake 2016; Wright et al. 2018; Lehtinen et al. 2020). In particular, to derive the Rossby number, these study employ a simple parameterization of the convective turnover time suggested by stellar interior models. It was found that for evolved stars, the Rossby-independent parameterizations break down in the rotation–activity relation (Lehtinen et al. 2020). This constitutes a strong argument in favor of the turbulent dynamo paradigm suggesting a common role of the turbulent process in dynamos operating in these stars.

8.5 Analogy of Mean-Field α and the Chiral Magnetic Effect

The α effect in mean-field dynamo theory is an effect that emerges after averaging over the scale of several turbulent eddies. We know already that turbulent diffusion is somewhat analogous to microphysical diffusion, which also emerges after averaging, but here after averaging over atomic scales. Interestingly, even for the α effect there can be an effect on atomic and subatomic scales, because fermions, such as electrons, are chiral. The spin of an electron emerging from the decay of a neutron is anti-aligned with its momentum vector, so their dot product is a negative pseudo-scalar, called the chirality. Positrons have positive chirality. In the presence of an ambient magnetic field, the spins align, but electrons and positrons move in opposite directions, causing therefore an electric current. This constitutes a microscopic α effect (Rogachevskii et al. 2017; Brandenburg et al. 2017c),

$$\alpha_{\text{micro}} \equiv \mu_5 \eta = 24\alpha_{\text{fine}}(n_{\text{L}} - n_{\text{R}})(\hbar c/k_{\text{B}}T)^2, \tag{65}$$

where μ_5 is the normalized chiral chemical potential (with units of inverse length), η is the microscopic magnetic diffusivity, $\alpha_{\text{fine}} \approx 1/137$ is the fine structure constant (quantifying the strength of electromagnetic interaction between charged particles), n_{L} and n_{R} are the number densities of left- and right-handed fermions, $\hbar \approx 10^{-27}$ erg s is the reduced Planck constant, $c \approx 3 \times 10^{10}$ cm s^{-1} is the speed of light, $k_{\text{B}} \approx 10^{-16}$ erg K^{-1} is the Boltzmann constant, and T is the temperature.

The applications of chiral MHD are manifold and range from condensed matter systems and heavy ion collisions to neutron stars and the early Universe; see Kharzeev (2014) for a review. Interestingly, because this microscopic α effect produces helical magnetic fields, and because the total chirality is conserved (Boyarsky et al. 2012), this effect does not last forever, but is being quenched in a form analogous to the catastrophic quenching formula, which takes the form (Rogachevskii et al. 2017)

$$\frac{\partial \mu_5}{\partial t} = -\lambda\eta\left(\mu_5\overline{\boldsymbol{B}}^2 - \eta_t\mu_0\overline{\boldsymbol{J}} \cdot \overline{\boldsymbol{B}}\right) - \Gamma_{\text{flip}}\mu_5, \tag{66}$$

where λ is a coupling constant which, in the catastrophic quenching formalism, is related to $2\eta_t k_{\text{f}}^2/B_{\text{eq}}^2$, and Γ_{flip} is a spin-flipping parameter, which is related to $2\eta k_{\text{f}}^2$ in the catastrophic quenching formalism (see, e.g., Field and Blackman 2002; Blackman and Brandenburg 2002).

There is a vast range of recent work in this field, which goes well beyond the scope of the present paper. We just mention here the paper of Masada et al. (2018), who studied chiral magnetohydrodynamic turbulence in core-collapse supernovae. They found that the inverse cascade related to the chiral effects impacts the magnetohydrodynamic evolution in the supernova core toward explosion.

8.6 Direct Statistical Simulations

An alternative or extension to mean-field theory in the usual sense is to solve the time-dependent system of one-point and two-point correlation functions. This is what is now known as Direct Statistical Simulations (Tobias et al. 2011; Tobias and Marston 2013, 2017) and has primarily been applied to two-dimensional turbulent shear flow problems. The dimensionality of the two-point correlation function doubles for those directions over which homogeneity cannot be assumed. On the other hand, the dynamics of the low order statistics is usually slower than that of the original equations. In addition, it is possible to reduce the complexity of the problem by employing Proper Orthogonal Decomposition (Allawala and Marston 2016; Allawala et al. 2020). This approach has not yet been applied to magneto-hydrodynamics and the dynamo problem. Such an approach would be able to incorporate both small-scale and large-scale dynamos at the same time. The small-scale dynamo problem would be solved through correlation functions, as has been done in Brandenburg and Subramanian (2000), for example. But their simulations were isotropic and did not include anisotropic dependencies on position. Nevertheless, this approach has the potential of being a strong competitor in addressing the high Reynolds number dynamics of problems of astro-physical and geophysical relevance; see Tobias (2021) for a recent review touching on these aspects.

9 Looking Forward

In this review, we have provided some insight into recent developments in our understanding of the generation of astrophysical large-scale magnetic fields. The current development of mean-field theory allows us to go beyond some of the original restrictions that were related to the assumption of large scale-separation and the inappropriate neglect of nonlinear effects due to higher order correlations in contributions to the mean turbulent electromotive force. A big portion of the progress comes from the development in the DNS of astrophysical tur-bulence. Noteworthy, the classical mean-field theory is based on the fundamental equations of electrodynamics and has well-known limits. With the new steps forward, we can take into account results of the DNS, e.g., the spectral kernels, and treat them as experimental facts. The necessity of some phenomenological additions to classical mean-field theory are motivated both by DNS and observations of the magnetic activity in astrophysical systems, such as those in our Sun and other stars. In this way, mean-field models become a valuable tool to understand the real and virtual worlds of the dynamo in stars and in DNS.

Acknowledgements We thank Hantao Ji for important discussions on the reversed field pinch in laboratory plasmas and dynamos in magnetically dominated settings. We are also indebted to the two reviewers for their constructive remarks that have led to improvements in the presentation. Support through the grant 2019-04234 from the Swedish Research Council (Vetenskapsrådet) (AB) is gratefully acknowledged. We thank for the allocation of computing resources provided by the Swedish National Infrastructure for Computing (SNIC) at the PDC Center for High Performance Computing Stockholm and Linköping. VP acknowledges financial support of the Ministry of Science and Higher Education (Subsidy No. 075-GZ/C3569/278). YM has been supported by MEXT/JSPS KAKENHI Grant numbers JP18H01212, JP18K03700, JP21K03612, JP23H01199; funding from Fukuoka University (Grant No. 171042, 177103 and GR2302).

Funding Open access funding provided by Stockholm University.

Declarations

References

Allawala A, Marston JB (2016) Statistics of the stochastically forced Lorenz attractor by the Fokker-Planck equation and cumulant expansions. Phys Rev E 94(5):052218. https://doi.org/10.1103/PhysRevE.94. 052218. 1604.00867

Allawala A, Tobias SM, Marston JB (2020) Dimensional reduction of direct statistical simulation. J Fluid Mech 898:A21. https://doi.org/10.1017/jfm.2020.382. 1708.07805

Altrock R, Howe R, Ulrich R (2008) Solar torsional oscillations and their relationship to coronal activity. In: Howe R, Komm RW, Balasubramaniam KS, Petrie GJD (eds) Subsurface and atmospheric influences on solar activity. ASP Conf Ser, vol 383, p 335

Augustson K, Brun AS, Miesch M, Toomre J (2015) Grand minima and equatorward propagation in a cycling stellar convective dynamo. Astrophys J 809(2):149. https://doi.org/10.1088/0004-637X/809/2/149. 1410.6547

Bagashvili SR, Shergelashvili BM, Japaridze DR, Chargeishvili BB, Kosovichev AG, Kukhianidze V, Ramishvili G, Zaqarashvili TV, Poedts S, Khodachenko ML, De Causmaecker P (2017) Statistical properties of coronal hole rotation rates: are they linked to the solar interior? Astron Astrophys 603:A134. https://doi.org/10.1051/0004-6361/201630377. 1706.04464

Baliunas SL, Donahue RA, Soon WH, Horne JH, Frazer J, Woodard-Eklund L, Bradford M, Rao LM, Wilson OC, Zhang Q, Bennett W, Briggs J, Carroll SM, Duncan DK, Figueroa D, Lanning HH, Misch T, Mueller J, Noyes RW, Poppe D, Porter AC, Robinson CR, Russell J, Shelton JC, Soyumer T, Vaughan AH, Whitney JH (1995) Chromospheric variations in main-sequence stars. Astrophys J 438:269–287. https://doi.org/10.1086/175072

Baryshnikova I, Shukurov A (1987) Oscillatory α^2-dynamo: numerical investigation. Astron Nachr 308(2):89–100. https://doi.org/10.1002/asna.2113080202

Beaudoin P, Charbonneau P, Racine E, Smolarkiewicz PK (2013) Torsional oscillations in a global solar dynamo. Sol Phys 282(2):335–360. https://doi.org/10.1007/s11207-012-0150-2. 1210.1209

Berdyugina SV (2005) Starspots: a key to the stellar dynamo. Living Rev Sol Phys 2(1):8. https://doi.org/10. 12942/lrsp-2005-8

Biskamp D, Müller WC (1999) Decay laws for three-dimensional magnetohydrodynamic turbulence. Phys Rev Lett 83(11):2195–2198. https://doi.org/10.1103/PhysRevLett.83.2195. physics/9903028

Blackman EG, Brandenburg A (2002) Dynamic nonlinearity in large-scale dynamos with shear. Astrophys J 579(1):359–373. https://doi.org/10.1086/342705. astro-ph/0204497

Blackman EG, Ji H (2006) Laboratory plasma dynamos, astrophysical dynamos and magnetic helicity evolution. Mon Not R Astron Soc 369(4):1837–1848. https://doi.org/10.1111/j.1365-2966.2006.10431.x. astro-ph/0604221

Bonanno A, Corsaro E (2022) On the origin of the dichotomy of stellar activity cycles. Astrophys J Lett 939(2):L26. https://doi.org/10.3847/2041-8213/ac9c05. 2210.11305

Boro Saikia S, Marvin CJ, Jeffers SV, Reiners A, Cameron R, Marsden SC, Petit P, Warnecke J, Yadav AP (2018) Chromospheric activity catalogue of 4454 cool stars. Questioning the active branch of stellar activity cycles. Astron Astrophys 616:A108. https://doi.org/10.1051/0004-6361/201629518. 1803.11123

Boyarsky A, Fröhlich J, Ruchayskiy O (2012) Self-consistent evolution of magnetic fields and chiral asymmetry in the early universe. Phys Rev Lett 108(3):031301. https://doi.org/10.1103/PhysRevLett.108. 031301. 1109.3350

Brandenburg A (1998) Disc turbulence and viscosity. In: Abramowicz MA, Björnsson G, Pringle JE (eds) Theory of black hole accretion disks, pp 61–90

Brandenburg A (2001) The inverse cascade and nonlinear alpha-effect in simulations of isotropic helical hydromagnetic turbulence. Astrophys J 550(2):824–840. https://doi.org/10.1086/319783. astro-ph/0006186

Brandenburg A (2008) The dual role of shear in large-scale dynamos. Astron Nachr 329(7):725. https://doi.org/10.1002/asna.200811027. 0808.0959

Brandenburg A (2016) Stellar mixing length theory with entropy rain. Astrophys J 832(1):6. https://doi.org/10.3847/0004-637X/832/1/6. 1504.03189

Brandenburg A (2017) Analytic solution of an oscillatory migratory α^2 stellar dynamo. Astron Astrophys 598:A117. https://doi.org/10.1051/0004-6361/201630033. 1611.02671

Brandenburg A (2018a) Advances in mean-field dynamo theory and applications to astrophysical turbulence. J Plasma Phys 84(4):735840404. https://doi.org/10.1017/S0022377818000806. 1801.05384

Brandenburg A (2018b) Magnetic helicity and fluxes in an inhomogeneous α^2 dynamo. Astron Nachr 339(631):631–640. https://doi.org/10.1002/asna.201913604

Brandenburg A, Chatterjee P (2018) Strong nonlocality variations in a spherical mean-field dynamo. Astron Nachr 339:118–126. https://doi.org/10.1002/asna.201813472. 1802.04231

Brandenburg A, Chen L (2020) The nature of mean-field generation in three classes of optimal dynamos. J Plasma Phys 86(1):905860110. https://doi.org/10.1017/S0022377820000082. 1911.01712

Brandenburg A, Dobler W (2001) Large scale dynamos with helicity loss through boundaries. Astron Astrophys 369:329–338. https://doi.org/10.1051/0004-6361:20010123. astro-ph/0012472

Brandenburg A, Käpylä PJ (2007) Magnetic helicity effects in astrophysical and laboratory dynamos. New J Phys 9(8):305. https://doi.org/10.1088/1367-2630/9/8/305. 0705.3507

Brandenburg A, Larsson G (2023) Turbulence with magnetic helicity that is absent on average. Atmosphere 14(6):932. https://doi.org/10.3390/atmos14060932. 2305.08769

Brandenburg A, Ntormousi E (2023) Galactic dynamos. Annu Rev Astron Astrophys 61(1):561–606. https://doi.org/10.1146/annurev-astro-071221-052807

Brandenburg A, Rädler KH (2013) Yoshizawa's cross-helicity effect and its quenching. Geophys Astrophys Fluid Dyn 107(1–2):207–217. https://doi.org/10.1080/03091929.2012.681307. 1112.1237

Brandenburg A, Sokoloff D (2002) Local and nonlocal magnetic diffusion and alpha-effect tensors in shear flow turbulence. Geophys Astrophys Fluid Dyn 96(4):319–344. https://doi.org/10.1080/03091920290032974. astro-ph/0111568

Brandenburg A, Subramanian K (2000) Large scale dynamos with ambipolar diffusion nonlinearity. Astron Astrophys 361:L33–L36. https://doi.org/10.48550/arXiv.astro-ph/0007450. astro-ph/0007450

Brandenburg A, Subramanian K (2005a) Astrophysical magnetic fields and nonlinear dynamo theory. Phys Rep 417:1–209. https://doi.org/10.1016/j.physrep.2005.06.005. arXiv:astro-ph/0405052

Brandenburg A, Subramanian K (2005b) Minimal tau approximation and simulations of the alpha effect. Astron Astrophys 439(3):835–843. https://doi.org/10.1051/0004-6361:20053221. astro-ph/0504222

Brandenburg A, Subramanian K (2005c) Strong mean field dynamos require supercritical helicity fluxes. Astron Nachr 326:400–408. https://doi.org/10.1002/asna.200510362. astro-ph/0505457

Brandenburg A, Subramanian K (2007) Simulations of the anisotropic kinetic and magnetic alpha effects. Astron Nachr 328(6):507. https://doi.org/10.1002/asna.200710772. 0705.3508

Brandenburg A, Tuominen I (1988) Variation of magnetic fields and flows during the solar cycle. Adv Space Res 8(7):185–189. https://doi.org/10.1016/0273-1177(88)90190-1

Brandenburg A, Tuominen I, Nordlund A, Pulkkinen P, Stein RF (1990) 3-D simulation of turbulent cyclonic magneto-convection. Astron Astrophys 232:277–291

Brandenburg A, Moss D, Tuominen I (1992) Stratification and thermodynamics in mean-field dynamos. Astron Astrophys 265:328–344

Brandenburg A, Jennings RL, Nordlund Å, Rieutord M, Stein RF, Tuominen I (1996) Magnetic structures in a dynamo simulation. J Fluid Mech 306:325–352. https://doi.org/10.1017/S0022112096001322

Brandenburg A, Saar SH, Turpin CR (1998) Time evolution of the magnetic activity cycle period. Astrophys J Lett 498(1):L51–L54. https://doi.org/10.1086/311297

Brandenburg A, Rädler KH, Schrinner M (2008) Scale dependence of alpha effect and turbulent diffusivity. Astron Astrophys 482(3):739–746. https://doi.org/10.1051/0004-6361:200809365. 0801.1320

Brandenburg A, Candelaresi S, Chatterjee P (2009) Small-scale magnetic helicity losses from a mean-field dynamo. Mon Not R Astron Soc 398(3):1414–1422. https://doi.org/10.1111/j.1365-2966.2009.15188.x. 0905.0242

Brandenburg A, Kleeorin N, Rogachevskii I (2010) Large-scale magnetic flux concentrations from turbulent stresses. Astron Nachr 331(1):5. https://doi.org/10.1002/asna.200911311. 0910.1835

Brandenburg A, Kemel K, Kleeorin N, Mitra D, Rogachevskii I (2011) Detection of negative effective magnetic pressure instability in turbulence simulations. Astrophys J Lett 740(2):L50. https://doi.org/10.1088/2041-8205/740/2/L50. 1109.1270

Brandenburg A, Rädler KH, Kemel K (2012) Mean-field transport in stratified and/or rotating turbulence. Astron Astrophys 539:A35. https://doi.org/10.1051/0004-6361/201117871. 1108.2264

Brandenburg A, Kleeorin N, Rogachevskii I (2013) Self-assembly of shallow magnetic spots through strongly stratified turbulence. Astrophys J Lett 776(2):L23. https://doi.org/10.1088/2041-8205/776/2/L23. 1306.4915

Brandenburg A, Gressel O, Jabbari S, Kleeorin N, Rogachevskii I (2014) Mean-field and direct numerical simulations of magnetic flux concentrations from vertical field. Astron Astrophys 562:A53. https://doi.org/10.1051/0004-6361/201322681. 1309.3547

Brandenburg A, Hubbard A, Käpylä PJ (2015) Dynamical quenching with non-local α and downward pumping. Astron Nachr 336(1):91–96. https://doi.org/10.1002/asna.201412141. 1412.0997

Brandenburg A, Ashurova MB, Jabbari S (2017a) Compensating Faraday depolarization by magnetic helicity in the solar corona. Astrophys J Lett 845(2):L15. https://doi.org/10.3847/2041-8213/aa844b. 1706.09540

Brandenburg A, Mathur S, Metcalfe TS (2017b) Evolution of co-existing long and short period stellar activity cycles. Astrophys J 845(1):79. https://doi.org/10.3847/1538-4357/aa7cfa. 1704.09009

Brandenburg A, Schober J, Rogachevskii I, Kahniashvili T, Boyarsky A, Fröhlich J, Ruchayskiy O, Kleeorin N (2017c) The turbulent chiral magnetic cascade in the early universe. Astrophys J Lett 845(2):L21. https://doi.org/10.3847/2041-8213/aa855d. 1707.03385

Browning MK, Miesch MS, Brun AS, Toomre J (2006) Dynamo action in the solar convection zone and tachocline: pumping and organization of toroidal fields. Astrophys J Lett 648(2):L157–L160. https://doi.org/10.1086/507869. astro-ph/0609153

Brummell NH, Hurlburt NE, Toomre J (1998) Turbulent compressible convection with rotation. II. Mean flows and differential rotation. Astrophys J 493(2):955–969. https://doi.org/10.1086/305137

Brummell NH, Clune TL, Toomre J (2002) Penetration and overshooting in turbulent compressible convection. Astrophys J 570(2):825–854. https://doi.org/10.1086/339626

Bushby PJ, Tobias SM (2007) On predicting the solar cycle using mean-field models. Astrophys J 661:1289–1296. https://doi.org/10.1086/516628. 0704.2345

Bushby PJ, Käpylä PJ, Masada Y, Brandenburg A, Favier B, Guervilly C, Käpylä MJ (2018) Large-scale dynamos in rapidly rotating plane layer convection. Astron Astrophys 612:A97. https://doi.org/10.1051/0004-6361/201732066. 1710.03174

Cameron R, Schüssler M (2015) The crucial role of surface magnetic fields for the solar dynamo. Science 347(6228):1333–1335. https://doi.org/10.1126/science.1261470. 1503.08469

Cameron R, Schüssler M (2023) Observationally guided models for the solar dynamo and the role of the surface field. Space Sci Rev 219. arXiv:2305.02253

Candelaresi S, Brandenburg A (2013) Kinetic helicity needed to drive large-scale dynamos. Phys Rev E 87:043104. https://doi.org/10.1103/PhysRevE.87.043104. 1208.4529

Cattaneo F, Hughes DW (2006) Dynamo action in a rotating convective layer. J Fluid Mech 553:401–418. https://doi.org/10.1017/S0022112006009165

Chabrier G, Küker M (2006) Large-scale α^2-dynamo in low-mass stars and brown dwarfs. Astron Astrophys 446(3):1027–1037. https://doi.org/10.1051/0004-6361:20042475. astro-ph/0510075

Charbonneau P (2014) Solar dynamo theory. Annu Rev Astron Astrophys 52:251–290. https://doi.org/10.1146/annurev-astro-081913-040012

Charbonneau P (2020) Dynamo models of the solar cycle. Living Rev Sol Phys 17(1):4. https://doi.org/10.1007/s41116-020-00025-6

Chatterjee P, Brandenburg A, Guerrero G (2010) Can catastrophic quenching be alleviated by separating shear and α effect? Geophys Astrophys Fluid Dyn 104(5):591–599. https://doi.org/10.1080/03091929.2010.504185. 1005.5708

Chatterjee P, Guerrero G, Brandenburg A (2011) Magnetic helicity fluxes in interface and flux transport dynamos. Astron Astrophys 525:A5. https://doi.org/10.1051/0004-6361/201015073. 1005.5335

Choudhuri AR, Chatterjee P, Jiang J (2007) Predicting solar cycle 24 with a solar dynamo model. Phys Rev Lett 98(13):131103. https://doi.org/10.1103/PhysRevLett.98.131103. arXiv:astro-ph/0701527

Covas E, Tavakol R, Moss D, Tworkowski A (2000) Torsional oscillations in the solar convection zone. Astron Astrophys 360:L21–L24. astro-ph/0010323

Cowling TG (1933) The magnetic field of sunspots. Mon Not R Astron Soc 94:39–48. https://doi.org/10.1093/mnras/94.1.39

Davidson PA (2000) Was Loitsyansky correct? A review of the arguments. J Turbul 1(1):6. https://doi.org/10.1088/1468-5248/1/1/006

Deardorff JW (1972) Theoretical expression for the countergradient vertical heat flux. J Geophys Res 77(30):5900–5904. https://doi.org/10.1029/JC077i030p05900

Del Sordo F, Guerrero G, Brandenburg A (2013) Turbulent dynamos with advective magnetic helicity flux. Mon Not R Astron Soc 429(2):1686–1694. https://doi.org/10.1093/mnras/sts398. 1205.3502

Donati JF, Landstreet JD (2009) Magnetic fields of nondegenerate stars. Annu Rev Astron Astrophys 47:333–370. https://doi.org/10.1146/annurev-astro-082509-101833. 0904.1938

Egorov P, Rüdiger G, Ziegler U (2004) Vorticity and helicity of the solar supergranulation flow-field. Astron Astrophys 425:725–728. https://doi.org/10.1051/0004-6361:20040531

Elstner D, Rüdiger G (2007) How can α^2-dynamos generate axisymmetric magnetic fields? Astron Nachr 328(10):1130–1132. https://doi.org/10.1002/asna.200710864

Elstner D, Fournier Y, Arlt R (2020) Various scenarios for the equatorward migration of sunspots. In: Kosovichev A, Strassmeier S, Jardine M (eds) Solar and stellar magnetic fields: origins and manifestations, vol 354, pp 134–137. https://doi.org/10.1017/S1743921319009888. 2003.08131

Fan Y, Fang F (2014) A simulation of convective dynamo in the solar convective envelope: maintenance of the solar-like differential rotation and emerging flux. Astrophys J 789(1):35. https://doi.org/10.1088/0004-637X/789/1/35. 1405.3926

Favier B, Bushby PJ (2013) On the problem of large-scale magnetic field generation in rotating compressible convection. J Fluid Mech 723:529–555. https://doi.org/10.1017/jfm.2013.132. 1302.7243

Ferriere K (1993) The full alpha-tensor due to supernova explosions and superbubbles in the galactic disk. Astrophys J 404:162. https://doi.org/10.1086/172266

Field GB, Blackman EG (2002) Dynamical quenching of the α^2 dynamo. Astrophys J 572(1):685–692. https://doi.org/10.1086/340233. astro-ph/0111470

Getling AV, Kosovichev AG, Zhao J (2021) Evolution of subsurface zonal and meridional flows in solar cycle 24 from helioseismological data. Astrophys J Lett 908(2):L50. https://doi.org/10.3847/2041-8213/abe45a. 2012.15555

Ghizaru M, Charbonneau P, Smolarkiewicz PK (2010) Magnetic cycles in global large-eddy simulations of solar convection. Astrophys J Lett 715(2):L133–L137. https://doi.org/10.1088/2041-8205/715/2/L133

Gopalakrishnan K, Subramanian K (2023) Magnetic helicity fluxes from triple correlators. Astrophys J 943(1):66. https://doi.org/10.3847/1538-4357/aca808. 2209.14810

Gressel O (2010) A mean-field approach to the propagation of field patterns in stratified magnetorotational turbulence. Mon Not R Astron Soc 405(1):41–48. https://doi.org/10.1111/j.1365-2966.2010.16440.x. 1001.5250

Gruzinov AV, Diamond PH (1994) Self-consistent theory of mean-field electrodynamics. Phys Rev Lett 72(11):1651–1653. https://doi.org/10.1103/PhysRevLett.72.1651

Guerrero G, Smolarkiewicz PK, de Gouveia Dal Pino EM, Kosovichev AG, Mansour NN (2016) Understanding solar torsional oscillations from global dynamo models. Astrophys J Lett 828:L3. https://doi.org/10.3847/2041-8205/828/1/L3. 1608.02278

Hanasoge S, Gizon L, Sreenivasan KR (2016) Seismic sounding of convection in the sun. Annu Rev Fluid Mech 48(1):191–217. https://doi.org/10.1146/annurev-fluid-122414-034534. 1503.07961

Hathaway DH (2012) Supergranules as probes of the sun's meridional circulation. Astrophys J 760:84. https://doi.org/10.1088/0004-637X/760/1/84. 1210.3343

Hathaway DH (2015) The solar cycle. Living Rev Sol Phys 12(1):4. https://doi.org/10.1007/lrsp-2015-4. 1502.07020

Hatori T (1984) Kolmogorov-style argument for the decaying homogeneous MHD turbulence. J Phys Soc Jpn 53(8):2539. https://doi.org/10.1143/JPSJ.53.2539

Hazra G, Choudhuri AR (2017) A theoretical model of the variation of the meridional circulation with the solar cycle. Mon Not R Astron Soc 472(3):2728–2741. https://doi.org/10.1093/mnras/stx2152. 1708.05204

Hazra G, Jiang J, Karak BB, Kitchatinov L (2019) Exploring the cycle period and parity of stellar magnetic activity with dynamo modeling. Astrophys J 884(1):35. https://doi.org/10.3847/1538-4357/ab4128. 1909.01286

Hazra G, Nandy D, Kitchatinov L, Choudhuri AR (2023) Mean field models of flux transport dynamo and meridional circulation in the sun and stars. Space Sci Rev 219(5):39. https://doi.org/10.1007/s11214-023-00982-y. 2302.09390

Hotta H, Kusano K (2021) Solar differential rotation reproduced with high-resolution simulation. Nat Astron 5:1100–1102. https://doi.org/10.1038/s41550-021-01459-0. 2109.06280

Hotta H, Rempel M, Yokoyama T (2016) Large-scale magnetic fields at high Reynolds numbers in magnetohydrodynamic simulations. Science 351(6280):1427–1430. https://doi.org/10.1126/science.aad1893

Howard R, Labonte BJ (1980) The sun is observed to be a torsional oscillator with a period of 11 years. Astrophys J Lett 239:L33–L36. https://doi.org/10.1086/183286

Howe R, Hill F, Komm R, Christensen-Dalsgaard J, Larson TP, Schou J, Thompson MJ, Ulrich R (2011) The torsional oscillation and the new solar cycle. J Phys Conf Ser 271(1):012074. https://doi.org/10.1088/1742-6596/271/1/012074

Hubbard A, Brandenburg A (2009) Memory effects in turbulent transport. Astrophys J 706(1):712–726. https://doi.org/10.1088/0004-637X/706/1/712. 0811.2561

Hubbard A, Brandenburg A (2012) Catastrophic quenching in $\alpha\Omega$ dynamos revisited. Astrophys J 748:51. https://doi.org/10.1088/0004-637X/748/1/51. 1107.0238

Hubbard A, Del Sordo F, Käpylä PJ, Brandenburg A (2009) The α effect with imposed and dynamo-generated magnetic fields. Mon Not R Astron Soc 398(4):1891–1899. https://doi.org/10.1111/j.1365-2966.2009.15108.x. 0904.2773

Jabbari S, Brandenburg A, Losada IR, Kleeorin N, Rogachevskii I (2014) Magnetic flux concentrations from dynamo-generated fields. Astron Astrophys 568:A112. https://doi.org/10.1051/0004-6361/201423499. 1401.6107

Jabbari S, Brandenburg A, Mitra D, Kleeorin N, Rogachevskii I (2016) Turbulent reconnection of magnetic bipoles in stratified turbulence. Mon Not R Astron Soc 459(4):4046–4056. https://doi.org/10.1093/mnras/stw888. 1601.08167

Ji H (1999) Turbulent dynamos and magnetic helicity. Phys Rev Lett 83(16):3198–3201. https://doi.org/10.1103/PhysRevLett.83.3198. astro-ph/0102321

Ji H, Prager SC (2002) The α dynamo effects in laboratory plasmas. Magnetohydrodynamics 38:191–210. https://doi.org/10.22364/mhd.38.1-2.15. astro-ph/0110352

Ji H, Prager SC, Sarff JS (1995) Conservation of magnetic helicity during plasma relaxation. Phys Rev Lett 74(15):2945–2948. https://doi.org/10.1103/PhysRevLett.74.2945

Kaneda Y, Ishihara T, Yokokawa M, Itakura K, Uno A (2003) Energy dissipation rate and energy spectrum in high resolution direct numerical simulations of turbulence in a periodic box. Phys Fluids 15(2):L21–L24. https://doi.org/10.1063/1.1539855

Käpylä PJ, Korpi MJ, Tuominen I (2004) Local models of stellar convection: Reynolds stresses and turbulent heat transport. Astron Astrophys 422:793–816. https://doi.org/10.1051/0004-6361:20035874. astro-ph/0312376

Käpylä PJ, Korpi MJ, Ossendrijver M, Stix M (2006a) Magnetoconvection and dynamo coefficients. III. α-effect and magnetic pumping in the rapid rotation regime. Astron Astrophys 455(2):401–412. https://doi.org/10.1051/0004-6361:20064972. astro-ph/0602111

Käpylä PJ, Korpi MJ, Tuominen I (2006b) Solar dynamo models with α-effect and turbulent pumping from local 3D convection calculations. Astron Nachr 327(9):884. https://doi.org/10.1002/asna.200610636. astro-ph/0606089

Käpylä PJ, Korpi MJ, Brandenburg A (2009a) Alpha effect and turbulent diffusion from convection. Astron Astrophys 500(2):633–646. https://doi.org/10.1051/0004-6361/200811498. 0812.1792

Käpylä PJ, Korpi MJ, Brandenburg A (2009b) Large-scale dynamos in rigidly rotating turbulent convection. Astrophys J 697(2):1153–1163. https://doi.org/10.1088/0004-637X/697/2/1153. 0812.3958

Käpylä PJ, Mantere MJ, Brandenburg A (2012) Cyclic magnetic activity due to turbulent convection in spherical wedge geometry. Astrophys J Lett 755(1):L22. https://doi.org/10.1088/2041-8205/755/1/L22. 1205.4719

Käpylä PJ, Mantere MJ, Brandenburg A (2013) Oscillatory large-scale dynamos from Cartesian convection simulations. Geophys Astrophys Fluid Dyn 107(1–2):244–257. https://doi.org/10.1080/03091929.2012.715158. 1111.6894

Käpylä PJ, Brandenburg A, Kleeorin N, Käpylä MJ, Rogachevskii I (2016) Magnetic flux concentrations from turbulent stratified convection. Astron Astrophys 588:A150. https://doi.org/10.1051/0004-6361/201527731. 1511.03718

Käpylä MJ, Käpylä PJ, Olspert N, Brandenburg A, Warnecke J, Karak BB, Pelt J (2016) Multiple dynamo modes as a mechanism for long-term solar activity variations. Astron Astrophys 589:A56. https://doi.org/10.1051/0004-6361/201527002. 1507.05417

Käpylä MJ, Rheinhardt M, Brandenburg A (2022) Compressible test-field method and its application to shear dynamos. Astrophys J 932(1):8. https://doi.org/10.3847/1538-4357/ac5b78. 2106.01107

Käpylä PJ, Browning MK, Brun AS, Guerrero G, Warnecke J (2023) Simulations of solar and stellar dynamos and their theoretical interpretation. Space Sci Rev 219. 2305.16790

Karak BB (2023) Models for the long-term variations of solar activity. Living Rev Sol Phys 20(1):3. https://doi.org/10.1007/s41116-023-00037-y. 2305.17188

Karak BB, Rheinhardt M, Brandenburg A, Käpylä PJ, Käpylä MJ (2014) Quenching and anisotropy of hydromagnetic turbulent transport. Astrophys J 795(1):16. https://doi.org/10.1088/0004-637X/795/1/16. 1406.4521

Katsova MM, Obridko VN, Sokoloff DD, Livshits IM (2021) Estimating the energy of solar and stellar superflares. Geomagn Aeron 61(7):1063–1068. https://doi.org/10.1134/S0016793221070094

Kemel K, Brandenburg A, Ji H (2011) Model of driven and decaying magnetic turbulence in a cylinder. Phys Rev E 84(5):056407. https://doi.org/10.1103/PhysRevE.84.056407. 1106.1129

Kharzeev DE (2014) The chiral magnetic effect and anomaly-induced transport. Prog Part Nucl Phys 75:133–151. https://doi.org/10.1016/j.ppnp.2014.01.002. 1312.3348

Kitchatinov LL (2013) Theory of differential rotation and meridional circulation. In: Kosovichev AG, de Gouveia Dal Pino E, Yan Y (eds) IAU symposium. IAU symposium, vol 294, pp 399–410. https://doi.org/10.1017/S1743921313002834. 1210.7041

Kitchatinov LL, Olemskoy SV (2011) Alleviation of catastrophic quenching in solar dynamo model with nonlocal alpha-effect. Astron Nachr 332(5):496. https://doi.org/10.1002/asna.201011549. 1101.3115

Kitchatinov LL, Pipin VV, Rüdiger G (1994a) Turbulent viscosity, magnetic diffusivity, and heat conductivity under the influence of rotation and magnetic field. Astron Nachr 315:157–170

Kitchatinov LL, Rüdiger G, Küker M (1994b) Lambda-quenching as the nonlinearity in stellar-turbulence dynamos. Astron Astrophys 292:125–132

Kleeorin N, Rogachevskii I (1999) Magnetic helicity tensor for an anisotropic turbulence. Phys Rev E 59:6724–6729

Kleeorin N, Rogachevskii I (2022) Turbulent magnetic helicity fluxes in solar convective zone. Mon Not R Astron Soc 515(4):5437–5448. https://doi.org/10.1093/mnras/stac2141. 2206.14152

Kleeorin NI, Ruzmaikin AA (1982) Dynamics of the average turbulent helicity in a magnetic field. Magnetohydrodynamics 18:116–122

Kleeorin NI, Ruzmaikin AA (1991) Large-scale flows excited by magnetic fields in the solar convective zone. Sol Phys 131(2):211–230. https://doi.org/10.1007/BF00151634

Kleeorin NI, Rogachevskii IV, Ruzmaikin AA (1989) Negative magnetic pressure as a trigger of largescale magnetic instability in the solar convective zone. Sov Astron Lett 15:274

Kleeorin N, Mond M, Rogachevskii I (1996) Magnetohydrodynamic turbulence in the solar convective zone as a source of oscillations and sunspots formation. Astron Astrophys 307:293–309

Kleeorin N, Moss D, Rogachevskii I, Sokoloff D (2000) Helicity balance and steady-state strength of the dynamo generated galactic magnetic field. Astron Astrophys 361:L5–L8. arXiv:astro-ph/0205266

Kleeorin N, Safiullin N, Kuzanyan K, Rogachevskii I, Tlatov A, Porshnev S (2020) The mean tilt of sunspot bipolar regions: theory, simulations and comparison with observations. Mon Not R Astron Soc 495(1):238–248. https://doi.org/10.1093/mnras/staa1047. 2001.01932

Kochukhov O (2021) Magnetic fields of M dwarfs. Astron Astrophys Rev 29(1):1. https://doi.org/10.1007/s00159-020-00130-3. 2011.01781

Köhler H (1973) The solar dynamo and estimate of the magnetic diffusivity and the α-effect. Astron Astrophys 25:467

Kosovichev AG, Pipin VV (2019) Dynamo wave patterns inside of the sun revealed by torsional oscillations. Astrophys J Lett 871(2):L20. https://doi.org/10.3847/2041-8213/aafe82

Kosovichev AG, Schou J, Scherrer PH, Bogart RS, Bush RI, Hoeksema JT, Aloise J, Bacon L, Burnette A, de Forest C, Giles PM, Leibrand K, Nigam R, Rubin M, Scott K, Williams SD, Basu S, Christensen-Dalsgaard J, Dappen W, Rhodes EJ Jr, Duvall TL Jr, Howe R, Thompson MJ, Gough DO, Sekii T, Toomre J, Tarbell TD, Title AM, Mathur D, Morrison M, Saba JLR, Wolfson CJ, Zayer I, Milford PN (1997) Structure and rotation of the solar interior: initial results from the MDI medium-l program. Sol Phys 170:43–61. https://doi.org/10.1023/A:1004949311268

Krause F, Rädler KH (1980) Mean-field magnetohydrodynamics and dynamo theory. Pergamon Press (also Akademie-Verlag: Berlin), Oxford

Küker M, Rüdiger G, Pipin VV (1996) Solar torsional oscillations due to the magnetic quenching of the Reynolds stress. Astron Astrophys 312:615–623

Kulsrud RM, Zweibel EG (2008) On the origin of cosmic magnetic fields. Rep Prog Phys 71(4):046901. https://doi.org/10.1088/0034-4885/71/4/046901. 0707.2783

Lehtinen J, Jetsu L, Hackman T, Kajatkari P, Henry GW (2016) Activity trends in young solar-type stars. Astron Astrophys 588:A38. https://doi.org/10.1051/0004-6361/201527420. 1509.06606

Lehtinen JJ, Spada F, Käpylä MJ, Olspert N, Käpylä PJ (2020) Common dynamo scaling in slowly rotating young and evolved stars. Nat Astron 4:658–662. https://doi.org/10.1038/s41550-020-1039-x. 2003.08997

Leighton RB (1969) A magneto-kinematic model of the solar cycle. Astrophys J 156:1. https://doi.org/10.1086/149943

Losada IR, Brandenburg A, Kleeorin N, Mitra D, Rogachevskii I (2012) Rotational effects on the negative magnetic pressure instability. Astron Astrophys 548:A49. https://doi.org/10.1051/0004-6361/201220078. 1207.5392

Malkus WVR, Proctor MRE (1975) The macrodynamics of alpha-effect dynamos in rotating fluids. J Fluid Mech 67:417–443. https://doi.org/10.1017/S0022112075000390

Masada Y, Sano T (2014a) Long-term evolution of large-scale magnetic fields in rotating stratified convection. Publ Astron Soc Jpn 66:S2. https://doi.org/10.1093/pasj/psu081. 1403.6221

Masada Y, Sano T (2014b) Mean-field modeling of an α^2 dynamo coupled with direct numerical simulations of rigidly rotating convection. Astrophys J Lett 794(1):L6. https://doi.org/10.1088/2041-8205/794/1/L6. 1409.3256

Masada Y, Sano T (2016) Spontaneous formation of surface magnetic structure from large-scale dynamo in strongly stratified convection. Astrophys J Lett 822(2):L22. https://doi.org/10.3847/2041-8205/822/2/L22. 1604.05374

Masada Y, Sano T (2022) Rotational dependence of large-scale dynamo in strongly-stratified convection: what causes it? 2206.06566

Masada Y, Yamada K, Kageyama A (2013) Effects of penetrative convection on solar dynamo. Astrophys J 778(1):11. https://doi.org/10.1088/0004-637X/778/1/11. 1304.1252

Masada Y, Kotake K, Takiwaki T, Yamamoto N (2018) Chiral magnetohydrodynamic turbulence in core-collapse supernovae. Phys Rev D 98(8):083018. https://doi.org/10.1103/PhysRevD.98.083018. 1805.10419

Matthaeus WH, Pouquet A, Mininni PD, Dmitruk P, Breech B (2008) Rapid alignment of velocity and magnetic field in magnetohydrodynamic turbulence. Phys Rev Lett 100(8):085003. https://doi.org/10.1103/PhysRevLett.100.085003. 0708.0801

Miesch MS, Toomre J (2009) Turbulence, magnetism, and shear in stellar interiors. Annu Rev Fluid Mech 41(1):317–345. https://doi.org/10.1146/annurev.fluid.010908.165215

Mitra D, Candelaresi S, Chatterjee P, Tavakol R, Brandenburg A (2010a) Equatorial magnetic helicity flux in simulations with different gauges. Astron Nachr 331:130. https://doi.org/10.1002/asna.200911308. 0911.0969

Mitra D, Tavakol R, Käpylä PJ, Brandenburg A (2010b) Oscillatory migrating magnetic fields in helical turbulence in spherical domains. Astrophys J Lett 719(1):L1–L4. https://doi.org/10.1088/2041-8205/719/1/L1. 0901.2364

Mitra D, Brandenburg A, Kleeorin N, Rogachevskii I (2014) Intense bipolar structures from stratified helical dynamos. Mon Not R Astron Soc 445(1):761–769. https://doi.org/10.1093/mnras/stu1755. 1404.3194

Moffatt HK (1978) Magnetic field generation in electrically conducting fluids. Cambridge University Press, Cambridge

Moss D, Brandenburg A (1992) The influence of boundary conditions on the excitation of disk dynamo modes. Astron Astrophys 256:371–374

Noraz Q, Brun AS, Strugarek A, Depambour G (2022) Impact of anti-solar differential rotation in mean-field solar-type dynamos. Exploring possible magnetic cycles in slowly rotating stars. Astron Astrophys 658:A144. https://doi.org/10.1051/0004-6361/202141946. 2111.12722

Nordlund A, Brandenburg A, Jennings RL, Rieutord M, Ruokolainen J, Stein RF, Tuominen I (1992) Dynamo action in stratified convection with overshoot. Astrophys J 392:647. https://doi.org/10.1086/171465

Noyes RW, Weiss NO, Vaughan AH (1984) The relation between stellar rotation rate and activity cycle periods. Astrophys J 287:769–773. https://doi.org/10.1086/162735

Obridko VN, Pipin VV, Sokoloff D, Shibalova AS (2021) Solar large-scale magnetic field and cycle patterns in solar dynamo. Mon Not R Astron Soc 504(4):4990–5000. https://doi.org/10.1093/mnras/stab1062. 2104.06808

Olspert N, Lehtinen JJ, Käpylä MJ, Pelt J, Grigorievskiy A (2018) Estimating activity cycles with probabilistic methods. II. The Mount Wilson Ca H&K data. Astron Astrophys 619:A6. https://doi.org/10.1051/0004-6361/201732525. 1712.08240

Ossendrijver M (2003) The solar dynamo. Astron Astrophys Rev 11(4):287–367. https://doi.org/10.1007/s00159-003-0019-3

Ossendrijver M, Stix M, Brandenburg A (2001) Magnetoconvection and dynamo coefficients: Dependence of the alpha effect on rotation and magnetic field. Astron Astrophys 376:713–726. https://doi.org/10.1051/0004-6361:20011041. astro-ph/0108274

Parker E (1955) Hydromagnetic dynamo models. Astrophys J 122:293–314

Parker EN (1967) The dynamical state of the interstellar gas and field. III. Turbulence and enhanced diffusion. Astrophys J 149:535. https://doi.org/10.1086/149283

Parker EN (1979) Cosmical magnetic fields: their origin and their activity. Clarendon Press, Oxford

Parker EN (1993) A solar dynamo surface wave at the interface between convection and nonuniform rotation. Astrophys J 408:707. https://doi.org/10.1086/172631

Paxton B, Marchant P, Schwab J, Bauer EB, Bildsten L, Cantiello M, Dessart L, Farmer R, Hu H, Langer N, Townsend RHD, Townsley DM, Timmes FX (2015) Modules for experiments in stellar astrophysics (mesa): binaries, pulsations, and explosions. Astrophys J Suppl Ser 220:15. https://doi.org/10.1088/0067-0049/220/1/15. 1506.03146

Pipin VV (2008) The mean electro-motive force and current helicity under the influence of rotation, magnetic field and shear. Geophys Astrophys Fluid Dyn 102:21–49. arXiv:astro-ph/0606265

Pipin VV (2015) Dependence of magnetic cycle parameters on period of rotation in non-linear solar-type dynamos. Mon Not R Astron Soc 451:1528–1539. https://doi.org/10.1093/mnras/stv1026. 1412.5284

Pipin VV (2018) Nonkinematic solar dynamo models with double-cell meridional circulation. J Atmos Sol-Terr Phys 179:185–201. https://doi.org/10.1016/j.jastp.2018.07.010. 1803.09459

Pipin VV (2021) Solar dynamo cycle variations with a rotational period. Mon Not R Astron Soc 502(2):2565–2581. https://doi.org/10.1093/mnras/stab033. 2008.05083

Pipin VV (2022) On the effect of surface bipolar magnetic regions on the convection zone dynamo. Mon Not R Astron Soc 514(1):1522–1534. https://doi.org/10.1093/mnras/stac1434. 2112.09460

Pipin VV (2023) Spatio-temporal non-localities in a solar-like mean-field dynamo. Mon Not R Astron Soc 522(2):2919–2927. https://doi.org/10.1093/mnras/stad1150. 2302.11176

Pipin VV, Kosovichev AG (2011) The subsurface-shear-shaped solar $\alpha\Omega$ dynamo. Astrophys J Lett 727:L45–L48. https://doi.org/10.1088/2041-8205/727/2/L45. 1011.4276

Pipin VV, Kosovichev AG (2016) Dependence of stellar magnetic activity cycles on rotational period in a nonlinear solar-type dynamo. Astrophys J 823:133. https://doi.org/10.3847/0004-637X/823/2/133. 1602.07815

Pipin VV, Kosovichev AG (2018) On the origin of the double-cell meridional circulation in the solar convection zone. Astrophys J 854:67. https://doi.org/10.3847/1538-4357/aaa759. 1708.03073

Pipin VV, Kosovichev AG (2019) On the origin of solar torsional oscillations and extended solar cycle. Astrophys J 887(2):215. https://doi.org/10.3847/1538-4357/ab5952

Pipin VV, Seehafer N (2009) Stellar dynamos with $\Omega \times j$ effect. Astron Astrophys 493:819–828. https://doi.org/10.1051/0004-6361:200810766. 0811.4225

Pipin VV, Yokoi N (2018) Generation of a large-scale magnetic field in a convective full-sphere cross-helicity dynamo. Astrophys J 859(1):18. https://doi.org/10.3847/1538-4357/aabae6. 1712.01527

Pipin VV, Sokoloff DD, Zhang H, Kuzanyan KM (2013) Helicity conservation in nonlinear mean-field solar dynamo. Astrophys J 768:46. https://doi.org/10.1088/0004-637X/768/1/46. 1211.2420

Pipin VV, Kosovichev AG, Tomin VE (2023) Effects of emerging bipolar magnetic regions in mean-field dynamo model of solar cycles 23 and 24. Astrophys J 949(1):7. https://doi.org/10.3847/1538-4357/acaf69. 2210.08764

Pouquet A, Frisch U, Leorat J (1976) Strong MHD helical turbulence and the nonlinear dynamo effect. J Fluid Mech 77:321–354. https://doi.org/10.1017/S0022112076002140

Racine É, Charbonneau P, Ghizaru M, Bouchat A, Smolarkiewicz PK (2011) On the mode of dynamo action in a global large-eddy simulation of solar convection. Astrophys J 735(1):46. https://doi.org/10.1088/0004-637X/735/1/46

Rädler KH (1969) On the electrodynamics of turbulent fields under the influence of Coriolis forces. Monats Dt Akad Wiss 11:194–201

Rädler KH, Kleeorin N, Rogachevskii I (2003) The mean electromotive force for MHD turbulence: the case of a weak mean magnetic field and slow rotation. Geophys Astrophys Fluid Dyn 97:249–269

Rädler KH, Brandenburg A, Del Sordo F, Rheinhardt M (2011) Mean-field diffusivities in passive scalar and magnetic transport in irrotational flows. Phys Rev E 84(4):046321. https://doi.org/10.1103/PhysRevE.84.046321. 1104.1613

Rempel M (2007) Origin of solar torsional oscillations. Astrophys J 655(1):651–659. https://doi.org/10.1086/509866. astro-ph/0610221

Rempel M, Bhatia T, Bellot Rubio L, Korpi-Lagg MJ (2023) Small-scale dynamos: from idealized models to solar and stellar application. Space Sci Rev 219(5):36. 2305.02787

Rheinhardt M, Brandenburg A (2010) Test-field method for mean-field coefficients with MHD background. Astron Astrophys 520:A28. https://doi.org/10.1051/0004-6361/201014700. 1004.0689

Rheinhardt M, Brandenburg A (2012) Modeling spatio-temporal nonlocality in mean-field dynamos. Astron Nachr 333:71–77. https://doi.org/10.1002/asna.201111625. 1110.2891

Rheinhardt M, Devlen E, Rädler KH, Brandenburg A (2014) Mean-field dynamo action from delayed transport. Mon Not R Astron Soc 441:116–126. https://doi.org/10.1093/mnras/stu438. 1401.5026

Rincon F (2021) Helical turbulent nonlinear dynamo at large magnetic Reynolds numbers. Phys Rev Fluids 6(12):L121701. https://doi.org/10.1103/PhysRevFluids.6.L121701. 2108.12037

Roberts GO (1972) Dynamo action of fluid motions with two-dimensional periodicity. Philos Trans R Soc Lond, Ser A 271(1216):411–454. https://doi.org/10.1098/rsta.1972.0015

Rogachevskii I, Ruchayskiy O, Boyarsky A, Fröhlich J, Kleeorin N, Brandenburg A, Schober J (2017) Laminar and turbulent dynamos in chiral magnetohydrodynamics. I. Theory. Astrophys J 846(2):153. https://doi.org/10.3847/1538-4357/aa886b. 1705.00378

Rüdiger G, Kichatinov LL (1990) The turbulent stresses in the theory of the solar torsional oscillations. Astron Astrophys 236(2):503–508

Rüdiger G, Kichatinov LL (1993) Alpha-effect and alpha-quenching. Astron Astrophys 269(1–2):581–588

Rüdiger G, Pipin VV (2000) Viscosity-alpha and dynamo-alpha for magnetically driven compressible turbulence in Kepler disks. Astron Astrophys 362:756–761

Rüdiger G, Kitchatinov LL, Brandenburg A (2011) Cross helicity and turbulent magnetic diffusivity in the solar convection zone. Sol Phys 269(1):3–12. https://doi.org/10.1007/s11207-010-9683-4. 1004.4881

Rüdiger G, Kitchatinov LL, Schultz M (2012) Suppression of the large-scale Lorentz force by turbulence. Astron Nachr 333(1):84–91. https://doi.org/10.1002/asna.201111635. 1109.3345

Ruzmaikin AA (1981) The solar cycle as a strange attractor. Comments Astrophys 9(2):85–93

Schrinner M (2011) Global dynamo models from direct numerical simulations and their mean-field counterparts. Astron Astrophys 533:A108. https://doi.org/10.1051/0004-6361/201116642. 1105.2912

Schrinner M, Rädler KH, Schmitt D, Rheinhardt M, Christensen U (2005) Mean-field view on rotating magnetoconvection and a geodynamo model. Astron Nachr 326(3):245–249. https://doi.org/10.1002/asna.200410384

Schrinner M, Rädler KH, Schmitt D, Rheinhardt M, Christensen UR (2007) Mean-field concept and direct numerical simulations of rotating magnetoconvection and the geodynamo. Geophys Astrophys Fluid Dyn 101(2):81–116. https://doi.org/10.1080/03091920701345707. astro-ph/0609752

Schrinner M, Petitdemange L, Dormy E (2011) Oscillatory dynamos and their induction mechanisms. Astron Astrophys 530:A140. https://doi.org/10.1051/0004-6361/201016372. 1101.1837

Schüssler M (1981) The solar torsional oscillation and dynamo models of the solar cycle. Astron Astrophys 94(2):L17

Schüssler M (1983) Stellar dynamo theory. In: Stenflo JO (ed) Solar and stellar magnetic fields: origins and coronal effects, IAU Symposium, vol 102. Reidel, Dordrecht, pp 213–236. https://doi.org/10.1017/S0074180900029880

See V, Jardine M, Vidotto AA, Donati JF, Boro Saikia S, Bouvier J, Fares R, Folsom CP, Gregory SG, Hussain G, Jeffers SV, Marsden SC, Morin J, Moutou C, do Nascimento JD, Petit P, Waite IA (2016) The connection between stellar activity cycles and magnetic field topology. Mon Not R Astron Soc 462:4442–4450. https://doi.org/10.1093/mnras/stw2010. 1610.03737

Seehafer N, Pipin VV (2009) An advective solar-type dynamo without the α effect. Astron Astrophys 508:9–16. https://doi.org/10.1051/0004-6361/200912614. 0910.2614

Shimada R, Hotta H, Yokoyama T (2022) Mean-field analysis on large-scale magnetic fields at high Reynolds numbers. Astrophys J 935(1):55. https://doi.org/10.3847/1538-4357/ac7e43. 2207.01639

Simard C, Charbonneau P, Bouchat A (2013) Magnetohydrodynamic simulation-driven kinematic mean field model of the solar cycle. Astrophys J 768(1):16. https://doi.org/10.1088/0004-637X/768/1/16

Simard C, Charbonneau P, Dubé C (2016) Characterisation of the turbulent electromotive force and its magnetically-mediated quenching in a global EULAG-MHD simulation of solar convection. Adv Space Res 58(8):1522–1537. https://doi.org/10.1016/j.asr.2016.03.041. 1604.01533

Snodgrass HB, Howard R (1985) Torsional oscillations of low mode. Sol Phys 95:221–228. https://doi.org/10.1007/BF00152399

Sokoloff DD, Shibalova AS, Obridko VN, Pipin VV (2020) Shape of solar cycles and mid-term solar activity oscillations. Mon Not R Astron Soc 497(4):4376–4383. https://doi.org/10.1093/mnras/staa2279. 2007.14779

Spruit HC (2003) Origin of the torsional oscillation pattern of solar rotation. Sol Phys 213:1–21. https://doi.org/10.1023/A:1023202605379. astro-ph/0209146

Steenbeck M, Krause F (1969) On the dynamo theory of stellar and planetary magnetic fields. I. AC dynamos of solar type. Astron Nachr 291:49–84. https://doi.org/10.1002/asna.19692910201

Steenbeck M, Krause F, Rädler KH (1966) Berechnung der mittleren Lorentz-Feldstärke für ein elektrisch leitendes Medium in turbulenter, durch Coriolis-Kräfte beeinflußter Bewegung. Z Naturforsch Teil A 21:369. https://doi.org/10.1515/zna-1966-0401

Stein RF, Nordlund Å (2012) On the formation of active regions. Astrophys J Lett 753(1):L13. https://doi.org/10.1088/2041-8205/753/1/L13. 1207.4248

Stejko AM, Kosovichev AG, Pipin VV (2021) Forward modeling helioseismic signatures of one- and two-cell meridional circulation. Astrophys J 911(2):90. https://doi.org/10.3847/1538-4357/abec70. 2101.01220

Stenflo JO (1992) Comments on the concept of an "extended solar cycle". In: Harvey KL (ed) The solar cycle. ASP Conf Ser, vol 27, p 421

Stepanov R, Bondar' NI, Katsova MM, Sokoloff D, Frick P (2020) Wavelet analysis of the long-term activity of V833 Tau. Mon Not R Astron Soc 495(4):3788–3794. https://doi.org/10.1093/mnras/staa1458. 2005.11136

Stix M (1974) Comments on the solar dynamo. Astron Astrophys 37(1):121–133

Stix M (1976) Differential rotation and the solar dynamo. Astron Astrophys 47:243–254

Stix M (1977) Coronal holes and the large-scale solar magnetic field. Astron Astrophys 59:73–78

Strugarek A, Beaudoin P, Charbonneau P, Brun AS, do Nascimento JD (2017) Reconciling solar and stellar magnetic cycles with nonlinear dynamo simulations. Science 357:185–187. https://doi.org/10.1126/science.aal3999. 1707.04335

Sur S, Brandenburg A (2009) The role of the Yoshizawa effect in the Archontis dynamo. Mon Not R Astron Soc 399(1):273–280. https://doi.org/10.1111/j.1365-2966.2009.15254.x. 0902.2394

Taylor JB (1974) Relaxation of toroidal plasma and generation of reverse magnetic fields. Phys Rev Lett 33(19):1139–1141. https://doi.org/10.1103/PhysRevLett.33.1139

Taylor JB (1986) Relaxation and magnetic reconnection in plasmas. Rev Mod Phys 58(3):741–763. https://doi.org/10.1103/RevModPhys.58.741

Timothy AF, Krieger AS, Vaiana GS (1975) The structure and evolution of coronal holes. Sol Phys 42(1):135–156. https://doi.org/10.1007/BF00153291

Tobias SM (1996) Grand minima in nonlinear dynamos. Astron Astrophys 307:L21

Tobias SM (2021) The turbulent dynamo. J Fluid Mech 912:P1. https://doi.org/10.1017/jfm.2020.1055. 1907.03685

Tobias SM, Marston JB (2013) Direct statistical simulation of out-of-equilibrium jets. Phys Rev Lett 110(10):104502. https://doi.org/10.1103/PhysRevLett.110.104502. 1209.3862

Tobias SM, Marston JB (2017) Three-dimensional rotating Couette flow via the generalised quasilinear approximation. J Fluid Mech 810:412–428. https://doi.org/10.1017/jfm.2016.727. 1605.07410

Tobias S, Weiss N (2007) The solar dynamo and the tachocline. In: Hughes DW, Rosner R, Weiss NO (eds) The solar tachocline, p 319

Tobias SM, Cattaneo F, Brummell NH (2008) Convective dynamos with penetration, rotation, and shear. Astrophys J 685(1):596–605. https://doi.org/10.1086/590422

Tobias SM, Dagon K, Marston JB (2011) Astrophysical fluid dynamics via direct statistical simulation. Astrophys J 727(2):127. https://doi.org/10.1088/0004-637X/727/2/127. 1009.2684

Ulrich RK (2001) Very long lived wave patterns detected in the solar surface velocity signal. Astrophys J 560:466–475. https://doi.org/10.1086/322524

Vainshtein SI, Cattaneo F (1992) Nonlinear restrictions on dynamo action. Astrophys J 393:165. https://doi.org/10.1086/171494

Vainshtein SI, Zeldovich IB, Ruzmaikin AA (1980) The turbulent dynamo in astrophysics. Izdatel Nauka, Moscow

Vidotto AA, Lehmann LT, Jardine M, Pevtsov AA (2018) The magnetic field vector of the sun-as-a-star – II. Evolution of the large-scale vector field through activity cycle 24. Mon Not R Astron Soc 480:477–487. https://doi.org/10.1093/mnras/sty1926. 1807.06334

Vishniac ET, Cho J (2001) Magnetic helicity conservation and astrophysical dynamos. Astrophys J 550:752–760

Viviani M, Warnecke J, Käpylä MJ, Käpylä PJ, Olspert N, Cole-Kodikara EM, Lehtinen JJ, Brandenburg A (2018) Transition from axi- to nonaxisymmetric dynamo modes in spherical convection models of solar-like stars. Astron Astrophys 616:A160. https://doi.org/10.1051/0004-6361/201732191. 1710.10222

Warnecke J (2018) Dynamo cycles in global convection simulations of solar-like stars. Astron Astrophys 616:A72. https://doi.org/10.1051/0004-6361/201732413. 1712.01248

Warnecke J, Brandenburg A, Mitra D (2011) Dynamo-driven plasmoid ejections above a spherical surface. Astron Astrophys 534:A11. https://doi.org/10.1051/0004-6361/201117023. 1104.0664

Warnecke J, Brandenburg A, Mitra D (2012) Magnetic twist: a source and property of space weather. J Space Weather Space Clim 2:A11. https://doi.org/10.1051/swsc/2012011. 1203.0959

Warnecke J, Losada IR, Brandenburg A, Kleeorin N, Rogachevskii I (2013) Bipolar magnetic structures driven by stratified turbulence with a coronal envelope. Astrophys J Lett 777(2):L37. https://doi.org/10.1088/2041-8205/777/2/L37. 1308.1080

Warnecke J, Käpylä PJ, Käpylä MJ, Brandenburg A (2016) Influence of a coronal envelope as a free boundary to global convective dynamo simulations. Astron Astrophys 596:A115. https://doi.org/10.1051/0004-6361/201526131. 1503.05251

Warnecke J, Rheinhardt M, Tuomisto S, Käpylä PJ, Käpylä MJ, Brandenburg A (2018) Turbulent transport coefficients in spherical wedge dynamo simulations of solar-like stars. Astron Astrophys 609:A51. https://doi.org/10.1051/0004-6361/201628136. 1601.03730

Warnecke J, Rheinhardt M, Viviani M, Gent FA, Tuomisto S, Käpylä MJ (2021) Investigating global convective dynamos with mean-field models: full spectrum of turbulent effects required. Astrophys J Lett 919(2):L13. https://doi.org/10.3847/2041-8213/ac1db5. 2105.07708

Willis AP (2012) Optimization of the magnetic dynamo. Phys Rev Lett 109(25):251101. https://doi.org/10.1103/PhysRevLett.109.251101. 1209.1559

Wilson PR, Altrocki RC, Harvey KL, Martin SF, Snodgrass HB (1988) The extended solar activity cycle. Nature 333:748–750. https://doi.org/10.1038/333748a0

Wright NJ, Drake JJ (2016) Solar-type dynamo behaviour in fully convective stars without a tachocline. Nature 535(7613):526–528. https://doi.org/10.1038/nature18638. 1607.07870

Wright NJ, Newton ER, Williams PKG, Drake JJ, Yadav RK (2018) The stellar rotation-activity relationship in fully convective M dwarfs. Mon Not R Astron Soc 479(2):2351–2360. https://doi.org/10.1093/mnras/sty1670. 1807.03304

Yokoi N (2013) Cross helicity and related dynamo. Geophys Astrophys Fluid Dyn 107:114–184. https://doi.org/10.1080/03091929.2012.754022. 1306.6348

Yokoi N, Schmitt D, Pipin V, Hamba F (2016) A new simple dynamo model for stellar activity cycle. Astrophys J 824(2):67. https://doi.org/10.3847/0004-637X/824/2/67. 1601.06348

Yoshimura H (1978) Nonlinear astrophysical dynamos - multiple-period dynamo wave oscillations and long-term modulations of the 22 year solar cycle. Astrophys J 226:706–719. https://doi.org/10.1086/156653

Yoshimura H (1981) Solar cycle Lorentz force waves and the torsional oscillations of the sun. Astrophys J 247:1102–1112. https://doi.org/10.1086/159120

Yoshizawa A, Yokoi N (1993) Turbulent magnetohydrodynamic dynamo for accretion disks using the cross-helicity effect. Astrophys J 407:540. https://doi.org/10.1086/172535

Yoshizawa A, Kato H, Yokoi N (2000) Mean field theory interpretation of solar polarity reversal. Astrophys J 537(2):1039–1053. https://doi.org/10.1086/309057

Zeldovich YB, Ruzmaikin AA, Sokoloff DD (1983) Magnetic fields in astrophysics. Gordon and Breach, New York

Zhou H, Blackman EG, Chamandy L (2018) Derivation and precision of mean field electrodynamics with mesoscale fluctuations. J Plasma Phys 84(3):735840302. https://doi.org/10.1017/S0022377818000375. 1710.04064

Publisher's Note Springer Nature remains neutral with regard to jurisdictional claims in published maps and institutional affiliations.

Authors and Affiliations

Axel Brandenburg[1,2,3,4] · Detlef Elstner[5] · Youhei Masada[6] · Valery Pipin[7]

✉ A. Brandenburg
brandenb@nordita.org

D. Elstner
delstner@aip.de

Y. Masada
ymasada@fukuoka-u.ac.jp

V. Pipin
pip@iszf.irk.ru

[1] Nordita, KTH Royal Institute of Technology and Stockholm University, Hannes Alfvéns väg 12, 10691 Stockholm, Sweden

[2] The Oskar Klein Centre, Department of Astronomy, 10691 Stockholm, Sweden

[3] School of Natural Sciences and Medicine, Ilia State University, 0194 Tbilisi, Georgia

[4] McWilliams Center for Cosmology and Department of Physics, Carnegie Mellon University, Pittsburgh, Pennsylvania 15213, USA

[5] Leibniz-Institut für Astrophysik Potsdam (AIP), An der Sternwarte 16, 14482 Potsdam, Germany

[6] Department of Applied Physics, Faculty of Science, Fukuoka University, Fukuoka 814-0180, Japan

[7] Institute of Solar-Terrestrial Physics, Russian Academy of Sciences, Irkutsk, 664033, Russia

Space Science Reviews (2023) 219:39
https://doi.org/10.1007/s11214-023-00982-y

Mean Field Models of Flux Transport Dynamo and Meridional Circulation in the Sun and Stars

Gopal Hazra[1,2] · Dibyendu Nandy[3,4] · Leonid Kitchatinov[5] · Arnab Rai Choudhuri[6,7]

Received: 14 February 2023 / Accepted: 22 June 2023 / Published online: 13 July 2023
© The Author(s) 2023

Abstract

The most widely accepted model of the solar cycle is the flux transport dynamo model. This model evolved out of the traditional $\alpha\Omega$ dynamo model which was first developed at a time when the existence of the Sun's meridional circulation was not known. In these models the toroidal magnetic field (which gives rise to sunspots) is generated by the stretching of the poloidal field by solar differential rotation. The primary source of the poloidal field in the flux transport models is attributed to the Babcock–Leighton mechanism, in contrast to the mean-field α-effect used in earlier models. With the realization that the Sun has a meridional circulation, which is poleward at the surface and is expected to be equatorward at the bottom of the convection zone, its importance for transporting the magnetic fields in the dynamo process was recognized. Much of our understanding about the physics of both the meridional circulation and the flux transport dynamo has come from the mean field theory obtained by averaging the equations of MHD over turbulent fluctuations. The mean field theory of meridional circulation makes clear how it arises out of an interplay between the centrifugal and thermal wind terms. We provide a broad review of mean field theories for solar magnetic fields and flows, the flux transport dynamo modelling paradigm and highlight some of their applications to solar and stellar magnetic cycles. We also discuss how the dynamo-generated magnetic field acts on the meridional circulation of the Sun and how the fluctuations in the meridional circulation, in turn, affect the solar dynamo. We conclude with some remarks on how the synergy of mean field theories, flux transport dynamo models and direct numerical simulations can inspire the future of this field.

Keywords Sun: dynamo · Sun: meridional circulation · Sun: magnetic topology · Stars: late-type · Stars: magnetic field

1 Introduction

The turbulent convection zones of the Sun and other stars host a magnetohydrodynamic dynamo mechanism which involves interactions between the velocity and the magnetic fields. When this interaction takes place within a rotating astrophysical object, it leads to the possibility of a large-scale magnetic field emerging out of such interactions (Parker 1955a;

Solar and Stellar Dynamos: A New Era
Edited by Manfred Schüssler, Robert H. Cameron, Paul Charbonneau, Mausumi Dikpati,
Hideyuki Hotta and Leonid Kitchatinov

Extended author information available on the last page of the article

Steenbeck et al. 1966; Moffatt 1978; Parker 1979). Historically this subject developed by solving the mean field equations which arise by averaging over turbulence at small scales. The challenge of the subject comes from the fact that physics at small scales may profoundly influence what is happening at the large scales. The physics of small scales is captured in the mean field equations through a set of parameters – the α-effect, turbulent diffusion, Reynolds stresses (including what is called the Λ-effect), and turbulent pumping. We shall collectively refer to them as 'turbulence parameters' (Moffatt 1978). The mean field theory has two aspects. (i) We have to estimate the turbulence parameters by some means. (ii) We have to solve the mean field equations in which these turbulence parameters appear. In the early years of research, turbulence parameters would be calculated analytically by making some suitable assumptions about the small-scale turbulence (e.g., Choudhuri 1998). Sometimes, observational data could be used to put important constraints on these parameters (e.g., Chae et al. 2008; Hazra and Miesch 2018). Within the last few years, it has been possible to calculate the turbulence parameters from numerical simulations of turbulence (e.g., Käpylä et al. 2006; Simard et al. 2016; Shimada et al. 2022; Warnecke et al. 2021). We expect more inputs from simulations in the coming years to put the mean field models on a firmer footing. There is a common consensus that the mean field models played a historically important role in the development of the subject. However, with increasingly complex and computationally intensive full magnetohydrodynamic (MHD) simulations being done by various groups around the world, are mean field models still relevant?

The mean field modelling approach is computationally less demanding and it is possible to make more extensive parameter space studies with them. However, there is a deeper reason why the mean field models continue to remain so relevant. Even the most ambitious numerical simulations undertaken at the present time have fluid and magnetic Reynolds numbers many orders of magnitude smaller than what they are for the Sun and other stars. They are still very far from producing sufficiently realistic results which can be compared with observational data in detail. On the other hand, by adjusting various parameters of mean field models suitably, it is often possible to achieve remarkable agreements with observations. This approach may justifiably be criticized as ad hoc. However, an understanding of what needs to be done in the mean field models to achieve convergence with observational data often provides great insights into various physical processes. Mean field models are expected to remain an active research area for many years to come.

The mean field theory of large-scale magnetic fields is easily adapted to an approach known as kinematic dynamo theory. In this approach, we have to specify the large-scale flows – such as the differential rotation and the meridional circulation – apart from the turbulence parameters, and then the evolution of the mean magnetic field is calculated – an approach pioneered by Parker (1955a) and Steenbeck et al. (1966). While the kinematic dynamo theory was developing, important developments also took place in the mean-field theory of large-scale flows – the initial impetus coming from efforts to explain the differential rotation of the Sun. For a few decades, the kinematic dynamo theory and the mean field theory of large-scale flows developed almost independently of each other. These two theories have come together in the last few years with the blossoming of the field of solar and stellar dynamos. Kinematic models of the solar dynamo could rely on the observations of the differential rotation and the meridional circulation of the Sun. We do not have similar detailed observational data of other stars. In order to build mean field models of stellar dynamos, the kinematic dynamo theory and the mean field theory of large-scale flows have to be combined together. In the case of the Sun also, as we have become aware of various feedback processes between the solar magnetic cycle and the large-scale flows, it has become essential to combine the two theories of the kinematic dynamo and large-scale flows –

to model such phenomena as torsional oscillations and variations of the meridional circulation.

The first models of the solar dynamo were constructed at a time when the only available knowledge about large-scale flows was the existence of differential rotation at the solar surface. Nothing was known about the meridional circulation of the Sun or the distribution of differential rotation underneath the solar surface. With the discovery of the meridional circulation and the realization that it is likely to play an important role in the dynamo process, a new type of dynamo model – the flux transport dynamo model – came into being. There are now efforts to apply the flux transport dynamo model to other stars. Since the meridional circulation plays a crucial role in the flux transport dynamo, the temporal variation of this circulation in the Sun may have a profound effect on the solar cycle. Some of the irregularities of the solar cycle may arise from fluctuations in the meridional circulation (Karak and Choudhuri 2011; Choudhuri and Karak 2012).

The solar dynamo has been the subject of several reviews (Choudhuri 2011; Charbonneau 2014; Karak et al. 2014a; Charbonneau 2020). The prospect for extrapolating the solar dynamo models to stars has been reviewed by Choudhuri (2017). We also refer to recent reviews of the meridional circulation (Choudhuri 2021b; Hanasoge 2022).

The mean field models of large-scale magnetic and velocity fields are described in the next section. Then Sect. 3 describes how large-scale flows in the Sun and stars – especially the meridional circulation – can be computed from the mean field model of large-scale flows. How the flux transport dynamo model arose for explaining the solar cycle will be discussed in the Sect. 4, with Sect. 5 summarizing its applications to other stars. Then Sect. 6 will point out the key role of meridional circulation variations in explaining the solar cycle irregularities. In the Sect. 7, we survey the current status of the important subject of computing the turbulent parameters from numerical simulations, which may provide important inputs to mean field models, and conclude our review.

2 Mean Field Theory of Large-Scale Magnetic and Velocity Fields

Magnetic and velocity fields of the Sun behave differently on large and small spatial scales. The fields of the scale comparable to the solar radius show repeatable – though not strictly periodic – evolution of their patterns in the course of 11-year solar cycles (Hathaway 2015). Cells of granular or supergranular solar convection, on the other hand, reconfigure themselves irregularly on a much shorter time scale. This leads to the basic idea of mean-field magnetohydrodynamics (MHD) that not detailed structures but mean statistical properties only of the small-scale turbulent magnetic (b) and velocity (v) fields are relevant to the dynamics of their large-scale magnetic (B) and velocity (V) counterparts. Different scales are separated by temporal or spatial averaging, which leaves the large-scale field unchanged but nullifies the turbulent fields: $\langle b \rangle = 0$, $\langle v \rangle = 0$, where the angular brackets signify the averaging.

Formulation of the mean-field MHD is a formidable task that is not completed up to now. Nevertheless, the main effects and methods of the mean-field theory were systematised already about forty years ago in the monographs by Moffatt (1978), Parker (1979) and Krause and Rädler (1980). The mean field induction equation of the theory,

$$\frac{\partial B}{\partial t} = \nabla \times (\mathcal{E} + V \times B - \eta \nabla \times B),$$ \hfill (1)

includes the small-scale turbulent fields via the mean electromotive force (EMF) $\mathcal{E} = \langle \mathbf{v} \times \mathbf{b} \rangle$. In what follows, we use the following basic expression for the EMF

$$\mathcal{E} = \alpha \mathbf{B} + \mathbf{v}^{\mathrm{dia}} \times \mathbf{B} - \eta_{\mathrm{T}} \nabla \times \mathbf{B}, \tag{2}$$

which includes so-called α-effect, diamagnetic pumping and eddy diffusion, respectively, in its right-hand side.

Among the three effects, diamagnetic pumping seems to be the least known one. The diamagnetic effect of turbulent conducting fluids consists in expulsion of magnetic fields from the regions of relatively high turbulent intensity (see Kitchatinov and Olemskoy 2012b, for pictorial explanation). The effect was predicted by Zeldovich (1957) and first derived by Rädler (1968) where the expression for the effective velocity

$$\mathbf{v}^{\mathrm{dia}} = -\frac{1}{2} \nabla \eta_{\mathrm{T}} \tag{3}$$

can be found (see also Eq. (3.10) in Kichatinov and Rüdiger 1992). The diamagnetic pumping has been detected in MHD laboratory experiment (Spence et al. 2007) and in direct numerical simulations (Tobias et al. 1998; Dorch and Nordlund 2001; Ossendrijver et al. 2002). If included in a dynamo model, the downward diamagnetic pumping with the effective velocity of Eq. (3) can concentrate magnetic fields at the bottom of the convection zone thus realising an overshoot dynamo even in distributed-type models (Kitchatinov and Olemskoy 2012b).

The coefficient α appearing in Eq. (2), which becomes a rank-2 tensor in the completely general situation, arises from helical turbulent motions and is crucial for the dynamo generation of the magnetic field. It was first evaluated by Parker (1955a) and Steenbeck et al. (1966). For isotropic turbulence, suitable assumptions lead to the expression

$$\alpha = -\tau \langle \mathbf{v} \cdot (\nabla \times \mathbf{v}) \rangle / 3, \tag{4}$$

where τ is the correlation time of turbulence (Choudhuri 1998). As we shall point out in Sect. 4, the flux transport dynamo involves a different mechanism for magnetic field generation: the Babcock–Leighton mechanism. This mechanism also can be represented by the coefficient α appearing in Eq. (2). However, α corresponding to this mechanism is not given by Eq. (4).

It has been found relatively recently that the conservation of magnetic helicity leads to a catastrophic quenching of the α-effect (Gruzinov and Diamond 1994; Brandenburg and Subramanian 2005), switching off the local α-effect of Eq. (2) in the case of large magnetic Reynolds number. It can be shown that non-local α-effect of Babcock–Leighton type is not subject to the catastrophic quenching when combined with downward diamagnetic pumping (Kitchatinov and Olemskoy 2011a; Brandenburg et al. 2015). The physical reason for alleviation of the catastrophic quenching in this case is spatial separation of the (near-bottom) toroidal field and the (near-surface) poloidal field generated from the toroidal field by non-local α-effect. The large-scale magnetic helicity is not produced in this case and helicity conservation does not demand an emergence of equal amount of opposite in sign small-scale helicity which is an essential part of the catastrophic quenching mechanism (Brandenburg and Subramanian 2005).

We now turn to the mean field theory of large-scale flows. The mean-field equation of motion is

$$\frac{\partial \mathbf{V}}{\partial t} + (\mathbf{V} \cdot \nabla)\mathbf{V} = \frac{1}{\mu\rho}(\mathbf{B} \cdot \nabla)\mathbf{B} - \frac{1}{\rho}\nabla\left(P + \frac{B^2}{2\mu}\right) + \mathbf{g} + \frac{1}{\rho}\nabla \cdot \mathcal{S}, \tag{5}$$

where P is pressure, $\boldsymbol{g}$ is gravity and

$$S_{ij} = -\rho\langle v_i v_j\rangle + \mu^{-1}\langle b_i b_j - \frac{1}{2}\delta_{ij}b^2\rangle \tag{6}$$

is the turbulence stress tensor combining the Reynolds and Maxwell stress of small-scale fields.

Mean field MHD has to express the turbulent stress in terms of large-scale fields. Galilean invariance demands the mean velocity to contribute via its spatial derivatives only:

$$S_{ij} = \rho N_{ijkl}\nabla_k V_l \tag{7}$$

(repetition of subscripts means summation). This linear relation can be seen as a formal definition of the eddy viscosity tensor N_{ijkl}. Enhanced dissipation of large-scale flows is a well-known effect of turbulence. This is however not all what convective turbulence in stars can do. Convection is driven by buoyancy forces which point upward or downward in radius. This imparts *anisotropy* with different intensities of radial and horizontal turbulent mixing. It is known since Lebedinsky (1941) that the eddy viscosity tensor for anisotropic turbulence does not satisfy the Onsager symmetry rule $N_{ijkl} = N_{klij}$. The rule has to be satisfied for true viscosity decreasing kinetic energy of mean flow (see Sect. I.9 in Lifshitz and Pitaevskii 1981). Violation of the rule signals that the anisotropic turbulence does not necessarily dissipate but can excite some kind of large-scale flow. An important application of this excitation effect was found in the theory of stellar differential rotation where it is known as the Λ-effect (Rüdiger 1989). The stress tensor of Eq. (7) for anisotropic turbulence does not vanish for rigid rotation,

$$S_{ij}^{\Lambda} = \rho N_{ijkl}\varepsilon_{klm}\Omega_m, \tag{8}$$

where Ω is the angular velocity and ε_{klm} is the fully antisymmetric tensor. The components $S_{r\phi}^{\Lambda}$ and $S_{\theta\phi}^{\Lambda}$ in spherical coordinates stand for the angular momentum fluxes by turbulence, which are the principal drivers of stellar differential rotation. Details of the Λ-effect theory can be found in Rüdiger (1989) and Rüdiger et al. (2013). Separation of the Λ-effect from true viscosity changes Eq. (7) to

$$S_{ij} = S_{ij}^{\Lambda} + \rho\mathcal{N}_{ijkl}\nabla_k V_l, \tag{9}$$

where $\mathcal{N}_{ijkl}$ is the true viscosity tensor with positive definite coefficients and symmetry, $\mathcal{N}_{ijkl} = \mathcal{N}_{klij}$, ensuring dissipation of large-scale flows. Derivation of the viscosity tensor for rotating turbulence can be found in Kitchatinov et al. (1994).

It may be noted that several effects in excess of the Λ-effect and eddy viscosity of Eq. (9) have been found in the extensive literature on turbulent stress. These include turbulent pressure, a slight modification of the large-scale Lorentz force (Kleeorin and Rogachevskii 1994; Rüdiger et al. 2012), and the anisotropic kinetic alpha-effect (Frisch et al. 1987). Equation (9) includes what matters for stellar applications only. A similar comment applies to the EMF of Eq. (2). The equation displays its three basic contributions in the simplest form. Rotationally induced anisotropy complicates them (Pipin 2008; Kitchatinov et al. 1994) so that, e.g., the eddy magnetic diffusivities for the directions along and across the rotation axis differ. The rotational anisotropy however is of modifying rather than principal nature for stellar dynamo modelling.

3 Meridional Circulation in the Sun and Stars

The global flow in the sun is known to vary little in course of the activity cycle. The flow is not magnetic by origin. We consider first hydrodynamics of the meridional flow and discuss its magnetic modification afterwards.

3.1 Meridional Flow Origin and Structure

Momentum density in the stellar convection zones is divergence-free, $\nabla \cdot (\rho \boldsymbol{u}) = 0$, to a good approximation (Lantz and Fan 1999). In this case, the meridional flow is a fluid circulation over *closed* stream-lines.

The circulation proceeds in a turbulent convection zone where the eddy viscosity resists the flow. Some forces supporting the flow against the viscous decay should therefore be present. Only non-conservative forces can transmit energy to a circulatory flow. Motion equation (5) can be curled to filter-out irrelevant conservative forces: see, for example, Choudhuri (2021b). This leads to a meridional flow description in terms of the azimuthal vorticity $\omega = (\nabla \times \boldsymbol{V})_\phi$:

$$\frac{\partial \omega}{\partial t} + r \sin\theta\, \nabla \cdot \left(\boldsymbol{V} \frac{\omega}{r \sin\theta} \right) + \mathcal{D}(\omega) = r \sin\theta \frac{\partial \Omega^2}{\partial z} - \frac{g}{r c_{\mathrm{p}}} \frac{\partial S}{\partial \theta}. \tag{10}$$

In this equation, $z = r \cos\theta$ is the (signed) distance from the equatorial plane, S is the specific entropy, and $\mathcal{D}$ accounts for the viscous dissipation of the meridional flow. The symbolic representation for the dissipation term is justified by complexity of its explicit formulation (Kitchatinov and Olemskoy 2011b). The dissipation term acts to decrease the meridional flow energy.

Two terms in the right-hand side of Eq. (10) stand for two principal drivers of the meridional flow. The first term includes driving by the centrifugal force. The force is conservative for cylinder-shaped (z-independent) rotation. Accordingly, the first term in the right-hand side of (10) accounts for the non-conservative part of the centrifugal force. The second term involves the non-conservative buoyancy (baroclinic) force.

Figure 1 illustrates the two drivers. If the angular velocity decreases with distance from the equatorial plane, as it does in the sun (Schou et al. 1998), a torque by the centrifugal force tends to drive anti-clockwise circulation (in the north-west quadrant of the convection zone). The baroclinic driving is proportional to the temperature variation with latitude inside the convection zone. The 'differential temperature' results from rotationally-induced anisotropy of the convective heat transport (Rüdiger et al. 2005). If the temperature increases with latitude, as it probably does in the Sun (Miesch et al. 2006; Kitchatinov and Olemskoy 2011b), a slightly cooler fluid at low latitudes tends to sink down and the warmer polar fluid tends to rise up and spread over the surface to drive a clockwise circulation (Fig. 1).

The two drivers of the meridional flow counteract each other in the sun. The counteraction probably is the general case with solar-type stars. This can be evidenced by normalizing Eq. (10) to dimensionless units. Measuring time in its viscous scale R^2/ν_{T} and multiplying Eq. (10) by this scale squared, gives the first and the second terms in the right-hand side of the normalised equation the coefficients of the Taylor (Ta) and Grashof (Gr) numbers

$$\mathrm{Ta} = \frac{4\Omega^2 R^4}{\nu_{\mathrm{T}}^2}, \quad \mathrm{Gr} = \frac{g R^3}{\nu_{\mathrm{T}}^2} \frac{\delta T}{T} \tag{11}$$

Fig. 1 Illustration of the centrifugal (left) and baroclinic (right) driving of the meridional flow (see text)

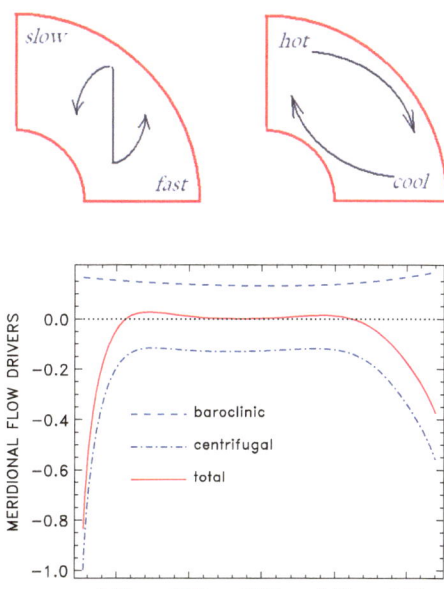

Fig. 2 Depth profiles of the baroclinic and centrifugal driving terms of the meridional flow equation (10) for the 45° latitude computed with the mean-field model by Kitchatinov and Olemskoy (2011b). The driving terms are normalised to the maximum absolute value one and their sum is shown by the red line

respectively, where δT is the differential temperature. Direct numerical simulations (Miesch et al. 2006) and mean-field models (Kitchatinov and Olemskoy 2011b) of the solar differential rotation give the value $\delta T / T \sim 10^{-5}$ for the normalised differential temperature varying moderately with depth. This leads to large characteristic values of Gr $\sim$ Ta $\sim 10^7$ for the standard mixing-length estimation $\nu_T \approx 5 \times 10^8$ m^2 s^{-1} of the eddy viscosity. Each term in the left side of Eq. (10) scales to a much smaller value. There is no other way to satisfy this equation but the two terms on its right-hand side almost balance each other. This leads to the balance equation

$$r \sin \theta \frac{\partial \Omega^2}{\partial z} - \frac{g}{r c_{\mathrm{p}}} \frac{\partial S}{\partial \theta} = 0. \qquad (12)$$

Equation (10) shows that the meridional flow results from a slight deviation from the thermo-rotational balance of Eq. (12). The vorticity equation also informs on how the balance is maintained. Every term in the right-hand side of this equation alone can drive a meridional flow of order one kilometer per second (Durney 1996). A considerable deviation from the balance would drive a fast meridional flow, which reacts back on the differential rotation and temperature to restore the balance. The meridional flow results from deviations from the thermo-rotational balance and also controls that the deviations are small. This consideration shows that a reasonable model for the meridional flow alone is not possible. A realistic model has to solve consistently for the meridional flow, differential rotation and heat transport.

Figure 2 shows the depth profiles of the meridional flow drivers computed with a mean-field model. The sum of the baroclinic and centrifugal drivers is close to zero in the bulk of the convection zone, which is therefore close to the thermo-rotational balance of Eq. (12). The balance is however violated near the top and bottom boundaries. This is because of the stress-free boundary conditions employed in the model. This condition of zero surface den-

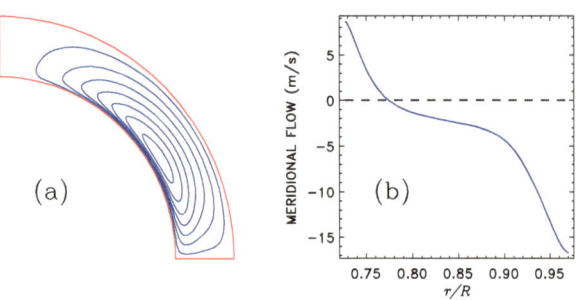

Fig. 3 Meridional flow stream-lines (a) and depth profile of the meridional velocity for the 45° latitude (b) from the same mean-field model as Fig. 2

sity of external forces ensures that the meridional flow is controlled by 'internal' processes inside the convection zone, not imposed externally.

The stress-free condition uniquely defines rotational gradient near the boundaries (Kitchatinov 2016) regardless of the balance condition of Eq. (12). As a result, deviation from cylinder-shaped rotation increases near the boundaries. This is probably why the centrifugal driving in Fig. 2 amplifies near the boundaries. On the contrary, the spherically symmetric heat flux at the bottom and free radiation on the top, prescribed in the model of Fig. 2, do not control the latitudinal entropy gradient and the near-surface baroclinic driving is not disturbed.

The stress-free condition together with zero radial velocity constitute a complete set of hydrodynamical boundary conditions. The extra condition of the thermo-rotational balance cannot be satisfied near the boundary. Thin boundary layers form where the balance is violated. Deviation from the balance excites the meridional flow inside the boundary layers.

The flow of Fig. 3 computed with the same mean-field model as Fig. 2 attains its largest velocity on the boundaries and decreases inside the convection zone. The flow of this Figure is in at least qualitative agreement with the recent seismological detection (Rajaguru and Antia 2015; Gizon et al. 2020; Hanasoge 2022). The surface flow velocity of Fig. 3 is close to solar observations. The flow profile and its bottom value are less certain. The turning point at about $r/R = 0.77$ and the bottom velocity of 8 to 9 m s^{-1} are close to Fig. 4 by Rajaguru and Antia (2015) but differ somewhat from $r/R \approx 0.8$ and bottom velocity of about 5 m s^{-1} found by Gizon et al. (2020). It may be noted that the computed bottom flow slows down and the turning point shifts to larger r if the model of Fig. 3 is modified by reducing mixing-length scale near the bottom (cf. Fig. 2 in Kitchatinov and Nepomnyashchikh 2017).

An unsettled issue is how deep the equatorward meridional counter flow penetrates within the solar interior. Based on a kinematic dynamo modelling approach Nandy and Choudhuri (2002) argued that a single cell flow penetrating below the convection zone into the stable, overshoot layer is important for explaining the low-mid latitude appearance of sunspots. While subsequent arguments have been made both against and for such a possibility (Gilman and Miesch 2004; Garaud and Brummell 2008), we note that the most recent observations do not rule out a deep meridional counter flow (Gizon et al. 2020).

Since the convection cells become much smaller at the top of the solar convection zone where the various scale heights are much smaller compared to the interior, the nature of convection clearly changes in a top layer and this complicates the issue of a boundary layer there. Observationally, helioseismic maps of differential rotation show a near-surface shear layer at the top of the solar convection zone. Recently Choudhuri (2021a) and Jha and Choudhuri (2021) have argued that this shear layer arises from the changed nature of convection rather than from a violation of Eq. (12).

The boundary layers in the solar model are not very thin (Fig. 2). Their thickness $D_E \sim \sqrt{\nu_T/\Omega}$ decreases with rotation rate. For faster rotation, the meridional flow retreats to increasingly thin boundary layers and weakens inside the convection zone (Kitchatinov and Olemskoy 2012a). Simultaneously, the differential rotation changes from the conical shape to cylinder-shaped pattern reflecting a faster increase of the Taylor number of Eq. (11) with rotation rate compared to the Grashof number.

3.2 Magnetic Modifications

The meridional flow equation (Eq. (10)) is modified with allowance for the large-scale axisymmetric magnetic field $\boldsymbol{B}$:

$$\frac{\partial \omega}{\partial t} + r \sin\theta \, \nabla \cdot \left(\frac{\boldsymbol{V}\omega - \boldsymbol{V}_A\omega_A}{r \sin\theta} \right) + \mathcal{D}(\omega)$$

$$= r \sin\theta \frac{\partial (\Omega^2 - \Omega_A^2)}{\partial z} - \frac{g}{r c_p} \frac{\partial S}{\partial \theta} - \frac{g\rho}{2r\gamma P} \frac{\partial V_A^2}{\partial \theta}, \tag{13}$$

where meridional flow driving terms are again collected in the right-hand side of the equation. The magnetic terms in Eq. (13) are formulated in terms of the Alfven velocity $\boldsymbol{V}_A = \boldsymbol{B}/\sqrt{\mu\rho}$ and the Alfven angular frequency Ω_A for the toroidal field $B_\phi = \sqrt{\mu\rho}\, r \sin\theta\, \Omega_A$; $\omega_A = (\nabla \times \boldsymbol{V}_A)_\phi$ is the magnetic vorticity and $\gamma = c_p/c_v$ is the adiabaticity index.

The first term on the right-hand side of Eq. (13) includes the non-conservative magnetic tension by the toroidal field. The minus sign in the contribution means that the tension force points towards the rotation axis, opposite to the centrifugal force. The last term on the right-hand side stands for the baroclinic driving by magnetic pressure. It was accounted for when deriving this term that the density varies much stronger in radius than in latitude and the convection zone stratification is close to adiabaticity.

It can be seen that magnetic contribution in the left-hand side of Eq. (13) includes the poloidal field only. This field is weak and this contribution is negligible for the sun. For stars with deep convection zones, the poloidal field can be strong (Gregory et al. 2012) and the magnetic advection of vorticity can be significant.

Assuming that the mean field in the deep convection zone of the sun is of order 1 Tesla, we could see that the magnetic terms in Eq. (13) are about two orders of magnitude smaller compared to the centrifugal driving. The magnetic terms are nevertheless large compared to each term on the left side of the equation. As in the hydrodynamical case, the meridional flow results from a disbalance of the driving terms in the right-hand side of Eq. (13) but the flow reacts back to ensure that the deviation from the balance remains small.

The magnetically modified thermo-rotational balance is global by nature. This in particular means that rotation law variation in torsional oscillations may not spatially coincide with the location of the magnetic fields producing the oscillations (Pipin and Kosovichev 2020).

The magnetic field can also affect the meridional flow indirectly by modifying the differential temperature (Spruit 2003; Hanasoge 2022) or differential rotation.

4 Modelling the Solar Cycle: The Paradigm Shift from the $\alpha\Omega$ Dynamo to the Flux Transport Dynamo

In the mean-field model, the magnetic field is assumed to be axisymmetric and can be written as

$$\boldsymbol{B} = B_\phi(r, \theta, t)\, \mathbf{e}_\phi + \nabla \times [A(r, \theta, t)\, \mathbf{e}_\phi], \tag{14}$$

where $B_\phi(r, \theta)\, \mathbf{e}_\phi$ is referred to as the toroidal field and $\nabla \times [A(r, \theta)\, \mathbf{e}_\phi] = \mathbf{B}_p$ gives the poloidal field. The velocity field associated with large-scale flows can be written as

$$\mathbf{V} = \mathbf{V}_m + r \sin\theta\, \Omega(r, \theta)\, \mathbf{e}_\phi, \tag{15}$$

where $\Omega(r, \theta)$ is the angular velocity in the interior of the Sun and $\mathbf{V}_m$ is the meridional circulation having components V_r and V_θ. On substituting Eq. (14) and Eq. (15) into Eq. (1) with $\mathcal{E}$ given by Eq. (2), some reasonable assumptions lead to the following coupled equations for the poloidal and the toroidal fields

$$\frac{\partial A}{\partial t} + \frac{1}{s}(\mathbf{V}_m.\nabla)(sA) = \eta_T \left(\nabla^2 - \frac{1}{s^2} \right) A + \alpha B, \tag{16}$$

$$\frac{\partial B}{\partial t} + \frac{1}{r}\left[\frac{\partial}{\partial r}(rV_r B) + \frac{\partial}{\partial \theta}(V_\theta B) \right] = \eta_T \left(\nabla^2 - \frac{1}{s^2} \right) B + s(\mathbf{B}_p.\nabla)\Omega + \frac{1}{r}\frac{d\eta_T}{dr}\frac{\partial}{\partial r}(rB), \tag{17}$$

where $s = r \sin\theta$. Note that we are not including the diamagnetic pumping term in this discussion.

When the first efforts were made to construct mean field models of the solar dynamo (Parker 1955a; Steenbeck et al. 1966), the existence of the meridional circulation was not yet known. The early models which took $\mathbf{V}_m = 0$ are now known as $\alpha\Omega$ dynamo models. In such models, the generation of the poloidal field involves the α-effect according to Eq. (16) and the generation of the toroidal field is due to differential rotation involving Ω according to Eq. (17). A remarkable result was that the $\alpha\Omega$ dynamo models could give periodic dynamo waves under certain circumstances (Parker 1955a; Steenbeck and Krause 1969). This raised the possibility of explaining the solar cycle with this model. In order to model the butterfly diagram of sunspots, we need to have the dynamo wave propagate in the equatorial direction. The condition for this was found to be

$$\alpha \frac{\partial \Omega}{\partial r} < 0 \tag{18}$$

in the northern hemisphere of the Sun. This is often referred to as the Parker–Yoshimura sign rule (Parker 1955a; Yoshimura 1975). In the 1970s when nothing was known about the nature of the differential rotation underneath the solar surface, many models of the solar dynamo were constructed by prescribing α and Ω in such a manner that the Eq. (18) was satisfied. Many of these models matched different aspects of the observational data of solar cycles reasonably well and it seemed that the subject was progressing in the right direction.

Several difficulties with the $\alpha\Omega$ dynamo models started becoming apparent by the late 1980s. Firstly, as helioseismology started producing the first maps of the angular velocity distribution inside the Sun, it was found to be completely different from what was being assumed in various $\alpha\Omega$ dynamo models: see, for example, Sect. 4 of Roberts (1972) and Sect. II of Stix (1976) for discussions of the types of differential rotation which they used. Secondly, it was established that the poloidal field of the Sun at the surface propagates poleward with the progress of the solar cycle, in contrast to the sunspots (forming from the toroidal field) which appear closer to the equator as the cycle progresses (Wang et al. 1989). In the simplest kinds of $\alpha\Omega$ dynamo models without meridional circulation, the poloidal and toroidal fields remain coupled to each other, and it is not possible to make them move in opposite directions. Thirdly and lastly, simulations of sunspot formation indicated that the toroidal field must be much stronger than what used to be assumed. Bipolar sunspots form

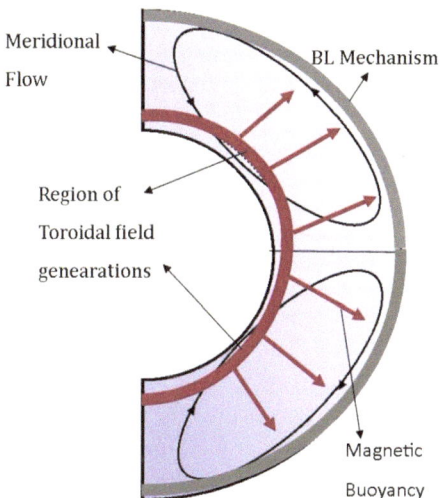

Fig. 4 A cartoon indicating the essential ingredients of the flux transport dynamo model. Taken from the PhD thesis (Hazra 2018)

when parts of the toroidal field rise through the convection zone due to magnetic buoyancy (Parker 1955b). Detailed simulations of this process based on the thin flux tube equation (Spruit 1981; Choudhuri 1990) showed that the Coriolis force due to the solar rotation tries to divert the rising flux tubes towards high latitudes (Choudhuri and Gilman 1987; Choudhuri 1989). Only if the magnetic field inside the flux tubes is sufficiently strong, it is able to counter the Coriolis force in such a manner that there is a match with the observational data (D'Silva and Choudhuri 1993; Fan et al. 1993; Caligari et al. 1995).

The flux transport dynamo model arose in response to these difficulties with the $\alpha\Omega$ dynamo models. Bipolar sunspot pairs on the solar surface appear with a tilt (Hale et al. 1919) – due to the action of the Coriolis force (D'Silva and Choudhuri 1993). Babcock (1961) and Leighton (1969) realized that the decay of such tilted bipolar sunspot pairs would give rise to a poloidal field. The flux transport dynamo model invokes this Babcock–Leighton mechanism for the generation of the poloidal field. Unlike the canonical α-effect, this mechanism does not require the toroidal field to be sufficiently weak. However, when the Babcock–Leighton mechanism is combined with the differential rotation given by helioseismology in a minimalistic dynamo model, the Parker–Yoshimura condition given by Eq. (18) is not satisfied at the low latitudes and dynamo waves are found to propagate in the poleward direction implying that sunspots would appear at higher latitudes with the progress of the solar cycle (Choudhuri et al. 1995), in contradiction with observations. We certainly need something else to turn things around. The meridional circulation was the first proposition which provides a way out of this conundrum.

Figure 4 is a cartoon summarizing how the flux transport dynamo model works. The dark red region at the bottom of the convection zone is where helioseismology has discovered a strong layer of differential rotation which overlaps with the stable overshoot layer beneath the convection zone. Dynamo models which incorporate direct helioseismic observations indicate that the toroidal field begins to be inducted in the convection zone (Muñoz-Jaramillo et al. 2009) and is subsequently amplified, stored and transported equatorward in the tachocline region (Nandy and Choudhuri 2002). However, see Spruit (2012) for various arguments against toroidal flux generation and storage in the tachocline region. Also, as pointed out by Muñoz-Jaramillo et al. (2009) that toroidal field generation is possible in the bulk of the convection zone due to latitudinal shear but an interesting question remains whether the toroidal field can be stored there long enough to be amplified to high values against magnetic buoyancy. A downward pumping can help in this context as studied by Guerrero and de Gouveia Dal Pino (2008) and Zhang and Jiang (2022) leading to a negligible role of tachocline in the FTD model. The stored toroidal field in the tachocline region that escapes out of the tachocline into the convection zone rises to form sunspots due to magnetic buoyancy indicated by the dark red arrows. The decay of sunspots near the surface indicated by the greyish color gives rise to the poloidal field by the Babcock–Leighton mechanism. The meridional circulation is shown by the black contours. It is equatorward at the bottom of the convection zone so that the toroidal field generated there is advected equatorward, producing sunspots closer to the equator with the progress of the solar cycle. On the other hand, the meridional circulation is poleward near the surface so that the poloidal field generated there is advected poleward.

The first 2D axisymmetric models of the flux transport dynamo were constructed in the mid-1990s (Choudhuri et al. 1995; Durney 1995), although some of the basic ideas were put forth on the basis of a 1D model in an earlier paper by Wang et al. (1991). That the meridional circulation can reverse the direction of the dynamo wave was demonstrated convincingly by Choudhuri et al. (1995) and paved the way for the formulation of the flux transport dynamo model. Within the next few years, different groups studied different aspects of the model (Durney 1997; Dikpati and Charbonneau 1999; Nandy and Choudhuri 2001; Küker et al. 2001; Nandy and Choudhuri 2002; Bonanno et al. 2002; Guerrero and Muñoz 2004; Chatterjee et al. 2004; Choudhuri et al. 2004). This model could explain various aspects of observational data pertaining to the solar cycle, especially the butterfly diagram of sunspots along with the time-latitude distribution of the poloidal field at the surface (Chatterjee et al. 2004).

A majority of the flux transport dynamo calculations assumed such a single-cell meridional flow as shown in Fig. 4. The nature of the meridional circulation deeper down in the convection zone remained uncertain till fairly recently and some groups claimed a more complicated, multi-cellular profile (Zhao et al. 2013) from helioseismology. Global 3D HD and MHD convection simulations (e.g., Passos et al. 2014; Featherstone and Miesch 2015) also reported a multi-cellular meridional flow profile for the Sun. Passos et al. (2015) reported a multi-cell meridional circulation with an equatorward return flow near the base of the convection zone from the mid-latitude. Featherstone and Miesch (2015) found a multi-cell meridional flow for the solar-like differential rotation but a single-cell profile for the slowly rotating stars with anti-solar differential rotation. Subsequent research has shown that under certain circumstances the flux transport paradigm can work even in the presence of complex, multi-cellular meridional flow (Jouve and Brun (2007), Hazra et al. (2014), Hazra and Nandy (2016)). However, helioseismology results from different groups are now converging on a single-cell flow pattern (Rajaguru and Antia 2015; Gizon et al. 2020) in agreement with what had been assumed in the majority of flux transport dynamo calculations, strongly validating the flux transport dynamo model.

The period of the flux transport dynamo is essentially set by the time scale of the meridional circulation. When other parameters are held fixed, the period T and the amplitude v_0 of the meridional circulation are found to obey the approximate relation

$$T \propto v_0^{-\gamma}. \tag{19}$$

The index γ is found to have a value close to 1 in different models of the flux transport dynamo (Dikpati and Charbonneau 1999; Yeates et al. 2008).

We note as a caveat that flux transport dynamo models incorporating both radial and latitudinal turbulent pumping as gleaned from magnetoconvection simulations can explain many of the observed features of the solar cycle, even in the absence of meridional circulation – circumventing the constraint of the Parker-Yoshimura sign rule (Hazra et al. 2019). However, unlike meridional circulation, the turbulent pumping profile in the solar convection zone remains completely unconstrained by observations.

One limitation of the 2D axisymmetric models of the flux transport dynamo is that the Babcock–Leighton mechanism is intrinsically a 3D mechanism and can be treated in 2D models only by making drastically simplifying assumptions. In fact, there has been a debate about the best way of treating this mechanism in 2D models (Durney 1997; Nandy and Choudhuri 2001; Muñoz-Jaramillo et al. 2010). One possible approach of handling this mechanism more realistically is to develop 3D kinematic/non-kinematic models in which the magnetic field is treated in 3D so that the dynamics of tilted bipolar sunspots can be computed explicitly (Yeates and Muñoz-Jaramillo 2013; Miesch and Dikpati 2014; Hazra et al. 2017; Hazra and Miesch 2018; Pipin 2022; Bekki and Cameron 2023). An important recent development in this context is a 3D kinematic Babcock–Leighton flux transport dynamo where the buoyant emergence of flux tubes is treated as a dynamic, magnetic field dependent process in a self-consistent manner (Kumar et al. 2019). All the recent developments in the 3D kinematic dynamo models are reviewed by Hazra (2021).

5 Extrapolation to Stellar Dynamos

Magnetic field and sun-like magnetic cycle have been observed in many solar-type stars with outer convection zone (e.g., Wilson 1978; Noyes et al. 1984a; Baliunas et al. 1995; Donati et al. 1997). Unlike the Sun, for which we have a lot of detailed observational data available, observation of surface magnetic field for other stars is quite limited. The observational estimate of magnetic activity for other stars mostly comes from indirect proxies of the magnetic field such as measurements of chromospheric Ca II H & K lines (Wilson 1978; Noyes et al. 1984a; Baliunas et al. 1995) and Coronal X-ray emission (Wright et al. 2011; Wright and Drake 2016). Also, one of the major difficulties in measuring stellar activity is that we need a long-term programme for monitoring stars as their cycle period will likely to be commensurate with the 11-year solar cycle period. Thanks to Mount Wilson observatory monitoring program (Wilson 1978), we have long-term data of Ca II H & K flux for 111 stars from spectral type F2-M2 on or near main sequence. Using this data, Noyes et al. (1984a) found that the magnetic activity of stars increases with the rotation rate. Actually, the magnetic activity better correlates with Rossby number, which is a ratio of the rotation period to the convective turnover time. In Fig. 8 of Noyes et al. (1984a), it is shown how the magnetic activity varies with the Rossby number. The magnetic activity first increases rapidly with increasing rotation rate (or decreasing Rossby number), and then it increases very slowly or even seems to be independent of Rossby number for rapidly rotating stars. This result was corroborated by other independent studies from coronal X-ray emission (e.g., Hempelmann et al. 1995; Wright et al. 2011). Zeeman Doppler Imaging (ZDI) technique also (Donati et al. 1997) emerges as a promising way of reconstructing surface magnetic field from other stars. Using this method, Vidotto et al. (2014a) analysed 73 late-F, G, K and M dwarf stars and reported a similar rotation-activity relation. Recently an extensive study by Reiners et al. (2022) of 292 M-dwarfs from CARMENES high resolution spectra also found the same rotation-activity relation.

While stellar activity follows a clear dependency on the rotation rate of the stars, the dependence of the stellar cycle period on rotation is somewhat complicated (Vaughan and Preston 1980; Noyes et al. 1984b). Mount Wilson sample of Ca II H & K shows two distinct

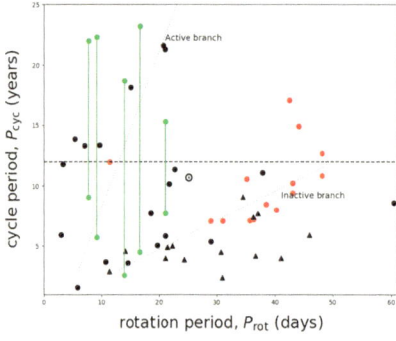 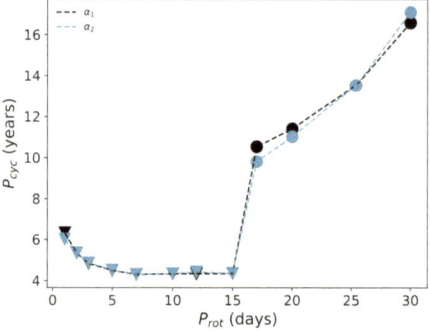

Fig. 5 Left: Rotational dependency of cycle period from observations - P_{cyc} vs P_{rot} plot (Taken from Boro Saikia et al. (2018)). The red symbols are stars with well defined activity cycles, green symbols are stars with multiple activity cycles, and black symbols are for stars with unconfirmed activity cycles. Mount Wilson stars are denoted as filled circles and triangles represent HARPS stars. The active and inactive branches from Böhm-Vitense (2007) are shown in dashed black lines. The Sun is shown as $\odot$. Right: P_{cyc} vs P_{rot} plot from theoretical model of Hazra et al. (2019). The black and blue colors represent two types of treatment in their Babcock–Leighton α effect. Stars with quadrupolar and dipolar parities are shown in triangular and circular symbols respectively

branches the active, young one and old slowly rotating one with a gap between them known as Vaughan & Preston gap (Vaughan and Preston 1980). It has been found that the cycle period decreases with decreasing rotation period of the stars in both of the branches. This data was further analysed carefully by many others (Saar and Brandenburg 1999; Saar 2002; Böhm-Vitense 2007) reporting similar trends. Figure 1 from Böhm-Vitense (2007) shows the clear trend of decreasing cycle period with increasing rotation rate along the inactive and active branches. A recent analysis of a larger sample (4454 Cool stars) shows that the Vaughan and Preston gap might be a result of a lack of data in the Wilson sample (Boro Saikia et al. 2018). The left panel of Fig. 5 (taken from Boro Saikia et al. (2018)) shows the P_{cyc}-P_{rot} diagram for the stars with observed cycle period. The stars with well-defined cycles in their sample show an increasing trend of cycle period with rotation period and there is no clear gap between inactive and active branches of stars. However, as it is clear from the figure, the uncertainty lies in the fast-rotating active branch. Similar results are also reported by Olspert et al. (2018) by an individual probabilistic analysis of the Ca II H & K data.

Many theoretical efforts have been made to understand the relation of magnetic activity and cycle period with rotation period of the stars (e.g., Durney and Robinson 1982; Robinson and Durney 1982; Brandenburg et al. 1994; Nandy 2004; Kitchatinov and Olemskoy 2015; Jouve et al. 2010; Karak et al. 2014b; Strugarek et al. 2017; Warnecke 2018; Hazra et al. 2019). There were some early efforts from traditional $\alpha\Omega$ mean-field dynamo (Durney and Robinson 1982; Robinson and Durney 1982; Brandenburg et al. 1994), before the importance of meridional circulation was properly recognized in dynamo theory (see Sect. 4 for details), to understand the observational behavior of stars. The observed dependence of magnetic activity on the rotation rate of the star is naturally explained from the mean-field $\alpha\Omega$ dynamo theory. The Ω-effect depends on the differential rotation which is connected to the rotation rate of the star. Also, the α-effect which is a measure of helical turbulence naturally relates to the rotation rate of the star. However, as the mean-field dynamo model was not able to explain all the properties of cyclic magnetic activity of the Sun, we need more observational constraints on mean-field dynamo parameters for explaining properties of the stellar magnetic activity from the $\alpha\Omega$ dynamo model. For kinematic dynamo, in the linear regime, the dynamo can sustain if the dynamo number $D = \frac{\alpha R^3}{\eta} \frac{1}{r} \frac{\partial \Omega}{\partial r}$ (η is the co-efficient of turbulent diffusivity and R is the outer stellar radius) exceeds a critical value D_c. In that case, the period of the dynamo cycle $P_{cyc} \propto D^{-1/2}$ (Noyes et al. 1984b). Hence $P_{cyc} \propto \Omega^{-1}$, which is in agreement with the stellar observation. However, in the non-linear regime, where the magnetic field grows until the Lorentz force alters the velocity field to permit some equilibrium, the α-effect or velocity shear is reduced as the field strength increases. As a result, the dynamo number gets reduced until a steady state is achieved and the cycle period has the approximately same value as it had for $D = D_c$. The quenching of dynamo action gives a cycle period almost independent of rotation Ω.

Meanwhile, the importance of meridional circulation in the solar dynamo theory, hence the Flux Transport Dynamo (FTD) theory (see Sect. 4 for details) was established to explain many properties of the solar magnetic field (Choudhuri et al. 1995; Chatterjee et al. 2004). The first comprehensive model of FTD for solar-like stars was carried out by Jouve et al. (2010). Two main ingredients of the FTD model differential rotation and meridional circulation were obtained from 3D hydrodynamic simulations as the observational data for them is not available for other stars. The 3D hydrodynamic simulations result a slower meridional circulation with an increasing rotation rate. In FTD models, as the cycle period is inversely proportional to the speed of meridional circulation (Dikpati and Charbonneau 1999; Chatterjee et al. 2004), the computed cycle periods with different rotation rates from these models are not compatible with observations. Similar results were reported from the scaling relation of stellar dynamo (Nandy 2004). Karak et al. (2014b) also constructed a theoretical model for stellar dynamo based on FTD model. They used the differential rotation and meridional circulation for stars with rotation periods of 1 day to 30 days from mean-field hydrodynamic models as presented in Sect. 2. They also reported an increase in the cycle period with increasing rotation rate, as the amplitude of meridional circulation from the mean-field hydrodynamic model decreases with the increasing rotation rate of stars.

Recently, Hazra et al. (2019) extended the study of Karak et al. (2014b) by incorporating radial turbulent pumping. Turbulent pumping was found to be unavoidable in a stratified stellar convection zone due to the topological asymmetric convective flows (Tobias et al. 1998; Käpylä et al. 2006; Miesch and Hindman 2011). A few previous studies in a solar context already showed that pumping is important in transporting poloidal field from the surface to the deeper convection zone and to match the results of FTD models with observed surface magnetic field (Guerrero and de Gouveia Dal Pino 2008; Cameron et al. 2012; Karak and Nandy 2012; Karak and Cameron 2016; Hazra and Nandy 2016). The inclusion of turbulent pumping suppresses the diffusion of the horizontal field and makes the behavior of magnetic field different than the traditional flux transport dynamo model. In addition to explaining the increasing magnetic activity with the rotation, the model of Hazra et al. (2019) can explain the decreasing trend of the cycle period with the increasing rotation rate of stars for the inactive branch of slowly rotating stars. In the right panel of Fig. 5, the dependence of cycle period with the rotation period of the stars is shown for two types of treatment of Babcock–Leighton α effect with rotation (see Sect. 2.3 in Hazra et al. (2019) for details). A direct comparison of their result (right panel of Fig. 5) with observation of cycle period dependency on rotation (left panel of Fig. 5) shows that the observational trend for the slowly rotating branch upto rotation period of 15 days is reproduced qualitatively well. However, the change of global parity from dipolar to quadrupolar reported in their simulation for the stars with rotation faster than the rotation period of 17 days has no counterpart in the observation. This is because there is no observational study yet regarding parity of the solar-like stars. Also the weak increasing trend of cycle period with faster rotation for stars with rotation period less than 15 days has not been quantified from observation, and hence needs further careful investigation. The global magnetic field distribution in the Sun and stars, its parity, and the structure of coronal magnetic fields are governed by the dynamo mechanism (Dash et al. 2023) and surface emergence and evolution of magnetic flux (Nandy et al. 2018; Kavanagh et al. 2021). The magnetic field topology in turn determines the global stellar magnetosphere and magnetized stellar wind (Réville et al. 2015; Vidotto et al. 2014b) that play critical roles in star-planet interactions (Das et al. 2019; Basak and Nandy 2021; Carolan et al. 2021) and the forcing of (exo)planetary space environments (Nandy et al. 2021; Hazra et al. 2022). Also, the stellar magnetic cycle alters the total X-ray and EUV (XUV) radiation from host stars affecting exoplanetary atmospheres (Hazra et al. 2020).

There is another idea that the observed dependency of stellar activity cycles on rotation rates might be a manifestation of the dependence on the effective temperature of stars (Kitchatinov 2022). By combining models of differential rotation and dynamo together for stars with different masses, Kitchatinov (2022) found shorter cycles for hotter stars. Also, note that the hotter stars rotate faster on average. Hence computed shorter cycles for hotter stars are basically for fast rotators.

The flux transport dynamo paradigm can in fact be elegantly captured via a mathematical formulation based on time-delay differential equations (Wilmot-Smith et al. 2006). Tripathi et al. (2021) show that such a truncated Babcock–Leighton model imbibing the effects of fluctuations and noise can simultaneously explain the observed bimodal distribution of long-term sunspot time series, the breakdown of gyrochronology relations in middle aged solar-type stars and the relative low activity of the Sun compared to other Sun-like stars. This lends further credence to the philosophy that the basic ideas of flux transport dynamo theory gleaned in the context of the Sun may apply to other solar-like stars and across a substantial phase of solar evolution.

Many 3D global convective simulations of the stellar dynamo are able to reproduce magnetic activity cycle and the solutions of these simulations are close to many features of the solar magnetic field evolution (e.g., Ghizaru et al. 2010; Augustson et al. 2015; Käpylä et al. 2016; Brun et al. 2022). However, there are limited studies (Strugarek et al. 2017; Viviani et al. 2018; Warnecke 2018; Brun et al. 2022) to understand the observed rotational dependencies on magnetic cycles using global dynamo simulations due to computational challenges. Strugarek et al. (2017) explored the $P_{cyc} - P_{rot}$ relation with a limited sample of rotation rates and found that the cycle period is almost inversely proportional to the Rossby numbers. Other simulation by Viviani et al. (2018) found no clear dependency of cycle period with increasing rotation because of the strong non-axisymmetric magnetic field in their simulations. Warnecke (2018) investigated the rotational dependency of magnetic cycles with rotation rates varying by a factor of 30 times. For moderately and rapidly rotating cases, they found well defined cycles but for slowly rotating runs, the magnetic cycles were mostly irregular with longer periods. For moderately and rapidly rotating stars, the cycle period increases weakly with rotation in their simulations contrary to the what have been seen in observations. The global convective simulations produces different results than the mean-field FTD theory possibly because the meridional circulation amplitude has almost no effect in setting up the cycle period, contrary to the FTD models.

6 Temporal Variations of the Meridional Circulation and Solar Cycle Fluctuations

Since the period of the flux transport dynamo depends on the strength of the meridional circulation, as indicated in Eq. (19), it is obvious that fluctuations in the meridional circulation would have an effect on the dynamo. We now discuss what we know about the temporal variations of the meridional circulation and how they may affect the dynamo.

6.1 Evidence for Meridional Circulation Variations

A variation of the meridional circulation with the solar cycle has been inferred both from helioseismology (e.g., Chou and Dai 2001; Beck et al. 2002; Basu and Antia 2010; Komm et al. 2015) and from the tracking of surface markers (Hathaway and Rightmire 2010; Mahajan et al. 2021). It has been found that the meridional circulation becomes weaker at the time of sunspot maximum. From GONG full-disk Dopplergrams and HMI instrument on SDO, Komm et al. (2015) computed the temporal variation of the amplitude of meridional circulation near the surface at three depths of 2.0 Mm, 7.1 Mm and 11.6 Mm over latitudes. In the left panel of Fig. 6, we show the observed meridional flow over time at the depth of 2.0 Mm (taken from Komm et al. (2015)). It is clear from the figure that the amplitude of the meridional flow became weaker near solar maxima around the years 2002 and 2014 and stronger at solar minimum around 2009. This is presumably caused by the back-reaction of the dynamo-generated magnetic field on the large-scale flows. Note that observations of the variation of the meridional flow over the solar cycle could be affected by the inflows towards active regions and they can mimic changes of the overall meridional flow speed (e.g., Brun and Rempel 2009; Cameron and Schüssler 2010). Some effects of the cyclic variation of the meridional circulation can be studied by introducing a simple quenching by the magnetic field (Karak and Choudhuri 2012). However, a proper theoretical understanding requires the solving of Eq. (10) simultaneously with the dynamo equations (Eq. (16) and Eq. (17)). Hazra and Choudhuri (2017) developed a perturbation approach to study this problem. Their result is shown in the right panel of Fig. 6 compares favorably with the observational data as shown in the left panel of Fig. 6. In this context, an independent study by Nandy et al. (2011) claims that a relatively faster meridional flow in the rising phase of the cycle followed by a slower flow in the declining phase (on average throughout the meridional flow loop in the convection zone) can explain the occurrence of unusually deep minima between solar activity cycles; the results of Hazra and Choudhuri (2017) does not appear to be inconsistent with this finding. Another numerical study by Saha et al. (2022) – based on a flux transport dynamo model – indicates that meridional circulation can reproduce the (observed) cyclic modulation of weakened activity during grand minima phases such as the Maunder minimum. However, very long-term observations of surface flow variations do not exist and current capabilities do not allow setting strong constraints on deep flow variations.

It may be noted that the back-reaction of the magnetic field also causes periodic variations in the differential rotation, the so-called torsional oscillations. There have been efforts to model this also within the framework of the flux transport dynamo (e.g., Chakraborty et al. 2009).

We are interested here in the question of whether there are more random, non-periodic variations in the meridional circulation. Since we have reliable observational data about the meridional circulation only for about a quarter century, this question cannot be answered on the basis of direct observations. However, the periods of the cycles depend on the strength of the meridional circulation, as indicated in Eq. (19). A reasonable extrapolation from this suggests that we may use the data about the durations of past cycles to draw inferences about the variations of the meridional circulation (Karak and Choudhuri 2011). In Fig. 7 by plotting the durations of various past cycles, we see that cycles 10 to 14 had an almost constant period somewhat longer than 11 yr, suggesting that the meridional circulation was probably weaker in that era. Then cycles 15 to 19 had an almost constant period somewhat shorter

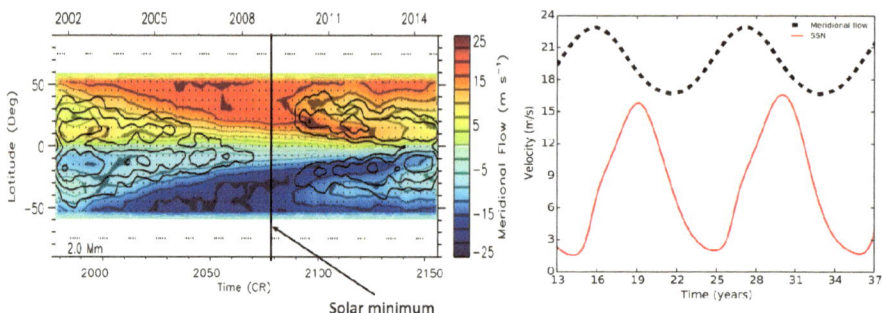

Fig. 6 Left panel: Observational estimate of meridional circulation as a function of time and latitude derived from GONG Dopplergrams at depth 2.0 Mm (Taken from Komm et al. (2015)) The black contours show the magnetic activity of magnitude 5, 10, and 20 G. The black vertical line shows the solar minimum. Right panel: Theoretical estimate of variation of meridional circulation with the solar cycle from the model of Hazra and Choudhuri (2017). The black dashed line shows the theoretically computed amplitude of meridional circulation and the red solid line shows two synthetic solar cycles. Adapted from Hazra and Choudhuri (2017)

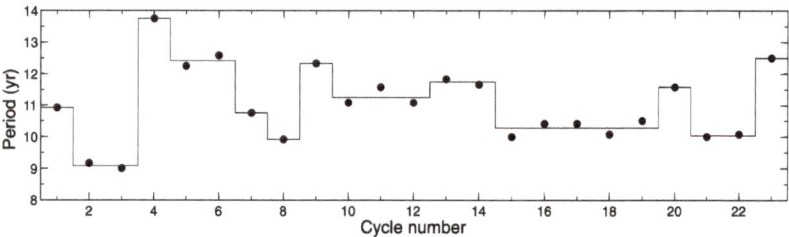

Fig. 7 Durations of various solar cycles beginning with the solar cycle 1. The solid filled circles show the observed period of the last 23 cycles. The solid line is for guiding the eye to discern the patterns in the variations of the solar cycle durations. Adapted from Karak and Choudhuri (2011)

than 11 yr, suggesting a stronger meridional circulation at that time. Based on such considerations, Karak and Choudhuri (2011) concluded that the meridional circulation had some random fluctuations with a coherence time of a few decades – perhaps in the range between 20 and 50 yr. Such fluctuations are expected to be a major cause behind the irregularities of the solar cycle.

6.2 Possible Causes Behind the Irregularities of the Solar Cycle

The earliest idea for explaining solar cycle irregularities was that this is a manifestation of nonlinear chaos (Weiss et al. 1984). Although the dynamo process certainly involves various kinds of nonlinearities, the most obvious nonlinearities are found not to produce any sustained chaotic behaviour and the various random fluctuations associated with the dynamo may be the more likely candidates for producing the cycle irregularities (Choudhuri 1992). However, there is one kind of observation that is presumably a signature of chaos: the Gnevyshev-Ohl effect obeyed over many cycles that the even cycle was stronger than the previous odd cycle. This is presumably due to period doubling just beyond bifurcation (Charbonneau et al. 2005).

We now try to identify the possible sources of fluctuations in the flux transport dynamo model. The Babcock–Leighton process depends on the tilts of active regions. We see a scatter in the tilt angles (Stenflo and Kosovichev 2012), presumably caused by the turbulent buffeting of flux tubes rising through the convection zone (Longcope and Choudhuri 2002). Choudhuri et al. (2007) proposed that the scatter in tilts gives rise to random fluctuations in the Babcock–Leighton process. This idea enabled them to make the first successful dynamo-based prediction of a solar cycle (Choudhuri et al. 2007; Jiang et al. 2007). More support for this idea has come from observational data (Dasi-Espuig et al. 2010) and simulations (Karak and Miesch 2017). Fluctuations in the Babcock–Leighton process have also been invoked to model the hemispheric asymmetry of sunspot cycles (Goel and Choudhuri 2009) and the Maunder minimum (Choudhuri and Karak 2009).

Since the Babcock–Leighton mechanism eventually builds up the polar field at the end of a cycle, the strength of the polar field gives an indication of the nature of fluctuations in the Babcock–Leighton mechanism during the cycle and is regularly used for modelling actual solar cycles. As we have systematic data of polar fields for not more than half a century, it is important to consider proxies of this field, such as geomagnetic indices (Wang and Sheeley 2009) and polar faculae (Muñoz-Jaramillo et al. 2013), for earlier times. Although the building up of the polar field is a collective process arising from many active regions, there are some indications that large "rogue" sunspot pairs (not satisfying Hale's polarity law may have a large effect (Jiang et al. 2015; Hazra et al. 2017). Stochastic fluctuations in the source terms for the poloidal field, both in the context of the mean-field α-effect and the Babcock–Leighton mechanism have been utilized within the flux transport dynamo paradigm to demonstrate the importance of these fluctuations in the occurrence and recovery from grand minima episodes (Hazra et al. 2014; Passos et al. 2014) and in the genesis of hemispheric decoupling and parity modulation in the sunspot cycle (Hazra and Nandy 2019).

One major limitation of using fluctuations in the Babcock–Leighton process alone for explaining the irregularities in the solar cycle is that these fluctuations cannot produce much variations in cycle durations (see however Kitchatinov et al. 2018). To explain the observed variations in the cycle periods, we need something else like the fluctuations in the meridional circulation or other transport coefficients such as turbulent pumping. We now turn to a discussion of the effects that such fluctuations would produce on the dynamo.

6.3 The Effects of Random Fluctuations in the Meridional Circulation

Karak (2010) varied the meridional circulation to match the periods of various solar cycles in the twentieth century and found that even the amplitudes of the cycle got matched to a certain extent. This was a clear indication that one of the causes behind the irregularities of the solar cycle was the fluctuations in the meridional circulation.

Suppose the meridional circulation has slowed down due to fluctuations, which will make the cycles longer. Diffusion will have more time to act and will try to make the cycles weaker. This would cause an anti-correlation between the strength of the cycle and its duration. A consequence of stronger cycles having shorter duration is that they should rise faster. The anti-correlation between the rise time and the cycle strength has been known for a long time and is called the Waldmeier effect. Karak and Choudhuri (2011) succeeded in explaining the Waldmeier effect by incorporating fluctuations in the meridional circulation in their dynamo model.

Choudhuri and Karak (2012) developed a comprehensive model of grand minima by including fluctuations in both the Babcock–Leighton process and in the meridional circulation in their dynamo simulations. By analyzing polar ice cores (^{14}C data), Usoskin et al. (2007) arrived at the result that there were about 27 grand minima in the last 11,000 yr. The results of Choudhuri and Karak (2012) are in broad agreement with this.

6.4 Solar Cycle Fluctuations and Cycle Forecasts

Can we utilize our understanding of solar cycle fluctuations to predict future sunspot cycles. Observations indicate that the polar field is a good precursor of the following sunspot cycle and a dynamo basis for this was already alluded to early on (Schatten et al. 1978). The first suggestion of using dynamo models for forecasting future solar cycle amplitudes – using poloidal field as inputs – was made by Nandy (2002). Subsequently detailed models based on the flux transport paradigm were worked out. It is noteworthy that while a variety of prediction techniques exist in the literature (Petrovay 2020), predictions based on the Babcock–Leighton paradigm and data driven flux transport dynamo models appear to now provide consistent results (Nandy 2021).

The theoretical explanation of why the polar field is such a good precursor was provided by Jiang et al. (2007), who pointed out that the polar field captures the essential outcome of the fluctuations in the Babcock–Leighton process. A series of papers utilizing the flux transport paradigm and stochastic fluctuations in the Babcock–Leighton source term – established the importance of cycle memory, i.e., the propagation of information of past polar fields to future sunspot cycles (Yeates et al. 2008; Karak and Nandy 2012). However, these studies did not look at aspects related to meridional flow variations.

If fluctuations in the meridional circulation are also important, can we find a precursor to capture the effects of that? It turns out that there is a time lag between the meridional circulation and its effect on the dynamo. The strength of a cycle does not depend on the value of the meridional circulation at the cycle maximum, but on its value a few years earlier – when the previous cycle was decaying. As a result, the decay rate of the previous cycle has a correlation with the strength of the next cycle and provides the appropriate precursor which encapsulates the effect of fluctuations in the meridional circulation (Hazra et al. 2015).

Hazra and Choudhuri (2019) realized that the polar field P at the end of the previous cycle and the decay rate R at that time can be the two precursors for predicting the next cycle, corresponding respectively to fluctuations in the Babcock–Leighton mechanism and fluctuations in the meridional circulation. From the data of past cycles, Hazra and Choudhuri (2019) found that an appropriate combination of P and R like PR or $\sqrt{PR}$ may be a better predictor for the next cycle than P or R alone. Figures 8(a) and (b) show the correlation of peak sunspot number of the next cycle with individual precursors P and R respectively. The correlations of combined precursors $\sqrt{PR}$ and PR with a peak sunspot number of the next cycle are also shown in Figs. 8(c) and (d) respectively. As we see in Fig. 8, the combined precursors give a better correlation than the individual ones. In an era when there had not been a significant fluctuation in the meridional circulation, the polar field P alone may be a good enough predictor for the next cycle. However, a combination of P and R may give a more complete formula for predicting the next cycle under more general circumstances.

7 The Future: Towards Bridging Mean Field Approaches, Flux Transport Dynamos and Full Magnetohydrodynamic Simulations

This review, although somewhat limited in scope due to space constraints, reinforces the view that mean field models and the flux transport dynamo paradigm have been very useful in explaining many of the observed properties of the solar cycle, including but not limited to, the latitudinal distribution and equatorward propagation of the sunspot belt, solar cycle fluctuations, parity modulation, and have played a critical role in devising data driven models for solar cycle predictions. The mathematical structure of these solar cycle models

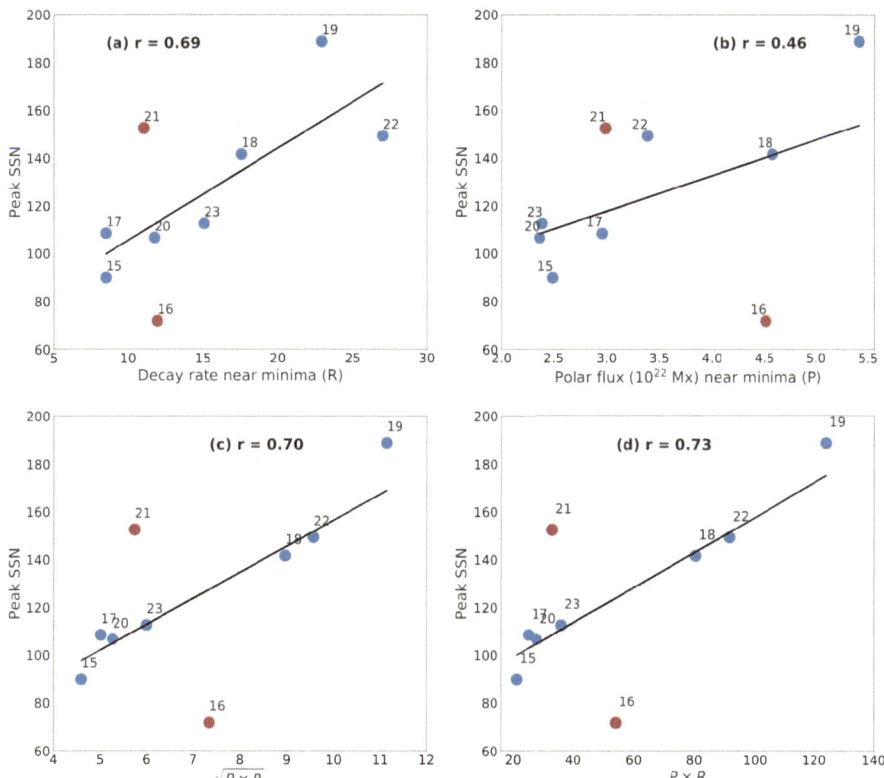

Fig. 8 The correlation plots of various precursors with the amplitude of the next cycle. Correlations of next cycle amplitude with (a) the decay rate at the late phase of the cycle R, (b) the polar field near minima of the cycle P (P is polar flux in Mx divided by 10^{22}), (c) the combined new precursor $\sqrt{PR}$, and (d) $P \times R$. Taken from Hazra and Choudhuri (2019)

is based on the canonical $\alpha\Omega$ dynamo equations, although in the Babcock–Leighton models, the poloidal source term is motivated from a fundamentally different perspective, or often explicitly added in an ad hoc manner to mimic the buoyant emergence of flux tubes. Moreover, these models rely significantly on a priori prescribed transport coefficients and large-scale flow profiles in stark contrast to full MHD models and magnetoconvection simulations. These appear to be orthogonal approaches and indeed, often these diverse communities have worked in silos. However, rich dividends and transformative progress may result from bridging these approaches and making use of observational constraints, when available.

The mean field, kinematic or flux transport dynamo models rely on multiple processes. The source of the toroidal field, differential rotation, is rather well constrained by helioseismic observations (Howe 2009). The Babcock–Leighton poloidal source is well constrained by near surface observations and is now thought to be the dominant driver of cycle to cycle variability over at least centennial time-scales (Dasi-Espuig et al. 2010; Cameron and Schüssler 2015; Bhowmik and Nandy 2018); these can be adequately captured in data driven surface flux transport models or dynamo models. The meridional circulation is well observed on the solar surface and results for the solar interior are now beginning to converge as already discussed indicating a largely single cell flow threading the solar convection zone

(Rajaguru and Antia 2015; Gizon et al. 2020) in keeping with the typical profile used in flux transport dynamo models.

The origin of the meridional flow seems to be well understood in the mean field theory and the models based on the theory agree closely with the seismologically detected single-cell circulation. The origin of the observed variability in the meridional flow is less certain however. Apart from direct modification by the Lorentz force, meridional flow of Eq. (13) is sensitive to variations in the differential rotation (Rempel 2005) and differential temperature (Spruit 2003). The dominant mechanism for the variability remains to be identified.

Many of the important ingredients which play a crucial role in the kinematic, flux transport dynamo modeling approach are in fact well constrained and we posit this is perhaps one of the underlying reasons for its success. As already argued, such models have in fact been the first to point out the importance of single-cell meridional circulation threading the convection zone, recovering which still remains a challenge for full MHD models, although there is progress towards that direction (Featherstone and Miesch 2015). In addition, the meridional circulation does not play any big role in setting up the cycle period in the 3D MHD model. This is just one of the examples of how the flux transport paradigm may serve as a useful guide for full MHD numerical simulations.

The reverse is also true. There is much that can be gleaned from mean field models of helical turbulent convection and full MHD simulations that are useful inputs for the flux transport models. For example, one of the widely utilized and popular sources of the poloidal field, the mean field α-effect cannot be directly observed in action. Although challenging and fraught with uncertainties on how to extract these transport coefficients, there are attempts to utilize full MHD models for constraining the mean field poloidal source (Käpylä et al. 2006; Simard et al. 2016; Warnecke et al. 2018; Shimada et al. 2022). Another case in point is the diamagnetic pumping (already discussed in this review) and turbulent pumping of magnetic fields. Full MHD simulations point out that turbulent pumping amplitudes can be effectively comparable or faster than meridional circulation (Käpylä et al. 2006). This has resulted in the construction of flux transport dynamo models that imbibe the physics of turbulent pumping of magnetic flux. Another outstanding issue is the amplitude of the effective turbulent magnetic diffusivity that is utilized in mean field or flux transport dynamo models. While this naturally arises out of turbulent convection driven by convective heat flux, its amplitude in the solar interior remains uncertain. Mixing length theory suggests strong turbulent diffusivity on the order of 10^{12}–10^{13} cm^2 s^{-1}; this sometimes introduces a problem in sustaining dynamo action, although, with low diffusion near the base of the convection zone and diamagnetic pumping, dynamo sustains (Kitchatinov and Olemskoy 2012b). There are also some recent efforts (Karak and Cameron 2016; Hazra et al. 2019) with high turbulent diffusivity $\sim 10^{12}$ cm^2 s^{-1}, which are able to produce magnetic cycles with added turbulent pumping in their model. The magnetic quenching of turbulent diffusivity in flux transport models has been demonstrated to be useful in sustaining magnetic cycles in this context too (Muñoz-Jaramillo et al. 2011). This is another example where ideas from mean field theory, magnetoconvection, and flux transport models come together to provide useful insight.

One fundamental challenge remains. Can one bring out the essence of the Babcock–Leighton mechanism – so successfully utilized in kinematic flux transport dynamo models and now proven to drive cycle to cycle variability – in direct numerical simulations of the solar magnetic cycle? Perhaps this is where the future lies, where ideas gleaned from all these diverse approaches may converge together.

Acknowledgements We would like to thank two anonymous referees for their valuable comments that helped to improve the review. GH acknowledges IIT Kanpur Initiation Grant (IITK/PHY/2022386) and ARIEL fellowship for financial support. DN acknowledges support for the Center of Excellence in Space Sciences

India at IISER Kolkata from the Ministry of Education, Government of India and multiple past students for sharing his journey of discovery. LK acknowledges financial support from the Ministry of Science and High Education of the Russian Federation. The research of ARC is supported by an Honorary Professorship from the Indian Institute of Science. All authors also thank International Space Science Institute for supporting the workshop "Solar and Stellar Dynamos: A New Era" where this review originated.

Author Contribution Second, third and fourth authors contributed equally to this work.

Declarations

Competing Interests The authors declare no competing interests.

References

Augustson K, Brun AS, Miesch M et al (2015) Grand minima and equatorward propagation in a cycling stellar convective dynamo. Astrophys J 809(2):149. https://doi.org/10.1088/0004-637X/809/2/149. arXiv: 1410.6547 [astro-ph.SR]

Babcock HW (1961) The topology of the Sun's magnetic field and the 22-YEAR cycle. Astrophys J 133:572. https://doi.org/10.1086/147060

Baliunas SL, Donahue RA, Soon WH et al (1995) Chromospheric variations in main-sequence stars. Astrophys J 438:269–287. https://doi.org/10.1086/175072

Basak A, Nandy D (2021) Modelling the imposed magnetospheres of Mars-like exoplanets: star-planet interactions and atmospheric losses. Mon Not R Astron Soc 502(3):3569–3581. https://doi.org/10.1093/mnras/stab225

Basu S, Antia HM (2010) Characteristics of solar meridional flows during solar cycle 23. Astrophys J 717(1):488–495. https://doi.org/10.1088/0004-637X/717/1/488. arXiv:1005.3031 [astro-ph.SR]

Beck JG, Gizon L, Duvall JTL (2002) A new component of solar dynamics: North-South diverging flows migrating toward the equator with an 11 year period. Astrophys J Lett 575(1):L47–L50. https://doi.org/10.1086/342636

Bekki Y, Cameron RH (2023) Three-dimensional non-kinematic simulation of the post-emergence evolution of bipolar magnetic regions and the Babcock–Leighton dynamo of the Sun. Astron Astrophys 670:A101. https://doi.org/10.1051/0004-6361/202244990. arXiv:2209.08178 [astro-ph.SR]

Bhowmik P, Nandy D (2018) Prediction of the strength and timing of sunspot cycle 25 reveal decadal-scale space environmental conditions. Nat Commun 9:5209. https://doi.org/10.1038/s41467-018-07690-0

Böhm-Vitense E (2007) Chromospheric activity in G and K main-sequence stars, and what it tells us about stellar dynamos. Astrophys J 657:486–493. https://doi.org/10.1086/510482

Bonanno A, Elstner D, Rüdiger G et al (2002) Parity properties of an advection-dominated solar $\alpha^2\Omega$-dynamo. Astron Astrophys 390:673–680. https://doi.org/10.1051/0004-6361:20020590. arXiv:astro-ph/0204308 [astro-ph]

Boro Saikia S, Marvin CJ, Jeffers SV et al (2018) Chromospheric activity catalogue of 4454 cool stars. Questioning the active branch of stellar activity cycles. Astron Astrophys 616:A108. https://doi.org/10.1051/0004-6361/201629518. arXiv:1803.11123 [astro-ph.SR]

Brandenburg A, Subramanian K (2005) Astrophysical magnetic fields and nonlinear dynamo theory. Phys Rep 417(1–4):1–209. https://doi.org/10.1016/j.physrep.2005.06.005

Brandenburg A, Charbonneau P, Kitchatinov LL et al (1994) Stellar dynamos: the Rossby number dependence. In: Caillault JP (ed) Cool stars, stellar systems, and the Sun, p 354

Brandenburg A, Hubbard A, Käpylä PJ (2015) Dynamical quenching with non-local α and downward pumping. Astron Nachr 336(1):91–96. https://doi.org/10.1002/asna.201412141

Brun AS, Rempel M (2009) Large scale flows in the solar convection zone. Space Sci Rev 144(1–4):151–173. https://doi.org/10.1007/s11214-008-9454-9

Brun AS, Strugarek A, Noraz Q et al (2022) Powering stellar magnetism: energy transfers in cyclic dynamos of Sun-like stars. Astrophys J 926(1):21. https://doi.org/10.3847/1538-4357/ac469b. arXiv:2201.13218 [astro-ph.SR]

Caligari P, Moreno-Insertis F, Schussler M (1995) Emerging flux tubes in the solar convection zone. I. Asymmetry, tilt, and emergence latitude. Astrophys J 441:886. https://doi.org/10.1086/175410

Cameron RH, Schüssler M (2010) Changes of the solar meridional velocity profile during cycle 23 explained by flows toward the activity belts. Astrophys J 720(2):1030–1032. https://doi.org/10.1088/0004-637X/720/2/1030. arXiv:1007.2548 [astro-ph.SR]

Cameron R, Schüssler M (2015) The crucial role of surface magnetic fields for the solar dynamo. Science 347(6228):1333–1335. https://doi.org/10.1126/science.1261470. arXiv:1503.08469 [astro-ph.SR]

Cameron RH, Schmitt D, Jiang J et al (2012) Surface flux evolution constraints for flux transport dynamos. Astron Astrophys 542:A127. https://doi.org/10.1051/0004-6361/201218906. arXiv:1205.1136 [astro-ph.SR]

Carolan S, Vidotto AA, Hazra G et al (2021) The effects of magnetic fields on observational signatures of atmospheric escape in exoplanets: double tail structures. Mon Not R Astron Soc 508(4):6001–6012. https://doi.org/10.1093/mnras/stab2947. arXiv:2110.05200 [astro-ph.EP]

Chae J, Litvinenko YE, Sakurai T (2008) Determination of magnetic diffusivity from high-resolution solar magnetograms. Astrophys J 683(2):1153–1159. https://doi.org/10.1086/590074

Chakraborty S, Choudhuri AR, Chatterjee P (2009) Why does the Sun's torsional oscillation begin before the sunspot cycle? Phys Rev Lett 102(4):041102. https://doi.org/10.1103/PhysRevLett.102.041102. arXiv:0907.4842 [astro-ph.SR]

Charbonneau P (2014) Solar dynamo theory. Annu Rev Astron Astrophys 52:251–290. https://doi.org/10.1146/annurev-astro-081913-040012

Charbonneau P (2020) Dynamo models of the solar cycle. Living Rev Sol Phys 17(1):4. https://doi.org/10.1007/s41116-020-00025-6

Charbonneau P, St-Jean C, Zacharias P (2005) Fluctuations in Babcock–Leighton dynamos. I. Period doubling and transition to chaos. Astrophys J 619(1):613–622. https://doi.org/10.1086/426385

Chatterjee P, Nandy D, Choudhuri AR (2004) Full-sphere simulations of a circulation-dominated solar dynamo: exploring the parity issue. Astron Astrophys 427:1019–1030. https://doi.org/10.1051/0004-6361:20041199. arXiv:astro-ph/0405027

Chou DY, Dai DC (2001) Solar cycle variations of subsurface meridional flows in the Sun. Astrophys J Lett 559(2):L175–L178. https://doi.org/10.1086/323724

Choudhuri AR (1989) The evolution of loop structures in flux rings within the solar convection zone. Sol Phys 123:217–239. https://doi.org/10.1007/BF00149104

Choudhuri AR (1990) A correction to Spruit's equation for the dynamics of thin flux tubes. Astron Astrophys 239(1–2):335–339

Choudhuri AR (1992) Stochastic fluctuations of the solar dynamo. Astron Astrophys 253:277–285

Choudhuri AR (1998) The physics of fluids and plasmas: an introduction for astrophysicists. Cambridge University Press, Cambridge

Choudhuri AR (2011) The origin of the solar magnetic cycle. Pramana 77(1):77–96. https://doi.org/10.1007/s12043-011-0113-4. arXiv:1103.3385 [astro-ph.SR]

Choudhuri AR (2017) Starspots, stellar cycles and stellar flares: lessons from solar dynamo models. Sci China, Phys Mech Astron 60(1):19601. https://doi.org/10.1007/s11433-016-0413-7. arXiv:1612.02544 [astro-ph.SR]

Choudhuri AR (2021a) A theoretical estimate of the pole-equator temperature difference and a possible origin of the near-surface shear layer. Sol Phys 296(2):37. https://doi.org/10.1007/s11207-021-01784-7. arXiv:2008.02983 [astro-ph.SR]

Choudhuri AR (2021b) The meridional circulation of the Sun: observations, theory and connections with the solar dynamo. Sci China, Phys Mech Astron 64(3):239601. https://doi.org/10.1007/s11433-020-1628-1. arXiv:2008.09347 [astro-ph.SR]

Choudhuri AR, Gilman PA (1987) The influence of the Coriolis force on flux tubes rising through the solar convection zone. Astrophys J 316:788–800. https://doi.org/10.1086/165243

Choudhuri AR, Karak BB (2009) A possible explanation of the Maunder minimum from a flux transport dynamo model. Res Astron Astrophys 9(9):953–958. https://doi.org/10.1088/1674-4527/9/9/001. arXiv:0907.3106 [astro-ph.SR]

Choudhuri AR, Karak BB (2012) Origin of grand minima in sunspot cycles. Phys Rev Lett 109:171103. https://doi.org/10.1103/PhysRevLett.109.171103. arXiv:1208.3947 [astro-ph.SR]

Choudhuri AR, Schussler M, Dikpati M (1995) The solar dynamo with meridional circulation. Astron Astrophys 303:L29

Choudhuri AR, Chatterjee P, Nandy D (2004) Helicity of solar active regions from a dynamo model. Astrophys J Lett 615(1):L57–L60. https://doi.org/10.1086/426054

Choudhuri AR, Chatterjee P, Jiang J (2007) Predicting solar cycle 24 with a solar dynamo model. Phys Rev Lett 98:131103. https://doi.org/10.1103/PhysRevLett.98.131103. arXiv:astro-ph/0701527

Das SB, Basak A, Nandy D et al (2019) Modeling star-planet interactions in far-out planetary and exoplanetary systems. Astrophys J 877(2):80. https://doi.org/10.3847/1538-4357/ab18ad. arXiv:1812.07767 [astro-ph.EP]

Dash S, Nandy D, Usoskin I (2023) Long-term forcing of Sun's coronal field, open flux and cosmic ray modulation potential during grand minima, maxima and regular activity phases by the solar dynamo mechanism. arXiv e-prints. https://doi.org/10.48550/arXiv.2208.12103. arXiv:2208.12103 [astro-ph.SR]

Dasi-Espuig M, Solanki SK, Krivova NA et al (2010) Sunspot group tilt angles and the strength of the solar cycle. Astron Astrophys 518:A7. https://doi.org/10.1051/0004-6361/201014301. arXiv:1005.1774 [astro-ph.SR]

Dikpati M, Charbonneau P (1999) A Babcock–Leighton flux transport dynamo with solar-like differential rotation. Astrophys J 518(1):508–520. https://doi.org/10.1086/307269

Donati JF, Semel M, Carter BD et al (1997) Spectropolarimetric observations of active stars. Mon Not R Astron Soc 291(4):658–682. https://doi.org/10.1093/mnras/291.4.658

Dorch SBF, Nordlund Å (2001) On the transport of magnetic fields by solar-like stratified convection. Astron Astrophys 365:562–570. https://doi.org/10.1051/0004-6361:20000141

D'Silva S, Choudhuri AR (1993) A theoretical model for tilts of bipolar magnetic regions. Astron Astrophys 272:621

Durney BR (1995) On a Babcock–Leighton dynamo model with a deep-seated generating layer for the toroidal magnetic field. Sol Phys 160:213–235. https://doi.org/10.1007/BF00732805

Durney BR (1996) On the influence of gradients in the angular velocity on the solar meridional motions. Sol Phys 169(1):1–32. https://doi.org/10.1007/BF00153830

Durney BR (1997) On a Babcock–Leighton solar dynamo model with a deep-seated generating layer for the toroidal magnetic field. IV. Astrophys J 486(2):1065–1077. https://doi.org/10.1086/304546

Durney BR, Robinson RD (1982) On an estimate of the dynamo-generated magnetic fields in late-type stars. Astrophys J 253:290–297. https://doi.org/10.1086/159633

Fan Y, Fisher GH, Deluca EE (1993) The origin of morphological asymmetries in bipolar active regions. Astrophys J 405:390. https://doi.org/10.1086/172370

Featherstone NA, Miesch MS (2015) Meridional circulation in solar and stellar convection zones. Astrophys J 804(1):67. https://doi.org/10.1088/0004-637X/804/1/67. arXiv:1501.06501 [astro-ph.SR]

Frisch U, She ZS, Sulem PL (1987) Large-scale flow driven by the anisotropic kinetic alpha effect. Phys D, Nonlinear Phenom 28(3):382–392. https://doi.org/10.1016/0167-2789(87)90026-1

Garaud P, Brummell NH (2008) On the penetration of meridional circulation below the solar convection zone. Astrophys J 674(1):498. https://doi.org/10.1086/524837

Ghizaru M, Charbonneau P, Smolarkiewicz PK (2010) Magnetic cycles in global large-eddy simulations of solar convection. Astrophys J Lett 715(2):L133–L137. https://doi.org/10.1088/2041-8205/715/2/L133

Gilman PA, Miesch MS (2004) Limits to penetration of meridional circulation below the solar convection zone. Astrophys J 611(1):568. https://doi.org/10.1086/421899

Gizon L, Cameron RH, Pourabdian M et al (2020) Meridional flow in the Sun's convection zone is a single cell in each hemisphere. Science 368(6498):1469–1472. https://doi.org/10.1126/science.aaz7119

Goel A, Choudhuri AR (2009) The hemispheric asymmetry of solar activity during the last century and the solar dynamo. Res Astron Astrophys 9(1):115–126. https://doi.org/10.1088/1674-4527/9/1/010. arXiv:0712.3988 [astro-ph]

Gregory SG, Donati JF, Morin J et al (2012) Can we predict the global magnetic topology of a pre-main-sequence star from its position in the Hertzsprung-Russell diagram? Astrophys J 755(2):97. https://doi.org/10.1088/0004-637X/755/2/97

Gruzinov AV, Diamond PH (1994) Self-consistent theory of mean-field electrodynamics. Phys Rev Lett 72(11):1651–1653. https://doi.org/10.1103/PhysRevLett.72.1651

Guerrero G, de Gouveia Dal Pino EM (2008) Turbulent magnetic pumping in a Babcock–Leighton solar dynamo model. Astron Astrophys 485:267–273. https://doi.org/10.1051/0004-6361:200809351. arXiv:0803.3466

Guerrero GA, Muñoz JD (2004) Kinematic solar dynamo models with a deep meridional flow. Mon Not R Astron Soc 350(1):317–322. https://doi.org/10.1111/j.1365-2966.2004.07655.x. arXiv:astro-ph/0402097 [astro-ph]

Hale GE, Ellerman F, Nicholson SB et al (1919) The magnetic polarity of Sun-spots. Astrophys J 49:153. https://doi.org/10.1086/142452

Hanasoge SM (2022) Surface and interior meridional circulation in the Sun. Living Rev Sol Phys 19(1):3. https://doi.org/10.1007/s41116-022-00034-7

Hathaway DH (2015) The solar cycle. Living Rev Sol Phys 12(1):4. https://doi.org/10.1007/lrsp-2015-4

Hathaway DH, Rightmire L (2010) Variations in the Sun's meridional flow over a solar cycle. Science 327(5971):1350. https://doi.org/10.1126/science.1181990

Hazra G (2018) Understanding the behavior of the Sun's large scale magnetic field and its relation with the meridional flow. PhD thesis, Indian Institute of Science, Bangalore

Hazra G (2021) Recent advances in the 3D kinematic Babcock–Leighton solar dynamo modeling. J Astrophys Astron 42(2):22. https://doi.org/10.1007/s12036-021-09738-y. arXiv:2009.03810 [astro-ph.SR]

Hazra G, Choudhuri AR (2017) A theoretical model of the variation of the meridional circulation with the solar cycle. Mon Not R Astron Soc 472(3):2728–2741. https://doi.org/10.1093/mnras/stx2152. arXiv:1708.05204 [astro-ph.SR]

Hazra G, Choudhuri AR (2019) A new formula for predicting solar cycles. Astrophys J 880(2):113. https://doi.org/10.3847/1538-4357/ab2718. arXiv:1811.01363 [astro-ph.SR]

Hazra G, Miesch MS (2018) Incorporating surface convection into a 3D Babcock–Leighton solar dynamo model. Astrophys J 864(2):110. https://doi.org/10.3847/1538-4357/aad556. arXiv:1804.03100 [astro-ph.SR]

Hazra S, Nandy D (2016) A proposed paradigm for solar cycle dynamics mediated via turbulent pumping of magnetic flux in Babcock–Leighton-type solar dynamos. Astrophys J 832(1):9. https://doi.org/10.3847/0004-637X/832/1/9. arXiv:1608.08167 [astro-ph.SR]

Hazra S, Nandy D (2019) The origin of parity changes in the solar cycle. Mon Not R Astron Soc 489(3):4329–4337. https://doi.org/10.1093/mnras/stz2476. arXiv:1906.06780 [astro-ph.SR]

Hazra G, Karak BB, Choudhuri AR (2014) Is a deep one-cell meridional circulation essential for the flux transport solar dynamo? Astrophys J 782(2):93. https://doi.org/10.1088/0004-637X/782/2/93. arXiv:1309.2838 [astro-ph.SR]

Hazra S, Passos D, Nandy D (2014) A stochastically forced time delay solar dynamo model: self-consistent recovery from a Maunder-like grand minimum necessitates a mean-field alpha effect. Astrophys J 789(1):5. https://doi.org/10.1088/0004-637X/789/1/5. arXiv:1307.5751 [astro-ph.SR]

Hazra G, Karak BB, Banerjee D et al (2015) Correlation between decay rate and amplitude of solar cycles as revealed from observations and dynamo theory. Sol Phys 290(6):1851–1870. https://doi.org/10.1007/s11207-015-0718-8. arXiv:1410.8641 [astro-ph.SR]

Hazra G, Choudhuri AR, Miesch MS (2017) A theoretical study of the build-up of the Sun's polar magnetic field by using a 3D kinematic dynamo model. Astrophys J 835(1):39. https://doi.org/10.3847/1538-4357/835/1/39. arXiv:1610.02726 [astro-ph.SR]

Hazra G, Jiang J, Karak BB et al (2019) Exploring the cycle period and parity of stellar magnetic activity with dynamo modeling. Astrophys J 884(1):35. https://doi.org/10.3847/1538-4357/ab4128. arXiv:1909.01286 [astro-ph.SR]

Hazra G, Vidotto AA, D'Angelo CV (2020) Influence of the Sun-like magnetic cycle on exoplanetary atmospheric escape. Mon Not R Astron Soc 496(3):4017–4031. https://doi.org/10.1093/mnras/staa1815. arXiv:2006.10634 [astro-ph.SR]

Hazra G, Vidotto AA, Carolan S et al (2022) The impact of coronal mass ejections and flares on the atmosphere of the hot Jupiter HD189733b. Mon Not R Astron Soc 509(4):5858–5871. https://doi.org/10.1093/mnras/stab3271. arXiv:2111.04531 [astro-ph.EP]

Hempelmann A, Schmitt JHMM, Schultz M et al (1995) Coronal X-ray emission and rotation of cool main-sequence stars. Astron Astrophys 294:515–524

Howe R (2009) Solar interior rotation and its variation. Living Rev Sol Phys 6(1):1. https://doi.org/10.12942/lrsp-2009-1. arXiv:0902.2406 [astro-ph.SR]

Jha BK, Choudhuri AR (2021) A theoretical model of the near-surface shear layer of the Sun. Mon Not R Astron Soc 506(2):2189–2198. https://doi.org/10.1093/mnras/stab1717. arXiv:2105.14266 [astro-ph.SR]

Jiang J, Chatterjee P, Choudhuri AR (2007) Solar activity forecast with a dynamo model. Mon Not R Astron Soc 381(4):1527–1542. https://doi.org/10.1111/j.1365-2966.2007.12267.x. arXiv:0707.2258 [astro-ph]

Jiang J, Cameron RH, Schüssler M (2015) The cause of the weak solar cycle 24. Astrophys J Lett 808(1):L28. https://doi.org/10.1088/2041-8205/808/1/L28. arXiv:1507.01764 [astro-ph.SR]

Jouve L, Brun AS (2007) On the role of meridional flows in flux transport dynamo models. Astron Astrophys 474(1):239–250. https://doi.org/10.1051/0004-6361:20077070. arXiv:0712.3200 [astro-ph]

Jouve L, Brown BP, Brun AS (2010) Exploring the P_{cyc} vs. P_{rot} relation with flux transport dynamo models of solar-like stars. Astron Astrophys 509:A32. https://doi.org/10.1051/0004-6361/200913103. arXiv:0911.1947 [astro-ph.SR]

Käpylä PJ, Korpi MJ, Ossendrijver M et al (2006) Magnetoconvection and dynamo coefficients. III. α-effect and magnetic pumping in the rapid rotation regime. Astron Astrophys 455:401–412. https://doi.org/10.1051/0004-6361:20064972. arXiv:astro-ph/0602111

Käpylä MJ, Käpylä PJ, Olspert N et al (2016) Multiple dynamo modes as a mechanism for long-term solar activity variations. Astron Astrophys 589:A56. https://doi.org/10.1051/0004-6361/201527002. arXiv:1507.05417 [astro-ph.SR]

Karak BB (2010) Importance of meridional circulation in flux transport dynamo: the possibility of a Maunder-like grand minimum. Astrophys J 724:1021–1029. https://doi.org/10.1088/0004-637X/724/2/1021. arXiv:1009.2479 [astro-ph.SR]

Karak BB, Cameron R (2016) Babcock–Leighton solar dynamo: the role of downward pumping and the equatorward propagation of activity. Astrophys J 832:94. https://doi.org/10.3847/0004-637X/832/1/94. arXiv:1605.06224 [astro-ph.SR]

Karak BB, Choudhuri AR (2011) The Waldmeier effect and the flux transport solar dynamo. Mon Not R Astron Soc 410(3):1503–1512. https://doi.org/10.1111/j.1365-2966.2010.17531.x. arXiv:1008.0824 [astro-ph.SR]

Karak BB, Choudhuri AR (2012) Quenching of meridional circulation in flux transport dynamo models. Sol Phys 278(1):137–148. https://doi.org/10.1007/s11207-012-9928-5. arXiv:1111.1540 [astro-ph.SR]

Karak BB, Miesch M (2017) Solar cycle variability induced by tilt angle scatter in a Babcock–Leighton solar dynamo model. Astrophys J 847(1):69. https://doi.org/10.3847/1538-4357/aa8636. arXiv:1706.08933 [astro-ph.SR]

Karak BB, Nandy D (2012) Turbulent pumping of magnetic flux reduces solar cycle memory and thus impacts predictability of the Sun's activity. Astrophys J Lett 761:L13. https://doi.org/10.1088/2041-8205/761/1/L13. arXiv:1206.2106 [astro-ph.SR]

Karak BB, Jiang J, Miesch MS et al (2014a) Flux transport dynamos: from kinematics to dynamics. Space Sci Rev 186(1–4):561–602. https://doi.org/10.1007/s11214-014-0099-6

Karak BB, Kitchatinov LL, Choudhuri AR (2014b) A dynamo model of magnetic activity in solar-like stars with different rotational velocities. Astrophys J 791:59. https://doi.org/10.1088/0004-637X/791/1/59. arXiv:1402.1874 [astro-ph.SR]

Kavanagh RD, Vidotto AA, Klein B et al (2021) Planet-induced radio emission from the coronae of M dwarfs: the case of Prox Cen and AU. Mon Not R Astron Soc 504(1):1511–1518. https://doi.org/10.1093/mnras/stab929. arXiv:2103.16318 [astro-ph.SR]

Kichatinov LL, Rüdiger G (1992) Magnetic-field advection in inhomogeneous turbulence. Astron Astrophys 260(1–2):494–498

Kitchatinov LL (2016) Rotational shear near the solar surface as a probe for subphotospheric magnetic fields. Astron Lett 42(5):339–345. https://doi.org/10.1134/S1063773716050054

Kitchatinov L (2022) The dependence of stellar activity cycles on effective temperature. Res Astron Astrophys 22(12):125006. https://doi.org/10.1088/1674-4527/ac9780. arXiv:2205.09952 [astro-ph.SR]

Kitchatinov LL, Nepomnyashchikh AA (2017) A joined model for solar dynamo and differential rotation. Astron Lett 43:332–343. https://doi.org/10.1134/S106377371704003X

Kitchatinov LL, Olemskoy SV (2011a) Alleviation of catastrophic quenching in solar dynamo model with nonlocal alpha-effect. Astron Nachr 332(5):496–501. https://doi.org/10.1002/asna.201011549

Kitchatinov LL, Olemskoy SV (2011b) Differential rotation of main-sequence dwarfs and its dynamo efficiency. Mon Not R Astron Soc 411:1059–1066. https://doi.org/10.1111/j.1365-2966.2010.17737.x

Kitchatinov LL, Olemskoy SV (2012a) Differential rotation of main-sequence dwarfs: predicting the dependence on surface temperature and rotation rate. Mon Not R Astron Soc 423(4):3344–3351. https://doi.org/10.1111/j.1365-2966.2012.21126.x

Kitchatinov LL, Olemskoy SV (2012b) Solar dynamo model with diamagnetic pumping and nonlocal α-effect. Sol Phys 276(1–2):3–17. https://doi.org/10.1007/s11207-011-9887-2

Kitchatinov LL, Olemskoy SV (2015) Dynamo saturation in rapidly rotating solar-type stars. Res Astron Astrophys 15:1801. https://doi.org/10.1088/1674-4527/15/11/003. arXiv:1503.07956 [astro-ph.SR]

Kitchatinov LL, Pipin VV, Rüdiger G (1994) Turbulent viscosity, magnetic diffusivity, and heat conductivity under the influence of rotation and magnetic field. Astron Nachr 315(2):157–170. https://doi.org/10.1002/asna.2103150205

Kitchatinov LL, Mordvinov AV, Nepomnyashchikh AA (2018) Modelling variability of solar activity cycles. Astron Astrophys 615:A38. https://doi.org/10.1051/0004-6361/201732549. arXiv:1804.02833 [astro-ph.SR]

Kleeorin N, Rogachevskii I (1994) Effective Ampère force in developed magnetohydrodynamic turbulence. Phys Rev E 50(4):2716–2730. https://doi.org/10.1103/PhysRevE.50.2716

Komm R, González Hernández I, Howe R et al (2015) Solar-cycle variation of subsurface meridional flow derived with ring-diagram analysis. Sol Phys 290(11):3113–3136. https://doi.org/10.1007/s11207-015-0729-5

Krause F, Rädler KH (1980) Mean-field magnetohydrodynamics and dynamo theory. Pergamon, Oxford

Küker M, Rüdiger G, Schultz M (2001) Circulation-dominated solar shell dynamo models with positive alpha-effect. Astron Astrophys 374:301–308. https://doi.org/10.1051/0004-6361:20010686

Kumar R, Jouve L, Nandy D (2019) A 3D kinematic Babcock Leighton solar dynamo model sustained by dynamic magnetic buoyancy and flux transport processes. Astron Astrophys 623:A54. https://doi.org/10.1051/0004-6361/201834705. arXiv:1901.04251 [astro-ph.SR]

Lantz SR, Fan Y (1999) Anelastic magnetohydrodynamic equations for modeling solar and stellar convection zones. Astrophys J Suppl Ser 121(1):247–264. https://doi.org/10.1086/313187

Lebedinsky AI (1941) Rotation of the Sun. Astron J (USSR) 18(1):10–25

Leighton RB (1969) A magneto-kinematic model of the solar cycle. Astrophys J 156:1. https://doi.org/10.1086/149943

Lifshitz EM, Pitaevskii LP (1981) Physical kinetics: Landau and Lifshitz course of theoretical physics, vol 10. Pergamon, Oxford

Longcope D, Choudhuri AR (2002) The orientational relaxation of bipolar active regions. Sol Phys 205:63–92. https://doi.org/10.1023/A:1013896013842

Mahajan SS, Hathaway DH, Muñoz-Jaramillo A et al (2021) Improved measurements of the Sun's meridional flow and torsional oscillation from correlation tracking on MDI and HMI magnetograms. Astrophys J 917(2):100. https://doi.org/10.3847/1538-4357/ac0a80. arXiv:2107.07731 [astro-ph.SR]

Miesch MS, Dikpati M (2014) A three-dimensional Babcock–Leighton solar dynamo model. Astrophys J Lett 785(1):L8. https://doi.org/10.1088/2041-8205/785/1/L8. arXiv:1401.6557 [astro-ph.SR]

Miesch MS, Hindman BW (2011) Gyroscopic pumping in the solar near-surface shear layer. Astrophys J 743(1):79. https://doi.org/10.1088/0004-637X/743/1/79. arXiv:1106.4107 [astro-ph.SR]

Miesch MS, Brun AS, Toomre J (2006) Solar differential rotation influenced by latitudinal entropy variations in the tachocline. Astrophys J 641(1):618–625. https://doi.org/10.1086/499621

Moffatt HK (1978) Magnetic field generation in electrically conducting fluids. Cambridge University Press, Cambridge

Muñoz-Jaramillo A, Nandy D, Martens PCH (2009) Helioseismic data inclusion in solar dynamo models. Astrophys J 698(1):461–478. https://doi.org/10.1088/0004-637X/698/1/461. arXiv:0811.3441 [astro-ph]

Muñoz-Jaramillo A, Nandy D, Martens PCH et al (2010) A double-ring algorithm for modeling solar active regions: unifying kinematic dynamo models and surface flux-transport simulations. Astrophys J Lett 720(1):L20–L25. https://doi.org/10.1088/2041-8205/720/1/L20. arXiv:1006.4346 [astro-ph.SR]

Muñoz-Jaramillo A, Nandy D, Martens PCH (2011) Magnetic quenching of turbulent diffusivity: reconciling mixing-length theory estimates with kinematic dynamo models of the solar cycle. Astrophys J Lett 727(1):L23. https://doi.org/10.1088/2041-8205/727/1/L23. arXiv:1007.1262 [astro-ph.SR]

Muñoz-Jaramillo A, Dasi-Espuig M, Balmaceda LA et al (2013) Solar cycle propagation, memory, and prediction: insights from a century of magnetic proxies. Astrophys J Lett 767(2):L25. https://doi.org/10.1088/2041-8205/767/2/L25. arXiv:1304.3151 [astro-ph.SR]

Nandy D (2002) Can theoretical solar dynamo models predict future solar activity? In: 34th COSPAR scientific assembly. p 53

Nandy D (2004) Exploring magnetic activity from the Sun to the stars. Sol Phys 224:161–169. https://doi.org/10.1007/s11207-005-4990-x

Nandy D (2021) Progress in solar cycle predictions: sunspot cycles 24-25 in perspective. Sol Phys 296(3):54. https://doi.org/10.1007/s11207-021-01797-2. arXiv:2009.01908 [astro-ph.SR]

Nandy D, Choudhuri AR (2001) Toward a mean field formulation of the Babcock–Leighton type solar dynamo. I. α-coefficient versus Durney's double-ring approach. Astrophys J 551(1):576–585. https://doi.org/10.1086/320057. arXiv:astro-ph/0107466 [astro-ph]

Nandy D, Choudhuri AR (2002) Explaining the latitudinal distribution of sunspots with deep meridional flow. Science 296(5573):1671–1673. https://doi.org/10.1126/science.1070955

Nandy D, Muñoz-Jaramillo A, Martens PCH (2011) The unusual minimum of sunspot cycle 23 caused by meridional plasma flow variations. Nature 471(7336):80–82. https://doi.org/10.1038/nature09786. arXiv:1303.0349 [astro-ph.SR]

Nandy D, Bhowmik P, Yeates AR et al (2018) The large-scale coronal structure of the 2017 August 21 great American eclipse: an assessment of solar surface flux transport model enabled predictions and observations. Astrophys J 853(1):72. https://doi.org/10.3847/1538-4357/aaa1eb

Nandy D, Martens PCH, Obridko V et al (2021) Solar evolution and extrema: current state of understanding of long-term solar variability and its planetary impacts. Prog Earth Planet Sci 8(1):40. https://doi.org/10.1186/s40645-021-00430-x

Noyes RW, Hartmann LW, Baliunas SL et al (1984a) Rotation, convection, and magnetic activity in lower main-sequence stars. Astrophys J 279:763–777. https://doi.org/10.1086/161945

Noyes RW, Weiss NO, Vaughan AH (1984b) The relation between stellar rotation rate and activity cycle periods. Astrophys J 287:769–773. https://doi.org/10.1086/162735

Olspert N, Lehtinen JJ, Käpylä MJ et al (2018) Estimating activity cycles with probabilistic methods. II. The Mount Wilson Ca H&K data. Astron Astrophys 619:A6. https://doi.org/10.1051/0004-6361/201732525. arXiv:1712.08240 [astro-ph.SR]

Ossendrijver M, Stix M, Brandenburg A et al (2002) Magnetoconvection and dynamo coefficients. II. Field-direction dependent pumping of magnetic field. Astron Astrophys 394:735–745. https://doi.org/10.1051/0004-6361:20021224

Parker EN (1955a) Hydromagnetic dynamo models. Astrophys J 122:293–314. https://doi.org/10.1086/146087

Parker EN (1955b) The formation of sunspots from the solar toroidal field. Astrophys J 121:491. https://doi.org/10.1086/146010

Parker EN (1979) Cosmical magnetic fields: their origin and their activity. Clarendon, Oxford

Passos D, Nandy D, Hazra S et al (2014) A solar dynamo model driven by mean-field alpha and Babcock–Leighton sources: fluctuations, grand-minima-maxima, and hemispheric asymmetry in sunspot cycles. Astron Astrophys 563:A18. https://doi.org/10.1051/0004-6361/201322635. arXiv:1309.2186 [astro-ph.SR]

Passos D, Charbonneau P, Miesch M (2015) Meridional circulation dynamics from 3D magnetohydrodynamic global simulations of solar convection. Astrophys J Lett 800(1):L18. https://doi.org/10.1088/2041-8205/800/1/L18. arXiv:1502.01154 [astro-ph.SR]

Petrovay K (2020) Solar cycle prediction. Living Rev Sol Phys 17(1):2. https://doi.org/10.1007/s41116-020-0022-z. arXiv:1907.02107 [astro-ph.SR]

Pipin VV (2008) The mean electro-motive force and current helicity under the influence of rotation, magnetic field and shear. Geophys Astrophys Fluid Dyn 102:21–49. https://doi.org/10.1080/03091920701374772

Pipin VV (2022) On the effect of surface bipolar magnetic regions on the convection zone dynamo. Mon Not R Astron Soc 514(1):1522–1534. https://doi.org/10.1093/mnras/stac1434. arXiv:2112.09460 [astro-ph.SR]

Pipin VV, Kosovichev AG (2020) Torsional oscillations in dynamo models with fluctuations and potential for helioseismic predictions of the solar cycles. Astrophys J 900(1):26. https://doi.org/10.3847/1538-4357/aba4ad

Rädler KH (1968) On the electrodynamics of conducting fluids in turbulent motion. II. Turbulent conductivity and turbulent permeability. Z Naturforsch Teil A 23:1851–1860. https://doi.org/10.1515/zna-1968-1124

Rajaguru SP, Antia HM (2015) Meridional circulation in the solar convection zone: time-distance helioseismic inferences from four years of HMI/SDO observations. Astrophys J 813(2):114. https://doi.org/10.1088/0004-637X/813/2/114

Reiners A, Shulyak D, Käpylä PJ et al (2022) Magnetism, rotation, and nonthermal emission in cool stars. Average magnetic field measurements in 292 M dwarfs. Astron Astrophys 662:A41. https://doi.org/10.1051/0004-6361/202243251. arXiv:2204.00342 [astro-ph.SR]

Rempel M (2005) Influence of random fluctuations in the Λ-effect on meridional flow and differential rotation. Astrophys J 631(2):1286–1292. https://doi.org/10.1086/432610

Réville V, Brun AS, Matt SP et al (2015) The effect of magnetic topology on thermally driven wind: toward a general formulation of the braking law. Astrophys J 798(2):116. https://doi.org/10.1088/0004-637X/798/2/116. arXiv:1410.8746 [astro-ph.SR]

Roberts PH (1972) Kinematic dynamo models. Philos Trans R Soc Lond Ser A 272(1230):663–698. https://doi.org/10.1098/rsta.1972.0074

Robinson RD, Durney BR (1982) On the generation of magnetic fields in late-type stars – a local time-dependent dynamo model. Astron Astrophys 108:322–325

Rüdiger G (1989) Differential rotation and stellar convection. Sun and the solar stars. Akademie Verlag, Berlin

Rüdiger G, Egorov P, Kitchatinov LL et al (2005) The eddy heat-flux in rotating turbulent convection. Astron Astrophys 431:345–352. https://doi.org/10.1051/0004-6361:20041670

Rüdiger G, Kitchatinov LL, Schultz M (2012) Suppression of the large-scale Lorentz force by turbulence. Astron Nachr 333(1):84–91. https://doi.org/10.1002/asna.201111635

Rüdiger G, Kitchatinov LL, Hollerbach R (2013) Magnetic processes in astrophysics: theory, simulations, experiments. Wiley-VCH, Weinheim

Saar S (2002) Stellar dynamos: scaling laws and coronal connections. In: Favata F, Drake JJ (eds) Stellar Coronae in the Chandra and XMM-NEWTON Era, p 311

Saar SH, Brandenburg A (1999) Time evolution of the magnetic activity cycle period. II. Results for an expanded stellar sample. Astrophys J 524:295–310. https://doi.org/10.1086/307794

Saha C, Chandra S, Nandy D (2022) Evidence of persistence of weak magnetic cycles driven by meridional plasma flows during solar grand minima phases. Mon Not R Astron Soc 517(1):L36–L40. https://doi.org/10.1093/mnrasl/slac104. arXiv:2209.14651 [astro-ph.SR]

Schatten KH, Scherrer PH, Svalgaard L et al (1978) Using dynamo theory to predict the sunspot number during solar cycle 21. Geophys Res Lett 5(5):411–414. https://doi.org/10.1029/GL005i005p00411

Schou J, Antia HM, Basu S et al (1998) Helioseismic studies of differential rotation in the solar envelope by the solar oscillations investigation using the Michelson Doppler imager. Astrophys J 505:390–417. https://doi.org/10.1086/306146

Shimada R, Hotta H, Yokoyama T (2022) Mean-field analysis on large-scale magnetic fields at high Reynolds numbers. Astrophys J 935(1):55. https://doi.org/10.3847/1538-4357/ac7e43. arXiv:2207.01639 [astro-ph.SR]

Simard C, Charbonneau P, Dubé C (2016) Characterisation of the turbulent electromotive force and its magnetically-mediated quenching in a global EULAG-MHD simulation of solar convection. Adv Space Res 58(8):1522–1537. https://doi.org/10.1016/j.asr.2016.03.041. arXiv:1604.01533 [astro-ph.SR]

Spence EJ, Nornberg MD, Jacobson CM et al (2007) Turbulent diamagnetism in flowing liquid sodium. Phys Rev Lett 98(16):164503. https://doi.org/10.1103/PhysRevLett.98.164503

Spruit HC (1981) Motion of magnetic flux tubes in the solar convection zone and chromosphere. Astron Astrophys 98:155–160

Spruit HC (2003) Origin of the torsional oscillation pattern of solar rotation. Sol Phys 213(1):1–21. https://doi.org/10.1023/A:1023202605379

Spruit H (2012) Theories of the solar cycle and its effect on climate. Prog Theor Phys Suppl 195:185–200. https://doi.org/10.1143/PTPS.195.185

Steenbeck M, Krause F (1969) On the dynamo theory of stellar and planetary magnetic fields. I. AC dynamos of solar type. Astron Nachr 291:49–84. https://doi.org/10.1002/asna.19692910201

Steenbeck M, Krause F, Rädler KH (1966) Berechnung der mittleren Lorentz-Feldstärke v X B für ein elektrisch leitendes Medium in turbulenter, durch Coriolis-Kräfte beeinflußter Bewegung (A calculation of the mean electromotive force in an electrically conducting fluid in turbulent motion, under the influence of Coriolis forces). Z Naturforsch Teil A 21:369–376. https://doi.org/10.1515/zna-1966-0401

Stenflo JO, Kosovichev AG (2012) Bipolar magnetic regions on the sun: global analysis of the SOHO/MDI data set. Astrophys J 745:129. https://doi.org/10.1088/0004-637X/745/2/129. arXiv:1112.5226 [astro-ph.SR]

Stix M (1976) Differential rotation and the solar dynamo. Astron Astrophys 47(2):243–254

Strugarek A, Beaudoin P, Charbonneau P et al (2017) Reconciling solar and stellar magnetic cycles with nonlinear dynamo simulations. Science 357(6347):185–187. https://doi.org/10.1126/science.aal3999. arXiv:1707.04335 [astro-ph.SR]

Tobias SM, Brummell NH, Clune TL et al (1998) Pumping of magnetic fields by turbulent penetrative convection. Astrophys J Lett 502(2):L177–L180. https://doi.org/10.1086/311501

Tripathi B, Nandy D, Banerjee S (2021) Stellar mid-life crisis: subcritical magnetic dynamos of solar-like stars and the breakdown of gyrochronology. Mon Not R Astron Soc 506(1):L50–L54. https://doi.org/10.1093/mnrasl/slab035. arXiv:1812.05533 [astro-ph.SR]

Usoskin IG, Solanki SK, Kovaltsov GA (2007) Grand minima and maxima of solar activity: new observational constraints. Astron Astrophys 471(1):301–309. https://doi.org/10.1051/0004-6361:20077704. arXiv:0706.0385 [astro-ph]

Vaughan AH, Preston GW (1980) A survey of chromospheric CA II H and K emission in field stars of the solar neighborhood. Publ Astron Soc Pac 92:385–391. https://doi.org/10.1086/130683

Vidotto AA, Gregory SG, Jardine M et al (2014a) Stellar magnetism: empirical trends with age and rotation. Mon Not R Astron Soc 441:2361–2374. https://doi.org/10.1093/mnras/stu728. arXiv:1404.2733 [astro-ph.SR]

Vidotto AA, Jardine M, Morin J et al (2014b) M-dwarf stellar winds: the effects of realistic magnetic geometry on rotational evolution and planets. Mon Not R Astron Soc 438(2):1162–1175. https://doi.org/10.1093/mnras/stt2265. arXiv:1311.5063 [astro-ph.SR]

Viviani M, Warnecke J, Käpylä MJ et al (2018) Transition from axi- to nonaxisymmetric dynamo modes in spherical convection models of solar-like stars. Astron Astrophys 616:A160. https://doi.org/10.1051/0004-6361/201732191. arXiv:1710.10222 [astro-ph.SR]

Wang YM, Sheeley NR (2009) Understanding the geomagnetic precursor of the solar cycle. Astrophys J Lett 694(1):L11–L15. https://doi.org/10.1088/0004-637X/694/1/L11

Wang YM, Nash AG, Sheeley NR, Jr (1989) Evolution of the Sun's polar fields during sunspot cycle 21: poleward surges and long-term behavior. Astrophys J 347:529. https://doi.org/10.1086/168143

Wang YM, Sheeley NR, Jr, Nash AG (1991) A new solar cycle model including meridional circulation. Astrophys J 383:431. https://doi.org/10.1086/170800

Warnecke J (2018) Dynamo cycles in global convection simulations of solar-like stars. Astron Astrophys 616:A72. https://doi.org/10.1051/0004-6361/201732413. arXiv:1712.01248 [astro-ph.SR]

Warnecke J, Rheinhardt M, Tuomisto S et al (2018) Turbulent transport coefficients in spherical wedge dynamo simulations of solar-like stars. Astron Astrophys 609:A51. https://doi.org/10.1051/0004-6361/201628136. arXiv:1601.03730 [astro-ph.SR]

Warnecke J, Rheinhardt M, Viviani M et al (2021) Investigating global convective dynamos with mean-field models: full spectrum of turbulent effects required. Astrophys J Lett 919(2):L13. https://doi.org/10.3847/2041-8213/ac1db5. arXiv:2105.07708 [astro-ph.SR]

Weiss NO, Cattaneo F, Jones CA (1984) Periodic and aperiodic dynamo waves. Geophys Astrophys Fluid Dyn 30(4):305–341. https://doi.org/10.1080/03091928408219262

Wilmot-Smith AL, Nandy D, Hornig G et al (2006) A time delay model for solar and stellar dynamos. Astrophys J 652(1):696–708. https://doi.org/10.1086/508013

Wilson OC (1978) Chromospheric variations in main-sequence stars. Astrophys J 226:379–396. https://doi.org/10.1086/156618

Wright NJ, Drake JJ (2016) Solar-type dynamo behaviour in fully convective stars without a tachocline. Nature 535:526–528. https://doi.org/10.1038/nature18638. arXiv:1607.07870 [astro-ph.SR]

Wright NJ, Drake JJ, Mamajek EE et al (2011) The stellar-activity-rotation relationship and the evolution of stellar dynamos. Astrophys J 743:48. https://doi.org/10.1088/0004-637X/743/1/48. arXiv:1109.4634 [astro-ph.SR]

Yeates AR, Muñoz-Jaramillo A (2013) Kinematic active region formation in a three-dimensional solar dynamo model. Mon Not R Astron Soc 436(4):3366–3379. https://doi.org/10.1093/mnras/stt1818. arXiv:1309.6342 [astro-ph.SR]

Yeates AR, Nandy D, Mackay DH (2008) Exploring the physical basis of solar cycle predictions: flux transport dynamics and persistence of memory in advection- versus diffusion-dominated solar convection zones. Astrophys J 673(1):544–556. https://doi.org/10.1086/524352. arXiv:0709.1046 [astro-ph]

Yoshimura H (1975) Solar-cycle dynamo wave propagation. Astrophys J 201:740–748. https://doi.org/10.1086/153940

Zeldovich YB (1957) Magnetic field in two-dimensional turbulence of conducting fluid. J Exp Theor Phys 4:460–462

Zhang Z, Jiang J (2022) A Babcock–Leighton-type solar dynamo operating in the bulk of the convection zone. Astrophys J 930(1):30. https://doi.org/10.3847/1538-4357/ac6177. arXiv:2204.14077 [astro-ph.SR]

Zhao J, Bogart RS, Kosovichev AG et al (2013) Detection of equatorward meridional flow and evidence of double-cell meridional circulation inside the Sun. Astrophys J Lett 774(2):L29. https://doi.org/10.1088/2041-8205/774/2/L29. arXiv:1307.8422 [astro-ph.SR]

Authors and Affiliations

Gopal Hazra[1,2]  **· Dibyendu Nandy[3,4] · Leonid Kitchatinov[5] · Arnab Rai Choudhuri[6,7]**

✉ G. Hazra
hazra@iitk.ac.in

D. Nandy
dnandi@iiserkol.ac.in

L. Kitchatinov
kit@iszf.irk.ru

A.R. Choudhuri
arnab@iisc.ac.in

[1] Dept. of Physics, Indian Institute of Technology, Kanpur, Kalyanpur, Kanpur, 208060, Uttar Pradesh, India

[2] Dept. of Astrophysics, University of Vienna, Türkenschanzstraße 17, Vienna, 1180, Austria

[3] Department of Physical Sciences, Indian Institute of Science Education and Research, Kolkata, Mohanpur, Kolkata, 741 246, WB, India

[4] Center of Excellence in Space Sciences India, Indian Institute of Science Education and Research, Kolkata, Mohanpur, Kolkata, 741 246, WB, India

[5] Institute of Solar-Terrestrial Physics SB RAS, Lermontov Str. 126a, Irkutsk, 664033, Russia

[6] Dept. of Physics, Indian Institute of Science, C V Raman Avenue, Bengaluru, 560012, Karnataka, India

[7] Max Planck Institute for the History of Science, Boltzmannstrasse 22, Berlin, 14195, Germany

Space Science Reviews (2023) 219:60
https://doi.org/10.1007/s11214-023-01004-7

Observationally Guided Models for the Solar Dynamo and the Role of the Surface Field

Robert H. Cameron[1]  · Manfred Schüssler[1]

Received: 1 May 2023 / Accepted: 13 September 2023 / Published online: 12 October 2023
© The Author(s) 2023

Abstract

Theoretical models for the solar dynamo range from simple low-dimensional "toy models" to complex 3D-MHD simulations. Here we mainly discuss appproaches that are motivated and guided by solar (and stellar) observations. We give a brief overview of the evolution of solar dynamo models since 1950s, focussing upon the development of the Babcock–Leighton approach between its introduction in the 1960s and its revival in the 1990s after being long overshadowed by mean-field turbulent dynamo theory. We summarize observations and simple theoretical deliberations that demonstrate the crucial role of the surface fields in the dynamo process and give quantitative analyses of the generation and loss of toroidal flux in the convection zone as well as of the production of poloidal field resulting from flux emergence at the surface. Furthermore, we discuss possible nonlinearities in the dynamo process suggested by observational results and present models for the long-term variability of solar activity motivated by observations of magnetically active stars and the inherent randomness of the dynamo process.

Keywords Solar activity · Solar cycle · Dynamo

1 Introduction

Studies of solar and stellar dynamos face a problem of utter complexity, i.e., the interaction of turbulent convection with rotation and magnetic field in a highly stratified medium, covering wide ranges of spatial and temporal scales. Attempts to directly attack this problem with numerical 3D-MHD simulations have made significant progress in recent years (e.g., Charbonneau 2020; Käpylä et al. 2023), but are still severely limited in spatial resolution, so that a faithful representation of the solar cycle has not been achieved so far. All other approaches to the dynamo problem resort to simplifications in order to obtain a tractable task. They largely rely on the surprising amount of regularity shown by the solar cycle,

Solar and Stellar Dynamos: A New Era
Edited by Manfred Schüssler, Robert H. Cameron, Paul Charbonneau, Mausumi Dikpati, Hideyuki Hotta and Leonid Kitchatinov

✉ M. Schüssler
schuessler@mps.mpg.de

[1] Max Planck Institute for Solar System Research, Justus-von-Liebig-Weg 3, 37077 Göttingen, Germany

e.g., Hale's polarity rules and the systematic tilt of sunspot groups (Joy's law), the butterfly diagram, and the regular reversals of the global dipole field (Hathaway 2015). These regularities suggest simplified concepts, such as the generation of azimuthal (toroidal) magnetic flux through winding of meridional (poloidal) magnetic field lines by differential rotation (Cowling 1953) or the formation of bipolar sunspot groups by the emergence of buoyantly rising magnetic flux tubes (Parker 1955b).

A variety of models addressing various aspects of the dynamo problem has been developed. These range from rather ad-hoc "toy models" over sophisticated two-scale approaches, pioneered by the model of Parker (1955a) and by mean-field theory of turbulent MHD flows (see review by Brandenburg et al. 2023) to 2D/3D flux-transport dynamo models including various physical processes considered to be relevant for the dynamo (see reviews by Charbonneau 2020; Hazra et al. 2023).

The simplest "model" in the literature is probably due to Barnes et al. (1980). These authors considered a digital narrow-band filtering of Gaussian white noise, corresponding to a randomly disturbed periodic signal that could be illustrated by a "pendulumn pelted with peas" (Yule 1927). The results of their short computer program (15 lines!) exhibit a very similar kind of variability as shown by the sunspot record, including the occurence of extended periods of low activity (grand minima). The important lesson from this result for models of the solar dynamo is that even a striking similarity of their output with records of solar activity alone does not necessarily imply the validity of a model.

In this paper, we focus upon models of the solar dynamo as well as for the cyclic generation and removal of magnetic flux that are guided by solar (and also stellar) observations. The motivation for such approaches comes from the fact that the very complex interactions in the convection zone can lead to comparably simple results, such as the stable differential rotation in latitude, the meridional circulation, the polarity rules of sunspot groups, or the butterfly diagram of magnetic flux emergence. This might perhaps be compared to the flow of water in a watermill: although the detailed turbulent flow pattern is utterly complex, the general result is simple: water flows down the potential well and faithfully drives the wheel. The paradigm of observationally motivated models is the scenario of Babcock (1961), which was later extended and put in a mathematical form by Leighton (1969). These seminal papers paved the way for what are now generally called Babcock–Leighton-type models of the solar dynamo.

The plan of this paper is as follows. In Sect. 2 we describe the principles of the Babock-Leighton scenario and review the evolution of dynamo models until the end of the 1980s, when mounting empirical evidence led to the renaissance of Babcock–Leighton dynamo models after an extended period of near oblivion. The generation and removal of toroidal and poloidal magnetic flux in the Sun and the crucial role of the observable surface field are discussed in Sect. 3. Nonlinearity of the dynamo process, predictability, and models for the long-term variability of solar activity are considered in Sect. 4. Section 5 gives a brief outlook.

2 The Babcock–Leighton Approach and the Evolution of Models for the Solar Dynamo

The seminal paper of Babcock (1961) lays out a purely observationally based scenario for the 11/22-year solar cycle. In a first step, the poloidal magnetic flux connected to the global dipole field is wound up by latitudinal differential rotation in the convection zone. This generates oppositely directed subsurface belts of toroidal field in both hemispheres. Subsequent instability, buoyant rise, and emergence of toroidal flux at the surface produces bipolar magnetic regions (BMRs) and sunspot groups in accordance with Hale's polarity rules. These regions are observed to be systematically tilted in the sense that the magnetic polarity leading in the direction of rotation is nearer to the equator than the following polarity. This tilt, together with the dispersal of the BMRs in the course of time, leads to preferred transport of leading-polarity flux over the equator, so that a surplus of the respective following-polarity flux builds up in both hemispheres. Poleward transport of this flux by convection and meridional circulation leads to the reversal of the polar fields and the buildup of an oppositely directed dipole field, a process already suggested by Babcock and Babcock (1955). This entails the generation and emergence of a reversed toroidal field followed by another reversal of the dipole field, thus giving rise to the 22-year magnetic cycle.

Babcock (1961) already noted that the systematic tilt of the sunspot groups could be "... the result of Coriolis forces, which induce a vorticity in the whole of the fluid associated with a BMR when it rises to the surface." The Coriolis effect is also invoked in the concept of "cyclonic convection" of Parker (1955a) and in the mean-field scheme based on turbulence theory pioneered by Steenbeck and Krause (1966) and Steenbeck et al. (1966). Possibly since both these approaches focus on the collective effect of a small-scale process (in the sense of a two-scale approach), neither Parker nor Steenbeck and coworkers apparently realized that the systematic tilt provides direct observational support for the action of the Coriolis effect as well a quantitative measure.

Leighton (1964) added a key component to Babcock's scenario in the form of a random-walk model for the quasi-diffusive transport of magnetic flux on the solar surface by supergranular convection. Subsequently, he put the Babcock model in the mathematical form of a one-dimensional dynamo equation (Leighton 1969). It is not obvious why this Babcock–Leighton (BL) model fell into near oblivion for about two decades thereafter. In order to provide some understanding for this development, a brief sketch of the evolution of solar dynamo studies from the early 1970s onward is given in what follows.

Perhaps since the mathematically involved mean-field theory was considered to provide a strong theoretical footing, models for the solar dynamo until the 1990s mostly relied on a mean-field turbulent "α-effect" based on correlations between small-scale fields operating within the convection zone. To explain the equatorward-directed migration of the activity belts shown by the butterfly diagram, these models generally required an inwardly increasing rotation rate, $d\Omega/dr < 0$, according to the Parker-Yoshimura sign rule (Parker 1955a; Köhler 1973; Yoshimura 1975). The effect of the observed strong latitudinal differential rotation was thought to be of secondary importance. Consequently, both key processes for the turbulent dynamo, radial differential rotation and α-effect, were assumed to operate in the solar interior, thus inaccessable to observation at that time. Therefore, the corresponding model parameters were largely unconstrained. The observable evolution of the surface field was seen as a mere epiphenomenon of a dynamo process operating deeply hidden in the convection zone and served solely as an observational constraint for the models.

The foundations of this approach were undermined when helioseismology made it possible to measure the differential rotation in the convection zone. It turned out that (except for a shallow near-surface shear layer) the increase of the rotation rate with depth required by these models is absent in the bulk of the convection zone (see review by Howe 2009). On the other hand, the rotation rate at the bottom of the convection zone showed a steep radial gradient across the "tachocline" (Brown et al. 1989), such that $d\Omega/dr < 0$ in solar latitudes ≥ 45 deg and $d\Omega/dr > 0$ in lower latitudes. Some years before, Galloway and Weiss (1981) already had suggested that the solar dynamo should work in a stably stratified layer of overshooting convection below the bottom of the convection zone in order to avoid the rapid buoyant loss of toroidal magnetic flux inferred by Parker (1975). These results led to the concept of turbulent dynamo action within the overshoot layer/tachocline (e.g., Rüdiger and Brandenburg 1995) or near its interface with the convection zone proper (Parker 1993), where a properly chosen combination of the signs of α-effect and radial gradient of the rotation rate could provide the correct conditions for equatorward propagating dynamo waves and activity belts (e.g., Tobias 1996; Charbonneau and MacGregor 1997). This approach was complemented by studies of equilibrium, stability, and dynamics of thin magnetic flux tubes starting with Spruit and van Ballegooijen (1982), Choudhuri and Gilman (1987), and Moreno-Insertis et al. (1992). Such studies (see reviews by Fan 2021; Işik et al. 2023) indicated that the toroidal field in the overshoot region must be amplified to about 10^5 G before becoming unstable (Ferriz-Mas and Schüssler 1993, 1995) and rising to the surface to form bipolar magnetic regions within the latitude range shown by the butterfly diagram (Schüssler et al. 1994). Twisting of the rising flux tubes due to the Coriolis force then leads to tilt angles consistent with observation (D'Silva and Choudhuri 1993; Fan et al. 1994; Caligari et al. 1995).

Although the combination of tachocline/overshoot layer dynamos and the dynamics of thin flux tubes offered a comprehensive picture from the generation of magnetic flux up to its emergence at the surface, these models relied on a number of untested assumptions and simplifications, so that their predictive power remained rather limited. In the course of time, a number of theoretical considerations and observational results cast severe doubt upon the validity of this aproach:

- The total energy of a toroidal field of 10^5 Gauss at the bottom of the convection zone would be comparable to the energy in the differential rotation of the tachocline (Rempel 2006). However, a corresponding strong variation of the differential rotation in the lower convection zone and tachocline in the course of the 11-year activity cycle is not observed (Basu and Antia 2019). Moreover, the interface between radiative core and convection zone cannot support much shear stress (Spruit 2011). Consequently, the radial differential rotation in the tachocline (mainly reflecting the transition from latitudinally differential rotation in the convection zone to nearly rigid rotation below) cannot generate a sizeable amount of toroidal magnetic flux unless being maintained by a very powerful downward transport of angular momentum.
- The latitudinal differential rotation is sufficient to create the total toroidal flux covered by the emergence of bipolar magnetic regions and sunspot groups. At the same time, the contribution of the observed radial differential rotation in the convection zone is only a few percent of that of the latitudinal differential rotation (Cameron and Schüssler 2015). Moreover, the radial shear in the tachocline is strong in high latitudes and weak in low latitudes, thus unfavorable for toroidal flux generation in low latitudes.
- Helioseismology indicates that the overshoot layer is much more strongly subadiabatic than assumed in the models for the storage and stability of toroidal flux (Christensen-Dalsgaard et al. 2011). On the other hand, part of the lower convection zone could be subadiabatically stratified (e.g., Spruit 1997; Hotta 2017).
- Observations of magnetically active stars show that partly and fully convective stars follow the same activity-rotation law (Wright and Drake 2016; Reiners et al. 2022), thus questioning the relevance of convective overshoot and the existence of a tachocline for the dynamo process. This is also supported by 3D-MHD simulations of partially and fully convective M dwarfs (Bice and Toomre 2023). Furthermore, even very cool, fully convective dwarfs (beyond spectral type M7) can exhibit activity cycles (Route 2016). Recently, Lu et al. (2023) found that the spin-down rates of M-type stars appear to sharply rise

across the transition between partly and fully convective stars, indicating more organized large-scale surface fields of fully convective stars.

- 3D-MHD simulations demonstrate the formation of super-equipartion magnetic flux concentrations and buoyantly rising flux loops within a simulated convection zone without overshoot and tachocline (Fan and Fang 2014; Chen et al. 2017). Nelson et al. (2014) obtained similar results, albeit with a simulation assuming three times the solar rotation rate.
- If emerged magnetic structures were "anchored" near to the bottom of the convection zone, they should show slower rotation than that of the surface plasma below about ± 30 deg latitude according to the helioseismically determined rotation profile. However, magnetic structures are observed rotate faster than the plasma at the surface (e.g., Howard 1996).

A new twist came when observations of a systematic poleward-directed meridional flow at the surface pioneered by Duvall (1979) and Howard (1979) became established during the 1980s (see review by Hanasoge 2022). This led to the suggestion that the associated deep return flow within the convection zone could transport toroidal flux equatorward. Models of flux-transport dynamos (FTDs) that rely upon this concept can provide a butterfly diagram consistent with observations regardless of the sign of the radial gradient of the rotation rate (Wang and Sheeley 1991; Durney 1995; Choudhuri et al. 1995). A recent review of FTD models has been provided by Hazra et al. (2023).

The confirmation of the systematic poleward surface flow also sparked the development of simulation models for the transport of magnetic flux at the solar surface on solar-cycle time scales, pioneered by the NRL group (DeVore et al. 1984; Wang et al. 1989b). Such Surface Flux Transport (SFT) models (see reviews by Mackay and Yeates 2012; Yeates et al. 2023) assume passive transport of vertically (radially) orientated magnetic flux by surface flows. Comparison of SFT simulations with observed synoptic magnetograms shows that the evolution of the surface flux can be faithfully described by flux emergence in systematically tilted bipolar magnetic regions followed by passive flux transport by differential rotation, meridional flow, and supergranular flows (the latter mostly being treated as a diffusion-like random walk, cf. Leighton 1964), and flux cancellation. In particular, the models confirm the buildup of polar fields due to the preferred transport of following-polarity flux toward the poles (Wang et al. 1989a).

The success of the SFT simulations led to the re-appraisal of the BL approach for the (re)generation of the poloidal field in the course of the dynamo process (Giovanelli 1985; Wang and Sheeley 1991), typically in connection with equatorward flux transport by meridional flow in the convection zone as Flux Transport Dynamo (FTD) models (Wang et al. 1991; Charbonneau 2020; Hazra et al. 2023). In the course of time, further lines of evidence for the validity of the BL approach and the crucial role of the surface flux for the solar dynamo became apparent:

- SFT models reproduce the evolution of the surface field, particularly the poleward drift of the following-polarity surface field leading to the buildup of the polar fields and the axial dipole (e.g., Wang et al. 2002; Baumann et al. 2004; Jiang et al. 2014b; Upton and Hathaway 2014; Whitbread et al. 2017).
- The polar field strength around cycle minimum is strongly correlated with the amplitude of the subsequent activity maximum (Legrand and Simon 1981; Wang and Sheeley 2009; Kitchatinov and Olemskoy 2011b; Hathaway and Upton 2016), thus providing a faithful predictor of cycle strength (Schatten et al. 1978; Petrovay 2020; Kumar et al. 2021; Bhowmik et al. 2023). Recent work by Kumar et al. (2022) and Biswas et al. (2023b)

suggests that an earlier prediction is possible by considering the rise rate of the polar field a few years before minimum.

- Cameron and Schüssler (2015) showed that the above correlation in fact reflects a causation: the net hemispheric toroidal flux generated during a half cycle results from the action of the latitudinal differential rotation on the poloidal flux connected to the polar fields (see Sect. 3.1 below).
- The observed azimuthal surface field (a proxy for flux emergence) evolves in accordance with an updated BL model (Cameron et al. 2018).

The timescale for the rise of flux loops formed by the toroidal field and the subsequent flux emergence is generally considered to be short compared to the cycle time scale. Therefore, BL dynamo models typically do not explicitly include these processes but rather incorporate a source term near the surface that is related to the deep-seated toroidal field in the convection zone (for a discussion of various approaches, see Choudhuri and Hazra 2016). A few examples of two-dimensional axisymmetric dynamo models with different prescriptions for the BL source term are the studies of Durney (1997), Dikpati and Charbonneau (1999), Nandy and Choudhuri (2001), Chatterjee et al. (2004), and Muñoz-Jaramillo et al. (2010). More complete overviews have been provided by Charbonneau (2020) and Hazra et al. (2023).

FTD/BL models mostly rely on radial differential rotation in the tachocline and often also assume penetration of the meridional flow into the stably stratified interior in order to "store" the toroidal magnetic flux at the bottom of the convection zone (e.g., Nandy and Choudhuri 2002), which seems to be difficult to achieve (cf. Gilman and Miesch 2004). In fact, Muñoz-Jaramillo et al. (2009) showed that no such penetration is necessary for FTD models to reproduce the basic features of the solar cycle. These authors also showed that the helioseismically determined differential rotation in the convection zone leads to a dominant production of toroidal flux by the latitudinal differential rotation (see also Guerrero and de Gouveia Dal Pino 2007). Recently, Zhang and Jiang (2022) presented a comprehensive dynamo model including the helioseismically determined differential rotation (including the near-surface shear layer), a one-cell meridional circulation, radial pumping keeping the surface field vertical and inhibiting diffusive loss of the toroidal field, and a BL source term. The latitudinal propagation of the toroidal flux belts in this model is provided by a combination of flux transport by the equatorward meridional return flow and the latitude dependence of the latitudinal rotational shear generating toroidal magnetic flux, the latter as already envisaged by Babcock (1961). Test cases with removal of the radial shear in the tachocline or putting the numerical boundary above the tachocline does not significantly change the results of the model, thus strongly suggesting that the tachocline shear is indeed largely irrelevant for the dynamo (cf. Spruit 2011; Brandenburg 2005; Cameron and Schüssler 2015).

Typically, FT/BL dynamo models are 2D (axisymmetric) or even 3D and time-dependent. Although they exhibit solar-like solutions for properly chosen values of the parameters (e.g., turbulent diffusivity, geometry and speed of the meridional flow, dynamo excitation, etc.), computational limitations do not permit a complete coverage of the parameter space. In order to systematically investigate the parameter ranges for solar-like solutions, Cameron and Schüssler (2017a) updated the 1D, two-layer model of Leighton (1969). Model quantities are the azimuthally averaged radial field at the surface and the radially integrated toroidal flux, both as a function of latitude. Poleward meridional flow at the surface, an equatorward return flow somewhere in the convection zone, latitudinal differential rotation as well as radial differential rotation in the near-surface shear layer are included. Furthermore, downward convective pumping of the near-surface horizontal field and turbulent diffusion of the

Fig. 1 Sketch of the updated Babcock–Leighton model of Cameron and Schüssler (2017a)

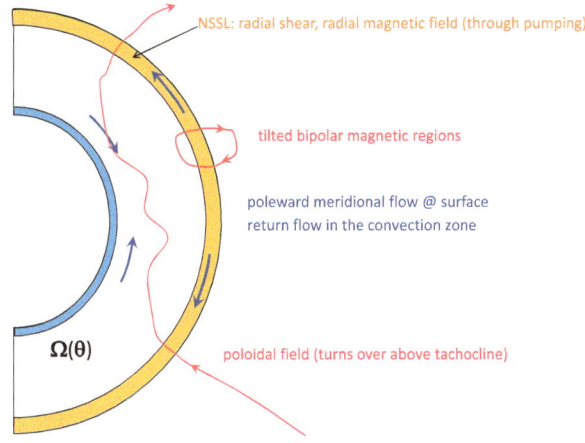

NSSL: radial shear, radial magnetic field (through pumping)

tilted bipolar magnetic regions

poleward meridional flow @ surface
return flow in the convection zone

$\Omega(\theta)$

poloidal field (turns over above tachocline)

radial surface field and the toroidal field are considered. Using the radially integrated magnetic flux has the advantage that the model neither depends on the radial distribution of the toroidal field in the convection zone nor on the depth location of the meridional return flow. The solutions of this linear model depend on four parameters which represent the driving of the dynamo (by emergence of tilted bipolar regions), the effective speed of the meridional return flow, the turbulent diffusivity affecting the toroidal field in the convection zone, and the effective radial shear below the near-surface shear layer. All other parameters (such as meridional flow and diffusivity at the surface) are taken from observations. The simplicity of the model permits the exploration of the full four-dimensional parameter space. Relevant ranges of parameters are identified by requiring that the model results meet observational constraints: positive growth rate, dipole parity, 22-year magnetic period, flux emergence concentrated in low latitudes, and a 90 deg phase difference between the maximum of flux emergence and maximum strength of the polar field. It turns out that these requirements strongly constrain the parameters, yielding about 2 m s^{-1} for the speed of the return flow, a turbulent diffusivity of about 80 km^2 s^{-1}, toroidal flux generation dominated by latitudinal differential rotation, and weak dynamo excitation not far above the threshold. A sketch of the model is shown in Fig. 1 and an example of the results in Fig. 2.

The model results indicate that the solar dynamo most probably is a flux-transport dynamo near marginal excitation, operating in the high-diffusivity regime (see also Yeates et al. 2008). The inferred speed of the equatorward flow transporting the toroidal flux is consistent with the latitudinal drift rate of the activity belts. The relevant values of the magnetic diffusivity affecting the toroidal flux are also consistent with the observed evolution of the solar activity belts (Cameron and Schüssler 2016). Dominance of latitudinal differential rotation entails a natural explanation for the concentration of flux emergence in low latitudes: the latitudinal rotational shear peaks at mid latitudes and the deep meridional return flow transports the generated toroidal flux equatorward. Consequently, it is not necessary to assume a threshold value of the toroidal field for the initiation of flux emergence, e.g., in the sense of a buoyancy instability.

Further development of BL dynamo models was provided through a closer connection between the near-surface evolution and the interior evolution of the magnetic field. Cameron et al. (2012) found that downward convective pumping (leading to predominantly radial field near the surface) is required to bring a 1D SFT model into accordance with the results of a 2D axisymmetric FTD model. Furthermore, FTD dynamos with convective pumping provide

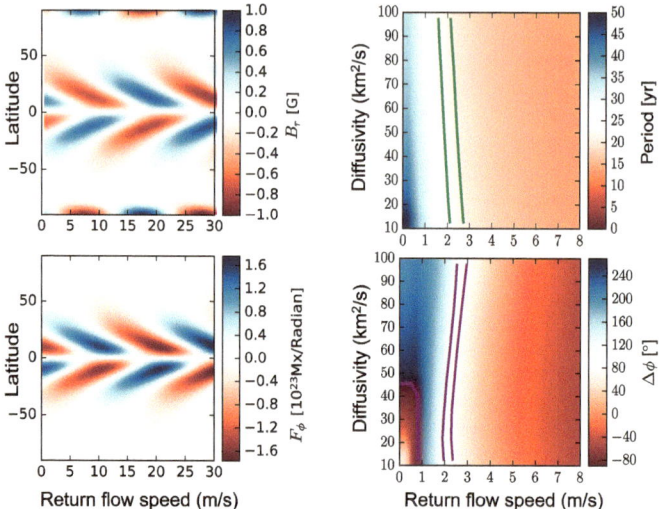

Fig. 2 Results from the updated Babcock–Leighton model of Cameron and Schüssler (2017a). *Left panels:* Radial surface field (upper panel) and depth-integrated toroidal flux (lower panel) as a function of latitude and time for a solution consistent with the properties of the solar cycle. *Right panels:* Properties of model solutions as functions of the amplitude of the effective speed of the equatorward return flow and of the magnetic diffusivity in the convection zone. Colour shadings indicate the dynamo period (upper panel) and the phase difference between the maximum of the polar field and the maximum rate of flux emergence (lower panel). The regions between the coloured lines indicate "solar-like" models: period in the range 21–23 yr and phase difference between 80 and 100 degrees. These results indicate a rather tight constraint for the speed of the return flow of about 2–3 m s^{-1}

"solar-like" results also with high turbulent diffusivity in the convective interior of the order of 10^{12} cm^2 s^{-1} consistent with mixing-length arguments (e.g., Karak and Cameron 2016; Hazra et al. 2019). Pumping also significantly reduces the loss of magnetic flux through the surface, so that the "restart" of the normal dynamo mode after a grand minimum can be achieved in a BL dynamo with the emergence of a few bipolar regions, without the need of an additional generation process (Karak and Miesch 2018).

Bhowmik and Nandy (2018) used the surface distribution of magnetic flux during activity minima resulting from a data-driven SFT simulation as input for a FTD model. A fully coupled 2D × 2D model was developed by Lemerle et al. (2015) and Lemerle and Charbonneau (2017). These authors connected an observationally calibrated SFT model in latitude and longitude with an axisymmetric FTD model operating in the meridional (radius-latitude) plane. The FTD model provides the source for the SFT model by stochastic flux emergence in tilted bipolar magnetic regions. In turn, the SFT model feeds back upon the FTD model in the form of a BL-like source term near the upper boundary of the dynamo model, resulting in a self-consistent coupling of the two models. Figure 3 illustrates that properly calibrated results from this model are consistent with the characteristic features of the corresponding solar observations. A similar approach was presented by Miesch and Dikpati (2014).

The BL source in such models is typically provided by "deposition" of tilted bipolar regions in the SFT part of the model, non-locally depending on the (latitude-dependent) strength of the deep-seated toroidal field. For the time delay between sunspot emergences, a probability distribution function obtained from sunspot observations can be assumed (e.g., Miesch and Dikpati 2014; Hazra et al. 2017). Such ad-hoc procedures ignore the details of the formation and rise of flux loops through the convection zone, in particular the devel-

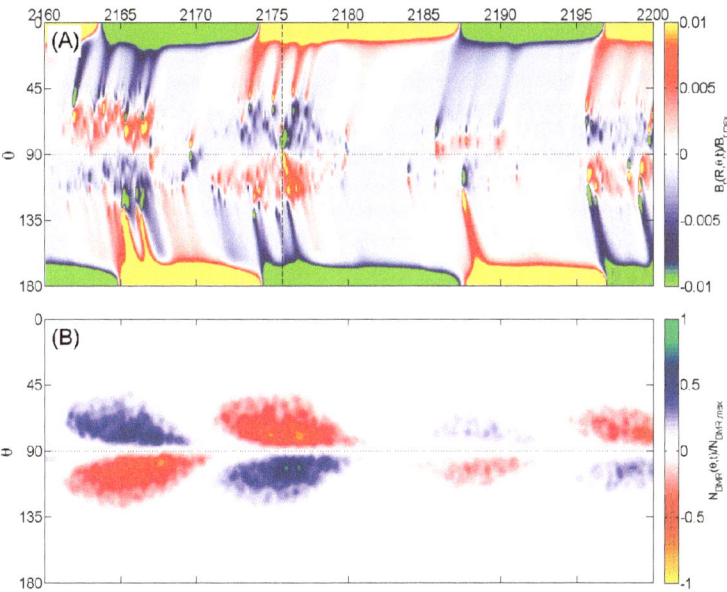

Fig. 3 Example result obtained by Nagy et al. (2017, part of their Fig. 1) using the 2D × 2D dynamo model developed by Lemerle and Charbonneau (2017). Shown are time-latitude diagrams of the radial field at the surface (upper panel) and of the number of flux emergences (butterly diagram, lower panel)

opment of the tilt angle. They also implicitly assumes a dynamical disconnection of the emerged bipolar region from its subsurface roots. Bekki and Cameron (2023) considered the post-emergence evolution of deposited bipolar regions in the framework of a nonlinear 3D simulation in a spherical shell, including the Lorentz force as well as solar-like differential rotation and meridional flow driven by a mean-field prescription. These authors found that the evolution depends sensitively on the initial shape and depth of the injected bipolar regions. When initialized with zero tilt, the Lorentz force in combination with the Coriolis force tends to produce negative tilt angles. The simulated BMRs also develop an systematic asymmetry between the strength of leading and following spots, which is consistent with observations.

Yeates and Muñoz-Jaramillo (2013) suggested a more consistent treatment of flux emergence by assuming helical upflows that transport flux loops through the convection zone, thus capturing aspects of buoyancy and advection by convective flows as well as the connection between the surface and interior field. A similar treatment was proposed by Pipin (2022). The idea of this approach is consistent with the results of detailed observations of the properties of emerging flux, which favor passive flux transport by convective upflows (Birch et al. 2016). On the other hand, comparison with well-calibrated SFT models (Whitbread et al. 2019) shows that after emergence the surface flux needs to be dynamically disconnected from its deep toroidal roots as suggested by Schüssler and Rempel (2005). Moreover, Schunker et al. (2019) and Schunker et al. (2020) showed that the tilt of bipolar magnetic regions develops only *after* emergence, possibly related to the extended flows converging towards the bipolar magnetic regions (Martin-Belda and Cameron 2016; Gottschling et al. 2021). In view of these results it seems fair to say that the formation of rising flux loops, their transfer through the convection zone, the actual emergence process, and the subsequent early evolution of the emerged magnetic flux are still poorly understood (see also reviews by

Fan 2021; Işik et al. 2023), so that these processes cause a significant amount of uncertainty in all FTD/BL models presented so far.

Among the first 3D studies of FTD/BL models for the solar dynamo are the models of Hazra et al. (2017) and Karak and Miesch (2017). While these authors used the deposition approach to represent flux emergence, Kumar et al. (2019) implemented the more consistent recipe of Yeates and Muñoz-Jaramillo (2013) in a kinematic 3D dynamo code. Another step towards a full 3D treatment was taken by Hazra and Miesch (2018), who introduced a 3D, stationary, convection-like flow field in the upper part of the convection zone. This serves to replace the diffusion term in standard SFT models by a more realistic explicit transport process. The assumed flow pattern was based upon an observational power spectrum of the surface flows and a downward extrapolation under the assumption of zero divergence of the mass flux and an imposed radial profile. The kinematic model ran into problems owing to unlimited small-scale dynamo action driven by the imposed stationary flow. Further results obtained with kinematic 3D models of BL dynamos have been reviewed by Hazra (2021). Bekki and Cameron (2023) used their 3D model to perform nonlinear BL dynamo simulations with the near-surface deposition of properly tilted bipolar regions. The Lorentz force provides nonlinear saturation of the dynamo amplitude. This feedback drives non-axisymmetric flows as well as solar-like zonal flows superposed upon the differential rotation and also modifies the meridional circulation. Other nonlinear effects (see also Sect. 4.1 that could lead to dynamo saturation and determine the cycle amplitude have been studied by Jiang (2020), Jiao et al. (2021), and Talafha et al. (2022).

3 Generation and Removal of Magnetic Flux

3.1 Toroidal Flux

Toroidal magnetic field (azimuthal field in the case of an average over longitude) is considered to be generated in the Sun through the action of differential rotation on a poloidal magnetic field. Part of the generated toroidal flux emerges when radial motions carry loops of magnetic flux through the solar surface. Flux emergence leads to the formation of two surface areas with oppositely-directed radial field components. These bipolar magnetic regions (BMRs) have been studied extensively (e.g. Hale et al. 1919; Harvey et al. 1975; Martin and Harvey 1979; Wilson et al. 1988; Harvey 1993; McClintock and Norton 2016; Schunker et al. 2016). A map showing the latitudes at which larger BMRs with sunspots emerge in time (known as the butterfly diagram) is shown in the upper panel of Fig. 4.

BMRs show various systematic properties. One example is Hale's law (Hale et al. 1919), which states that the leading polarity (with respect to the direction of rotation) of BMRs in each hemisphere has the same sign in most (up to 95%) of the cases during an activity cycle. The leading polarity is opposite in the two hemispheres and switches from cycle to cycle. These properties imply a systematic East-West component of the horizontal field during emergence, which flips sign between each hemisphere and from cycle to cycle. This can be illustrated in a time-latitude map of the longitudinally averaged azimuthal component of the surface field covering four activity cycles, which is shown in the middle panel of Fig. 4 (cf. Cameron et al. 2018; Liu and Scherrer 2022). The latitudes where sunspot groups emerge correspond to locations where the surface azimuthal field is strong. Weaker azimuthal field corresponds to flux emergence of smaller BMRs (ephemeral regions, Martin and Harvey 1979).

Fig. 4 Observationally based time-latitude diagrams of sunspots (top panel, data from Royal Greenwich Observatory/NOAA: http://solarcyclescience.com/AR_Database), longitudinally averaged azimuthal surface field (middle panel, data from Wilcox Solar Observatory, WSO), and longitudinally averaged radial field (lower panel, data from WSO)

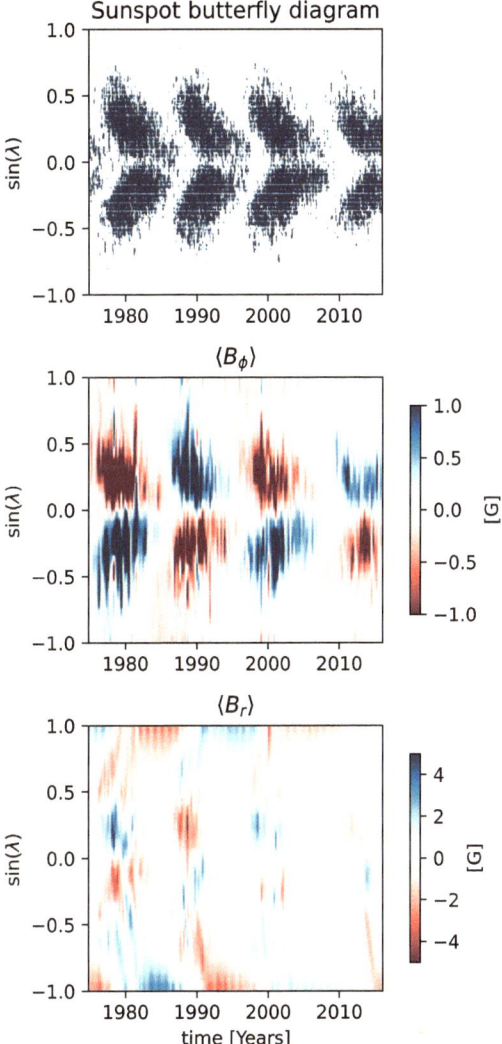

Another systematic property of BMRs is Joy's law (Hale et al. 1919). It refers to the tendency for the leading part of a BMR or sunspot group to be closer to the equator. Because of Hale's law, the leading parts mostly have the same polarity, so that there is a systematic tendency in both hemispheres that more radial magnetic flux of one polarity appears closer to the equator. The radial field of the leading parts in the other hemisphere has the opposite polarity, and the polarities switch from cycle to cycle. These properties are illustrated in the time-latitude diagram for the longitudinally averaged radial surface field shown in the bottom panel of Fig. 4. The locations of flux emergence can be clearly seen, with the leading polarity dominating on the equatorward side of the "butterfly wings". Plumes of magnetic field with widths of a few months connect the butterfly wings to the polar regions, where the field switches polarity with the period of the solar cycle.

An important feature of the time-latiude diagram of the toroidal surface field (middle panel of Fig. 4) is that during times of activity maximum (when there are many sunspots)

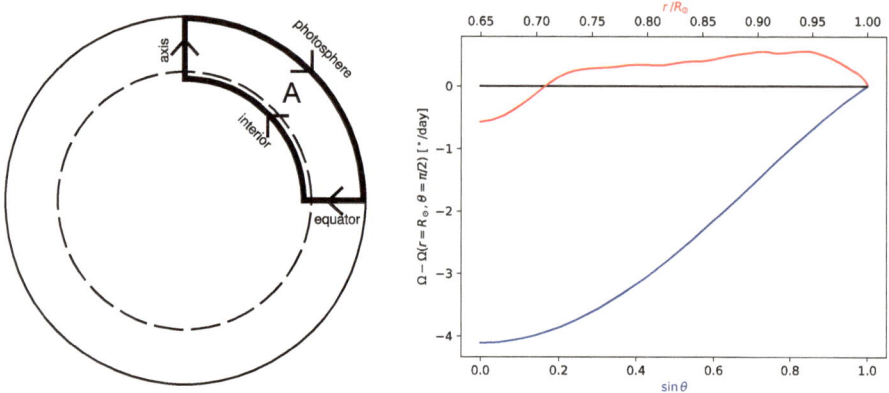

Fig. 5 *Left panel:* Area and contour used for calculating the net axisymmetric toroidal flux in a hemisphere. The bottom part of the contour is placed at a radius R_{interior}, where the penetration of the 22 year cyclic component of the magnetic field is negligible. *Right panel:* Solar differential rotation along segments of the contour in the left panel. Shown are the radial dependence of Ω on the equatorial plane (red curve, upper scale) and the (co)latitudinal dependence at the surface (blue curve, lower scale), both relative to the equatorial surface rotation rate. The data for this plot was obtained from Larson and Schou (2018)

the toroidal field at every latitude in each hemisphere is of the same sign. This suggests to consider the evolution of the net subsurface toroidal flux in each hemisphere. For this we start from the induction equation,

$$\frac{\partial \mathbf{B}}{\partial t} = \nabla \times (\mathbf{U} \times \mathbf{B} - \eta \nabla \times \mathbf{B}), \tag{1}$$

where $\mathbf{B}$ is the magnetic field, $\mathbf{U}$ the flow velocity, and η the (molecular) magnetic diffusivity. Following Cameron and Schüssler (2015), Cameron and Schüssler (2020), and Jeffers et al. (2022), we now consider the area A that is enclosed by the thick outline in the left panel of Fig. 5 and apply Stokes' theorem to the induction equation, viz.

$$\frac{\partial \iint_A \mathbf{B} \cdot d\mathbf{A}}{\partial t} = \oint_{\delta A} (\mathbf{U} \times \mathbf{B} - \eta \nabla \times \mathbf{B}) \cdot d\mathbf{l}, \tag{2}$$

where δA indicates the boundary of A and $d\mathbf{l}$ is the corresponding line element. We now take the azimuthal average of Eq. (2) and neglect the diffusive term in the contour integral since the magnetic Reynolds number is very large. This leads to

$$\frac{\partial \iint_A \langle \mathbf{B} \rangle \cdot d\mathbf{A}}{\partial t} = \oint_{\delta A} (\langle \mathbf{U} \rangle \times \langle \mathbf{B} \rangle + \langle \mathbf{u} \times \mathbf{b} \rangle) \cdot d\mathbf{l}, \tag{3}$$

where $\langle \ldots \rangle$ indicates the azimuthal average. This equation represents the temporal change of net toroidal flux in the Northern hemisphere resulting from the inductive action of the flow field on the magnetic field. The quantities $\mathbf{b} = \mathbf{B} - \langle \mathbf{B} \rangle$ and $\mathbf{u} = \mathbf{U} - \langle \mathbf{U} \rangle$ denote, respectively, the fluctuating components of magnetic field and flow velocity with respect to the corresponding azimuthal averages.

Since the Sun's radial differential rotation in the equatorial plane is small compared to its latitudinal differential rotation (see right panel of Fig. 5) it is convenient to use a frame of reference rotating with the surface equatorial rate. In this frame, we have

$$\oint_{\delta A} (\langle \mathbf{U} \rangle \times \langle \mathbf{B} \rangle) \cdot d\mathbf{l} = \int_0^{\pi/2} \left(\Omega(R_\odot, \theta) - \Omega(R_\odot, \frac{\pi}{2}) \right) \sin\theta \langle B_r \rangle R_\odot^2 d\theta$$

$$- \int_{R_{\text{interior}}}^{R_\odot} \left(\Omega(r, \frac{\pi}{2}) - \Omega(R_\odot, \frac{\pi}{2}) \right) \langle B_\theta \rangle r \, dr \,, \tag{4}$$

where θ is the colatitude. Only the surface part of the contour integral (first term on right-hand side) and the part in the equatorial plane (second term) contribute to the contour integral. The bottom part in the interior vanishes since $\mathbf{B}$ is zero there and the part along the rotation axis vanishes since we consider azimuthal averages of all quantities. The solenoidality of the magnetic field implies

$$\int_0^{\pi/2} \langle B_r \rangle R_\odot^2 \sin\theta d\theta = - \int_{R_{\text{interior}}}^{R_\odot} \langle B_\theta \rangle r \, dr, \tag{5}$$

which means that the net magnetic flux through the equatorial plane beneath the surface is equal in magnitude to the net flux through the solar surface in each hemisphere.

Owing to the weak radial differential rotation in the equatorial plan), the net toroidal flux generation is strongly dominated by the surface part of the contour integral, i.e., the first term on the right-hand side of Eq. (4). This can be directly evaluated from synoptic observations of the radial field on the surface and the Sun's differential rotation, with a net hemispheric toroidal flux of $5\text{–}8 \times 10^{23}$ Mx per cycle being thus generated. The integrand for the surface part in the contour integral (Eq. (4)) is strongly dominated by the polar caps (see bottom left panel of Fig. 6), so that the flux associated with the polar dipole field represents the relevant poloidal source for the generation of net toroidal flux. Other poloidal flux that is contained in the convection zone and does not cross the surface leads to equal amounts of East-West and West-East orientated toroidal flux in a hemisphere and thus does not contribute to the net toroidal flux required by Hale's polarity laws.

The term in Eq. (3) involving the fluctuating components, $\oint_{\delta A} \langle \mathbf{u} \times \mathbf{b} \rangle \cdot d\mathbf{l}$, is dominated by (turbulent) diffusive fluxes across the axis and the equatorial plane. Some of the other terms, such as those owing to the turbulent electromotoric force (α-effect), are expected to vanish due to symmetries – along the axis owing to the conservation of helicity and at the equator owing to the expected vanishing of the kinetic helicity, which in its simplest form scales like $\cos\theta$.

Another contribution from the fluctuating terms is the change of the net toroidal flux by flux emergence and submergence through the surface, which is described by

$$\left(\frac{d\Phi}{dt} \right)_{\text{em,subm}} = \int_{\pi/2}^0 \langle u_r b_\phi \rangle |_{R_\odot} R_\odot d\theta \,. \tag{6}$$

Flux loss through the surface was part of the original Babcock–Leighton model (Leighton 1969), but later fell out of favour since it was thought that the emerged flux is mostly retracted back through the surface (e.g., Wallenhorst and Topka 1982). This process can take place in flux cancellation events when loops of emerged flux become sufficiently narrow, so that the tension force dominates. Some retraction of flux is possibly required to account for the amount of flux brought to the surface by repeated emergences in so-called "nests of activity" (e.g., Gaizauskas et al. 1983).

Parker (1984) argued that a net loss of toroidal flux from the Sun would require an organized sequence of reconnection events between narrowly-spaced bipolar regions, thus

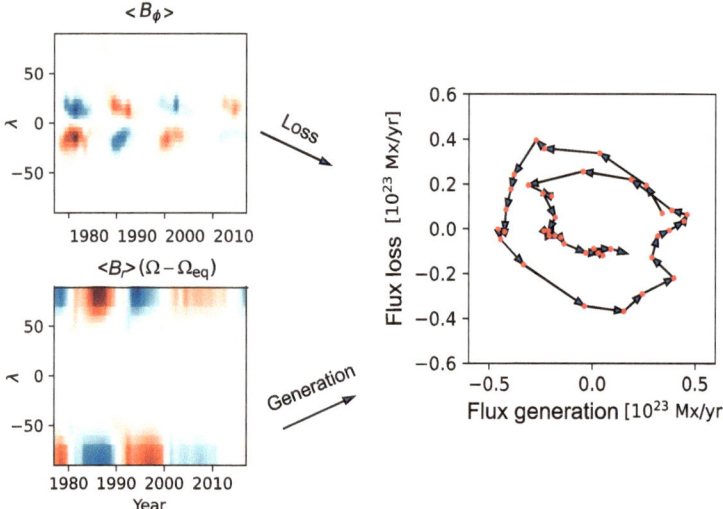

Fig. 6 Generation and loss of net unsigned toroidal flux in the Northern hemisphere, covering four activity cycles on the basis of synoptic magnetograms from the Wilcox Solar Observatory. The time-latitude diagrams on the left side show the observational input for evaluating the dominating surface part of the contour integral (Eqs. (4) and 6). Toroidal flux is generated proportional to the product of the azimuthally averaged surface radial field and the surface rotation rate relative to the equatorial rate (lower left panel). Toroidal flux is lost through flux emergence at a rate proportional to the product of $\langle B_\phi \rangle$ at the surface (upper left panel) and the radial emergence velocity estimated as $v_{em} = 200$ m s^{-1} (Centeno 2012). The panel to the right shows a phase plot of the yearly integrated values of the generated and lost toroidal flux indicated by red dots. The connecting line segments with arrows illustrate the change from one year to the next

effectively shedding the mass from the toroidal field lines and allowing them to freely escape with the solar wind. However, observations show that the bipolar regions generally are much too widely spaced for this process to operate on the Sun, so that Parker concluded that in dynamo models the solar photosphere should be approximated by an impenetrable boundary.

Parker's argument, however, ignores that what is relevant for the dynamo models is the effect of the emergence on the *mean* (azimuthally averaged) toroidal field. Cameron and Schüssler (2020) showed that the amount of the azimuthally averaged toroidal field lost due to the emergence of a bipolar region is time-independent, irrespective of the further evolution of its magnetic flux. While cancellation/retraction of most of the field may occur, at the same time the remaining surface flux spreads out in longitude (transported by near-surface flows) to eventually fully encircle the Sun. In this way, toroidal field detached from the solar interior is formed which can be carrried away by the solar wind and the same amount of flux is lost from the mean toroidal field in the interior. A quantitative analysis of the net azimuthally averaged toroidal flux lost during a cycle arrives at 3.3×10^{23} Mx/hemisphere/cycle (Cameron and Schüssler 2020). A similar estimate of 5×10^{23} was obtained using in situ measurements of the magnetic field and flows in the solar wind (Bieber and Rust 1995). The observations thus indicate that, integrated over a cycle, the amount of the toroidal magnetic flux generated by the axisymmetric flows and fields $(\oint_{\delta A} (\langle \mathbf{U} \rangle \times \langle \mathbf{B} \rangle) \cdot \mathbf{dl})$ is similar to the amount lost through flux emergence $(\oint_{\delta A} \langle \mathbf{u} \times \mathbf{b} \rangle \cdot \mathbf{dl})$.

The toroidal flux budget shown in Fig. 6 shows the approximate balance between the amount of toroidal magnetic flux lost throughout a cycle due to flux emergence and the amount produced by the winding up of poloidal flux threading the solar surface, both being determined by surface observations of the magnetic field (B_ϕ and B_r). In hindsight, this result reinforces the justification for neglecting the radial shear in the tachocline. There is a phase shift between the production rate and loss of about 90°, which is consistent with the loss through flux emergence being proportional to the total subsurface toroidal flux. The phase diagram is based on Eqs. (4) and (6), which confirms the concept that the toroidal flux is mainly generated by the action of latitudinal differential rotation on the poloidal flux threading the solar surface in the polar dipole field.

3.2 Poloidal Flux

The previous section has shown that the application of Stokes' theorem to the induction equation over a properly defined meridional area illuminates the essence of the generation and loss of net toroidal flux during solar activity cycles. A similar procedure can be used to determine the evolution of the net radial flux threading a hemisphere of the Sun (cf. Durrant et al. 2004; Pipin and Kosovichev 2023). To this end, we integrate the radial component of the azimuthally averaged induction equation over the photospheric surface area, A, of the Northern hemisphere, NH, and apply Stokes' theorem, which yields

$$\frac{\partial \Phi_{\text{pol,NH}}}{\partial t} = \frac{\partial}{\partial t} \oint_A B_r dA = 2\pi R_\odot \left(\langle u_r b_\theta \rangle - \langle u_\theta b_r \rangle \right) , \tag{7}$$

where the contour integral is taken over the boundary of A, i.e., the equator. The two terms on the r.h.s. correspond to observable quantities. The term $\langle u_r b_\theta \rangle$ quantifies the effect of flux emerging across the equator. This term is important when large and highly tilted active regions emerge with polarities on either side of the equator (e.g., Cameron et al. 2013). The term $\langle u_\theta b_r \rangle$ corresponds to convective motions carrying radial flux across the equator and can be approximated by a "turbulent" diffusion term, viz.

$$\langle u_\theta b_r \rangle = \eta_T R_\odot \frac{\partial B_r}{\partial \theta} . \tag{8}$$

There is a systematic component to $\partial B_r / \partial \theta$ corresponding to the systematic tilt angle of bipolar regions (see the time-latitude diagram of the radial magnetic field in Fig. 4). A random component is introduced by large, highly tilted active regions which emerge near to the equator (Cameron et al. 2013), also called "rogue active regions" (Nagy et al. 2017). Such regions, together with "Anti-Hale" active regions with reversed polarity sequence (e.g., Li 2018; Muñoz-Jaramillo et al. 2021) can introduce significant randomness into the net flux in each hemisphere and the axial dipole moment (Cameron et al. 2014; Hazra et al. 2017; Pal et al. 2023). This randomness thus carries over into the production of the toroidal flux emerging in the subsequent cycle. Jiang et al. (2015) and Whitbread et al. (2018) showed that the low amplitude of solar cycle 24 can be understood in terms of rogue active regions that emerged during cycle 23.

Figure 7 illustrates the budget of poloidal surface flux for four activity cycles on the basis of synoptic magnetograms from the Wilcox Solar Observatory. The poloidal flux is determined by integration of the longitudinally averaged radial field over the solar surface, viz.

$$\Phi_{\text{pol}} = \pi R_\odot^2 \int_0^\pi \langle B_r \rangle \text{sign}(\pi/2 - \theta) \sin\theta d\theta , \tag{9}$$

where the results for the two hemispheres are averaged.

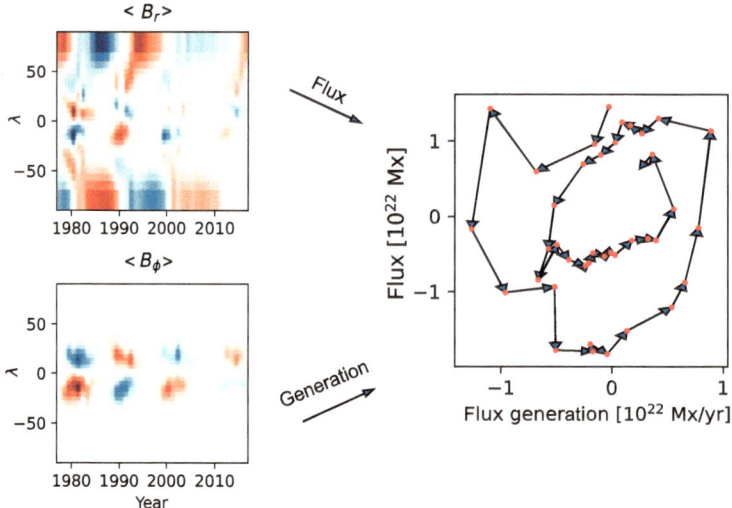

Fig. 7 Systematic part of the generation rate and amount of net poloidal flux in the course of four activity cycles. The left panels illustrate the observational input: the azimuthally averaged radial surface field (upper left panel) determines the total net poloidal surface flux (cf. Eq. (9)) while the azimuthally averaged azimuthal surface field (lower left panel) represents flux emergence in tilted bipolar regions and thus determines the rate of generation of poloidal flux via flux transported over the equator (cf. Eq. (15)). The right panel shows a phase plot of both quantities, with each red dot representing the average value for one year. The line segments with arrows represent the change from one year to the next

The generation term for the poloidal flux results from summing the contributions of all bipolar regions to the amount of magnetic flux which is carried across the equator. The contribution of a single bipolar region (i) depends on the magnetic flux of each polarity, Φ_i, the angular separation of the two polarities in latitude, $\delta_{\lambda,i}$, the latitude of emergence, λ_i, and a parameter characterizing the surface evolution of the magnetic flux near the equator, λ_R (cf. Petrovay et al. 2020). This parameter depends on the latitudinal derivative of the surface meridional flow speed, V, at the equator ($\lambda = 0$), and the turbulent diffusivity at the surface, η_T, viz.

$$\lambda_R = \sqrt{\frac{\eta_T}{R_\odot (dV/d\lambda)_{\lambda=0}}} \,. \tag{10}$$

Petrovay et al. (2020) estimated the time-integrated amount of magnetic flux crossing the equator (and thus the contribution of the bipolar region to the buildup of the poloidal field) by the leading term of a Taylor expansion as

$$\Phi_{\text{pol},i} = \frac{\Phi_i \delta_{\lambda,i}}{\sqrt{2\pi}\lambda_R} e^{-\lambda_i^2/(2\lambda_R^2)} \,. \tag{11}$$

During its emergence, the flux tube forming the bipolar region contributes to the longitudinally averaged toroidal surface field, $\langle B_\phi \rangle$, by the amount $\langle B_{\phi,i} \rangle$ according to

$$\Phi_i \frac{\delta_{\phi,i}}{2\pi} = \langle B_{\phi,i} \rangle v_{\text{em}} \Delta t \, R_\odot \, \Delta\lambda \,, \tag{12}$$

where $\delta_{\phi,i}$ is the longitudinal angular extent of the bipolar region, Δt is the time over which the emergence takes place, $\Delta\lambda$ is the latitudinal width of the leading polarity of the emerging region, and v_{em} is the mean emergence speed (assumed to be the same for all emergences). The latitudinal and longitudinal extents are related by $\delta_{\lambda,i} = \tan(\alpha_i)\cos(\lambda_i)\delta_{\phi,i}$, where α_i is tilt angle. We thus obtain

$$\Phi_i\delta_{\lambda,i} = \langle B_{\phi,i}\rangle\, v_{\text{em}}\, \Delta t\, R_\odot\, \Delta\lambda\, 2\pi\, \tan(\alpha_i)\cos(\lambda_i)\,. \tag{13}$$

Inserting this result into Eq. (11), we obtain

$$\Phi_{\text{pol},i} = \frac{\langle B_{\phi,i}\rangle\, v_{\text{em},i}\, \Delta t\, R_\odot\, \Delta\lambda\, \sqrt{2\pi}\, \tan(\alpha_i)\, \cos(\lambda_i)}{\lambda_R}\, e^{-\lambda_i^2/(2\lambda_R^2)} \tag{14}$$

For the tilt angle we consider Joy's law without scatter, so that $\alpha_i = \alpha(\lambda_i)$. We further assume that the emergence velocity is the same for all bipolar regions with a value of $v_{\text{em}} = 200\ \text{m s}^{-1}$ (Centeno 2012). Adding together the contributions of the individual bipolar regions to obtain the rate of generation of poloidal flux then amounts to the latitude integral

$$\frac{d\Phi_{\text{pol}}}{dt} = \frac{\sqrt{2\pi}\, v_{\text{em}}}{\lambda_R} \int_{-\pi/2}^{\pi/2} \langle B_\phi\rangle\, R_\odot\, \tan[\alpha(\lambda)]\, \cos(\lambda)\, e^{-\lambda^2/(2\lambda_R^2)} d\lambda \tag{15}$$

We choose the parameters as in Jiang et al. (2014a) with $\eta_T = 250\ \text{km}^2\ \text{s}^{-1}$, surface meridional flow $V(\lambda) = 11\sin(180\lambda/75)\ \text{m s}^{-1}$, and Joy's law for a cycle of intermediate strength in the form $\alpha(\lambda) = 0.7 \times 1.3\sqrt{|\lambda|}\,\text{sign}(\lambda)$.

The poloidal flux budget resulting from using the observed longitudinally averaged radial and azimuthal surface fields in Eqs. (9) and (15) is illustrated by the phase diagram on the right side of Fig. 7. It demonstrates that poloidal flux and poloidal flux generation as determined through Eqs. (9) and (15) are consistent with each other. This confirms the basic concept that the poloidal field is being generated by the joint contributions of tilted bipolar regions. Compared to the toroidal flux budget (Fig. 6), the phase diagram is somewhat less smooth. This is a consequence of the significant random component affecting the generation of poloidal flux. This component is only included in the poloidal flux but not in the rate of poloidal flux generation, where we have assumed Joy's law without scatter.

3.3 Flux Transport

The net toroidal and poloidal flux budgets discussed in the previous sections show that the observable magnetic flux at the surface plays a crucial role in the solar dynamo, highlighting the role of differential rotation, flux emergence, and Joy's law. This section briefly discusses additional processes which modifiy the spatial distribution of the fluxes. In terms of the poloidal flux, the budget shown in Fig. 7 demonstrates that flux carried across the equator is essential. Furthermore, the toroidal flux budget is dominated by the polar fields, which means that the poleward transport of flux is an important ingredient of the dynamo process: surface flux which crosses the equator is transported to the poles by a combination of the large-scale meridional flow and small-scale flows associated with convection. These processes are captured by the surface flux transport model (Jiang et al. 2014b; Yeates et al. 2023).

The butterfly diagram of flux emergence demonstrates that there is a similar requirement for the equatorward transport of the subsurface toroidal flux from latitudes of about 55°, where the toroidal flux is most efficiently produced (cf. Spruit 2011), towards the equator.

Low-latitude flux emergence in accordance with Joy's law then leads to the cross-equatorial transport of poloidal field. The nature of this transport of toroidal flux is not directly constrained, so that dynamo waves, turbulent pumping, and meridional circulation are all viable candidates. Gizon et al. (2020) showed that the speed of the deep equatorward meridional flow inferred from helioseismology is consistent with the observed migration of the activity belts if the toroidal flux is distributed over the lower half of the convection zone. This lends credibility to the concept of flux transport dynamos (recently reviewed by Hazra et al. 2023).

4 Nonlinearity, Predictability, and Long-Term Variability

4.1 Nonlinear Effects

An excited hydromagnetic dynamo leads to exponential growth of an initially weak magnetic field until further amplification becomes limited by the action of the Lorentz force, which introduces a nonlinearity into the system. The nonlinearity can affect large-scale flows (e.g, differential rotation and meridional flow), modify turbulence effects (e.g., the mean-field α-term, turbulent diffusivity, and turbulent pumping), or change the properties of flux emergence. A nonlinearity often invoked in mean-field models is "α-quenching" (Steenbeck and Krause 1969; Stix 1972), a parameterization of the decreasing efficiency of turbulence to produce poloidal field from the toroidal field as the field amplitude grows. In the case of a system with a high magnetic Reynolds number, R_m, such as the Sun, this nonlinearity is potentially catastrophic since the ratio of the small-scale field and the mean field scales as $\sqrt{R_m}$ (Cattaneo and Vainshtein 1991). This causes saturation of the dynamo at field amplitudes many orders of magnitude smaller than observed on the Sun. However, catastrophic quenching may be alleviated by removal of magnetic helicity from the system (Kleeorin et al. 2000; Hubbard and Brandenburg 2012).

In the Babcock–Leighton framework, the poloidal field generation results from the systematic tilt angle of bipolar magnetic regions. While there is observational evidence that the average tilt angle decreases with increasing cycle strength (Dasi-Espuig et al. 2010; McClintock and Norton 2013; Jiao et al. 2021) as well as with increasing maximum field strength of bipolar magnetic regions (Jha et al. 2020), catastrophic quenching is not expected in this case (Kitchatinov and Olemskoy 2011a). Sufficiently strong magnetic field can also reduce the turbulent magnetic diffusivity (e.g., Kleeorin and Rogachevskii 2007; Guerrero et al. 2009). This affects the operation of a Babcock–Leighton dynamo and may, in the case of fluctuating sources, lead to subcritical dynamo action (Vashishth et al. 2021).

Magnetically induced changes of the large-scale differential rotation (zonal flows, cf. Labonte and Howard 1982) and of the meridional circulation (systematic inflows towards active regions, cf. Gizon et al. 2001) have been observed (see also Hathaway et al. 2022). The zonal flows are probably too weak to have a substantial effect on the dynamo mechanism. The inflows towards active regions explain part of the cyclic variation of the meridional flow (Cameron and Schüssler 2012; Mahajan et al. 2023). Their effects have been studied in the framework of surface flux transport and Babcock–Leighton models (Martin-Belda and Cameron 2017; Nagy et al. 2020). It was found that, in principle, the resulting changes to the meridional flow can provide nonlinear saturation of the dynamo.

Another nonlinearity which has been studied in the Babcock–Leighton framework is 'latitudinal quenching' (Jiang 2020; Talafha et al. 2022). This effect is related to the observation that the average latitudes of sunspots are located more poleward in strong cycles as compared to weak cycles (Waldmeier 1955; Solanki et al. 2008; Hathaway 2015). Since bipolar

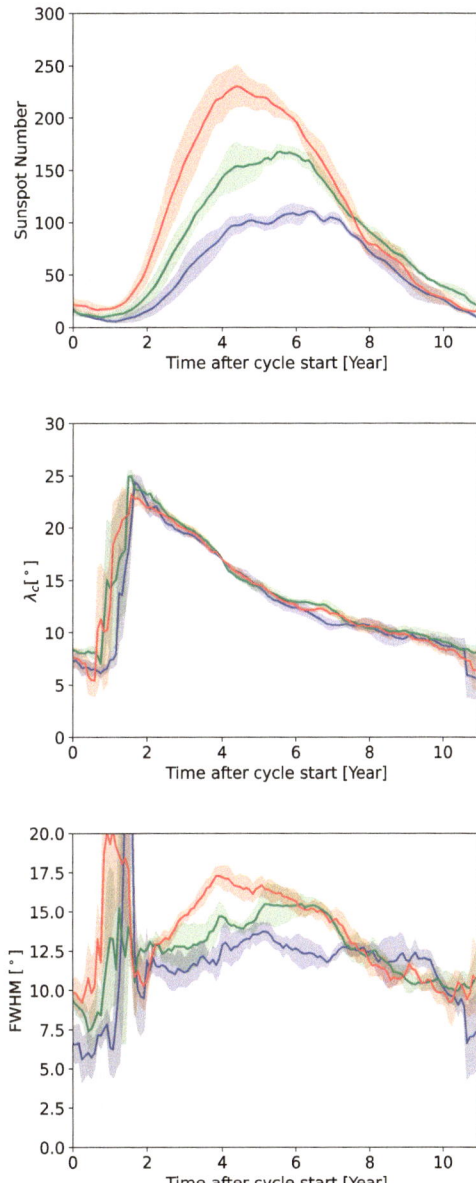

Fig. 8 Properties of the sunspot zones as function of time from cycle start, based upon the historical sunspot record. Shown are the sunspot number (top panel), central latitude (middle panel), and full width at half maximum (lower panel) of the sunspot zones. The quantities are averaged over weak cycles ($n = 12, 14, 24, 16$, blue lines), intermediate cycles ($n = 15, 20, 17, 23$, green lines), and strong cycles ($n = 18, 21, 22, 19$, red lines). Shading indicates the range covered by the ± 1 stderr, calculated using the scatter between the four cycles in each group

regions at higher latitudes contribute less to the flux crossing the equator and thus to the buildup of the poloidal field (Jiang et al. 2014a), this effect provides an amplitude-limiting nonlinearity as demonstrated by (Karak 2020) using a 3D BL dynamo model..

A closer look at the properties of flux emergence depending on cycle strength yields further insight into the nature of latitude quenching. In Fig. 8, we used 13-month smoothed sunspot sunspot numbers (Version 2) between solar cycles 12 and 24 from the SILSO data base (https://www.sidc.be/silso/datafiles) and 12-month averages of observed times, latitudes, and areas of sunspots given in the Royal Greenwich Observatory and USAF/NOAA data bases (downloaded from http://solarcyclescience.com/activeregions.html) to plot the sunspot number, the central latitude, and the full width at half maximum (FWHM) of the sunspot zones ("butterfly wings"), respectively, as functions of time since the start of a cycle. Depending on their peak sunspot number, four cycles each were put into groups of strong, medium, and weak cycles. Central latitude and FWHM of the sunspot zones were determined from Gaussian fits of the sunspot data with unsigned latitudes, thus merging both hemispheres.

Since the solar cycle is not perfectly periodic, comparing the evolution of different cycles requires a definition of a reference time for each cycle. One possibility is to fit the shape of the time evolution of the sunspot number to a given functional relationship. Using such a procedure, Hathaway (2011) found that the central latitude of the sunspot belts propagates equatorward in the same way for all cycles (see also Waldmeier 1939). We thus take the reference time as the instant at which the central latitude is at $19°$ and define the start of the cycle to be 4 years before that instant. These particular choices have no significant impact on the analysis and the results.

The figure confirms earlier results of Waldmeier (1955): (i) the activity of stronger cycles rises faster and peaks earlier than that of weaker cycles (often referred to as "Waldmeier effect") while the declining phase is independent of cycle strength, and (ii) the time profile of the propagation of the sunspot zones towards the equator is independent of cycle strength (see also Hathaway 2011). Furthermore, the full wdth at half maximum of the sunspot zones in the declining phase is also independent of cycle strength (Cameron and Schüssler 2016), while stronger cycles show broader sunspot zones (see also Mandal et al. 2017; Biswas et al. 2022). These properties reveal three aspects of latitude quenching, namely

1. The Waldmeier effect has the consequence that flux emergence around cycle maxima on average occurs in higher latitudes for stronger cycles than for weaker cycles, thus being less effective for the buildup of the polar field. This corresponds to a negative feedback.
2. The broader wings of the sunspot zones during the maximum phases of stronger cycles have the opposite effect since more flux emerges in lower latitudes.
3. The fact that all three properties, sunspot number as well as central latitude and width of the sunspot zones, behave independently of cycle strength in the declining phase of the cycles means that, during this most critical phase for the buildup of the poloidal field, the amount of magnetic flux transferred across the equator is independent of cycle strength, thus corresponding to negative feedback.

These three aspects reveal the action of an underlying nonlinearity connected to flux emergence. It is plausible that the cycle strength reflects the amount of toroidal magnetic flux in the convection zone which is available for flux emergence. In strong cycles, latitudinal differential rotation, which is steepest in mid latitudes, acts upon a stronger poloidal field and thus produces more and stronger toroidal magnetic flux. A nonlinearity matching the cycle properties discussed above could be that flux emergence occurs when a critical field strength of the order of the equipartition field strength is exceeded. While the latitude drift of the sunspot zones is independent of cycle strength (possibly being determined by a deep meridional flow towards the equator, which is largely unaffected by the magnetic field), the critical field strength is reached earlier in strong cycles, meaning that more flux emerges at higher latitudes (point 1 of the list above). At the same time, the critical field strength is exceeded in a broader range of latitudes (point 2). Consequently, stronger cycles have lost a bigger part of their available toroidal flux earlier in the cycle compared to weaker cycles, so that in the later phases flux emergence and width of the sunspot zones become independent of cycle strength. Biswas et al. (2022) confirmed this conjecture using a Babcock–Leighton

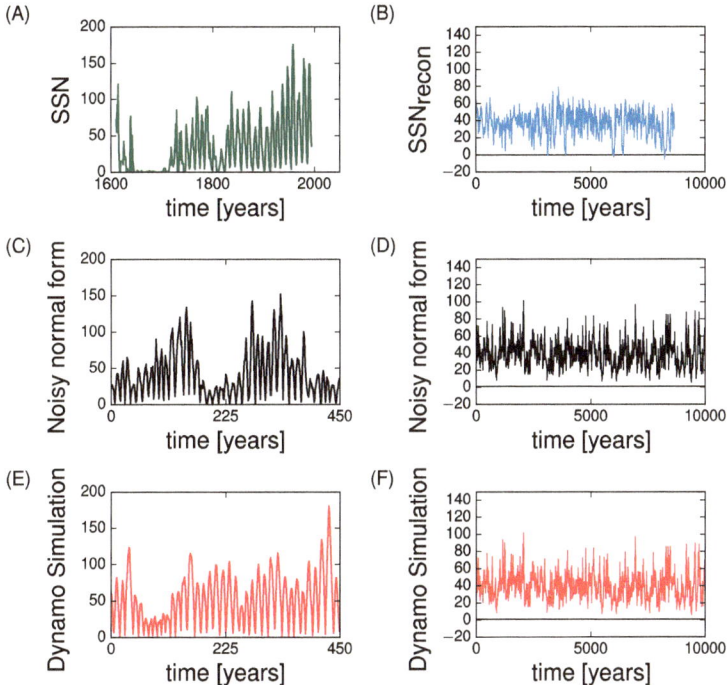

Fig. 9 Observed and simulated records of solar activity. The historical sunspot record (sunspot group number, panel A) and the level of solar activity reconstructed from cosmogenic isotopes samples (Usoskin et al. 2016) are shown in panels A and B, respectively. Panels C and D in the middle show random realizations of a normal-form model (near a Hopf bifurcation) model with noise, covering the same lengths of time as panels A and B. Panels E and F at the bottom shows realizations from a simple Babcock–Leighton dynamo model. Figure reprinted with permission from Cameron and Schüssler (2017b), copyright by AAS

flux-transport dynamo model. They suggested that, in the declining phase of all cycles, flux emergence compensates the increase of the toroidal field strength due to the pileup of flux near the stagnation point of the equatorward meridional flow. Consequently, the mean field strength remains near to the critical field strength for flux emergence and all cycles decline in the same way.

All mechanisms discussed in this section are plausible explanations for the nonlinearity limiting the amplitude of the solar dynamo. Distinguishing which combination of them acts on the Sun is an open observational challenge.

4.2 Aspects of Long-Term Variability

The sunspot cycle shows variability on a wide range of timescales (see reviews by Usoskin 2023; Biswas et al. 2023a). Panel A of Fig. 9 shows the historical sunspot record from telescopic observations since the beginning of the 17th century. It reveals significant cycle-to-cycle fluctuations of cycle strength together with longer-term variability. Particularly conspicuous is the period of very low sunspot activity between 1645 to 1715 and the period of high average sunspot activity between about 1940 and 2006. Other such "grand minima" and "grand maxima" are found in reconstructions of solar activity during the past millenia (albeit at a coarser time resolution) on the basis of various records of cosmogenic isotopes (e.g., Solanki et al. 2004; Usoskin et al. 2016). An example of such a reconstruction is shown in panel B of Fig. 9.

Observational studies of the rotational evolution (gyrochronology) of solar-type stars (e.g., Metcalfe et al. 2016; van Saders et al. 2016; David et al. 2022; Metcalfe et al. 2022, 2023) indicate that magnetic braking by a stellar wind is significantly reduced for stars near or beyond the solar age, which suggests a decline of their large-scale magnetic fields (for a different view, see Kotorashvili et al. 2023). This indicates that, at its current rotation rate, the Sun could be approaching a transition point where its gobal dynamo switches off, so that at present the excitation of the solar dynamo is only weakly supercritical (or even subcritical, e.g., Karak et al. 2015; Tripathi et al. 2021; Vashishth et al. 2021). Weak excitation of the solar dynamo is also supported by the existence of grand minima as shown by the activity records since highly supercritical dynamos do not exhibit such episodes (e.g., Kitchatinov and Olemskoy 2010; Cameron and Schüssler 2017b; Vashishth et al. 2023).

The observationally well-studied rotation-activity relation for magnetically active stars (e.g., Brun and Browning 2017) suggests the rotation rate as the relevant control parameter for dynamo excitation. For most dynamo models, the transition from decaying field to excited oscillatory dynamo action corresponds to a supercritical Hopf bifurcation (e.g., Tobias et al. 1995), when a fixed point (equilibrium) spawns a limit cycle (oscillatory solution). The behaviour of a weakly excited nonlinear system near a Hopf bifurcation is fully described by a generic normal-form model that is independent of the specific properties of the system, including also the nature of its nonlinearity (e.g., Guckenheimer and Holmes 1983).

Cameron and Schüssler (2017b) applied this concept to the solar dynamo and showed that the observed power spectrum of the solar cycle (from the sunspot record and from the reconstruction based on cosmogenic isotopes) is consistent with a noisy normal-form model whose parameters are completely determined by observations. The noise in the system results from the large scatter in the observed tilt angles of active regions, which is possibly associated with the interaction of magnetic flux and convective motions (Longcope and Fisher 1996). As discussed in Sects. 3.1 and 3.2, the tilt angle scatter leads to randomness in the amount of flux transported across the equator, and hence to randomness in the amount of toroidal field generated for the subsequent cycle.

Panels C and D of Fig. 9 show an example realization of the normal-form model covering 10,000 years. Panel B of Fig. 10 gives the corresponding temporal power spectrum, which is consistent with the observed spectrum shown in panel A of Fig. 10. Moreover, the model also exhibits extended periods of low activity (grand minima) whose statistical properties in terms of the distributions of their lengths and the waiting times between grand minima is consistent with the reconstructed record of solar activity by Usoskin et al. (2016).

Likewise, the updated 1D Babcock–Leighton dynamo model of Cameron and Schüssler (2017a) with slightly supercritical excitation and including noise in the source term for the poloidal field also yields time series that are statistically similar to the empirical series (see panels E and F of Fig. 9). The corresponding power spectra in panels C and D of Fig. 10) show a good match to the observed spectra (panel A of Fig. 10). A similar approach was taken by Kitchatinov and Nepomnyashchikh (2017) and Kitchatinov et al. (2018), who studied a more comprehensive dynamo model.

Cameron and Schüssler (2019) carried out a detailed analysis of power pectra obtained with the generic noisy normal-form model (incorporating only the 11/22 year base period) in order to evaluate the statistical significance of periodicities inferred from the record of reconstructed solar activity on the basis of cosmogenic isotopes. They showed that power spectra from realizations covering 10,000 years of simulated time (matching the length of the reconstructed solar data) typically exhibit spectral peaks at various periods that are qualitatively similar to those found in the solar data. Such peaks, which result from the stochastic

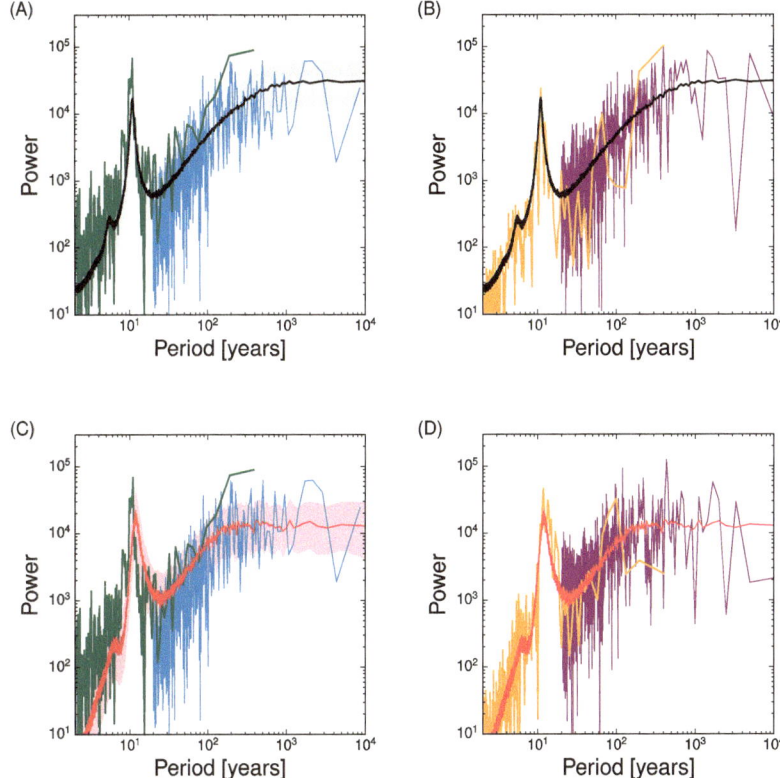

Fig. 10 Power specta of empirical solar data and models. (A) Sunspot group number (green), and reconstruction from cosmogenic isotopes (blue), together with the expectation value from 10.000 realizations of the normal form model (black curve) and the corresponding 25% to 75% quartiles (grey shading). (B) One realization of the normal form model. (C) Same as (A), but compared to the expectation value and quartiles from realizations of an updated Babcock–Leighton model (red curve and pink shading). (D) One relization of the updated Babcock–Leighton model. Figure reprinted with permission from Cameron and Schüssler (2017b), copyright by AAS

noise in the dynamo excitation, can reach significance levels of 3σ. These results cast doubt on the proposition that seemingly significant periodicities such as the 90-year Gleissberg and the 210-year de Vries "cycles" are intrinsic to the solar dynamo and not just statistical fluctuations. In fact, the sharpness of the corresponding peaks in the power spectrum indicates a random origin since spectral peaks representing intrinsic dynamo periodicities tend to be broadened owing to the damping inherent to the dynamo process.

Apart from stochastic forcing as discussed above, long-term variability in the dynamo process can also arise from magnetic feedback on the flow and from time delays in the dynamo process. Comprehensive reviews of models for the long-term variability are provided by Karak (2023) and Biswas et al. (2023a).

Let us finally briefly discuss the operation of the dynamo during grand minima and the recovery from such episodes. In both, the noisy normal form model and dynamo simulations shown in Fig. 9, dynamo excitation is weakly supercritical, so that the system can transition into grand-minimum phases by the observed level of stochastic noise. During such weak-field episodes, the system is essentially kinematic and the dynamo recovers on the kinematic growth time. Such a scenario is supported by cosmogenic isotope data showing continued 11-year periodicity during the main phase of the Maunder minimum between the years 1645 to 1700 (Beer et al. 1998). Alternatively, the dynamo excitation may require a minimum field amplitude. Dynamo action would then switch off once the threshold is fallen below, but can be restarted by sufficiently strong stochastic fluctuations, thus leading to on-off intermittency (e.g., Schmitt et al. 1996).

In the BL framework, both possibilities could arise. If the systematic tilt angle of bipolar magnetic regions (Joy's law), which is essential for the BL dynamo process, is restricted to active regions with a magnetic flux exceeding, say, 10^{20} Mx, this could involve a threshold effect if the underlying flux loops originate from an undular instability of toroidal field stored in the overshoot region. Studies in the framework of the thin-flux-tube approximation (see review by Fan 2021) show that this instability requires that the field strength exceeds a critical value for the formation of buoyantly rising loops. These loops then acquire a systematic tilt by the action of the Coriolis force. This could mean that the BL dynamo switches off when its amplitude falls below a threshold value, because the emergence of tilted active regions then ceases. The dynamo would then not work in the kinematic regime and would need an additional process (e.g., a distributed turbulent α-effect) in order to restart (e.g., Hazra et al. 2014).

On the other hand, if the small ephemeral regions, which have a statistical tendency to obey Hale's law (Martin and Harvey 1979; Harvey 1992), should also have a tendency to obey Joy's law, then the BL dynamo can operate during grand minima when sunspots and big active regions are absent. In fact, there is an observed phase shift between sunspot activity and cosmogenic isotope concentration during grand minima: the two are in phase except during grand minima when they are out of phase (Beer et al. 1998; Usoskin et al. 2001). Wang and Sheeley (2013) argued that this shift is because the axial dipole moment dominates the signal in cosmogenic isotopes during periods of low activity while the equatorial dipole moment (reflecting low-latitude active region emergences) dominates outside grand minima. This indicates that even during grand minima, the polar fields were behaving in a manner similar that of the BL dynamo, albeit based on ephemeral regions rather than on larger active regions.

4.3 Notes on Predictability

Based on sufficient understanding of the global solar dynamo, the observed state of the solar magnetic field can be used to predict its future evolution (for comprehensive reviews, see Petrovay 2020, Dhowmik et al. 2023). In the Babcock–Leighton model, the cycle-to-cycle variability is directly related to the amount of magnetic flux that gets across the equator. At the end of a cycle, all this flux ends up in the polar regions and the amplitude of the polar field is strongly correlated with the strength of the subsequent cycle (e.g., Kumar et al. 2021). The physical basis for this correlation is explained in Sect. 3.1. Predicting the strength of a cycle thus becomes a matter of predicting the polar field strength at the end of a cycle (Schatten et al. 1978), which is almost equivalent to predicting how much flux is transported across the equator (see Sect. 3.2). This can be achieved by performing surface-flux-transport simulations using the actual observations of latitude, magnetic flux, and tilt angle of each emerging active region. Such simulations rather accurately reproduce the amount of flux in each hemisphere and the resulting axial dipole moment as a predictor for the subsequent cycle (Jiang et al. 2015; Yeates et al. 2023). A prediction for the next cycle during an ongoing cycle (i.e., before all the active regions of the cycle have emerged) can be made by including all active regions which have been observed and performing Monte-Carlo simulations of the effect of the active regions which have not yet emerged, using their average properties (Cameron et al. 2016). The result of this approach is the prediction of the polar field at the end of the cycle with error estimates. These can then be converted into predictions for the strength of the subsequent cycle.

In any case, all predictions are limited by the inherent randomness of the dynamo process (Jiang et al. 2018; Kitchatinov et al. 2018). For instance, individual "rogue active regions" can have a strong effect and may potentially even shut down the large-scale dynamo and initiate a grand minimum episode (Nagy et al. 2017).

5 Outlook

Considering the enormous range of scales and the complexity of physical processes governing the interaction of turbulent convection, differential rotation, and magnetic field in the solar convection zone, it is rather surprising that a comparatively simple approach such as the Babcock–Leighton scenario seems to provide such a successful description of the solar dynamo process. An important factor here is that key ingredients of the model, such as the properties of bipolar magnetic regions (latitude range, tilt, flux distribution, etc.) and surface flows, can be obtained by observations. To a large extent, such observations are not available for other stars, so that basic input is missing for the application of the model. Furthermore, the underlying processes leading to flux emergence, i.e., the roles of convective flows, magnetic buoyancy, and instabilities deep in the convection zone, are largely not understood. In the absence of direct observational evidence, 3D MHD simulations seem to be the only possibility here. Considering the impressive progress that simulations have seen during the last decade (recently reviewed by Käpylä et al. 2023), there is hope that the complexity of these processes will be better understood in the not-too-far future, providing an even better basis for simplified models such as the Babcock–Leighton approach. Still, we need to exercise caution and be aware of the severe limitations of all our efforts to understand the solar dynamo as lucidly expounded by Parker (2009) and Spruit (2011, 2012).

Funding Open Access funding enabled and organized by Projekt DEAL.

Data Availability Sunspot data were obtained from the World Data Center SILSO, Royal Observatory of Belgium, Brussels. The data for the aa-index were obtained from https://isgi.unistra.fr/indices_aa.php.

Declarations

References

Babcock HW (1961) The topology of the Sun's magnetic field and the 22-year cycle. Astrophys J 133:572. https://doi.org/10.1086/147060

Babcock HW, Babcock HD (1955) The Sun's magnetic field, 1952–1954. Astrophys J 121:349. https://doi.org/10.1086/145994

Barnes JA, Tryon PV, Sargent HH (1980) Sunspot cycle simulation using random noise. In: Pepin RO, Eddy JA, Merrill RB (eds) The Ancient Sun: fossil record in the Earth, moon and meteorites, pp 159–163

Basu S, Antia HM (2019) Changes in solar rotation over two solar cycles. Astrophys J 883(1):93. https://doi.org/10.3847/1538-4357/ab3b57

Baumann I, Schmitt D, Schüssler M et al (2004) Evolution of the large-scale magnetic field on the solar surface: a parameter study. Astron Astrophys 426:1075–1091. https://doi.org/10.1051/0004-6361:20048024

Beer J, Tobias S, Weiss N (1998) An active Sun throughout the Maunder minimum. Sol Phys 181(1):237–249. https://doi.org/10.1023/A:1005026001784

Bekki Y, Cameron RH (2023) Three-dimensional non-kinematic simulation of post-emergence evolution of bipolar magnetic regions and Babcock-Leighton dynamo of the Sun. Astron Astrophys 670:A101. https://doi.org/10.1051/0004-6361/202244990

Bhowmik P, Nandy D (2018) Prediction of the strength and timing of sunspot cycle 25 reveal decadal-scale space environmental conditions. Nat Commun 9:5209. https://doi.org/10.1038/s41467-018-07690-0

Bhowmik P, Jiang J, Upton L et al (2023) Physical models for solar cycle predictions. Space Sci Rev 219:40. https://doi.org/10.1007/s11214-023-00983-x.

Bice CP, Toomre J (2023) Effects of full-sphere convection on M-dwarf dynamo action, flux emergence, and spin-down. Astrophys J 951(1):79. https://doi.org/10.3847/1538-4357/acd2db

Bieber JW, Rust DM (1995) The escape of magnetic flux from the Sun. Astrophys J 453:911. https://doi.org/10.1086/176451

Birch AC, Schunker H, Braun DC et al (2016) A low upper limit on the subsurface rise speed of solar active regions. Sci Adv 2(7):e1600,557–e1600,557. https://doi.org/10.1126/sciadv.1600557

Biswas A, Karak BB, Cameron R (2022) Toroidal flux loss due to flux emergence explains why solar cycles rise differently but decay in a similar way. Phys Rev Lett 129(24):241102. https://doi.org/10.1103/PhysRevLett.129.241102

Biswas A, Karak BB, Kumar P (2023b) Exploring the reliability of polar field rise rate as a precursor for an early prediction of solar cycle. arXiv e-prints arXiv:2308.01155

Biswas A, Karak B, Usoskin I et al (2023a) Long-term modulation of solar cycles. Space Sci Rev 219:19. https://doi.org/10.1007/s11214-023-00968-w.

Brandenburg A (2005) The case for a distributed solar dynamo shaped by near-surface shear. Astrophys J 625(1):539–547. https://doi.org/10.1086/429584

Brandenburg A, Elstner D, Masada Y et al (2023) Turbulent processes and mean-field dynamo. Space Sci Rev 219:55. https://doi.org/10.1007/s11214-023-00999-3

Brown TM, Christensen-Dalsgaard J, Dziembowski WA et al (1989) Inferring the Sun's internal angular velocity from observed p-mode frequency splittings. Astrophys J 343:526. https://doi.org/10.1086/167727

Brun AS, Browning MK (2017) Magnetism, dynamo action and the solar-stellar connection. Living Rev Sol Phys 14:4. https://doi.org/10.1007/s41116-017-0007-8

Caligari P, Moreno-Insertis F, Schüssler M (1995) Emerging flux tubes in the solar convection zone. I. Asymmetry, tilt, and emergence latitude. Astrophys J 441:886. https://doi.org/10.1086/175410

Cameron RH, Schüssler M (2012) Are the strengths of solar cycles determined by converging flows towards the activity belts? Astron Astrophys 548:A57. https://doi.org/10.1051/0004-6361/201219914

Cameron R, Schüssler M (2015) The crucial role of surface magnetic fields for the solar dynamo. Science 347(6228):1333–1335. https://doi.org/10.1126/science.1261470

Cameron RH, Schüssler M (2016) The turbulent diffusion of toroidal magnetic flux as inferred from properties of the sunspot butterfly diagram. Astron Astrophys 591:A46. https://doi.org/10.1051/0004-6361/201527284

Cameron RH, Schüssler M (2017a) An update of Leighton's solar dynamo model. Astron Astrophys 599:A52. https://doi.org/10.1051/0004-6361/201629746

Cameron RH, Schüssler M (2017b) Understanding solar cycle variability. Astrophys J 843(2):111. https://doi.org/10.3847/1538-4357/aa767a

Cameron RH, Schüssler M (2019) Solar activity: periodicities beyond 11 years are consistent with random forcing. Astron Astrophys 625:A28. https://doi.org/10.1051/0004-6361/201935290

Cameron RH, Schüssler M (2020) Loss of toroidal magnetic flux by emergence of bipolar magnetic regions. Astron Astrophys 636:A7. https://doi.org/10.1051/0004-6361/201937281

Cameron RH, Schmitt D, Jiang J et al (2012) Surface flux evolution constraints for flux transport dynamos. Astron Astrophys 542:A127. https://doi.org/10.1051/0004-6361/201218906

Cameron RH, Dasi-Espuig M, Jiang J et al (2013) Limits to solar cycle predictability: cross-equatorial flux plumes. Astron Astrophys 557:A141. https://doi.org/10.1051/0004-6361/201321981

Cameron RH, Jiang J, Schüssler M et al (2014) Physical causes of solar cycle amplitude variability. J Geophys Res 119(2):680–688. https://doi.org/10.1002/2013JA019498

Cameron RH, Jiang J, Schüssler M (2016) Solar cycle 25: another moderate cycle? Astrophys J Lett 823(2):L22. https://doi.org/10.3847/2041-8205/823/2/L22

Cameron RH, Duvall TL, Schüssler M et al (2018) Observing and modeling the poloidal and toroidal fields of the solar dynamo. Astron Astrophys 609:A56. https://doi.org/10.1051/0004-6361/201731481

Cattaneo F, Vainshtein SI (1991) Suppression of turbulent transport by a weak magnetic field. Astrophys J Lett 376:L21. https://doi.org/10.1086/186093

Centeno R (2012) The naked emergence of solar active regions observed with SDO/HMI. Astrophys J 759(1):72. https://doi.org/10.1088/0004-637X/759/1/72

Charbonneau P (2020) Dynamo models of the solar cycle. Living Rev Sol Phys 17(1):4. https://doi.org/10.1007/s41116-020-00025-6

Charbonneau P, MacGregor KB (1997) Solar interface dynamos. II. Linear, kinematic models in spherical geometry. Astrophys J 486(1):502–520. https://doi.org/10.1086/304485

Chatterjee P, Nandy D, Choudhuri AR (2004) Full-sphere simulations of a circulation-dominated solar dynamo: exploring the parity issue. Astron Astrophys 427:1019–1030. https://doi.org/10.1051/0004-6361:20041199

Chen F, Rempel M, Fan Y (2017) Emergence of magnetic flux generated in a solar convective dynamo. I. The formation of sunspots and active regions, and the origin of their asymmetries. Astrophys J 846(2):149. https://doi.org/10.3847/1538-4357/aa85a0

Choudhuri AR, Gilman PA (1987) The influence of the Coriolis force on flux tubes rising through the solar convection zone. Astrophys J 316:788. https://doi.org/10.1086/165243

Choudhuri AR, Hazra G (2016) The treatment of magnetic buoyancy in flux transport dynamo models. Adv Space Res 58(8):1560–1570. https://doi.org/10.1016/j.asr.2016.03.015

Choudhuri AR, Schüssler M, Dikpati M (1995) The solar dynamo with meridional circulation. Astron Astrophys 303:L29

Christensen-Dalsgaard J, Monteiro MJPFG, Rempel M et al (2011) A more realistic representation of overshoot at the base of the solar convective envelope as seen by helioseismology. Mon Not R Astron Soc 414(2):1158–1174. https://doi.org/10.1111/j.1365-2966.2011.18460.x

Cowling TG (1953) Solar electrodynamics. In: Kuiper GP (ed) The Sun. University of Chicago Press, Chicago, p 532

Dasi-Espuig M, Solanki SK, Krivova NA et al (2010) Sunspot group tilt angles and the strength of the solar cycle. Astron Astrophys 518:A7. https://doi.org/10.1051/0004-6361/201014301

David TJ, Angus R, Curtis JL et al (2022) Further evidence of modified spin-down in Sun-like stars: pileups in the temperature-period distribution. Astrophys J 933(1):114. https://doi.org/10.3847/1538-4357/ac6dd3

DeVore CR, Boris JP, Sheeley NR (1984) The concentration of the large-scale solar magnetic field by a meridional surface flow. Sol Phys 92(1–2):1–14. https://doi.org/10.1007/BF00157230

Dikpati M, Charbonneau P (1999) A Babcock-Leighton flux transport dynamo with solar-like differential rotation. Astrophys J 518(1):508–520. https://doi.org/10.1086/307269

D'Silva S, Choudhuri AR (1993) A theoretical model for tilts of bipolar magnetic regions. Astron Astrophys 272:621

Durney BR (1995) On a Babcock-Leighton dynamo model with a deep-seated generating layer for the toroidal magnetic field. Sol Phys 160(2):213–235. https://doi.org/10.1007/BF00732805

Durney BR (1997) On a Babcock-Leighton solar dynamo model with a deep-seated generating layer for the toroidal magnetic field. IV. Astrophys J 486(2):1065–1077. https://doi.org/10.1086/304546

Durrant CJ, Turner JPR, Wilson PR (2004) The mechanism involved in the reversals of the Sun's polar magnetic fields. Sol Phys 222(2):345–362. https://doi.org/10.1023/B:SOLA.0000043577.33961.82

Duvall JTL (1979) Large-scale solar velocity fields. Sol Phys 63(1):3–15. https://doi.org/10.1007/BF00155690

Fan Y (2021) Magnetic fields in the solar convection zone. Living Rev Sol Phys 18(1):5. https://doi.org/10.1007/s41116-021-00031-2

Fan Y, Fang F (2014) A simulation of convective dynamo in the solar convective envelope: maintenance of the solar-like differential rotation and emerging flux. Astrophys J 789(1):35. https://doi.org/10.1088/0004-637X/789/1/35

Fan Y, Fisher GH, McClymont AN (1994) Dynamics of emerging active region flux loops. Astrophys J 436:907. https://doi.org/10.1086/174967

Ferriz-Mas A, Schüssler M (1993) Instabilities of magnetic flux tubes in a stellar convection zone I. Equatorial flux rings in differentially rotating stars. Geophys Astrophys Fluid Dyn 72(1):209–247. https://doi.org/10.1080/03091929308203613

Ferriz-Mas A, Schüssler M (1995) Instabilities of magnetic flux tubes in a stellar convection zone II. Flux rings outside the equatorial plane. Geophys Astrophys Fluid Dyn 81(3):233–265

Gaizauskas V, Harvey KL, Harvey JW et al (1983) Large-scale patterns formed by solar active regions during the ascending phase of cycle 21. Astrophys J 265:1056–1065. https://doi.org/10.1086/160747

Galloway DJ, Weiss NO (1981) Convection and magnetic fields in stars. Astrophys J 243:945–953. https://doi.org/10.1086/158659

Gilman PA, Miesch MS (2004) Limits to penetration of meridional circulation below the solar convection zone. Astrophys J 611(1):568–574. https://doi.org/10.1086/421899

Giovanelli RG (1985) The sunspot cycle and solar magnetic fields. I – The mechanism as inferred from observation. II – The interaction of flux tubes with the convection zone. Aust J Phys 38:1045–1089. https://doi.org/10.1071/PH851045

Gizon L, Duvall JTL, Larsen RM (2001) Probing surface flows and magnetic activity with time-distance helioseismology. In: Brekke P, Fleck B, Gurman JB (eds) Recent insights into the physics of the Sun and heliosphere: highlights from SOHO and other space mission. Astronomical Society of the Pacific, p 189

Gizon L, Cameron R, Pourabdian M et al (2020) Meridional flow in the Sun's convection zone is a single cell in each hemisphere. Science 368(6498):1469–1472. https://doi.org/10.1126/science.aaz7119

Gottschling N, Schunker H, Birch AC et al (2021) Evolution of solar surface inflows around emerging active regions. Astron Astrophys 652:A148. https://doi.org/10.1051/0004-6361/202140324

Guckenheimer J, Holmes P (1983) Nonlinear oscillations, dynamical systems, and bifurcations of vector fields. Applied mathematical sciences. Springer, New York

Guerrero G, de Gouveia Dal Pino EM (2007) How does the shape and thickness of the tachocline affect the distribution of the toroidal magnetic fields in the solar dynamo? Astron Astrophys 464(1):341–349. https://doi.org/10.1051/0004-6361:20065834

Guerrero G, Dikpati M, de Gouveia Dal Pino EM (2009) The role of diffusivity quenching in flux-transport dynamo models. Astrophys J 701(1):725–736. https://doi.org/10.1088/0004-637X/701/1/725

Hale GE, Ellerman F, Nicholson SB et al (1919) The magnetic polarity of Sun-spots. Astrophys J 49:153. https://doi.org/10.1086/142452

Hanasoge SM (2022) Surface and interior meridional circulation in the Sun. Living Rev Sol Phys 19(1):3. https://doi.org/10.1007/s41116-022-00034-7

Harvey KL (1992) The cyclic behavior of solar activity. In: Harvey KL (ed) The solar cycle. ASP Conference Series, vol 27. Astronomical Society of the Pacific, San Francisco, p 335

Harvey KL (1993) Magnetic bipoles on the Sun. PhD thesis, University of Utrecht, Netherlands

Harvey KL, Harvey JW, Martin SF (1975) Ephemeral active regions in 1970 and 1973. Sol Phys 40(1):87–102. https://doi.org/10.1007/BF00183154

Hathaway DH (2011) A standard law for the equatorward drift of the sunspot zones. Sol Phys 273(1):221–230. https://doi.org/10.1007/s11207-011-9837-z

Hathaway DH (2015) The solar cycle. Living Rev Sol Phys 12(1):4. https://doi.org/10.1007/lrsp-2015-4

Hathaway DH, Upton LA (2016) Predicting the amplitude and hemispheric asymmetry of solar cycle 25 with surface flux transport. J Geophys Res Space Phys 121(11):10,744–10,753. https://doi.org/10.1002/2016JA023190

Hathaway DH, Upton LA, Mahajan SS (2022) Variations in differential rotation and meridional flow within the Sun's surface shear layer 1996–2022. Front Astron Space Sci 9:419. https://doi.org/10.3389/fspas.2022.1007290

Hazra G (2021) Recent advances in the 3D kinematic Babcock-Leighton solar dynamo modeling. J Astrophys Astron 42(2):22. https://doi.org/10.1007/s12036-021-09738-y

Hazra G, Miesch MS (2018) Incorporating surface convection into a 3D Babcock-Leighton solar dynamo model. Astrophys J 864(2):110. https://doi.org/10.3847/1538-4357/aad556

Hazra S, Passos D, Nandy D (2014) A stochastically forced time delay solar dynamo model: self-consistent recovery from a Maunder-like grand minimum necessitates a mean-field alpha effect. Astrophys J 789(1):5. https://doi.org/10.1088/0004-637X/789/1/5

Hazra G, Choudhuri AR, Miesch MS (2017) A theoretical study of the build-up of the sun's polar magnetic field by using a 3D kinematic dynamo model. Astrophys J 835(1):39. https://doi.org/10.3847/1538-4357/835/1/39

Hazra G, Jiang J, Karak BB et al (2019) Exploring the cycle period and parity of stellar magnetic activity with dynamo modeling. Astrophys J 884(1):35. https://doi.org/10.3847/1538-4357/ab4128

Hazra G, Nandy D, Kitchatinov L et al (2023) Mean field models of flux transport dynamo and meridional circulation in the Sun and stars. Space Sci Rev 219:39. https://doi.org/10.1007/s11214-023-00982-y

Hotta H (2017) Solar overshoot region and small-scale dynamo with realistic energy flux. Astrophys J 843(1):52. https://doi.org/10.3847/1538-4357/aa784b

Howard R (1979) Evidence for large-scale velocity features on the sun. Astrophys J Lett 228:L45–L50. https://doi.org/10.1086/182900

Howard RF (1996) Solar active regions as diagnostics of subsurface conditions. Annu Rev Astron Astrophys 34:75–110. https://doi.org/10.1146/annurev.astro.34.1.75

Howe R (2009) Solar interior rotation and its variation. Living Rev Sol Phys 6(1):1. https://doi.org/10.12942/lrsp-2009-1

Hubbard A, Brandenburg A (2012) Catastrophic quenching in $\alpha\Omega$ dynamos revisited. Astrophys J 748(1):51. https://doi.org/10.1088/0004-637X/748/1/51

Işik E, van Saders J, Reiners R, Metcalfe T (2023) Scaling and evolution of stellar magnetic activity. Space Sci Rev 219

Jeffers SV, Cameron RH, Marsden SC et al (2022) The crucial role of surface magnetic fields for stellar dynamos: ϵ Eridani, 61 Cygni A, and the Sun. Astron Astrophys 661:A152. https://doi.org/10.1051/0004-6361/202142202

Jha BK, Karak BB, Mandal S et al (2020) Magnetic field dependence of bipolar magnetic region tilts on the Sun: indication of tilt quenching. Astrophys J Lett 889(1):L19. https://doi.org/10.3847/2041-8213/ab665c

Jiang J (2020) Nonlinear mechanisms that regulate the solar cycle amplitude. Astrophys J 900(1):19. https://doi.org/10.3847/1538-4357/abaa4b

Jiang J, Cameron RH, Schüssler M (2014a) Effects of the scatter in sunspot group tilt angles on the large-scale magnetic field at the solar surface. Astrophys J 791(1):5. https://doi.org/10.1088/0004-637X/791/1/5

Jiang J, Hathaway DH, Cameron RH et al (2014b) Magnetic flux transport at the solar surface. Space Sci Rev 186(1–4):491–523. https://doi.org/10.1007/s11214-014-0083-1

Jiang J, Cameron RH, Schüssler M (2015) The cause of the weak solar cycle 24. Astrophys J Lett 808(1):L28. https://doi.org/10.1088/2041-8205/808/1/L28

Jiang J, Wang JX, Jiao QR et al (2018) Predictability of the solar cycle over one cycle. Astrophys J 863(2):159. https://doi.org/10.3847/1538-4357/aad197

Jiao Q, Jiang J, Wang ZF (2021) Sunspot tilt angles revisited: dependence on the solar cycle strength. Astron Astrophys 653:A27. https://doi.org/10.1051/0004-6361/202141215

Käpylä PJ, Browning MK, Brun AS et al (2023) Simulations of solar and stellar dynamos and their theoretical interpretation. Space Sci Rev 219:58. https://doi.org/10.1007/s11214-023-01005-6 arXiv:2305.16790

Karak BB (2020) Dynamo saturation through the latitudinal variation of bipolar magnetic regions in the Sun. Astrophys J Lett 901(2):L35. https://doi.org/10.3847/2041-8213/abb93f

Karak B (2023) Models for the long-term variations of solar activity. Living Rev Sol Phys 20:3. https://doi.org/10.1007/s41116-023-00037-y

Karak BB, Cameron R (2016) Babcock-Leighton solar dynamo: the role of downward pumping and the equatorward propagation of activity. Astrophys J 832(1):94. https://doi.org/10.3847/0004-637X/832/1/94

Karak BB, Miesch M (2017) Solar cycle variability induced by tilt angle scatter in a Babcock-Leighton solar dynamo model. Astrophys J 847(1):69. https://doi.org/10.3847/1538-4357/aa8636

Karak BB, Miesch M (2018) Recovery from Maunder-like grand minima in a Babcock-Leighton solar dynamo model. Astrophys J Lett 860(2):L26. https://doi.org/10.3847/2041-8213/aaca97

Karak BB, Kitchatinov LL, Brandenburg A (2015) Hysteresis between distinct modes of turbulent dynamos. Astrophys J 803(2):95. https://doi.org/10.1088/0004-637X/803/2/95

Kitchatinov L, Nepomnyashchikh A (2017) How supercritical are stellar dynamos, or why do old main-sequence dwarfs not obey gyrochronology? Mon Not R Astron Soc 470(3):3124–3130. https://doi.org/10.1093/mnras/stx1473

Kitchatinov LL, Olemskoy SV (2010) Dynamo hysteresis and grand minima of solar activity. Astron Lett 36(4):292–296. https://doi.org/10.1134/S1063773710040079

Kitchatinov LL, Olemskoy SV (2011a) Alleviation of catastrophic quenching in solar dynamo model with nonlocal alpha-effect. Astron Nachr 332(5):496. https://doi.org/10.1002/asna.201011549

Kitchatinov LL, Olemskoy SV (2011b) Does the Babcock-Leighton mechanism operate on the Sun? Astron Lett 37(9):656–658. https://doi.org/10.1134/S0320010811080031

Kitchatinov LL, Mordvinov AV, Nepomnyashchikh AA (2018) Modelling variability of solar activity cycles. Astron Astrophys 615:A38. https://doi.org/10.1051/0004-6361/201732549

Kleeorin N, Rogachevskii I (2007) Nonlinear turbulent magnetic diffusion and effective drift velocity of a large-scale magnetic field in two-dimensional magnetohydrodynamic turbulence. Phys Rev E 75(6):066315. https://doi.org/10.1103/PhysRevE.75.066315

Kleeorin N, Moss D, Rogachevskii I et al (2000) Helicity balance and steady-state strength of the dynamo generated galactic magnetic field. Astron Astrophys 361:L5–L8. arXiv:astro-ph/0205266

Köhler H (1973) The solar dynamo and estimate of the magnetic diffusivity and the α-effect. Astron Astrophys 25:467

Kotorashvili K, Blackman EG, Owen JE (2023) Why the observed spin evolution of older-than-solar like stars might not require a dynamo mode change. Mon Not R Astron Soc 522(1):1583–1590. https://doi.org/10.1093/mnras/stad981

Kumar R, Jouve L, Nandy D (2019) A 3D kinematic Babcock Leighton solar dynamo model sustained by dynamic magnetic buoyancy and flux transport processes. Astron Astrophys 623:A54. https://doi.org/10.1051/0004-6361/201834705

Kumar P, Nagy M, Lemerle A et al (2021) The polar precursor method for solar cycle prediction: comparison of predictors and their temporal range. Astrophys J 909(1):87. https://doi.org/10.3847/1538-4357/abdbb4

Kumar P, Biswas A, Karak BB (2022) Physical link of the polar field buildup with the Waldmeier effect broadens the scope of early solar cycle prediction: Cycle 25 is likely to be slightly stronger than Cycle 24. Mon Not R Astron Soc 513(1):L112–L116. https://doi.org/10.1093/mnrasl/slac043

Labonte BJ, Howard R (1982) Torsional waves on the Sun and the activity cycle. Sol Phys 75(1–2):161–178. https://doi.org/10.1007/BF00153469

Larson TP, Schou J (2018) Global-mode analysis of full-disk data from the Michelson Doppler imager and the helioseismic and magnetic imager. Sol Phys 293(2):29. https://doi.org/10.1007/s11207-017-1201-5

Legrand JP, Simon PA (1981) Ten cycles of solar and geomagnetic activity. Sol Phys 70(1):173–195. https://doi.org/10.1007/BF00154399

Leighton RB (1964) Transport of magnetic fields on the Sun. Astrophys J 140:1547. https://doi.org/10.1086/148058

Leighton RB (1969) A magneto-kinematic model of the solar cycle. Astrophys J 156:1. https://doi.org/10.1086/149943

Lemerle A, Charbonneau P (2017) A coupled 2 × 2D Babcock-Leighton solar dynamo model. II. Reference dynamo solutions. Astrophys J 834(2):133. https://doi.org/10.3847/1538-4357/834/2/133

Lemerle A, Charbonneau P, Carignan-Dugas A (2015) A coupled 2 × 2D Babcock-Leighton solar dynamo model. I. Surface magnetic flux evolution. Astrophys J 810(1):78. https://doi.org/10.1088/0004-637X/810/1/78

Li J (2018) A systematic study of Hale and anti-Hale sunspot physical parameters. Astrophys J 867(2):89. https://doi.org/10.3847/1538-4357/aae31a

Liu AL, Scherrer PH (2022) Solar toroidal field evolution spanning four sunspot cycles seen by the Wilcox Solar Observatory, the Solar and Heliospheric Observatory/Michelson Doppler Imager, and the Solar Dynamics Observatory/Helioseismic and Magnetic Imager. Astrophys J Lett 927(1):L2. https://doi.org/10.3847/2041-8213/ac52ae

Longcope DW, Fisher GH (1996) The effects of convection zone turbulence on the tilt angles of magnetic bipoles. Astrophys J 458:380. https://doi.org/10.1086/176821

Lu Y, See V, Amard L et al (2023) An abrupt change in the stellar spin-down law at the fully convective boundary. arXiv e-prints arXiv:2306.09119

Mackay DH, Yeates AR (2012) The Sun's global photospheric and coronal magnetic fields: observations and models. Living Rev Sol Phys 9(1):6. https://doi.org/10.12942/lrsp-2012-6

Mahajan SS, Sun X, Zhao J (2023) Removal of active region inflows reveals a weak solar cycle scale trend in the near-surface meridional flow. Astrophys J 950(1):63. https://doi.org/10.3847/1538-4357/acc839

Mandal S, Karak B, Banerjee D (2017) Latitude distribution of sunspots: analysis using sunspot data and a dynamo model. Astrophys J 851(1):70. https://doi.org/10.3847/1538-4357/aa97dc

Martin SF, Harvey KH (1979) Ephemeral active regions during solar minimum. Sol Phys 64(1):93–108. https://doi.org/10.1007/BF00151118

Martin-Belda D, Cameron RH (2016) Surface flux transport simulations: effect of inflows toward active regions and random velocities on the evolution of the Sun's large-scale magnetic field. Astron Astrophys 586:A73. https://doi.org/10.1051/0004-6361/201527213

Martin-Belda D, Cameron RH (2017) Inflows towards active regions and the modulation of the solar cycle: a parameter study. Astron Astrophys 597:A21. https://doi.org/10.1051/0004-6361/201629061

McClintock BH, Norton AA (2013) Recovering Joy's law as a function of solar cycle, hemisphere, and longitude. Sol Phys 287(1–2):215–227. https://doi.org/10.1007/s11207-013-0338-0

McClintock BH, Norton AA (2016) Tilt angle and footpoint separation of small and large bipolar sunspot regions observed with HMI. Astrophys J 818(1):7. https://doi.org/10.3847/0004-637X/818/1/7

Metcalfe TS, Egeland R, van Saders J (2016) Stellar evidence that the solar dynamo may be in transition. Astrophys J Lett 826(1):L2. https://doi.org/10.3847/2041-8205/826/1/L2

Metcalfe TS, Finley AJ, Kochukhov O et al (2022) The origin of weakened magnetic braking in old solar analogs. Astrophys J Lett 933(1):L17. https://doi.org/10.3847/2041-8213/ac794d

Metcalfe TS, Strassmeier KG, Ilyin IV et al (2023) Constraints on magnetic braking from the G8 Dwarf Stars 61 UMa and τ Cet. Astrophys J Lett 948:L6. https://doi.org/10.3847/2041-8213/acce38. arXiv e-prints arXiv:2304.09896

Miesch MS, Dikpati M (2014) A three-dimensional Babcock-Leighton solar dynamo model. Astrophys J Lett 785(1):L8. https://doi.org/10.1088/2041-8205/785/1/L8

Moreno-Insertis F, Schüssler M, Ferriz-Mas A (1992) Storage of magnetic flux tubes in a convective over-shoot region. Astron Astrophys 264(2):686–700

Muñoz-Jaramillo A, Nandy D, Martens PCH (2009) Helioseismic data inclusion in solar dynamo models. Astrophys J 698(1):461–478. https://doi.org/10.1088/0004-637X/698/1/461

Muñoz-Jaramillo A, Nandy D, Martens PCH et al (2010) A double-ring algorithm for modeling solar active regions: unifying kinematic dynamo models and surface flux-transport simulations. Astrophys J Lett 720(1):L20–L25. https://doi.org/10.1088/2041-8205/720/1/L20

Muñoz-Jaramillo A, Navarrete B, Campusano LE (2021) Solar anti-Hale bipolar magnetic regions: a distinct population with systematic properties. Astrophys J 920(1):31. https://doi.org/10.3847/1538-4357/ac133b

Nagy M, Lemerle A, Labonville F et al (2017) The effect of "rogue" active regions on the solar cycle. Sol Phys 292(11):167. https://doi.org/10.1007/s11207-017-1194-0

Nagy M, Lemerle A, Charbonneau P (2020) Impact of nonlinear surface inflows into activity belts on the solar dynamo. J Space Weather Space Clim 10:62. https://doi.org/10.1051/swsc/2020064

Nandy D, Choudhuri AR (2001) Toward a mean field formulation of the Babcock-Leighton type solar dynamo. I. α-coefficient versus Durney's double-ring approach. Astrophys J 551(1):576–585. https://doi.org/10.1086/320057

Nandy D, Choudhuri AR (2002) Explaining the latitudinal distribution of sunspots with deep meridional flow. Science 296(5573):1671–1673. https://doi.org/10.1126/science.1070955

Nelson NJ, Brown BP, Sacha Brun A et al (2014) Buoyant magnetic loops generated by global convective dynamo action. Sol Phys 289(2):441–458. https://doi.org/10.1007/s11207-012-0221-4

Pal S, Bhowmik P, Mahajan SS et al (2023) Impact of anomalous active regions on the large-scale magnetic field of the Sun. Astrophys J 953(1):51. https://doi.org/10.3847/1538-4357/acd77e

Parker EN (1955a) Hydromagnetic dynamo models. Astrophys J 122:293. https://doi.org/10.1086/146087

Parker EN (1955b) The formation of sunspots from the solar toroidal field. Astrophys J 121:491. https://doi.org/10.1086/146010

Parker EN (1975) The generation of magnetic fields in astrophysical bodies. X. Magnetic buoyancy and the solar dynamo. Astrophys J 198:205–209. https://doi.org/10.1086/153593

Parker EN (1984) Magnetic buoyancy and the escape of magnetic fields from stars. Astrophys J 281:839–845. https://doi.org/10.1086/162163

Parker EN (1993) A solar dynamo surface wave at the interface between convection and nonuniform rotation. Astrophys J 408:707. https://doi.org/10.1086/172631

Parker EN (2009) Solar magnetism: the state of our knowledge and ignorance. Space Sci Rev 144(1–4):15–24. https://doi.org/10.1007/s11214-008-9445-x

Petrovay K (2020) Solar cycle prediction. Living Rev Sol Phys 17(1):2. https://doi.org/10.1007/s41116-020-0022-z

Petrovay K, Nagy M, Yeates AR (2020) Towards an algebraic method of solar cycle prediction. I. Calculating the ultimate dipole contributions of individual active regions. J Space Weather Space Clim 10:50. https://doi.org/10.1051/swsc/2020050

Pipin VV (2022) On the effect of surface bipolar magnetic regions on the convection zone dynamo. Mon Not R Astron Soc 514(1):1522–1534. https://doi.org/10.1093/mnras/stac1434

Pipin VV, Kosovichev AG (2023) Magnetic flux budget in mean-field dynamo model of solar cycles 23 and 24. arXiv:2306.04124

Reiners A, Shulyak D, Käpylä PJ et al (2022) Magnetism, rotation, and nonthermal emission in cool stars. Average magnetic field measurements in 292 M dwarfs. Astron Astrophys 662:A41. https://doi.org/10.1051/0004-6361/202243251

Rempel M (2006) Flux-transport dynamos with Lorentz force feedback on differential rotation and meridional flow: saturation mechanism and torsional oscillations. Astrophys J 647(1):662–675. https://doi.org/10.1086/505170

Route M (2016) The discovery of solar-like activity cycles beyond the end of the main sequence? Astrophys J Lett 830(2):L27. https://doi.org/10.3847/2041-8205/830/2/L27

Rüdiger G, Brandenburg A (1995) A solar dynamo in the overshoot layer: cycle period and butterfly diagram. Astron Astrophys 296:557

Schatten KH, Scherrer PH, Svalgaard L et al (1978) Using dynamo theory to predict the sunspot number during solar cycle 21. Geophys Res Lett 5(5):411–414. https://doi.org/10.1029/GL005i005p00411

Schmitt D, Schuessler M, Ferriz-Mas A (1996) Intermittent solar activity by an on-off dynamo. Astron Astrophys 311:L1–L4

Schunker H, Braun DC, Birch AC et al (2016) SDO/HMI survey of emerging active regions for helioseismology. Astron Astrophys 595:A107. https://doi.org/10.1051/0004-6361/201628388

Schunker H, Birch AC, Cameron RH et al (2019) Average motion of emerging solar active region polarities. I. Two phases of emergence. Astron Astrophys 625:A53. https://doi.org/10.1051/0004-6361/201834627

Schunker H, Baumgartner C, Birch AC et al (2020) Average motion of emerging solar active region polarities. II. Joy's law. Astron Astrophys 640:A116. https://doi.org/10.1051/0004-6361/201937322

Schüssler M, Rempel M (2005) The dynamical disconnection of sunspots from their magnetic roots. Astron Astrophys 441(1):337–346. https://doi.org/10.1051/0004-6361:20052962

Schüssler M, Caligari P, Ferriz-Mas A et al (1994) Instability and eruption of magnetic flux tubes in the solar convection zone. Astron Astrophys 281:L69–L72

Solanki SK, Usoskin IG, Kromer B et al (2004) Unusual activity of the Sun during recent decades compared to the previous 11,000 years. Nature 431(7012):1084–1087. https://doi.org/10.1038/nature02995

Solanki SK, Wenzler T, Schmitt D (2008) Moments of the latitudinal dependence of the sunspot cycle: a new diagnostic of dynamo models. Astron Astrophys 483(2):623–632. https://doi.org/10.1051/0004-6361:20054282

Spruit HC (1997) Convection in stellar envelopes: a changing paradigm. Mem Soc Astron Ital 68:397–413. arXiv:astro-ph/9605020

Spruit HC (2011) Theories of the solar cycle: a critical view. In: Miralles MP, Sánchez Almeida J (eds) The Sun, the solar wind, and the heliosphere. IAGA Special Sopron Book Series, vol 4. Springer, Dordrecht, p 39. https://doi.org/10.1007/978-90-481-9787-3_5

Spruit HC (2012) Theories of the solar cycle and its effect on climate. Prog Theor Phys Suppl 195:185–200. https://doi.org/10.1143/PTPS.195.185

Spruit HC, van Ballegooijen AA (1982) Stability of toroidal flux tubes in stars. Astron Astrophys 106(1):58–66

Steenbeck M, Krause F (1966) Erklärung stellarer und planetarer Magnetfelder durch einen turbulenzbedingten Dynamomechanismus. Z Naturforsch A 21:1285. https://doi.org/10.1515/zna-1966-0813

Steenbeck M, Krause F (1969) On the dynamo theory of stellar and planetary magnetic fields. I. AC dynamos of solar type. Astron Nachr 291:49–84. https://doi.org/10.1002/asna.19692910201

Steenbeck M, Krause F, Rädler KH (1966) Berechnung der mittleren Lorentz-Feldstärke für ein elektrisch leitendes Medium in turbulenter, durch Coriolis-Kräfte beeinflußter Bewegung. Z Naturforsch Teil A 21:369. https://doi.org/10.1515/zna-1966-0401

Stix M (1972) Non-linear dynamo waves. Astron Astrophys 20:9

Talafha M, Nagy M, Lemerle A et al (2022) Role of observable nonlinearities in solar cycle modulation. Astron Astrophys 660:A92. https://doi.org/10.1051/0004-6361/202142572

Tobias SM (1996) Diffusivity quenching as a mechanism for Parker's surface dynamo. Astrophys J 467:870. https://doi.org/10.1086/177661

Tobias SM, Weiss NO, Kirk V (1995) Chaotically modulated stellar dynamos. Mon Not R Astron Soc 273(4):1150–1166. https://doi.org/10.1093/mnras/273.4.1150

Tripathi B, Nandy D, Banerjee S (2021) Stellar mid-life crisis: subcritical magnetic dynamos of solar-like stars and the breakdown of gyrochronology. Mon Not R Astron Soc 506(1):L50–L54. https://doi.org/10.1093/mnrasl/slab035

Upton L, Hathaway DH (2014) Predicting the Sun's polar magnetic fields with a surface flux transport model. Astrophys J 780(1):5. https://doi.org/10.1088/0004-637X/780/1/5

Usoskin IG (2023) A history of solar activity over millennia. Living Rev Sol Phys 20:2. https://doi.org/10.1007/s41116-023-00036-z

Usoskin IG, Mursula K, Kovaltsov GA (2001) Heliospheric modulation of cosmic rays and solar activity during the Maunder minimum. J Geophys Res 106(A8):16,039–16,046. https://doi.org/10.1029/2000JA000105

Usoskin IG, Gallet Y, Lopes F et al (2016) Solar activity during the Holocene: the Hallstatt cycle and its consequence for grand minima and maxima. Astron Astrophys 587:A150. https://doi.org/10.1051/0004-6361/201527295

van Saders JL, Ceillier T, Metcalfe TS et al (2016) Weakened magnetic braking as the origin of anomalously rapid rotation in old field stars. Nature 529(7585):181. https://doi.org/10.1038/nature16168

Vashishth V, Karak BB, Kitchatinov L (2021) Subcritical dynamo and hysteresis in a Babcock-Leighton type kinematic dynamo model. Res Astron Astrophys 21(10):266. https://doi.org/10.1088/1674-4527/21/10/266

Vashishth V, Karak BB, Kitchatinov L (2023) Dynamo modelling for cycle variability and occurrence of grand minima in Sun-like stars: rotation rate dependence. Mon Not R Astron Soc 522(2):2601–2610. https://doi.org/10.1093/mnras/stad1105

Waldmeier M (1939) Die Zonenwanderung der Sonnenflecken. Astron Mitt Eidgenöss Sternwarte Zür 14:470–481

Waldmeier M (1955) Ergebnisse und Probleme der Sonnenforschung. Geest & Portig, Leipzig

Wallenhorst SG, Topka KP (1982) On the disappearance of a small sunspot group. Sol Phys 81(1):33–46. https://doi.org/10.1007/BF00151977

Wang YM, Sheeley NR (1991) Magnetic flux transport and the Sun's dipole moment: new twists to the Babcock-Leighton model. Astrophys J 375:761. https://doi.org/10.1086/170240

Wang YM, Sheeley NR (2009) Understanding the geomagnetic precursor of the solar cycle. Astrophys J Lett 694(1):L11–L15. https://doi.org/10.1088/0004-637X/694/1/L11

Wang YM, Sheeley NR (2013) The solar wind and interplanetary field during very low amplitude sunspot cycles. Astrophys J 764(1):90. https://doi.org/10.1088/0004-637X/764/1/90

Wang YM, Nash AG, Sheeley NR (1989a) Evolution of the Sun's polar fields during sunspot cycle 21: poleward surges and long-term behavior. Astrophys J 347:529. https://doi.org/10.1086/168143

Wang YM, Nash AG, Sheeley NR (1989b) Magnetic flux transport on the Sun. Science 245(4919):712–718. https://doi.org/10.1126/science.245.4919.712

Wang YM, Sheeley NR, Nash AG (1991) A new solar cycle model including meridional circulation. Astrophys J 383:431. https://doi.org/10.1086/170800

Wang YM, Sheeley NR, Lean J (2002) Meridional flow and the solar cycle variation of the sun's open magnetic flux. Astrophys J 580(2):1188–1196. https://doi.org/10.1086/343845

Whitbread T, Yeates AR, Muñoz-Jaramillo A et al (2017) Parameter optimization for surface flux transport models. Astron Astrophys 607:A76. https://doi.org/10.1051/0004-6361/201730689

Whitbread T, Yeates AR, Muñoz-Jaramillo A (2018) How many active regions are necessary to predict the solar dipole moment? Astrophys J 863(2):116. https://doi.org/10.3847/1538-4357/aad17e

Whitbread T, Yeates AR, Muñoz-Jaramillo A (2019) The need for active region disconnection in 3D kinematic dynamo simulations. Astron Astrophys 627:A168. https://doi.org/10.1051/0004-6361/201935986

Wilson PR, Altrocki RC, Harvey KL et al (1988) The extended solar activity cycle. Nature 333(6175):748–750. https://doi.org/10.1038/333748a0

Wright NJ, Drake JJ (2016) Solar-type dynamo behaviour in fully convective stars without a tachocline. Nature 535(7613):526–528. https://doi.org/10.1038/nature18638

Yeates AR, Muñoz-Jaramillo A (2013) Kinematic active region formation in a three-dimensional solar dynamo model. Mon Not R Astron Soc 436(4):3366–3379. https://doi.org/10.1093/mnras/stt1818

Yeates AR, Nandy D, Mackay DH (2008) Exploring the physical basis of solar cycle predictions: flux transport dynamics and persistence of memory in advection- versus diffusion-dominated solar convection zones. Astrophys J 673(1):544–556. https://doi.org/10.1086/524352

Yeates AR, Cheung MCM, Jiang J et al (2023) Surface flux transport on the Sun. Space Sci Rev 219:31. https://doi.org/10.1007/s11214-023-00978-8

Yoshimura H (1975) Solar-cycle dynamo wave propagation. Astrophys J 201:740–748. https://doi.org/10.1086/153940

Yule GU (1927) On a method of investigating periodicities in disturbed series, with special reference to wolfer's sunspot numbers. Philos Trans R Soc Lond A 226:267–298

Zhang Z, Jiang J (2022) A Babcock-Leighton-type solar dynamo operating in the bulk of the convection zone. Astrophys J 930(1):30. https://doi.org/10.3847/1538-4357/ac6177

Publisher's Note Springer Nature remains neutral with regard to jurisdictional claims in published maps and institutional affiliations.

Space Science Reviews (2023) 219:40
https://doi.org/10.1007/s11214-023-00983-x

Physical Models for Solar Cycle Predictions

Prantika Bhowmik[1] · Jie Jiang[2,3] · Lisa Upton[4] · Alexandre Lemerle[5,6] ·
Dibyendu Nandy[7,8]

Received: 22 March 2023 / Accepted: 22 June 2023 / Published online: 14 July 2023
© The Author(s) 2023

Abstract
The dynamic activity of stars such as the Sun influences (exo)planetary space environments
through modulation of stellar radiation, plasma wind, particle and magnetic fluxes. Energetic solar-stellar phenomena such as flares and coronal mass ejections act as transient perturbations giving rise to hazardous space weather. Magnetic fields – the primary driver of
solar-stellar activity – are created via a magnetohydrodynamic dynamo mechanism within
stellar convection zones. The dynamo mechanism in our host star – the Sun – is manifest
in the cyclic appearance of magnetized sunspots on the solar surface. While sunspots have
been directly observed for over four centuries, and theories of the origin of solar-stellar
magnetism have been explored for over half a century, the inability to converge on the exact mechanism(s) governing cycle to cycle fluctuations and inconsistent predictions for the
strength of future sunspot cycles have been challenging for models of the solar cycles. This
review discusses observational constraints on the solar magnetic cycle with a focus on those
relevant for cycle forecasting, elucidates recent physical insights which aid in understanding solar cycle variability, and presents advances in solar cycle predictions achieved via
data-driven, physics-based models. The most successful prediction approaches support the
Babcock-Leighton solar dynamo mechanism as the primary driver of solar cycle variability
and reinforce the flux transport paradigm as a useful tool for modelling solar-stellar magnetism.

Keywords Solar magnetic fields · Sunspots · Solar dynamo · Solar cycle predictions ·
Magnetohydrodynamics

1 Introduction

The Sun's magnetic field is the primary determinant of the electromagnetic and particulate
environment around our planet as well as the heliosphere. Solar magnetic field variability
is manifested through different spatial and temporal scales: from long-term decadal-scale
variations in open magnetic flux, 10.7 cm radio flux, and total solar irradiance to short-term
sporadic energetic events such as flares and coronal mass ejections (CMEs). While longer-
term variations depend on the distribution and evolution of the large-scale global magnetic

---

Solar and Stellar Dynamos: A New Era
Edited by Manfred Schüssler, Robert H. Cameron, Paul Charbonneau, Mausumi Dikpati,
Hideyuki Hotta and Leonid Kitchatinov

Extended author information available on the last page of the article

field of the Sun, short-term perturbations like CMEs or flares originate primarily from localised magnetic structures of relatively smaller spatial scales. Despite the differences, solar magnetic variability of different spatial and temporal scales can intermingle and influence each other since all follow the same laws of physics. Ultimately, solar magnetic fields therefore are responsible for shaping space weather and space climate (Nandy et al. 2023).

High energy radiation and particle fluxes originating from extreme space weather events (flares and CMEs) can damage satellites orbiting the Earth and are hazardous to astronaut health. The impact of such events can harm critical infrastructures on the ground, resulting in direct or cascading failures across vital services such as communications and navigational networks, electric power grids, water supply, healthcare, transportation services etc. (Schrijver et al. 2015).

Flares and CMEs are linked to the complex magnetic field distribution on the solar surface, which is dictated by the emergence of sunspots and the subsequent evolution of the active region associated magnetic flux. Thus the frequency of short-lived energetic events depends on the number of sunspots emerging within a solar cycle. Simultaneously, the slower and longer-term evolution of the magnetic field of sunspots determines the amplitude of open magnetic flux and the speed and structure of solar wind emanating from the Sun, in effect, defining the space climate. It has direct consequences on the forcing of planetary atmospheres, the life span of orbiting satellites and planning of future space missions. Thus understanding and predicting phenomena which governs space weather and space climate is a scientific pursuit with immense societal relevance – in which solar cycle predictions occupy a central role (NRC 2013; Schrijver et al. 2015; 2017; NSWSAP 2019, 2022).

The methodologies for predicting different aspects of space weather and space climate are diverse but broadly relies upon observations, empirical methods, computational models and consideration of the physics of the system. Here we focus primarily on the last theme, i.e., developing our understanding of solar variability using the laws of physics towards attaining the goal of solar cycle predictions. Now physics-based prediction on different time scales itself is an extensive topic, and a complete narrative is beyond the scope of this review. Instead, we limit ourselves to decadal-centennial scale variability associated with the sunspot cycle. We emphasize that physical understanding gleaned from successful solar cycle prediction models also apply to other Sun-like stars with similar dynamo mechanisms.

Sunspots are understood to be the product of a dynamo mechanism operating within the Sun's convection zone (SCZ, hereafter) dictated by the laws of magnetohydrodynamics (Charbonneau 2020). In the SCZ, the kinetic energy stored in the ionised plasma converts to the magnetic energy primarily stored in the toroidal component of the magnetic field. The toroidal field, following significant amplification, rises through the SCZ due to magnetic buoyancy (Parker 1955) and emerges on the solar surface as strong localised magnetic field concentrations, forming Bipolar Magnetic Regions (BMRs, primarily), of which the larger ones are optically identified as sunspots. One of the mechanisms that contribute to poloidal field generation is the mean-field α-effect which relies on helical turbulent convection twisting rising toroidal fields whose net impact is to produce a finite poloidal component. On the surface, observed small and large-scale plasma motion redistributes the magnetic field associated with the BMRs, resulting in the reversal and growth of the existing global magnetic dipole moment (poloidal component) of the Sun. This process – termed as the Babcock-Leighton (B-L, hereafter) mechanism (Babcock 1961; Leighton 1969) – is another means to generate the poloidal component.

The strength of the magnetic dipole at the end of a solar cycle is found to be one of the best precursors for predicting the amplitude of the following cycle. This in itself is related to the stretching of the poloidal field through the deterministic process of differential rotation.

However, observations, analytic theory and data-driven models of decadal-centennial scale variability in the solar dynamo mechanism indicates that the B-L mechanism is the primary source of variability in the poloidal field and hence in the sunspot cycle (Cameron and Schüssler 2015; Bhowmik and Nandy 2018).

Any physics-based model aiming for solar cycle predictions must be dictated by the laws of magnetohydrodynamics, contain the essence of the dynamo mechanism, and be constrained by observed plasma and magnetic field properties in the solar surface and the interior. Some recent studies (Petrovay 2020; Nandy 2021) have explored the diversity of methods employed for sunspot cycle predictions and their credibility. Compared to these studies, this review will primarily focus on physics-based predictions of the sunspot cycle. Also see Jiang et al. (2023) for a comparison of cycle 25 predictions using physics-based models.

In the following sections, we begin with a brief account of the observed distribution of magnetic field (primarily on the solar surface) and plasma flows which serve as building blocks and constraints for computational models (Sect. 2). This is followed by a short description of the computational models of magnetic field evolution on the solar surface and interior which has shown great promise as predictive models of the solar cycle (Sect. 3). Physical insights on sources of irregularities in the strength of the solar cycle and amplitude modulation mechanisms – which are gleaned from simulations and attempts to match observations – are discussed in Sect. 4. In Sect. 5 we present a review of physics-based solar cycle predictions limiting ourselves to data-driven modelling approaches. We conclude in Sect. 6 with a discussion on the relevance of solar cycle predictions models for theories of solar-stellar magnetism and end with some remarks on future prospects in the field of solar cycle predictability.

2 Constraints from Solar Observation

Solar magnetic field observations are primarily limited to the visible layer of the Sun, i.e., the photosphere. Plasma flows are observed both on the surface as well as in the interior through inference using tools of helioseismology. Computational models utilize these information for purposes of constraining and calibrating the models. The main goals driving this synergy between observations and models are to achieve a better understanding of the physical processes ongoing in the Sun and develop predictive models of solar activity. In this section, we focus on observations which are relevant to surface flux transport (SFT) and dynamo models of the solar magnetic field. For a detailed account of solar observations, see Norton et al. (2023), this collection.

2.1 Sunspot Number

Although sunspots have been observed systematically through telescopes from the early 1600's, in the early 1800's solar astronomer Samuel Heinrich Schwabe began plotting the number of sunspots as a function of time and discovered that their appearance was cyclic with a period of about eleven years (Schwabe 1844). As new phenomena (e.g., flares, CMEs, etc.) on the Sun were discovered, it was found that they too varied along with the sunspot number. Solar activity is often characterized by the Monthly Sunspot Number, a count of the number of Sunspots or Active Regions observed each month as a function of time. The official count is maintained by the Sunspot Index and Long-term Solar Observations (SILSO) at

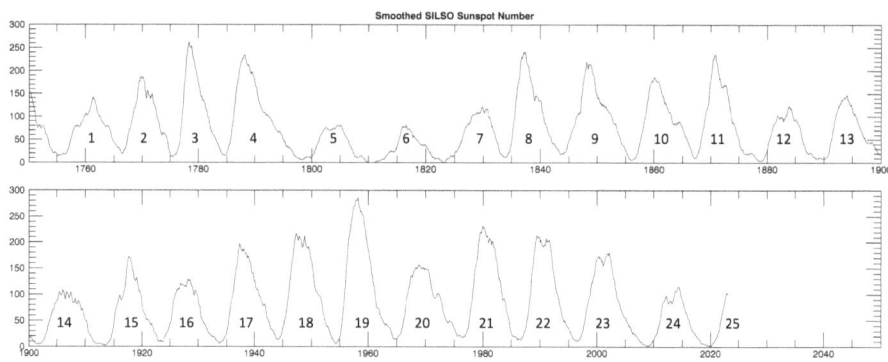

Fig. 1 The SILSO Sunspot Number. The smoothed SILSO Monthly mean total sunspot number (v2.0), smoothed illustrates the rise and fall of solar activity from 1750 to the present (marked with solar cycle numbers)

the Royal Observatory of Belgium, Brussels.[1] In 2015, the sunspot number data were revised to version 2.0, to account for changes in observers and provide a more consistent data series throughout the historical record (Clette et al. 2014). Version 2.0 of the SILSO Monthly mean total sunspot number, smoothed with a 13 month Gaussian filter, is illustrated in Fig. 1.

The SILSO data series now shows nearly 25 solar cycles. Each solar cycle has a period of about 11 years and an average amplitude (v2.0) of 180, with a range of about 80 (e.g., Solar Cycle 5) to 280 (e.g., Solar Cycle 19). The length of the cycle correlates with the amplitude of the cycle such that bigger cycles tend to be shorter in duration and weaker cycles tend to be longer (Waldmeier 1935). The shape of the solar cycle approximately appears as an asymmetric Gaussian function, with a rapid rising phase and a longer decaying phase. However, for a better fitting with the observed sunspot cycle, combinations of power law and exponential functions have also been utilized (Hathaway et al. 1994; Jiang et al. 2018). Shorter term variability in the cycle, on the order of about 2 years, causes many cycles to have two or more peaks, which are often more pronounced in weaker cycles (Karak et al. 2018). Sunspot cycles have other irregularities too, for example, the Sun entered into a prolonged near-minimum state during 1645–1715 (this was pointed out by G. Spörer and E. W. Maunder in the 1890s). This phase, known as the Maunder Minimum (Eddy 1976), is a period where the solar activity cycle was operating at an exceptionally weak state for several decades.

2.2 Magnetic Field Observations Relevant for Solar Cycle Models

Perhaps one of the most significant solar discoveries in the twentieth century, was the realization that sunspots are magnetic in nature (Hale 1908b,a). Sunspots are now known to host very strong magnetic fields on the order of 1000 G and often appear as a collection of spots. The magnetic counterparts of sunspots correspond to Active Regions. An Active Region, in general, appears as a pair of spots (with opposite polarities) which are referred to as BMR. Sunspots are the optical counterparts of active regions. Besides the spots, the background surface field has a strength of only a few Gauss. The dynamics of the Sun's surface magnetic field is captured by the 'Magnetic Butterfly Diagram', see Fig. 2. This figure is created by plotting the longitudinally average radial magnetic field on the Sun as a function of latitude

---

[1]https://www.sidc.be/silso/home.

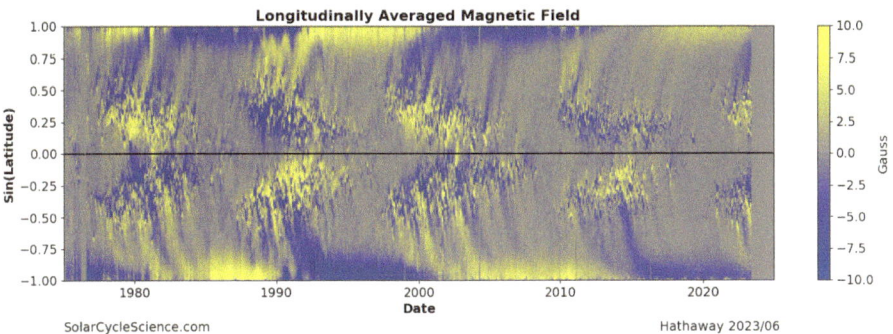

Fig. 2 The magnetic butterfly diagram illustrates the evolution of magnetic flux on the Sun over several cycles, providing observational constraints that govern models used to make solar cycle predictions. The SOLIS/MDI/HMI magnetic field data averaged over each Carrington Rotation is plotted as a function of latitude and time, with the positive (negative) polarity shown in yellow (blue). Figure courtesy D.H. Hathaway via www.solarcyclescience.com

and time. This figure illustrates several properties that serve as constraints for the SFT and dynamo models and which are tests for predictive models. The properties include:

- Sporer's Law: Active Regions begin emerging at mid-latitudes (∼ 30 degrees). As the cycle progresses, active regions form bands of flux that moves equator-ward (Carrington 1858).
- Joy's Law for Tilt Angles: Active Regions tend to have a characteristic tilt such that the angle between the local parallel of latitude and the line joining the leading polarity spot (which appears closer to the equator) and the following polarity spot statistically increases with increasing latitude (Hale et al. 1919).
- Hale's Polarity Law: The relative polarity of Active Regions (between the leading and following spots) is opposite across the equator, and this polarity changes sign from one cycle to the next (Hale et al. 1919). Thus a complete magnetic cycle has a periodicity of 22 years.
- Polar Fields: In addition to the flux in emerging Active Regions, the Sun possesses unipolar concentrations of magnetic flux near both of the poles. The polarity is opposite across hemispheres, reverses polarity at about the time of solar cycle maximum (Babcock and Livingston 1958) and attains its maximum amplitude around cycle minimum. This large-scale field is known as the 'Polar field', and its evolution plays an important role in solar cycle predictions. However, due to projection effects, polar field measurements suffer from erroneous values, thus prompting the need for off-ecliptic space missions focusing on the Sun's poles (Nandy et al. 2022).
- Magnetic Field Surges: Streams of weak polarity flux are carried from the active latitude to the poles in about 2 – 3 years. These streams are responsible for the build-up and reversal of the polar fields (Babcock 1961). The strength of these surges reaching the poles can vary significantly based on the associated active regions. The emergence latitudes, tilt and flux of the active region, and frequency of active region emergence are all important factors in determining how much a given active region will contribute to the polar field evolution. The high latitudes emergence with large tilt (Yeates et al. 2015) and long-lasting activity complexes (Wang et al. 2020) tend to generate prominent poleward surges. However, the surges resulting from high latitude emergence usually only bring the transient influence of polar field (Jiang et al. 2014a; Yeates et al. 2015). On the contrary, 'rogue region' (e.g., with a large area and high tilt angle) emerging at low latitudes

usually cannot generate prominent surges but can influence the polar field development significantly (Nagy et al. 2017).

The redistribution of the active regions' magnetic flux across the solar surface and the interior convection zone happens through the collective effect of small and large-scale plasma motions which provide additional constraints on models.

2.3 Plasma Flows

Plasma flows in the solar convection zone may be divided into three categories based on the physical role they play in the solar dynamo mechanism: convective flows, differential rotation, and meridional circulation. The thermal flux through the solar convection zone and consequent temperature gradient causes the plasma within the solar convection zone to rise to upper layers, transfer or radiate their energy away and sink back down after cooling. As a result, convective cells with a spectrum of different scales (Hathaway et al. 2015) are formed ranging from granules (diameter ~ 1 Mm) with lifetimes of minutes to hours, to supergranules (diameter ~ 30 Mm) with lifetimes of days, and to the largest convective structures (diameter ~ 200 Mm) with lifetimes of months. These convective motions are turbulent in nature and effectively distribute the magnetic field in the entire convection zone, including the solar surface, similar to a diffusive process.

The Sun rotates differentially which was first found by tracking sunspots on the solar surface (Adams 1911; Belopolsky 1933; Howard 1984). This differential rotation at the surface is such that the large-scale plasma flow speed along the direction of solar rotation varies latitudinally with a faster-rotating equator than the poles. Later, helioseismology (Schou et al. 1998; Basu 2016) was utilized to obtain the structure and spatial variation of rotation rate inside the solar convection zone. The radiative zone rotates as a solid body resulting in a strong radial shear within the tachocline which is thought to encompass the (stable) overshoot layer at the base of the convection zone. The differential rotation plays a crucial role in the generation and amplification of the toroidal component of the Sun's large-scale magnetic field (see Sect. 3).

Another large-scale subsurface plasma flow known as the meridional circulation (Hanasoge 2022) which in near-surface layers carries plasma from the equatorial region to the poles (in both hemispheres) with a varying speed dependent on latitude. The flow speed becomes zero at the equator and the poles, and the circulation attains its peak speed $(10 - 20 \text{ m s}^{-1}$, about 1% of the mean solar rotation rate) near mid-latitude. The law of mass conservation dictates an equator-ward return flow of plasma deeper inside the solar convection zone, which, however, remained hard to map using helioseismic observations due to its small amplitude. While some recent studies (Rajaguru and Antia 2015; Liang et al. 2018) have suggested that meridional circulation is a single-cell flow where the return flow is at the depth of the solar convection zone (depth $< 0.8 \ R_\odot$), others (Hathaway 2012; Zhao et al. 2013; Hathaway et al. 2022) suggest that it may be multi-cellular in depth and or latitude. The shape of the meridional profile in latitude and radius is crucial in determining the various properties of the sunspot cycles, including cycle duration. Early flux transport dynamo models suggest that a deep single-cell meridional flow threading the convection zone is necessary to match solar cycle observations (Nandy and Choudhuri 2002; Hazra et al. 2014). However, the recent Babcock-Leighton-type dynamo models (Zhang and Jiang 2022; Zhang et al. 2022) indicate that the deep single-cell meridional flow is not crucial to match solar cycle observations.

Note that both small-scale and large-scale plasma flows are not static. Helioseismic observation shows that the Sun's differential rotation varies with time in an oscillatory fashion and is correlated with the solar activity cycle – it is known as the solar torsional oscillation (Zhao and Kosovichev 2004; Howe 2009). Meridional circulation also exhibits cycle-dependent temporal variation in its peak speed, with reduced amplitude during cycle maximum compared to the minimum (Hathaway and Rightmire 2010; Hathaway and Upton 2014; Hathaway et al. 2022). However, for both large-scale plasma flows, such variations constitute less than 20% of the average profiles. Thus, computational models with time-independent plasma flow profiles can reproduce the majority of the observed large-scale magnetic field variability.

2.4 Polar Fields as Precursors of the Strength of Sunspot Cycles

The temporal evolution of the averaged polar field has a $\pi/2$ phase difference with the sunspot cycle. As mentioned earlier, the average polar field strength at cycle minimum serves as an important element in predicting sunspot cycles. Although direct observation of the polar fields became available only in the 1970s, indirect measures of polar flux exist based on proxies. Polar flux evolution derived from polar faculae observations (Muñoz-Jaramillo et al. 2012) cover a period of 100 years (during cycles $14-24$). Note that the average polar flux during cycle minimum is a close representation of the Sun's magnetic dipole (axial) moment – which acts as a seed to generate the following solar cycle. In fact, the average polar flux at the nth cycle minimum has the maximum positive correlation with the amplitude of the $n+1^{th}$ cycle [see, Fig. 3]. The correlation decreases drastically for the amplitude of cycles n^{th}, $n+2^{th}$ and $n+3^{th}$ as depicted in Fig. 3. Figure 3 reflects on two crucial aspects of solar cycle predictability: first, a strong solar cycle does not result in a strong polar field generation at that cycle minimum [see, Fig. 3(a)] and the memory of the polar field (of nth cycle) diminishes beyond the next $(n+1^{th})$ cycle [see, Fig. 3(c) and (d)]. It is important to note here that these empirical observational evidences especially the decreasing correlation beyond the immediate cycle, were preceded by flux transport dynamo models exploring the memory issue which predicted that the sunspot cycle memory is limited primarily to one cycle alone in certain parameter regimes related to flux transport timescales (Yeates et al. 2008; Karak and Nandy 2012).

3 Physical Modeling Approaches

In an astrophysical magnetised plasma system like our Sun, we expect the following properties of the plasma will be satisfied: the velocity is non-relativistic, the collisional mean free path of the atomic or molecular constituents of the plasma is much shorter than competing plasma length scales, and the plasma is electrically neutral and non-degenerate. In such a system, the evolution of the magnetic field is dictated by the magnetohydrodynamic (MHD) induction equation, which is a combination of Ohm's law, Ampère's law and Maxwell's equations:

$$\frac{\partial \mathbf{B}}{\partial t} = \nabla \times (\mathbf{u} \times \mathbf{B} - \eta \nabla \times \mathbf{B}). \tag{1}$$

Here $\mathbf{u}$ and $\mathbf{B}$ are the plasma velocity and magnetic fields, respectively, and $\eta = 1/\mu\sigma$ is the magnetic diffusivity, with μ the magnetic permeability and σ the electric conductivity. Additionally, the magnetic field satisfies the divergence-free condition, $\nabla \cdot \mathbf{B} = 0$. The spatio-temporal evolution of the plasma flow is dictated by Navier–Stokes equation,

$$\frac{\partial \mathbf{u}}{\partial t} + (\mathbf{u} \cdot \nabla)\mathbf{u} = -\frac{1}{\rho}\nabla P + \mathbf{g} + \frac{1}{\rho}(\mathbf{J} \times \mathbf{B}) + \nu\nabla^2\mathbf{u}, \tag{2}$$

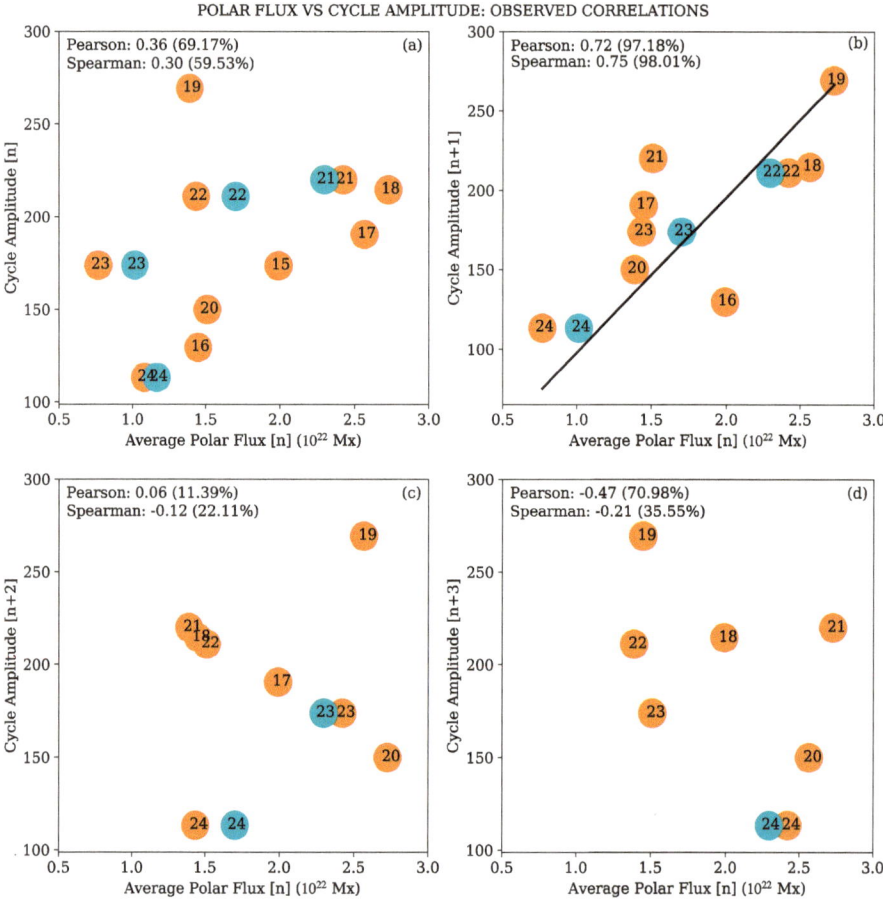

Fig. 3 Observed cycle-to-cycle correlations between the polar flux at cycle minima (say, [n]) and the cycle amplitude of different cycles, namely (a) cycle [n], (b) cycle [n+1], (c) cycle [n+2], and (d) cycle [n+3]. The numbers inside the circles indicate the associated solar cycle numbers. The colors of the circles differ based on the source of polar flux data, orange: averaged polar flux obtained from polar faculae count, cyan: the average dipole moment (scaled appropriately to place them in the figure). Image reproduced with permission from Nandy (2021) copyright by Springer Link

where ρ is plasma density, P is plasma pressure, **g** is the gravitational acceleration, $\mathbf{J} = (\nabla \times \mathbf{B})/\mu$ is the electric current, and ν is kinematic viscosity. Additionally, the plasma flow obeys mass conservation through the continuity equation. Along with equations (1) and (2), one should take into account the conservation of energy and equation of states involving the pressure and plasma density. In an isolated system, where we can ignore the Poynting flux, the mechanical energy stored in the flow (**u**) acting in the opposite direction of the Lorentz force ($\mathbf{J} \times \mathbf{B}$) is converted to magnetic energy. This magnetic energy can decay through the dissipation of the electrical currents supporting the magnetic field.

Thus the sustainability of a dynamo depends on the relative strength between the induction effect controlled by the velocity field (the first term on the R.H.S in equation (1) and the Ohmic dissipation (the second term on the R.H.S in equation (1). The ratio of these two terms is known as the magnetic Reynolds number, $R_m = uL/\eta$, where L is the length scale determining whether inductive effect overcomes the dissipative processes. In most astrophysical systems, a very large L ensures a very high R_m, which is crucial for the survival and growth of the dynamo.

In an ideal scenario, solving the complete set of MHD equations associated with the conservation of mass, momentum, energy, and magnetic flux including the magnetic induction equation in the SCZ should provide the Sun-like spatio-temporal evolution of the velocity and magnetic field with the given Sun-like plasma properties. However, this requires the numerical models to be capable of comprising a wide range of spatial and temporal scales characterizing fluid turbulence at high viscous and magnetic Reynolds number medium – which is quite challenging from the computational point of view. While with increasing computational power and improved algorithms, full MHD models are becoming more realistic, the parameter regimes are still nowhere near the real solar convection zone. Moreover, all the existing MHD models operate with enhanced dissipation, much stronger than the characteristic dissipation in the solar interior. A comprehensive account of the MHD simulations of solar dynamos is presented in Käpylä et al. (2023), this collection, thus, we restrain ourselves from going into further details.

The scope of the growth of the dynamo is encapsulated within the advective part of the induction equation $[\nabla \times (\mathbf{u} \times \mathbf{B})]$ in Eq. (1), where any pre-existing magnetic field ($\mathbf{B}$) is amplified by the plasma flow through the shearing term $[\mathbf{B} \cdot \nabla(\mathbf{u})]$, compression and rarefication $[\mathbf{B}(\nabla \cdot \mathbf{u})]$, and advection $[(\mathbf{u} \cdot \nabla)\mathbf{B}]$. While any positive gradient in the plasma flow ensures growth of $\mathbf{B}$, the dynamo-generated magnetic field should have the following observed characteristics (see Sect. 2) in the solar context:

- The large-scale magnetic field (the dipole component) should reverse on a decadal scale.
- The sunspot generating field component should have a $\pi/2$ phase difference with the dipole component, should exhibit an equator-ward migration, and the associated polarity should be anti-symmetric across the equator.
- On the solar surface, the dynamo model is expected to reproduce observed features of sunspots and the associated flux evolution, which include pole-ward migration of the diffused field and generation of observed polar field (as seen in Fig. 2).
- Moreover, the solar dynamo models should result in amplitude fluctuations in both the sunspot-generating component and the large-scale dipole component, along with observed empirical patterns and correlations between them.

Reproducing all these intricate details of the observed solar magnetic field and the velocity field while solving the full set of MHD equations in the turbulent convection zone indeed becomes a challenging problem. Thus one major and very successful alternative approach in the dynamo community has been to focus on the evolution of the magnetic field only by solving the induction equation (1) while utilizing prescribed plasma flow derived from observation (Charbonneau 2020). These are often termed as kinematic dynamo models. Another modelling approach, namely Surface Flux Transport (SFT) models, simulate only one half of the solar cycle, namely, the evolution of magnetic fields which have emerged on the solar surface mediated via prescribed flow parameters which are constrained by observations. We discuss them briefly below.

3.1 Solar Surface Flux Transport Models as Tools for Polar Field Predictions

The genesis of solar surface magnetic field evolution models, as well as the Babcock-Leighton mechanism for polar field generation can be traced to the heuristic ideas first proposed by Babcock (1961). Babcock attempted to explain the behavior of the large-scale solar magnetic fields through a phenomenological description of the topology of the Sun's magnetic field and its evolution which was related to the emergence of systematically tilted BMRs and the subsequent diffusion of their flux, cross-equatorial cancellation and migration to the poles – culminating in the large-scale dipolar field reversal. This process was envisaged to be complemented by the stretching of the submerged large-scale flux systems by differential rotation to produce the sunspot forming toroidal field. Later, Leighton (1964) put these ideas on a firmer theoretical foundation. He suggested that the radial magnetic field at the surface of the Sun is advected and diffused kinematically like a passive scalar field.

The computational models capturing the evolution of this radial magnetic field $[B_r(\theta, \phi, R_\odot)]$ associated with the active regions are known as Surface Flux Transport (SFT) models. The temporal evolution of the longitudinal averaged radial field obtained from such simulations should have the distinct features observed in the magnetic butterfly diagram (Fig. 2). Then the SFT mechanism may also be derived from the MHD induction equation (1) as the time evolution of the radial component of the magnetic field B_r, evaluated at $r = R_\odot$, as:

$$\frac{\partial B_r}{\partial t} = -\frac{1}{r \sin \theta} \frac{\partial}{\partial \theta} \left(u_\theta B_r \sin \theta\right) - \frac{1}{r \sin \theta} \frac{\partial}{\partial \phi} \left(u_\phi B_r\right) + \eta_T \nabla^2 B_r. \tag{3}$$

Here, u_θ and u_ϕ denote two large-scale plasma flows on the solar surface: meridional circulation and differential rotation, respectively. The diffusivity η_T represents a simplification of the effect of turbulent convective motions of the plasma on B_r. Note that we haven't considered the contribution from radial diffusion, which appears in the mean-field MHD formulation of SFT (Yeates et al. 2023). To the linear formulation above 3, a source term must be added to account for the additional influx of magnetic field associated with the emergence of active regions, which are primarily BMRs. For a detailed description of SFT models and their theoretical considerations, refer to Jiang et al. (2014b), Yeates et al. (2023).

3.1.1 Genesis of Surface Flux Transport Simulations

Following the pioneering work by Babcock (1961) and Leighton (1964) describing the evolution of B_r on the solar surface, DeVore et al. (1984) created the first SFT model of the Sun. Their SFT model was originally used to constrain meridional flow at the surface, which was difficult to measure and very uncertain at that time. To mimic the emergence of active regions on the solar surface, Sheeley et al. (1985) included bipolar active region sources based on observed statistics. Wang et al. (1989) explored the role of surface flux transport and dissipation processes such as differential rotation, meridional flow, and diffusion to investigate their role in the reversal and build-up of the polar fields. They found that a) differential rotation was essential for separating the leading and following polarity flux in bipolar active regions, b) diffusion played a crucial role in cross-equatorial flux cancellation of the leading polarities and b) meridional flow was essential for transporting the following polarity flux to the poles aiding in polar field reversal and build-up. The primary physical ingredients of the surface processes resulting in the observed solar cycle associated polar field dynamics were now in place.

3.1.2 Evolution Towards Data Driven Surface Flux Transport Models

More evolved SFT models now have the ability to incorporate the observed flows (static and time-evolving) and assimilate data for realistic simulations of solar surface field evolution and polar field predictions. These models are also paving the way for realistic coronal field and heliospheric modeling by providing time-dependent, data assimilated boundary conditions at the solar photosphere.

While many modern SFT models continue to parameterize the small-scale convective motions with a diffusivity coefficient, a novel class of magnetoconvective SFT (mSFT) models have been developed which emulate not only the spreading motions of convection outflows, but also the concentration of the magnetic network formed at the boundaries of convective structures. The first attempt at this was achieved by introducing a random attractor matrix (Worden and Harvey 2000) to replace the diffusivity. The attractor method was later adapted by the Air Force Data-Assimilative Photospheric Flux Transport model (ADAPT) (Arge et al. 2010; Hickmann et al. 2015). Another SFT model invoked a collision and fragmentation algorithm (Schrijver 2001). An alternative approach known as the Advective Flux Transport (AFT) (Upton and Hathaway 2014) model has been developed which mimics surface convection through spherical harmonics to generate an evolving velocity flow field that reproduces the size, velocities, and lifetimes of the observed convective spectrum (Hathaway et al. 2015).

Another major advancement in SFT models, brought about by the space-based Doppler-Magnetographs, is the availability of high cadence and high-resolution magnetograms. Schrijver and De Rosa (2003) were one of the firsts to directly assimilate magnetogram data into the SFT model by incorporating SOHO/MDI magnetic field observations within $60°$ of the disk center. Using their SFT maps as an inner boundary condition to a PFSS model, they were able to create an accurate reconstruction of the interplanetary magnetic field (IMF). More formal data assimilation processes (e.g., Kalman filtering) require that the observed data be merged with the simulated data in a way that accounts for the uncertainties. ADAPT (Hickmann et al. 2015) and AFT (Upton and Hathaway 2014) type SFT models employ Kalman filtering. The ADAPT model is used in conjunction with WSA-ENLIL model to aid in Space Weather Predictions. Since the surface field distribution drives the coronal magnetic field, SFT (and AFT) models have shown great capabilities for coronal field simulations and predictions (Nandy et al. 2018; Dash et al. 2020; Mikić et al. 2018; Mackay and Upton 2022; Yeates and Bhowmik 2022).

Studies are illuminating the influence that flow variations have on polar field dynamics. Hathaway and Upton (2014) found that variations in the meridional flow had a significant impact ($\sim 20\%$) on the polar field strength. Cross equatorial flow (Komm 2022) significantly influenced the residual magnetic flux transported to the poles. These simulations clearly show that despite being relatively weak, the shape and amplitude of the meridional circulation is a crucial element shaping the solar cycle dynamics.

Incorporating the observed sunspots statistics (Jiang et al. 2011) on the solar surface is crucial – where the active region's emergence latitude and associated tilt angle and magnetic flux become major deciding factors to the final contribution to the dipole moment evolution (Petrovay et al. 2020a; Wang et al. 2021). It may appear that the final dipole moment at the end of a cycle (which acts as a seed to the following cycle) will then be deterministic – a strong cycle producing a strong poloidal field at cycle minimum. However, observation suggests otherwise [see Fig. 3(a)]: saturation in the final dipole moment, which is linked with two factors: tilt-quenching (Dasi-Espuig et al. 2010; Jiang et al. 2011) and latitude-quenching (Jiang 2020) (see Sect. 4.3.2 for more details). Moreover, scatter in the active region tilt angles (in addition to the systematic tilt according to Joy's law) introduces substantial uncertainty in the final dipole moment at cycle minimum (Jiang et al. 2014a; Bhowmik 2019). For example, a few big rogue active regions emerging at low latitudes with a "wrong" (i.e., opposite to the majority for this cycle) tilt angles can reduce the dipole moment amplitude, thus weakening the seed for the following cycle (Jiang et al. 2015; Nagy et al. 2017).

Observationally constrained and flux calibrated SFT models can now match decadal to centennial-scale solar surface magnetic field dynamics and are being used for predicting the polar field amplitude based on synthetic data inputs of the declining phase of the cycle – with a reasonably good degree of success (Cameron et al. 2010; Upton and Hathaway 2014; Cameron et al. 2016; Bhowmik and Nandy 2018; Wang et al. 2002). These models have become useful tools for understanding solar surface flux transport dynamics, exploring the nuances of the Babcock-Leighton mechanism for solar poloidal field generation, and are being coupled to data-driven dynamo models for predicting the strength of the sunspot cycle.

3.2 Flux Transport Dynamo Models as a Tool for Sunspot Cycle Predictions

Kinematic or flux transport dynamo models have shown exceptional fidelity for being used as tools for solar cycle predictions. The utility of these models are due to the possibility of prescribing observationally constrained, or theoretically "expected" velocity profiles ($\mathbf{u}$) and assimilating observations of the poloidal field to obtain the spatio-temporal evolution of the solar magnetic field ($\mathbf{B}$) by using Eq. (1). These models use two large-scale time-independent velocity profiles to incorporate the observed differential rotation and meridional circulation (see Sect. 2).

Based on the observed properties of the surface field, the large-scale solar magnetic field at cycle minimum can be reasonably approximated to be axisymmetric (independent of ϕ) and antisymmetric across the equatorial plane. This simplifies the kinematic dynamo problem further. Thus in spherical polar coordinates (r, θ, ϕ), the magnetic field ($\mathbf{B}$) can be expressed as,

$$\mathbf{B}(r, \theta, t) = \nabla \times \mathcal{A}(r, \theta, t)\,\hat{\mathbf{e}}_\phi + \mathcal{B}(r, \theta, t)\,\hat{\mathbf{e}}_\phi. \tag{4}$$

The first term in the R.H.S. of the above equation is the poloidal component ($\mathbf{B_P}$, hereafter) in the meridional plane expressed through a vector potential ($\mathcal{A}$) and the second term ($\mathcal{B}$) corresponds to the toroidal component ($\mathbf{B_T}$, hereafter). The velocity can also be expressed similarly as a combination of the poloidal (meridional circulation) and toroidal (differential rotation) components. All these simplifications lead us to two separate but coupled equations for $\mathbf{B_P}$ and $\mathbf{B_T}$, where the first corresponds to the axial dipole moment (or averaged polar field), and the latter is related to the sunspot-generating strong magnetic field.

The solution to the set of equations produces $\mathbf{B_P}$ and $\mathbf{B_T}$ with a $\pi/2$ phase difference, both having roughly decadal-scale periodicity (considering amplitude only). It is reflected in Fig. 4 showing the observed evolution of averaged polar field and sunspot cycles over four decades. A cycle begins with the strongest $\mathbf{B_P}$ and the weakest $\mathbf{B_T}$. In the rising phase of a cycle, with increasing $\mathbf{B_T}$ (i.e., more sunspots of opposite polarity), $\mathbf{B_P}$ (i.e., average polar field) weakens gradually through the B-L mechanism. $\mathbf{B_P}$ changes its polarity during the cycle maximum and progresses towards a new maximum value (with opposite polarity) during the declining phase of the cycle while $\mathbf{B_T}$ continues to decrease till the cycle minimum. Polarity-wise, $\mathbf{B_P}$ and $\mathbf{B_T}$ have a 22-year-long periodicity, which is also evinced through Hale's polarity law (as discussed in Sect. 2). In the following section, we describe how the generation process of the two components of the Sun's magnetic field rely on each other.

3.2.1 Poloidal to Toroidal Field

The induction equation for the toroidal component ($\mathbf{B_T}$) includes a source term originating from the differential rotation [$\Omega(r, \theta)$] in the SCZ, compared to which the sink term due

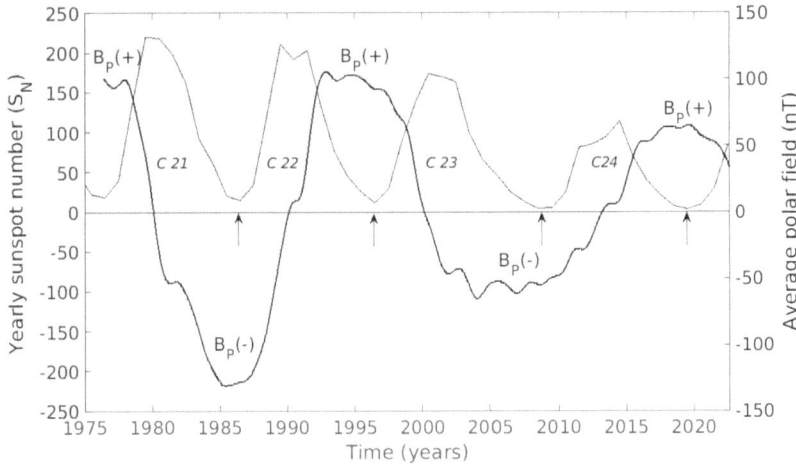

Fig. 4 The temporal evolution of yearly sunspot number (source: WDC-SILSO, Royal Observatory of Belgium, Brussels) and average polar field (source: Wilcox Solar Observatory) during sunspot cycles: 21 – 24. The arrows denote the epochs of cycle minima approximately. The sign of the average polar field corresponds to the sign of $B_P(+/-)$, and the sunspot number is related to the amplitude of $\mathbf{B_T}$

to diffusive decay is negligible. Thus any pre-existing $\mathbf{B_P}$ will be amplified through shearing along the azimuthal direction (ϕ) by $\mathbf{B_P} \cdot \nabla\Omega$ and generate new $\mathbf{B_T}$. The differential rotation in the solar convection zone and the stable overshoot layer at its base (coinciding with the tachocline where turbulent diffusivity is suppressed) plays important roles in the amplification and storage of $\mathbf{B_T}$ (Muñoz-Jaramillo et al. 2009).

Following sufficient amplification by differential rotation, $\mathbf{B_T}$ satisfies the magnetic buoyancy condition (Jouve et al. 2018; Fan 2021). Any perturbed part of the toroidal field rises as a flux rope through the SCZ, where it encounters the Coriolis force and turbulent diffusivity in the medium (Weber et al. 2011). The flux tube which eventually emerges through the solar surface creates a pair of spots (in general, following Hale's polarity rule) with a certain latitude-dependent tilt angle (following Joy's law), a somewhat fragmented structure and a reduced strength compared to its initial amplitude. A more detailed account of active region emergence has been discussed in Weber et al. (2023).

3.2.2 Toroidal to Poloidal Field Generation

Strictly axisymmetric flows and fields cannot sustain the dynamo process (Cowling's theorem). Thus to sustain a dynamo, a non-axisymmetric process must be invoked. Elaborate studies on kinematic dynamo models have utilized different mechanisms to convert $\mathbf{B_T}$ to $\mathbf{B_P}$ (Cameron et al. 2017; Charbonneau 2020) by utilizing intrinsically non-axisymmetric processes which are parameterized in the dynamo equations. We present below a very brief narrative of such approaches.

Turbulence and Mean-Field Electrodynamics Approach The thermally driven environment in the SCZ results in turbulent motion of the plasma, which therefore has a mean large-scale flow along with a fluctuating component [$\mathbf{u} = \langle \mathbf{u} \rangle + \mathbf{u}'$]. While the mean component, $\langle \mathbf{u} \rangle$, corresponds to the standard axisymmetric large-scale plasma velocity (differential rotation and meridional circulation), the fluctuating component, $\mathbf{u}'$, vanishes when averaged in the azimuthal direction. The magnetic field can be decomposed in a similar fashion: $\mathbf{B} = \langle \mathbf{B} \rangle + \mathbf{B}'$. Although the fluctuating parts of the velocity and the magnetic field vanish individually when averaged azimuthally, their product, $\mathcal{E} = \langle \mathbf{u}' \times \mathbf{B}' \rangle$ will sustain and overcome the restriction set by Cowling's theorem. $\mathcal{E}$ is known as the mean turbulent electromotive force, and part of it serves as a source term (the α-effect) in the induction equation of $\mathbf{B_P}$. From a physical point of view, it can be linked to the helical twisting of the toroidal field component ($\mathbf{B_T}$) by helical turbulent convection. For a thorough description, please refer to Brandenburg et al. (2023), Hazra et al. (2023), this collection.

The Babcock-Leighton Mechanism The magnetic axis connecting the opposite polarities of active regions has a certain tilt with respect to the east-west (toroidal) direction which arises due to the action of the Coriolis force on buoyantly rising toroidal flux tubes (an inherently non-axisymmetric process, see Fisher et al. (2000) and references within). Thus, all active regions have non-zero components of magnetic moments along the north-south (poloidal) direction – which collectively contributes to the axial dipole moment generation and evolution (Petrovay et al. 2020a). Section 3.1 describes how the magnetic flux initially concentrated within tilted active regions decay and redistributes across the solar surface to generate the large-scale magnetic field. Thus is the so called Babcock-Leighton mechanism which converts $\mathbf{B_T}$ to $\mathbf{B_P}$. Observational evidence not only strongly supports the B-L mechanism, they also help constrain data-driven predictive SFT and dynamo models (Passos et al. 2014; Cameron and Schüssler 2015; Bhowmik and Nandy 2018; Bhowmik 2019). For a more detailed account readers may consult Cameron and Schüssler (2023), this collection.

One critical aspect of B-L type dynamos is the spatial dissociation between the source regions of $\mathbf{B_P}$ (on the solar surface) and $\mathbf{B_T}$ (in the deep SCZ). For the B-L dynamo to function effectively, the spatially segregated layers must be connected to complete the dynamo loop. The transport of $\mathbf{B_P}$ generated on the solar surface to the deeper layers of SCZ can occur through various processes. These include meridional circulation (Wang et al. 1991; Choudhuri et al. 1995; Nandy and Choudhuri 2002) which acts as a conveyor belt connecting surface layers to the deep convection zone, turbulent diffusion (Yeates et al. 2008) as well as turbulent pumping Hazra and Nandy (2016). All these processes are an integral part of any flux-transport dynamo model, irrespective of whether the dominant poloidal field generation process is the mean-field or the B-L mechanism.

A new approach towards predictive solar cycle modeling is the coupling of a 2D SFT model to an internal 2D dynamo model – where the output from the first model serves as an upper boundary condition of the second one (Lemerle and Charbonneau 2017). Subsequently, the internal distribution of the toroidal magnetic field ($\mathbf{B_T}$) in the dynamo model generates synthetic sunspots emerging in the SFT model. This model, therefore, has the advantage of incorporating a full non-axisymmetric representation of the solar surface at a much lower numerical cost than the 3D models. The primary weakness of this $2 \times 2D$ model is its obligation to tackle the different spatial resolutions in the SFT and the dynamo components.

Presenting an elaborate account of all important works on SFT and solar dynamo modelling approaches is beyond the scope of this review; instead we have elaborated only on the primary physical mechanisms that are at the heart of the solar dynamo mechanism. We now turn our focus to processes that are at the basis of solar cycle fluctuations, understanding which is important from the perspective of solar cycle predictions.

4 Physical Processes Influencing Solar Cycle Predictability

As shown in Sect. 2, apart from its about 11-year periodicity, a prominent property of the solar activity record is the strong variability of the cycle amplitudes, including extended intervals of very low activity, e.g., Maunder minimum, or particularly high activity, e.g., modern maximum (Usoskin 2017, Biswas et al. 2023, this collection). Stochastic perturbations inherent in the turbulent solar-stellar convection zones and nonlinearities are two viable candidates for explaining solar cycle fluctuations. Understanding what drives these fluctuations and our ability to account for them either through first principles or observational data assimilation paves the way towards physics-based solar cycle predictions.

4.1 Numerical Weather Forecasts and Nonlinear Time Series Analysis of Solar Activity Proxies

Insights into the development of numerical weather or climate forecasting models over half a century serve as an useful analogy for physics-based solar cycle predictions (Wiin-Nielsen 1991) and could inspire the progress in physics-based solar cycle predictions. Numerical weather forecasts correspond to applying physical laws to the atmosphere, solving mathematical equations associated with these laws, and generating reliable forecasts within a certain timescale. The breakthrough from the 1930s to the 1950s can be classified into two groups. One is the physics of atmospheric dynamics. Vilhelm Bjerknes formulated the atmospheric prediction problem. C. G. Rossby (1939) derived the barotropic vorticity equation and proposed the first theory of the atmospheric long waves, i.e., Rossby waves. J. Charney (1948) developed the quasi-geostrophic theory for calculating the large-scale motions of planetary-scale waves. The second is the genesis of numerical calculation methods and the application of the computational method led by Lewis Fry Richardson and John von Neumann. From 1955 onwards, numerical forecasts generated by computers were issued regularly. People mainly concentrate on four domains to increase the performance of predictions (Kalnay 2003); improve the representation of small-scale physical processes, utilize more comprehensive (spatial and temporal) observational data, use more accurate methods of data assimilation, and utilize more and more powerful supercomputers.

Edward Lorenz (1963) opened the doors of physics based weather forecasting by establishing the importance of nonlinear dynamics in the context of convecting systems and meteorology. In the subsequent remarkable papers (Lorenz 1965, 1969), Lorenz made a fundamental discovery related to the predictability of weather arguing that nonlinearity leads to chaotic dynamics making long-range forecasts impossible. We now know that the chaotic nature of the atmosphere imposes a limit of about two weeks in weather forecasts even with ideal models and perfect observations.

Advances in time-series analysis of non-linear dynamics since the 1980s have made it possible to distinguish between stochastic behavior and deterministic chaos in principle. Strange attractor reconstruction based on correlation integral and embedding dimension (Takens 1981; Grassberger and Procaccia 1983) and the method of surrogate data (Theiler et al. 1992; Paluš and Novotná 1999) are the most widely used methods to look for chaotic behavior. Numerous attempts in this field have been invoked in the literature by analyzing different time series of solar activity proxies, e.g., sunspot number data (Price et al. 1992), sunspot area data (Carbonell et al. 1994), cosmogenic data (Hanslmeier and Brajša 2010), polar faculae (Deng et al. 2016), and so on. However, these studies show highly diverging results. For example, some studies (Mundt et al. 1991; Rozelot 1995; Hanslmeier and Brajša 2010; Deng et al. 2016) report evidence for the presence of low-dimensional deterministic chaos in solar cycles. On the other hand, others (Price et al. 1992; Carbonell et al. 1994; Mininni et al. 2000) find no evidence that sunspots are generated by a low-dimensional chaotic process. Even in studies showing evidence of chaos, divergent values of the system parameters (e.g., maximum Lyapunov exponent) were estimated – indicating divergent prediction time scales. It is suggested that results claiming the existence of chaos were derived from short scaling regions obtained using very low time delays in the computations for the correlation dimension (Carbonell et al. 1994). Furthermore, the ways to filter or smooth solar activity data also strongly impact the results (Price et al. 1992; Petrovay 2020).

In brief, despite the intensive investigation of solar activity data, there is no consensus on whether chaos or inherent stochastic perturbations or noise, or a combination of these processes drive solar cycle fluctuations. The insufficient length of the solar activity data, that is sparsity of data in phase space, compromises statistical sampling. Clearly distinguishing between stochastic modulation and deterministic chaos remains an outstanding issue. However, driving predictive physical models with observational data of the poloidal component of Sun's magnetic field (which is primarily subject to stochasticity and nonlinear effects) provides a way out of this conundrum.

4.2 Low-Order Models of the Solar Cycle

Model building aims to use our understanding of a physical system by establishing dynamical equations explaining a physical phenomena and furthermore, aid in the interpretation of observational data. One such approach – low-order dynamo models – usually approximate physical processes that occur in a dynamo through truncated equations. Such models have the advantage of exploring a wide variety of solar behavior that is governed by the same underlying mathematical structure, without studying the dynamo process in detail, or making other modeling assumptions. Sections 3.5 and 3.6 of Petrovay (2020) give an overview of this topic classified by two types of low-order models: truncated models and generic normal-form models. See also the review by Lopes et al. (2014). Here we present the progress in this field by classifying them based on different physical processes influencing solar cycle variability and predictability. Although there is no conclusive evidence of the presence or absence of chaos, most recent studies suggest that the irregular component in the variation of solar activity is dominated by stochastic mechanisms.

4.2.1 Deterministic Chaos Subject to Weak Stochastic Perturbations

Such studies assume the non-linear solar dynamo is a chaotic oscillator, subject only to weak stochastic perturbations. The generic normal-form equations are investigated utilizing the theory of non-linear dynamics by bifurcation analysis. The bifurcation sequences are robust. Although the approach has no actual predictive power, they provide an understanding of generic properties and explain the origin of assumed chaotic behavior. Tobias et al. (1995) used a Poincare-Birkhoff normal form for a saddle-node or Hopf bifurcation. Their results show that stellar dynamos are governed by equations that possess the bifurcation structure. Modulation of the basic cycle and chaos are found to be a natural consequence of the successive transitions from a non-magnetic state to periodic cyclic activity and then to periodically modulated cyclic activity followed by chaotically modulated cycles. This behaviour can be further complicated by symmetry-breaking bifurcations that lead to mixed-mode modulation of cyclic behaviour (Knobloch et al. 1998). Trajectories in the phase space spend most of the time in the dipole subspace, displaying modulated cyclic activity, but occasionally flip during a deep grand minimum. Wilmot-Smith et al. (2005) extended the model of Tobias et al. (1995) to include an axisymmetry-breaking term. The model is able to reproduce some properties found in observations of solar-type stars. Their solution also exhibits clustering of minima, together with periods of reduced and enhanced magnetic activity.

There are also some studies using truncated dynamo models to investigate chaotic behaviour. For example, some argue that mixed modes of symmetry can only appear as a result of symmetry-breaking bifurcations in the nonlinear domain based on a constructed minimal nonlinear α-Ω dynamo (Jennings and Weiss 1991). Tobias (1996) show that grand minima naturally occur in their low-order non-linear α-Ω dynamo if the magnetic Prandtl number is small. The pattern of magnetic activity during grand minima can be contrasted both with sunspot observations and with the cosmogenic record.

Yoshimura (1978) suggested that a time-delay mechanism is intrinsic to the feedback action of a magnetic field on the dynamo process. Wilmot-Smith et al. (2006) constructed a truncated dynamo model to mimic the generation of field components in spatially segregated layers and their communication was mimicked through the use of time delays in a dynamo model involving delay differential equations. A variety of dynamic behaviors including periodic and aperiodic oscillations similar to solar variability arise as a direct consequence of the introduction of time delays in the system. Hazra et al. (2014) extended the model of Wilmot-Smith et al. (2006) by introducing stochastic fluctuations to investigate the solar behaviour during a grand minimum. Recently Tripathi et al. (2021) apply the model with an additive noise to understand the breakdown of stellar gyrochronology relations at about the age of the Sun (van Saders JL et al. 2016). The one-dimensional iterative map is an effective and classical method to investigate the dynamics of a system. Using this method, studies (Durney 2000; Charbonneau 2001) explored the dynamical consequences of the time delay in the B-L type dynamo. As the dynamo number increases beyond criticality, the system exhibits a classical transition to chaos through successive period doubling bifurcations.

4.2.2 Weakly Nonlinear Limit Cycle Affected by Random Noise

Since the non-stationary nature of solar convection is an intimate part of the solar dynamo, a rich body of literature regards that solar variability is largely governed by stochastic perturbations. Random noise has been used to fully mimic the behaviour of the solar cycle (Barnes et al. 1980) while others describe the global behavior of the solar cycle in terms of a Van der Pol oscillator (Mininni et al. 2001); a stochastic parameter corresponding to a stochastic mechanism in the dynamo process was introduced in the Van der Pol equations to model irregularities in the solar cycle were modeled. The mean values and deviations obtained for the periods, rise times, and peak values, were in good agreement with the values obtained from the sunspot time series. Another example is a low-order dynamo model with a stochastic α-effect (Passos and Lopes 2011) in which grand minima episodes manifested; this model is characterized by a non-linear oscillator whose coefficients retain most of the physics behind dynamo theory.

While most low-order models have a loose connection with observations, Cameron and Schüssler (2017) developed a generic normal-form model, whose parameters are all constrained by observations. They introduce multiplicative noise to the generic normal-form model of a weakly nonlinear system near a Hopf bifurcation. Their model reproduces the characteristics of the solar cycle variability on timescales between decades and millennia, including the properties of grand minima, which suggest that the variability of the solar cycle can be understood in terms of a weakly nonlinear limit cycle affected by random noise. In addition, they argue that no intrinsic periodicities apart from the 11-year cycle are required to understand the variability.

4.3 Babcock-Leighton-Type Kinematic Dynamo Models

Over the past two decades – supported by advances in flux transport models and observational evidence – Babcock-Leighton type solar dynamo models have become the mainstream approach to solar cycle modeling. The B-L mechanism imbibes the processes of emergence of toroidal fields through the convection zone and their subsequent decay and transport by supergranular diffusion and large-scale surface flow fields over the surface, i.e., the SFT processes discussed in Sect. 3.1. Since the SFT processes are directly observed, these act as a source of data assimilation in the B-L dynamo paving the way towards data-driven predictions – akin to what has been achieved in weather models.

4.3.1 Effects of the Meridional Flow and the Time Delay

The meridional flow plays an essential role in the B-L type flux transport dynamo (FTD). The flow strength can modulate not only the cycle strength but also the cycle period. There exists a rich literature describing the effects of the meridional flow on modulation of solar cycles based on FTD models (Yeates et al. 2008; Karak 2010; Nandy et al. 2011; Karak and Choudhuri 2013). Bushby and Tobias (2007) introduced weak stochastic perturbations in the penetration depth of the meridional flow and the results showed significant modulation in the activity cycle; while they argue that this modulation leads to a loss of predictability. Nandy (2021) provides counter arguments pointing out short-term prediction up to one cycle is possible due to the inherent one-cycle memory in the sunspot cycle; see also Yeates et al. (2008), Karak and Nandy (2012), Hazra et al. (2020), Kumar et al. (2021). We note that recently Zhang and Jiang (2022) developed a B-L type dynamo working in the bulk of the convection zone. The model has a much weaker dependence on the flow. Only the flow at the solar surface plays a key role in the polar field generation as in the SFT models implying that the flux transport paradigm is not hostage to meridional circulation being the primary transporter of magnetic flux within the SCZ – as also argued in Hazra and Nandy (2016), Karak and Cameron (2016).

In Sect. 4.2 we have shown numerous attempts at the analysis of the dynamical system using low-order models. For the first time, Charbonneau et al. (2005) presented a series of numerical simulations of the PDE-based 2D B-L type dynamo model incorporating amplitude-limiting quenching nonlinearity. The solutions show a well-defined transition to chaos via a sequence of period-doubling bifurcations as the dynamo numbers $C_S = s_0 R_\odot / \eta_t$ (s_0: strength of the source term, η_t is the turbulent magnetic diffusivity in the Sun's convective envelope) is increased. The results are presented in Fig. 5. Hence they suggest that the time delay inherent to the B-L type dynamo process, acting in conjunction with a simple amplitude-quenching algebraic-type nonlinearity could naturally lead to the observed fluctuations in the amplitude of the solar cycle. The time delay was regarded as the third class of fluctuation mechanisms by Charbonneau et al. (2005). The method was further extended (Charbonneau et al. 2007) to investigate the odd-even pattern in sunspot cycle peak amplitudes. Indeed, it is now being recognized that time delays introduced into the dynamo system due to the finite time necessary for flux transport processes to bridge the source layers of the poloidal and toroidal fields across the convection zone introduces a memory into the system which makes solar cycle predictions a realistic possibility (Nandy 2021).

4.3.2 Observable Nonlinear and Stochastic Mechanisms in the Source Term

Within the framework of the B-L type dynamo, the emergence and decay of tilted bipolar sunspots give rise to the poloidal field. The amount of poloidal field depends on the sunspot

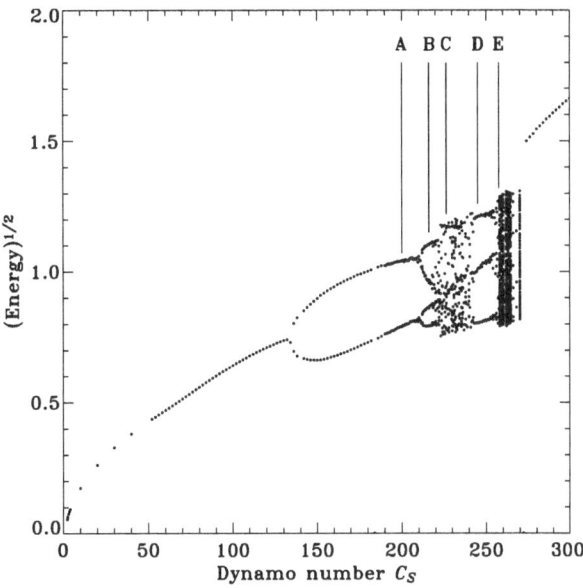

Fig. 5 Bifurcation diagram reconstructed from a sequence of numerical 2D B-L type dynamo solutions with increasing dynamo numbers C_S. Vertical lines labeled 'A' to 'E' correspond to representative 2-periodic, 4-periodic, 5-periodic, 3-periodic, and chaotic solutions, respectively. Image reproduced with permission from Charbonneau et al. (2005), copyright by the American Astronomical Society

properties, e.g., the tilt angle of the bipolar sunspots, which show both the systematic property resulting from Joy's law and the stochastic one due to the tilt scatter. Hence, the B-L mechanism has an inherent randomness. Studies like Charbonneau and Dikpati (2000), Passos et al. (2014), Kitchatinov et al. (2018), Hazra and Nandy (2019a), Saha et al. (2022) took the stochastic fluctuation in the poloidal field source term as a free parameter and investigated their possible effects on the cycle modulation. Based on a B-L type solar dynamo with an additional mean-field α-effect, Sanchez et al. (2014) quantified the intrinsic limit of predictability, i.e., e-folding time τ, which is the equivalent of the two weeks for the weather forecast. As expected, the e-folding time is shown to decrease corresponding to a short forecast horizon, with the increase of the α-effect.

Studies (Kitchatinov and Olemskoy 2011; Choudhuri and Karak 2012; Olemskoy et al. 2013; Karak and Miesch 2017) attempted to estimate the parameters of the B-L mechanism and their fluctuations using historical sunspot data. Jiang et al. (2014a) measured the tilt-angle scatter using the observed tilt-angle data and quantified the effects of this scatter on the evolution of the solar surface field using SFT simulations with flux input based upon the recorded sunspot groups. The result showed that the effect of the scatter on the evolution of the large-scale magnetic field at the solar surface reaches a level of over 30%. When a BMR with area A and tilt angle α emerges, it has the (initial) axial dipole field strength $D_i \propto A^{1.5} \sin \alpha$. We define the final contribution of a BMR to the axial dipole field as the final axial dipole field strength D_f. Jiang et al. (2014a) show that D_f has the Gaussian latitudinal dependence. The result was confirmed by others (Nagy et al. 2017; Whitbread et al. 2018; Petrovay et al. 2020b), that is

$$D_f = D_i \exp(-\lambda^2/\lambda_R^2), \qquad (5)$$

where λ_R is determined by the ratio of equatorial flow divergence to diffusivity. Wang et al. (2021) further generalized the result to ARs with realistic configuration, which usually show large differences in evolution from the idealized BMR approximation for δ-type ARs (Jiang et al. 2019; Yeates 2020). Hence big ARs emerging close to the equator could have big effects on the polar field evolution, and on the subsequent cycle evolution based on the correlation between the polar field at cycle minimum and the subsequent cycle strength. These are referred to as rogue ARs by Nagy et al. (2017). Jiang et al. (2015) demonstrated that these low-latitude regions with abnormal polarity could indeed be the cause of the weak polar field at the end of cycle 23, hence the low amplitude of cycle 24. Simulations by Nagy et al. (2017) indicate that in the most extreme case, such an event could lead to a grand minimum; they argue that the emergence of rogue ARs in the late phase of a cycle may limit the scope of predicting the dipole moment (and polar field amplitude) at the minimum of a cycle. However, it is likely that the impact of such rogue regions may be estimated through the ensemble prediction approach (Cameron et al. 2016; Bhowmik and Nandy 2018; Jiang et al. 2018).

The stochastic properties of the BMRs emergence mentioned above provide the observable stochastic mechanisms in solar cycle modulation. The systematic properties of the BMRs emergence have recently been suggested to be observable nonlinearities. Historical data show that the cycle amplitude has an anti-correlation with the mean tilt angle of BMRs (Dasi-Espuig et al. 2010; Jiao et al. 2021) and a positive correlation with the mean emergence latitudes (Solanki et al. 2008; Jiang et al. 2011). Jiang (2020) investigated the effects of the latitude and tilt's properties on the solar cycle, which are referred to as latitudinal quenching and tilt quenching, respectively as depicted in Fig. 6 (also see Jha et al. 2020). They defined the final total dipole moment, which is the total dipole moment generated by all sunspot emergence during a whole cycle. For a stronger cycle, more ARs emerge at higher latitudes, limiting the amount of flux cancellation between the leading spots of opposite polarities through cross-equatorial interactions. Effectively, this causes a weaker impact on the dipole moment evolution as magnetic flux from both leading and following spots of opposite polarities advect towards the poles. Additionally, ARs from stronger cycles are likely to have smaller mean tilt angles, thus collectively contributing less to the dipole moment. In comparison, in a weaker solar cycle, more ARs emerging at lower latitudes ensure enhanced cross-equatorial flux cancellation, thereby, a larger effect on the dipole moment. Weaker cycles also tend to have ARs with higher mean tilt angles, which again will be more impactful in determining the dipole moment evolution. Thus, both forms of quenching lead to the expected final total dipolar moment being enhanced for weak cycles and saturated to a nearly constant value for normal and strong cycles. This could be an explanation of the observed long-term solar cycle variability, e.g., the odd-even rule. Karak (2020) verified that latitudinal quenching is a potential mechanism for limiting the magnetic field growth in the Sun using a three-dimensional B-L type dynamo model. Talafha et al. (2022) systematically explored the relative importance played by these two forms of quenching in the solar dynamo showing that this is governed by λ_R.

5 Physics-Based Solar Cycle Predictions

The importance of the dipole moment (or the average polar field) in solar cycle predictions is established through observation and dynamo theory. The most successful empirical method for solar cycle predictions based on the polar field precursor (Schatten et al. 1978) in fact predated the solar dynamo model based predictions. Thus, physics-based predictions of the solar cycle, in general, are either based on SFT simulations aiming to estimate the dipole moment (related to polar flux) at the cycle minimum or involve dynamo simulations with modified poloidal field source term according to the observed (or simulated) dipole moment at the cycle minimum. Nandy (2002) first alluded to the possibility of developing data-driven

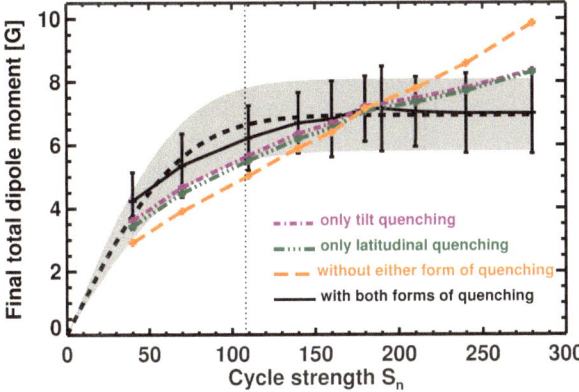

Fig. 6 Effects of observable nonlinear and stochastic mechanisms in the source term on the poloidal field generation. The black solid curve indicates the expected values from 100 SFT simulations using random sunspot group realizations including latitudinal and tilt quenching. Error bars correspond to the 1σ standard deviation, caused by the randomness in the properties of sunspot groups. Green dashed–triple-dotted and purple dashed–dotted curves show the expected values for SFT simulations with only the latitudinal and tilt quenching, respectively. The orange dashed curve shows the expected value of SFT simulations without latitudinal or tilt quenching. Image reproduced with permission from Jiang (2020), copyright by the American Astronomical Society

predictive solar dynamo models by utilizing the observed poloidal field as inputs; although this particular branch of solar cycle predictions is relatively new, significant progress has been achieved through contributions from multiple works predicting the past cycle 24 and present cycle 25 using physics-based models.

5.1 Role of SFT Models in Solar Cycle Predictions

Despite the dissimilarities among different SFT models regarding their treatments of the emerged sunspot statistics and observed transport processes on the photosphere, SFT models have played a major role in physics-based solar cycle predictions, especially of cycle 25. The idea lies in the fact that the Sun's magnetic axial dipole moment at the end of a solar cycle is strongly correlated with the following cycle's peak amplitude. This positive correlation is found from observation spanning multiple cycles (Muñoz-Jaramillo et al. 2013) and is also supported by the principles of the solar dynamo mechanism (see Sect. 3.2 for more details).

Using SFT simulations, Cameron et al. (2016) presented the first prediction of cycle 25 about four years before the cycle minimum (which occurred at the end of 2019). Their simulation started with the observed synoptic magnetogram at the end of 2015. The sunspot emergence statistics in the declining phase (years: 2016 – 2020) of cycle 24 were generated using 50 randomizations which included uncertainties associated with the emergence timing, latitude-longitude position, tilt angle and magnetic flux of the sunspots. They provided a possible range of the axial dipole moment at the end of cycle 24. Based on the positive correlation between dipole moment and the following cycle amplitude, they predicted a moderately strong sunspot cycle 25. Hathaway and Upton (2016), Upton and Hathaway (2018) took a similar approach to estimate the axial dipole moment at the end of cycle 24 using their AFT simulations. However, the sunspot statistics corresponding to the declining phase of cycle 24 were taken from the declining phase of solar cycle 14. The uncertainties in predicting the dipole moment were realised by considering stochastic variations in the convective motion details, sunspot tilt angles, and changes in the meridional flow profile. Their predicted dipole moment amplitude suggested that cycle 25 would be a weak to moderate cycle. Iijima et al. (2017) argued that the axial dipole moment does not vary significantly in the last three years of the declining phase of any sunspot cycle. Thus to predict the dipole moment at the end of cycle 24, they initiated their SFT simulation in 2017 with an observed synoptic magnetogram and continued till the end of 2019 without assimilating any sunspots. Their prediction suggested a weak solar cycle 25. The importance of correctly simulating the surface magnetic field distribution and their consecutive inclusion in an interior dynamo model was extensively utilized by other studies (Labonville et al. 2019; Bhowmik and Nandy 2018) for cycle predictions.

As addressed in previous sections, the randomness in sunspot emergence properties plays an important role in the final amplitude of the axial dipole moment at cycle minimum. Assimilating the uncertainties associated with the sunspots and the observed magnetogram used as the initial condition in the SFT simulations, Jiang et al. (2018) demonstrated a generalized scheme to investigate the predictability of the solar cycle over one cycle. The scheme requires at least three years of sunspot observation for the ongoing cycle to predict the statistical properties of the sunspots during the remaining phase. Their prediction of cycle 25 based on this scheme suggests a weak-to-moderate cycle peak.

5.2 Dynamo-Based Solar Cycle Predictions with Data Assimilation

The B-L type 2D flux transport dynamo models were utilized for the first time to predict cycle 24 during the mid-2000s (Dikpati et al. 2006; Choudhuri et al. 2007). However, the only two physics-based predictions of cycle 24 diverged significantly from each other (with a difference of ~ 100 sunspots during the maximum). Despite using similar dynamo models, such divergence can arise from two aspects: differences in the dominating flux transport processes (Yeates et al. 2008) and how the B-L source term is designed according to the observed magnetic field. Exploring these two points is crucial for understanding the physics behind sunspot cycle predictions as well as providing realistic forecasts.

All kinematic flux transport dynamo models consider the following transport processes at the least: differential rotation, meridional circulation and magnetic diffusion. They use analytical functions corresponding to the observed differential rotation (Howe 2009). The observed meridional flow on the solar surface sets constraints for the meridional circulation profile used within the SCZ. The exact structuring of this flow at different layers of the SCZ still requires further observational support (Hanasoge 2022), but recent helioseismic studies suggest a one-cell meridional circulation (Rajaguru and Antia 2015; Liang et al. 2018). Nonetheless, these models assume an equatorward branch of the meridional flow near the tachocline which ensures observed latitudinal propagation of sunspot activity belts in both hemispheres and an appropriate cycle duration (Choudhuri et al. 1995; Dikpati and Charbonneau 1999; Nandy and Choudhuri 2002; Chatterjee et al. 2004; Hazra et al. 2014; Hazra and Nandy 2019b). Furthermore, the amplitude and profiles of magnetic diffusivity are also based on analytical functions, which only vary with the depth of the SCZ (Muñoz-Jaramillo et al. 2011).

However, based on the strength of diffusivity, flux transport dynamo models behave differently and can be categorized into two major classes: advection dominated (diffusivity order, $\eta \sim 10^{10}$ cm^2 s^{-1}, see Dikpati and Charbonneau 1999) and diffusion dominated ($\eta \sim 10^{12}$ cm^2 s^{-1}, see Choudhuri et al. 2007). The strength of the diffusivity decides which transport mechanism between meridional circulation and magnetic diffusivity will be more effective for convecting $\mathbf{B_P}$ to the deeper layers of the SCZ (Yeates et al. 2008). It also determines whether $\mathbf{B_P}$ associated with multiple past solar cycles can survive in the SCZ at the prescribed diffusivity and contribute simultaneously to the generation of new $\mathbf{B_T}$ of the following cycle (Yeates et al. 2008; Jiang et al. 2007). However, the inclusion of turbulent pumping (Karak and Nandy 2012; Hazra and Nandy 2016) as an additional transport process in flux transport dynamo diminishes the difference between the advection-dominated and diffusion-dominated regimes. All these results are crucial for estimating the dynamical memory of the solar dynamo models and their ability to accurately predict the future solar cycle amplitude. Dynamical memory is a measure of determining the range of temporal association of the poloidal field ($\mathbf{B_P}$) of a certain sunspot cycle (say, n^{th}) with the toroidal field ($\mathbf{B_T}$) of following cycles (say, $n+1^{th}$, $n+2^{th}$, $n+3^{th}$, etc.). Note that for advection-dominated dynamo models, the dynamical memory is about two solar cycles, whereas it's about half a solar cycle for diffusion-dominated dynamo models or in models with turbulent pumping (Yeates et al. 2008; Karak and Nandy 2012).

Besides the transport parameters, how we assimilate observational data to model the poloidal field ($\mathbf{B_P}$) source will influence the successive generation of the toroidal field ($\mathbf{B_T}$), thus is crucial for solar cycle predictions. Below, we discuss this aspect of 'data-driven' dynamo models in the context of solar cycle prediction.

As mentioned in Sect. 3.2, surface flux transport processes acting on emerged active regions produces the large-scale photospheric field and serves as a reliable means for poloidal field generation. Multiple efforts have been made to assimilate the observed surface data in flux transport dynamo models. Dikpati and Gilman (2006) included observed sunspot group areas to formulate the $\mathbf{B_P}$ source while assuming all spots of any solar cycle are distributed in the same latitudinal belt (between 5° and 35°) and have similar tilt angles. However, their data-driven model failed to correctly predict the solar cycle 24 peak (Dikpati et al. 2006). The primary reasons for this disparity were that the idealized realization of the sunspots (fixed latitude and tilt angle) results in a poloidal source at the minimum (of nth cycle) directly proportional to the preceding cycle (n^{th}) amplitude and that the low magnetic diffusivity in their flux transport dynamo model increased the dynamical memory to more than two solar cycles. Thus according to their model, not only does a strong solar cycle (n^{th}) produce a strong poloidal source, its strength influences several following solar cycles ($n+1^{th}$, $n+2^{th}$ and $n+3^{th}$).

In contrast, Choudhuri et al. (2007) and Jiang et al. (2007) used a diffusion-dominated flux transport dynamo model to predict sunspot cycle 24. For modeling the $\mathbf{B_P}$ source, they relied on observed large-scale surface magnetic field distribution (for example, the axial dipole moment) during the solar cycle 23 minimum. Their prediction was a good match to the observed peak of cycle 24. The positive correlation between $\mathbf{B_P}(n)$ and the $\mathbf{B_T}(n+1)$ using a diffusion-dominated dynamo model where the dynamic memory is half a solar cycle ensured the success of their prediction. Guo et al. (2021) took a similar approach by combining observed axial dipole moment to a diffusion-dominated dynamo model to predict cycle 25. In a recent review, Nandy (2021) discusses in details how observations and stochastically forced dynamo simulations support only a short half- to one-cycle memory in the solar cycle, suggesting that the latter class of dynamo models are the right approach to take for solar cycle predictions.

Recently, Bhowmik and Nandy (2018) assimilated a series of surface magnetic field distribution at eight successive solar minima (cycles 16−23 minima) in a flux transport dynamo model (diffusion-dominated). The surface maps were obtained from their calibrated century-scale SFT simulation, which assimilates the observed statistics of emerging bipolar sunspot pairs during that period. Their coupled SFT-dynamo simulations reproduced past solar cycles (17−23) with reasonable accuracy (except cycle 19). The same methodology was utilized to provide an ensemble forecast for cycle 25 while assimilating the predicted

Fig. 7 Comparison of physical model-based predictions of solar cycle 25 peak amplitude. The average of the six predictions is 107.75 sunspots (with $\pm 1\sigma = 17.15$). Details on each of these predictions are described in Sects. 5.1 and 5.2

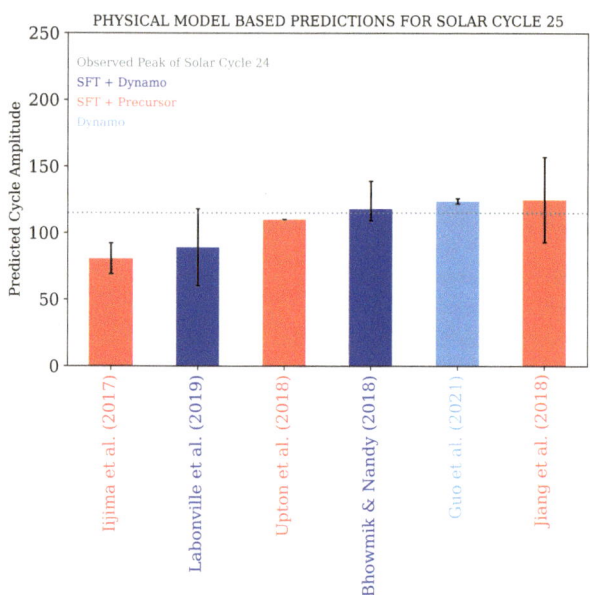

surface field distributions at cycle 24 minimum from their SFT simulations. Their prediction indicates a weak cycle 25 (with a peak SSN of 118 and a range from 109–139), similar to, or slightly stronger than cycle 24. Note that the upper bound of the Bhowmik and Nandy (2018) prediction as reported in Nandy (2021) was misreported as 155 but should have been 139. The 2 × 2D SFT-dynamo model by Lemerle and Charbonneau (2017) is another example of coupling the surface magnetic field to the internal dynamo, which occurs more intimately in their model. Labonville et al. (2019) utilized the same model to assimilate the series of BMRs observed during cycles 23 and 24 (from Yeates et al. 2007) into the SFT part of the simulations. They first calibrated the model by including cycle 23 data to produce an ensemble forecast for cycle 24 and subsequently comparing it with observation. They then assimilated the BMRs series for cycles 23 and 24 to present an ensemble forecast for cycle 25, including its amplitude (weaker than cycle 24), rising and declining phases and northern and southern asymmetry.

All physics-based predictions for cycle 25 are depicted in Fig. 7.

5.3 Comparison with Other Prediction Methods

The methodologies of predictions are not restricted to physical modeling only. They can be based on (a) precursor technique-based forecasts, (b) non-linear mathematical modeling, (c) statistical forecasts, (d) machine learning and neural network, (e) spectral methods etc. (Petrovay 2020; Nandy 2021). Note that most of the precursor-based forecasts consider the physics of solar dynamo and cycle predictability to some extent without performing any computational modeling. For example, based on semi-empirical and semi-physical approaches Hazra and Choudhuri (2019) and Kumar et al. (2022) predicted solar cycle 25 amplitude, where the latter group claimed that polar field evolution after the polarity reversal exhibits similar features like the Waldmeier effect and can be utilized for cycle predictions. Nonetheless, the forecasts based on diverging techniques show a wide variation in predicted cycle amplitudes: with peak sunspot numbers ranging between 50 and 250 for solar cycle

25 (see Fig. 3 of Nandy 2021). In that regard, physics-based predictions of cycle 25 have reached a consensus with an average of 107.75 sunspots (with $\pm 1\sigma = 17.15$). In contrast, for cycle 24 physics-based predictions, the difference between the two predicted peaks was more than 100 sunspots.

6 Summary and Future Prospects

It is noteworthy that while the only two physics-based predictions of solar cycle 24 diverged significantly from each other, physics-based predictions of cycle 25 show significantly more convergence. This is possibly an indication of increasing understanding of the solar dynamo process – as argued in Nandy (2002) – and advances in assimilating observational data in the computational models used for predictions. However, there are significant improvements that are still necessary in the adopted modeling approaches.

While all physical models of solar cycle predictions have been 2D in nature, global 3D dynamo models have the promise of capturing the surface flux transport dynamics and internal magnetic field evolution self-consistently. Some recent works have solved the induction equation in three spatial dimensions within a dynamo framework thus going beyond the 2D axisymmetric models (Yeates and Muñoz-Jaramillo 2013; Miesch and Dikpati 2014; Hazra et al. 2017; Karak and Miesch 2017; Kumar et al. 2019; Whitbread et al. 2019). These are B-L type dynamos working in a kinematic mode with modules to incorporate realistic sunspot emergence and decay of the associated flux. These models provide the opportunity of further development towards dynamical models, imbibing in-built axisymmetric and non-axisymmetric feedback mechanisms (Nagy et al. 2020), thus slowly closing the gap between these phenomenological B-L type dynamo models and the full MHD dynamo models. However, such models still have two significant discrepancies: the polar field generated on the surface is much stronger than the observed order of magnitude, and sunspot emergence at higher latitudes is prevented artificially. Nonetheless, these models hold the promise of self-consistently imbibing elements of the surface flux transport dynamics leading to polar field reversal and build-up as well as solar internal magnetic field evolution, which 2D models cannot.

Finally, we end by posing the following provocative questions. From a purely utilitarian view, are dynamo models at all necessary for solar cycle predictions and can they provide higher accuracy or advantage compared to empirical polar field precursor based techniques? There is no doubt that physical approaches based on surface flux transport models and dynamo models have significantly advanced our understanding of solar cycle predictability; however, the resolution of the above questions are fundamental to sustained growth prospects of this field of research.

Acknowledgements All authors acknowledge support from the International Space Science Institute (Bern) for facilitating enriching interactions which led to the planning of this review.

Author Contribution D.N. coordinated and planned the writing of this review in consultation with all authors. P.B. managed and led the manuscript writing process to which all authors contributed.

Funding D.N. acknowledges financial support for the Center of Excellence in Space Sciences India from the Ministry of Education, Government of India. P.B. acknowledges support from the project ST/W00108X/1 funded by the Science and Technology Facilities Council (STFC), United Kingdom. J.J. was supported by the National Natural Science Foundation of China grant Nos. 12173005 and 11873023. L.U. was supported by NASA Heliophysics Living With a Star grants NNH16ZDA010N-LWS and NNH18ZDA001N-LWS and by NASA grant NNH18ZDA001N-DRIVE to the COFFIES DRIVE Center managed by Stanford University. A.L. acknowledges support from the Fonds de Recherche du Québec – Nature et Technologie (Programme de recherche collégiale).

Declarations

Competing Interests The authors declare no conflicts of interest.

References

Adams WS (1911) An investigation of the rotation period of the sun by spectroscopic methods. Publ Carnegie Inst Wash 138:1–132

Arge CN, Henney CJ, Koller J et al (2010) Air Force Data Assimilative Photospheric Flux Transport (ADAPT) model. In: Maksimovic M, Issautier K, Meyer-Vernet N et al (eds) Twelfth international solar wind conference. AIP Conference Proceedings, vol 1216. pp 343–346. https://doi.org/10.1063/1.3395870

Babcock HW (1961) The topology of the Sun's magnetic field and the 22-year cycle. Astrophys J 133:572. https://doi.org/10.1086/147060

Babcock HD, Livingston WC (1958) Changes in the Sun's polar magnetic field. Science 127:1058

Barnes JA, Sargent HH III, Tryon PV (1980) Sunspot cycle simulation using random noise. In: Pepin RO, Eddy JA, Merrill RB (eds) The ancient Sun: fossil record in the Earth, Moon and meteorites, pp 159–163

Basu S (2016) Global seismology of the Sun. Living Rev Sol Phys 13(1):2. https://doi.org/10.1007/s41116-016-0003-4. arXiv:1606.07071 [astro-ph.SR]

Belopolsky A (1933) Bestimmung der Sonnenrotation auf spektroskopischem Wege in den Jahren 1931, 1932 und 1933 in Pulkovo. Z Astrophys 7:357

Bhowmik P (2019) Polar flux imbalance at the sunspot cycle minimum governs hemispheric asymmetry in the following cycle. Astron Astrophys 632:A117. https://doi.org/10.1051/0004-6361/201834425

Bhowmik P, Nandy D (2018) Prediction of the strength and timing of sunspot cycle 25 reveal decadal-scale space environmental conditions. Nat Commun 9:5209. https://doi.org/10.1038/s41467-018-07690-0. arXiv:1909.04537 [astro-ph.SR]

Biswas A, Karak BB, Usoskin I et al (2023) Long-term modulation of solar cycles. Space Sci Rev 219(3):19. https://doi.org/10.1007/s11214-023-00968-w. arXiv:2302.14845 [astro-ph.SR]

Brandenburg A, Elstner D, Masada Y et al (2023) Turbulent processes and mean-field dynamo. Space Sci Rev 219. arXiv:2303.12425 [astro-ph.SR]

Bushby PJ, Tobias SM (2007) On predicting the solar cycle using mean-field models. Astrophys J 661(2):1289–1296. https://doi.org/10.1086/516628. arXiv:0704.2345 [astro-ph]

Cameron RH, Schüssler M (2015) The crucial role of surface magnetic fields for the solar dynamo. Science 347(6228):1333–1335. https://doi.org/10.1126/science.1261470. arXiv:1503.08469 [astro-ph.SR]

Cameron RH, Schüssler M (2017) Understanding solar cycle variability. Astrophys J 843(2):111. https://doi.org/10.3847/1538-4357/aa767a. arXiv:1705.10746 [astro-ph.SR]

Cameron R, Schüssler M (2023) Observationally guided models for the solar dynamo and the role of the surface field. Space Sci Rev 219. arXiv:2305.02253 [astro-ph.SR]

Cameron RH, Jiang J, Schmitt D et al (2010) Surface flux transport modeling for solar cycles 15-21: effects of cycle-dependent tilt angles of sunspot groups. Astrophys J 719(1):264–270. https://doi.org/10.1088/0004-637X/719/1/264. arXiv:1006.3061 [astro-ph.SR]

Cameron RH, Jiang J, Schüssler M (2016) Solar cycle 25: another moderate cycle? Astrophys J Lett 823(2):L22. https://doi.org/10.3847/2041-8205/823/2/L22. arXiv:1604.05405 [astro-ph.SR]

Cameron RH, Dikpati M, Brandenburg A (2017) The global solar dynamo. Space Sci Rev 210(1–4):367–395. https://doi.org/10.1007/s11214-015-0230-3. arXiv:1602.01754 [astro-ph.SR]

Carbonell M, Oliver R, Ballester JL (1994) A search for chaotic behaviour in solar activity. Astron Astrophys 290:983–994

Carrington RC (1858) On the distribution of the solar spots in latitudes since the beginning of the year 1854, with a map. Mon Not R Astron Soc 19:1–3. https://doi.org/10.1093/mnras/19.1.1

Charbonneau P (2001) Multiperiodicity, chaos, and intermittency in a reduced model of the solar cycle. Sol Phys. 199(2): 385–404. https://doi.org/10.1023/A:1010387509792

Charbonneau P (2020) Dynamo models of the solar cycle. Living Rev Sol Phys 17(1):4. https://doi.org/10.1007/s41116-020-00025-6

Charbonneau P, Dikpati M (2000) Stochastic fluctuations in a Babcock-Leighton model of the solar cycle. Astrophys J 543(2):1027–1043. https://doi.org/10.1086/317142

Charbonneau P, St-Jean C, Zacharias P (2005) Fluctuations in Babcock-Leighton dynamos. I. Period doubling and transition to chaos. Astrophys J 619(1):613–622. https://doi.org/10.1086/426385

Charbonneau P, Beaubien G, St-Jean C (2007) Fluctuations in Babcock-Leighton dynamos. II. Revisiting the Gnevyshev-Ohl rule. Astrophys J 658(1):657–662. https://doi.org/10.1086/511177

Charney J (1948) On the scale of atmospheric motions. Geofys Publ 17:1–17

Chatterjee P, Nandy D, Choudhuri AR (2004) Full-sphere simulations of a circulation-dominated solar dynamo: exploring the parity issue. Astron Astrophys 427:1019–1030. https://doi.org/10.1051/0004-6361:20041199. arXiv:astro-ph/0405027 [astro-ph]

Choudhuri AR, Karak BB (2012) Origin of grand minima in sunspot cycles. Phys Rev Lett 109(17):171103. https://doi.org/10.1103/PhysRevLett.109.171103. arXiv:1208.3947 [astro-ph.SR]

Choudhuri AR, Schussler M, Dikpati M (1995) The solar dynamo with meridional circulation. Astron Astrophys 303:L29

Choudhuri AR, Chatterjee P, Jiang J (2007) Predicting solar cycle 24 with a solar dynamo model. Phys Rev Lett 98(13):131103. https://doi.org/10.1103/PhysRevLett.98.131103. arXiv:astro-ph/0701527 [astro-ph]

Clette F, Svalgaard L, Vaquero JM et al (2014) Revisiting the sunspot number. Space Sci Rev 186: 35–103. https://doi.org/10.1007/s11214-014-0074-2

Dash S, Bhowmik P, Athira BS et al (2020) Prediction of the Sun's coronal magnetic field and forward-modeled polarization characteristics for the 2019 July 2 total solar eclipse. Astrophys J 890(1):37. https://doi.org/10.3847/1538-4357/ab6a91. arXiv:1906.10201 [astro-ph.SR]

Dasi-Espuig M, Solanki SK, Krivova NA et al (2010) Sunspot group tilt angles and the strength of the solar cycle. Astron Astrophys 518:A7. https://doi.org/10.1051/0004-6361/201014301. arXiv:1005.1774 [astro-ph.SR]

Deng LH, Li B, Xiang YY et al (2016) Comparison of chaotic and fractal properties of polar faculae with sunspot activity. Astron J 151(1):2. https://doi.org/10.3847/0004-6256/151/1/2

DeVore CR, Boris JP, Sheeley NR Jr (1984) The concentration of the large-scale solar magnetic field by a meridional surface flow. Sol Phys 92(1–2):1–14. https://doi.org/10.1007/BF00157230

Dikpati M, Charbonneau P (1999) A Babcock-Leighton flux transport dynamo with solar-like differential rotation. Astrophys J 518(1):508–520. https://doi.org/10.1086/307269

Dikpati M, Gilman PA (2006) Simulating and predicting solar cycles using a flux-transport dynamo. Astrophys J 649(1):498–514. https://doi.org/10.1086/506314

Dikpati M, de Toma G, Gilman PA (2006) Predicting the strength of solar cycle 24 using a flux-transport dynamo-based tool. Geophys Res Lett 33(5):L05102. https://doi.org/10.1029/2005GL025221

Durney BR (2000) On the differences between odd and even solar cycles. Sol Phys 196(2):421–426. https://doi.org/10.1023/A:1005285315323

Eddy JA (1976) The Maunder minimum. Science 192:1189–1202. https://doi.org/10.1126/science.192.4245.1189

Fan Y (2021) Magnetic fields in the solar convection zone. Living Rev Sol Phys 18(1):5. https://doi.org/10.1007/s41116-021-00031-2

Fisher GH, Fan Y, Longcope DW et al (2000) The solar dynamo and emerging flux – (invited review). Sol Phys 192:119–139. https://doi.org/10.1023/A:1005286516009

Grassberger P, Procaccia I (1983) Characterization of strange attractors. Phys Rev Lett 50(5):346–349. https://doi.org/10.1103/PhysRevLett.50.346

Guo W, Jiang J, Wang JX (2021) A dynamo-based prediction of solar cycle 25. Sol Phys 296(9):136. https://doi.org/10.1007/s11207-021-01878-2. arXiv:2108.01412 [astro-ph.SR]

Hale GE (1908a) On the probable existence of a magnetic field in Sun-spots. Astrophys J 28:315. https://doi.org/10.1086/141602

Hale GE (1908b) The Zeeman effect in the Sun. Publ Astron Soc Pac 20(123):287. https://doi.org/10.1086/121847

Hale GE, Ellerman F, Nicholson SB et al (1919) The magnetic polarity of Sun-spots. Astrophys J 49:153. https://doi.org/10.1086/142452

Hanasoge SM (2022) Surface and interior meridional circulation in the Sun. Living Rev Sol Phys 19(1):3. https://doi.org/10.1007/s41116-022-00034-7

Hanslmeier A, Brajša R (2010) The chaotic solar cycle. I. Analysis of cosmogenic ^{14}C-data. Astron Astrophys 509:A5. https://doi.org/10.1051/0004-6361/200913095

Hathaway DH (2012) Supergranules as probes of the Sun's meridional circulation. Astrophys J 760(1):84. https://doi.org/10.1088/0004-637X/760/1/84. arXiv:1210.3343 [astro-ph.SR]

Hathaway DH, Rightmire L (2010) Variations in the Sun's meridional flow over a solar cycle. Science 327(5971):1350. https://doi.org/10.1126/science.1181990

Hathaway DH, Upton L (2014) The solar meridional circulation and sunspot cycle variability. J Geophys Res Space Phys 119(5):3316–3324. https://doi.org/10.1002/2013JA019432. arXiv:1404.5893 [astro-ph.SR]

Hathaway DH, Upton LA (2016) Predicting the amplitude and hemispheric asymmetry of solar cycle 25 with surface flux transport. J Geophys Res Space Phys 121(11):10,744–10,753. https://doi.org/10.1002/2016JA023190. arXiv:1611.05106 [astro-ph.SR]

Hathaway DH, Wilson RM, Reichmann EJ (1994) The shape of the sunspot cycle. Sol Phys 151(1):177–190. https://doi.org/10.1007/BF00654090

Hathaway DH, Teil T, Norton AA et al (2015) The Sun's photospheric convection spectrum. Astrophys J 811(2):105. https://doi.org/10.1088/0004-637X/811/2/105. arXiv:1508.03022 [astro-ph.SR]

Hathaway DH, Upton LA, Mahajan SS (2022) Variations in differential rotation and meridional flow within the Sun's surface shear layer 1996–2022. Front Astron Space Sci 9:1007290. https://doi.org/10.3389/fspas.2022.1007290. arXiv:2212.10619 [astro-ph.SR]

Hazra G, Choudhuri AR (2019) A new formula for predicting solar cycles. Astrophys J 880(2):113. https://doi.org/10.3847/1538-4357/ab2718. arXiv:1811.01363 [astro-ph.SR]

Hazra S, Nandy D (2016) A proposed paradigm for solar cycle dynamics mediated via turbulent pumping of magnetic flux in Babcock-Leighton-type solar dynamos. Astrophys J 832(1):9. https://doi.org/10.3847/0004-637X/832/1/9. arXiv:1608.08167 [astro-ph.SR]

Hazra S, Nandy D (2019a) The origin of parity changes in the solar cycle. Mon Not R Astron Soc 489(3):4329–4337. https://doi.org/10.1093/mnras/stz2476. arXiv:1906.06780 [astro-ph.SR]

Hazra S, Nandy D (2019b) The origin of parity changes in the solar cycle. Mon Not R Astron Soc 489(3):4329–4337. https://doi.org/10.1093/mnras/stz2476. arXiv:1906.06780 [astro-ph.SR]

Hazra G, Karak BB, Choudhuri AR (2014) Is a deep one-cell meridional circulation essential for the flux transport solar dynamo? Astrophys J 782(2):93. https://doi.org/10.1088/0004-637X/782/2/93. arXiv:1309.2838 [astro-ph.SR]

Hazra S, Passos D, Nandy D (2014) A stochastically forced time delay solar dynamo model: self-consistent recovery from a Maunder-like grand minimum necessitates a mean-field alpha effect. Astrophys J 789(1):5. https://doi.org/10.1088/0004-637X/789/1/5. arXiv:1307.5751 [astro-ph.SR]

Hazra G, Choudhuri AR, Miesch MS (2017) A theoretical study of the build-up of the Sun's polar magnetic field by using a 3D kinematic dynamo model. Astrophys J 835(1):39. https://doi.org/10.3847/1538-4357/835/1/39. arXiv:1610.02726 [astro-ph.SR]

Hazra S, Brun AS, Nandy D (2020) Does the mean-field α effect have any impact on the memory of the solar cycle? Astron Astrophys 642:A51. https://doi.org/10.1051/0004-6361/201937287. arXiv:2003.02776 [astro-ph.SR]

Hazra G, Nandy D, Kitchatinov L et al (2023) Mean field models of flux transport dynamo and meridional circulation in the Sun and stars. Space Sci Rev 219. arXiv:2302.09390 [astro-ph.SR]

Hickmann KS, Godinez HC, Henney CJ et al (2015) Data assimilation in the ADAPT photospheric flux transport model. Sol Phys 290(4):1105–1118. https://doi.org/10.1007/s11207-015-0666-3. arXiv:1410.6185 [math-ph]

Howard R (1984) Solar rotation. Annu Rev Astron Astrophys 22:131–155. https://doi.org/10.1146/annurev.aa.22.090184.001023

Howe R (2009) Solar interior rotation and its variation. Living Rev Sol Phys 6(1):1. https://doi.org/10.12942/lrsp-2009-1. arXiv:0902.2406 [astro-ph.SR]

Iijima H, Hotta H, Imada S et al (2017) Improvement of solar-cycle prediction: plateau of solar axial dipole moment. Astron Astrophys 607:L2. https://doi.org/10.1051/0004-6361/201731813. arXiv:1710.06528 [astro-ph.SR]

Jennings RL, Weiss NO (1991) Symmetry breaking in stellar dynamos. Mon Not R Astron Soc 252:249–260. https://doi.org/10.1093/mnras/252.2.249

Jha BK, Karak BB, Mandal S et al (2020) Magnetic field dependence of bipolar magnetic region tilts on the Sun: indication of tilt quenching. Astrophys J Lett 889(1):L19. https://doi.org/10.3847/2041-8213/ab665c. arXiv:1912.13223 [astro-ph.SR]

Jiang J (2020) Nonlinear mechanisms that regulate the solar cycle amplitude. Astrophys J 900(1):19. https://doi.org/10.3847/1538-4357/abaa4b. arXiv:2007.07069 [astro-ph.SR]

Jiang J, Chatterjee P, Choudhuri AR (2007) Solar activity forecast with a dynamo model. Mon Not R Astron Soc 381(4):1527–1542. https://doi.org/10.1111/j.1365-2966.2007.12267.x. arXiv:0707.2258 [astro-ph]

Jiang J, Cameron RH, Schmitt D et al (2011) The solar magnetic field since 1700. I. Characteristics of sunspot group emergence and reconstruction of the butterfly diagram. Astron Astrophys 528:A82. https://doi.org/10.1051/0004-6361/201016167. arXiv:1102.1266 [astro-ph.SR]

Jiang J, Cameron RH, Schüssler M (2014a) Effects of the scatter in sunspot group tilt angles on the large-scale magnetic field at the solar surface. Astrophys J 791(1):5. https://doi.org/10.1088/0004-637X/791/1/5. arXiv:1406.5564 [astro-ph.SR]

Jiang J, Hathaway DH, Cameron RH et al (2014b) Magnetic flux transport at the solar surface. Space Sci Rev 186(1–4):491–523. https://doi.org/10.1007/s11214-014-0083-1. arXiv:1408.3186 [astro-ph.SR]

Jiang J, Cameron RH, Schüssler M (2015) The cause of the weak solar cycle 24. Astrophys J Lett 808(1):L28. https://doi.org/10.1088/2041-8205/808/1/L28. arXiv:1507.01764 [astro-ph.SR]

Jiang J, Wang JX, Jiao QR et al (2018) Predictability of the solar cycle over one cycle. Astrophys J 863(2):159. https://doi.org/10.3847/1538-4357/aad197. arXiv:1807.01543 [astro-ph.SR]

Jiang J, Song Q, Wang JX et al (2019) Different contributions to space weather and space climate from different big solar active regions. Astrophys J 871(1):16. https://doi.org/10.3847/1538-4357/aaf64a. arXiv:1901.00116 [astro-ph.SR]

Jiang J, Zhang Z, Petrovay K (2023) Comparison of physics-based prediction models of solar cycle 25. J Atmos Sol-Terr Phys 243:106018. https://doi.org/10.1016/j.jastp.2023.106018. arXiv:2212.01158 [astro-ph.SR]

Jiao Q, Jiang J, Wang ZF (2021) Sunspot tilt angles revisited: dependence on the solar cycle strength. Astron Astrophys 653:A27. https://doi.org/10.1051/0004-6361/202141215. arXiv:2106.11615 [astro-ph.SR]

Jouve L, Brun AS, Aulanier G (2018) Interactions of twisted Ω-loops in a model solar convection zone. Astrophys J 857(2):83. https://doi.org/10.3847/1538-4357/aab5b6. arXiv:1803.04709 [astro-ph.SR]

Kalnay E (2003) Atmospheric modeling, data assimilation and predictability. Cambridge University Press, Cambridge

Käpylä PJ, Browning MK, Brun AS et al (2023) Simulations of solar and stellar dynamos and their theoretical interpretation. Space Sci Rev 219. arXiv:2305.16790 [astro-ph.SR]

Karak BB (2010) Importance of meridional circulation in flux transport dynamo: the possibility of a Maunder-like grand minimum. Astrophys J 724(2):1021–1029. https://doi.org/10.1088/0004-637X/724/2/1021. arXiv:1009.2479 [astro-ph.SR]

Karak BB (2020) Dynamo saturation through the latitudinal variation of bipolar magnetic regions in the Sun. Astrophys J Lett 901(2):L35. https://doi.org/10.3847/2041-8213/abb93f. arXiv:2009.06969 [astro-ph.SR]

Karak BB, Cameron R (2016) Babcock-Leighton solar dynamo: the role of downward pumping and the equatorward propagation of activity. Astrophys J 832(1):94. https://doi.org/10.3847/0004-637X/832/1/94. arXiv:1605.06224 [astro-ph.SR]

Karak BB, Choudhuri AR (2013) Studies of grand minima in sunspot cycles by using a flux transport solar dynamo model. Res Astron Astrophys 13(11):1339–1357. https://doi.org/10.1088/1674-4527/13/11/005. arXiv:1306.5438 [astro-ph.SR]

Karak BB, Miesch M (2017) Solar cycle variability induced by tilt angle scatter in a Babcock-Leighton solar dynamo model. Astrophys J 847(1):69. https://doi.org/10.3847/1538-4357/aa8636. arXiv:1706.08933 [astro-ph.SR]

Karak BB, Nandy D (2012) Turbulent pumping of magnetic flux reduces solar cycle memory and thus impacts predictability of the Sun's activity. Astrophys J Lett 761(1):L13. https://doi.org/10.1088/2041-8205/761/1/L13. arXiv:1206.2106 [astro-ph.SR]

Karak BB, Mandal S, Banerjee D (2018) Double peaks of the solar cycle: an explanation from a dynamo model. Astrophys J 866(1):17. https://doi.org/10.3847/1538-4357/aada0d. arXiv:1808.03922 [astro-ph.SR]

Kitchatinov LL, Olemskoy SV (2011) Does the Babcock-Leighton mechanism operate on the Sun? Astron Lett 37(9):656–658. https://doi.org/10.1134/S0320010811080031. arXiv:1109.1351 [astro-ph.SR]

Kitchatinov LL, Mordvinov AV, Nepomnyashchikh AA (2018) Modelling variability of solar activity cycles. Astron Astrophys 615:A38. https://doi.org/10.1051/0004-6361/201732549. arXiv:1804.02833 [astro-ph.SR]

Knobloch E, Tobias SM, Weiss NO (1998) Modulation and symmetry changes in stellar dynamos. Mon Not R Astron Soc 297(4):1123–1138. https://doi.org/10.1046/j.1365-8711.1998.01572.x

Komm R (2022) Is the subsurface meridional flow zero at the equator? Sol Phys 297(7):99. https://doi.org/10.1007/s11207-022-02027-z

Kumar R, Jouve L, Nandy D (2019) A 3D kinematic Babcock Leighton solar dynamo model sustained by dynamic magnetic buoyancy and flux transport processes. Astron Astrophys 623:A54. https://doi.org/10.1051/0004-6361/201834705. arXiv:1901.04251 [astro-ph.SR]

Kumar P, Karak BB, Vashishth V (2021) Supercriticality of the dynamo limits the memory of the polar field to one cycle. Astrophys J 913(1):65. https://doi.org/10.3847/1538-4357/abf0a1. arXiv:2103.11754 [astro-ph.SR]

Kumar P, Biswas A, Karak BB (2022) Physical link of the polar field buildup with the Waldmeier effect broadens the scope of early solar cycle prediction: cycle 25 is likely to be slightly stronger than cycle 24.

Mon Not R Astron Soc 513(1):L112–L116. https://doi.org/10.1093/mnrasl/slac043. arXiv:2203.11494 [astro-ph.SR]

Labonville F, Charbonneau P, Lemerle A (2019) A dynamo-based forecast of solar cycle 25. Sol Phys 294(6):82. https://doi.org/10.1007/s11207-019-1480-0

Leighton RB (1964) Transport of magnetic fields on the Sun. Astrophys J 140:1547. https://doi.org/10.1086/148058

Leighton RB (1969) A magneto-kinematic model of the solar cycle. Astrophys J 156:1. https://doi.org/10.1086/149943

Lemerle A, Charbonneau P (2017) A coupled 2 × 2D Babcock-Leighton solar dynamo model. II. Reference dynamo solutions. Astrophys J 834(2):133. https://doi.org/10.3847/1538-4357/834/2/133. arXiv:1606.07375 [astro-ph.SR]

Liang ZC, Gizon L, Birch AC et al (2018) Solar meridional circulation from twenty-one years of SOHO/MDI and SDO/HMI observations. Helioseismic travel times and forward modeling in the ray approximation. Astron Astrophys 619:A99. https://doi.org/10.1051/0004-6361/201833673. arXiv:1808.08874 [astro-ph.SR]

Lopes I, Passos D, Nagy M et al (2014) Oscillator models of the solar cycle. Towards the development of inversion methods. Space Sci Rev 186(1–4):535–559. https://doi.org/10.1007/s11214-014-0066-2. arXiv:1407.4918 [astro-ph.SR]

Lorenz EN (1963) Deterministic nonperiodic flow. J Atmos Sci 20(2):130–148. https://doi.org/10.1175/1520-0469(1963)020<0130:DNF>2.0.CO;2

Lorenz EN (1965) A study of the predictability of a 28-variable atmospheric model. Tellus 17(3):321–333. https://doi.org/10.1111/j.2153-3490.1965.tb01424.x. https://onlinelibrary.wiley.com/doi/abs/10.1111/j.2153-3490.1965.tb01424.x

Lorenz EN (1969) The predictability of a flow which possesses many scales of motion. Tellus 21(3):289–307. https://doi.org/10.1111/j.2153-3490.1969.tb00444.x. https://onlinelibrary.wiley.com/doi/abs/10.1111/j.2153-3490.1969.tb00444.x

Mackay DH, Upton LA (2022) A comparison of global magnetofrictional simulations of the 2015 March 20 solar eclipse. Astrophys J 939(1):9. https://doi.org/10.3847/1538-4357/ac94c7

Miesch MS, Dikpati M (2014) A three-dimensional Babcock-Leighton solar dynamo model. Astrophys J Lett 785(1):L8. https://doi.org/10.1088/2041-8205/785/1/L8. arXiv:1401.6557 [astro-ph.SR]

Mikić Z, Downs C et al (2018) Predicting the corona for the 21 August 2017 total solar eclipse. Nat Astron 2:913–921. https://doi.org/10.1038/s41550-018-0562-5

Mininni PD, Gómez DO, Mindlin GB (2000) Stochastic relaxation oscillator model for the solar cycle. Phys Rev Lett 85(25):5476–5479. https://doi.org/10.1103/PhysRevLett.85.5476

Mininni PD, Gomez DO, Mindlin GB (2001) Simple model of a stochastically excited solar dynamo. Sol Phys 201(2):203–223. https://doi.org/10.1023/A:1017515709106

Mundt MD, Bruce MIW, Chase RRP (1991) Chaos in the sunspot cycle: analysis and prediction. J Geophys Res 96(A2):1705–1716. https://doi.org/10.1029/90JA02150

Muñoz-Jaramillo A, Nandy D, Martens PCH (2009) Helioseismic data inclusion in solar dynamo models. Astrophys J 698(1):461–478. https://doi.org/10.1088/0004-637X/698/1/461. arXiv:0811.3441 [astro-ph]

Muñoz-Jaramillo A, Nandy D, Martens PCH (2011) Magnetic quenching of turbulent diffusivity: reconciling mixing-length theory estimates with kinematic dynamo models of the solar cycle. Astrophys J Lett 727(1):L23. https://doi.org/10.1088/2041-8205/727/1/L23. arXiv:1007.1262 [astro-ph.SR]

Muñoz-Jaramillo A, Sheeley NR, Zhang J et al (2012) Calibrating 100 years of polar faculae measurements: implications for the evolution of the heliospheric magnetic field. Astrophys J 753(2):146. https://doi.org/10.1088/0004-637X/753/2/146. arXiv:1303.0345 [astro-ph.SR]

Muñoz-Jaramillo A, Dasi-Espuig M, Balmaceda LA et al (2013) Solar cycle propagation, memory, and prediction: insights from a century of magnetic proxies. Astrophys J Lett 767(2):L25. https://doi.org/10.1088/2041-8205/767/2/L25. arXiv:1304.3151 [astro-ph.SR]

Nagy M, Lemerle A, Labonville F et al (2017) The effect of "rogue" active regions on the solar cycle. Sol Phys 292(11):167. https://doi.org/10.1007/s11207-017-1194-0. arXiv:1712.02185 [astro-ph.SR]

Nagy M, Lemerle A, Charbonneau P (2020) Impact of nonlinear surface inflows into activity belts on the solar dynamo. J Space Weather Space Clim 10:62. https://doi.org/10.1051/swsc/2020064

Nandy D (2002) Can theoretical solar dynamo models predict future solar activity? In: 34th COSPAR scientific assembly, p 53

Nandy D (2021) Progress in solar cycle predictions: sunspot cycles 24-25 in perspective. Sol Phys 296(3):54. https://doi.org/10.1007/s11207-021-01797-2. arXiv:2009.01908 [astro-ph.SR]

Nandy D, Choudhuri AR (2002) Explaining the latitudinal distribution of sunspots with deep meridional flow. Science 296(5573):1671–1673. https://doi.org/10.1126/science.1070955

Nandy D, Muñoz-Jaramillo A, Martens PCH (2011) The unusual minimum of sunspot cycle 23 caused by meridional plasma flow variations. Nature 471(7336):80–82. https://doi.org/10.1038/nature09786. arXiv:1303.0349 [astro-ph.SR]

Nandy D, Bhowmik P, Yeates AR et al (2018) The large-scale coronal structure of the 2017 August 21 great American eclipse: an assessment of solar surface flux transport model enabled predictions and observations. Astrophys J 853(1):72. https://doi.org/10.3847/1538-4357/aaa1eb

Nandy D, Banerjee D, Bhowmik P et al (2022) Exploring the solar poles: the last great frontier of the Sun. arXiv e-prints. https://doi.org/10.48550/arXiv.2301.00010. arXiv:2301.00010 [astro-ph.IM]

Nandy D, Baruah Y, Bhowmik P et al (2023) Causality in heliophysics: magnetic fields as a bridge between the Sun's interior and the Earth's space environment. J Atmos Sol-Terr Phys 248:106081. https://doi.org/10.1016/j.jastp.2023.106081

Norton A, Howe R, Upton L et al (2023) Solar cycle observations. Space Sci Rev 219. arXiv:2305.19803 [astro-ph.SR]

NRC (2013) Solar and space physics: a science for a technological society. The National Academies Press, Washington. https://doi.org/10.17226/13060. https://nap.nationalacademies.org/catalog/13060/solar-and-space-physics-a-science-for-a-technological-society

NSWSAP (2019) National space weather strategy and action plan. The National Academies Press, Washington. https://trumpwhitehouse.archives.gov/wp-content/uploads/2019/03/National-Space-Weather-Strategy-and-Action-Plan-2019.pdf

NSWSAP (2022) Space weather research-to-operations and operations-to-research framework. The National Academies Press, Washington. https://www.whitehouse.gov/wp-content/uploads/2022/03/03-2022-Space-Weather-R2O2R-Framework.pdf

Olemskoy SV, Choudhuri AR, Kitchatinov LL (2013) Fluctuations in the alpha-effect and grand solar minima. Astron Rep 57(6):458–468. https://doi.org/10.1134/S1063772913050065. arXiv:1305.2660 [astro-ph.SR]

Paluš M, Novotná D (1999) Sunspot cycle: a driven nonlinear oscillator? Phys Rev Lett 83(17):3406–3409. https://doi.org/10.1103/PhysRevLett.83.3406

Parker EN (1955) The formation of sunspots from the solar toroidal field. Astrophys J 121:491. https://doi.org/10.1086/146010

Passos D, Lopes I (2011) Grand minima under the light of a low order dynamo model. J Atmos Sol-Terr Phys 73(2–3):191–197. https://doi.org/10.1016/j.jastp.2009.12.019

Passos D, Nandy D, Hazra S et al (2014) A solar dynamo model driven by mean-field alpha and Babcock-Leighton sources: fluctuations, grand-minima-maxima, and hemispheric asymmetry in sunspot cycles. Astron Astrophys 563:A18. https://doi.org/10.1051/0004-6361/201322635. arXiv:1309.2186 [astro-ph.SR]

Petrovay K (2020) Solar cycle prediction. Living Rev Sol Phys 17(1):2. https://doi.org/10.1007/s41116-020-0022-z. arXiv:1907.02107 [astro-ph.SR]

Petrovay K, Nagy M, Yeates AR (2020a) Towards an algebraic method of solar cycle prediction. I. Calculating the ultimate dipole contributions of individual active regions. J Space Weather Space Clim 10:50. https://doi.org/10.1051/swsc/2020050. arXiv:2009.02299 [astro-ph.SR]

Petrovay K, Nagy M, Yeates AR (2020b) Towards an algebraic method of solar cycle prediction. I. Calculating the ultimate dipole contributions of individual active regions. J Space Weather Space Clim 10:50. https://doi.org/10.1051/swsc/2020050. arXiv:2009.02299 [astro-ph.SR]

Price CP, Prichard D, Hogenson EA (1992) Do the sunspot numbers form a "chaotic" set? J Geophys Res 97(A12):19,113–19,120. https://doi.org/10.1029/92JA01459

Rajaguru SP, Antia HM (2015) Meridional circulation in the solar convection zone: time-distance helioseismic inferences from four years of HMI/SDO observations. Astrophys J 813(2):114. https://doi.org/10.1088/0004-637X/813/2/114. arXiv:1510.01843 [astro-ph.SR]

Rossby CG (1939) Relation between variations in the intensity of the zonal circulation of the atmosphere and the displacements of the semi-permanent centers of action. J Mar Res 2:38–55. https://doi.org/10.1357/002224039806649023

Rozelot JP (1995) On the chaotic behaviour of the solar activity. Astron Astrophys 297:L45

Saha C, Chandra S, Nandy D (2022) Evidence of persistence of weak magnetic cycles driven by meridional plasma flows during solar grand minima phases. Mon Not R Astron Soc 517(1):L36–L40. https://doi.org/10.1093/mnrasl/slac104. arXiv:2209.14651 [astro-ph.SR]

Sanchez S, Fournier A, Aubert J (2014) The predictability of advection-dominated flux-transport solar dynamo models. Astrophys J 781(1):8. https://doi.org/10.1088/0004-637X/781/1/8

Schatten KH, Scherrer PH, Svalgaard L et al (1978) Using dynamo theory to predict the sunspot number during solar cycle 21. Geophys Res Lett 5(5):411–414. https://doi.org/10.1029/GL005i005p00411

Schou J, Antia HM, Basu S et al (1998) Helioseismic studies of differential rotation in the solar envelope by the solar oscillations investigation using the Michelson Doppler imager. Astrophys J 505(1):390–417. https://doi.org/10.1086/306146

Schrijver CJ (2001) Simulations of the photospheric magnetic activity and outer atmospheric radiative losses of cool stars based on characteristics of the solar magnetic field. Astrophys J 547(1):475–490. https://doi.org/10.1086/318333

Schrijver CJ, De Rosa ML (2003) Photospheric and heliospheric magnetic fields. Sol Phys 212(1):165–200. https://doi.org/10.1023/A:1022908504100

Schrijver CJ, Kauristie K, Aylward AD et al (2015) Understanding space weather to shield society: a global road map for 2015-2025 commissioned by COSPAR and ILWS. Adv Space Res 55(12):2745–2807. https://doi.org/10.1016/j.asr.2015.03.023. arXiv:1503.06135 [physics.space-ph]

Schwabe H (1844) Sonnenbeobachtungen im Jahre 1843. Astron Nachr 21(15):233. https://doi.org/10.1002/asna.18440211505

Sheeley NR Jr, DeVore CR, Boris JP (1985) Simulations of the mean solar magnetic field during sunspot cycle-21. Sol Phys 98(2):219–239. https://doi.org/10.1007/BF00152457

Solanki SK, Wenzler T, Schmitt D (2008) Moments of the latitudinal dependence of the sunspot cycle: a new diagnostic of dynamo models. Astron Astrophys 483(2):623–632. https://doi.org/10.1051/0004-6361:20054282

Takens F (1981) Detecting strange attractors in turbulence. Lecture notes in mathematics, vol 898. Springer, Berlin, p 366. https://doi.org/10.1007/BFb0091924

Talafha M, Nagy M, Lemerle A et al (2022) Role of observable nonlinearities in solar cycle modulation. Astron Astrophys 660:A92. https://doi.org/10.1051/0004-6361/202142572. arXiv:2112.14465 [astro-ph.SR]

Theiler J, Eubank S, Longtin A et al (1992) Testing for nonlinearity in time series: the method of surrogate data. Phys D, Nonlinear Phenom 58(1–4):77–94. https://doi.org/10.1016/0167-2789(92)90102-S

Tobias SM (1996) Grand minimia in nonlinear dynamos. Astron Astrophys 307:L21

Tobias SM, Weiss NO, Kirk V (1995) Chaotically modulated stellar dynamos. Mon Not R Astron Soc 273(4):1150–1166. https://doi.org/10.1093/mnras/273.4.1150

Tripathi B, Nandy D, Banerjee S (2021) Stellar mid-life crisis: subcritical magnetic dynamos of solar-like stars and the breakdown of gyrochronology. Mon Not R Astron Soc 506(1):L50–L54. https://doi.org/10.1093/mnrasl/slab035. arXiv:1812.05533 [astro-ph.SR]

UN (2017) Space weather: Special report of the Inter-Agency Meeting on Outer Space Activities on developments within the United Nations system related to space weather. A/AC.105/1146. United Nations, New York. https://documents-dds-ny.un.org/doc/UNDOC/GEN/V17/027/62/PDF/V1702762.pdf

Upton L, Hathaway DH (2014) Predicting the Sun's polar magnetic fields with a surface flux transport model. Astrophys J 780(1):5. https://doi.org/10.1088/0004-637X/780/1/5. arXiv:1311.0844 [astro-ph.SR]

Upton LA, Hathaway DH (2018) An updated solar cycle 25 prediction with AFT: the modern minimum. Geophys Res Lett 45(16):8091–8095. https://doi.org/10.1029/2018GL078387. arXiv:1808.04868 [astro-ph.SR]

Usoskin IG (2017) A history of solar activity over millennia. Living Rev Sol Phys 14:3. https://doi.org/10.1007/s41116-017-0006-9

van Saders JL, Ceillier T, Metcalfe TS et al (2016) Weakened magnetic braking as the origin of anomalously rapid rotation in old field stars. Nature 529(7585):181–184. https://doi.org/10.1038/nature16168. arXiv:1601.02631 [astro-ph.SR]

Waldmeier M (1935) Neue Eigenschaften der Sonnenfleckenkurve. Astron. Mitt. Eidgenöss. Sternwarte Zür. 14:105–136

Wang YM, Nash AG, Sheeley NR Jr (1989) Magnetic flux transport on the Sun. Science 245(4919):712–718. https://doi.org/10.1126/science.245.4919.712

Wang YM, Sheeley NR Jr, Nash AG (1991) A new solar cycle model including meridional circulation. Astrophys J 383:431. https://doi.org/10.1086/170800

Wang YM, Lean J, Sheeley NR Jr (2002) Role of a variable meridional flow in the secular evolution of the Sun's polar fields and open flux. Astrophys J Lett 577(1):L53–L57. https://doi.org/10.1086/344196

Wang ZF, Jiang J, Zhang J et al (2020) Activity complexes and a prominent poleward surge during solar cycle 24. Astrophys J 904(1):62. https://doi.org/10.3847/1538-4357/abbc1e. arXiv:2009.12483 [astro-ph.SR]

Wang ZF, Jiang J, Wang JX (2021) Algebraic quantification of an active region contribution to the solar cycle. Astron Astrophys 650:A87. https://doi.org/10.1051/0004-6361/202140407. arXiv:2104.04307 [astro-ph.SR]

Weber MA, Fan Y, Miesch MS (2011) The rise of active region flux tubes in the turbulent solar convective envelope. Astrophys J 741:11. https://doi.org/10.1088/0004-637X/741/1/11. arXiv:1109.0240 [astro-ph.SR]

Weber MA, Schunker H, Jouve L et al (2023) Understanding active region emergence and origins on the Sun and other cool stars. Space Sci Rev 219. arXiv:2306.06536

Whitbread T, Yeates AR, Muñoz-Jaramillo A (2018) How many active regions are necessary to predict the solar dipole moment? Astrophys J 863(2):116. https://doi.org/10.3847/1538-4357/aad17e. arXiv:1807.01617 [astro-ph.SR]

Whitbread T, Yeates AR, Muñoz-Jaramillo A (2019) The need for active region disconnection in 3D kinematic dynamo simulations. Astron Astrophys 627:A168. https://doi.org/10.1051/0004-6361/201935986. arXiv:1907.02762 [astro-ph.SR]

Wiin-Nielsen A (1991) The birth of numerical weather prediction. Tellus, Ser A Dyn Meteorol Oceanogr 43(4):36–52. https://doi.org/10.3402/tellusa.v43i4.11937

Wilmot-Smith AL, Martens PCH, Nandy D et al (2005) Low-order stellar dynamo models. Mon Not R Astron Soc 363(4):1167–1172. https://doi.org/10.1111/j.1365-2966.2005.09514.x

Wilmot-Smith AL, Nandy D, Hornig G et al (2006) A time delay model for solar and stellar dynamos. Astrophys J 652(1):696–708. https://doi.org/10.1086/508013

Worden J, Harvey J (2000) An evolving synoptic magnetic flux map and implications for the distribution of photospheric magnetic flux. Sol Phys 195(2):247–268. https://doi.org/10.1023/A:1005272502885

Yeates AR (2020) How good is the bipolar approximation of active regions for surface flux transport? Sol Phys 295(9):119. https://doi.org/10.1007/s11207-020-01688-y. arXiv:2008.03203 [astro-ph.SR]

Yeates AR, Bhowmik P (2022) Automated driving for global nonpotential simulations of the solar corona. Astrophys J 935(1):13. https://doi.org/10.3847/1538-4357/ac7de4

Yeates AR, Muñoz-Jaramillo A (2013) Kinematic active region formation in a three-dimensional solar dynamo model. Mon Not R Astron Soc 436(4):3366–3379. https://doi.org/10.1093/mnras/stt1818. arXiv:1309.6342 [astro-ph.SR]

Yeates AR, Mackay DH, van Ballegooijen AA (2007) Modelling the global solar corona: filament chirality observations and surface simulations. Sol Phys 245:87–107. https://doi.org/10.1007/s11207-007-9013-7. arXiv:0707.3256

Yeates AR, Nandy D, Mackay DH (2008) Exploring the physical basis of solar cycle predictions: flux transport dynamics and persistence of memory in advection- versus diffusion-dominated solar convection zones. Astrophys J 673(1):544–556. https://doi.org/10.1086/524352. arXiv:0709.1046 [astro-ph]

Yeates AR, Baker D, van Driel-Gesztelyi L (2015) Source of a prominent poleward surge during solar cycle 24. Sol Phys 290(11):3189–3201. https://doi.org/10.1007/s11207-015-0660-9. arXiv:1502.04854 [astro-ph.SR]

Yeates AR, Cheung MCM, Jiang J et al (2023) Surface flux transport on the Sun. Space Sci Rev 219(4):31. https://doi.org/10.1007/s11214-023-00978-8. arXiv:2303.01209 [astro-ph.SR]

Yoshimura H (1978) Nonlinear astrophysical dynamos: multiple-period dynamo wave oscillations and long-term modulations of the 22 year solar cycle. Astrophys J 226:706–719. https://doi.org/10.1086/156653

Zhang Z, Jiang J (2022) A Babcock-Leighton-type solar dynamo operating in the bulk of the convection zone. Astrophys J 930(1):30. https://doi.org/10.3847/1538-4357/ac6177. arXiv:2204.14077 [astro-ph.SR]

Zhang Z, Jiang J, Zhang H (2022) A potential new mechanism for the butterfly diagram of the solar cycle: latitude-dependent radial flux transport. Astrophys J Lett 941(1):L3. https://doi.org/10.3847/2041-8213/aca47a. arXiv:2212.00948 [astro-ph.SR]

Zhao J, Kosovichev AG (2004) Torsional oscillation, meridional flows, and vorticity inferred in the upper convection zone of the Sun by time-distance helioseismology. Astrophys J 603(2):776

Zhao J, Bogart RS, Kosovichev AG et al (2013) Detection of equatorward meridional flow and evidence of double-cell meridional circulation inside the Sun. Astrophys J Lett 774(2):L29. https://doi.org/10.1088/2041-8205/774/2/L29. arXiv:1307.8422 [astro-ph.SR]

Publisher's Note Springer Nature remains neutral with regard to jurisdictional claims in published maps and institutional affiliations.

Authors and Affiliations

Prantika Bhowmik[1] ⓘ · **Jie Jiang[2,3]** ⓘ · **Lisa Upton[4]** ⓘ · **Alexandre Lemerle[5,6]** ⓘ · **Dibyendu Nandy[7,8]** ⓘ

✉ P. Bhowmik
prantika.bhowmik@durham.ac.uk

J. Jiang
jiejiang@buaa.edu.cn

L. Upton
lisa.upton@swri.org

A. Lemerle
alexandre.lemerle@umontreal.ca

D. Nandy
dnandi@iiserkol.ac.in

[1] Department of Mathematical Sciences, Durham University, Stockton Road, Durham, DH1 3LE, UK

[2] School of Space and Environment, Beihang University, Beijing, China

[3] Key Laboratory of Space Environment Monitoring and Information Processing of MIIT, Beijing, China

[4] Department of Solar and Heliospheric Physics, Southwest Research Institute, Boulder, 80302, CO, USA

[5] Département de Physique, Collège de Bois-de-Boulogne, Montréal, Québec, Canada

[6] Département de Physique, Université de Montréal, Montréal, Québec, Canada

[7] Center of Excellence in Space Sciences India, Indian Institute of Science Education and Research Kolkata, Mohanpur, 741246, West Bengal, India

[8] Department of Physical Sciences, Indian Institute of Science Education and Research Kolkata, Mohanpur, 741246, West Bengal, India